Calcium

Chemistry, Analysis, Function and Effects

Food and Nutritional Components in Focus

Series Editor:
Professor Victor R. Preedy, *School of Medicine, King's College London, UK*

Titles in the Series:
1: Vitamin A and Carotenoids: Chemistry, Analysis, Function and Effects
2: Caffeine: Chemistry, Analysis, Function and Effects
3: Dietary Sugars: Chemistry, Analysis, Function and Effects
4: B Vitamins and Folate: Chemistry, Analysis, Function and Effects
5: Isoflavones: Chemistry, Analysis, Function and Effects
6: Fluorine: Chemistry, Analysis, Function and Effects
7: Betaine: Chemistry, Analysis, Function and Effects
8: Imidazole Dipeptides: Chemistry, Analysis, Function and Effects
9: Selenium: Chemistry, Analysis, Function and Effects
10: Calcium: Chemistry, Analysis, Function and Effects

How to obtain future titles on publication:
A standing order plan is available for this series. A standing order will bring delivery of each new volume immediately on publication.

For further information please contact:
Book Sales Department, Royal Society of Chemistry, Thomas Graham House, Science Park, Milton Road, Cambridge, CB4 0WF, UK
Telephone: +44 (0)1223 420066, Fax: +44 (0)1223 420247
Email: booksales@rsc.org
Visit our website at www.rsc.org/books

Calcium

Chemistry, Analysis, Function and Effects

Edited by

Victor R. Preedy
School of Medicine, King's College London, UK
Email: editor@publicationeditor.org.uk

ROYAL SOCIETY OF CHEMISTRY

THE QUEEN'S AWARDS FOR ENTERPRISE: INTERNATIONAL TRADE 2013

Food and Nutritional Components in Focus No. 10

Print ISBN: 978-1-84973-887-3
PDF eISBN: 978-1-78262-213-0
ISSN: 2045-1695

A catalogue record for this book is available from the British Library

Published by The Royal Society of Chemistry,
Thomas Graham House, Science Park, Milton Road,
Cambridge CB4 0WF, UK

Registered Charity Number 207890

For further information see our web site at www.rsc.org

Printed and bound by CPI Group (UK) Ltd, Croydon, CR0 4YY

Preface

Recently, there have been major advances in our understanding of the chemistry and function of nutritional components. This has been enhanced by rapid developments in analytical techniques and instrumentation. Chemists, food scientists and nutritionists are, however, separated by divergent skills, and professional disciplines. Hitherto this transdisciplinary divide has been difficult to bridge.

The series *Food and Nutritional Components in Focus* aims to cover in a single volume the chemistry, analysis, function and effects of single components in the diet or its food matrix. Its aim is to link scientific disciplines so that information becomes more meaningful and applicable to health in general.

The series *Food and Nutritional Components in Focus* covers the latest knowledge base and has a structured format with major subsections covering

- Compounds in Context
- Chemistry
- Analysis
- Function and Effects

In some books the section on Chemistry is also linked with Biochemistry. Each chapter has a novel cohort of features namely by containing:

- *Summary Points*
- *Key Facts* (areas of focus explained for the lay person)
- *Definitions of Words and Terms*

The series covers numerous classes of dietary components including, for example, minerals, vitamins, food additives, and so on. The chapters are written by national or international experts, specialists and leaders in the field.

Food and Nutritional Components in Focus No. 10
Calcium: Chemistry, Analysis, Function and Effects
Edited by Victor R. Preedy

Published by the Royal Society of Chemistry, www.rsc.org

Calcium has the following material: Section 1 Calcium in Context has material on dietary sources and metabolism, ethnicity and geography, availability, milk and dairy products, legumes, vegetables, cereals, baked goods and meals. Section 2 Chemistry and Biochemistry has chapters covering chemistry, biological roles, analysis and vitamin D. Section 3 Analysis has material on ultrasonic-dialysis capillary electrophoresis, fluorescent polyanions, osteoclastic calcium resorption, X-ray diffraction patterns, nanocalcium, milk, food frequency questionnaires, calcium digestibility, *in vivo*, *in vitro* and *ex vivo* techniques, bioavailability and CaCo-2 cells. Section 4 Function and Effects has extensive coverage on calcium in relation to adolescents, dietary calcium, protein intake, bioaccessibility, legumes, rice calcium and phytic acid, acrylamide, gluten, breadmaking, soymilk, prebiotics, drinking water, saliva, intestinal absorption, calcium-sensing receptors, taste cells, calcium signalling, gene expression, mitochondria, pregnancy, lactation, bone health, hypertension, transporters, cholesterol metabolism, vitamin D, exercise, osteoprotegerin body fat, critical care, and hypercalcemia.

Calcium is specifically designed for chemists, analytical scientists, forensic scientists, food scientists, dieticians, nutritionists, food scientists, health professionals and research academics. The series is suitable for lecturers and teachers in food and nutritional sciences. Importantly, the series will be a valuable resource for college or university libraries as a reference guide.

Professor Victor R. Preedy
King's College London

Contents

Calcium in Context

Chapter 1 **Calcium in the Context of Dietary Sources and Metabolism** 3
Maciej S. Buchowski

1.1 Overview of Calcium and Its Physiological Functions 3
1.1.1 Sources of Calcium in the Diet 4
1.1.2 Calcium Absorption and Excretion 8
1.1.3 Calcium Homeostasis and Systemic Balance 9
1.1.4 Calcium Bioavailability and Dietary Factors Affecting Calcium Absorption 11
Summary Points 13
Key Facts 14
Definitions of Words and Terms 16
References 17

Chapter 2 **The Biological Roles of Calcium: Nutrition, Diseases and Analysis** 21
Leonardo M. Moreira, Raphael P. Araujo, Fernando P. Leonel, Henrique V. N. Machado, Alexandre O. Teixeira, Fabio V. Santos, Vanessa J. S. V. Santos, and Juliana P. Lyon

2.1 Introduction 21
2.1.1 Calcium in the Biological Medium 23
2.1.2 Calcium in the Diet 24
2.1.3 The Calcium Metabolism 25
2.1.4 The Calcium Action in Several Cardiovascular Diseases 25
2.1.5 Calcium and Osteoporosis 26

Food and Nutritional Components in Focus No. 10
Calcium: Chemistry, Analysis, Function and Effects
Edited by Victor R. Preedy

Published by the Royal Society of Chemistry, www.rsc.org

2.1.6 Methods of Analysis and Evaluation of Calcium in Feeds and Biological Tissues 26
2.1.7 Conclusions 27
Summary Points 27
Key Facts of Calcium in the Human Body 27
Definitions of Words and Terms 27
List of Abbreviations 28
References 28

Chapter 3 Food Sources of Calcium Vary by Ethnicity and Geography 30
Noreen Willows

3.1 Introduction 30
3.2 Milk and Dairy Foods Consumed Worldwide 30
3.2.1 Europe 31
3.2.2 Middle East 32
3.2.3 Canada, United States of America, and Australia 33
3.2.4 South Asia 33
3.2.5 East and Southeast Asia 34
3.2.6 Africa 36
3.3 Nondairy Sources of Calcium Consumed Worldwide 36
3.3.1 Asia 36
3.3.2 Africa 38
3.3.3 Mexico and Central America 39
Summary Points 39
Key Facts 39
Definitions of Words and Terms 40
List of Abbreviations 41
References 41

Chapter 4 Calcium Availability in Specific Foods: Milk and Dairy Products, Legumes, Vegetables, Cereals, Baked Goods and Cooked Meals 46
Mª Victorina Aguilar Vilas

4.1 Introduction 46
4.2 Bioavailability: Influencing Factors 47
4.2.1 Calcium Intake and Speciation 47
4.2.2 Vitamin D 47
4.2.3 Proteins or Specific Amino Acids 49
4.2.4 Carbohydrates 49
4.2.5 Fat 50
4.2.6 Minerals 50
4.2.7 Oxalic Acid 50
4.2.8 Phytic Acid 50

4.2.9 Fiber 50
4.2.10 Other Dietary Factors 51
4.3 Calcium Availability in Specific Foods 51
4.3.1 Milk and Dairy Products 51
4.3.2 Legumes 56
4.3.3 Vegetables 57
4.3.4 Cereals and their Derivatives 58
4.3.5 Baked Goods 59
4.3.6 Beverages 59
4.3.7 Cooked Meals 60
Summary Points 60
Key Facts of Availability 60
Definition of Words and Terms 61
List of Abbreviations 61
References 62

Chemistry and Biochemistry

Chapter 5 The Chemistry of Calcium 67
Daniel Perrone and Mariana Monteiro

5.1 Chemical Properties of Calcium 67
5.2 Calcium Isotopes in Nutrition Research 70
Summary Points 71
Keys Facts of Alkaline-Earth Metals 72
Definitions and Explanation of Key Terms 72
List of Abbreviations 73
References 73

Chapter 6 Vitamin D and Impact on Total-Body Calcium 75
Howard A. Morris

6.1 Introduction 75
6.2 Calcium is an Essential Nutrient 76
6.2.1 Recommended Daily Intakes for Calcium 76
6.2.2 Oestrogen, the Calcium Economy and Influence on Recommended Daily Intake 77
6.3 Vitamin D is an Essential Nutrient 78
6.3.1 Vitamin D Requirement and Metabolism 79
6.3.2 Vitamin D and Metabolic Bone Disorders 80
6.3.3 Regulation of Plasma 1,25D Levels 83
6.4 Vitamin D Activities and Regulation of Bone-Mineral Homeostasis 84
6.4.1 The Basic Multicellular Unit of Bone Remodeling 84
6.4.2 Vitamin D Activities within Osteoblasts and Osteocytes 86
6.5 Conclusions 87

Summary Points 87
Key Facts 88
Definitions and Explanations 88
List of Abbreviations 89
References 89

Analysis

Chapter 7 Ultrasonic-Dialysis Capillary Electrophoresis Inductively Coupled Plasma Optical Emission Spectrometry Analysis of Calcium Speciation in Red Blood Cells 95
Biyang Deng, Shuangjiao Sun, and Yingzi Wang

7.1 Introduction 95
7.2 Methodological Considerations 97
7.3 Effects of Various Factors 99
7.3.1 Effect of Buffer Concentration 99
7.3.2 Effect of Applied Voltage 100
7.4 Lysis of HRBC 100
7.5 Electrophoresis Separation of Calcium Species in HRBC Cytoplasm 100
7.6 Electrophoresis Separations of Calcium Species in HRBC 101
7.7 Quantification of Calcium-Containing Species in HRBC 102
7.8 Identification of Other Calcium Species Using Ultrasonic Dialysis 102
7.9 Identification of Free Ca^{2+} in HRBC Cytoplasm 104
7.10 Quantification, Detection Limit and Recovery of Free Ca^{2+} in HRBC Cytoplasm 104
Summary Points 104
Key Facts 105
Key Facts of Ultrasonic Dialysis 105
Key Facts of Capillary Electrophoresis Inductively Coupled Plasma Mass Spectrometry/Atomic Emission Spectrometry 105
Key Facts of Metallomics 106
Key Facts of Requirements for Reliable Elemental Speciation Analysis 106
Key Facts of Detection Systems for Elemental Speciation Analysis 106
Definitions of Words and Terms 107
List of Abbreviations 108
Acknowledgments 108
References 108

Chapter 8 Using Fluorescent Polyanions to Assay for Osteoclastic Calcium-Resorption Activity **111**
Tatsuya Miyazaki and Osamu Suzuki

8.1 Introduction 111
8.2 Current Methods of Evaluating Osteoclast Function 112
8.3 Fluorescent Polyanions to Assay for Osteoclastic Calcium-Resorption Activity 113
8.3.1 CaP (Hydroxyapatite) Plate 114
8.3.2 Coating of Fluorescein-Labeled Polyanion to CaP Plate 115
8.3.3 Bone Resorption Assay 116
8.3.4 Evaluation of Drugs for Osteoporosis 119
8.4 Concluding Remarks 120
Summary Points 120
Key Facts about Fluorescent Polyanions and CaP Resorption 121
Definitions of Words and Terms 122
List of Abbreviations 123
References 123

Chapter 9 Methods for the X-Ray Diffraction Patterns of Nanocalcium in Milk **126**
Ching-Hsiang Chen, Liang-Yih Chen, and Hsiao-Chien Chen

9.1 Overview of Calcium in Food Nanotechnology 126
9.1.1 Function of Calcium in Human Body 126
9.1.2 Food Nanotechnology 128
9.1.3 Analysis of Calcium 129
9.2 X-Ray Diffraction Technology 130
9.2.1 X-Ray Generation 131
9.2.2 Bravais Lattices 132
9.2.3 Bragg Spectrometer 132
9.3 Structure of Nanocalcium in Milk 136
9.4 Conclusion 138
Summary Points 140
Key Facts 140
Key Facts of Nanocalcium Additive and its Importance 140
Key Facts of Structural Identification by Using X-ray Diffraction 141
Definitions of Terms 141
List of Abbreviations 142
References 143

Chapter 10 Using Food-Frequency Questionnaires for Calcium Intakes **146**
B. Pampaloni and M. L. Brandi

10.1 Introduction 146
10.2 Using FFQs to Assess Dietary Habits in Epidemiological Studies and Calcium Intakes 148
10.2.1 Dietary Assessment in Different Ethnic Groups 150
10.2.2 Obesity 151
10.2.3 Diabetes Type 2 151
10.2.4 Stroke 152
10.2.5 Cardiometabolic Risk Factors 153
10.2.6 Cardiometabolic Risk Factors in Young People 154
10.2.7 Gastroesophageal Reflux Disease 154
10.2.8 Kidney Stones 154
10.2.9 Colorectal Cancer (CRC) 155
10.2.10 Calcium and Bone 156
10.3 Conclusion 159
Summary Points 159
Key Facts 160
Key Facts of Food-Frequency Questionnaires and their Use 160
List of Abbreviations 160
References 161

Chapter 11 Calcium Digestibility Using *In Vivo*, *In Vitro* and *Ex Vivo* Techniques **164**
F. Rossi

11.1 Introduction 164
11.2 Methods for Estimating Ca Digestibility 165
11.2.1 *In Vitro* Methods 165
11.2.2 *In Vivo* Studies 169
11.2.3 Cell Culture 173
Summary Points 174
Key Facts 174
List of Abbreviations 176
References 177

Chapter 12 Determining Calcium Bioavailability Using Caco-2 Cells **179**
Amparo Alegría, Guadalupe Garcia-Llatas, and Reyes Barberá

12.1 Introduction 179
12.2 Caco-2 Cells 181

12.2.1 Generalities 181
12.2.2 Methodological Aspects of Caco-2 Cell Assays 181
12.3 Calcium Absorption Mechanisms in Caco-2 Cells 184
12.4 Bioavailability of Calcium-Enriched Foods 184
12.5 Influence of Food Components and Processing upon Calcium Bioavailability 185
12.5.1 Fiber and Phytic Acid 185
12.5.2 Calcium–Iron Interaction 186
12.5.3 Maillard Reaction Products 186
12.5.4 Peptides and Proteins 187
12.5.5 Processing 188
Summary Points 196
Key facts 196
Caco-2 Cells as a Model to Study Calcium Bioavailability 196
Definitions of Words and Terms 196
List of Abbreviations 197
References 197

Function and Effects

Chapter 13 Adolescents and Dietary Calcium 203
Marta Mesías, Isabel Seiquer, and M. Pilar Navarro

13.1 The Role of Calcium in Adolescence 203
13.2 Recommended Calcium Intakes 204
13.3 Main Food Sources of the Daily Calcium Intake 206
13.4 Determinant Factors Affecting the Calcium Absorption during Adolescence 208
13.4.1 Hormonal Factors 208
13.4.2 Dietary Factors 209
13.5 Determinant Factors Affecting Calcium Retention during Adolescence 211
13.5.1 Hormonal Factors 213
13.5.2 Dietary Factors 214
13.6 Other Factors Affecting Bone-Mass Acquisition 216
13.7 Conclusions 216
Summary Points 217
Key Facts 217
Key Facts of Calcium 217
Key Facts of Adolescence 218
Key Facts of Osteoporosis 218
Definitions of Words and Terms 218
List of Abbreviations 219
Acknowledgements 219
References 219

Chapter 14 The Influence of Protein Intake on Calcium Balance **223**
E. Rouy and D. Tome

14.1 Introduction 223
14.2 Protein Intake and Calcium Metabolism: Interacting at Multiple Levels 224
14.2.1 Protein Intake Increases Urinary Calcium Excretion 224
14.2.2 Does Protein Intake Increase Calcium Absorption? 224
14.2.3 Is Protein-Induced Acidosis a Threat to Calcium Metabolism? 226
14.2.4 The Calcium-Sensing Receptor also Senses Amino Acids 227
14.2.5 Overall Effect of Protein Intake on Calcium Balance 228
14.3 Effects of Protein Intake on Calcium Metabolism and Implications for Health 229
14.3.1 Protein Intake and Bone: Anabolic or Catabolic? 229
14.3.2 Protein Intake, Urinary Calcium and Kidney-Stone Formation 230
14.4 Conclusions 231
Summary Points 231
Key Facts 232
Key Facts on Protein 232
Key Facts on Osteoporosis 232
Key Facts on Kidney Health 232
Definitions of Words and Terms 233
List of Abbreviations 234
References 234

Chapter 15 Bioaccessibility of Calcium in Legumes **237**
Mª Jesús Lagarda, Antonio Cilla, and Reyes Barberá

15.1 Introduction 237
15.2 Calcium Content in Legumes 238
15.3 Calcium Bioaccessibility Studies in Legumes 240
15.3.1 Dietetic Factors Affecting Calcium Bioaccessibility. Influence of Processing 242
Summary Points 252
Key Facts 252
In Vitro Bioavailability of Ca in Legumes 252
Definitions of Words and Terms 252
List of Abbreviations 253
References 253

Chapter 16 Calcium – Function and Effects: Rice Calcium and Phytic Acid Levels **256**
Jianfen Liang, Yifan He, Qian Gao, Xuan Wang, and M. J. Robert Nout

16.1 Introduction 256
16.2 Calcium in Rice 257
16.2.1 Calcium Contents in Rice Products 258
16.2.2 Availability of Calcium from Rice Products 259
16.3 Phytic Acid in Rice 259
16.3.1 Phytic Acid and its Chemical Determination 259
16.3.2 Phytic Acid Contents of Rice and Rice Products 262
16.4 Location of Calcium, Phosphorous and Phytic Acid in Rice Kernel 265
16.4.1 Location of Elements with X-Ray Microscope 266
16.4.2 Location of Phytic Acid by Abrasive Method 266
16.5 Conclusion 269
Summary Points 269
Definitions of Words and Terms 270
List of Abbreviations 271
Acknowledgements 271
References 271

Chapter 17 Adding Calcium to Foods and Effect on Acrylamide **274**
Neslihan Göncüoğlu Taş, Aytül Hamzalıoğlu, Tolgahan Kocadağlı, and Vural Gökmen

17.1 Introduction 274
17.2 Acrylamide Formation Mechanism 275
17.3 Mitigation Mechanism of Acrylamide Formation by Calcium 278
17.4 Mitigation of Acrylamide Formation in Foods by Calcium Salts 281
17.5 Effect of Calcium Addition on Sugar Degradation in Foods and Model Systems 283
Summary Points 285
Key Facts of Mitigation of Acrylamide Formation with Calcium 286
Definition of Words and Terms 286
List of Abbreviations 287
References 287

Chapter 18 Addition of Calcium to Gluten and Nongluten Flours for Breadmaking **291**
Urszula Krupa-Kozak

18.1 Introduction 291
18.2 Calcium – Role and Demands 292
18.2.1 Calcium in the Human Body 292
18.2.2 Recommendations and Guidelines 293
18.3 Gluten Sensitivity 294
18.4 Calcium in Traditional Breadmaking 296
18.4.1 Nutritional Value of Cereals 296
18.4.2 The Breadmaking Process 297
18.4.3 Cereals Processing *vs.* Calcium Content 299
18.4.4 Calcium Supplements and Additives in Breadmaking 300
18.5 Calcium in Gluten-Free Breadmaking 301
18.5.1 Technological and Nutritional Quality of Gluten-Free Baked Products 301
18.5.2 Calcium in Gluten-Free Baked Products 302
Summary Points 304
Key Facts of Breadmaking 304
Definitions and Explanations 305
List of Abbreviations 306
References 306

Chapter 19 Calcium-Fortified Soymilk: Function and Health Benefits **310**
Lily Stojanovska, Mutamed Ayyash, and Vasso Apostolopoulos

19.1 Introduction 310
19.2 Calcium Fortification and Bioavailability 312
19.3 Soy Isoflavones 313
19.4 Soymilk and the Immune Response 314
19.5 Soymilk and Bone Health 315
19.6 Soymilk Properties in Health and Disease 316
19.7 Effects of Soymilk in Cholesterol 316
19.8 Effects of Soymilk in Metabolic Syndrome Disorders 318
19.9 Effects of Soymilk in Cancer 318
19.10 Adverse Effects of Soymilk 319
19.11 Conclusion 320
Summary Points 320
Key Facts 321
Key Facts of Soymilk History 321
Key Facts of Calcium-Fortification Method 322
Key Facts of Osteoporosis 322

Definitions of Words and Terms 323
List of Abbreviations 324
References 324

Chapter 20 The Effect of Prebiotics on Calcium Absorption and Utilization 329
Steven Jakeman and Connie Weaver

20.1 Introduction 329
20.2 What are Prebiotics? 330
20.3 Calcium Absorption 331
20.4 Prebiotic Mediated Calcium Absorption 331
20.4.1 Changes in the Lumen Environment 331
20.4.2 Changes to the Large Intestine 332
20.4.3 Changes to the Gut-Microbial Population 333
20.5 Types of Prebiotics 334
20.5.1 Disaccharides 334
20.5.2 Oligosaccharides 334
20.5.3 Polysaccharides 336
20.5.4 Synbiotics 336
20.5.5 Mixed Prebiotics 336
20.6 Factors that Influence Prebiotic Mediated Calcium Absorption 337
20.6.1 Age 337
20.6.2 Intervention Length and Calcium Status 339
20.6.3 Dose Amount 340
20.6.4 Genetics and Pre-Existing Conditions 341
20.7 Prebiotics on Bone 341
20.8 Conclusion and Future Work 343
Summary Points 343
Key Facts 344
Key Features of Prebiotics 344
Definitions of Words and Terms 344
List of Abbreviations 345
References 345

Chapter 21 Health Aspects of Calcium in Drinking Water 349
A. Ata Alturfan and Ebru Emekli Alturfan

21.1 Drinking Water 349
21.2 Calcium and Water Hardness 350
21.3 Water as a Source of Calcium and Bioavailability of Calcium in Water 351
21.4 Calcium Content of Drinking Water in Different Regions 352

21.5 Potential Effects of Drinking-Water Hardness on Health 353
21.5.1 Effects of the Hardness of Drinking Water on Cardiovascular Diseases 353
21.5.2 Effects of Calcium in Water on Bone Metabolism 356
21.5.3 Effects of Calcium in Water on Kidney Stones 356
21.6 Conclusion 357
Summary Points 358
Key Facts 358
Key Facts of Potential Health Effects of Drinking-Water Hardness 358
Key Facts of Calcium and Water Hardness 358
Definitions of Words and Terms 359
List of Abbreviations 360
References 360

Chapter 22 Calcium in Saliva and Impact on Health 364
Aysen Yarat, Ebru Emekli Alturfan, and Serap Akyuz

22.1 Importance of Calcium in Biological Systems 364
22.2 Saliva Production, Composition and Salivary Calcium Concentration 365
22.3 The Role of Salivary Calcium in Maintenance of Tooth Integrity 369
22.4 Salivary Calcium and Periondontitis 373
22.5 Salivary Calcium and Dental Caries 374
22.6 Salivary Calcium and Diabetes 375
22.7 Salivary Calcium as a Diagnostic Tool 376
22.8 Conclusion 377
Summary Points 377
Key Facts of Salivary Calcium in Maintenance of Tooth Integrity 378
Definitions of Words and Terms 378
List of Abbreviations 379
References 379

Chapter 23 Intestinal Absorption of Calcium 384
Brian R. Stephens and James S. Jolliff

23.1 Introduction 384
23.2 Calcium Absorption 385
23.3 The Paracellular Pathway 386
23.4 The Transcellular Pathway 387
23.5 Facilitated Diffusion 388

23.6 Endoplasmic Reticulum-Mediated Transport of Calcium 388
23.7 Vesicular Transport 389
23.8 Calcium Homeostasis 389
23.9 Vitamin D 389
23.10 Other Factors Affecting Calcium Absorption 390
23.11 Studies from Other Animal Models 391
Summary Points 391
Key Facts of Calbindin 392
Definition of Words and Terms 392
List of Abbreviations 393
References 393

Chapter 24 The Calcium-Sensing Receptor in Intestinal Cells 396
Tohru Hira, Shingo Nakajima, and Hiroshi Hara

24.1 Introduction 396
24.2 Multiple Agonists for CaSR 397
24.3 CaSR in the Gastrointestinal Tract 398
24.3.1 Gut-Hormone Secretion 398
24.3.2 Gastrin and Gastric-Acid Secretion 398
24.3.3 CCK Secretion 399
24.3.4 GLP-1 Secretion 401
24.4 Colonic-Fluid Secretion 403
24.5 Epithelial Proliferation and Differentiation 403
24.6 Future Perspectives 404
Summary Points 406
Key Facts 406
CaSR 406
Enteroendocrine System 407
Nutrient-Sensing Receptors 407
Definitions of Words and Terms 407
List of Abbreviations 408
References 409

Chapter 25 Taste Cells and Calcium Signaling 413
Kathryn F. Medler

25.1 Introduction 413
25.2 Taste Cell Signaling 416
25.2.1 Ionic Stimuli 417
25.2.2 Chemically Complex Stimuli 419
25.3 Calcium Regulation in Taste Cells 421
25.4 Effects of Obesity on Taste Cell Signaling 422
25.5 Conclusions 425
Summary Points 425

Key Facts of Taste 425
Definitions of Words and Terms 426
List of Abbreviations 427
References 427

Chapter 26 Intracellular Calcium Modulation of Gene Expression 431
Mariana Casas and Enrique Jaimovich

26.1 Introduction 431
26.2 Muscle-Cell Adaptations to Training 432
26.3 Gene Transcription Regulation by Calcium in Skeletal Muscle 433
26.4 What We Do Know about the Mechanism from Work in Myotubes and Adult Muscle Fibers 434
26.5 Fiber-Type Diversity 435
26.5.1 Ca^{2+}-Dependent Kinases and Phosphatases Involved in Muscle Phenotype Specification 437
26.5.2 Ras-MAPK Pathway 440
26.5.3 IP_3-Dependent Ca^{2+} Signals and Muscle Plasticity 441
Summary Points 441
Key Facts for Muscle Calcium Signals and Gene Expression 442
Definitions of Words and Terms 443
References 444

Chapter 27 Mitochondrial Calcium Homeostasis and Implications for Human Health 448
Tito Calì, Denis Ottolini, and Marisa Brini

27.1 Introductory Elements of Calcium Signaling 448
27.2 Mitochondrial Calcium Handling 450
27.2.1 The Molecular Identification of the Players 452
27.3 Mitochondrial Calcium in Human Health 457
Summary Points 460
Key Facts 461
Key Facts of Calciums 461
Key Facts of Mitochondrial Ca^{2+} History 461
Key Facts of the Molecular Identification of the Mitochondrial Ca^{2+} Players 462
Key Facts of ATP 462
Key Facts of Apoptosis 462
Key Facts of Autophagy 463
Definitions of Words and Terms 463

List of Abbreviations 464
Acknowledgements 465
References 465

Chapter 28 Pregnancy and Calcium 468
Belal Alshaikh and Kamran Yusuf

28.1 Introduction 468
28.2 Maternal Adaptations during Pregnancy 469
28.2.1 Intestinal Calcium Absorption 469
28.2.2 Renal Excretion of Calcium 469
28.2.3 Skeletal Calcium Metabolism 470
28.3 Calcium Transfer from Mother to Fetus 470
28.4 Calcitropic Hormones – Changes during Pregnancy 472
28.4.1 Parathyroid Hormone (PTH) 472
28.4.2 Parathyroid Hormone-Related Protein (PTHrP) 472
28.4.3 1,25-Dihydroxyvitamin D_3 (Calcitriol) 472
28.4.4 Calcitonin 473
28.5 Maternal Calcium Metabolism and Fetal Health 473
28.6 Maternal Calcium Metabolism and Infant Growth 475
28.7 Impact of Pregnancy on Disorders of Calcium Metabolism 475
28.7.1 Pre-eclampsia 475
28.7.2 Gestational Diabetes 476
28.7.3 Hyperparathyroidism 476
28.7.4 Hypoparathyroidism 476
28.7.5 Pseudohypoparathyroidism 477
Summary Points 477
Key Features of Pregnancy and Calcium 478
Definitions of Key Terms 478
List of Abbreviations 480
References 480

Chapter 29 Calcium Supplementation during Pregnancy and Lactation: Implications for Maternal and Infant Bone Health 484
Flávia Fioruci Bezerra and Carmen Marino Donangelo

29.1 Introduction 484
29.2 Calcium Supplementation and Maternal Bone Outcomes 485
29.2.1 Effects of Calcium Supplementation during Pregnancy 486
29.2.2 Effects of Calcium Supplementation during Lactation 491

29.3 Calcium Supplementation and Fetal/Infant Bone Growth 495
29.3.1 Fetal Bone Growth 496
29.3.2 Evaluation of Fetal and Infant Bone Outcomes 496
29.3.3 Effects of Maternal Calcium Supplementation during Pregnancy on Fetal and Infant Bone Growth 500
29.3.4 Effects of Maternal Calcium Supplementation on Breast-Milk Calcium Concentration 502
29.4 Conclusions 503
Summary Points 503
Key Facts of Bone 504
Definitions of Key Terms 504
List of Abbreviations 505
References 505

Chapter 30 The Effects of Dietary Calcium on Hypertension 509
Filippo Rossi

30.1 Introduction 509
30.2 Does Calcium Really Affect Blood Pressure? 510
30.3 Ca in the Regulation of the Renin–Angiotensin System 513
30.4 Conclusion 514
Summary Points 515
Key Facts 515
Definition and Explanation of Key Terms 516
List of Abbreviations 517
References 517

Chapter 31 Interaction of Dietary Calcium with Genes of Transporters, Receptors and Enzymes Involved in Cholesterol Metabolism 519
Xiaobo Wang, Lin Lei, Yuwei Liu, Ka Ying Ma, Jingnan Chen, Rui Jiao, Yu Huang, and Zhen-Yu Chen

31.1 Introduction 519
31.2 Cholesterol Absorption 520
31.3 Cholesterol in Circulation 521
31.4 Cholesterol Homeostasis 523
31.5 Effect of Dietary Calcium on Plasma Lipoprotein in Ovariectomized Hamsters 523
31.6 Interaction of Dietary Calcium with Genes of Transporters, Receptors and Enzymes Involved in Cholesterol Metabolism 525
31.7 Conclusion 526

Summary Points 527
Key Facts of Cholesterol Absorption and Elimination 527
Definition of Words and Terms 527
List of Abbreviations 528
Acknowledgements 528
References 528

Chapter 32 Bone Health: The Independent and Combined Effects of Calcium, Vitamin D and Exercise in Children and Adolescents 530
Dimitris Vlachopoulos, Luis Gracia-Marco, Alan R. Barker, Inge Huybrechts, Luis A. Moreno, and Theodora Mouratidou

32.1 Introduction 530
32.2 Bone Mass Increase during Childhood and Adolescence 532
32.3 Calcium, Vitamin D and Bone Status in Children and Adolescents 533
32.4 Impact of Physical Activity on Bone Status during Growth 535
32.4.1 The Role of Physical Activity on Bone Mass 535
32.4.2 Weight-Bearing Exercise and Bone Mass 536
32.5 Associations of Calcium, Vitamin D and Exercise with Bone Mass of Children and Adolescents 538
32.6 Conclusions 540
Summary Points 540
Key Facts of Bone Health during Growth 541
Definitions and Explanations of Key Terms 541
List of Abbreviations 542
References 542

Chapter 33 Dietary Calcium and Osteoprotegerin 547
Dianjun Sun, Jun Yu, Yanhui Gao, and Yuanyuan Li

33.1 Osteoprotegerin in Health: A Brief Overview 547
33.1.1 Molecular Structure 548
33.1.2 Expression and Regulation 548
33.1.3 Function 549
33.2 Calcium Intake and Osteoprotegerin 550
33.3 Calcium and Osteoprotegerin and Bone Diseases 550
33.3.1 Calcium and Bone Diseases 550
33.3.2 Osteoprotegerin and Bone Diseases 552
33.4 Osteoprotegerin and Diabetes 554
33.5 Osteoprotegerin and Cardiovascular Diseases 555
33.6 Clinical Application of Osteoprotegerin 557
33.6.1 Osteoprotegerin and Hypercalcemia 557

33.6.2 Osteoprotegerin and Bone-Tissue Engineering 558
Summary Points 558
Key Facts 559
Definitions of Words and Terms 559
List of Abbreviations 559
References 560

Chapter 34 Inadequate Calcium Intake and Body Fat in Adults 565
Diane K. Tidwell and Matthew J. McAllister

34.1 Introduction 565
34.2 Randomized Clinical Trials Investigating Calcium and Body Fat in Adults 566
34.3 Observational Studies Investigating Calcium and Body Fat in Adults 568
34.4 The Biological Plausibility of Calcium's Effect on Body Fat 570
34.5 Summary 575
Summary Points 576
Key Facts 576
Definitions and Explanations of Key Terms 577
List of Abbreviations 577
References 578

Chapter 35 Calcium in Critical Care 582
Dima Youssef and Karin Amrein

35.1 Introduction 582
35.1.1 Recommended Intake 584
35.1.2 Sources of Calcium 584
35.1.3 Calcium Supplements 584
35.1.4 Calcium Absorption 585
35.1.5 Side Effects of Calcium Supplements 585
35.1.6 Special Situations 585
35.2 Hypocalcemia 585
35.2.1 Causes of Hypocalcemia 586
35.2.2 Clinical Manifestations of Hypocalcemia 589
35.2.3 Diagnostic Approach to Hypocalcemia 590
35.2.4 Treatment of Hypocalcemia 592
35.3 Hypercalcemia 593
35.3.1 Causes of Hypercalcemia 593
35.3.2 Clinical Manifestations of Hypercalcemia 596
35.3.3 Diagnostic Approach to Hypercalcemia 597
35.3.4 Treatment of Hypercalcemia 598

Summary Points 600
Key Facts for Calcium in Critical Care 601
Definitions of Words and Terms 601
List of Abbreviations 602
References 602

Chapter 36 Hypercalcemia: An Overview of its Pathology 605
Franco Lumachi and Stefano M. M. Basso

36.1 Calcium Distribution in the Human Body 605
36.2 Regulatory Mechanisms of Calcium Metabolism 607
36.3 Evaluation of Hypercalcemia 610
36.4 Primary Hyperparathyroidism and Other Benign Diseases 611
36.5 Malignancy-Associated Hypercalcemia 614
36.6 Conclusions 617
Summary Points 617
Key Facts 618
Key Facts of Calcium Functions and Control of Ionized Calcium 618
Key Facts of Parathyroid Hormone Actions 618
Key Facts of Benign Hypercalcemia 619
Malignancy-Associated Hypercalcemia 619
Definitions and Explanations of Key Terms 620
List of Abbreviations 621
References 622

Subject Index 625

Section I
Calcium in Context

CHAPTER 1

Calcium in the Context of Dietary Sources and Metabolism

MACIEJ S. BUCHOWSKI*[a]

[a]Division of Gastroenterology, Hepatology and Nutrition, Department of Medicine, Vanderbilt University, 1161 21st Avenue South, A4103 MCN, Nashville, Tennessee 37232, USA
*E-mail: maciej.buchowski@vanderbilt.edu

1.1 Overview of Calcium and Its Physiological Functions

Calcium is a divalent cation with an atomic weight of 40, one of the most abandoned elements in the Earth's biosphere and it is present in both solid matter and in aqueous solutions. A solid, calcium carbonate, occurs in marble, chalk, limestone, and calcite, calcium sulfate in anhydrite and gypsum, calcium fluoride in fluorspar or fluorite and calcium phosphate occurs in apatite. Calcium also occurs in numerous silicates and aluminosilicates. Many organisms concentrate calcium compounds in their shells or skeletons. For example, calcium carbonate is formed in the shells of oysters and in the skeletons of coral, which are often used as a calcium source in dietary supplements. In soil, calcium usually is present as a cation in colloids. In plants, calcium is present in the leaves, stems, roots, and seeds in

Food and Nutritional Components in Focus No. 10
Calcium: Chemistry, Analysis, Function and Effects
Edited by Victor R. Preedy

Published by the Royal Society of Chemistry, www.rsc.org

concentration ranging from 0.1% to almost 10%. In living cells, calcium is one of 21 elements occurring as mineral elements in biosphere and is essential for conducting cell functions. In mammals, calcium is present in all cells and accounts for up to 4% of total body weight.

In humans, it ranks fifth after oxygen, carbon, hydrogen, and nitrogen and it makes up 1.9% of the body by weight. Approximately 99% of calcium is contained in bones and teeth as calcium hydroxyapatite ($Ca_{10}[PO_4]_6[OH]_2$) and the remainder is inside the cells (0.9%) and extracellular fluid (0.1%). In the bone and teeth, calcium constitutes 25% of the dry weight and 40% of the ash weight. The extracellular fluid contains ionized calcium at concentrations of about 4.8 mg/100 mL (1.20 mmol L^{-1}) maintained by the parathyroid–vitamin D axis as well as complexed calcium at concentrations of about 1.6 mg/100 mL (0.4 mmol L^{-1}). The plasma contains a protein-bound calcium fraction at a concentration of 3.2 mg/100 mL (0.8 mmol L^{-1}). In the cellular compartment, the total calcium concentration is lower than in extracellular fluid by several orders of magnitude (Robertson and Marshall, 1981).

In bone and teeth, the most calcified structure in the animal and human body, the role of calcium is structural and mechanical, determining their hardness and strength (Abrams, 2011; Hill *et al.*, 2013). The second most calcified structure is the vasculature. Once considered a passive process, vascular calcification has emerged as an actively regulated form of tissue biomineralization, in which skeletal morphogens and osteochondrogenic transcription factors are expressed by cells within the vessel wall, regulating the deposition of vascular calcium (Bithika and Dwight, 2012). Another physiological role of calcium is to act as an activator for several key cellular enzymes such as pancreatic lipase, acid phosphatase, cholinesterase, ATPase, and succinic dehydrogenase (Nicholls, 2002; Brownlee *et al.*, 2010; Hung *et al.*, 2010; Glancy and Balaban, 2012; Tarasov *et al.*, 2012). Through its role in enzyme activation, calcium stimulates muscle contraction (*i.e.* promotes muscle tone and normal heartbeat) and regulates the transmission of nerve impulses from one cell to another through its control over acetylcholine production (Harnett and Biancani, 2003). Calcium is also essential for the normal clotting of blood, by stimulating the release of thromboplastin from the blood platelets (Østerud, 2010; Diamond, 2013). In conjunction with phospholipids, calcium plays a key role in the regulation of the permeability of cell membranes and consequently over the uptake of other nutrients by the cell (Brenner and Moulin, 2012; Kiselyov *et al.*, 2012). On a molecular level, calcium is an important second messenger participating in many activities. For example, when physicochemical insults deregulate calcium delicate homeostasis, it acts as an intrinsic stressor producing or increasing cell damage (Cerella *et al.*, 2010).

1.1.1 Sources of Calcium in the Diet

Dietary calcium comes from food sources associated with dairy products, other foods such as vegetables and cereals, foods fortified with inorganic or organic calcium, and from dietary supplements containing calcium.

Dairy foods are excellent sources of calcium and a major supplier of dietary calcium in the developed and majority of less developed countries (Table 1.1). For example, more than 40% of dietary calcium in North American and British diet come from milk, cheese, and yogurt and from foods to which dairy products have been added such as pizza, lasagna, and dairy desserts (Annonymous, 2011a).

A major nondairy source of calcium is green vegetables such as kale, turnip greens, bok choy, and Chinese cabbage, which provide approximately ~7% of dietary calcium (Table 1.2).

Other nondairy sources of calcium are grains, legumes, fruits, meat, poultry, fish (Tables 1.3 and 1.4), and eggs each providing 1% to 5% of calcium in a typical Western-style diet (Annonymous, 2009).

Other excellent sources of calcium are nuts (Table 1.5) and spices (Table 1.6).

In African diet, calcium-rich foods include crabs, edible caterpillars, locust beans, millet, and cowpea, baobab, and amaranth leaves (Annonymous, 2000).

Table 1.1 Calcium content of selected dairy food sources commonly present in diet. This table contains contents of calcium in dairy foods and eggs. The data is extracted from the USDA Nutrient Data Laboratory: http://fnic.nal.usda.gov/food-composition/usda-nutrient-data-laboratory.

Food source	Calcium (mg/100 g)
Milk, nonfat (with added vitamin A and D)	170
Milk, reduced-fat (2% milk fat)	140
Milk, whole (3.25% milk fat)	120
Yogurt, plain, low fat (with or without added fruit)	120–190
Milk, dry, nonfat, regular, without added vitamin A and D	1250
Cottage cheese, nonfat, 1% or 2% milk fat	60–85
Ice creams, milk based, all natural	130–170
Cheese (*e.g.* cheddar, Swiss, muenster, provolone, mozzarella)	400–800
Egg, whole; raw or cooked	56–62

Table 1.2 Calcium content of selected vegetables and fruits commonly present in diet. This table contains contents of calcium in vegetables and fruits. The data is extracted from the USDA Nutrient Data Laboratory: http://fnic.nal.usda.gov/food-composition/usda-nutrient-data-laboratory.

Food source	Calcium (mg/100 g)
Turnip greens, kale, collards; cooked or boiled and drained	130–150
Kale, arugula, turnip greens, soybeans, beet green; raw	120–200
Okra, Chinese cabbage, broccoli, spinach; raw	80–100
Carrots, tomatoes, brussels sprouts, endive, squash; raw	30–40
Onions, asparagus, lima beans, green peas; raw	15–30
Tangerines, blackcurrants, oranges; raw	40–60
Blackberries, kiwifruit, grapes; raw	27–37
Papaya, grapefruit, gooseberries; raw	20–25
Apples, pears, apricots, peaches; raw	4–8

Table 1.3 Calcium content of fish, seafood, and meats commonly present in diet. This table contains contents of calcium in selected fish, seafood, and meat. The data is extracted from the USDA Nutrient Data Laboratory: http://fnic.nal.usda.gov/food-composition/usda-nutrient-data-laboratory.

Food source	Calcium (mg/100 g)
Fish, salmon, pink, canned, drained solids	283
Mollusks, oyster, eastern, wild, cooked, moist heat	140
Fish: trout, herring, pike; cooked	50–90
Fish: grouper, ocean perch, tuna, skipjack; cooked	27–44
Chicken, turkey: all meats broiled cooked or roasted	10–30
Pork: loin, other cuts, separable lean and fat, cooked or broiled	30–70
Beef: tenderloin, steak, separable lean and fat, trimmed to 1/8″ fat, all grades, cooked, broiled	10–20
Veal, lamb: all grades, cooked, broiled	20–30
Game: all grades, cooked, broiled	10–15

Table 1.4 Calcium content of bread, bakery, cereal, and pasta products commonly present in diet. This table contains contents of calcium in selected bread, bakery, and pasta products and cereal. The data is extracted from the USDA Nutrient Data Laboratory: http://fnic.nal.usda.gov/food-composition/usda-nutrient-data-laboratory.

Food source	Calcium (mg/100 g)
Muffins, English muffins, mixed grains toasted	30–250
Bread, mixed grains, commercially prepared	50–600
Crackers, standard snack type	80–250
Pancakes, plain, standard commercial dry mix	70–220
Tortillas, ready-to-bake or -fry, corn	50–175
Cereals, ready-to-eat, single or mixed grains, regular	30–80
Pasta, noodles, macaroni, spaghetti (dry)	5–40

Table 1.5 Calcium content of nuts and seeds commonly present in diet. This table contains contents of calcium in selected nuts and seeds. The data is extracted from the USDA Nutrient Data Laboratory: http://fnic.nal.usda.gov/food-composition/usda-nutrient-data-laboratory.

Food source	Calcium (mg/100 g)
Almonds	260–340
Lotus, sesame seeds	140–160
Hazelnuts, filberts	117
Pumpkin, butternuts, sunflower seed kernels	52–87
Chestnuts raw, unpeeled	12–27

In several developed countries, an important source of dietary calcium are foods fortified with calcium, which do not naturally contain calcium such as orange juice, other beverages, soy milk and tofu and ready-to-eat cereals (Table 1.7) (Calvo *et al.*, 2004; Rafferty *et al.*, 2007; Poliquin *et al.*, 2009).

In recent decades, dietary supplements became an important source of dietary calcium. The use of vitamin and mineral supplements that include

Table 1.6 Calcium content of spices commonly present in diet. This table contains contents of calcium in selected dried and fresh spices. The data is extracted from the USDA Nutrient Data Laboratory: http://fnic.nal.usda.gov/food-composition/usda-nutrient-data-laboratory.

Food source	Calcium (mg/100 g)
Savory, dried	2100
Marjoram, dried	1990
Thyme, dried	1890
Tarragon, dried	1139
Cinnamon, dried	1002
Curry powder, dried	525
Caraway seed, dried	689
Dill weed, fresh	252
Bay leaf, dried	834

Table 1.7 Calcium content of selected fortified foods. This table contains contents of calcium in selected foods fortified with calcium either during the technological process (tofu) or to increase amount of calcium (orange juice, infant formula, ready-to-eat cereal, flour). The data is extracted from the USDA Nutrient Data Laboratory: http://fnic.nal.usda.gov/food-composition/usda-nutrient-data-laboratory.

Food source	Calcium (mg/100 g)
Tofu, raw, firm, prepared with calcium sulfate	680
Orange juice, chilled, includes from concentrate, fortified with calcium and vitamin D	140
Infant formula	125–432
Cereals, ready-to-eat, single or mixed grains, fortified	150–450
Corn or wheat flour, white, all-purpose, enriched, fortified	130–250

calcium becomes commonplace in several populations, especially in the developed countries. For example, among the United States (US) population based on a national survey, about 40% of adults, but almost 70% of older women reported calcium intake from supplements (Bailey *et al.*, 2010). Current estimates from the National Health and Nutrition Survey (NHANES) showed that in the US adult population between 2007 and 2010, dietary supplements users have approximately 10% higher calcium intake than nonusers (Wallace *et al.*, 2014).

The most common forms of supplemental calcium are calcium carbonate and calcium citrate. Generally, less calcium carbonate is required to achieve a given dose of elemental calcium because calcium carbonate provides 40% of elemental calcium, compared with 20% for calcium citrate. However, compared with calcium citrate, calcium carbonate is more often associated with gastrointestinal side effects, including constipation, flatulence, and bloating (Straub, 2007). In contrast, calcium citrate is less dependent than calcium carbonate on stomach acid for absorption (Recker, 1985) and thus, can be taken without food. Other forms of calcium in dietary supplements include calcium lactate, gluconate, glucoheptonate, and hydroxyapatite and their relevance for life stage groups may vary. The health benefits of calcium supplements are still debatable. For example, in a 2013 update, the *US Preventive Services Task Force* concludes that the current evidence is insufficient to assess the balance of the benefits and harms of combined vitamin D and calcium supplementation for the primary prevention of fractures in premenopausal women or in men (Moyer, 2013). Some recent studies have raised concern about an increased cardiovascular risk with the use of calcium supplements, but the findings are considered inconsistent and inconclusive (Xiao *et al.*, 2013).

1.1.2 Calcium Absorption and Excretion

Calcium is absorbed in the intestine by passive diffusion (paracellular) or by active transport (transcellular) across the intestinal mucosa (Bronner, 2009). The rate of paracellular calcium uptake is considered nonsaturable, while transcellular transport can be upregulated under conditions of dietary calcium constraints. The paracellular route is tied to a downhill concentration gradient between the luminal and the extracellular compartments throughout the entire intestine. Although canonically thought to be constant, the recent evidence suggests that paracellular calcium transport is regulated, at least in part, by $1,25(OH)_2$ vitamin D (Christakos, 2012).

Transcellular calcium absorption can also take place against an uphill gradient, but requires molecular machinery in the form of distinct calcium transport proteins and energy from hydrolyzable adenosine triphosphate (ATP) (Auchère *et al.*, 1998). The absorption occurs mostly in the duodenum and the jejunum (Pansu *et al.*, 1983) and the process is activated by calcitriol and is dependent on the intestinal vitamin D receptor (VDR) and physiologic factors such as the presence of calcium-regulating hormones and the life stage (Whiting, 2010; Gallagher, 2013). Since a concentration gradient is not a prerequisite for this process, transcellular transport accounts for most of the absorption of calcium at low and moderate intake levels (Table 1.8).

The solubility of calcium salts is increased in the acid environment of the stomach, but the dissolved calcium ions to some extent reassociate and precipitate in the jejunum and ileum where the pH is closer to neutral. Recent observations indicate that a reduction of gastric acidity may impair effective calcium uptake throughout the entire intestine (Kopic and Geibel, 2013). In the neutral environment, the absorbability of calcium is determined mainly by the presence of other food components such as lactose, glucose, fatty acids, phosphorus, and oxalate, which can bind to soluble calcium, are released resulting in complex luminal interactions. For example, absorption of calcium supplements, and especially those that are less soluble, is substantially better if they are taken with a meal perhaps by food-stimulated gastric secretion and delayed emptying allowing dispersion and dissolution

Table 1.8 Physiological and dietary factors affecting calcium absorption. This table lists the key factors that affect calcium absorption in humans and lists calcium roles in living cells, and describes calcium physiological role in the human body. Calcium absorption is a complex process influenced by several physiological and dietary factors either promoting or inhibiting this process.

Factors promoting absorption	Factors interfering with absorption
Physiological	
Growth-promoting hormones	Lack of stomach acid
Vitamin D_3	Diminishing absorption with aging
Optimal gastric acidity	Vitamin D deficiency
Dietary	
Lactose	High-fiber diet
Ingestion with a meal	Oxalates and phosphates
Phosphorus in optimal ratio with calcium	High protein intake

of calcium. In the gastrointestinal lumen, calcium can compete or interfere with the absorption of other minerals such as iron, zinc, and magnesium.

Calcium is excreted in urine, feces, and body tissues and fluids, such as sweat. Calcium excretion in the urine is a function of the balance between the calcium load filtered by the kidneys and the efficiency of reabsorption from the renal tubules. Most of the calcium (~98%) is reabsorbed by either passive or active processes occurring at four sites in the kidney, each contributing to maintaining neutral calcium balance. The majority of the filtered calcium (~70%) is reabsorbed passively in the proximal tubule and the remaining 30% actively in the ascending loop of Henle, the distal tubule, and collecting duct (Allen and Woods, 1994).

1.1.3 Calcium Homeostasis and Systemic Balance

Regulation of calcium homeostasis during a lifetime is a complex process reflecting a balance among intestinal calcium absorption, bone calcium influx and efflux, and renal calcium excretion. Maintaining the level of circulating ionized calcium within a narrow physiological range between 8.5 and 10.5 mg dL^{-1} (2.12 and 2.62 mmol L^{-1}) is critical for normal body function (Jeon, 2008). Homeostasis of serum calcium level is maintained through an endocrine system comprised of controlling factors, epithelial calcium channels, and feedback mechanisms that includes vitamin D metabolites, primarily calcitriol, and parathyroid hormone (PTH) (Peacock, 2010). Any perturbations in calcium homeostasis can result in hypocalcemia or hypercalcemia and adaptations in calcium handling must occur during a lifetime that include growth and aging (Felsenfeld *et al.*, 2013).

Calcium systemic balance is essential for a multitude of physiological processes, ranging from cell signaling to maintenance of bone health. A systemic calcium balance (positive, neutral, or negative) is the measure derived by taking the difference between the total intake and the sum of the urinary, fecal, and sweat calcium excretion. These measures have some limitations and are generally cross-sectional in nature, and their precision differs. Long-term balance studies for calcium are rarely carried out because of the difficult study protocol. Calcium balance can also be estimated by using stable isotopes to trace the amount of calcium absorbed from a single feeding. In general, a positive calcium balance is indicative of calcium accretion also termed net calcium retention, neutral balance suggests maintenance of bone, and a negative balance indicates bone loss.

The relevance of the calcium balance state varies depending upon the developmental stage. Infancy through late adolescence periods are characterized by positive calcium balance due to enhanced bone formation. In female adolescents and adults, even within the normal menstrual cycle, there are measurable fluctuations in calcium balance owing to the effects of fluctuating sex steroid levels and other factors on the basal rates of bone formation and resorption. Later in life, menopause and age-related bone loss lead to a net loss of calcium due to enhanced bone resorption.

In an average adult human, daily calcium intake is approximately 800–1000 mg per day (20–25 mmol). From this amount, about 25–50% is absorbed and passes into the exchangeable calcium pool (Figure 1.1). This pool consists of the small amount of calcium in the blood, lymph,

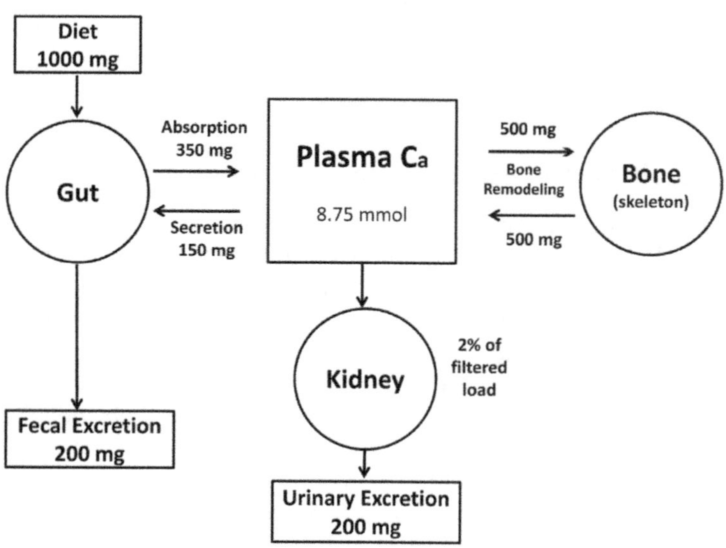

Figure 1.1 Calcium metabolism in an adult human under neutral balance (input = output). Outline of calcium absorption (mg per day) from diet by the gut, reabsorption by the kidney, turnover in bone, and urinary and fecal excretion.

and other body fluids, and accounts for 1% of the total body calcium. Calcium located in bones and teeth (99%) is inaccessible to most physiological processes. Approximately 150 mg per day (3.75 mmol) of calcium enter the intestinal lumen in intestinal secretions such as digestive enzymes and bile, but about 30% of this calcium is reabsorbed (Allen and Woods, 1994). The kidneys filter about 8.6 g per day (215 mmol) of calcium, almost all (~98%) of which is reabsorbed so that only 100 to 200 mg per day (2.5 to 5 mmol) is excreted in approximately equal amounts in the urine and stool. Calcium loss from the skin is about 15 mg per day (0.4 mmol) depending on sweating. In the adult human, the extracellular calcium pool turns over approximately 20 to 30 times per day, while the bone pool turns over every 5 to 6 years.

1.1.4 Calcium Bioavailability and Dietary Factors Affecting Calcium Absorption

Calcium availability from diet varies with form of calcium ingested. In general, bioavailability is increased when calcium is well solubilized and inhibited in the presence of agents that bind calcium or form insoluble calcium salts. The absorption of calcium is about 30% from dairy and fortified foods (*e.g.*, orange juice, tofu, and soymilk) and nearly twice as high from certain leafy green vegetables and calcium supplements (Table 1.9).

Dietary fiber has an adverse effect on calcium absorption in humans and can impair the calcium balance significantly. The majority of marked adverse effects of dietary fiber could be explained by the calcium-binding capacity of phytic acid. However, other constituents of dietary fiber also have the ability to bind calcium. For example, uronic acids present in hemicellulose can bind calcium strongly and may explain the inhibition of dietary fiber calcium absorption from cellulose-containing foods. Pectin present in dietary fiber especially in fruits and vegetables do not affect calcium absorption most likely because 80% of uronic acids in pectin are methylated and cannot bind calcium (Allen and Woods, 1994).

Table 1.9 Foods ranked according to absorbability of calcium. This table lists the key facts about foods in relation to their calcium absorbability divided into four categories.

Absorbability of calcium	Food
Excellent >50%	Kale, broccoli, turnip greens, brussels sprouts, rutabaga, mustard greens, bok choy, cauliflower, watercress
Good ~30%	Milk, dairy products, yogurt, soy milk, calcium-set soy tofu, soy isolates
Fair ~20%	Almonds, sesame seeds, pinto beans, sweet potatoes, nuts
Poor ~5%	Spinach, rhubarb, collard greens

Other food compounds such as oxalic acid that also bind calcium could significantly interfere with absorption and decrease calcium bioavailability. The poor absorption of calcium from spinach (5% compared to 27% absorption from milk) has been attributed to its binding to oxalic acid in spinach. However, other factors may also be involved because calcium absorption from calcium oxalate is twice as high as that from that the plant. It has been documented that during the absorption of calcium oxalate there is no tracer exchange, suggesting that absorption occurs without dissociation of the molecule (Heaney and Weaver, 1989). Similarly, solubility of salts such as citrate or citrate-malate is not related to their absorbability. Thus, the form in which the calcium approaches the mucosal brush border and in which it is transported by the paracellular mechanism may be a better predictor of its bioavailability than the solubility of the salt.

Some food constituents increase calcium bioavailability. For example, lactose enhances the absorption of calcium in animals and human infants. Lactose increases the diffusional component of calcium and perhaps of phosphorus, especially in the ileum, and probably acts osmotically to alter the junctions between the epithelial cells (Kobayashi *et al.*, 1975). Solubility of the dominant chemical form of calcium in specific foods, or of calcium supplements, has a negligible effect on calcium absorption after lactase treatment can be ascribed to the fact that most metabolizable sugars enhance calcium absorption. However, it is doubtful that lactose improves the absorbability of calcium from dairy products beyond infancy. For example, in adults the absorption of calcium from yogurt is the same as that from milk, even though the lactose in yogurt is hydrolyzed in the stomach by lactase originating from the bacteria in the yogurt (Smith *et al.*, 1985).

Dark green, leafy vegetables are often relatively high in calcium. Absorption from many of these is expected to be good, if they are low in oxalic acid (*e.g.*, kale, broccoli, turnip and mustard greens, and collard). For example, absorption from kale is as good as from milk. Other factors may also be involved because calcium absorption from calcium oxalate is twice as high as that from spinach (Heaney and Weaver, 1989). There is no interference of calcium oxalate with the absorption of calcium in milk when the two are consumed together.

Protein intake stimulates acid release in the stomach, and this, in turn, enhances calcium absorption. However, it has long been known that protein also increases urinary calcium excretion. The effect of protein on calcium retention and hence bone health has been controversial (Allen and Woods, 1994). Sodium and potassium in the diet may also affect the calcium balance. High intakes of sodium increase urinary calcium excretion. In contrast, adding more potassium to a high-sodium diet might help decrease calcium excretion, particularly in postmenopausal women (Sellmeyer *et al.*, 2002; Annonymous, 2011b). Phosphate in food is a mixture of inorganic and organic and similarly to calcium, the portion of phosphorus absorption is due to saturable, active transport facilitated by calcitriol. However, fractional phosphorus absorption is virtually constant across a broad range of intakes, suggesting that absorption occurs primarily by a passive, concentration-dependent process. Interestingly, several observational studies have suggested that the consumption of carbonated soft drinks with high levels of phosphate is associated with reduced bone mass and increased fracture risk. However, it is likely that the effect is due to replacing milk with soda, rather than to phosphorus itself (Heaney and Rafferty, 2001).

Alcohol intake can affect calcium nutriture by reducing calcium absorption although the amount of alcohol required to cause an effect and whether moderate alcohol consumption is helpful or harmful to bone are unknown (Hirsch and Peng, 1996). Caffeine from coffee and tea modestly increases calcium excretion and reduces absorption (Heaney and Recker, 1982). Other studies have indicated that caffeine intake from two to three cups of coffee per day might result in bone loss, but only in individuals with low milk or low total calcium intake (Harris and Dawson-Hughes, 1994).

However, the extent to which these food compounds affect calcium absorption varies, and food combinations affect overall absorption efficiency. For example, eating spinach with milk at the same time reduces the absorption of the calcium in the milk (Weaver and Heaney, 1991). In contrast, wheat products (with the exception of wheat bran) do not appear to have a negative impact on calcium absorption (Weaver *et al.*, 1991). Nevertheless, calcium from foods of plant origin is less bioavailable than calcium from foods of animal origin such as milk and dairy products (Weaver, 2009).

The calcium salts most commonly used for food fortification or as supplements exhibit similar absorbability when tested in pure chemical form (Rafferty *et al.*, 2007). In contrast, the absorbability of calcium from pharmaceutical preparations is usually lower than predictions from studies of pure salts (Weaver and Heaney, 2006). Calcium citrate appears to be better absorbed than calcium carbonate when they are taken with food (Harvey *et al.*, 1988). Other research suggests similar bioavailability of the forms of calcium carbonate and citrate (Heaney *et al.*, 1999). Another form of supplemental calcium, calcium formate, showed a better ability to deliver calcium to the bloodstream after oral administration than both calcium carbonate and calcium citrate (Hanzlik *et al.*, 2005).

Summary Points

- This chapter focuses on calcium in the context of dietary sources and providing bases of calcium metabolism in the human body.
- Calcium is an inorganic element essential to living cells present in the Earth's biosphere as a solid matter and aqueous solution.
- In humans, calcium is an essential constituent of bones and teeth, participates in vascular calcification, and is necessary for activation catalytic and mechanical properties of proteins in key enzymes.

- Dietary sources of calcium include dairy and nondairy foods, fortified foods, and supplements.
- Calcium is readily absorbed through the gastrointestinal tract (through vitamin D_3 action) and calcium absorption is facilitated by some food components by forming calcium complexes with some food components and by high gastric acidity through aiding solubilization of the calcium salts.
- The calcium balance is measured as the difference between calcium absorbed and excreted is essential for many physiological processes, ranging from cell signaling to maintenance of bone health.
- Regulation of calcium homeostasis is based on the interrelationship among intestinal calcium absorption, bone influx and efflux of calcium, and renal calcium excretion.

Key Facts

Key facts about calcium and its forms in biosphere, lists calcium roles in living cells, and describes calcium physiological role in the human body.

1. Calcium is a mineral widely abandoned in the biosphere and occurs mostly as calcium salts such as carbonate, sulfate, fluorite, and phosphate.
2. Calcium is present in all living cells including plant cells.
3. In humans, 99% of the body's calcium is stored in the bones and teeth where it supports their structure and functions.
4. The remaining 1% of supports critical metabolic functions such as vascular contraction and vasodilation, muscle function, nerve transmission, blood clotting, and intracellular signaling and hormonal secretion.

Key facts about calcium sources of dietary calcium. Key facts about major sources of calcium that include dairy and dairy products, nondairy foods such as vegetables, grains and soy foods, foods fortified with calcium such as juices and cereals, and calcium supplements such calcium carbonate and citrate.

1. The major source of calcium in the diet is dairy milk, milk products, and food sources associated with dairy products.
2. Major nondairy foods containing calcium include green leafy vegetables, grains, cereals, and legumes.
3. Other sources of calcium include foods fortified with inorganic or organic calcium such as fruit juices, beverages, and cereals.
4. Dietary supplements containing calcium are becoming important sources of dietary calcium especially in older adults.
5. Calcium carbonate and calcium citrate are major sources of calcium in dietary supplements.

Key facts about calcium absorption and excretion. Key facts about intestinal calcium absorption by two independent pathways and calcium excretion from the human body mainly in urine and stools.

1. Calcium absorption occurs across the intestinal mucosa *via* either passive nonsaturable diffusion (paracellular) or by active transport (transcellular) pathways.
2. Transcellular active transport takes place against an uphill gradient and requires calcium-transport proteins, energy from ATP. The process is activated by calcitriol, dependent on the intestinal vitamin D receptor (VDR) and the presence of calcium-regulating hormones.
3. The paracellular passive transport depends primarily on calcium quantity and availability in the diet.
4. The solubility of calcium salts is increased in the acid environment of the stomach and the intestine absorbs between 25 and 35% of the ingested calcium.
5. Calcium leaves the body mainly in urine and feces, but also in other body tissues and fluids, such as sweat.
6. Calcium excretion in the urine is a function of the balance between calcium load and reabsorption from the renal tubules (~98%).

Key facts about calcium homeostasis and systemic balance in the human body.

1. Regulation of calcium homeostasis during a lifetime is a complex process reflecting a balance among intestinal calcium absorption, bone calcium influx and efflux, and renal calcium excretion.
2. Homeostasis of serum calcium level is maintained through an endocrine system comprised of controlling factors, epithelial calcium channels, and feedback mechanisms that includes calcitriol and parathyroid hormone (PTH).
3. Exchangeable calcium pool accounts for ~1% of calcium in the human body and turnovers 20–30 times a day.
4. Systemic calcium balance (positive, neutral, or negative) is the measure derived by taking the difference between the total intake and the sum of the urinary, fecal, and sweat calcium excretion.

Key facts about calcium bioavailability from food sources and calcium salts and dietary factors affecting its absorption such as fiber, oxalic acid, lactose, and protein.

1. Calcium availability from diet varies with form of calcium ingested.
2. The absorption of calcium is about 30% from dairy and fortified foods (*e.g.*, orange juice, tofu, and soymilk) and nearly twice as high from certain leafy green vegetables and calcium supplements.

3. Dietary fiber has an adverse effect on calcium absorption in humans and can impair significantly calcium balance.
4. Other food compounds that bind calcium (*e.g.* oxalic acid) could significantly interfere with calcium absorption and decrease calcium bioavailability from some foods (*e.g.* spinach).
5. Lactose from milk increases calcium bioavailability especially in infants.
6. Other food components affecting calcium bioavailability include protein, sodium and potassium, phosphates, caffeine, and alcohol.
7. The calcium salts most commonly used for food fortification or as supplements exhibit similar absorbability.

Definitions of Words and Terms

Vitamin D_3. This is a fat-soluble vitamin that could be either provided with the diet or synthesized from 7-dehydrocholesterol with adequate sunlight exposure in humans.

Calcitriol. This is the hormonally active form of vitamin D, also termed 1,25-dihydroxyvitamin D_3, that increases the level of calcium (Ca^{2+}) in the blood by increasing the uptake of calcium from the gut into the blood. Calcitriol is used to treat and prevent low levels of calcium in the blood of patients whose kidneys or parathyroid glands are not working normally.

Parathyroid Hormone (PTH). This is an 84 amino acid peptide acting primarily in kidney. PTH major physiological functions to raise plasma calcium *via* bone resorption and renal calcium reabsorption and to stimulate the metabolism of vitamin D to its active hormonal form 1,25-dihydroxyvitamin D_3.

NHANES. The National Health and Nutrition Examination Survey (NHANES) is a program of studies designed to assess the health and nutritional status of adults and children in the United States since early 1960s. The survey is unique in that it combines interviews and physical examinations. Findings from NHANES are used to determine the prevalence of major diseases and risk factors for diseases.

Calcitonin. This is a hormone (32 amino acid peptide) produced by the thyroid gland under conditions of hypercalcemia that lowers the levels of calcium and phosphate in the blood and promotes bone formation. Calcitonin inhibits bone removal by the osteoclasts and at the same time promotes bone formation by the osteoblasts.

Bone remodeling. This is a process by which bone is renewed to maintain strength and mineral homeostasis. The bone remodeling unit is composed of a tightly coupled group of highly specialized osteoclasts and osteoblasts that sequentially carry out resorption of old bone and formation of new bone.

Osteoclasts. These are large bone-resorbing cells triggered by parathyroid hormone (PTH) in response to hypocalcemia. Osteoclasts are formed from the conjoining of several cells created by the bone marrow and travel in the circulatory system working in perfect synchronization with osteoblasts to maintain the skeletal system.

Osteoblasts. These are bone-forming connective tissue cells found at the bone surface that can be stimulated to proliferate and differentiate as osteocytes.

Osteocytes. These are cells enclosed in bone that manufacture type 1 collagen and other substances that make up the bone extracellular matrix.

References

Abrams, S. A., 2011. Calcium and vitamin D requirements for optimal bone mass during adolescence. *Current Opinion in Clinical Nutrition & Metabolic Care*. 14: 605–609.

Allen, L. and Woods, R., 1994. Calcium and phosphorus. In: Shills, M. E., Olson, J. and Shike, M. (ed.) *Modern Nutrition in Health and Disease*. Lea & Fibiger, Philadelphia, pp. 144–163.

Annonymous, 2000. Agriculture, food and nutrition for Africa: a resource book. In: Nations, Faaootu (ed.). FAO Corporate Document Repository.

Annonymous, 2009. *Nutrient Availability Data (2009)*. U.S. Department of Agriculture/Economic Research Service.

Annonymous, 2011a. *Dietary Reference Intakes for Calcium and Vitamin D*. National Academies Press Online, Washington, DC, 1132 pp.

Annonymous, 2011b. Dietary reference intakes for calcium and vitamin D. In: Ross, A. C., Taylor, C. L., Yaktine, A. L. and Valle, H. B. D. (ed.). The National Academies Press, Washington, DC.

Auchère, D., Tardivel, S., Gounelle, J., Drüeke, T. and Lacour, B., 1998. Role of transcellular pathway in ileal Ca^{2+} absorption: stimulation by low-Ca^{2+} diet. *American Journal of Physiology*. 275: G951–G956.

Bailey, R., Dodd, K., Goldman, J., Gahche, J., Dwyer, J., Moshfegh, A., Sempos, C. and Picciano, M., 2010. Estimation of total usual calcium and vitamin D intakes in the United States. *Journal of Nutrition*. 140: 817–822.

Bithika, T. and Dwight, A. T., 2012. Arterial calcification and bone physiology: role of the bone–vascular axis. *Nature Reviews Endocrinology*. 8: 529–543.

Brenner, C. and Moulin, M., 2012. Physiological roles of the permeability transition pore. *Circulation Research*. 111: 1237–1247.

Bronner, F., 2009. Recent developments in intestinal calcium absorption. *Nutrition Reviews*. 67: 109–113.

Brownlee, I. A., Forster, D. J., Wilcox, M. D., Dettmar, P. W., Seal, C. J. and Pearson, J. P., 2010. Physiological parameters governing the action of pancreatic lipase. *Nutrition Research Reviews*. 23: 146–154.

Calvo, M. S., Whiting, S. J. and Barton, C. N., 2004. Vitamin D fortification in the United States and Canada: current status and data needs. *American Journal of Clinical Nutrition*. 80: 1710S–1716S.

Cerella, C., Diederich, M. and Ghibelli, L., 2010. The dual role of calcium as messenger and stressor in cell damage, death, and survival. *International Journal of Cell Biology*. DOI: 10.1155/2010/546163.

Christakos, S., 2012. Recent advances in our understanding of 1,25-dihydroxyvitamin D_3 regulation of intestinal calcium absorption. *Archives of Biochemistry and Biophysics*. 523: 73–76.

Diamond, S. L., 2013. Systems biology of coagulation. *Journal of Thrombosis and Haemostasis*. 11: 224–232.

Felsenfeld, A., Rodriguez, M. and Levine, B., 2013. New insights in regulation of calcium homeostasis. *Current Opinion in Nephrology and Hypertension*. 22: 371–376.

Gallagher, J. C., 2013. Vitamin D and aging. *Endocrinology and Metabolism Clinics of North America*. 42: 319–332.

Glancy, B. and Balaban, R. S., 2012. Role of mitochondrial Ca^{2+} in the regulation of cellular energetics. *Biochemistry*. 51: 2959–2973.

Hanzlik, R. P., Fowler, S. C. and Fisher, D. H., 2005. Relative bioavailability of calcium from calcium formate, calcium citrate, and calcium carbonate. *Journal of Pharmacology and Experimental Therapeutics*. 313: 1217–1222.

Harnett, K. M. and Biancani, P., 2003. Calcium-dependent and calcium-independent contractions in smooth muscles. *American Journal of Medicine*. 115: 24–30.

Harris, S. S. and Dawson-Hughes, B., 1994. Caffeine and bone loss in healthy postmenopausal women. *American Journal of Clinical Nutrition*. 60: 573–578.

Harvey, J. A., Zobitz, M. M. and Pak, C. Y. C., 1988. Dose dependency of calcium absorption: a comparison of calcium carbonate and calcium citrate. *Journal of Bone and Mineral Research*. 3: 253–258.

Heaney, R. and Recker, R., 1982. Effects of nitrogen, phosphorus, and caffeine on calcium balance in women. *Journal of Laboratory and Clinical Medicine*. 99: 46–55.

Heaney, R. P., Dowell, M. S. and Barger-Lux, M. J., 1999. Absorption of calcium as the carbonate and citrate salts, with some observations on method. *Osteoporosis International*. 9: 19–23.

Heaney, R. P. and Rafferty, K., 2001. Carbonated beverages and urinary calcium excretion. *American Journal of Clinical Nutrition*. 74: 343–347.

Heaney, R. P. and Weaver, C. M., 1989. Oxalate: effect on calcium absorbability. *American Journal of Clinical Nutrition*. 50: 830–832.

Hill, T. R., Aspray, T. J. and Francis, R. M., 2013. Vitamin D and bone health outcomes in older age. *Proceedings of the Nutrition Society*. 72: 372–380.

Hirsch, P. and Peng, T., 1996. Effects of alcohol on calcium homeostasis and bone. In: Anderson, J. J. B. and Garner, S. C. (ed.) *Calcium and Phosphorus in Health and Disease*. CRC Press, Boca Raton, FL.

Hung, C. H.-L., Ho, Y.-S. and Chang, R. C.-C., 2010. Modulation of mitochondrial calcium as a pharmacological target for Alzheimer's disease. *Ageing Research Reviews*. 9: 447–456.

Jeon, U., 2008. Kidney and calcium homeostasis. *Electrolyte & Blood Pressure*. 6: 68–76.

Kiselyov, K. K., Ahuja, M., Rybalchenko, V., Patel, S. and Muallem, S., 2012. The intracellular Ca^{2+} channels of membrane traffic. *Channels*. 6: 344–351.

Kobayashi, A., Kawai, S., Obe, Y. and Nagashima, Y., 1975. Effects of dietary lactose and lactase preparation on the intestinal absorption of calcium and magnesium in normal infants. *American Journal of Clinical Nutrition*. 28: 681–683.

Kopic, S. and Geibel, J., 2013. Gastric acid, calcium absorption, and their impact on bone health. *Physiological Reviews*. 93: 189–268.

Moyer, V. A., 2013. Vitamin D and calcium supplementation to prevent fractures in adults: U.S. Preventive Services Task Force recommendation statement. *Annals of Internal Medicine*. 158: 691–696.

Nicholls, D. G., 2002. Mitochondrial function and dysfunction in the cell: its relevance to aging and aging-related disease. *International Journal of Biochemistry & Cell Biology*. 34: 1372–1381.

Østerud, B., 2010. Tissue factor expression in blood cells. *Thrombosis Research*. 125(suppl. 1): S31–S34.

Pansu, D., Bellaton, C., Roche, C. and Bronner, F., 1983. Duodenal and ileal calcium absorption in the rat and effects of vitamin D. 244: G695–G700.

Peacock, M., 2010. Calcium metabolism in health and disease. *Clinical Journal of the American Society of Nephrology*. 5: S23–S30.

Poliquin, S., Joseph, L. and Gray-Donald, K., 2009. Calcium and vitamin D intakes in an adult Canadian population. *Canadian Journal of Dietetic Practice and Research*. 70: 21–27.

Rafferty, K., Walters, G. and Heaney, R. P., 2007. Calcium fortificants: overview and strategies for improving calcium nutriture of the U.S. population. *Journal of Food Science*. 72: R152–R158.

Recker, R. R., 1985. Calcium absorption and achlorhydria. *New England Journal of Medicine*. 313: 70–73.

Robertson, W. and Marshall, R., 1981. Ionized calcium in body fluids. *Critical Reviews in Clinical Laboratory Sciences*. 15: 85–125.

Sellmeyer, D., Schloetter, M. and Sebastian, A., 2002. Potassium citrate prevents increased urine calcium excretion and bone resorption induced by a high sodium chloride diet. *Journal of Clinical Endocrinology and Metabolism*. 87: 2008–2012.

Smith, T. M., Kolars, J. C., Savaiano, D. A. and Levitt, M. D., 1985. Absorption of calcium from milk and yogurt. *American Journal of Clinical Nutrition*. 42: 1197–1200.

Straub, D. A., 2007. Calcium supplementation in clinical practice: a review of forms, doses, and indications. *Nutrition in Clinical Practice*. 22: 286–296.

Tarasov, A. I., Griffiths, E. J. and Rutter, G. A., 2012. Regulation of ATP production by mitochondrial Ca^{2+}. *Cell Calcium*. 52: 28–35.

Wallace, T. C., McBurney, M. and Fulgoni, V. L., 2014. Multivitamin/mineral supplement contribution to micronutrient intakes in the United States, 2007–2010. *Journal of the American College of Nutrition*. 33: 94–102.

Weaver, C. and Heaney, R. P., 2006. Calcium. In: Shils, M., Shike, M., Ross, A., Caballero, B. and Cousins, R. (ed.) *Modern Nutrition in Health and Disease*. Lippincott Williams & Wilkins, Baltimore, MD.

Weaver, C. M., 2009. Should dairy be recommended as part of a healthy vegetarian diet? Point. *American Journal of Clinical Nutrition*. 89: 1634S–1637S.

Whiting, S. J., 2010. Calcium: a nutrient deserving a special issue. *Nutrients*. 2: 1044–1047.

Xiao, Q., Murphy, R. A., Houston, D. K., Harris, T. B., Chow, W. and Park, Y., 2013. Dietary and supplemental calcium intake and cardiovascular disease mortality: the national institutes of health–AARP diet and health study. *JAMA Internal Medicine*. 173: 639–646.

CHAPTER 2

The Biological Roles of Calcium: Nutrition, Diseases and Analysis

LEONARDO M. MOREIRA*[a], RAPHAEL P. ARAUJO[a], FERNANDO P. LEONEL[a], HENRIQUE V. N. MACHADO[a], ALEXANDRE O. TEIXEIRA[a], FABIO V. SANTOS[b], VANESSA J. S. V. SANTOS[b], AND JULIANA P. LYON[c]

[a]Departamento de Zootecnia, Universidade Federal de São João Del Rei, Av. Visconde do Rio Preto s/n, Fábricas, São João Del Rei, Minas Gerais, Brazil; [b]Universidade Federal de São João Del Rei, Campus Centro Oeste D. Lindu, Rua Sebastião Gonçalves Coelho, 400, Chanadour, Divinópolis, Minas Gerais, Brazil; [c]Universidade Federal de São João Del Rei, Departamento de Ciências Naturais, Campus Dom Bosco, Praça Dom Helvécio, 74, Fábricas, Sala C 2.16 bloco C, São João Del Rei, Minas Gerais, Brazil
*E-mail: leonardomarmo@ufsj.edu.br

2.1 Introduction

Humphry Davy is considered to be the discoverer of calcium (1808), in spite of some previous experiments with animal bones involving calcium having been developed after the year 1700 (McDowell, 1992). Calcium is an alkaline-earth metal with atomic number 20, being its electronic distribution $[Ar]4s^2$ ($1s^22s^22p^63s^23p^64s^2$). Calcium presents 1.97 angstroms of metallic

Food and Nutritional Components in Focus No. 10
Calcium: Chemistry, Analysis, Function and Effects
Edited by Victor R. Preedy

Published by the Royal Society of Chemistry, www.rsc.org

radius (1.00 angstroms of ionic radius to Ca^{2+}), and a Pauling's electronegativity of 1.0 (Lee, 1996).

The alkaline-earth metals constitute the Group IIA in the periodic table. These elements present a well-defined behavior of highly reactive metals, although they are less reactive than the metals of the Group IA (Lee, 1996). In any case, they react easily with a great variability of nonmetals (Mahan and Myers, 1987), producing colorless ionic compounds, which present the alkaline-earth metal as a divalent cation (Lee, 1996). Indeed, the relative ease with which both s electrons are lost from these elements atoms leads to compounds in which only the +2 oxidation state is found (it occurs because M^+ is unstable with respect to disproportionation) (Huheey *et al.*, 1993). The metals of this group are less strongly reducing that the alkali metals (Group IA), but still must be considered strongly reducing (Huheey *et al.*, 1993). The metals of the group IIA (alkaline-earth metals) are significantly harder than the metals of the Group IA (Alkaline metals) (Mahan and Myers, 1987).

Calcium is the fifth most abundant element by mass in the Earth's crust (4.66%), being one component of several common mineral salts (Lee, 1996). It is important to notice that the calcium is not found free in nature (the elements of the Group IIA (Be; Mg; Ca; Sr; Ba; Ra) are not found in the metallic state in nature (Mahan and Myers, 1987)), in spite of the great number of compounds with calcium in their respective compositions (Lee, 1996). Some of the most common compounds encountered on Earth are limestone (calcium carbonate ($CaCO_3$)); gypsum (calcium sulfate ($CaSO_4{\cdot}2H_2O$)); fluorite (calcium fluorite (CaF_2)) and apatite (calcium fluorophosphates ($CaFO_3P$) or calcium chlorophosphate ($CaClO_3P$)) (Holden, 2001). Chloride and sulfate salts of calcium are water soluble; most other inorganic calcium salts are only slightly soluble in water (NRC, 2005).

Fairly pure deposits of calcium carbonate in the form of limestone are common and are a major source of calcium used in animal-diet formulation. Nearly pure calcium sulfate is also found in the form of gypsum that is used in Portland cement, drywall, and plaster production as well as animal feeds. Organic salts of calcium, such as calcium propionate, are very soluble and are increasingly being used in diet form (NRC, 2005).

In any case, the calcium in its metallic state is a silvery white metallic solid, which presents its surface covered with a thin layer of oxide that protects this metal from several types of chemical attacks by air, such as the oxidative attack.

The study focused on calcium is particularly relevant for some areas of interest, such as agriculture and soil science, since the levels of calcium in the soil constitute one important aspect related to the fertility of the soil, being a key nutrient to several cultures. Indeed, the reactivity of agricultural limestones and the respective relation between calcium and other metallic contents, such as the magnesium level, has been evaluated in order to optimize the soil acidity neutralization and the availability of mineral nutrients to plant-growing cultures (Bellingieri *et al.*, 1988).

2.1.1 Calcium in the Biological Medium

Calcium is a biologically relevant element to plants and animals. In plants, for example, the calcium ion presents a crucial regulator role of the growth and development through being associated to a great number of physiological processes (Hepler, 2005). In animals, the calcium ions are associated with various physiological and pathological processes, being focus of interest of a high number of research works. Extracellular calcium is essential for formation of skeletal tissues, transmission of nervous tissue impulses, excitation of skeletal and cardiac muscle contraction, blood clotting, eggshell formation, and as a component of milk. Intracellular calcium, while 1/10 000 the concentration of extracellular calcium, is involved in the activity of a wide array of enzymes and serves as an important "second messenger" conveying information from the surface of the cell to the interior of the cell (NCR, 2005).

The mineral salts constitute around 4% of the corporal composition of vertebrate animals, with the calcium and phosphorus representing more than 50% of this content (Underwood, 1981). In the biological medium, calcium presents great relevance, as its divalent cation, Ca^{2+} is associated with many biochemical roles, such as a messenger for hormonal action, a trigger for muscle contraction, in the initiation of blood clotting, and in the stabilization of protein structures (Shriver and Atkins, 1999). The amount of ionized calcium in normal serum is approximately 1.3 mM L^{-1} (Ettori and Scooggan, 1959). X-ray diffraction and NMR results illustrate how these functions are controlled by conformational changes induced by Ca^{2+}, when it binds to calmodulin, troponin C, and related proteins: these proteins are involved in the activation of membrane channels and receptors on cell surfaces (Shriver and Atkins, 1999).

The higher concentration of calcium in the animal organism is present in an inorganic form, as salts of hydroxyapatite in the bone matrix. This form of calcium represents 99% of the calcium that is present in several types of animals, such as mammals. The remaining 1% of all calcium found in the animal organism is divided between the vascular and intracellular environments, being associated with cell membranes and endoplasmic reticulum (Berchielli *et al.*, 2006).

In agreement with Shriver and Atkins, the many roles of the Ca^{2+} ion appear to arise from its affinity for the hard ligand oxygen in conjunction with the lability of its complexes, which fall between the alkali-metal ions and the d-metal ions, and the less-labile ions of its lighter congeners in Group 2 (Be^{2+} and Mg^{2+}) (Shriver and Atkins, 1999). In fact, the divalent cation of calcium (Ca^{2+}) is a Lewis's acid considered significantly "hard", in Pearson's concept, making this cation present a great affinity for "soft" Lewis's bases, in Pearson's classification, which is the case of the bases that presents an oxygen atom as a donor site. Indeed, as a consequence of its low selectivity, Ca^{2+} can bind neutral oxygen donor ligands (carbonyls and alcohols) in competition with water (Shriver and Atkins, 1999), implying a great potentiality to formations of chemical bonds with several relevant biological molecules.

Intracellular calcium plays a role in signaling and as a second messenger in many types of cells and its concentration is closely regulated in cells (Gunter and Gunter, 2001). In fact, techniques for measuring intracellular free calcium are using radiometric and nonradiometric fluorescence with significant success (Gunter and Gunter, 2001).

2.1.2 Calcium in the Diet

In total there are 20 minerals that are essential for maintenance and normal functioning of the body (De Groote *et al.*, 2002). Insufficient supply of these minerals results in deficiency symptoms leading to reduced performance. Therefore, sufficient amounts should be supplied. Conversely, an oversupply of minerals may harm the production efficiency together with a possible negative effect on the environment. Therefore, apart from the requirement for various minerals their availability should also be known to prevent oversupply. Different terms used to express the nutritive value of the minerals are described together with various factors that may affect their nutritive value (De Groote *et al.*, 2002).

The mineral relationships is related to the acid–base interaction, as the maintenance of physiological acid–base equilibrium requires the excretion of excessive dietary cations and anions. The consumption of excess mineral cations in comparison to the anions amount or exceeding the quantity of anions in comparison to the cations amount results in acid–base disturbances (Shohl, 1939), as the mineral interrelationships were found to affect numerous metabolic processes (Leach, 1979; Mongin, 1980).

In this context, the control of the dosages of calcium in the diet is a relevant subject to animal health. In fact, the absence or the insufficient quantities of this micronutrient affects the bones, being related to bone diseases, such as osteoporosis, include, increasing significantly the risk of fractures. On the other hand, the excessive ingestion of this element is also extremely dangerous to animal organisms. The high biological levels of calcium affects the absorption of other important divalent cations, such as iron(II), zinc(II) and Mn(II), which can provoke direct and indirect adverse consequences. An elevated ingestion of calcium (more than 2000 mg by day), for example, mainly calcium of mineral origin, especially when associated to high levels of vitamin D, can generate calculus in kidneys. Another adverse effect related to the high levels of calcium is intestinal constipation.

Ions such as, for example, Mg(II) can be dislocated and, for consequence, absorbed in insufficient quantities. In fact, Ca(II) and Mg(II) are alkaline-earth metals that present proximal affinities regarding Lewis acids, which is the case of several cations of biological interest. A particular topic related to pathological occurrences due to excess calcium is associated to circulatory affections, in which Ca(II) presents a central role. Indeed, in the animal body, the magnesium is deeply related to the calcium and the phosphorus, with respect to the distribution and metabolism (Maynard, 1979). The dietary cation–anion balance (DCAB) effect on the macromineral (calcium, phosphorus and magnesium) balance, urinary and fecal pH, serum concentration of calcium, phosphorus and magnesium has been evaluated in several animals, as the manipulation of the DCAB significantly affected the macromineral metabolism, mainly the calcium metabolism (Gomide *et al.*, 2004).

2.1.3 The Calcium Metabolism

The metabolism of calcium is associated with three basic factors: The absorption of calcium from the alimentation process; the calcium elimination by the urinary system; the fixation and/or release associated to bones and teeth.

The absorption of calcium happens, mainly, in the duodenum, involving passive and active absorption. When the level of calcium of the diet is relatively low, a large proportion of the calcium is absorbed by active transport. The absorption of calcium in the animal intestine occurs accordingly with necessity, *i.e.*, with the change of the levels of requirements (McDowell, 1992). The absorption of calcium is dependent on the medium pH, as the acid medium favors the solubility of the calcium ions and, consequently, increases its absorption that occurs, principally, in the acid medium of the duodenal environment (McDowell, 1992).

The presence of calcium in the blood is regulated by vitamin D and by the hormones calcitonin and parathormone. When the levels of calcium in blood are high, the thyroid gland secretes calcitonin, leading to the following major consequences: inhibition of vitamin D activation; impairment of the calcium absorption in the kidneys and intestine; and inhibition of the calcium liberation by the osteoclasts. In its turn, when the calcium levels in blood are lowered, the parathyroid gland secretes the parathormone, leading to vitamin D activation. Vitamin D, together with parathormone, stimulates calcium absorption by kidneys and intestine and stimulates osteoclasts to resorb bone tissues liberating calcium to the blood (Whitney and Rolfes, 2008) (Figure 2.1).

2.1.4 The Calcium Action in Several Cardiovascular Diseases

A group of academic and industry experts in the fields of nutrition, cardiology, epidemiology, food science, bone health, and integrative medicine examined the data on the relationship between calcium-supplement use and risk of cardiovascular events. The results of these studies indicated a small increase in the risk of adverse cardiovascular events (Heaney *et al.*, 2012). On the other hand, Jorde and Bonaa (2000) claim that calcium possesses a protective effect against hypertension. Research studies sponsored by the National Heart, Lung, and Blood Institute (NHLBI) led to the development of the DASH eating plan (dietary approach to stop hypertension), which consists in eating foods that are low in saturated and *trans* fats and rich in potassium, calcium, magnesium, fiber, and protein. Jacqmain *et al.* (2003) suggested that the calcium intake is related to less cholesterol, less diabetes and protection against colon cancer.

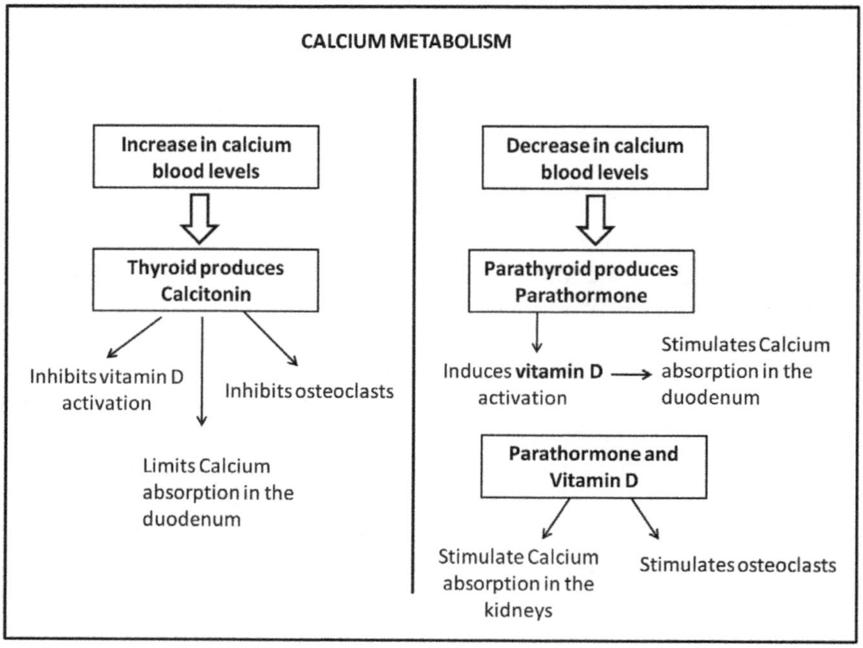

Figure 2.1 Schematic metabolism of calcium.

2.1.5 Calcium and Osteoporosis

Osteoporosis is a systemic skeletal disease characterized by a low bone mass. The disease is related to ageing and it is known to affect 44 million of people in the United States, mostly woman beyond 50 (Whitney and Rolfes, 2008). Osteoporosis is a silent disease and the bone mass loss is not shown in blood tests or by symptoms in the organism. There are several predisposing factors for the risk of developing osteoporosis, as we can mention: age, gender, genetic influence, hormones, physical activity level and diet.

According to Gennari (2001), reduced intake of calcium can be correlated to reduced bone mass and osteoporosis. Calcium intake should be appropriate, especially in the growth stage of development in order that a person achieves the ideal peak of bone mass at the age of 20 (Teegarden *et al.*, 1999).

2.1.6 Methods of Analysis and Evaluation of Calcium in Feeds and Biological Tissues

Determination of calcium in feeds and biological tissues is usually accomplished by wet or dry ashing of the sample followed by suspension of the ash in an acidic solution for analysis by atomic absorption spectrophotometry. Indeed, atomic absorption measurement is conducted at a wavelength of 422.7 nm and can detect as little as 0.01 mg Ca per L (NRC, 2005).

2.1.7 Conclusions

The biological action of the calcium ion and its derivative compounds constitute one of the more relevant studies related to the areas of bioinorganic chemistry (inorganic biochemistry), human and animal nutrition and medicinal chemistry.

Summary Points

- Calcium is an alkaline-earth metal with atomic number 20 and is the fifth most abundant element by mass in the Earth's crust, being one component of several common mineral salts.
- Calcium is important in plant physiology as well as to several metabolic and cellular functions of the human organism, as well as to the formation of bone tissues.
- Insufficient calcium intake may lead to the development of osteoporosis. On the other hand, the excessive ingestion of calcium increases the risk of renal calculus and intestinal constipation.
- The presence of calcium in the blood is regulated by vitamin D and by the hormones calcitonin and parathormone.
- The intake of calcium is important to prevent cardiovascular diseases.

Key Facts of Calcium in the Human Body

1. Calcium is the most abundant mineral in the human organism, as 99% of calcium is stored in bones and teeth.
2. Extracellular calcium is essential for formation of skeletal tissues, transmission of nervous tissue impulses, excitation of skeletal and cardiac muscle contraction, blood clotting, eggshell formation, and as a component of milk.
3. Diet is the main source of calcium to the organism.
4. Although dairy products are excellent sources of calcium, vegans may also obtain this nutrient from soy and its derivatives, beans and other grains as well as from green leaves such as broccoli, collards cabbage kale and mustard greens.
5. Calcium and phosphate are salivary buffers. Tooth decay occurs because of the demineralization promoted by bacterial acid products and the calcium present in saliva helps in the remineralization of teeth.

Definitions of Words and Terms

Calcitonin is a polypeptide hormone produced by the parafolicular cells of thyroid.

Osteoporosis progressive disease characterized by the decrease in bone mass, which can lead to an increase in the risk of bone fracturing.

Parathormone also known as parathyroid hormone, is a polypeptide hormone produced by the chief cells of the parathyroid gland.
Ratiometric fluorescence. Method based on the use of a ratio between two fluorescence intensities.
Nonradiometric fluorescence. Method that employs indicators with a shift in their fluorescence intensity. The cytosolic-free calcium concentration is related with fluorescence intensity.

List of Abbreviations

Ar	Argon
Ca^{2+}	Divalent cation of calcium
Be	Beryllium
Be^{2+}	Divalent cation of beryllium
Mg	Magnesium
Mg^{2+}	Divalent cation of magnesium
Sr	Strontium
Ba	Barium
Ra	Radium
$CaCO_3$	Calcium carbonate
CaF_2	Calcium fluoride
NCR	National Research Council of the National Academies
NMR	Nuclear magnetic resonance

References

Bellingieri, P. A., Alcarde, J. C. and Souza, E. C. A., 1988. Agricultural limestones reactivity and the relation between calcium and magnesium contents. *Anais da Escola Superior de Agricultura, "Luiz de Queiroz," Universidade de Sao Paulo*. 45: 499–515.

Berchielli, T. T., Pires, A. V. and de Oliveira, S. G., 2006. *Nutrição de Ruminantes*. Funep, 583 pp.

De Groote, G., Lippens, M., Jongbloed, A. W. and Meschy, F., 2002. *Study on the Bioavailability of Major and Trace Minerals*, International Association of the European (EU) Manufacturers of Major, Trace and Specific Feed Mineral Materials.

Ettori, J. and Scooggan, S. M., 1959. Ionized calcium in biological media. *Nature*. 184: 1315–1316.

Gennari, C., 2001. Calcium and vitamin D nutrition and bone disease of the elderly. *Public Health and Nutrition*. 4: 547–549.

Gomide, C. A., Zanetti, M. A., Penteado, M. V. C., Carrer, C. R. O., Del Claro, G. R. and Netto, A. S., 2004. Influence of the dietary cation–anion difference on calcium, phosphorus and magnesium balance in sheep. *Arquivo Brasileiro de Medicina Veterinaria e Zootecnia*. 56(3): 363–369.

Gunter, K. K. and Gunter, T. E., 2001. Measurements of intracellular free calcium concentration in biological systems. *Current Protocols in Toxicology*, Ch. 2.

Heaney, R. P., Kopecky, S., Maki, K. C., Hathcock, J., MacKay, D. and Wallace, T. C., 2012. A review of calcium supplements and cardiovascular disease risk. *Advances in Nutrition – An International Review Journal*. 3: 763–771.

Hepler, P. K., 2005. Calcium: a central regulator of plant growth and development. *Plant Cell*. 17: 2142–2155.

Holden, N. E., 2001. *History of the Origin of the Chemical Elements and Their Discoverers*.

Huheey, J. E., Keiter, E. A. and Keiter, R. L., 1993. *Inorganic Chemistry – Principles of Structure and Reactivity*. HarperCollinsCollegePublishers.

Jacqmain, M., *et al.*, 2003. Calcium intake, body composition and lipoprotein lipid concentrations. *American Journal of Clinical Nutrition*. 77: 1448–1452.

Jorde, R. and Bonaa, K. H., 2000. Calcium from dairy products, vitamin D and blood pressure: the Tromso study. *American Journal of Clinical Nutrition*. 71: 1530–1535.

Leach, R. M., 1979. Dietary electrolytes: story with many facets. *Feedstuffs*. 27.

Lee, J. D., *Concise Inorganic Chemistry*, Chapman & Hall, 1996.

Mahan, B. M. and Myers, R. J., 1987. *University Chemistry*. 4th edn. Benjamin-Cummings Publishing Company.

Maynard, L. A., Loosli, J. K., Hintz, H. F. and Warner, R. G., 1979. *Animal Nutrition*. 7th edn. McGraw-Hill, New York, 602 pp.

McDowell, L. R., 1992. Calcium and phosphorus. In: *Minerals in Animal and Human Nutrition. Chapter 2*. Academy Press, Inc., pp. 26–77.

Mongin, P., 1980. Electrolytes in nutrition: review of basic principles and practical application in poultry and swine. In: *Third Ann. Int. Mineral Conf. Orlando, FL*, p. 1.

National Research Council of the National Academies, 2005. *Mineral Tolerance of Animals*. 2nd edn.

Shohl, A. T., 1939. *Mineral Metabolism*. Reinhold Publishing Corp., New York.

Shriver, D. F. and Atkins, P. W., *Inorganic Chemistry*. 3rd edn. Oxford University Press, 1999.

Teegarden, D., *et al.*, 1999. Previous milk consumption is associated with greater bone density in young women. *American Journal of Clinical Nutrition*. 69: 1014–1017.

Underwood, J. E., 1981. *Los minerals en la nutricion del Ganado*. Acribia, Zaragoza, 210 pp.

Whitney, E. and Rolfes, S. R., 2008. *Nutriçao*. Ed Cengage Learning, São Paulo, 342 pp.

CHAPTER 3

Food Sources of Calcium Vary by Ethnicity and Geography

NOREEN WILLOWS*[a]

[a]Faculty of Agricultural, Life & Environmental Sciences, University of Alberta, 4-378 Edmonton Clinic Health Academy, Mailbox #54, 11405 87 Avenue, Edmonton, Alberta, Canada T6G 2P5
*E-mail: noreen.willows@ualberta.ca

3.1 Introduction

This chapter discusses food sources of calcium among populations worldwide. The focus is on the contribution of dairy, vegetables, cereal grains and legumes to calcium intake. The chapter also describes the contribution to calcium nutrition of less common but potentially important sources of calcium such as insects. Discussed in brief are the genetic, sociodemographic and religious reasons a calcium-contributing food is consumed. In the modern context, calcium supplements and foods intentionally fortified with calcium (*e.g.*, calcium-fortified orange juice) contribute to calcium intake; however, a discussion of these calcium sources is not included in this chapter.

3.2 Milk and Dairy Foods Consumed Worldwide

Dairy is a concentrated source of calcium; hence, populations that consume dairy have the highest calcium intakes (Prentice, 2014). For example, a 250 mL serving of milk or buttermilk contains 300 mg of calcium (Osteoporosis

Food and Nutritional Components in Focus No. 10
Calcium: Chemistry, Analysis, Function and Effects
Edited by Victor R. Preedy

Published by the Royal Society of Chemistry, www.rsc.org

Canada, 2014). Although global dairy production has increased in recent decades, especially in developing countries, there are still large between-country differences in dairy consumption (Prentice, 2014). Dairy is included in European, Middle Eastern, and South Asian cuisines; however, it is not a part of traditional Polynesian cuisines, nonpastoral African cuisines, the cuisines of the indigenous peoples of the Americas, or most East or Southeast Asian cuisines (Kittler and Sucher, 2008). Dairy consumption is greatest among cultures that historically maintained mammals for milking, and where the ability to hydrolyze lactose persists in the adult population. Lactase activity persists into adulthood at high frequency in Northern European and African pastoralist populations; at moderate frequency in Southern European and Middle Eastern populations; and, at low frequency in nonpastoral Asian, Pacific, indigenous American, and African populations (Ingram *et al.*, 2009; Prentice, 2014). Cheese, fermented milk beverages like kefir and fermented dairy products like yoghurt are lower in lactose than milk and therefore better tolerated in populations that produce low levels of lactase (Savaiano, 2010).

3.2.1 Europe

In countries predominantly populated by persons of Northern, Central European and Scandinavian ancestry, milk (including buttermilk and fermented milk) and dairy products in the form of hard cheese, cheese curd, whipped cream, ice cream and sour cream make a significant contribution to calcium intake. Dairy is used in sauces, soups, stews, noodle dishes, dumplings and baked products. Many adults in Southern Europe do not drink milk but do eat cheese (Barer-Stein, 1999; Kittler and Sucher, 2008).

Dairy is the main source of calcium for British adults aged 19 to 64 years old. Milk and milk products contribute to 43% of mean overall calcium intake, with semiskimmed milk and cheese being the main contributors, providing 17% and 11% of calcium, respectively (Henderson, 2003). Among 12 year old Northumbrian children in the United Kingdom milk contributes to 25% of calcium intake, more than any other food (Moynihan *et al.*, 1996). Among Flemish preschoolers 2.5 to 6.5 years old milk, sweetened milk drinks and cheese are the main sources of calcium in the diet (26, 25 and 11% of calcium, respectively) (Huybrechts *et al.*, 2011).

Dairy provides the majority of dietary calcium in Southern Europe countries. Among Spanish children 7 to 11 years old dairy provides 64.7% of calcium (Ortega *et al.*, 2012). Dairy provides 58.7% of calcium among Spanish adults (Estaire *et al.*, 2012). In central Italy the principal source of calcium among adults is hard cheese (parmigiano), since the quantity of milk consumed daily is small and yoghurt consumption is infrequent (Matteucci and Giampietro, 2008). Dairy is the primary source of calcium among adults in Greece (Magkos *et al.*, 2006). Milk is not a preferred beverage among Greek adults, whose dairy-product choices are yoghurt and hard and soft cheeses, some like feta that are preserved in salt brine (Barer-Stein, 1999).

Dietary practices of members of the Eastern Christian Orthodox Church influence dairy consumption. Russian, Greek, Serbian, Romanian, Bulgarian, Ukrainian, and Moldavian ethnic groups commonly practice this religion. Dietary rules of this religion require vegetarianism during religious fasts through the avoidance of most animal foods including milk and all dairy products. Adherents of these dietary rules avoid dairy every Wednesday and Friday throughout the year and during the Christmas, Lent and Assumption fasts, resulting in a significant reduction in calcium intake (Kittler and Sucher, 2008; Lazarou and Matalas, 2010).

3.2.2 Middle East

In the Middle East fresh milk is seldom consumed as a beverage, in part because the hot climate causes milk to spoil quickly. One study of 316 Lebanese adults 30 to 50 years old found that only 15.9% drank a daily cup of milk, while 53.1% never drank milk (Gannagé-Yared *et al.*, 2005). Milk in the Middle East is used in puddings and custards. In Jewish, Iraqi, Jordanian, Lebanese and Syrian cuisines cultured (soured) milk such as leben (buttermilk made from soured milk) and yoghurt are eaten with bread, pilafs, meats, or fruits. Cheeses from camel, goat, or sheep milk may be preserved in brine or olive oil. Labneh is consumed. This fresh cheese is prepared by straining yoghurt. In Israel, cottage cheese is eaten for breakfast and incorporated into dishes, particularly filled pastries (Barer-Stein, 1999; Kittler and Sucher, 2008).

Dairy consumed in Iran includes milk, cheese, yoghurt, ice cream, cocoa milk, a dairy product make from fermented and dried sheep milk (kashk), and a beverage made by diluting yoghurt with water (dough) (Karandish *et al.*, 2005). A study of postmenopausal Iranian women found that dairy contributed to 61 ± 19% of dietary calcium (Karandish and Naghashpoor, 2010). A study of pregnant Iranian women found that dairy provided 49% of dietary calcium; however, 43% of women consumed <1 serving of dairy per day. Milk and yoghurt were the main dairy products consumed by pregnant women, followed by cheese (Karandish *et al.*, 2005).

In Israel, the consumption by Jews of dairy is dictated by religious dietary laws (Stern, 2004). Food is kosher, meaning acceptable to eat, when prepared according to these laws. Keeping kosher concerning dairy applies also to individuals living outside of Israel who practice the Jewish religion. To be kosher, only the milk of a cow, goat or sheep can be consumed. Jewish religious observances require the strict separation of dairy from meat and poultry. These foods cannot be cooked together, eaten at the same meal, or eaten close together in time to ensure that dairy does not mix with meat or poultry in the digestive tract. The requirement for separation means that cheese made from animal-derived rennet is forbidden (Stern, 2004). Following an ultra-Orthodox Jewish lifestyle may therefore result in fewer opportunities to consume dairy and consequently a lower calcium intake. Indeed, a large survey of Jewish adults in Israel examined dietary intake in relation to religious observance. It found higher dairy product consumption among less religiously adherent individuals (Shmueli and Tamir, 2007).

3.2.3 Canada, United States of America, and Australia

A large proportion of the population of Canada, the United States of America and Australia have European heritage, hence, high dairy consumption. Results from a large national survey in the United States indicated the highest ranked sources of calcium for adults were milk (22.5% of calcium) and cheese (21.6% of calcium) (O'Neil *et al.*, 2012). Results from the same national survey indicated that among children the two top sources of calcium were also milk (33.2% of calcium) and cheese (19.4% of calcium) (Keast *et al.*, 2013). In the United States, African-Americans of most age groups consume significantly less dairy overall, milk and cheese compared with their non-African-American counterparts. Yoghurt consumption is negligible among African-Americans (Fulgoni *et al.*, 2007). The Food Habits of Canadians study found that among participants aged 18 to 65 years old milk contributed to about 35% of calcium consumption whereas cheese contributed to 12 to 19% of calcium consumption (Johnson-Down *et al.*, 2006). Most (84–98%) Australian children aged 2 to 16 years old participating in a 2007 national survey consumed dairy. Milk was the predominant dairy food consumed by 58–88% of children, followed by cheese (consumed by 36–53% of children) and yoghurt (consumed by 12–39% of children). Children typically consumed 243–384 g per day of milk, which was the primary source of calcium (Baird *et al.*, 2012).

3.2.4 South Asia

Dairy is an ingredient of most South Asian cuisines. The consumption of milk, buttermilk, ghee (clarified butter), curds, and fermented milk products like yoghurt is common. Milk may be drunk fresh or as a diluted yoghurt drink known as lassi. Although in South Asia milk from the cow is common, milk from goats, sheep, buffalo, mares and camels is also used (Kilara and Iya, 1992; Kittler and Sucher, 2008). In India, milk (whether unpasteurized or pasteurized) is often boiled before its consumption to prevent spoilage. Sugar and/or condiments are sometimes added to improve its flavor and perceived digestibility. Dilution of milk is also a common practice in many Indian households, which significantly diminishes its nutritional profile (Bahman *et al.*, 2012). In traditional South Asian cooking hard cheese is not used, rather a fresh cheese known as paneer is incorporated into many dishes. Desserts made with milk are common including sweet milk and rice pudding (khir), a fudge made with condensed milk or cream and sugar (barfi), a sweetened frozen dairy product made with sweetened evaporated milk (kulfi), and dumplings made of thickened or reduced milk, soaked in rose-flavored sugar syrup (gulab jamun). Tea (chai) is made with boiled milk and heavily sweetened (Barer-Stein, 1999; Kilara and Iya, 1992; Kittler and Sucher, 2008).

Dairy consumption in South Asia is driven by religion, socioeconomic status, social practices, and availability. Hinduism, one of the major religions of India, considers milk and dairy products from the cow to be spiritually pure. Consuming milk, dairy products and ghee are all important to maintaining ritual purity in Hinduism (Kilara and Iya, 1992; Kittler and Sucher, 2008). Food availability and economic status exert an important impact on food intake. While impoverished individuals often have very low dairy intake, due to lack of availability dairy consumption can even be low among individuals from better off socioeconomic groups. Girls may be deprived of nourishing foods like dairy when it is in limited supply so that boys can eat (Gupta, 1987).

Two hundred adolescents from lower and upper economic stratum in India were included in a study of calcium intake and sources of calcium. The contribution of dairy products to daily calcium intake was the highest in upper economic strata boys (271 mg, 219–382 mg) and the lowest in lower economic strata girls (43 mg, 14–66 mg) (data presented as median, 25th and 75th percentiles). The main source of calcium was dairy products in upper economic strata adolescents while it was dark green leafy vegetables in lower economic strata adolescents. Girls from both groups had less access to dairy products than boys (Sanwalka *et al.*, 2010). Dairy consumption may be lowest in rural areas. One study of healthy adults in south India found that milk and milk products contributed to 15% of energy intake of urban living individuals but to 5% of energy intake of rural-living individuals. Calcium intake was significantly lower in rural than urban individuals. It ranged from 262 mg per day for rural woman to 323 mg per day for urban men (Kumar, 2007).

A study of impoverished nonpregnant and nonlactating women between the ages of 30 and 60 years residing in a large urban slum in India found that consumption of milk and milk products was only 62 g per day. Considering few other sources of calcium in the diet, and the low overall intake of food among women, total calcium intake was very low (270 mg per day) (Shatrugna *et al.*, 2005). In a study in Bangladesh, very low income women consumed, on average, 3.6 g per day of dairy whereas high-income women consumed, on average, 89.3 g per day. Milk was the second highest source of calcium among higher-income women, contributing to more than 19% of calcium. In contrast, among low-income women milk supplied only 3% of calcium. For both groups of women low consumption of dairy was the result of a lack of availability of milk and dairy products in the region coupled with their high cost (Islam *et al.*, 2003).

3.2.5 East and Southeast Asia

The diet of East and Southeast Asia is traditionally nondairy based; consequently, most nutrition surveys in Asia indicate low dairy consumption. An exception is a long history of milk production and consumption in some areas of China, namely Inner Mongolia, Shanghai, Beijing and Tianjin (Fuller *et al.*, 2007). Among postmenopausal women in Vietnam included in a nutrition study in 2004 only 0.38% of calcium consumed was from cow's milk and milk powder combined (Khan *et al.*, 2008). Results of the 2008–2009 Korea National Health and Nutrition Examination Survey indicated that adults consume milk and dairy products less than once per day (4.1 times per week for men and 4.7 times per week for women) (Yoon *et al.*, 2012). Data from the 2005–2008 National Nutrition Surveys in Taiwan found that 45.4% of respondents ate no dairy. Forty seven percent of first-grade children consumed a serving or more of dairy daily, while 37% of 6th graders did. Only 20% of nonelderly adults ate at least a daily serving of dairy. Physiological limitations (diarrhea, 22%; bloating, 2.9%; constipation, 1.3%), some seemingly due to lactose intolerance, accounted for about 25% of dairy avoidance. Most (66.7%) avoidance was due to dietary habits or dislike. Only 3% of respondents avoided dairy because they could not afford it (Lee *et al.*, 2009).

As Asian diets and consumption patterns become more westernized, the consumption of dairy is increasing. The rapidly developing dairy industry in China has become an important source of dairy for China and other countries (Fuller *et al.*, 2007). Even small amounts of dairy can make a significant contribution to calcium intake. In Japan, dairy was the predominant source of calcium for all age groups in a 2002 national nutrition survey (Sato *et al.*, 2005). Data from the 2001–2002 Nutrition and Health Survey in Taiwan Elementary School Children showed that children 6–12 years old had a very low calcium intake and a low intake of dairy, yet dairy (fresh milk, yogurt, cheese and other dairy products) was the primary source of calcium. Dairy provided children with 164 mg calcium per day, on average. The mean daily intake of dairy in children 6–9 years old was 0.9 servings, whereas it was 0.6 servings in children 10–12 years old (Wu *et al.*, 2007). Data from the Korean National Health and Nutrition Examination Survey 2007–2010 showed that among children and adolescents 1 to 18 years of age the highest ranked source of calcium was dairy (milk, modified milk powder, formula, breast milk, goat's milk, yoghurt, ice cream, and cheese). Dairy contributed to 35% of calcium intake followed by vegetables (17.3%), grains (11.3%) and seafood (9.9%). Dairy contributed to 57% of calcium intake at age 1–2 years and to 24% of calcium intake at age 15–18 years (Im *et al.*, 2013). Milk consumption in Singapore was determined in 2009 from a random sample of children. The mean weekday consumption of milk for 3 to 6 year-old children was 635 mL and the mean intake of calcium from all sources was 1032 mg per day. The mean weekday consumption of milk for 7 to 10 year-old children was 359 mL and the mean intake of calcium from all sources was 733 mg per day (Goh and Jacob, 2011). Milk contributed the majority of calcium in Singapore children's diets.

In China, there is great variation in milk consumption across regions and between rural and urban areas. There have been increases in milk consumption, largely in urban areas, and school programs promote milk consumption among children (Fuller *et al.*, 2007). The 2007 Chinese Food Guide Pagoda recommends the daily consumption of 300 g of milk and milk products (Ge, 2011). A survey of households in Beijing, Shanghai, and Guangzhou in the fall of 2001 focused on dairy consumption. Survey data revealed that all but five of 300 households reported purchasing one or more dairy products. About 91% of the households purchased milk, but less than 60% purchased yogurt or ice cream (Fuller *et al.*, 2007). Data from the 2004 China Economic, Population, Nutrition and Health Survey showed that adults aged 18–45 years consumed, on average, 12 g per day of milk and dairy products (Zhai *et al.*, 2009). A survey of Chinese adolescents living in Guangzhou city in 2006–2007 demonstrated that nearly 70% consumed less than the recommended daily 300 g intake of dairy. The average intake of dairy (sources included whole milk, whole milk powder, skim/low-fat milk, skim/low-fat milk powder, yoghurt, milk tea, cheese, and ice cream) among adolescents was 261 g per day (Zhang *et al.*, 2012).

3.2.6 Africa

Intake of milk and dairy is low among ethnic Africans except for among herding populations (Prentice, 2014). The pastoral Maasai of East Africa rely on meat, milk and blood from cattle for sustenance. Cow's milk is consumed fresh, boiled and mixed with blood and meat in stews, and made into sour milk and yoghurt. Goat milk is also consumed. Wild herbs high in calcium are added to fresh or boiled milk and soups (Oiye *et al.*, 2009). The adult diet of the pastoral Fulani of Nigeria is rich in dairy, in particular milk, cheese, butter oil, and yogurt. The vast majority of adults drink milk (Glew *et al.*, 2001). Dairy products are losing their dominance in the diets of the Fulani of Central Africa who have become nonnomadic. The need for cash income to meet the requirements of sedentary living requires the Fulani to sell most of their dairy and milk products. These factors have resulted in less availability of milk and dairy products for Fulani children (Ekpo *et al.*, 2008).

3.3 Nondairy Sources of Calcium Consumed Worldwide

The remainder of this chapter will focus on nondairy sources of calcium in regions that are not large dairy consumers.

3.3.1 Asia

In Asia, traditional food sources of calcium are grains, vegetables, legumes, seafood and the soft bones of small fish. The Chinese Food Pagoda 2007 depicts eight food categories for dietary adequacy with recommended daily intakes for each one: 250–400 g for cereals, 300–500 g for vegetables, 200–400 g for fruits, 50–75 g for meat, 70–100 g for fish and shrimps, 25–50 g for eggs, 30–50 g for soybean products and nuts, and 300 g for dairy (Ge, 2011). Apart from dairy, in China food categories that contribute to calcium nutrition are cereals (*e.g.*, white rice, porridge, noodles, bread, cake and biscuits); vegetables (*e.g.*, dark green leafy vegetables, cruciferous vegetables such as gai lan and bok choy, melon, radish, pepper, carrot, tomato, starchy tubers, fresh corn, fresh beans, allium – onion, garlic, leek, and chives, mushroom and fungi); fish and shrimp (*e.g.*, freshwater and saltwater fish, canned fish, salted fish, crab, shrimp, prawn, squid, cuttlefish, scallops, mussel and whelk); and, soybean products (*e.g.*, hard tofu, fried tofu pop, soft tofu, soy products such as tofu curd and vegetarian chicken, soy drink, bean curd pudding, fresh and dried soybean) (Zhang *et al.*, 2012).

Among Korean children and adolescents participating in the Korea National Health and Nutrition Examination Survey 2007–2010 the highest ranking source of calcium after dairy was vegetables (*e.g.*, kimchi – pickled vegetables, radish leaves, welsh onion, onion, radish, soybean sprouts, carrot, spinach) (contributing 17.3% of calcium). Two plants very high in calcium (mg per standard serving) were radish leaves (149 mg/60 g) and mugwort (103.5 mg/45 g) (Im *et al.*, 2013). The next highest ranking sources of calcium were grains (*e.g.*, rice, barley, wheat and their products) (11.3% of calcium) and seafood (*e.g.*, boiled and dried anchovy, fish paste, dried and raw shrimp, loach mudfish, squid, opossum shrimp, mackerel) (9.9% of calcium). Soy (*e.g.*, soybeans, tofu, fried tofu, bean-curd dregs, soya milk, bean flour) provided 6.4% of calcium (Im *et al.*, 2013). Data from the Nutrition and Health Survey in Taiwan Elementary School Children 2001–2002 showed that after dairy, the main sources of calcium in the diet of children were green and yellow vegetables (81.4 mg calcium per day), soybean and soybean products (42.2 mg calcium per day), and fish and seafood (not including seaweed) (38.7 mg calcium per day) (Wu *et al.*, 2007).

Whole small fish are an excellent calcium source. Japanese Dietary Guidelines recommend individuals to drink milk and eat green/yellow vegetables, beans, and small fish to get a sufficient amount of calcium (Japan Dietetic Association, 2014). In Japan, whole fish are eaten, as well as cakes made from whole fish, including the bones. In the traditional Korean diet fish high in calcium (mg per standard serving) are dried icefish (98.2 mg/10 g) and dried anchovy (64.5 mg/5 g) (Im *et al.*, 2013). In low-income countries such as Bangladesh, small fish play an important role in the diet since they are often used in sauces and curries, which are eaten with the main staple. Among some Bangladeshi women fish are the main source of calcium (Islam *et al.*, 2003).

The soybean (or soya bean) is a legume native to East Asia that has been cultivated for millennia. In China, soybeans are referred to as "poor man's cow" as they are high in protein and are made into products resembling milk and cheese (Kittler and Sucher, 2008). Buddhism has been influential in the development of soy foods in China and other Asian countries where it is practised. This is because soy foods can substitute for animal foods making a vegetarian diet possible (Huang and Ang, 1992). Soybeans in Asia are consumed as sprouts, boiled beans and boiled bean pods (edamame). They are processed to make soymilk, soybean curd (tofu), fermented tofu, fermented cooked soybeans (*e.g.*, black beans, natto, tempeh), fermented soybean paste (*e.g.*, miso), soy flour and condiments (*e.g.*, soya sauce). Cooked soybeans provide 85 mg of calcium per 125 mL. To make tofu, ground soybeans are soaked in water to which a coagulant, either calcium or magnesium sulfate, is added. Made with calcium sulfate, 125 g of tofu provides 193 mg of calcium (Osteoporosis Canada, 2014). In Korea, a typical serving of tofu is 60 g and provides 75.6 mg of calcium (Im *et al.*, 2013). In Japan, a medium serving of tofu (100 g) provides 90 mg of calcium, whereas a medium serving (60 g) of deep-fried tofu provides 160 mg of calcium (Sato *et al.*, 2005).

Potential sources of calcium in the Asian diet are bones and bird egg-shells cooked or pickled in vinegar. In China, postpartum mothers eat whole eggs that have been placed into a vinegar solution that dissolves the shell (Newman, 2004). In southern China, postpartum mothers eat a special mixture called *geung cho* of ginger, pig's feet, and boiled eggs with the shells on cooked in black vinegar (Fong, 2001). One preservation technique to make duck or chicken eggs taste sour involves immersing eggs in vinegar and salt until the shells soften. The eggs are called *zao dan*. Some eggs are kept in the solution so long that the shells dissolve (Newman, 2004).

The contribution of preserved and pickled eggs to calcium nutrition could be substantial. Eggshells are about 95% calcium carbonate and a single chicken eggshell contains up to 3 g (3000 mg) calcium (Roberts, 2010). The softened shells of pickled eggs are edible (McGee, 2007). Calcium carbonate from the eggshell might penetrate the pickled egg and be ingested. The acid in vinegar may dissolve calcium from bone and eggshells when soup is made using it; however, it is unknown how much calcium is provided by *geung cho*. Pork bones cooked in vinegar are stated to provide 130 mg calcium per 110 g (as cited in Kuhnlein, 1984).

3.3.2 Africa

Both wild and cultivated dark green leafy vegetables are an important part of the traditional starch-based African diet. Wild greens eaten throughout Africa often have more calcium than cultivated greens (Kinabo *et al.*, 2006; Van Der Walt *et al.*, 2009). Among two- to five-year-old children in KwaZulu-Natal in South Africa dark-green leafy vegetables such as Swiss chard and imifino (a collective term for various dark-green leaves) contribute 21 to 39% of total dietary calcium (Faber *et al.*, 2007).

Insects are a neglected but potentially important source of calcium in Africa and elsewhere. Humans eat over 1600 species of insects deliberately, although cultures are highly variable in their consumption of insects (Raubenheimer and Rothman, 2013). Insects are consumed in central and southern Africa, Asia, Australia (by the indigenous Aborigines) and Latin America. The calcium content of five species of insects consumed by the Luo of Kenya varies from 33 mg/100 g dry matter for *Onyoso mammon* (ant) to 341 mg/100 g dry matter for *Onjiri mammon* (cricket) (Christensen *et al.*, 2006). The calcium content of 16 of 22 insect species eaten in Nigeria has been reported. Two insects relatively high in calcium are *A. trifasciata* (stem girdler) containing 61.3 mg of calcium per 100 g and *Z variegatus* (grasshopper) containing 42.4 mg of calcium per 100 g (Alamu *et al.*, 2013). To fully appreciate the contribution of insects to nutritional status there is the need for additional standardized data on both the types and quantities of insects consumed in Africa and elsewhere and their macro- and micronutrient composition (Raubenheimer and Rothman, 2013).

3.3.3 Mexico and Central America

In Mexico and some Central American countries (*e.g.*, Guatemala) maize (corn) tortillas are daily dietary staples and a major source of calcium for persons whose intake of dairy is limited. To make tortillas using the traditional method maize undergoes a process called nixtamalization, or liming, whereby the kernels are cooked and soaked overnight in a lime (calcium hydroxide) water solution before being ground to form dough. Commercial corn flour is also available for the preparation of tortillas; it is usually lime treated for only 1 h. Tortillas prepared using lime-treated, home-prepared corn flour have 1.7 times more calcium than tortillas prepared from commercial lime-treated corn flour (Rosado *et al.*, 2005). In Mexico it is common to consume six tortillas daily with a combined weight of 180 g. When made from commercial flour six tortillas provide 187.6 mg of calcium; when made with homemade flour six tortillas provide 326.8 mg of calcium (Rosado *et al.*, 2005).

Summary Points

- This chapter focuses on variations in food sources of calcium worldwide.
- Genetic, cultural, sociodemographic and religious reasons help to explain differences in the calcium-containing foods that are consumed worldwide.
- The diets of ethnic Europeans are relatively high in dairy sources of calcium, whereas the diets of ethnic Africans and individuals living in East and Southeast Asia are low in dairy.
- Global dairy production has intensifying in recent decades increasing the amount of milk and dairy products consumed in regions where they were not traditionally incorporated into the diet.
- Even where dairy is consumed, nondairy sources of calcium must be ingested to meet calcium adequacy.
- Nondairy sources of calcium are fish with edible bones, leafy greens, legumes such as the soybean and soybean derivatives like tofu made with calcium sulfate, grains such as maize treated with calcium hydroxide used to prepare tortillas, and insects.
- To widen our understanding of food sources of calcium increased knowledge of the calcium content of edible insects, wild greens, pickled eggs and Asian bone soups is required.

Key Facts

Key facts about the digestion of milk and dairy foods in adulthood, in particular the requirement for the persistent production of the enzyme lactase by the intestinal epithelium.

1. The main carbohydrate in milk is lactose.
2. Lactase is the enzyme produced by the intestinal epithelium that hydrolyzes lactose to glucose and galactose.
3. Lactase production decreases after the weaning phase in most humans.
4. It has been estimated that 65% or more of the human adult population are lactase nonpersistent.
5. Lactase activity persists into adulthood at high frequency in Northern European dairying and African pastoralist populations; at moderate frequency in Southern European and Middle Eastern populations; and at low frequency in nonpastoral Asian, Pacific, indigenous American, and African populations.
6. When lactase production is nonpersistent into adulthood, drinking milk can cause significant gastrointestinal discomfort such as diarrhea, cramps, bloating and flatulence.
7. Cheese and fermented milk products like yoghurt are lower in lactose than milk and therefore better tolerated by individuals who produce low levels of lactase.
8. Differences in lactase persistence help to explain variations in milk and dairy product consumption within and between populations.

Definitions of Words and Terms

Cuisine. A cuisine is a characteristic style or method of cooking often named after the geographic region from which it originated.

Dairy. Dairy pertains to milk and milk products.

Dairy product. A dairy product is a food or beverage produced from the milk of mammals such as cows, goats, sheep, yaks, horses, and camels.

Ethnicity. An ethnicity is a social group of people who identify with each other based on common ancestral, religious, linguistic, social, cultural, or national experience.

Lactase. Lactase is an enzyme produced by the epithelial cells of the small intestine that hydrolyzes lactose into glucose and galactose.

Lactase persistence. Lactase persistence is the continued production of the enzyme lactase in adulthood.

Lactose. Lactose is the primary carbohydrate in milk. This disaccharide must be hydrolyzed to its two constituent monosaccharides, glucose and galactose, to be transported across the intestinal epithelium.

Lactose intolerance. Lactose intolerance is a group of symptoms (cramps, diarrhea, flatulence, gurgling, bloating, nausea, and combinations thereof) that are caused by the inability to enzymatically cleave lactose into its two constituent sugars.

Nomad. A pastoral group that moves their herds seasonally to avoid depleting pasturage.

Pastoralist. Pastoralists are nomads who raise lifestock such as camels, goats, sheep and cattle. Many of these animals are milk producing.

List of Abbreviations

g Grams
mg Milligrams

References

Alamu, O. T., Amao, A. O., Nwokedi, C. I., Oke, O. A. and Lawa, I. O., 2013. Diversity and nutritional status of edible insects in Nigeria: a review. *International Journal of Biodiversity and Conservation*. 5(4): 215–222.

Bahman, S., Yadav, N., Kumar, A., Ganguly, S., Garg, V. and Marwaha, S. S., 2012. Impact of household practices on the nutritional profile of milk. *Indian Journal of Public Health*. 56(1): 82–87.

Baird, D. L., Syrette, J., Hendrie, G. A., Riley, M. D., Bowen, J. and Noakes, M., 2012. Dairy food intake of Australian children and adolescents 2–16 years of age: 2007 Australian National Children's Nutrition and Physical Activity Survey. *Public Health Nutrition*. 15(11): 2060–2073.

Barer-Stein, T., 1999. *You Eat What You Are: People, Culture and Food Traditions*. 2nd edn, Firefly Books, Richmond Hill, Ontario, 544 pp.

Christensen, D. L., Orech, F. O., Mungai, M. N., Larsen, T., Friis, H. and Aagaard-Hansen, J., 2006. Entomophagy among the Luo of Kenya: a potential mineral source? *International Journal of Food Sciences and Nutrition*. 57: 198–203.

Ekpo, U. F., Omotayo, A. M. and Dipeolu, M. A., 2008. Prevalence of malnutrition among settled pastoral Fulani children in Southwest Nigeria. *BMC Research Notes*. 1(1): 7.

Estaire, P., González-Rodríguez, L. G., López-Sobaler, A. M. and Ortega, R. M., 2012. Food sources and intake of calcium in a representative sample of Spanish adults. *Food and Nutrition Sciences*. 3: 1269–1276.

Faber, M., Van Jaarsveld, P. J. and Laubscher, R., 2007. The contribution of dark-green leafy vegetables to total micronutrient intake of two- to five-year-old children in a rural setting. *Water SA*. 33(3): 407–412.

Fong, S., 2001. Celebrating baby beginnings. *Flavor and Fortune*. 8: 13–14.

Fulgoni, III, V., Nicholls, J., Reed, A., Buckley, R., Kafer, K., Huth, P., DiRienzo, D. and Miller, G. D., 2007. Dairy consumption and related nutrient intake in African-American adults and children in the United States: continuing survey of food intakes by individuals 1994–1996, 1998, and the National Health and Nutrition Examination Survey 1999–2000. *Journal of the American Dietetic Association*. 107(2): 256–264.

Fuller, F., Beghin, J. and Rozelle, S., 2007. Consumption of dairy products in urban China: results from Beijing, Shangai and Guangzhou. *Australian Journal of Agricultural and Resource Economics*. 51(4): 459–474.

Gannagé-Yared, M. H., Chemali, R., Sfeir, C., Maalouf, G. and Halaby, G., 2005. Dietary calcium and vitamin D intake in an adult Middle Eastern population: food sources and relation to lifestyle and PTH. *International Journal for Vitamin and Nutrition Research*. 75(4): 281–289.

Ge, K., 2011. The transition of Chinese dietary guidelines and the food guide program. *Asia Pacific Journal of Clinical Nutrition*. 20(3): 439–446.

Glew, R. H., Williams, M., Conn, C. A., Cadena, S. M., Crossey, M., Okolo, S. N. and VanderJagt, D. J., 2001. Cardiovascular disease risk factors and diet of Fulani pastoralists of northern Nigeria. *American Journal of Clinical Nutrition*. 74(6): 730–736.

Goh, D. Y. T. and Jacob, A., 2011. Children's consumption of beverages in Singapore: knowledge, attitudes and practice. *Journal of Paediatrics and Child Health*. 47(7): 465–472.

Gupta, M. D., 1987. Selective discrimination against female children in rural Punjab, India. *Population and Development Review*. 13: 77–100.

Henderson, L., 2003. The national diet & nutrition survey: adults aged 19 to 64 years. Available at: http://multimedia.food.gov.uk/multimedia/pdfs/ndnsv3.pdf.

Huang, Y. and Ang, C. Y. W., 1992. Vegetarian foods for Chinese Buddhists. *Food Technology*. 46: 105–108.

Huybrechts, I., Lin, Y., De Keyzer, W., Sioen, I., Mouratidou, T., Moreno, L. A., Slimani, N., Jenab, M., Vandevijvere, S., De Backer, G. and De Henauw, S., 2011. Dietary sources and sociodemographic and economic factors affecting vitamin D and calcium intakes in Flemish preschoolers. *European Journal of Clinical Nutrition*. 65(9): 1039–1047.

Im, J. G., Kim, S. H., Lee, G. Y., Joung, H. and Park, M. J., 2013. Inadequate calcium intake is highly prevalent in Korean children and adolescents: the Korea National Health and Nutrition Examination Survey (KNHANES) 2007–2010. *Public Health Nutrition*. 1–7.

Ingram, C. J., Mulcare, C. A., Itan, Y., Thomas, M. G. and Swallow, D. M., 2009. Lactose digestion and the evolutionary genetics of lactase persistence. *Human Genetics*. 124(6): 579–591.

Islam, M. Z., Lamberg-Allardt, C., Kärkkäinen, M. and Ali, S. M. K., 2003. Dietary calcium intake in premenopausal Bangladeshi women: do socio-economic or physiological factors play a role? *European Journal of Clinical Nutrition*. 57(5): 674–680.

Japan Dietetic Association, 2014. Japan's dietary guidelines. Available at: http://www.dietitian.or.jp/english/newsletter/jpndiet.html.

Johnson-Down, L., Ritter, H., Starkey, L. J. and Gray-Donald, K., 2006. Primary food sources of nutrients in the diet of Canadian adults. *Canadian Journal of Dietetic Practice and Research*. 67(1): 7–13.

Karandish, M., Mohammadpour-Ahranjani, B., Rashidi, A., Maddah, M., Vafa, M. R. and Neyestani, T. R., 2005. Inadequate intake of calcium and dairy products among pregnant women in Ahwaz city, Iran. *Malaysian Journal of Nutrition*. 11: 111–120.

Karandish, M. and Naghashpoor, M., 2010. Calcium intake pattern among postmenopausal women in Ahwaz, Iran. *Journal of Biological Sciences*. 10: 251–254.

Keast, D. R., Fulgoni, V. L., Nicklas, T. A. and O'Neil, C. E., 2013. Food sources of energy and nutrients among children in the United States: National Health and Nutrition Examination Survey 2003–2006. *Nutrients*. 5(1): 283–301.

Khan, N. C., Mai, L. B., Hien, V. T. T., Lam, N. T., Hoa, V. O., Phuong, T. M., Nhung, B. T., Nakamori, M., Shimizu, Y. and Yamamoto, S., 2008. Development and validation of food frequency questionnaire to assess calcium intake in postmenopausal Vietnamese women. *Journal of Nutritional Science and Vitaminology*. 54(2): 124–129.

Kilara, A. and Iya, K. K., 1992. Food and dietary habits of the Hindu. *Food Technology*. 46: 94–104.

Kinabo, J., Mnkeni, A. P., Nyaruhucha, C. N. M., Msuya, J., Haug, A. and Ishengoma, J., 2006. Feeding frequency and nutrient content of foods commonly consumed in the iringa and morogoro regions in Tanzania. *International Journal of Food Sciences and Nutrition*. 57(1–2): 9–17.

Kittler, P. G. and Sucher, K., 2008. *Food and Culture*. Thompson Wadsworth, Canada, 5th edn, 544 pp.

Kuhnlein, H. V., 1984. Nutritional value of traditional food practices. *Research in Food Science and Nutrition*. 4: 63–71.

Kumar, E. G. T., 2007. High prevalence of low dietary calcium, high phytate consumption, and vitamin D deficiency in healthy south Indians. *American Journal of Clinical Nutrition*. 85(4): 1062–1067.

Lazarou, C. and Matalas, A. L., 2010. A critical review of current evidence, perspectives and research implications of diet-related traditions of the Eastern Christian Orthodox Church on dietary intakes and health consequences. *International Journal of Food Sciences and Nutrition*. 61(7): 739–758.

Lee, M. S., Huang, L. Y., Chen, M. C. and Wahlqvist, M. L., 2009. The demography of food in health security: current experience with dairy consumption in Taiwan. *Asia Pacific Journal of Clinical Nutrition*. 18(4): 585–589.

Magkos, F., Manios, Y., Babaroutsi, E. and Sidossis, L. S., 2006. Development and validation of a food frequency questionnaire for assessing dietary calcium intake in the general population. *Osteoporosis International*. 17(2): 304–312.

Matteucci, E. and Giampietro, O., 2008. Current dietary pattern in central Italy. *Open Nutrition Journal*. 2: 15–22.

McGee, H., 2007. *On Food and Cooking: The Science and Lore of the Kitchen*. Simon and Schuster Canada Inc, 896 pp.

Moynihan, P., Adamson, A., Rugg, A., Appleton, D. and Butler, T., 1996. Dietary sources of calcium and the contribution of flour fortification to total calcium intake in the diets of Northumbrian adolescents. *British Journal of Nutrition*. 75: 495–505.

Newman, J. M., 2004. Eggs: fresh and preserved, Chinese style. *Flavor and Fortune*. Winter Volume 11: 27, 29, 30, 32. Available at: http://www.flavorandfortune.com/dataaccess/article.php?ID=491. Accessed 19 April 2014.

Oiye, S., Simel, J. O., Oniang'o, R. and Johns, T., 2009. The Maasai food system and food and nutrition security. In: Kuhnlein, H. V., Erasmus, B. and Spigelski, D. (ed.) *Indigenous Peoples' Food Systems: The Many Dimensions of Culture, Diversity and Environment for Nutrition and Health*. Food and Agriculture Organization of the United Nations (FAO), pp. 231–249.

O'Neil, C. E., Keast, D. R., Fulgoni, V. L. and Nicklas, T. A., 2012. Food sources of energy and nutrients among adults in the US: NHANES 2003–2006. *Nutrients*. 4(12): 2097–2120.

Ortega, R. M., López-Sobaler, A. M., Jiménez, O. A., Navia, L. B., Ruiz-Roso, C. D. M. B., Rodríguez-Rodríguez, E. and López, P. B., 2012. Food sources and average intake of calcium in a representative sample of Spanish schoolchildren. *Nutricion Hospitalaria*. 27(3): 715–723.

Osteoporosis Canada, 2014. Calcium: an important nutrient that builds stronger bones. Available at: http://www.osteoporosis.ca/osteoporosis-and-you/nutrition/calcium-requirements/. Accessed 28 May 2014.

Prentice, A. M., 2014. Dairy products in global public health. *American Journal of Clinical Nutrition*. 99(suppl. 5): 1212S–1216S.

Raubenheimer, D. and Rothman, J. M., 2013. Nutritional ecology of entomophagy in humans and other primates. *Annual Review of Entomology*. 58: 141–160.

Roberts, J. R., 2010. Factors affecting egg shell and internal egg quality. In: Proceedings of the 18th Annual ASAIM SE Asian Feed Technology and Nutrition Workshop. May 24–27, 2010 Le Meridien Siem Reap, Cambodia, pp. 1–9. Available at: http://www.asaimsea.com/pdfs/FTNW10-Factors_Roberts%20paper.pdf. Accessed 19 April 2014.

Rosado, J. L., Díaz, M., Rosas, A., Griffit, I. and García, O. P., 2005. Calcium absorption from corn tortilla is relatively high and is dependent upon calcium content and liming in Mexican women. *Journal of Nutrition*. 135(11): 2578–2581.

Sanwalka, N. J., Khadilkar, A. V., Mughal, M. Z., Sayyad, M. G., Khadilkar, V. V., Shirole, S. C., Divate, U. P. and Bhandari, D. R., 2010. A study of calcium intake and sources of calcium in adolescent boys and girls from two socioeconomic strata, in Pune, India. *Asia Pacific Journal of Clinical Nutrition*. 19(3): 324–329.

Sato, Y., Tamaki, J., Kitayama, F., Kusaka, Y., Kodera, Y., Koutani, A. and Iki, M., 2005. Development of a food-frequency questionnaire to measure the dietary calcium intake of adult Japanese women. *Tohoku Journal of Experimental Medicine*. 207(3): 217–222.

Savaiano, D., 2010. Lactose intolerance: an unnecessary risk for low bone density. *Nestle Nutrition Workshop Series. Paediatric Programme*. 67: 161–171.

Shatrugna, V., Kulkarni, B., Kumar, P. A., Rani, K. U. and Balakrishna, N., 2005. Bone status of Indian women from a low-income group and its relationship to the nutritional status. *Osteoporosis International*. 16(12): 1827–1835.

Shmueli, A. and Tamir, D., 2007. Health behavior and religiosity among Israeli Jews. *Israel Medical Association Journal*. 9(10): 703–707.

Stern, L., 2004. *How to Keep Kosher*. 1st edn. HarperCollins, 303 pp.

Van Der Walt, A. M., Loots, D. T., Ibrahim, M. I. M. and Bezuidenhout, C. C., 2009. Minerals, trace elements and antioxidant phytochemicals in wild African dark-green leafy vegetables (morogo). *South African Journal of Science*. 105(11–12): 444–448.

Wu, S. J., Pan, W. H., Yeh, N. H. and Chang, H. Y., 2007. Dietary nutrient intake and major food sources: the Nutrition and Health Survey of Taiwan Elementary School Children 2001–2002. *Asia Pacific Journal of Clinical Nutrition*. 16(suppl. 2): 518–533.

Yoon, E. H., Noh, H., Lee, H. M., Hwang, H. S., Park, H. K. and Park, Y. S., 2012. Bone mineral density and food-frequency in Korean adults: the 2008 and 2009 Korea National Health and Nutrition Examination Survey. *Korean Journal of Family Medicine*. 33(5): 287–295.

Zhai, F., Wang, H., Du, S., He, Y., Wang, Z., Ge, K. and Popkin, B. M., 2009. Prospective study on nutrition transition in China. *Nutrition Reviews*. 67(s1): S56–S61.

Zhang, C. X., Chen, Y. M., Chen, W. Q., Su, Y. X., Wang, C. L. and Wu, J. N., 2012. Food group intake among adolescents in Guangzhou city compared with the Chinese dietary guidelines. *Asia Pacific Journal of Clinical Nutrition*. 21(3): 450–456.

CHAPTER 4

Calcium Availability in Specific Foods: Milk and Dairy Products, Legumes, Vegetables, Cereals, Baked Goods and Cooked Meals

Mª VICTORINA AGUILAR VILAS*[a]

[a]Department of Biomedical Sciences, Alcala University, 28871 Alcalá de Henares, Madrid, Spain
*E-mail: mvictorina.aguilar@uah.es

4.1 Introduction

Calcium is an essential nutrient for cell viability that can only be obtained from food. Its biological importance stems firstly from its structural function, providing integrity to the skeleton and maintaining intracellular structures and biological membrane integrity and permeability, and secondly, from the essential role of calcium ions in biochemical processes such as extra- and intracellular signaling, blood coagulation, membrane permeability, *etc.* (Peacock, 2010).

Unlike many other nutrients, calcium remains stable over time throughout the food production chain. However, intake is usually deficient among

Food and Nutritional Components in Focus No. 10
Calcium: Chemistry, Analysis, Function and Effects
Edited by Victor R. Preedy

Published by the Royal Society of Chemistry, www.rsc.org

most of the world's population, partly because not all ingested calcium is absorbed.

Due to the consequences for health and the economy of a deficient calcium intake, it is essential to formulate strategies for optimizing its absorption, such as changing recommended dietary intakes, establishing recommendations for supplementing or enriching foods and dietary products, or proposing technological processes of varying degrees of complexity. However, such strategies require knowledge of the bioavailability of calcium in different foods and, above all, of the different factors that influence this parameter.

4.2 Bioavailability: Influencing Factors

Bioavailability can be defined as the efficiency with which a nutrient is used by the body, and is usually expressed as a percentage. Since there are several methods for determining bioavailability, all of which may yield different results, the comparison of different foods and/or food components presented in both this and the following section is based on data obtained using the same method.

Between 20% and 60% of ingested calcium is absorbed and used; however, various factors may modify these figures, including the interaction of calcium with other food components, pathophysiological factors and even lifestyles. Of these, only dietary factors will be discussed here.

4.2.1 Calcium Intake and Speciation

The bioavailability of calcium depends on the chemical form in which it appears. Thus, of the various salts used as supplements, it is calcium formate that yields the highest absorption rates due to its low molecular weight, high calcium content and water solubility (Figure 4.1) (Hanzlik *et al.*, 2005; Heaney *et al.*, 2005).

There are conflicting opinions concerning the amount that should be ingested; the traditional recommendation was to take calcium in small doses to increase absorption, whereas these days the recommendation is to take higher amounts of calcium to increase its bioavailability.

4.2.2 Vitamin D

This acts at the levels of absorption and excretion alike. In calcium absorption, calcitriol acts on the transcellular pathway, inducing two different types of response. The first of these is fast acting and produces a concentration of presynthesized calcium channels (TRPV6) in the apical membrane, whilst the second is slow acting and induces increased synthesis of calcium channels, calbindin and calcium transporters and pumps such as the PMCA or the Na/Ca exchanger (NCX), responsible for calcium extrusion from the basolateral membrane (Figure 4.2).

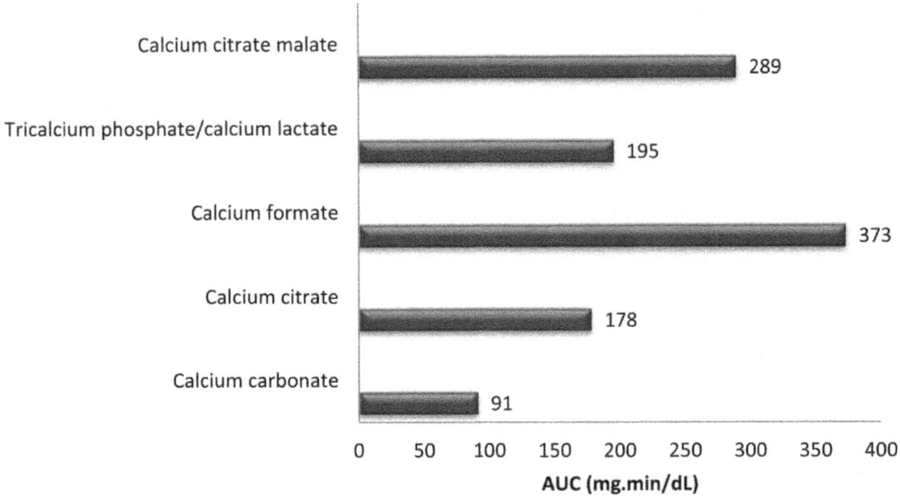

Figure 4.1 Increment in area under the curve for the serum calcium for five calcium compounds. Different calcium compounds have different ΔAUCs, indicative of increases in serum calcium. These increases are dependent on different factors such as the physical and chemical characteristics of each of them. Data that has been used in this figure comes from Hanzlik *et al.* (2005) and Heaney *et al.* (2005).

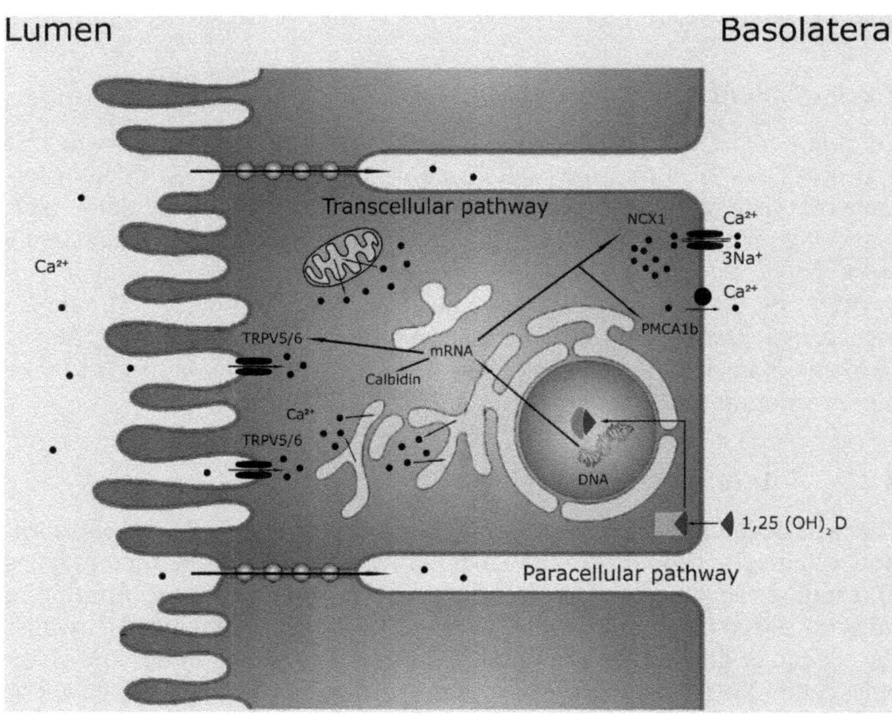

Figure 4.2 Vitamin D effect on calcium absorption. The calcitriol binds to the VDR that induces the synthesis of proteins such as TRPV6, calbindin.

4.2.3 Proteins or Specific Amino Acids

The effect of proteins on the bioavailability of calcium is controversial because although protein increases gastric acidity and thus enhances the degree of calcium absorption, it also intensifies urinary calcium loss. Possible mechanisms of action include the existence of an intrinsic relationship between dietary protein and calcium metabolism through modulation of growth factor IGF-1 levels (Tala de Souza *et al.*, 2002), or an action mediated by PTH and calcitriol (Theobald, 2005). However, the results of observational and clinical studies differ. Thus, in some studies such as those carried out by Shapses and Sukumar (2010), high protein diets were positively associated with bone density changes in the femoral neck and spine, whereas some epidemiological studies have suggested that protein induces a reduction in bone mass, a finding attributed to a protein-induced acid overload that must be mitigated by the release of bone calcium and increased urinary calcium excretion. In a meta-analysis conducted in 2009, Darling *et al.* concluded that although protein may exert a slight beneficial effect on bone, in the long term this does not result in a lower risk of fractures (Darling *et al.*, 2009).

This action differs according to the type of protein; even different varieties of casein have different effects, as will be discussed later.

Amino acids such as lysine and arginine also enhance calcium absorption by increasing intestinal permeability and producing N, respectively.

Some protein derivatives, such as phosphopeptides resulting from casein proteolysis, favor absorption due to the formation of soluble complexes with calcium at luminal pH, which act at the level of the cell membrane (Perales *et al.*, 2005) or by preventing interaction between calcium and other minerals such as iron.

4.2.4 Carbohydrates

In general, carbohydrates increase absorption through several mechanisms; for example, fructose forms soluble complexes with calcium. The majority of studies have focused on lactose, and several mechanisms have been proposed to explain its action:

- Acidification of the intestinal environment produced by fermentation of lactose, which increases soluble calcium and thus absorption.
- Activation of passive and active transport pathways.
- Hyperplastic action, stimulating colon development and consequently increasing the surface area for absorption. This latter hypothesis has been rejected because the positive effect of lactose remains the same after colectomy (Wasserman, 2012).

Findings on the effect of other sugars such as arabinose have been contradictory; some studies have found a decreased efficiency of calcium absorption (Dierpenmaat-Wolters and Diepenmaat, 1993), perhaps due to the formation of nonabsorbable complexes, whereas others have reported a slight increase in the absorption of this ion.

4.2.5 Fat

Fat decreases calcium absorption, but this effect is only significant in the case of lipid malabsorption (*e.g.* in cases of steatorrhoea, coeliac disease, short-bowel syndrome and pancreatitis), due to the formation of insoluble, nonabsorbable calcium soaps with fatty acids, especially long-chain saturated fatty acids.

4.2.6 Minerals

Iron, magnesium and phosphorus reduce calcium absorption because they share the same absorption mechanism. In the case of phosphorus, this compensates for, or at least minimizes, hypercalciuria induced by high protein diets. Other minerals such as sodium increase urinary excretion.

4.2.7 Oxalic Acid

Oxalic acid, present in vegetables and legumes, binds to calcium to form a compound that is not easily absorbed in the intestine; however, the absorption of calcium from other foods consumed in the same meal is not affected. Nevertheless, foods containing oxalic acid are harmful when eaten in large quantities. In fact, the bioavailability of calcium is determined by the oxalic acid (g kg^{-1})/calcium (g kg^{-1}) ratio. Any foodstuff with a ratio higher than 2.25 is a poor source of calcium and removes this ion.

4.2.8 Phytic Acid

As with oxalic acid, this compound, found in cereals and legumes, decreases the bioavailability of calcium, in this case because the six phosphates that comprise its molecule have a high tendency to bind to the calcium cation in the gastrointestinal tract, forming calcium phytate, which cannot be absorbed due to its insolubility. Both inositol hexaphosphoric acid (IP6) and some of its derivatives, such as IP5, present this complexing action. A phytate : calcium ratio of <2 is used to estimate the relative bioavailability of calcium, and reabsorption of endogenously excreted calcium is also affected.

Phytates bind to calcium from other foodstuffs eaten at the same meal, preventing its absorption.

4.2.9 Fiber

Different fiber components such as uronic acids, cellulose and hemicellulose reduce the efficiency of calcium absorption by binding and transfer to distal regions of the intestine, preventing absorption, or by forming insoluble complexes with calcium. Not all components of fiber act *via* the same mechanism, since cellulose appears to adsorb calcium and does not bind it to the hydroxyl groups in its polymer chains. However, other fiber components such as fructo-oligosaccharides (FOS), resistant starch or gum, may promote the absorption of calcium since the short chain fatty acids formed by their colonic fermentation:

- facilitate its passage through the membrane by passive diffusion in the form of calcium acetate (Lopez *et al.*, 2006);
- reduce pH, favoring absorption by passive diffusion through increasing its solubility;
- modify exchangers (Ca—H) for the transcellular pathway and may also induce the synthesis of calcium-binding proteins (calbindin D9k) that accelerate the absorption of calcium *via* the transcellular pathway (Lopez *et al.*, 2006).

FOS can even inhibit the negative action of phytic acid, since colonic fermentation enhanced by FOS promotes better phytic acid hydrolysis.

According to some long-term studies, an adaptation to the repeated consumption of fiber may occur in such a way that after 3–4 weeks, it no longer affects the bioavailability of minerals (Frølich, 1995).

4.2.10 Other Dietary Factors

Low molecular weight organic acids such as citric or malic acid form soluble complexes with calcium, favoring absorption. Alcohol and caffeine also affect absorption of this mineral; alcohol reduces intestinal absorption by primary or secondary malnutrition or by inhibiting certain liver enzymes that activate vitamin D. Meanwhile, caffeine appears to have a slightly negative effect on both absorption and urinary excretion of calcium because it induces an increase in urinary calcium (Theobald, 2005).

4.3 Calcium Availability in Specific Foods

This section will deal exclusively with foodstuffs that provide a high amount of calcium in the diet.

4.3.1 Milk and Dairy Products

4.3.1.1 Milk

Milk and dairy products are the main source of calcium, providing 48% of the recommended intake in Spain, 80% in North America and Canada and 100% in infants.

Human milk is the only strictly natural food we consume in our lives. Although it possesses a calcium content of 34 ± 2.9 g/100 mL, the degree of absorption differs according to the literature sources consulted. Thus, the ESPGHAN cites a figure of 75%, whereas substantially lower values (13%) have been reported in other studies (Roig *et al.*, 1999), differences that may be due to the methodology employed.

Since human milk is only consumed for a short period of time and is then replaced, usually by bovine milk, we shall compare bioavailability in both milks. In studies using similar methodologies, it has been observed that calcium bioavailability is greater in bovine milk than in human milk (20–30% *vs.* 13%) (Roig *et al.*, 1999), due to the components of bovine milk and the distribution of calcium in its different phases. Thus, bovine milk has higher calcium content (Table 4.1), and the mineral is usually bound to casein, of which bovine milk contains a high proportion, especially α-casein. Casein forms micelles containing calcium (57%), both as colloidal calcium phosphate and bound to serine and threonine residues located near the N terminus. Human milk contains less casein and furthermore, the predominant form is β-casein, which presents a lower tendency to bind to calcium (12%). The phosphopeptides formed during technological or digestive processing also contribute to the high bioavailability of calcium in milk, since they maintain calcium in a soluble form in the intestinal lumen, favoring the intraluminal stage of calcium absorption (Abrams *et al.*, 1990).

Serum proteins affect the speed of absorption rather than bioavailability *per se*, although they generally exert a lower effect on the rate of milk absorption than casein (Soral-Smietana *et al.*, 2013). However, some milk components, such as α-lactalbumin in human milk, increase calcium absorption due to the generation of peptides that facilitate the absorption of divalent cations such as calcium (Lönnerdal, 2003). Other milk components such as oligosaccharides (lactose, lactulose) favor the absorption of calcium in a dose-dependent manner due to the increase in luminal fluid they induce in order to maintain isotonicity. In turn, this increase in fluid may lead to an increase in intestinal distension, permeability of the junctions between enterocytes and passive calcium transport (Brommage *et al.*, 1993).

As discussed above, fats have an adverse effect on calcium bioavailability due to the formation of insoluble calcium soaps in the intestine, which may eventually cause intestinal obstruction. However, their effect differs according to the type of milk considered, since fatty acids in bovine milk are

Table 4.1 This table shows different dates of calcium availability depending on the species of animal origin. There is a huge variability in the data of the same species due to the methodology used.

Milk	Total calcium (mg)/100 mL	% Calcium availability
Human	34.0	13–75
Cow	125–150	32.1
Buffalo	180–200	48
Goat	126.8	18.3 ± 2.4
Sheep	184.0	18.2 ± 0.5

esterified at carbons 1 and 3. Lipases act on these ester bonds, freeing acids such as palmitic acid that bind to calcium to form calcium palmitate, which is then excreted in feces. This does not occur in human milk since 70% of the fatty acids are located in position 2; thus, they are not hydrolyzed by pancreatic lipase but remain bound to glycerol.

The Ca/P ratio is another key factor in the absorption of calcium, and is more appropriate in human milk (2 : 4) than in bovine milk (1 : 3): as a result, breastfed babies have higher plasma levels of calcium than babies fed bovine milk, who may run the risk of hypocalcaemia.

Prebiotics such as FOS are sometimes added to bovine milk; FOS increase the absorption of calcium, especially those that consist of three monomers such as 1-kestose, which forms the substrate of lactobacillus and bifidobacterium species for the production of short-chain fatty acids.

The bioavailability of calcium has also been observed to be greater in buffalo, goat and sheep milk (Table 4.2), a finding attributed to their higher calcium content (Shen *et al.*, 1995; Ahmad *et al.*, 2013). The high bioavailability of calcium in buffalo milk is explained by its Ca/P ratio of 1.6 : 4, which is similar to breast milk, and because approximately two thirds of the calcium is bound to casein, adopting a dispersed colloidal configuration that is easily absorbed, whereas only 10% is in the form of ionized calcium.

Table 4.2 This table shows the differences between cow's and human milk components. The presence, structure and quantity of some of these are factors that influence the bioavailability of the calcium.[a]

Components	Human milk (100 mL)	Cow's milk (100 mL)
Water	87	79
Energy (kcal)	62–70	68
Proteins (g)	0.89–1.1	3.5
Casein (g)	0.3–0.5 (40)	2–3 (82)
Whey protein (g) (%)	0.4–0.6 (60)	0.6–0.7 (18)
α-Lactalbumin (g)	0.2–0.3	0.12
β-Lactoglobulin (g)	—	0.30
Serum albumin (g)	0.03	0.03
Immunoglobulins (g)	0.06–0.1	0.07
Lactoferrin (g)	0.1–0.3	Traces
Lysozyme (g)	0.05	Traces
Carbohydrate (g)	7.7	6.1
Lactose (g)	6–6.5	5
Oligosaccharides (g)	1–1.2	0.1
Total fat (g)	3.5–4	3.8
Linoleic acid (% total fat)	12.5	2
Linolenic acid (% total fat)	1.01	0.5
Arachidonic acid (%total fat)	0.6	0.3
Docosahexaenoic acid (% total fat)	0.3	ND
Phosphorus	15	96

[a]ND: not detected.

4.3.1.2 *Dairy Products*

Technological processes influence the bioavailability of calcium in dairy products. Thus, the skimming process yields greater bioavailability, 24.5 to 31.5%, due to the partial removal of fat (Theobald, 2005). The same applies to the production of other dairy products such as cheese or evaporated milk, which contain more calcium in colloidal form than the milk they are made from, and so a greater proportion of calcium is absorbed. The fermentation process involved in the production of yoghourt causes a reduction in micellar calcium and an increase in the amount of serum calcium in the form of nonsedimenting calcium ions, which are also easily absorbed (De la Fuente *et al.*, 2003).

Various supplements such as calcium compounds or fiber are sometimes added to dairy products for nutritional purposes. In the case of fiber, it has been shown that any type of added fiber, whether insoluble such as wheat or bamboo or soluble such as inulin, pectin or plantago psyllium, reduces the bioavailability of calcium in yoghourt by between 7% and 21% (Rodriguez *et al.*, 2008).

Bipolar emulsifiers may also be added to dairy products, and these attract divalent cations such as calcium, lowering its bioavailability (Perales *et al.*, 2005).

Nowadays, a wide range of milks to which fruit juices and cereals have been added are marketed as functional foods; however, these have a lower total calcium content and calcium solubility than milk. Furthermore, their low casein content and the cereal fiber they contain, which forms insoluble compounds, affect the enhancing action of fructose and citric and malic acids (Perales *et al.*, 2005).

4.3.1.3 *Infant Formulas*

There are also many circumstances in which breastfeeding is not possible or must be supplemented. In these cases, infant formula milk is employed (infant formula, follow-on, toddler) or sometimes, special formulas (antireflux, lactose-free, anticonstipation, protein hydrolyzate, soy-based formulas, *etc.*) with a modified carbohydrate, protein or mineral composition, intended for feeding infants presenting some kind of pathology.

The aim of adapted milks is not so much to reproduce the exact characteristics of human milk but rather to develop a formula in which the bioavailability of nutrients is similar. The bioavailability of calcium in infant formula milk varies widely, ranging from 13% to 31% (Roig *et al.*, 1999; Perales *et al.*, 2005). The factor that exerts the greatest influence on this parameter is the calcium content of formulas: the higher the content, the greater the bioavailability (González-Bermúdez *et al.*, 2011). Nevertheless, the serum protein/casein ratio also plays an important role, because as mentioned above, higher casein fraction values are associated with greater calcium bioavailability. In fact, formulas made with hydrolyzed serum protein yield poorer results (13.3 ± 1.2%) than those obtained for casein-based formulas (21.2 ± 0.6%) (Bosscher *et al.*, 2001).

Iron is sometimes added to formulas and affects the absorption of calcium both by impairing casein hydrolysis (Drago and Valencia, 2004) and through the direct interaction that occurs between these two (Jovani *et al.*, 2000; Perales *et al.*, 2005).

The fats, usually vegetable, which are added to infant formulas also influence bioavailability. In formulas with a linoleate : stearate ratio similar to that of breast milk, calcium absorption is 38% of the intake, *i.e.* higher than when linoleate or oleate are added (Barnes *et al.*, 1974).

4.3.1.4 *Special Formulas*

Lactose-free formulas: Increasingly sought after by parents as a means to avoid lactose intolerance, these formulas present lower calcium absorption rates than those that do contain lactose (Table 4.3), calling into question the use of these formulas given their consequences.

Soy-based formulas: These possess a higher protein content than bovine milk (2.25 compared to 1.8 g/100 kcal) and are supplemented with calcium and phosphorous (Ca/P ratio = 1.2 : 2) to compensate for the high phytate content. Thus, the multiple factors that inhibit calcium absorption are taken into account during manufacture so that bioavailability is similar to that of formulas produced from bovine milk (Table 4.3).

The same could be said of lactose-free, soy-based formulas, which contain more calcium to compensate for the presence of soy proteins and the absence of lactose.

Antireflux formulas: These contain nondigestible carbohydrates that reduce intestinal absorption of minerals such as calcium, iron and zinc due to their chelating properties. They may also contain thickening agents such as alginates or guar gum, which reduce the absorption of calcium because they bind to this ion and form nonabsorbable compounds (Bosscher *et al.*, 2001).

Anticonstipation formulas: These contain modified triglycerides so that palmitic acid occupies the beta position. Thus, they contain 45% palmitic acid

Table 4.3 The availability expressed as dialysis percentage obtained in different infant formulas, with different compositions, is shown in this table.

Samples	% Dialyzed calcium
Adapted	21.9–31.6
Without lactose	15.3
Soy based	31.2
Whey based	13.3
Casein based	21.2
Soy based supplemented with calcium	46.4
Adapted supplemented with ferrous sulfate	30.8
Adapted supplemented with ferrous lactate	28.7
Follow on	24.1–31.0

at beta position, in an attempt to reproduce the composition of breast milk. They also have a Ca/P ratio of 2, similar to that of human milk.

The technological processes involved in the manufacture of infant formulas affect the bioavailability of calcium. For example, the Maillard reaction that occurs during the preparation of these formulas has a greater effect on the bioavailability of calcium in liquid formulas than in powder ones, because the compounds produced during the early stages of nonenzymatic browning form absorbable complexes with Ca. Furthermore, the addition of polar emulsifiers induces a reaction with divalent cations, decreasing their bioavailability, whilst the addition of prebiotics that increase the solubility of calcium or exert trophic effects on the intestinal mucosa increases the absorption of this ion (Hicks *et al.*, 2012).

4.3.2 Legumes

Legumes are a good source of proteins, complex carbohydrates and water-soluble vitamins and minerals, and have a low lipid content. However, the presence of antinutrients such as phytate or oxalate decreases their nutritional value.

Dialyzable calcium in legumes ranges from 2.45% in raw lentils to 26.7% in pinto beans (Table 4.4), although these values may vary according to the method used to quantify bioavailability (Viadel *et al.*, 2006). The bioavailability of calcium in soybeans is similar to that in breast milk, despite the presence of phytic and oxalic acid (Heaney, 1991).

The negative effects of these antinutrients can be reduced by technological processes or food-preparation methods such as germination, soaking, fermentation or phytase treatment, all of which break down the inositol hexaphosphoric molecule, increasing the amount of dialyzable calcium. The same does not occur with boiling, since in this case IP5 is formed, sequestering the calcium cation and affecting bioavailability. However, in the case of oxalic acid, boiling is responsible for the loss of soluble oxalate: 59.6% in white beans, 19.5% in chickpeas and 42.6% in lentils, which may help explain the greater bioavailability of calcium in cooked legumes (Viadel *et al.*, 2006).

Table 4.4 This table shows the dialyzable calcium obtained by several authors and the content of phytic acid and fiber dietary from legumes most consumed among the Western population.

Legumes	Calcium total content (mg/100 g)	% Dialyzable calcium	Phytic acid (g/100 g)	Fiber (g/100 g)
Beans white	103	21.8	0.6–1.8	25.4
Beans pinto	52	26.7	1.1–1.2	25.4
Beans red	24	24.4	1.3	25.4
Chickpeas	145	14.6	0.6–1	11.7–15
Lentils	56	2.45	0.8–1	11.7–15
Soybean seeds	139	11.1	1.3–2.2	22

Legumes also have a high fiber content (Table 4.4), and some of their components, such as pectins (especially in beans), located in the cotyledons of legumes, have chains that bind to calcium through nonmethylate free carboxylic acid groups or uronic acid, reducing the bioavailability of calcium. Calcium absorption is higher in ready-to-eat lentils, because these have been soaked in a solution of EDTA, which reacts with phytic acid and reduces its chelating action. Furthermore, ascorbic and citric acids are usually added, enhancing calcium absorption and offsetting the negative effect of phytic acid (Viadel *et al.*, 2006).

Although the bioavailability of calcium in soy products depends on the manufacturing process employed, for example, the type of coagulant used to precipitate the soy protein in the production of tofu, values are similar to those of bovine milk (31%).

4.3.3 Vegetables

Vegetables, especially leafy greens, have a high calcium concentration (Table 4.5), even higher than in dairy products, but bioavailability is much lower due to the presence of calcium absorption inhibitors.

The antinutrient present in these foodstuffs that produces the most drastic reduction in the bioavailability of calcium is oxalic acid. In fact, studies on bioavailability usually classify foods according to their oxalate content. Thus, it has been shown that bioavailability is low in oxalate-rich (>10 mg/portion) foodstuffs such as spinach and rhubarb, whereas others with a low content of this acid, such as those belonging to the Brassica family (*e.g.* broccoli and kale) are good sources of calcium (Table 4.5). However, differences

Table 4.5 Antinutrients as dietary fiber, oxalic acid and phytate affect to the calcium availability, especially in green leafy vegetables.

Samples	Total calcium content (mg/100 g)	% Calcium available	Fiber (g/100 g)	Phytic acid (g/100 g)	Oxalic acid (mg/100 g)
Indian mulberry	238	17.8	7.9	7.8	291
Pak-paw	370	6.6	9.2	11.3	853
Tree kale blanched	134	34.4	4.1	3.8	2.4
Pak-chee-lao blanched	206	32.7	4.0	15	100
Kale blanched	132	38.9	3.0	7.9	1.3
Celery blanched	138	36.9	1.8	5.2	8.4
Chinese cabbage	94	32.7	2.2	10.7	5.7
Sesbania	112	14.8	5.0	14.2	580
Wild betel	229	2.5	5.4	5.8	700
Spinach	92	5.1	6.3	0.22	870
Chinese spinach	215	8.4	10.1	10.4	853
Broccoli	35	22.9	2.4	0.16	Traces
Cauliflower	22	23.4	2.1	Traces	1.0
Cabbage	40	24.8	3.3	Traces	Traces
Rhubarb	139.2	8.5	1.8	Traces	805

in bioavailability are not only due to the species or growing conditions in question but also to the geographical origin usually associated with the variety. The calcium content in some of these plants is equal to that of milk, but bioavailability is greater. These findings are important for Oriental populations that consume virtually no milk and whose main sources of calcium are vegetables, especially brassicas.

The negative effect of oxalic acid can be reduced by subjecting the food to different cooking processes, particularly boiling, because some of the soluble oxalate is dissolved in water.

Phytic acid and fiber also reduce calcium uptake from vegetables, although their inhibitory effect is less pronounced than that of oxalate. The effect of fiber depends on the component concerned, since some, such as the pentosans and hexosans, which are partially soluble, nondigestible hemicelluloses, can increase calcium absorption when fermented by colonic bacteria and may be why kale, with a uronic acid content similar to that of spinach, presents greater calcium bioavailability (Kamchan *et al.*, 2004).

Lastly, plants also contain substances such as malic or citric acid, which enhance the bioavailability of calcium.

4.3.4 Cereals and their Derivatives

Cereals provide more than 50% of the energy consumed in developed countries, but their content in calcium and antinutrients varies; consequently, the bioavailability of this mineral in each also differs (Table 4.6). Although it is the presence of phytates (0.5–6% wt) that is usually associated with low calcium absorption, in some cereals such as wheat, which contributes 4% to calcium intake, it is the fiber that most reduces bioavailability of this ion (Aguilar

Table 4.6 The processing of cereals decreases the content of antinutrients that is responsible for the higher bioavailability of calcium as, for example, in bread or breakfast cereals.

Samples	% Available calcium
Rice	16
Wheat	17
Wheat flour	14.9
Whole-wheat flour	13.6
White bread	23
Whole-wheat bread	10.6
Muffin	7.2
Breakfast cereals	0.59–26.2
Infant cereals	
Eight cereals–honey[a]	17.2
Gluten-free cereals[a]	19.2
Multicereals[a]	18.4
Rice cream[a]	16.7
Wheat[a]	19.9

[a]Reconstitution with follow-on-formula.

et al., 2012). In cereals, this parameter depends on the part of the grain in question. Thus, calcium bioavailability is only slightly reduced in wheat flour due to the latter's low phytate and fiber content, whereas a greater reduction is observed in wheat bran, associated not with the inhibitory capacity of phytate, which is virtually null regardless of the phytate/calcium molar ratio, but with the high fiber content (especially cellulose and lignin), since fiber has a nonsaturable, linear capacity to bind calcium (Weaver *et al.*, 1996). The low effect of phytate on calcium absorption in wheat may be related to the higher presence of phytase in this cereal.

Normally, cereals are consumed in their processed state, and the technological processes involved modify the presence of antinutrients. For example, breakfast cereals have a lower phytate or fiber content than the raw cereals from which they are made as a result of extrusion, fermentation, milling or roasting. Due to this reduction in antinutrients and the addition of other ingredients that may include components that enhance calcium absorption, the amount of calcium dialysate present in cereals such as those made from rice and malt is comparable to that found in bovine milk (26.2%) (Aguilar *et al.*, 2012).

Bioavailability of calcium is higher in white bread than in wheat flour, since the baking process reduces the amount of phytate by activating endogenous wheat and yeast phytases, although its content remains sufficiently high so as to compromise the bioavailability of this mineral. The increase in the dialyzable calcium fraction that occurs when baking white bread is not seen in wholemeal bread, perhaps because the phytate/calcium molar ratio is not modified (Frontela *et al.*, 2011).

4.3.5 Baked Goods

Bioavailability is higher in baked goods such as biscuits or sponge cakes than in the wheat flour from which they are made. Thus, the percentage of calcium dialysate in sponge cake (Frontela *et al.*, 2011) is 5.27 ± 0.47% compared with 1.13 ± 0.41% in wheat flour. This increase in calcium dialysate may be due both to the reduction in phytates that occurs during fermentation and depends on factors such as degree of flour extraction, phytase activity, dough pH or addition of calcium salts, and to the contribution of the other ingredients used to make the cake.

4.3.6 Beverages

Calcium bioavailability in mineral water ranges from 23.8% to 47.5%, values that are similar or even superior to those for bovine milk and basically depend on the water's calcium salt content (Bohmer *et al.*, 2000).

As for coffee, it has been estimated that a cup of coffee causes a loss of 2–3 mg of calcium that is easily compensated for by adding 1 tablespoon of milk. Moderate consumption of caffeine (1 or 2 cups of coffee a day) has very few negative effects provided that calcium intake is sufficient. Beverages such as mate contain approximately 25% calcium dialysate. Bioavailability in artificial juices can reach 30% due to their citric and ascorbic acid content (Binaghi *et al.*, 2007).

Carbonated drinks, and especially cola, can present special problems due to their high phosphoric acid and caffeine content, and may cause calcium deficiency because in addition to disrupting the calcium balance, children often drink them instead of milk (Kristensen *et al.*, 2005).

4.3.7 Cooked Meals

In a study conducted in Spain of 30 commonly consumed dishes in school canteens that incorporated different food groups such as meat, fish, eggs, vegetables, legumes, cereals and derivatives, calcium dialyzability ranged from 0.75% to 61.3%, with vegetables being the poorest source of calcium. In fact, the dish with the highest total calcium content, spinach omelette, presented the lowest percentage of dialyzability due to its high oxalic acid content. The addition of pasta to vegetable dishes, for example macaroni with a tomato sauce, reduced the percentage of calcium dialysate due to the presence of fiber and phytates that, as mentioned earlier, negatively affect the bioavailability of calcium (Cámara *et al.*, 2005).

Summary Points

- This chapter is concerned with calcium availability.
- Calcium is an essential nutrient related to the prevention to several chronic diseases.
- Calcium intake is usually deficient in much of the world's population, especially among adolescents and older women.
- Not all ingested calcium is used by the body, for this reason, it is necessary to know the calcium bioavailability to establish nutritional strategies.
- Several factors affect calcium availability, for example: phytate, oxalic acid, protein or mineral contents.
- Milk is the food that has a greater calcium bioavailability.
- Food processing improves the calcium bioavailability to reduce the content of antinutrients.

Key Facts of Availability

1. Bioavailability allows estimation of the portion of ingested nutrient that is used by the body. Different methods have been used to determine the mineral bioavailability.
2. Calcium is a mineral essential for cellular viability and for human health.
3. The most used method is based on simulated gastrointestinal digestion and transportation with Caco-2 cells.

4. Mineral availability varies between foods due to the presence of various compounds.
5. It is necessary to know the mineral's bioavailability in order to establish the recommended dietary allowances (RDA), or supplementation and food-fortification guidelines.

Definition of Words and Terms

Bioavailability. Efficiency with which a nutrient is used through normal metabolic pathways.
Cellular signaling. This is a term referring to the complex system of cellular communication that controls cellular activities and coordinates cell actions.
Coeliac disease. This is an autoimmune disorder of the small intestine that occurs in genetically predisposed people of all ages and it is associated with a reaction by ingesting gluten, which is found in wheat, barley, rye and possibly oats.
Food fermentation. This is a desirable conversion of food carbohydrates to alcohols and carbon dioxide or organic acids using micro-organism as yeasts, bacteria, or a mixture, under anaerobic conditions.
Malting. Controlled cereal germination, followed by disruption of this natural process, drying grain by means of heat.
Pancreatitis. Inflammation of the pancreas.
Phytate (salt form). Phytic acid or inositol hexakisphosphate is a saturated cyclic acid that is the principal storage form of phosphorus in many plants tissues. This compound binds minerals, reducing their ability to be absorbed.
Ready-to-eat lentils. Industrially processed lentils that do not require soaking. Depending on the brand, they only need to be steamed in the pouch in vegetable stock.
Short-bowel syndrome. This is a disorder in the absorption of nutrients due to the loss of anatomical or functional of a large part of the small intestine
Steatorrhoea. This is a disease characterized by abnormal accumulation of lipid in the hepatic cells.

List of Abbreviations

TRPV6	Transient receptor potential cation channel, subfamily V, member 6. It is a membrane calcium channel involved in the first step in calcium absorption in the intestine
PMCA	Plasma membrane Ca^{2+} ATPase
NCX	Sodium–calcium exchanger
IGF-1	Insulin-like growth factor 1 (IGF-1)
PTH	Parathyroid hormone

FOS	Fructo-oligosaccharides
ESPGHAN	European Society of Paediatric Gastroenterology, Hepatology and Nutrition.
EDTA	Ethylenediaminetetraacetic acid
Wt	Weight
RDA	Recommended dietary allowances
ΔAUCs	Increment in area under the curve
VDR	Vitamin D receptor

References

Abrams, S. A., Vieira, N. E. and Yergey, A. L., 1990. Unequal distribution of a stable isotope calcium tracer between casein and whey formulas, human milk and cow's. *Journal of Nutrition*. 120: 1672–1676.

Aguilar, M. V., Mateos, C., Meseguer, I. and Martínez Para, M. C., 2012. Calcium availability in breakfast cereals: effect of other food components. *European Food Research and Technology*. 235: 480–495.

Ahmad, S., Anjum, F., Huma, N., Sameen, A. and Zahoor, T., 2013. Composition and physico-chemical characteristics of buffalo milk with particular emphasis on lipids, proteins, minerals, enzymes and vitamins. *Journal of Animal and Plant Sciences*. 23: 62–74.

Barnes, L. A., Morrow, G., Silverio, J., Finnegan, L. P. and Heitman, S. E., 1974. Calcium and fat absorption from infant formulas with different fat blends. *Pediatrics*. 54: 217–221.

Binaghi, J. M., Lopez, L., Ronayne, P. and Valencia, M., 2007. Evaluación de la influencia de distintos componentes de la dieta sobre la biodisponibilidad potencial de minerales en alimentos complementarios. *Revista Chilena de Nutrición*. 34: 56–60.

Bohmer, H., Muller, H. and Resch, K. L., 2000. Calcium supplementation with calcium-rich mineral waters: a systematic review and meta-analysis of its bioavailability. *Osteoporosis International*. 11: 938–943.

Bosscher, D., Van Caillie-Bertrand, M., Robberecht, H., Van Dyck, K., Van Cauwenbergh, R. and Deelstra, H., 2001. In vitro availability of calcium, iron, and zinc from first-age infant formulae and human milk. *Journal of Pediatric Gastroenterology and Nutrition*. 32: 54–58.

Brommage, R., Binacãoea, C. and Carrie, S. A., 1993. Intestinal calcium absorption in rats is stimulated by dietary lactulose and other resistant sugars. *Journal of Nutrition*. 123: 2186–2194.

Cámara, F., Amaroa, M. A., Barberá, R. and Clementec, G., 2005. Bioaccessibility of minerals in school meals. Comparison between dialysis and solubility methods. *Food Chemistry*. 92: 481–489.

Darling, A. L., Millward, D. J., Torgerson, D. J., Hewitt, C. E. and Lanham-New, S. A., 2009. Dietary protein and bone health: a systematic review and meta-analysis. *American Journal of Clinical Nutrition*. 90: 1674–1692.

De la Fuente, M. A., Montes, F., Guerrero, G. and Juárez, M., 2003. Total and soluble contents of calcium, magnesium, phosphorus and zinc in yoghurts. *Food Chemistry*. 80: 573–578.

Dierpenmaat-Wolters, M. G. E. and Diepenmaat, H. B., 1993. A mathematical model for investigation of the relation between the in vitro availability of minerals and trace elements and the composition of food products. In: Schlemmer, U. (ed.) *Bioavailability'93. Nutritional, Chemical and Food Processing Implications of Nutrient Availability. Part 2*. Bundesforschungsanstalt für Ernährung. Karlsruhe. pp. 93–97.

Drago, S. R. and Valencia, M. E., 2004. Influence of components of infant formulas on in vitro iron, zinc, and calcium availability. *Journal of Agricultural and Food Chemistry*. 52: 3202–3207.

Frølich, W., 1995. Bioavailability of micronutrients in a fiber-rich diet, especially related to minerals. *European Journal of Clinical Nutrition*. 49: S116–S122.

Frontela, C., Ros, G. and Martínez, C., 2011. Phytic acid content and in vitro iron, calcium and zinc bioavailability in bakery products: the effect of processing. *Journal of Cereal Science*. 54: 173–179.

González-Bermúdez, C. A., Frontela-Saseta, C., Peso-Echarri, P., López-Nicolás, R. and Martínez-Graciá, C., 2011. Use of anti-regurgitant infant formulas in unweaned babies. Effect on mineral availability. *Revista Chilena de Nutrición*. 38: 482–490.

Hanzlik, R. P., Fowler, S. C. and Fisher, D. H., 2005. Relative bioavailability of calcium from calcium formate, calcium citrate, and calcium carbonate. *Journal of Pharmacology and Experimental Therapeutics*. 313: 1217–1222.

Heaney, R. P., Rafferty, K., Dowell, M. S. and Bierman, J., 2005. Calcium fortification systems differ in bioavailability. *Journal of the American Dietetic Association*. 105: 807–809.

Heaney, R. P., Weaver, C. M. and Fitzsimmons, M. L., 1991. Soybean phytate content: effect on calcium absorption. *American Journal of Clinical Nutrition*. 53: 745–747.

Hicks, P. D., Hawthorne, K. M., Berseth, C. L., Marunycz, J. D., Heubi, J. E. and Abrams, S. A., 2012. Total calcium absorption is similar from infant formulas with and without prebiotics and exceeds that in human milk-fed infants. *BMC Pediatrics*. 12: 118.

Jovani, M., le Masle, M. and Alegria, A., 2000. In vitro interactions between calcium, zinc, copper and iron in milk- and soy-based infant formulas. *Food Science and Technology International*. 6: 25–31.

Kamchan, A., Puwastien, P., Sirichakwal, P. P. and Kongkachuichai, R., 2004. In vitro calcium bioavailability of vegetables, legumes and seeds. *Journal of Food Composition and Analysis*. 17: 311–320.

Kristensen, M., Jensen, M., Kudsk, J., Henriksen, M. and Molgaard, C., 2005. Short-term effects on bone turnover of replacing milk with cola beverages: a 10-day interventional study in young men. *Osteoporosis International*. 16: 1803–1808.

Lönnerdal, B., 2003. Nutritional and physiologic significance of human milk proteins. *American Journal of Clinical Nutrition*. 77: 1537S–1543S.

Lopez, E., Teucher, B., Boza, J. J., Martínez, A. A., Majsak, G., Baró, L., Carrero, J. J., González, M., Fonollá, J. and Fairweather-Tait, S., 2006. Absorption of calcium from milks enriched with fructooligosaccharides, caseino-phosphopeptides, tricalcium phosphate, and milk solids. *American Journal of Clinical Nutrition*. 83: 310–316.

Peacock, M., 2010. Calcium metabolism in health and disease. *Clinical Journal of the American Society of Nephrology*. 5: S23–S30.

Perales, S., Barberá, R., Lagarda, M. J. and Farre, R., 2005. Bioavailability of calcium from milk-based formulas and fruit juices containing milk and cereals estimated by in vitro methods solubility, dialyzability, and uptake and transport by Caco-2 cells. *Journal of Agricultural and Food Chemistry*. 53: 3721–3726.

Rodriguez, M. S., Montero, M., Staffolo, M. D., Martino, M., Bevilacqua, A. and Albertengo, L., 2008. Chitosan influence on glucose and calcium availability from yogurt in vitro comparative study with plants fibre. *Carbohydrate Polymers*. 4: 797–801.

Roig, M. J., Alegria, A., Barbera, R., Farre, R. and Lagarda, M. J., 1999. Calcium dialysability as an estimation of bioavailability in human milk, cow milk and infant formulas. *Food Chemistry*. 64: 403–409.

Shapses, S. A. and Sukumar, D., 2010. Protein intake during weight loss: effects on bone. In: Burckhardt, P., Dawson-Hughes, B. and Weaver, C. M. (ed.) *Nutritional Aspects of Osteoporosis*. Springer, London, UK, pp. 9–16.

Shen, L., Robberecht, H., Van Dael, P. and Deelstra, H., 1995. Estimation of the bioavailability of zinc and calcium from human, cow's, goat, and sheep milk by an in vitro method. *Biological Trace Element Research*. 49: 107–118.

Soral-Smietana, M., Zdunczyk, Z., Wronkowska, M., Juskiewicz, J. and Zander, L., 2013. Mineral composition and bioavailability of calcium and phosphorus from acid whey concentrated by various membrane processes. *Journal of Elementology*. 18: 115–125.

Tala de Souza, E. M., de Sousa, L. M., Fernandes Arruda, S. and de Almeida Siqueira, E. M., 2002. Protein improves the bioavailability of calcium and phosphorus from an alternative dietary supplement in rats. *Nutrition Research*. 22: 945–955.

Theobald, H. E., 2005. Dietary calcium and health. Briefing paper. *British Nutrition Foundation. Nutrition Bulletin*. 30: 237–277.

Viadel, B., Barberá, R. and Farré, R., 2006. Calcium, iron and zinc uptakes by Caco-2 cells from white beans and effect of cooking. *International Journal of Food Sciences and Nutrition*. 57: 190–197.

Wasserman, R., 2012. *The Transfer of Calcium and Strontium Across Biological Membranes*. Elsevier Science, New York, 464 pp.

Weaver, C. M., Heaney, R. P., Teegarden, D. and Hinders, S. M., 1996. Wheat bran abolishes the inverse relationship between calcium load size and absorption fraction in women. *Journal of Nutrition*. 126: 303–307.

Section II
Chemistry and Biochemistry

CHAPTER 5

The Chemistry of Calcium

DANIEL PERRONE*[a] AND MARIANA MONTEIRO[b]

[a]Laboratório de Bioquímica Nutricional e de Alimentos, Departamento de Bioquímica, Instituto de Química, Universidade Federal do Rio de Janeiro, Av. Athos da Silveira Ramos 149, CT, Bloco A, sala 528A, 21941-909, Rio de Janeiro, Brazil; [b]Departamento de Nutrição Básica e Experimental, Instituto de Nutrição Josué de Castro, Universidade Federal do Rio de Janeiro, Av. Carlos Chagas Filho, 373, CCS, Bloco J, 2° andar, sala 16, 21941-902, Rio de Janeiro, Brazil
*E-mail: danielperrone@iq.ufrj.br

5.1 Chemical Properties of Calcium

Calcium plays a myriad of key roles in cellular biochemistry and in the physiology of living organisms. Calcium participates as a secondary messenger in signal-transduction pathways and is involved with the contraction of muscle cells and the release of neurotransmitters by nervous cells. Calcium is also the cofactor of many enzymes, and is essential for the maintenance of the electrical potential difference across cellular membranes. Finally, 70% of bone is made up of hydroxyapatite, a mineral composed of different calcium salts.

Calcium is an alkaline-earth metal with symbol Ca and atomic number 20, located in the fourth period of the periodic table in Group 2. Calcium is the fifth most abundant element and the third most abundant metal in the Earth's crust by weight. However, since calcium easily reacts with oxygen and water it is not found in the metallic form in nature. The name calcium derives from the Latin word for "lime", *calx*. Humans have been using

Food and Nutritional Components in Focus No. 10
Calcium: Chemistry, Analysis, Function and Effects
Edited by Victor R. Preedy

Published by the Royal Society of Chemistry, www.rsc.org

calcium-containing compounds since antiquity. The Great Pyramid of Giza in Egypt has an outside cover made entirely from limestone ($CaCO_3$, calcium carbonate). In the first century, the Romans prepared concrete by mixing lime (CaO, calcium oxide) with volcanic rocks. In the 10th century, a material known as plaster of Paris or gypsum plaster ($CaSO_4$, calcium sulfate) was used to set broken bones. Metallic calcium was isolated for the first time by Sir Humphry Davy in 1808 by the hydrolysis of a mixture of lime (CaO) and mercury oxide (HgO). Calcium atomic and physical properties are listed in Table 5.1.

Pure calcium has two allotropes; a low-temperature face-centered cubic form that transforms into a high-temperature body centered cubic form at 443 °C (Wale and Pelton, 1987). Calcium electronic configuration ([Ar] $4s^2$) is characterized by two electrons in the valence shell, which are easily lost, increasing atomic stability and allowing the formation of ionic bonds with anions.

Ca^{2+} has a radius of 99 pm and forms colorless inorganic salts that are usually water soluble, with the exception of $CaSO_4$, $CaCO_3$ and $Ca_3(PO_4)_2$ (calcium phosphate). The solubility of calcium complexes or chelates may be very different from solubilities of inorganic salts (Table 5.2) and are

Table 5.1 Atomic and physical properties of calcium.

Atomic number	20
Atomic weight	40.078(4) Da
Electronic configuration	[Ar] $4s^2$
Atomic radius	197 pm
Covalent radius	176 ± 10 pm
Van der Waals radius	231 pm
Oxidation states (common in bold)	**+2**, +1, **0**
Ionization potential	1st: 589.8 kJ mol^{-1}
	2nd: 1145.4 kJ mol^{-1}
	3rd: 4912.4 kJ mol^{-1}
Electron affinity	−0.02455 eV
Electronegativity (Pauling's scale)	1.00
Melting point	1115 K, 842 °C
Boiling point	1757 K, 1484 °C

Table 5.2 Solubility (mg mL^{-1}) of calcium inorganic salts and chelates.

Salt/chelate	Aqueous solubility (mg mL^{-1})	pH of water solution
Calcium phosphate	0.0012–0.30	6.3
Calcium oxalate	0.0051–0.0061	8.5
Calcium carbonate	0.0066–0.014	8.5
Calcium phytate	0.01	5.9
Hydroxyapatite	0.04	7.0
Calcium citrate	0.20–3.6	5.6
Calcium sulfate	2.55	5.5
Calcium glycerophosphate	7.2	8.8
Calcium citrate malate	29.1	5.5
Bisglycinocalcium	282.3	5.4

highly dependent on pH. Considering that one of the most important factors regarding the bioavailability of calcium (and minerals in general) is its solubility (Heaney *et al.*, 1990) and that the pH of the gastrointestinal tract changes with time and region, varying from very acidic in the stomach to less acidic or nearly neutral in the intestine, pH has a great impact on the absorption of calcium. Goss *et al.* (2007) evaluated the equilibrium solubility of five calcium salts (calcium oxalate, calcium phosphate, calcium citrate, calcium glycerophosphate, calcium carbonate) at various pH values (Figure 5.1) and observed a decrease in solubility when the pH was increased. These results suggest that while most calcium salts, with the exception of calcium oxalate, are expected to dissolve in the stomach, calcium probably does not stay in solution in the intestine, its major site of absorption. The conditions of the small intestine are especially relevant, since insoluble calcium compounds cannot diffuse through the brush border membranes of enterocytes and therefore cannot be absorbed.

Calcium standard atomic mass is 40.078(4) Da and shows six naturally occurring isotopes, five of which are stable: ^{40}Ca, ^{42}Ca, ^{43}Ca, ^{44}Ca, ^{46}Ca. The other naturally occurring isotope, ^{48}Ca, is considered to be stable for practical reasons since it has a very long half-life (approximately 10^{18} years). Data regarding calcium naturally occurring isotopes are listed in Table 5.3.

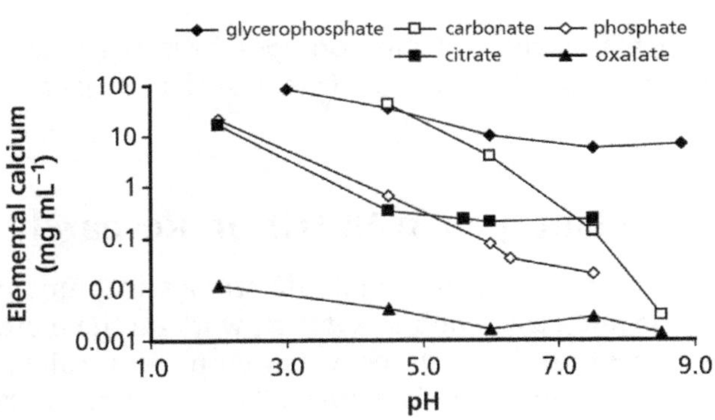

Figure 5.1 Log solubility plot of calcium salts (median value) in aqueous solutions open to the atmosphere (reproduced with permission from Goss *et al.*, 2007).

Table 5.3 Calcium naturally occurring isotopes data.

Isotope	Atomic mass (Da)	Representative isotopic composition (atom%)	Nuclear spin (I)
^{40}Ca	39.96259	96.94156	0
^{42}Ca	41.95862	0.647	0
^{43}Ca	42.95877	0.135	$7/2^{-}$
^{44}Ca	43.95548	2.086	0
^{46}Ca	45.95369	0.004	0
^{48}Ca	47.95253	0.187	0

Table 5.4 Calcium radioactive isotopes data.

Isotope	Atomic mass (Da)	Half-life	Mode of decay[a]
^{34}Ca	34.01412	<35 × 10^{-9} s	ND
^{35}Ca	35.00494	25.7 × 10^{-3} s	β^+ to ^{35}K
^{36}Ca	35.99309	102 × 10^{-3} s	β^+ to ^{36}K
^{37}Ca	36.98587	181.1 × 10^{-3} s	β^+ to ^{37}K
^{38}Ca	37.97632	440 × 10^{-3} s	β^+ to ^{38}K
^{39}Ca	38.97072	859.6 × 10^{-3} s	β^+ to ^{39}K
^{41}Ca	40.96259	102 000 years	EC to ^{41}K
^{45}Ca	44.95619	162.67 days	β^- to ^{45}Sc
^{47}Ca	46.95455	4.536 days	β^- to ^{47}Sc
^{49}Ca	48.95567	8.718 min	β^- to ^{49}Sc
^{50}Ca	49.95752	13.9 s	β^- to ^{50}Sc
^{51}Ca	50.96150	10.0 s	β^- to ^{51}Sc
^{52}Ca	51.96510	4.6 s	β^- to ^{52}Sc
^{53}Ca	52.97005	90 × 10^{-3} s	β^- to ^{53}Sc
^{54}Ca	53.97435	50 × 10^{-3} s	ND
^{55}Ca	54.98055	30 × 10^{-3} s	ND
^{56}Ca	55.98557	10 × 10^{-3} s	ND
^{57}Ca	56.99236	5 × 10^{-3} s	ND

[a] β^+ = beta plus decay (positron emission); β^- = beta minus decay (electron emission); EC = electron capture; ND = not determined.

So far, eighteen unstable calcium isotopes have been described (Audi *et al.*, 2003), showing half-lives ranging from less than 35 ns to 102 000 years (Table 5.4).

5.2 Calcium Isotopes in Nutrition Research

The use of calcium isotopes for metabolic studies in humans started in 1953, when Bellin and Laszlo published their work on the metabolism and removal of radioisotope ^{45}Ca. Other works employing calcium radioisotopes were performed until the late 1990s (Weaver *et al.*, 1996; Krall and Dawson-Hughes, 1999), but only with healthy adults or subjects in conditions of mineral deficiency. Although the use of radioisotopes has several advantages over stable isotopes, such as faster and simpler analysis and sample preparation and lower instrumentation cost, the main drawback of studies with radioisotopes is exposure to radiation. Therefore, most scientists turned to stable isotopes for investigating the metabolism of minerals in humans, especially in groups such as pregnant women or infants. Since stable isotopes do not decay, they can be used for long-term studies and the analysis of samples may be delayed. Moreover, one major advantage is the possibility of performing studies with multiple isotopes of multiple minerals without mutual interference.

Most studies with stable isotopes refer to mineral absorption or bioavailability. The simplest approach to evaluate mineral absorption is to orally administer a stable isotope and then collect and analyze the stool samples collected over a long period (up to 3 weeks) (Davidsson *et al.*, 1996). The administered isotope may be extrinsically added to a single or multiple meals or intrinsically incorporated into a food. The absorption extent is calculated by subtracting the amount of isotopes recovered in feces from the amount ingested orally. It is very important when using this technique that stool samples are completely collected, otherwise absorption will be overestimated. On the other hand, some nutrients, such as calcium, may undergo endogenous fecal excretion, that is they are excreted into the gastrointestinal tract, after its absorption, and not reabsorbed, leading to underestimated absorption results. Intravenously administering another calcium isotope and analyzing its content in the stools allows the determination of fecal excretion extent and correction of absorption results. This double stable isotope approach may also be used for determining absorption based on the analysis of plasma or urine samples at specific time points (Lee *et al.*, 1994; Moser-Veillon *et al.*, 2001; Ceglia *et al.*, 2010; Lee *et al.*, 2011), avoiding the need of complete feces collection, which is troublesome and often impractical. Stable isotope techniques are also used to study calcium metabolism, mainly bone calcium turnover (O'Brien *et al.*, 2006; Morgan *et al.*, 2012) and endogenous excretion (Abrams *et al.*, 1991; LeBlanc *et al.*, 1995).

Calcium stable isotopes were firstly analyzed by radiochemical neutron activation analysis (Janghorbani *et al.*, 1981), but mass-spectrometry techniques progressively became the method of choice due to their higher precision. In this field, fast atom bombardment mass spectrometry (Jiang and Smith, 1987) and thermal ionization mass spectrometry (Fairweather-Tait *et al.*, 1995) have been used, but presently, the most employed technique is inductively coupled plasma mass spectrometry (ICP-MS) (Patterson *et al.*, 1999). Although calcium isotope ratios measurements by ICP-MS using conventional sample introduction is subjected to interferences of several other elements, technological advances such as high-resolution inductively coupled plasma sector field mass spectrometry (HR-ICP-SFMS) (Field *et al.*, 2003) and multiple-collector inductively coupled plasma mass spectrometry (MC-ICP-MS) (Morgan *et al.*, 2011) have been developed to overcome such limitations.

Summary Points

- Calcium is an alkaline-earth metal with atomic number 20.
- Calcium is found as a divalent cation and usually forms water-soluble salts.
- The solubility of calcium salts and chelates is strongly dependent on pH affecting its absorption through the gastrointestinal tract.
- Calcium's atomic mass is 40.973 and it shows six stable and eighteen radioactive isotopes.
- Calcium stable isotopes are widely used for studies on the absorption, bioavailability and metabolism of calcium in humans.
- Inductively coupled plasma mass spectrometry is the most employed technique for calcium isotopic analysis.

Keys Facts of Alkaline-Earth Metals

1. Alkaline-earth metals are the elements located in Group 2 of the periodic table: beryllium (Be), magnesium (Mg), calcium (Ca), strontium (Sr), barium (Ba) and radium (Ra).
2. Alkaline-earth metals are soft, shinny, silvery-white metals that readily lose their two valence electrons to form doubly charged positive ions with oxidation number +2.
3. Cations of alkaline-earth metals are readily hydrated in solution, and the number of water molecules attached to the element increases with ionic radius due to steric factors (*e.g.* 4 for Be^{2+}, 6 for Mg^{2+} and 8 for Ca^{2+}).
4. Alkaline-earth metals are good electric conductors and crystallize in one or more of three closed-packed crystal forms (hexagonal close-packed, body-centered cubic and face-centered cubic).
5. All alkaline-earth metals react with hydrogen forming hydrides (*e.g.* CaH_2) and with halogens forming halides (*e.g.* $CaCl_2$).
6. With the exception of beryllium, alkaline-earth metals react with oxygen producing oxides (*e.g.* CaO) and with water producing hydroxides (*e.g.* $Ca(OH)_2$) and hydrogen gas.
7. Beryllium is mostly used in the defence industry, but it is also employed in electronics to produce semiconductors, electric insulators and heat conductors.
8. Magnesium and calcium have many commercial applications and both play significant roles in geological and biological processes.

Definitions and Explanation of Key Terms

Allotropy. Chemical property of some elements, especially those found naturally as solids, to be found in nature in two or more structurally different forms in the same physical state.

Beta decay. Physical phenomenon observed when the neutron to proton ratio is either too great (beta minus decay; electron emission) or too small (beta plus decay; positron emission).

Bioavailability. Fraction of a nutrient or non-nutrient present in food that is digested, absorbed, metabolized, transported and exerts its biological role.

Electron capture. Radioactive decay in which an electron is absorbed by the nucleus that simultaneously emits an electron neutrino. Observed in atom nuclei with an abundance of protons but not enough energy to undergo a positron emission.

Ionization energy. Corresponds to the minimum energy (usually in kJ mol^{-1}) needed to remove the outermost electron of an atom or molecule in the ground state.

Nuclear spin. Intrinsic quantum mechanical property that represents the total angular momentum of a nucleus.

Nucleophilicity. Ability of an electron-rich chemical species to donate two electrons forming a covalent bond with a positive nuclear charge of an electron-poor species, known as an electrophile.

Oxidation state. Number assigned to an element that represents the number of electrons that are gained, lost or shared when a chemical bond is formed with another element.

Radioisotopes. Isotopes that have an unstable nucleus and therefore emit particles and energy to reach a more stable balance between protons and neutrons.

Signal transduction. Cellular process, also known as cell signaling, in which extracellular signal activate a specific receptor located on the cell surface.

List of Abbreviations

ICP-MS	Inductively coupled plasma mass spectrometry
HR-ICP-SFMS	High-resolution inductively coupled plasma sector field mass spectrometry
MC-ICP-MS	Multiple-collector inductively coupled plasma mass spectrometry

References

Abrams, S. A., Esteban, N. V., Vieira, N. E. and Yergey, A., 1991. Dual tracer stable isotopic assessment of calcium absorption and endogenous fecal excretion in low birth weight infants. *Pediatric Research*. 29: 615–618.

Audi, G., Bersillon, O., Blachot, J. and Wapstra, A. H., 2003. The NUBASE evaluation of nuclear and decay properties. *Nuclear Physics A*. 729: 3–128.

Ceglia, L., Abrams, S. A., Harris, S. S., Rasmussen, H. M., Dallal, G. E. and Dawson-Hughes, B., 2010. A simple single serum method to measure fractional calcium absorption using dual stable isotopes. *Experimental and Clinical Endocrinology & Diabetes*. 118: 653–656.

Davidsson, B. L., Mackenzie, J., Kastenmayer, P., Aggett, P. J. and Hurrell, R. F., 1996. Zinc and calcium apparent absorption from an infant cereal: a stable isotope study in healthy infants. *British Journal of Nutrition*. 75: 291–300.

Fairweather-Tait, S., Prenctice, A., Heumann, K. G., Jarjou, L. M. A., Stirling, D. M., Wharf, S. G. and Turnlund, J. R., 1995. Effect of calcium supplements and stage of lactation on the calcium absorption efficiency of lactating women accustomed to low calcium intake. *American Journal of Clinical Nutrition*. 62: 1188–1192.

Field, M. P., Shapses, S., Cifuentes, M. and Sherrel, R. M., 2003. Precise and accurate determination of calcium isotope ratios in urine using HR-ICP-SFMS. *Journal of Analytical Atomic Spectrometry*. 18: 727–733.

Goss, S. L., Lemons, K. A., Kerstetter, J. E. and Bogner, R. H., 2007. Determination of calcium salt solubility with changes in pH and P_{CO_2}, simulating varying gastrointestinal environments. *Pharmacy and Pharmacology*. 59: 1485–1492.

Heaney, R. P., Recker, R. R. and Weaver, C. M., 1990. Absorbability of calcium sources: the limited role of solubility. *Calcified Tissue International*. 46: 300–304.

Janghorbani, M., Sundaresan, A. and Young, V. R., 1981. Accurate measurement of stable isotopes ^{46}Ca and ^{48}Ca in human feces, plasma, and urine in relation to human nutrition of calcium. *Clinica Chimica Acta*. 113: 267–280.

Jiang, X. and Smith, D. L., 1987. Quantitation of stable isotopic tracers of calcium by fast atom bombardment mass spectrometry. *Analytical Chemistry*. 59: 2570–2574.

Krall, E. and Dawson-Hughes, B., 1999. Smoking increases bone loss and decreases intestinal calcium absorption. *Journal of Bone and Mineral Research*. 14: 215–220.

LeBlanc, A., Schneider, V., Spector, E., Evans, H., Rowe, R., Lane, H., Demers, L. and Lipton, A., 1995. Calcium absorption, endogenous excretion, and endocrine changes during and after long-term bed rest. *Bone*. 16: 301S–304S.

Lee, B. W. T. K., Leung, S. S. F., Fairweather-Tait, S. J., Leung, D. M. Y., Tsang, H. S. Y., Eagles, J., Fox, T., Wang, S. H., Xu, Y. C., Zeng, W. P., Lau, J. and Masarei, J. R. L., 1994. True fractional calcium absorption in Chinese children measured with stable isotopes (^{42}Ca and ^{44}Ca). *British Journal of Nutrition*. 72: 883–897.

Lee, W., McCabe, G. P., Martin, B. R. and Weaver, C. M., 2011. Validation of a simple isotope method for estimating true calcium fractional absorption in adolescents. *Osteoporosis International*. 22: 159–166.

Morgan, J. L. L., Gordon, G. W., Arrua, R. C., Skulan, J. L., Anbar, A. D. and Bullen, T. D., 2011. High-precision measurement of variations in calcium isotope ratios in urine by multiple collector inductively coupled plasma mass spectrometry. *Analytical Chemistry*. 83: 6956–6962.

Morgan, J. L. L., Skulan, J. L., Gordon, G. W., Romaniello, S. J., Smith, S. M. and Anbar, A. D., 2012. Rapidly assessing changes in bone mineral balance using natural stable calcium isotopes. *Proceedings of the National Academy of Science of the United States of America*. 109: 9989–9994.

Moser-Veillon, P. B., Mangels, A. R., Vieira, N. E., Yergey, A. L., Patterson, K. Y., Hill, A. D. and Veillon, C., 2001. Calcium fractional absorption and metabolism assessed using stable isotopes differ between postpartum and never pregnant women. *Journal of Nutrition*. 131: 2295–2299.

O'Brien, K. O., Donangelo, C. M., Zapata, C. L. V., Abrams, S. A., Spencer, M. and King, J. C., 2006. Bone calcium turnover during pregnancy and lactation in women with low calcium diets is associated with calcium intake and circulating insulin-like growth factor 1 concentrations. *American Journal of Clinical Nutrition*. 83: 317–323.

Patterson, K. Y., Veillon, C., Hill, A. D., Moser-Veillon, P. B. and O'Haver, T. C., 1999. Measurement of calcium stable isotope tracers using cool plasma ICP-MS. *Journal of Analytical Atomic Spectrometry*. 14: 1673–1677.

Wale, C. W. and Pelton, A. D., 1987. The Ca–Li (calcium–lithium) system. *Bulletin of Alloy Phase Diagrams*. 8: 125–127.

Weaver, C. M., Heaney, R. P., Teegarden, D. and Hinders, S. M., 1996. Wheat bran abolishes the inverse relationship between calcium load size and absorption fraction in women. *Journal of Nutrition*. 126: 303–307.

CHAPTER 6

Vitamin D and Impact on Total-Body Calcium

HOWARD A. MORRIS*[a,b]

[a]School of Pharmacy and Medical Sciences, Division of Health Science, University of South Australia, Adelaide, South Australia 5000, Australia; [b]Chemical Pathology Directorate, SA Pathology, Box 14 Rundle Mall Post Office, Adelaide, South Australia 5000, Australia
*E-mail: howard.morris@health.sa.gov.au

6.1 Introduction

Ninety nine per cent of total-body calcium is localized to the skeleton. The impact of vitamin D on total-body calcium has marked effects on skeletal structure and bone mineral content. The skeleton is a biologically active organ providing structural support for the body, permitting movement and protecting vital organs such as the brain. However, its role as a store of calcium for the plasma compartment dominates over these other functions as it is the maintenance of plasma-calcium levels that is essential for a variety of activities acutely necessary for life. Daily dietary requirements for calcium and vitamin D for optimal health have been the subject of investigations for over 100 years and yet remain controversial today. If the daily intake of calcium is insufficient to meet obligatory losses then it is the skeletal compartment that must provide requirements for the plasma compartment. This loss of calcium from the skeleton can compromise its strength and increase the risk of fractures. Comparable data for vitamin D status is also becoming

Food and Nutritional Components in Focus No. 10
Calcium: Chemistry, Analysis, Function and Effects
Edited by Victor R. Preedy

Published by the Royal Society of Chemistry, www.rsc.org

available indicating that it is the combined requirement for vitamin D and calcium that is necessary for plasma-calcium homeostasis and optimal skeletal health. Current research continues to elucidate novel physiological activities for both of these nutrients, improving our understanding both on their biological roles and daily requirements for health.

6.2 Calcium is an Essential Nutrient

Calcium is an essential daily nutrient throughout life. Plasma-calcium levels are required to be maintained within narrow limits to ensure optimal neuromuscular activities through modulation of the excitability of neurons. Low plasma ionized calcium, defined as hypocalcaemia, increases neuronal excitability, while hypercalcaemia depresses neuromuscular activity. Plasma-calcium levels also effect the activities of ionized calcium as (1) a critical intracellular signaling entity through movement from one intracellular compartment to another, (2) as part of the intercellular matrix holding cells together at tight junctions and (3) as a cofactor in the blood coagulation cascade (Silverthorn, 2013). Maintenance of plasma-calcium homeostasis is under strict regulatory control to which calciotropic hormones and other mechanisms contribute.

The critical physiological property of calcium determining its daily requirement as an essential nutrient is the limitation of renal tubules to reabsorb 100% of calcium from renal glomerular filtrate. Consequently, calcium is subject to an obligatory loss in urine irrespective of the level of calcium in the diet. Obligatory urine calcium excretion has been demonstrated to exert biological effects on the total-body calcium economy through its effect on bone-mineral homeostasis to stimulate movement of calcium from the bone mineral compartment to the plasma compartment by bone resorption (Nordin *et al.*, 1991). Calcium is also subject to obligatory losses through the skin and the gut (Fleet and Schoch, 2010). During periods of fasting when dietary intakes are unavailable then skeletal calcium and phosphate act as the body store contributing these key mineral ions to the maintenance of plasma-calcium and phosphate homeostasis.

6.2.1 Recommended Daily Intakes for Calcium

The Institute of Medicine (IOM) updated the dietary reference intakes for calcium and vitamin D for North America in 2009 (Institute of Medicine (IOM) Report, 2011). The Recommended Dietary Allowance (RDA) is the intake that covers the requirement for 97.5% of the population for each specified group.

Skeletal bone mineral content increases rapidly during growth such that in adulthood the skeleton contains on average 1000 g of calcium. When growth has ceased but before age-related bone loss commences, net bone resorption is close to zero. However, obligatory losses of calcium in the urine, skin and digestive juices amount to a loss of about 200 mg per day (d). Another important factor determining the level of the RDA for calcium is that on

Table 6.1 Calcium dietary reference intakes by life stage (mg per day).

Life stage group	Recommended dietary allowance
Children 4–8 years	1000
Males 9–18 years	1300
Males 19–70 years	1000
Males > 70 years	1200
Females 9–18 years	1300
Females 19–50 years	1000
Females > 50 years	1200

average calcium absorption is only 25 percent of dietary consumption (Fleet and Schoch, 2010). The RDA for calcium at various life stages are presented in Table 6.1.

Intestinal calcium absorption causes small but physiologically significant increases in plasma-calcium levels detected by the extracellular calcium-sensing receptor (CaSR) (Brown *et al.*, 1993). The CaSR transfers signals of the extracellular ionized calcium level into the intracellular compartment. It is expressed in a wide variety of cells including bone-forming cells, osteoblasts, such that plasma ionized calcium through CaSR signaling is linked with vitamin D receptor (VDR)-activated pathways to promote osteoblast differentiation and formation of mineralized bone (Dvorak-Ewell *et al.*, 2011). These physiological mechanisms support epidemiological evidence that inadequate dietary calcium intakes contribute to a range of pathologies.

6.2.2 Oestrogen, the Calcium Economy and Influence on Recommended Daily Intake

No assessment of the impact of vitamin D on total-body calcium can be considered without understanding the impact of oestrogen and the interaction between total-body calcium and oestrogen status. In terms of clinical impact, oestrogen is the most important regulator of the calcium economy and bone-mineral homeostasis as a result of postmenopausal osteoporosis. Oestrogen affects calcium fluxes in men and women although the sharp fall of oestrogen status in women at the menopause confers a greater impact for the latter. At the menopause there is marked increase in bone resorption due to the loss of the effects of oestrogen on calcium fluxes at the intestine, kidney and bone (Prince, 1994). During a 5 year time interval around the menopause, bone-mineral density (BMD) decreases in the spine and hip by about 10 percent and total-body calcium decreases by 8 percent (Greendale *et al.*, 2012).

Randomized controlled trials in women following the menopause demonstrate that calcium supplementation significantly slows the rate of bone loss compared to placebo (Greendale *et al.*, 2012) and disruption to the calcium economy by oestrogen deficiency is physiologically significant for maintaining the bone-mineral status of the skeleton. This disruption can be at least partially ameliorated by increasing dietary calcium intakes.

When assessing the effects of dietary calcium intakes the major clinical issue is whether higher intakes reduce the risk of fracture. A meta-analysis of randomized controlled trials on the effects of calcium supplementation alone or calcium supplementation with vitamin D on the risk of fracture demonstrated a statistically significant 10% reduction in fractures when data from 6517 women were included with the optimal supplement at a level of 1200 mg calcium per day (Tang *et al.*, 2007). Calcium alone was as effective as calcium plus vitamin D. A further meta-analysis, this time including data from 68 517 patients examined the effects of vitamin D alone *versus* vitamin D plus calcium on fracture (DIPART, 2010). This analysis demonstrated that calcium and vitamin D reduced the risk of all fractures by 8% and risk of hip fracture by 17%, both statistically significant. Vitamin D supplementation alone had no effect. Other meta-analyses have demonstrated reduced risk of fracture of vitamin D and calcium supplementation largely in elderly, frail subjects only when the dose vitamin ranged between 800 and 2000 IU vitamin D per day (Bischoff-Ferrari *et al.*, 2012). All of these trials included a calcium supplement between 800 and 1200 mg per day.

6.3 Vitamin D is an Essential Nutrient

Adequate vitamin D status and calcium and phosphate intakes prevent the development of rickets in children and osteomalacia in adults. Current data from meta-analyses of randomized controlled clinical trials in largely elderly, frail subjects confirm that adequate vitamin D status and dietary calcium intake reduce the risk of osteoporosis including reducing the risk of falls, fractures and premature death (Murad *et al.*, 2011; Bischoff-Ferrari *et al.*, 2012; Bjelakovic *et al.*, 2014). Considerable controversy still exits with regard to the levels of vitamin D as assessed by serum 25-hydroxyvitamin D concentration required for musculoskeletal and other health benefits and the interaction with dietary calcium intake (Institute of Medicine (IOM) Report, 2011; Holick *et al.*, 2012).

1,25-Dihydroxyvitamin D (1,25D) is the active metabolite of vitamin D with the highest affinity for the vitamin D receptor (VDR), a protein with activity as a nuclear transcriptional regulator (DeLuca and Schnoes, 1976). These properties largely elucidated the endocrine mode of action of vitamin D. Numerous reports on vitamin D activities arose later that did not fit within the narrow paradigm of an endocrine mechanism for 1,25D acting through the VDR. During the first decade of the 21st century there has been a flowering of knowledge of novel aspects of vitamin D metabolism within a wide range of nonrenal tissues, activation of the VDR within the tissue of synthesis (thus autocrine and paracrine actions of 1,25D), multiple receptors for 1,25D and activities of VDR independent of 1,25D (reviewed in Ryan *et al.* (2013a)). Reports describing significant risks to health from an inadequate vitamin D status have generated considerable interest amongst the medical and lay communities alike in vitamin D status.

6.3.1 Vitamin D Requirement and Metabolism

Vitamin D_3 (also known as cholecalciferol) largely arises from sunlight exposure of the skin where UVB irradiation converts 7-dehydrocholesterol to previtamin D_3 that at body temperature isomerizes to vitamin D_3 (DeLuca and Schoes, 1976). Vitamin D can also be sourced from the diet from both animal (vitamin D_3) and plant origins (vitamin D_2, ergocalciferol). Dietary supplements are available in both of these forms. The IOM Report recommended for the general, healthy population aged 1–70 years 600 IU per day, while for individuals older than 70 years this was increased to 800 IU per day (Institute of Medicine (IOM) Report, 2011). A supplement of 800 IU per day is recommended to achieve serum 25D levels of 50 nmol L^{-1} or greater. Vitamin D toxicity does occur if doses larger than 20 000 IU daily are taken on a long-term basis leading the IOM to define a tolerable upper limit of 4000 IU per day. This may be lower in older, thin people, those with impaired renal function and in people that take large doses of calcium supplements.

In the circulation vitamin D and each of its metabolites are largely bound to the vitamin D binding protein (DBP). Vitamin D is activated by two consecutive hydroxylation reactions catalyzed by specific P450 enzymes (Figure 6.1). The enzyme vitamin D-25 hydroxylase (CYP2R1) synthesizes the prohormone 25-hydroxyvitamin D (25D). The liver activity appears to be the major site of synthesis for plasma 25D (reviewed in Ryan *et al.* (2013b)) but it is also expressed in a variety of tissues and synthesis of 25D in tissues other than the liver do not largely contribute to the plasma compartment of 25D. The measurement of the concentration of 25D in serum is the biomarker of vitamin D status because of the properties of 25D with regard to solubility and binding to vitamin D binding protein.

Synthesis of the biologically active metabolite, 1,25D (Omdahl *et al.*, 2002) is catalyzed by 25-hydroxyvitamin-D-1 alpha-hydroxylase (CYP27B1). During normal states of health renal synthesis is apparently the sole source of plasma 1,25D (Anderson *et al.*, 2005), while during pregnancy the placenta may make a contribution to plasma 1,25D levels. The CYP27B1 gene is expressed in many tissues, which under various physiological states can be higher than the level expressed in the kidney.

The enzyme 25-hydroxyvitamin D-24-hydroxylase (CYP24) is responsible for catabolism of vitamin D metabolites through its multicatalytic activity to cleave 4 hydrocarbons in the synthesis of calcitroic acid, a metabolite excreted *via* the kidney (Omdahl *et al.*, 2002). The combined activities of the renal CYP24 and renal CYP27B1 enzymes are responsible for determining plasma levels of 1,25D, while the renal CYP24 enzyme is responsible for the synthesis of plasma 24,25-dihydroxyvitamin D when 25D is the substrate and 1,24,25-trihydroxyvitamin D when 1,25D is the substrate. Biological activity has been proposed for 24,25-dihydroxyvitamin D through activation of a specific membrane-associated receptor involved in bone fracture repair (St-Arnaud and Naja, 2011).

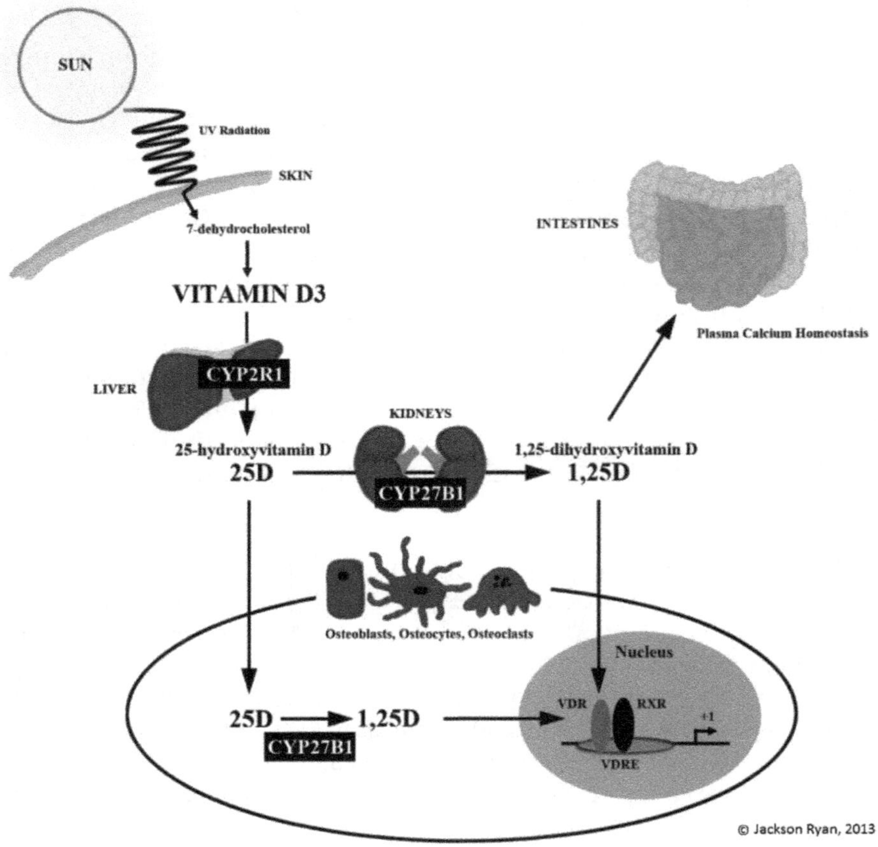

Figure 6.1 Endocrine and bone-cell metabolism of vitamin D. Circulating 25D is converted by the enzyme CYP27B1, to the active metabolite 1,25D in the kidney and bone cells. Kidney synthesis of 1,25D only contributes to the plasma level, which is responsible for its endocrine activities. Autocrine/paracrine activities of 1,25D within the three major bone cells regulate cell proliferation and maturation as well as transcription of genes regulating bone cell activities. Reproduced with permission of Jackson Ryan ©.

6.3.2 Vitamin D and Metabolic Bone Disorders

6.3.2.1 Nutritional Vitamin D Deficiency, Rickets and Osteomalacia

The disease of rickets was first described over four hundred years ago and sunlight was recognized as an antirachitic factor in the 19th century but it was not until the early 1900s that vitamin D was identified as the physiological antirachitic factor derived from sunlight (Shore and Chesney, 2013). Rickets is a softening of bones in children causing fractures and deformity involving the growth plate. It is characterized by increased levels of bone proteins or matrix relative to bone mineral as a result of a mineralization defect. Severe vitamin D deficiency is one of the causes of rickets in children. The disease is known as osteomalacia in adults. Under conditions of very low plasma 25D levels there is insufficient substrate for the renal CYP27B1 enzyme to generate adequate plasma 1,25D levels. Correction of hypocalcaemia and hypophosphatemia in humans and animal models is sufficient to prevent the onset of rickets or resolve this disease without necessarily correcting the vitamin D deficiency (reviewed in Ryan *et al.* (2013a)).

The most obvious phenotype of vitamin D deficiency in all mammals is disruption of plasma calcium and phosphate homeostasis (DeLuca and Schnoes, 1976). In the kidney CYP27B1 mRNA levels can be increased by some 40-fold under the combined influence of PTH, low 1,25D and hypocalcemia that is accompanied by a marked increase of CYP27B1 activity (Omdahl *et al.*, 2002). The increased renal CYP27B1 enzyme activity maintains production of 1,25D at relatively low plasma levels of its substrate 25D. Data from both clinical (Need *et al.*, 2008) and animal-model (Anderson *et al.*, 2008) studies demonstrate that adequate serum 1,25D levels can be maintained with serum 25D levels at 20 nmol L^{-1} or greater to stimulate intestinal calcium absorption to normalize plasma-calcium levels. Plasma 1,25D optimizes renal tubular reabsorption of calcium to maintain plasma-calcium homeostasis (Cochran *et al.*, 2005). Plasma 1,25D acts on bone cells to stimulate bone resorption and inhibit bone formation (Lieben *et al.*, 2012). Endocrine activities of 1,25D on the intestine, kidney and bone maintain plasma calcium and phosphate homeostasis by stimulating calcium fluxes across these organs in the required direction.

6.3.2.2 Genetic Disorders of Vitamin D Activity

Hereditary 1,25-dihydroxyvitamin-D-resistant rickets (HVDRR) arises from inactivation of the gene for the nuclear vitamin D receptor (reviewed by Ryan *et al.* (2013b)). HVDRR patients mainly present with hypocalcaemia, hypophosphatemia, and hyperparathyroidism with signs of rickets in childhood or osteomalacia in adulthood. Similar to nutritional deficiency of vitamin D, correction of plasma calcium and phosphate levels lowers PTH levels and resolves the mineralization defect of rickets or osteomalacia.

Genetically modified mouse models in which the gene for nuclear VDR has been ablated in the germ-line (global-VDRKO) mimic the pathology of HVDRR patients demonstrating a rachitic phenotype with hypocalcaemia, hypophosphatemia and hyperparathyroidism when fed a normal laboratory chow diet (reviewed in Ryan *et al.* (2013b)). When these mice are fed a "rescue diet" containing high levels of calcium and phosphate and lactose from weaning sufficient to normalize plasma calcium and phosphate and reduce PTH levels, the animal is protected from rickets. At 10 weeks of age their bone-mineral status including bone volume and strength are reported to be equal to wild-type mice fed the same diet. These data support the concept that vitamin D activities necessary to maintain bone-mineral status are limited to those required to maintain plasma calcium and phosphate homeostasis. Controversy in the field arose with data from older global VDRKO mice fed the rescue diet from weaning until 17 weeks of age with bone-mineral status significantly reduced compared with wild-type mice. Quantification of bone-cell number and activities identified that osteoblast number and mineral apposition rate were reduced in global VDRKO mice together with reduced osteoclast cell number and expression of the gene coding for receptor activator of nuclear factor-B ligand (RANKL) in bone tissue. The reduced bone volume was the result of reduced bone formation prompting the authors to state that "bone formation may be more vitamin D-dependent as the animals age" (reviewed in Ryan *et al.* (2013b)). Such data provides objective evidence that vitamin D activity through the nuclear VDR may enhance bone formation with a positive effect on bone-mineral status by a direct action on osteoblasts in older animals.

With the identification of 1,25D (calcitriol) as the biologically active metabolite of vitamin D and its availability as a therapeutic agent, another rare genetic cause of rickets was characterized; that arising from an inborn error of vitamin D metabolism such that 25D is unable to be converted to 1,25D. This disease is termed pseudo-vitamin D deficiency rickets (PDDR) (Glorieux *et al.*, 2011). Similar to nutritional vitamin D deficiency or HVDDR, PDDR patients usually present during their first year of life with signs of marked hypocalcemia including hypocalcemic seizures and radiological changes of rickets.

Similar to the experience with the global VDRKO mouse model, two mouse models in which the *Cyp27b1* gene is ablated in germ cells (global CYP27B1KO) were characterized models of PDDR in the early 21st century (Dardenne *et al.*, 2003). These mice demonstrate hypocalcaemia, hypophosphatemia, hyperparathyroidism and rickets when fed a diet with normal levels of calcium and phosphate while with a "rescue diet", sufficient to normalize the blood-mineral disturbances, the development of rickets was prevented. A difference observed between the global-VDRKO and global-CYP27B1KO mouse lines, which persisted even with the "rescue" diet, is that the growth plate in the CYP27B1KO mice remained enlarged and distorted in contrast to VDRKO mice in which it was normalized.

In summary, rickets in children and osteomalacia in adults can arise from either inactivation of the vitamin D system through ablation of either of the genes coding for the nuclear VDR or the enzyme synthesizing 1,25D or severe vitamin D deficiency and extremely low levels of serum 25D. This bone disease is resolved through normalization of plasma calcium and phosphate homeostasis not necessarily correcting vitamin D activity, indicating that vitamin D activity in the aetiology of rickets/osteomalacia is to maintain plasma calcium and phosphate homeostasis.

6.3.2.3 Vitamin D, Osteoporosis, Falls and Fractures

Osteoporosis is a distinct metabolic bone disease from rickets or osteomalacia. In osteoporosis the chemical composition of bone is normal and the extent of mineralization of the bone matrix is normal. The pathology arising from osteoporosis manifests as an increased risk of fracture due to the low level of bone mineral within the bone as an organ. Clinical evidence associates low vitamin D status with increased risk of hip fracture in the elderly (Lai *et al.*, 2010). The serum 25D level associated with increased risk of hip fracture is less severe than that identified for plasma-calcium homeostasis and osteomalacia; approximately 40 nmol L^{-1} and 20 nmol L^{-1}, respectively, (reviewed in Ryan *et al.* (2013b)).

The causal relationship between a low vitamin D status and hip fracture has been established through randomized, placebo-controlled clinical trials (RCTs) of vitamin D supplementation, usually in combination with a calcium supplement, to reduce the risk of fracture. Various meta-analyses of data from these RCTs have provided conflicting interpretations. However, when the dose of the vitamin D supplement is considered, a significant benefit of vitamin D at a dosage between 800 and 2000 IU per day combined with calcium at around 1000 mg per day has been consistently established (Bischoff-Ferrari *et al.*, 2012). Such analyses now include data from more than 60 000 subjects (DIPART, 2010). Vitamin D alone has no effect on the risk of fracture, neither does treatment with calcitriol (1,25D) (Avenell *et al.*, 2009).

Therefore, low serum 25D levels but not low serum 1,25D levels are associated with loss of bone mineral (osteoporosis) and increased risk of fractures without disruption of plasma calcium or phosphate homeostasis. Normalization of serum 25D levels and an adequate dietary calcium sufficient to attain calcium balance are both necessary to improve bone-mineral density and reduce the risk of fracture.

6.3.3 Regulation of Plasma 1,25D Levels

Plasma 1,25D is a key component of the endocrine factors regulating plasma-calcium homeostasis and, as such, levels of plasma 1,25D are tightly regulated. The renal metabolism of serum 25D by CYP27B1 and CYP24 enzymes is central to regulating plasma 1,25D that is largely maintained by regulation of renal expression of the genes coding for these enzymes (Anderson *et al.*, 2012). At least 85% of the variance in plasma 1,25D levels can be accounted for by variance in renal mRNA levels for CYP27B1 and CYP24 (Anderson *et al.*, 2005). Renal CYP27B1 expression, in turn, is regulated by each of the calcium- and phosphate-regulating hormones. Parathyroid hormone (PTH) is the major stimulatory signal for renal CYP27B1 expression.

The inhibition of renal CYP27B1 expression also plays a critical role in regulating plasma 1,25D levels. 1,25D is a major negative regulator of renal CYP27B1 expression. Fibroblast growth factor 23 (FGF23), a plasma hormone produced largely by mature osteocytes and a key regulator of plasma phosphate homeostasis is a potent inhibitor of CYP27B1 expression in the kidney, suppressing gene expression by some 3-fold in mice (Chanakul *et al.*, 2013). FGF23 has recently been identified to inhibit CYP27B1 expression in a number of extrarenal tissues including the heart, lung, spleen, aorta and testis.

The catabolism of 1,25D through the activity of the renal CYP24 enzyme activity in the renal proximal convoluted tubules is also critical for regulating plasma 1,25D levels. An important factor for the regulation of plasma

1,25D level is the inverse relationship between CYP24 and CYP27B1 expression in the kidney. This property is unusual and could be unique to renal tissue (Anderson *et al.*, 2005). This reciprocal relationship between mRNA levels of CYP27B1 and CYP24 is largely the product of reciprocal actions of the calciotropic hormones on the expression of these genes. For example, the major stimulator for CYP24 expression is 1,25D, which directly inhibits renal CYP27B1 expression.

High plasma PTH levels decrease CYP24 activity in kidney cells, while another calciotropic hormone, calcitonin, which is elevated by high plasma-calcium levels, increases CYP24 expression in human kidney cell line, although not demonstrated as yet *in vivo*. FGF23 also lowers plasma 1,25D levels by stimulating CYP24 expression in the kidney in addition to its inhibition of CYP27B1 expression (Anderson *et al.*, 2012).

6.4 Vitamin D Activities and Regulation of Bone-Mineral Homeostasis

Since 99% of body calcium is localized to the skeleton, activities of vitamin D that modulate bone cell activities and affect bone-mineral homeostasis are crucial to understanding the impact of vitamin D on total-body calcium.

6.4.1 The Basic Multicellular Unit of Bone Remodeling

Skeletal architecture is determined by the activities of the three major bone-cell types. Osteoblasts are bone-forming cells derived from a mesenchymal cell lineage. Osteocytes are the most numerous cell type, embedded in bone mineral and arise from the terminal-differentiation of osteoblasts. Osteoclasts are bone-resorbing cells and are derived from the haemopoietic cell lineage. Bone architecture is modified throughout adulthood by the coordinated activities of these three cell types located at a common anatomical site known as the basic multicellular unit of bone remodeling (Figure 6.2). Bone growth is also regulated by these cell types in addition to chrondrocytes, cells responsible for synthesis of the cartilaginous growth plate. Unlike bone remodeling, bone growth or modeling takes place with bone forming and bone resorbing activities taking place at different anatomical sites (McGee-Lawrence and Westendorf, 2011). This discussion will focus on bone remodeling as an example of vitamin D modulating bone-cell activities.

Each of these bone-cell types express the VDR and demonstrate vitamin D biological activities. Furthermore, each of the cell types express genes coding for the CYP27B1 and CYP24 enzymes, being capable of metabolizing vitamin D, particularly 25D to form the active metabolite 1,25D to exert biological activities (Atkins *et al.*, 2011). In bone tissue the local synthesis of 1,25D exerts paracrine or autocrine biological effects and does not contribute to plasma levels (Anderson *et al.*, 2005). Therefore, these cells are capable of exerting activities of vitamin D when plasma 25D levels are adequate and

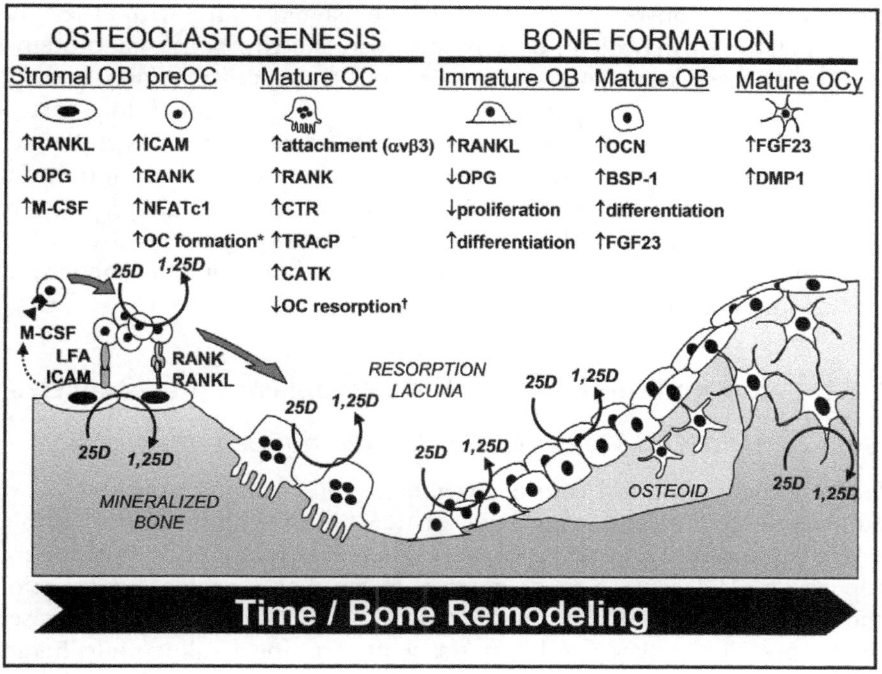

Figure 6.2 The basic muticellular unit of bone remodeling and vitamin D activities. The potential effects of 25-hydroxyvitamin D (25D) metabolism by bone cells. Cellular events in bone remodeling are presented with the potential role of metabolism of 25D into 1,25-dihydroxyvitamin D (1,25D) in osteoblasts (OB) and osteocytes (OCy), as well as in osteoclast (OC) lineage cells. Osteoblasts support osteoclast differentiation from pre-OC, which form bone-resorbing, mature OC. The resorbed site is then populated by immature OB, which proliferate and differentiate into mature OB. These synthesize an unmineralized bone matrix (osteoid). Some OBs remain in osteoid (pre-OCy) and differentiate into mature OCy, concomitant with bone mineralization. The activities of 1,25D are listed below each cell type. Reproduced from Atkins *et al.* (2011) with permission of the publishers. © Vitamin D Third Edition. Academic Press, San Diego, USA.

when plasma 1,25D levels are low, as occurs when dietary calcium intake is adequate. Although the best characterized direct activity of vitamin D on bone cells is the endocrine action of 1,25D on osteoblasts to induce RANKL stimulating osteoclastogenesis and bone resorption, vitamin D exerts a variety of activities on each of the cell types that vary amongst the cell types and even amongst osteoblasts depending on the stage of maturation (Figure 6.2).

Bone formation initially requires the proliferation of osteoblasts within the basic multicellular unit following the cessation of bone resorption by osteoclastic cells. Proliferation is followed by the coordinated synthesis of osteoid matrix by osteoblasts, which is required before mineral deposition as the calcium phosphate hydroxyapatite crystal along type-I collagen fibrils.

Type-I collagen constitute some 90% of the osteoid matrix with other proteins including osteocalcin, osteopontin, osteonectin, bone sialoprotein-1 and proteoglycans present. At this stage most of the osteoblasts undergo apoptosis. Those few embedded in the osteoid matrix mature into preosteocytes, a process that is followed by their transition into osteocytes embedded in mineralized tissue. A small number of osteoblasts transform into lining cells localized to the surface of the bone. Mineralization is tightly controlled by the synthesis of stimulating and inhibitory factors by osteoblasts, osteocytes and osteoclasts at various stages of maturation (McGee-Lawrence and Westendorf, 2011; Atkins *et al.*, 2011).

6.4.2 Vitamin D Activities within Osteoblasts and Osteocytes

6.4.2.1 Bone-Cell VDR: A Major Regulator of Cellular Activities

In vitro experimentation with primary bone cells isolated from rodents or humans demonstrate that inclusion of 1,25D in the culture media inhibits osteoblast proliferation and enhances osteoblast maturation and mineral deposition. The expression of many genes key to osteoblast maturation and mineral deposition are modulated by 1,25D. These include increased gene expression of type-I collagen; tissue nonspecific alkaline phosphatase (Tnap); matrix Gla protein, (an inhibitor of aberrant calcification of cartilage and arteries); osteopontin (Opn), (another inhibitor of mineralization); bone sialoprotein and osteocalcin (Ocn). Osteocytes in culture respond to 1,25D with induction of FGF23 and dentin matrix protein 1 (DMP1), a source of mineralization inhibitors. Direct genomic actions of 1,25D-VDR complex may modulate gene expression; however, the expression of other genes may be stimulated indirectly through an activity that enhances maturation of the osteoblast phenotype (Atkins *et al.*, 2011).

The stage of maturation is apparently a major regulator of transcriptional response to 1,25D within the osteoblast cell lineage. Induction of RANKL expression by 1,25D is increased in immature primary normal human bone-like cells compared to their more mature counterparts as well as in mouse osteoblasts derived from calvaria that demonstrate a more immature osteoblast phenotype. Mouse calvarial cells express RANKL some 1000-fold greater than mature osteoblast-like cells with physiological levels of 1,25D. When higher levels of 1,25D are added to the cell media, the immature calvarial cells increased RANKL expression a further 5-fold, while 1,25D did not increase RANKL expression in the more mature cells (Yang *et al.*, 2013).

The vitamin D system also mediates anabolic activities within bone under *in vivo* conditions when dietary calcium intake is adequate. Transgenic mice in which the human VDR gene is overexpressed only in mature osteoblast lineage cells including some osteocytes (OSVDR mice) have increased cortical and trabecular bone as a result of increased bone mineral apposition by osteoblasts and decreased bone resorption activity. This increased bone volume phenotype is lost when they are fed low dietary calcium (reviewed in Ryan *et al.* (2013a)).

6.4.2.2 Synthesis of 1,25D by Mesenchymal Stem Cells, Osteoblasts, Osteocytes and Osteoclasts

Each of the bone-cell types in both humans and rodents has the capability to metabolize vitamin D with particular emphasis on the conversion of 25D to 1,25D (Figure 6.1) (Ryan *et al.*, 2013a). Physiological levels of 25D in cell-culture media elicit similar biological effects as 1,25D, for which expression of CYP27B1 is essential (Atkins *et al.*, 2011). The expression of CYP27B1 is increased as osteoblast-like cells mature and when mice are fed adequate dietary calcium compared to lower dietary calcium (Anderson *et al.*, 2010). In contrast, the renal expression of CYP27B1 is downregulated with a higher dietary calcium intake associated with low plasma PTH levels.

Conversion of 25D to 1,25D and activation of VDR within bone cells stimulates maturation of mesenchymal stem cells, osteoblasts and osteoclasts and mineral deposition *in vitro*. It is unclear at this time whether the endogenous production of 1,25D in bone cells elicits any different actions compared to endocrine 1,25D from plasma. Preliminary data from a transgenic mouse model in which CYP27B1 expression is increased in osteoblasts and osteocytes indicate increased bone volumes due to a modest increase in bone formation rather than a reduction in osteoclast activity (reviewed in Ryan *et al.* (2013a)).

6.5 Conclusions

Vitamin D exhibits a duality of activities on bone tissue; it is capable of stimulating bone resorption and inhibiting bone formation or inhibiting bone resorption and stimulating bone formation. Each of these activities supports the flux of calcium and phosphate across bone cells, either out of bone tissue into the plasma compartment or from the plasma compartment into bone tissue. Evidence implicates dietary calcium intake and plasma 1,25D levels as the determinants of the direction of these mineral ion fluxes. Many aspects of the interactions between calcium and possibly phosphate signaling and other calciotropic hormones, particularly PTH, on vitamin D metabolism and activity within bone tissue remain to be elucidated. This knowledge provides an insight into the level of vitamin D metabolites and dietary calcium required for bone health. It is evident that these activities all contribute to regulation of plasma calcium and phosphate homeostasis with consequences for total-body calcium.

Summary Points

- The metabolism and activities of vitamin D with regard to total-body calcium are described.
- While 99% of total-body calcium is contained with the skeleton, the plasma ionized fraction is critical for neuromuscular activities essential for life.

- Vitamin D is metabolized into a hormone regulating plasma ionized calcium and phosphate homoeostasis that has effects on bone-mineral status.
- The active metabolite of vitamin D, 1,25-dihydroxyvitamin D (1,25D), exerts endocrine activities to regulate plasma calcium and phosphate homeostasis.
- 1,25D exerts autocrine or paracrine activities within bone cells modulating many biological activities.
- Vitamin D status impacts on total-body calcium by modulating bone-mineral homeostasis in the context of maintaining plasma calcium and phosphate homeostasis.

Key Facts

1. Calcium and vitamin D are essential nutrients for health.
2. The active form of vitamin D regulates plasma calcium and phosphate homeostasis.
3. Maintenance of plasma-calcium homeostasis is critical for maintenance of neuromuscular activity.
4. The skeleton is an essential store to support plasma calcium and phosphate homeostasis as well as providing a structural and mechanical role for the body.
5. The active vitamin D metabolite regulates a range of activities to modulate cell development as well as the direction and extent of calcium ion fluxes within bone tissue.
6. These activities modulate the structural properties of the skeleton and deficiencies in vitamin D status and dietary calcium intake contribute to the burden of metabolic bone disease in the community.

Definitions and Explanations

Plasma ionized calcium is a signaling ion contributing to the rate of neuronal firing and necessary for adequate function of muscles critical for life.

Rickets is a metabolic bone disease in children identified as a defect in mineralization of the skeleton and is associated with defective growth of the skeleton. In adults, when growth has ceased, this disease is know as *osteomalacia.*

Osteoporosis is a metabolic bone disease in which there is too little bone mineral in the bone as an organ. Consequently it has an increased risk of fracture. The chemical composition of bone is normal.

CYP27B1 (25-hydroxyvitamin D-1 alpha hydroxylase) is the enzyme that converts 25-hydroxyvitamin D to 1,25-dihydroxyvitamin D.

1,25-Dihydroxyvitamin D (1,25D) is the biologically active metabolite of vitamin D.

25-Hydroxyvitamin D is the prohormone of vitamin D whose level in serum indicates the level of vitamin D status.
Endocrine actions of 1,25D involve synthesis of 1,25D by the kidney, increasing plasma 1,25D levels and biological activities at organs distant from the kidney.
Autocrine actions of 1,25D involve synthesis of 1,25D in a range of tissues including bone cells and biological activities within the cell of synthesis without changing plasma 1,25D levels.
Paracrine actions of 1,25D involve synthesis of 1,25D in a range of tissues and biological activities within adjacent cells without changing plasma 1,25D levels.
Osteoblasts are bone cells responsible for bone formation.
Osteocytes are the most numerous cells in bone tissue, they are terminally differentiated osteoblasts and are responsible for detecting mechanical forces on bone tissue as well as other activities.
Osteoclasts are bone cells responsible for bone resorption.
PTH, parathyroid hormone, one of the calciotropic hormones secreted by the parathyroid gland when plasma ionized calcium levels are low.

List of Abbreviations

CYP27B1	25-Hydroxyvitamin D-1 alpha hydroxylase
CYP24	25-Hydroxyvitamin D-24-hydroxylase
VDR	Nuclear vitamin D receptor
PTH	Parathyroid hormone

References

Anderson, P. H., O'Loughlin, P. D., May, B. K. and Morris, H. A., 2005. Modulation of CYP27B1 and CYP24 mRNA expression in bone is independent of circulating 1,25(OH)2D3 levels. *Bone*. 36: 654–662.

Anderson, P. H., Sawyer, R. K., Moore, A. J., May, B. K., O'Loughlin, P. D. and Morris, H. A., 2008. Vitamin D depletion induces RANKL-mediated osteoclastogenesis and bone loss in a rodent model. *Journal of Bone Mineral Research*. 23: 1789–1797.

Anderson, P. H., Iida, S., Tyson, J. H., Turner, A. G. and Morris, H. A., 2010. Bone CYP27B1 gene expression is increased with high dietary calcium in mineralising osteoblasts. *Journal of Steroid Biochemistry Molecular Biology*. 121: 71–75.

Anderson, P. H., Turner, A. G. and Morris, H. A., 2012. Vitamin D actions to regulate calcium and skeletal homeostasis. *Clinical Biochemistry*. 45: 880–886.

Atkins, G. J., Findlay, D. M., Anderson, P. H. and Morris, H. A., 2011. Target genes; bone proteins. In: Feldman D., Pike J. W. and Adams J. S. (ed.) *Vitamin D*. 3rd edn. Academic Press, San Diego, USA, pp. 411–424.

Avenell, A., Gillespie, W. J., Gillespie, L. D. and O'Connell, D., 2009. Vitamin D and vitamin D analogues for preventing fractures associated with involutional and post-menopausal osteoporosis. *Cochrane Database Systematic Reviews*. 15: CD000227.

Bischoff-Ferrari, H. A., Willett, W. C., Orav, E. J., Lips, P., Meunier, P. J., Lyons, R. A., Flicker, L., Wark, J., Jackson, R. D., Cauley, J. A., Meyer, H. E., Pfeifer, M., Sanders, K. M., Stähelin, H. B., Theiler, R. and Dawson-Hughes, B., 2012. A pooled analysis of vitamin D dose requirements for fracture prevention. *New England Journal of Medicine*. 367: 40–49.

Bjelakovic, G., Gluud, L. L., Nikolova, D., Whitfield, K., Wetterslev, J., Simonetti, R. G., Bjelakovic, M. and Gluud, C., 2014. Vitamin D supplementation for prevention of mortality in adults. *Cochrane Database Systematic Reviews*. 1: CD007470.

Brown, E. M., Gamba, G., Riccardi, D., Lombardi, M., Butters, R., Kifor, O., Sun, A., Hediger, M. A., Lytton, J. and Hebert, S. C., 1993. Cloning and characterization of an extracellular Ca^{2+}-sensing receptor from bovine parathyroid. *Nature*. 366: 575–580.

Chanakul, A., Zhang, M. Y., Louw, A., Armbrecht, H. J., Miller, W. L., Portale, A. A. and Perwad, F., 2013. FGF-23 regulates CYP27B1 transcription in the kidney and in extra-renal tissues. *PLoS One*. 8: e72816.

Cochran, M., Coates, P. T. H. and Morris, H. A., 2005. The effect of calcitriol on fasting urine calcium loss and renal tubular reabsorption of calcium in patients with mild renal failure. Actions of a permissive hormone. *Clinical Nephrology*. 64: 98–102.

Dardenne, O., Prudhomme, J., Hacking, S. A., Glorieux, F. H. and St-Arnaud, R., 2003. Rescue of the pseudo-vitamin D deficiency rickets phenotype of CYP27B1-deficient mice by treatment with 1,25-dihydroxyvitamin D3: biochemical, histomorphometric, and biomechanical analyses. *Journal of Bone and Mineral Research*. 18: 637–643.

DeLuca, H. F. and Schnoes, H. K., 1976. Metabolism and mechanism of action of vitamin D. *Annual Review of Biochemistry*. 45: 631–666.

DIPART (Vitamin D Individual Patient Analysis of Randomized Trials) Group, 2010. Patient level pooled analysis of 68 500 patients from seven major vitamin D fracture trials in US and Europe. *British Medical Journal*. 340: b5463.

Dvorak-Ewell, M. M., Chen, T. H., Liang, N., Garvey, C., Liu, B., Tu, C., Chang, W., Bikle, D. D. and Shoback, D. M., 2011. Osteoblast extracellular Ca^{2+}-sensing receptor regulates bone development, mineralization, and turnover. *Journal of Bone and Mineral Research*. 26: 2935–2947.

Fleet, J. C. and Schoch, R. D., 2010. Molecular mechanisms for regulation of intestinal calcium absorption by vitamin D and other factors. *Critical Reviews in Clinical Laboratory Sciences*. 47: 181–195.

Glorieux, F. H., Edouard, T. and St-Arnaud, R., 2011. Pseudo-vitamin D deficiency. In: Feldman D., Pike J. W. and Adams J. S. (ed.) *Vitamin D*. 3rd edn. Academic Press, San Diego, USA, pp. 1187–1195.

Greendale, G. A., Sowers, M., Han, W., Huang, M. H., Finkelstein, J. S., Crandall, C. J., Lee, J. S. and Karlamangla, A. S., 2012. Bone mineral density loss in relation to the final menstrual period in a multiethnic cohort: results from the Study of Women's Health Across the Nation (SWAN). *Journal of Bone and Mineral Research*. 27: 111–118.

Holick, M. F., Binkley, N. C., Bischoff-Ferrari, H. A., Gordon, C. M., Hanley, D. A., Heaney, R. P., Murad, M. H. and Weaver, C. M., 2012. Guidelines for preventing and treating vitamin D deficiency and insufficiency revisited. *Journal of Clinical Endocrinology and Metabolism*. 97: 1153–1158.

Institute of Medicine (IOM) Report, 2011. In: Ross A. C., Taylor C. L., Yaktine A. L. and Del Valle H. B. (ed.) *Dietary Reference Intakes for Calcium and Vitamin D*. The National Academies Press, Washington, DC.

Lai, J. K., Lucas, R. M., Clements, M. S., Roddam, A. W. and Banks, E., 2010. Hip fracture risk in relation to vitamin D supplementation and serum 25-hydroxyvitamin D levels: a systematic review and meta-analysis of randomised controlled trials and observational studies. *BMC Public Health*. 10: 331.

Lieben, L., Masuyama, R., Torrekens, S., Van Looveren, R., Schrooten, J., Baatsen, P., Lafage-Proust, M. H., Dresselaers, T., Feng, J. Q., Bonewald, L. F., Meyer, M. B., Pike, J. W., Bouillon, R. and Carmeliet, G., 2012. Normocalcemia is maintained in mice under conditions of calcium malabsorption by vitamin D-induced inhibition of bone mineralization. *Journal of Clinical Investigation*. 122: 1803–1815.

McGee-Lawrence, M. E. and Westendorf, J. J., 2011. Histone deacetylases in skeletal development and bone mass maintenance. *Gene*. 474: 1–11.

Murad, M. H., Elamin, K. B., Abu Elnour, N. O., Elamin, M. B., Alkatib, A. A., Fatourechi, M. M., Almandoz, J. P., Mullan, R. J., Lane, M. A., Liu, H., Erwin, P. J., Hensrud, D. D. and Montori, V. M., 2011. Clinical review: the effect of vitamin D on falls: a systematic review and meta-analysis. *Journal of Clinical Endocrinology and Metabolism*. 96: 2997–3006.

Need, A. G., O'Loughlin, P. D., Morris, H. A., Coates, P. S., Horowitz, M. and Nordin, C. B., 2008. Vitamin D metabolites and calcium absorption in severe vitamin D deficiency. *Journal of Bone and Mineral Research*. 23: 1859–1863.

Nordin, B. E. C., Need, A. G., Morris, H. A., Horowitz, M. and Robertson, W. G., 1991. Evidence for a renal calcium leak in postmenopausal women. *Journal of Clinical Endocrinology and Metabolism*. 72: 401–407.

Omdahl, J. L., Morris, H. A. and May, B. K., 2002. Hydroxylase enzymes of the vitamin D pathway: expression, function, and regulation. *Annual Review of Nutrition*. 22: 139–166.

Prince, R. L., 1994. Counterpoint: estrogen effects on calciotropic hormones and calcium homeostasis. *Endocrine Reviews*. 15: 301–309.

Ryan, J. W., Reinke, D., Kogawa, M., Turner, A. G., Atkins, G. J., Anderson, P. H. and Morris, H. A., 2013a. Novel targets of vitamin D action in bone: action of the vitamin D receptor in osteoblasts, osteocytes and osteoclasts. *Current Drug Targets*. 14: 1683–1688.

Ryan, J. W., Anderson, P. H., Turner, A. G. and Morris, H. A., 2013b. Vitamin D activities and metabolic bone disease. *Clinica Chimica Acta*. 425: 148–152.

Shore, R. M. and Chesney, R. W., 2013. Rickets, Part I. *Pediatric Radiology*. 43: 140–151.

Silverthorn, D. E., 2013. *Human Physiology*. 6th edn. Pearson, Boston, USA, 890 pp.

St-Arnaud, R. and Naja, R. P., 2011. Vitamin D metabolism, cartilage and bone fracture repair. *Molecular and Cellular Endocrinology*. 347: 48–54.

Tang, B. M., Eslick, G. D., Nowson, C., Smith, C. and Bensoussan, A., 2007. Use of calcium or calcium in combination with vitamin D supplementation to prevent fractures and bone loss in people aged 50 years and older: a meta-analysis. *Lancet*. 370: 657–666.

Yang, D., Atkins, G. J., Turner, A. G., Anderson, P. H. and Morris, H. A., 2013. Differential effects of 1,25-dihydroxyvitamin D on mineralization and differentiation in two different types of osteoblast-like cultures. *Journal of Steroid Biochemistry and Molecular Biology*. 136: 166–170.

Section III
Analysis

CHAPTER 7

Ultrasonic-Dialysis Capillary Electrophoresis Inductively Coupled Plasma Optical Emission Spectrometry Analysis of Calcium Speciation in Red Blood Cells

BIYANG DENG*[a], SHUANGJIAO SUN[a], AND YINGZI WANG[a]

[a]Key Laboratory for the Chemistry and Molecular Engineering of Medicinal Resources (Ministry of Education of China), College of Chemistry and Chemical Engineering, Guangxi Normal University, Guilin 541004, China
*E-mail: bydeng@mailbox.gxnu.edu.cn

7.1 Introduction

Calcium is a very important macroelement for life and a necessary component of blood and other body fluids. In many essential cell processes, such as secretion, platelet aggregation, muscle contraction, gene transcription, and cell proliferation, calcium signaling plays a key role (Keel *et al.*, 2008; Smyth *et al.*, 2009). Calcium is also an important activator for many enzymatic reactions, including the transmission of nerve impulses, kidney function, and

Food and Nutritional Components in Focus No. 10
Calcium: Chemistry, Analysis, Function and Effects
Edited by Victor R. Preedy

Published by the Royal Society of Chemistry, www.rsc.org

breathing. The toxicity, bioavailability, and transport properties of an element are highly dependent on its chemical form (Rosen and Hieftje, 2004). Different species of the same element can have different chemical and biological behaviors (Finney and O'Hallaoran, 2003).

There are three forms of calcium in the human body: free Ca^{2+}, protein-bound, and anion chelated. Only changes in the free element of calcium are physiologically relevant because it is directly involved in the mechanisms of many biological processes (Bushinsky and Monk, 1998). Free Ca^{2+} links many physiological stimuli *via* their intracellular effectors through interacting with binding proteins, whose occupancy determines the cellular effect of stimulation (Friel and Chiel, 2008). Valeyev *et al.*, suggested a potential mechanism for selective and differential activation of Ca^{2+} – CaM targets by the same CaM molecules that are involved in a variety of intracellular functions (Valeyev *et al.*, 2008).

Trace elements in biofluids and tissues are very important to human and animal physiology (Deng *et al.*, 2008a; Deng *et al.*, 2010). Because the free calcium ion concentration is an important factor in physiological processes, researchers have developed several methods to determine the concentration of free Ca^{2+} in biological fluids. The developed analytical methods include liquid ion-exchange specific-ion electrodes (Ross, 1967), nuclear magnetic resonance (Murphy *et al.*, 1987), ion-selective electrodes (Fogh-Andersen *et al.*, 1978), fluorescent indicators (Blatter and Wier, 1990; Hyrc *et al.*, 2007), electrolyte analyzers (Unterer *et al.*, 2004), and radioisotope tracing of Ca (Hudec *et al.*, 2004). However, all these methods have their shortcomings and weaknesses. For example, the electrolyte analyzer and radioisotope tracing methods are difficult to perform and require precise control of experimental conditions and the fluorescent indicators bind not only to free Ca^{2+} but also to proteins, which may then alter their fluorescent properties (Gilchrist *et al.*, 1997). In addition, the results can be easily interfered by coexisting heavy-metal ions (*e.g.*, Mn^{2+}, Co^{2+}, Zn^{2+}, Cd^{2+}, Sr^{2+}, and Ba^{2+}) in the sample (Snitsarev *et al.*, 1996).

Hyphenated techniques have been applied to speciation analysis (Olesik *et al.*, 1995; Bacquart *et al.*, 2007). Since the 1990s, capillary electrophoresis (CE) and determination of ionic species have become popular methods in research and clinical testing (Dernovics and Lobinski, 2008). Coupling the capillary electrophoresis method with an analytical spectroscopic technique can provide a higher resolving power, offers rapid and efficient separations with minimal sample size and reagent consumption, and the possibility of separations with only minor disturbances in the existing equilibrium between the various species (Yan *et al.*, 2003; Deng *et al.*, 2010; Timerbaev, 2013). For example, coupling the CE technique to inductively coupled plasma-atomic emission spectrometry or mass-spectrometry techniques provides a powerful method for metal speciation because of the high-resolution separation efficiency of CE and the sensitive element-selective detection capability of ICP spectrometry technique (Richardson *et al.*, 2004; Deng *et al.*, 2007; Deng *et al.*, 2008a,b).

Traditional equilibrium dialysis has shown its shortcoming of long dialysis times, whereas ultrasonic dialysis technology has revealed advantages in many fields (Zhang *et al.*, 2007). Low-intensity ultrasound enhances the movement of liquids and favors mass transfer (Sinisterra, 1992). These properties enhance dialysis, resulting in rapid movement of small molecules through the dialysis membrane, and thus, equilibrium can be reached more rapidly.

Prototypical cells of HRBC demonstrate only basal Ca^{2+} transport, as their membranes do not possess receptor-mediated functions and have no apparent cytoplasmic organelles (Hudec *et al.*, 2004). A method was developed to determine free Ca^{2+} in HRBC cytoplasm using the capillary electrophoresis-inductively coupled plasma-optical emission spectrometry (CE-ICP-OES) technique (Deng *et al.*, 2010). Ultrasonic dialysis (UD) equipment has been constructed and used to help identify the range of molecular mass of the HRBC calcium-containing species detected using CE-ICP-OES technique. The combination of CE and ICP-OES is selected because the CE method requires only a small sample and produces high resolution results, whereas ICP provides selectivity and sensitivity as a specific calcium detector. Using this method, a total of nine calcium-containing species were found in HRBC, one of them was free Ca^{2+}.

7.2 Methodological Considerations

Reagents: The reagents used in application of CE-ICP-OES combined technique should all be analytical grade, which usually include: sodium chloride, potassium chloride, magnesium chloride, nitric acid (65%), hydrogen peroxide (30%), hydrochloric acid (37%), tris(hydroxymethyl)-aminomethane, and calcium standard solution (1 g L^{-1}). All reagents were dissolved or diluted in sub-boiled distilled water and filtered through a 0.45 μm membrane filter before use. Sub-boiled distilled water was used throughout the experiments.

Instrumentation: A commercial axial-viewed inductively coupled plasma-optical emission spectrometer with suitable instrument control software is used as the element-specific detector for calcium. The separation voltage is supplied by a high-voltage direct-current supply system. A fused-silica capillary with an inner diameter of 100 μm and outer diameter of 365 μm is used. The ultrasonic equipment includes an ultrasonic cleaner. Three dialysis membranes with molecular weights of 15, 50, and 100 kDa are used.

The interface between CE and ICP-OES consists of a glass tip, a stainless steel capillary, a platinum wire, epoxy glue, and a PTFE screw thread (Deng *et al.*, 2008b). This interface has many advantages, such as ease of building, ease of coupling to an ICP spectrometer, high nebulization efficiency, no sample dilution by sheath electrolyte, and long lifetime. The sample exit end acts as the cathode.

Ultrasonic Dialysis Device: The ultrasonic dialysis device is shown in Figure 7.1, consisting of three dialysis bags and a polypropylene dialysis pipe (inner diameter 3 cm with volume of approximately 5 mL) fixed on a stainless steel wire frame. The molecular weight cutoff of the inner dialysis tubing should be 100 kDa, the center one 50 kDa, and the outer tube 15 kDa. The sample

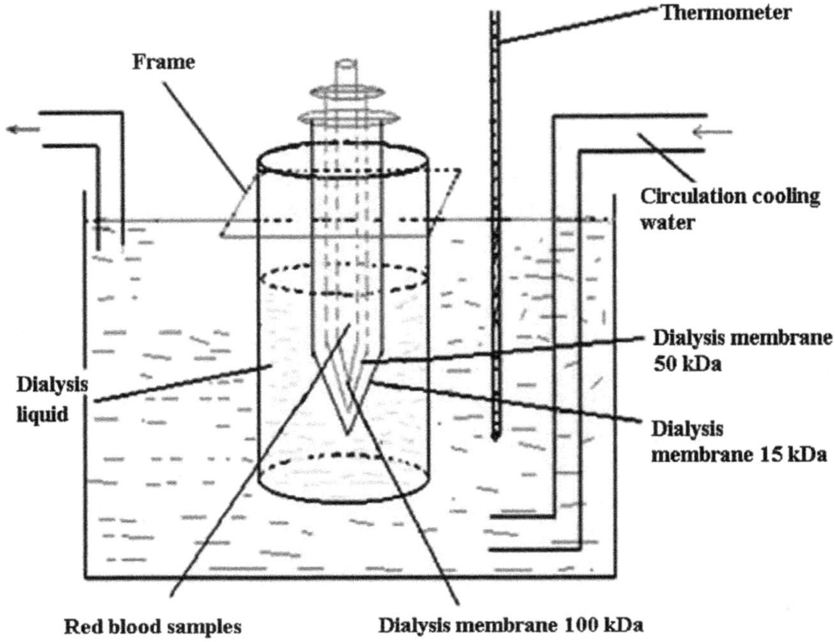

Figure 7.1 Ultrasonic dialysis equipment. The ultrasonic dialysis equipment was used in the study of the molecular weight range of calcium speciation in human red blood cell. Data from Deng *et al.* (2010) with permission from the Royal Society of Chemistry.

is placed in the innermost dialysis bag and temperature (set at 4 °C) is controlled using recirculated cooling water and a thermometer.

The dialysis bags are stoppered to prevent undesired transfer of fluid. All the dialysis bags should be submersed in dialyzate solution and the innermost dialysis bag is filled with HRBC lysate. The dialysis device is installed in the ultrasonic cleaner. The operating frequency of the ultrasonic cleaner is set at 59 kHz. The contents of each dialysis bag are directly detected with CE-ICP-OES technique after 5 h.

Operating Conditions for CE and ICP-OES: Before each run, the CE capillary should be purged with 0.1 mmol L^{-1} NaOH solution for 3 min, then with sub-boiled distilled water and the electrolyte buffer, each for 5 min, and finally with the electrolyte (Tris–HCl buffer, pH 7.40) for 10 min. Other operation conditions are listed in Table 7.1.

Sample Treatment: Human blood samples are collected from 10 healthy male volunteers (25–35 years age, 60–65 kg weight) at a hospital by professional phlebotomists according to institutional review board approved procedures. Samples of HRBC are obtained *via* transferring the blood samples into centrifuge tubes containing the anticoagulant heparin sodium, then immediately centrifuging at 4 °C and 3000*g* for 20 min. The buffer coat and the

Table 7.1 Operation conditions for calcium speciation in human blood cells. The electrolyte buffer and separation voltage are the best important parameters in capillary electrophoresis. Data from Deng *et al.* (2010) with permission from the Royal Society of Chemistry.

Capillary electrophoresis separation conditions		ICP-OES operation conditions	
Electrolyte buffer	40 mmol L^{-1} Tris–HCl, pH = 7.4	RF power	1000 W
CE capillary	110 cm × 100 μm i.d.	Coolant gas flow rate	15 L min^{-1}
Separation voltage	20 kV	Auxiliary gas flow rate	0.2 L min^{-1}
Sample injection	Hydrodynamic 0.04 MPa × 10 s	Carrier gas flow rate	0.8 L min^{-1}
		Calcium wavelength	393.366 nm

plasma should be discarded and the red cells be resuspended in a mixture of 140 mmol L^{-1} NaCl, 5 mmol L^{-1} KCl, 1 mmol L^{-1} $MgCl_2$, and 10 mmol L^{-1} Tris–HCl (pH 7.40) solutions. The samples are then centrifuged at 4 °C and 5000*g* for 15 min, and the superstratum solution and superstratum erythrocytes are discarded. The red cells should be washed three times following the same protocol and the last washing solution is analyzed with an ICP-OES technique.

The erythrocyte numbers are counted in 2 mL of the cell sample obtained from the above procedure, which in the experiment was typically observed as 9.0×10^{12} cells per L with an average red cell volume of 86.3 fL. Another 2 mL of the cell sample is prepared using the same method and calcium standard solution (20 μg L^{-1}) is added to this sample.

The HRBC samples are stored in polyethylene pipettes inside a Teflon flask stored at 4 °C before the analysis. All sample containers and glassware are soaked with 2 mol L^{-1} nitric acid solution and are thoroughly rinsed with sub-boiled distilled water before use.

Total Concentration of Calcium in HRBC: A 2 mL sample of the HRBC is transferred to a beaker and then 2.5 mL of HNO_3 and 1 mL of H_2O_2 solutions are carefully added when analyzing the total calcium content in HRBC. The HRBC sample is digested by heating with an electric heater at 120 °C until dried. After cooling, the sample should be redissolved in a mixture of 1 mL of sub-boiled distilled water and 1 mL of HCl solution, and finally be diluted to 50 mL with sub-boiled distilled water.

A reagent blank is prepared under the same procedure except that the original 2 mL HRBC sample is replaced by sub-boiled distilled water.

7.3 Effects of Various Factors

7.3.1 Effect of Buffer Concentration

In examination with buffer solutions of HAc–NaAc at pH 5.5 and Tris–HCl pH at 7.40, it was observed that both of these solutions can separate calcium species in HRBC, with consideration of physiological conditions

(Bushinsky and Monk, 1998). The optimal buffer concentration has also been investigated through testing of five different concentrations (20, 30, 40, 50, and 80 mmol L^{-1}) of Tris–HCl buffer solutions. In these investigations, improvement of the resolution was observed when the buffer concentration was increased from 20 to 40 mmol L^{-1}. However, the electrical current was unstable at high buffer concentrations (50 and 80 mmol L^{-1}). As a compromise between separation efficiency and the peak profile, a buffer concentration of 40 mmol L^{-1} can be chosen for further experiments.

7.3.2 Effect of Applied Voltage

The effect of applied voltage (varying from 16 to 22 kV) on the separation has also been investigated (Deng *et al.*, 2010). Increasing the applied voltage results in shorter migration times; however, higher applied voltages produce larger currents and increase the Joule heating effect. The optimum peak definition has been achieved with a voltage of 20 kV to limit this heating effect inside the capillary. With an applied voltage of 20 kV, the effect of injection time (8–15 s) on the separation of 20 μg L^{-1} Ca^{2+} has also been investigated. An injection time of longer than 10 s only slightly increases the peak height, but reduces the peak repeatability of migration time. Therefore, a 20 kV applied voltage and 10 s injection time may be considered as a compromise for a rapid and efficient separation with well-defined separation peaks and sensitivity.

7.4 Lysis of HRBC

Chemical treatment, ultrasonic treatment, and small-pore filtration are three methods used for lysis of HRBC. However, both filtration and chemical treatment have their disadvantages (Deng *et al.*, 2008b). For example, the filtration needed a long time and complex operation process; chemical reagent treatment would disturb calcium speciation and could lead to air-bubble formation and stoppage of the capillary electrophoresis current. Consequently, in the experiment, ultrasonic-wave treatment at room temperature for 30 min should be chosen for lysis of HRBC. The sonicated sample treatment is followed by centrifuging at 12 000*g* for 10 min and the upper-layer solution is analyzed using a CE-ICP-OES technique. With this lysing method, the electrophoresis current provides stable conditions with high reproducibility.

7.5 Electrophoresis Separation of Calcium Species in HRBC Cytoplasm

Mammalian mature red blood cells do not have intimae, nuclei or other organelles, consequently, during HRBC lysing, cytoplasm and inorganic salts are easily released and effectively separated from the cell membrane. The ultrasonic wave at room temperature has been used for 30 min for lysis of HRBC (Deng *et al.*, 2010). The sonication of a sample is followed by centrifuging at

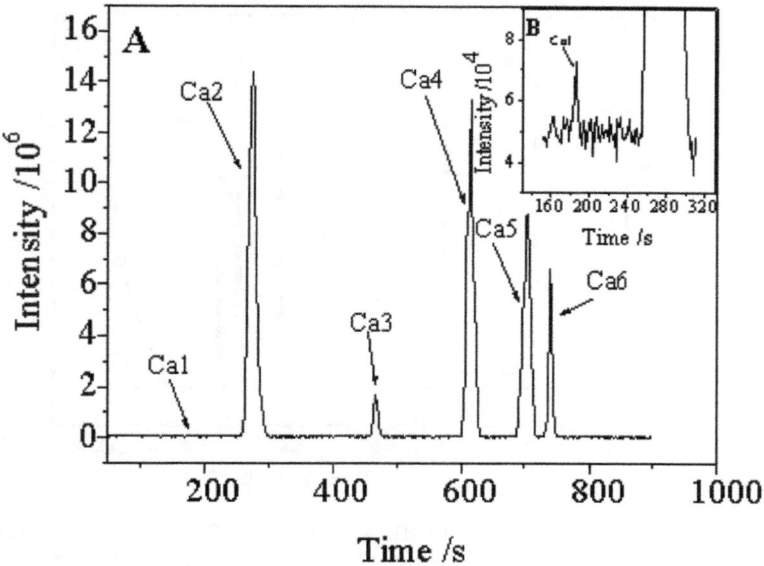

Figure 7.2 Speciation of calcium in human red blood cell cytoplasm using CE-ICP-OES technique. Separation conditions for the sample are described in the experimental procedures section. Data from Deng *et al.* (2010) with permission from the Royal Society of Chemistry.

12 000*g* for 10 min before filtration of the upper-layer solution through a 0.45 μm membrane filter. Under the optimized experimental conditions (20 kV applied voltage, 40 mmol L^{-1} pH 7.40 Tris–HCl buffer, hydrodynamic injection 10 s at 0.04 MPa), six primary calcium species have been observed from a normal person HRBC cytoplasm sample (Figure 7.2A). These species, labeled as Ca1 to Ca6, reveal migration times from 186 to 740 s. The insert (Figure 7.2B) shows the Ca1 peak (186 s) with expanded scale, which was suspected to be free Ca^{2+} ion. As the experimental observations show, the Ca contents in the other five species (Ca2 to Ca6) in erythrocyte cytoplasm are larger than that of Ca1.

7.6 Electrophoresis Separations of Calcium Species in HRBC

To prevent the HRBC samples from clogging the capillary because of their high viscosity, they are diluted two-fold with sub-boiled distilled water before injection into the capillary. Figure 7.3 represents calcium speciation for a normal HRBC sample. Although HRBC contains cytoplasma too, the Ca1 peak is not observed in the HRBC because of the lower free-calcium ion concentration. Therefore, there are eight species (Ca2 to Ca9) with migration times from 148 to 868 s.

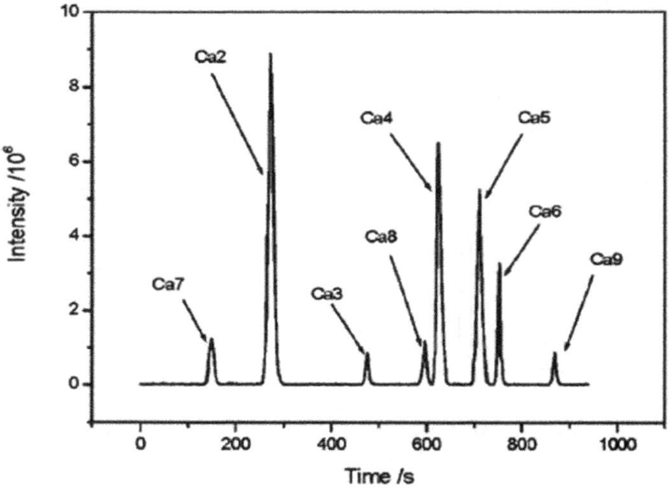

Figure 7.3 Speciation of calcium in HRBC using CE-ICP-OES technique (1 : 1 dilution with sub-boiled distilled water). Separation conditions for the sample are described in the experimental procedures Section. Data from Deng *et al.* (2010) with permission from the Royal Society of Chemistry.

7.7 Quantification of Calcium-Containing Species in HRBC

The Ca-containing species in HRBC samples are quantified using an external calibration curve against the peak area of free Ca^{2+} solution (1–30 mg L^{-1}). Under the optimized experimental conditions, the analytical results and the experimental uncertainty (expressed as RSD) of the determined concentrations and migration times for 10 replicate analyses of one sample are presented in Table 7.2. From these results, the total calcium content obtained using CE-ICP-OES technique is 59.08 mg L^{-1}. In comparison, the total concentration of calcium in the same HRBC samples is 62.38 mg L^{-1} when directly measured with an ICP-OES technique, indicating the recovery of Ca-containing species with the CE-ICP-OES method as 94.71%.

7.8 Identification of Other Calcium Species Using Ultrasonic Dialysis

The electropherogram of a dialyzed HRBC sample is presented in Figure 7.4. Traces of (a–d) in the diagram are the analytical results from the innermost dialysis bag (a) to the dialyzate (d). Peaks are labeled according to their migration times. Only one peak (Ca5) is observed in all traces with estimated molecular weight of smaller than 15 kDa. The Ca9 peak is present in all samples except the dialyzate and appears to have a molecular weight of 15–50 kDa.

Table 7.2 Analytical results of calcium speciation in HRBC (n = 10). RSD is an abbreviation of relative standard deviation. Data from Deng *et al.* (2010) with permission from the Royal Society of Chemistry.

Ca species	Migration time (s)	RSD (%)	Concentration (mg L^{-1})	RSD (%)
Ca2	273	1.3	21.80	4.1
Ca3	476	1.1	1.210	3.1
Ca4	622	1.0	14.65	4.1
Ca5	711	1.5	11.45	3.5
Ca6	750	1.8	4.090	3.9
Ca7	148	1.5	2.680	1.5
Ca8	596	1.4	1.950	2.5
Ca9	868	2.0	1.250	3.3

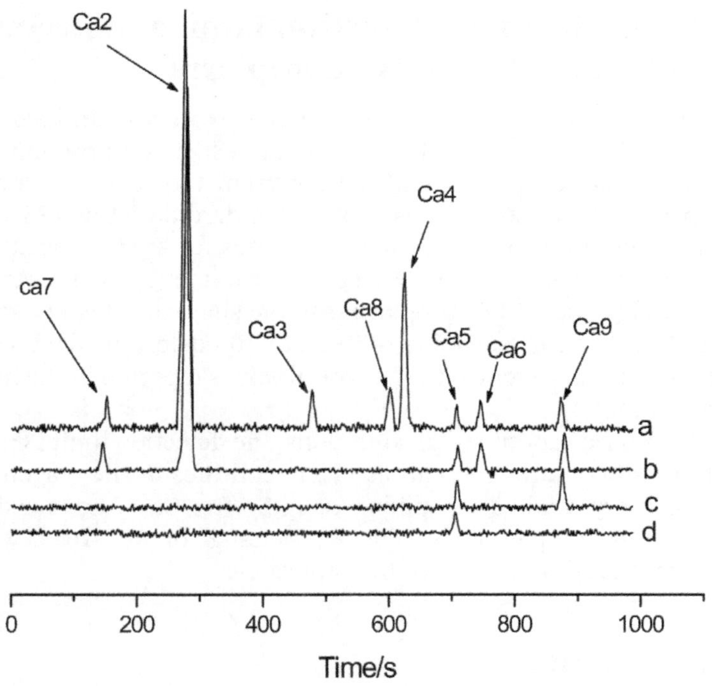

Figure 7.4 Electropherogram of a human red blood cell sample after dialysis. Separation conditions for the sample are described in the experimental procedures Section. Data from Deng *et al.* (2010) with permission from the Royal Society of Chemistry.

The molecular weights of Ca2, Ca6, and Ca7 are estimated in the range of 50–100 kDa, whereas the molecular weights of Ca3, Ca4, and Ca8 are estimated to be larger than 100 kDa. Comparing traces in Figures 7.4 and 7.2, Ca7, Ca8, and Ca9 peaks are identified as calcium-erythrocyte membrane species according to their migration times.

7.9 Identification of Free Ca^{2+} in HRBC Cytoplasm

To identify the Ca1 peak at 186 s, 20 μg L^{-1} of free Ca^{2+} should be spiked to the HRBC cytoplasm. Comparing with the unspiked sample (Figure 7.2), the peak height for Ca1 (at 188 s) substantially increases in the spiked sample and no significant change in peak intensity is observed for other peaks. The migration times for Ca1 seem to be nearly identical, suggesting that the peak labeled as Ca1 corresponds to free Ca^{2+} in the HRBC cytoplasm sample. To further confirm this assignment, an electropherogram of a standard solution of 20 μg L^{-1} free Ca^{2+} is produced that reveals a single peak at 183 s migration time. Based on these similar migration times, peak Ca1 can be assigned to free Ca^{2+} ion.

7.10 Quantification, Detection Limit and Recovery of Free Ca^{2+} in HRBC Cytoplasm

The concentration of free Ca^{2+} in HRBC cytoplasm can be quantified through an external calibration against the peak area in which a calibration curve is constructed with free Ca^{2+} standard solution from 4 to 50 μg L^{-1}. The experimental uncertainties, expressed as relative standard deviation (RSD), of the determined concentrations and migration times for seven replicate analyses of one sample are less than 2%. The measurement of concentration of free Ca^{2+} (Ca1) in the HRBC cytoplasm sample shows a value of 4.48 μg L^{-1} (112 nmol L^{-1}). The concentration of free Ca^{2+} ion determined with CE-ICP-OES technique is consistent with the typical values reported in the literature (Lindner *et al.*, 1993).

Under optimized experimental conditions, the detection limit (3σ) for free Ca^{2+} is 12 nmol L^{-1} through analysis of 11 replicates of the reagent blank. Free Ca^{2+} recovery, studied *via* spiking 20 μg L^{-1} free Ca^{2+} into four different HRBC cytoplasm samples show values in the range of 96.0–103% with RSD values between 1.4% and 4.4% for five replicates.

Summary Points

- A new method for analyses of calcium species in human red blood cells cytoplasm using ultrasonic dialysis and capillary electrophoresis-inductively coupled plasma-optical emission spectrometry techniques is reviewed.
- The optimum experimental conditions for calcium speciation are obtained as 20 kV applied voltage, 40 mmol L^{-1} pH 7.40 tris(hydroxymethyl)-aminomethane–HCl buffer solution, and hydrodynamic injection for 10 s at 0.04 MPa.
- Six primary calcium species are examined from a normal person red blood cells cytoplasm sample.
- Eight species are examined from the human red blood cells samples.

- Free Ca^{2+} in human red blood cells cytoplasm is identified from the electropherogram using a standard solution consisting of 20 μg L^{-1} free Ca^{2+}.
- The concentration of free Ca^{2+} in human red blood cells cytoplasm sample is obtained as 4.48 μg L^{-1} *via* an external calibration against the peak area.
- The recovery is obtained in the range of 96.0–103% with relative standard deviation values between 1.4% and 4.4% for five replicates.
- In human red blood cells, three species reveal molecular weights of greater than 100 kDa, three in the range of 50–100 kDa, one in the range of 15–50 kDa, and one of smaller than 15 kDa excepting free Ca^{2+}.
- The molecular weight of one of these calcium species located in the erythrocyte membrane is between 50 and 100 kDa.
- One species in human red blood cells shows an even faster migration time than free Ca^{2+}.

Key Facts

Key Facts of Ultrasonic Dialysis

1. Ultrasonic dialysis is separation of electrolytes through a membrane with ultrasonic dialysis equipment (Li *et al.*, 1996).
2. The diffusion rate of electrolyte through a membrane with ultrasonic irradiation is faster than that without ultrasonic irradiation.
3. The enhancement of particle permeation and diffusion coefficient in liquid is due to the ultrasonic vibration.
4. The membrane permeability increases with irradiation time.
5. The lifetime of membrane decreases with increasing sound intensity.
6. The degradation of permeability by ultrasound is due to ultrasonic cavitation.

Key Facts of Capillary Electrophoresis Inductively Coupled Plasma Mass Spectrometry/Atomic Emission Spectrometry

1. Capillary electrophoresis inductively coupled plasma mass spectrometry/atomic emission spectrometry is a combination of capillary electrophoresis with inductively coupled plasma mass spectrometry and inductively coupled plasma atomic emission spectrometry techniques.
2. In 1995, the combination of capillary electrophoresis with inductively coupled plasma-mass spectrometry and inductively coupled plasma-atomic emission spectrometry techniques was first developed to study rapid elemental speciation (Olesik *et al.*, 1995).
3. The analysis times with capillary electrophoresis-inductively coupled plasma-mass spectrometry combination technique are shortened by 2 min through combination.

4. The detection limits for capillary electrophoresis-inductively coupled plasma-mass spectrometry technique are generally 60 times larger than those typically obtained using inductively coupled plasma-mass spectrometry technique for elemental analysis.
5. The peak area and elution time reproducibility in capillary electrophoresis-inductively coupled plasma-mass spectrometry technique are typically better than 3% of relative standard deviation for successive injections.

Key Facts of Metallomics

1. Metallomics study was proposed by Haraguchi in 2004 (Haraguchi, 2004).
2. Metallomics is a scientific field in symbiosis with genomics and proteomics.
3. In metallomics, all metal-containing biomolecules are defined as "metallomes", in a similar manner to genomes in genomics and proteomes in proteomics.
4. Chemical speciation for specific identification of bioactive metallomes and the distributions of the elements in man, human blood serum and sea-water are the most important analytical technologies to establish metallomics as the integrated biometal science.
5. Metabolic functions and syntheses of proteins and genes cannot be performed without the aid of various metalloenzymes and metal ions.

Key Facts of Requirements for Reliable Elemental Speciation Analysis

1. A high separation power is essential.
2. The speciation of analytes can be identified.
3. It possesses a high detection capacity.
4. Speciation information is not lost during analysis.
5. Short analysis times are preferred.
6. It is suitable for complex matrices, such as environmental and biological samples.
7. The easy implementation, low running costs, small sample volumes, and high throughput should be considered in speciation analysis.

Key Facts of Detection Systems for Elemental Speciation Analysis

1. The inductively coupled plasma mass spectrometry technique provides extremely low detection limits.
2. In comparison with inductively coupled plasma mass spectrometry, the advantage of inductively coupled plasma atomic emission spectrometry

technique is the speciation analysis of lighter elements, such as calcium and magnesium.
3. Due to high operation costs of inductively coupled plasma-based techniques, atomic fluorescence and atomic absorption spectrometry are easier to assemble, maintain, and operate.
4. The molecular mass spectrometry is a useful tool because of delivering structural information.
5. Nonselective detection techniques are applied in speciation analysis, such as photometry, contactless conductivity, amperometry, and chemiluminescence.

Definitions of Words and Terms

Hyphenated technique. In order to get more information, two or more instruments are combined in use.
Capillary electrophoresis. A new type of liquid separation technology with a capillary as the separation channels and high voltage direct current as the driving force.
Inductively coupled plasma optical emission spectrometry. This technique is used for bulk elemental chemical analysis in which most elements with concentrations from large amounts to parts per billion, or in many cases parts per trillion, in solution can be measured.
Element speciation. The distribution of different chemical species of an element in a defined system.
Calcium species. Calcium species refers to calcium existing with different molecular structures in different forms of the same element.
Cytoplasm. The cytoplasm comprises cytosol and the organelles. All of the contents of the cells of prokaryote organisms are contained within the cytoplasm.
Red blood cell. A kind of cell with the largest number in the blood that is the main medium of blood to carry oxygen in the vertebrate body.
Detection limit. Minimum detectable concentration or amount.
Quantitation limit. This refers to the minimum detectable quantity when the accuracy and the precision of the method meet the requirements.
Dialysis. This is a purification technology that small molecules go through a semipermeable membrane to diffuse into water (or buffer) and separate from biological macromolecules.
Ultrasonic dialysis. Application of ultrasound to accelerate the dialysis separation of small molecules through a semipermeable membrane with ultrasonic dialysis equipment.
Calibration curve. The curve that shows the quantitative relation between the material concentration (or the quantity) and the instrument response values.
Relative standard deviation. The ratio of standard deviation and the mean value of calculation results.

List of Abbreviations

HRBC	Human red blood cells
CE	Capillary electrophoresis
CE-ICP-OES	Capillary electrophoresis-inductively coupled plasma-optical emission spectrometry
MS	Mass spectrometry
UD	Ultrasonic dialysis
Tris	Tris(hydroxymethyl)-aminomethane
RSD	Relative standard deviation
AES	Atomic emission spectrometry

Acknowledgments

This study was supported by the National Natural Science Foundation of China (no. 21365006, no. 20965001) and the Guangxi Science Foundation of China (no. 2010GXNSFA013051). The authors appreciate Dr Saeed Doroudiani for editing the manuscript and providing useful suggestions.

References

Bacquart, T., Devès, G., Carmona, A., Tucoulou, R., Bohic, S. and Ortega, R., 2007. Subcellular speciation analysis of trace element oxidation states using synchrotron radiation micro-X-ray absorption near-edge structure. *Analytical Chemistry*. 79: 7353–7359.

Blatter, L. A. and Wier, W. G., 1990. Intracellular diffusion, binding, and compartmentalization of the fluorescent calcium indicators indo-1 and fura-2. *Biophysical Journal*. 58: 1491–1499.

Bushinsky, D. A. and Monk, R. D., 1998. Calcium. *Lancet*. 352: 306–311.

Deng, B., Feng, J. and Meng, J., 2007. Speciation of inorganic selenium using capillary electrophoresis-inductively coupled plasma-atomic emission spectrometry with on-line hydride generation. *Analytica Chimica Acta*. 583: 92–97.

Deng, B., Li, X., Zhu, P., Xu, X., Xu, Q. and Kang, Y., 2008a. Speciation of magnesium in rat plasma using capillary electrophoresis-inductively coupled plasma-atomic emission spectrometry. *Electrophoresis*. 29: 1534–1539.

Deng, B., Zhu, P., Wang, Y., Feng, J., Li, X., Xu, X., Lu, H. and Xu, Q., 2008b. Determination of free calcium and calcium-containing species in human plasma by capillary electrophoresis-inductively coupled plasma optical emission spectrometry. *Analytical Chemistry*. 80: 5721–5726.

Deng, B., Wang, Y., Zhu, P., Ni, X., Lu, H. and Xu, X., 2010. Identification and analysis of calcium speciation in red blood cells by ultrasonic-dialysis capillary electrophoresis inductively coupled plasma optical emission spectrometry. *Journal of Analytical Atomic Spectrometry*. 25: 1859–1863.

Dernovics, M. and Lobinski, R., 2008. Speciation analysis of selenium metabolites in yeast-based food supplements by ICPMS-assisted hydrophilic interaction HPLC-hybrid linear ion trap/orbitrap MS. *Analytical Chemistry*. 80: 3975–3984.

Finney, L. A. and O'Hallaoran, T. V., 2003. Transition metal speciation in the cell: insights from the chemistry of metal ion receptors. *Science*. 300: 931–936.

Fogh-Andersen, N., Christiansen, T. F., Komarmy, L. and Siggaard-Andersen, O., 1978. Measurement of free calcium ion in capillary blood and serum. *Clinical Chemistry*. 24: 1545–1552.

Friel, D. D. and Chiel, H. J., 2008. Calcium dynamics: analyzing the Ca^{2+} regulatory network in intact cells. *Trends in Neurosciences*. 31: 8–19.

Gilchrist, J. S. C., Palahniuk, C. and Bose, R., 1997. Spectroscopic determination of sarcoplasmic reticulum Ca^{2+} uptake and Ca^{2+} release. *Molecular and Cellular Biochemistry*. 172: 159–170.

Haraguchi, H., 2004. Metallomics as integrated biometal science. *Journal of Analytical Atomic Spectrometry*. 19: 5–14.

Hudec, R., Lakatos, B., Orlicky, J. and Varecka, L., 2004. Reconstitution of the basal calcium transport in resealed human red blood cell ghosts. *Biochemical and Biophysical Research Communications*. 325: 1172–1179.

Hyrc, K. L., Rzeszotnik, Z., Kennedy, B. R. and Goldberg, M. P., 2007. Determining calcium concentration in heterogeneous model systems using multiple indicators. *Cell Calcium*. 42: 576–589.

Keel, S. B., Doty, R. T., Yang, Z., Quiqley, J. G., Chen, J., Knoblaugh, S., Kingsley, P. D., De Domennico, I., Vaughn, M. B., Kaplan, J., Palis, J. and Abkowitz, J. L., 2008. A heme export protein is required for red blood cell differentiation and iron homeostasis. *Science*. 319: 825–828.

Li, H., Ohdaira, E. and Ide, M., 1996. Effect of ultrasonic irradiation on permeability of dialysis membrane. *Japanese Journal of Applied Physics*. 35: 3255–3258.

Lindner, A., Hinds, T. R., Davidson, R. C. and Vincenzi, F. F., 1993. Increased cytosolic free calcium in red blood cells is associated with essential hypertension in humans. *American Journal of Hypertension*. 6: 771–779.

Murphy, E., Berkowitz, L. R., Orringer, E., Levy, L., Gabel, S. A. and London, R. E., 1987. Cytosolic free calcium levels in sickle red blood cells. *Blood*. 69: 1469–1474.

Olesik, J. W., Kinzer, J. A. and Olesik, S. V., 1995. Capillary electrophoresis inductively coupled plasma spectrometry for rapid elemental speciation. *Analytical Chemistry*. 67: 1–12.

Richardson, D. D., Kannamkumarath, S. S., Wuiloud, R. G. and Caruso, J. A., 2004. Hydride generation interface for speciation analysis coupling capillary electrophoresis to inductively coupled plasma mass spectrometry. *Analytical Chemistry*. 76: 7137–7142.

Rosen, A. and Hieftje, G. M., 2004. Inductively coupled plasma mass spectrometry and electrospray mass spectrometry for speciation analysis: applications and instrumentation. *Spectrochimica Acta, Part B: Atomic Spectroscopy*. 59: 135–146.

Ross, J. W., 1967. Calcium-selective electrode with liquid ion exchanger. *Science*. 156: 1378–1379.

Timerbaev, A. R., 2013. Element speciation analysis using capillary electrophoresis: twenty years of development and applications. *Chemical Reviews*. 113: 778–812.

Sinisterra, J. V., 1992. Application of ultrasound to biotechnology: an overview. *Ultrasonics*. 30: 180–185.

Smyth, J. T., Petranka, J. G., Boyles, R. R., DeHaven, W. I., Fukushima, M., Johnson, K. L., Williams, J. G. and Putney, Jr, J. M., 2009. Phosphorylation of STIM1 underlies suppression of store-operated calcium entry during mitosis. *Nature Cell Biology*. 11: 1465–1472.

Snitsarev, V. A., Mcnulty, T. J. and Taylor, C. W., 1996. Endogenous heavy metal ions perturb fura-2 measurements of basal and hormone-evoked Ca^{2+} signals. *Biophysical Journal*. 71: 1048–1056.

Unterer, S., Lutz, H., Gerber, B., Glaus, T. M., Hässig, M. and Reusch, C. E., 2004. Evaluation of an electrolyte analyzer for measurement of ionized calcium and magnesium concentrations in blood, plasma, and serum of dogs. *American Journal of Veterinary Research*. 65: 183–187.

Valeyev, N. V., Heslop-Harrison, P., Postlethwaite, I., Kotov, N. V. and Bates, D. G., 2008. Multiple calcium binding sites make calmodulin multifunctional. *Molecular BioSystems*. 4: 66–73.

Yan, X., Yin, X., Jiang, D. and He, X., 2003. Speciation of mercury by hydrostatically modified electroosmotic flow capillary electrophoresis coupled with volatile species generation atomic fluorescence spectrometry. *Analytical Chemistry*. 75: 1726–1732.

Zhang, P., Zhang, G. and Wang, W., 2007. Ultrasonic treatment of biological sludge: floc disintegration, cell lysis and inactivation. *Bioresource Technology*. 98: 207–210.

CHAPTER 8

Using Fluorescent Polyanions to Assay for Osteoclastic Calcium-Resorption Activity

TATSUYA MIYAZAKI[a] AND OSAMU SUZUKI*[a]

[a]Division of Craniofacial Function Engineering, Tohoku University Graduate School of Dentistry, 4-1 Seiryo-machi, Aoba-ku, Sendai 980-8575, Japan
*E-mail: suzuki-o@m.tohoku.ac.jp

8.1 Introduction

Bone tissue consists of approximately 70% mineral crystals, 20% organic matters, and 10% water (Rath *et al.*, 2000). The prototype of the mineral crystals is hydroxyapatite, which is one of the calcium phosphates and provides stiffness and compressional strength. Collagen is the major organic matrix that confers tensile strength to the bone. There are two kinds of cells contributing to bone metabolism, one is osteoblast and the other is osteoclast. Osteoblast synthesizes the bone matrix and osteoclast degrades it, resulting in the release of calcium ions (Glade and Krook, 1982). Osteoclast forms a tight attachment, so-called sealing zone, by rearranging actin cytoskeleton on the surface of the mineralized bone matrix. A structure called a ruffled border in osteoclast works to resolve the mineral by releasing protons and to degrade the matrix by synthesizing lysosomal proteases such as cathepsin K (Baron *et al.*, 1985; Stenbeck and Horton, 2000).

Food and Nutritional Components in Focus No. 10
Calcium: Chemistry, Analysis, Function and Effects
Edited by Victor R. Preedy

Published by the Royal Society of Chemistry, www.rsc.org

The imbalance of bone metabolism results in pathogenesis of bone diseases, such as osteoporosis, osteopetrosis and heterotopic ossification (Glade and Krook, 1982). Osteoporosis, a major metabolic bone disease, is thought to come from higher bone resorption rate by osteoclasts than bone formation rate by osteoblasts (Raisz, 2005). It is well known that osteoblast is coupled with osteoclast through a direct or indirect manner. The receptor activator of the NF-kappaB ligand (RANKL) signal from osteoblast and a soluble factor, such as the macrophage colony-stimulating factor (M-CSF), are involved in the formation of mature osteoclast from preosteoclasts (Quinn *et al.*, 1998; Yasuda *et al.*, 1998).

To elucidate the mechanism of bone metabolism and the pathogenesis of osteoporosis, animal models have been used together with the advance of recent molecular biological techniques (Panda *et al.*, 2001; Zhang *et al.*, 2005). However, there are a few reports about evaluating bone resorption activity *in vivo* or organ culture *in vitro*. Although the animal experiments are helpful to evaluate and analyze the mechanism of action of the drug, it seems to be undesirable from the point of view of animal welfare and cost of experiments. Therefore, the development of simple cell-culture assay systems *in vitro* may be required for screening the drugs or analysis of bone metabolism.

8.2 Current Methods of Evaluating Osteoclast Function

Several assay systems have been established for evaluating osteoclast differentiation. They include the analyses of tartrate-resistant acid phosphatase (TRAP) staining (Minkin, 1982), actin ring formation (Stenbeck and Horton, 2000), and expression of cathepsin K or matrix metalloproteinase (Delaissé *et al.*, 2003). However, these methods are not direct assays for osteoclast function, namely resorbing activity, and are not suitable for screening or evaluating many new drug candidates. To evaluate the osteoclast function, a simple method of measuring collagenase activity was developed recently (Greenwalt, 2008). This assay uses fluorescent-labeled collagen-coated plates in which the degradation activity of osteoclasts is evaluated by measuring the release of fluorescent-labeled collagen fragments. On the other hands, a pit assay, which measures bone resorption using dentin slices or calcium phosphate (CaP)-coated plates, has been frequently utilized (Tamura *et al.*, 1993). Although the pit assay is a principal and traditional evaluation system that makes it possible to measure CaP resorption activity by osteoclasts directly, it requires considerable time and effort to analyze pit areas or number by using image-analyzing software. If the area of pictures taken from one sample is small, it indicates a large variation. In some cases, it also needs toluidine blue or Von Kossa staining for the remaining CaP before capturing photographs.

8.3 Fluorescent Polyanions to Assay for Osteoclastic Calcium-Resorption Activity

The schematic mechanism and procedure for measuring bone resorption activity are illustrated in Figure 8.1. Beforehand of cell culture, a CaP plate labeled with fluorescent polyanion is prepared. Osteoclasts cultured on the CaP labeled with polyanion degrade and release it into culture medium. Bone resorption activity can be evaluated by simply measuring the fluorescence intensity of the culture medium after adjusting pH. The comparison of traditional pit assay and this resorption assay is summarized in Table 8.1.

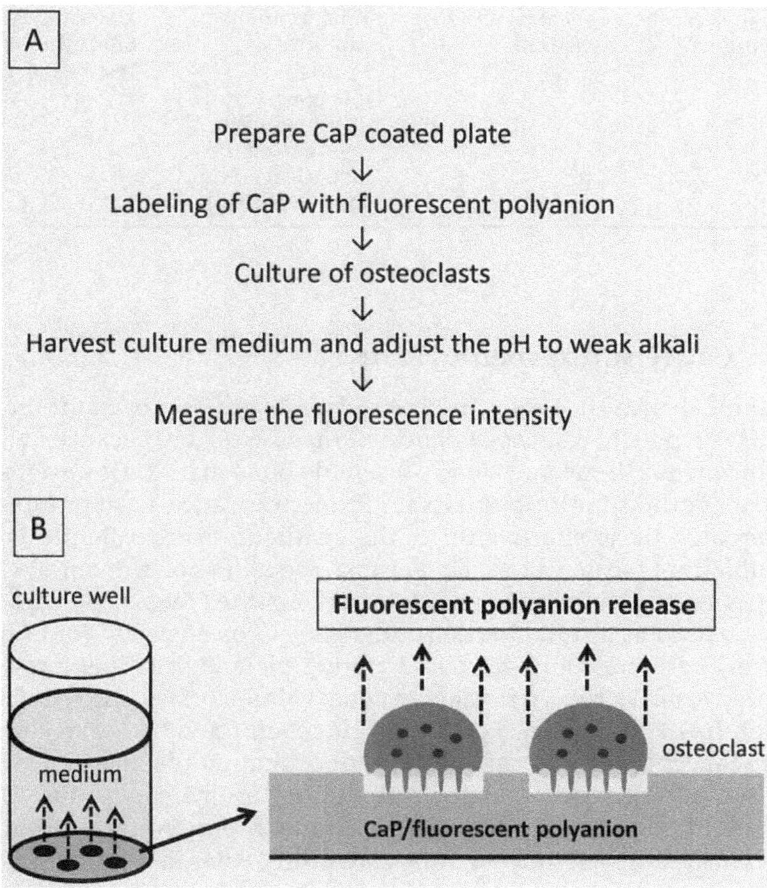

Figure 8.1 Schematic representation of method (A) and mechanism (B) for measuring bone-resorption activity using CaP labeled with fluorescent polyanions. A part of (B) was reproduced with modification from Figure 1A of Miyazaki *et al.*, 2011. *Analytical Biochemistry*. 410: 7–12 with permission of Elsevier.

Table 8.1 Comparison of the assays. The method of traditional pit assay and the resorption assay using fluorescent polyanions are compared by their merits and demerits.

Assay	Method	Merit	Demerit
Traditional pit assay	Measure the area or depth of pits by image-analyzing software	Dentin resembles to the native bone	Trained technicians are required to prepare dentin slices Time consuming to measure by image analysis
Assay of using CaP labeled with fluorescent polyanions	Measure the fluorescence intensity released into the medium	Simple method and a large number of specimens analysis is available Time-course measurement sampling from the identical culture well is available	Fluorescent-labeled polyanions have a possibility to affect the resorption activity

8.3.1 CaP (Hydroxyapatite) Plate

In general, dentin slices or synthetic hydroxyapatite are used for pit assay *in vitro*. The polished slices of dentin of tusks are used, because dentin is a homogeneous tissue and closely resemble bone matrix. However, acquisition of dentin is limiting and its uniform preparation is labor intensive. To overcome these shortcomings, the synthetic hydroxyapatite is used as a substitute for dentin slices. Because the main component of CaP in bone tissue is calcium-deficient carbonate apatite (Tas, 2003), carbonate apatite is used as a suitable resorption assay of osteoclasts. For example, preparing carbonate apatite-coated culture plate is described previously (Miyazaki *et al.*, 2010). Preincubated equal volumes of Na_2HPO_4 and $CaCl_2$ solution (pH 7.4) were mixed in a CO_2 incubator at 37 °C. The supersaturated concentration of calcium and phosphate under neutral pH was used to induce spontaneous precipitation of hydroxyapatite (Eanes *et al.*, 1965). After washing the CaP slurry with pure water, the slurry is poured into each well of 24- or 48-well culture plates and then dried. It is important to adjust the volume of the CaP coating. A thick CaP coating interferes with osteoclastogenesis, whereas a thin CaP coating is insufficient to evaluate resorption activity in the medium. Although sometimes sintered CaP is used for pit assay, the nonsintered CaP may be suitable for this assay because of high capacity of carrying fluorescein-labeled polyanions.

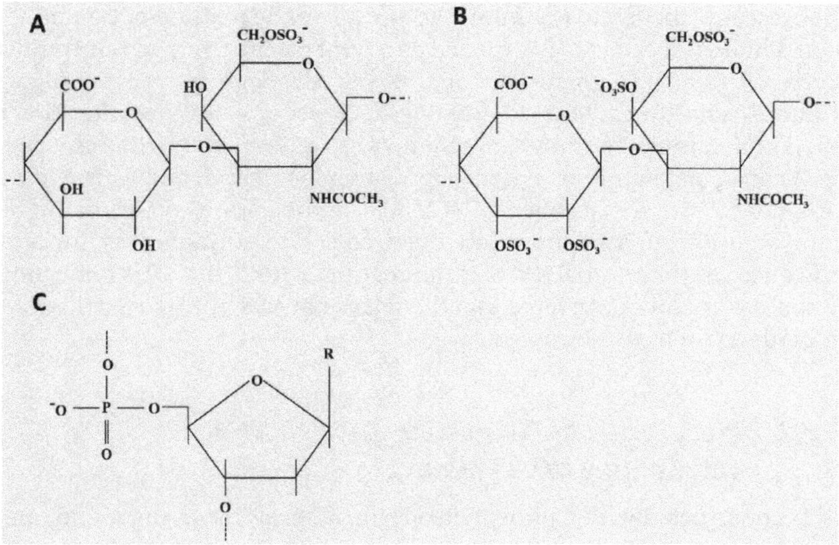

Figure 8.2 Structure of polyanions using this assay. Structure of chondroitin sulfate C (A), chondroitin polysulfate (B) and deoxyribonucleic acid (C). R indicates nucleobase.

8.3.2 Coating of Fluorescein-Labeled Polyanion to CaP Plate

As a matter of course, polyanions using in this assay system should not be toxic against cultured cells. It also needs a strong binding affinity to the CaP. Polyanions such as sulfated glycosaminoglycan, *e.g.* chondroitin sulfate (CS), or deoxyribonucleic acid (DNA), are distributed widely as a natural product, and they reveal low toxicity to the cells. The sulfate groups in CS or phosphate group in DNA bond strongly with calcium ion in CaP crystal (Figure 8.2). Both the long polymer and strong negative charge are needed to bind tightly with coated CaP.

8.3.2.1 Preparation of Fluorescein-Labeled CS and Coating to CaP Plate

CS is constructed with glucuronic acid and *N*-acetyl-galactosamine and has some structural isomer variants modified with sulfate groups at different positions (Seldin *et al.*, 1984). We usually use a synthetic chondroitin polysulfate (CPS), one of sulfated glycosaminoglycans for this assay (Figure 8.2B). In addition to the strong binding affinity to the CaP, CPS is resistant to the degradation enzymes of CS, such as chondroitinase, hyaluronidase or some other glycosaminoglycanases. These properties of CPS reduce the noise and results in providing the stable assay system.

Fluorescent labeling to CS is done according to the method of Ogamo *et al.* (1982). Fluoresceinamine (FA) is used as a fluorescent group and introduced to carboxyl groups in glucuronic acid of CS. Although free FA reveals very low fluorescence, it converts to a strong fluorescent substance after the formation of an amide bond by the amino group of FA. So, the fluorescence of free FA can be negligible even if the amide bond is cleaved and FA is released. Coating of CaP with FA-labeled CS (FACS) is attained only by incubating in a neutral solution such as phosphate buffered saline solution. It is important to add an excess amount of FACS. If the coating of FACS to CaP is not enough, CaP spaces remain noncovered and the released FACS by osteoclastic resorption binds again to those spaces.

8.3.2.2 Preparation of Fluorescein-Labeled DNA and Coating to CaP Plate

DNA is constructed with a phosphate group, a deoxyribose sugar and one of four nucleobase (Figure 8.2C). We usually use a calf thymus DNA for the assay due to its huge molecular weight property. Because the production of abundant fluorescein-labeled DNA at low cost has not been established, two-step coating described below is adopted. First, a CaP plate is coated with DNA by the incubation in DNA solution. After washing the plate, the coated DNA is labeled with fluorescent dye. There are many fluorescent dyes to bind DNA. Hoechst 33258, one of the major blue fluorescent dyes, binds to the minor groove of double-stranded DNA. Hoechst 33258 dye should not be added to the DNA solution directly before coating to CaP, because DNA becomes insoluble.

As a matter of course, it is possible to measure the resorption activity using an unlabeled DNA-coated CaP plate. The released DNA content in the medium can be measured by the DNA quantification methods such as using Hoechst 33258 dye (Kim *et al.*, 1988). The released DNA by osteoclastic resorption is abundant, and the DNA resulting from the culture cell is negligible. Generally, as the fluorescent dyes bonded to DNA are potentially mutagenic and carcinogenic, it needs care in handling.

8.3.3 Bone Resorption Assay

Osteoclasts are large multinucleated cells formed by the fusion of hematopoietic, mononuclear progenitors of the monocyte/macrophage lineage (Väänänen and Laitala-Leinonen, 2008). Osteoclasts are usually few in number in the bone and are difficult to isolate and induce proliferation. A murine macrophage cell line RAW264 or RAW264.7 (RAW cells) is frequently used for the study of osteoclast differentiation and to pursue the cellular function. RAW cells differentiate into TRAP positive mononuclear cells and then fuse into multinucleated cells by only stimulating with RANKL (Matsumoto *et al.*, 2000). Another cell source using this assay is mononuclear cells derived from bone marrow. As there are many kinds of cell populations in bone marrow, the cells to differentiate into osteoclasts need to be selected. One of the methods to prepare preosteoclast from bone marrow cells is harvesting mononuclear cells using M-CSF, and then the osteoclast differentiation is induced with RANKL on the assay plate. When the contaminant cell that does not possess the ability to differentiate into osteoclast is abundant in the assay well, they interfere with the fluorescent release from the CaP layer by osteoclastic resorption.

In this assay, preosteoclast are inoculated on a CaP plate coated with fluorescein-labeled polyanion and cultured in the medium containing fetal bovine serum and the stimulators such as M-CSF and RANKL. It is important to use phenol red-free basal medium and avoid using strong fluorescence substances. Bisphosphonate, β-estradiol or other drugs for regulating osteoclast differentiation are added at the same time. After five or six days, multinuclear osteoclasts are formed. The culture supernatant is harvested into a 96-well plate and mixed with an equal volume of alkalescent buffer because the fluorescence intensity of FA changes depending on the pH of medium. The fluorescence intensity is measured using a fluorescence plate reader with an excitation wavelength (Ex) of 485 nm and an emission wavelength (Em) of 535 nm for FACS, Ex: 360 nm and Em: 460 nm for Hoechst 33258.

After finishing the culture, observation of the resorption area (pit formation) is available using a microscope. The cells in the plate should be removed by treating with sodium hypochlorite solution and washed with tap water and then dried. The pit area can be observed without any staining when using the CaP plate described in the Section 8.3.1. Photographs are taken in several different regions in the dark field, and the pit areas are measured with image-analyzing software if needed (Miyazaki *et al.*, 2011). Figure 8.3 shows the formation of pits and fluorescent polyanion release. With RANKL stimulation, the increase of fluorescence intensity was observed as well as the formation of pit area.

In addition to fluorescent polyanions such as FACS and DNA/Hoechst 33258, calcein is also thought to be a candidate in this assay. Calcein is well known as a calcium-binding fluorescent molecule, and is used for analyzing the bone metabolism *in vivo* (Hori *et al.*, 1985) and staining the pit area *in vitro* (Coxon *et al.*, 2008). The fluorescence intensity in the medium released by the osteoclastic resorption was compared using the above fluorescent substances (Figure 8.4). The increase of fluorescent intensity was observed in a dose-dependent manner with RANKL concentration when the three fluorescent-labeled polyanions were used. On the other hand, when calcein was used for CaP labeling, a spontaneous fluorescent release was observed, resulting in a lack of clarity for the detection of the RANKL-induced CaP resorption activity, although pit formation was obviously observed. This indicates that strong and tight binding to CaP, by means of a polyanion, is essential for this assay, which reduces spontaneous fluorescent release in the culture medium.

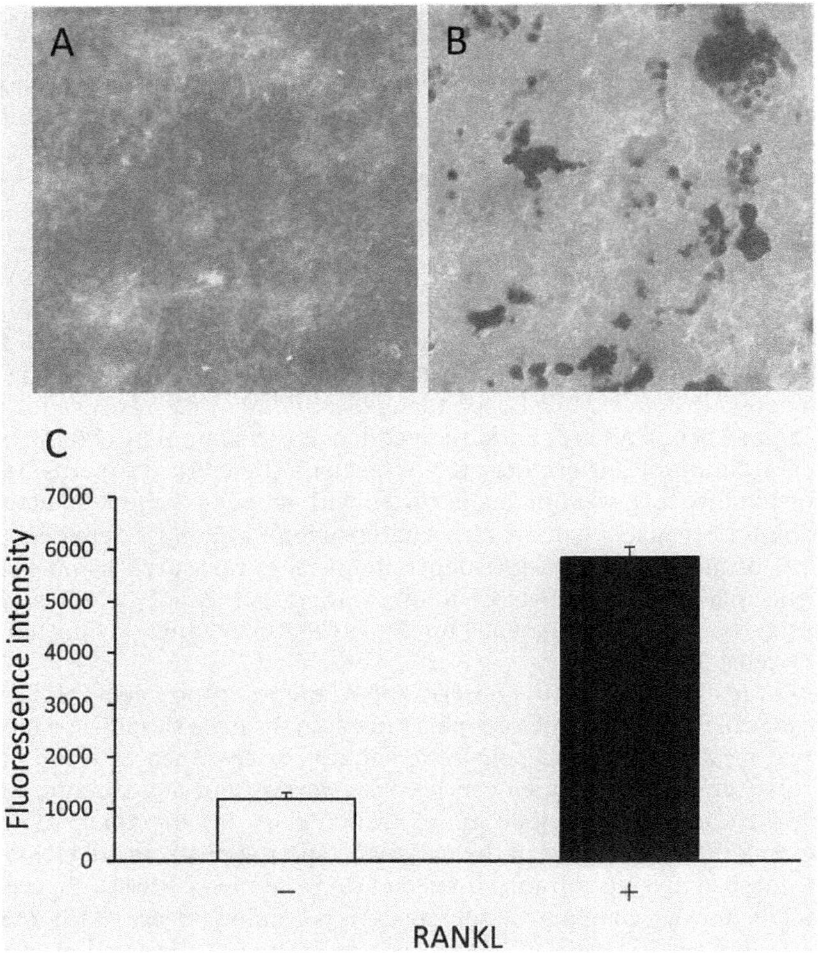

Figure 8.3 Bone resorption of differentiated osteoclast induced with RANKL. Photographs are the pit formation by RAW264 cells cultured in the absence (A) or presence (B) of RANKL (100 ng mL^{-1}) for six days using FACPS/CaP-coated plates. The black regions in B are the resorption areas. (C) The fluorescence intensity of the culture supernatant in the absence (−) or presence (+) of RANKL stimulation were measured (mean ± standard deviation, $n = 3$).

On the osteoclast differentiation, macrophage-derived cells first differentiate into mononuclear osteoclastic cells and then they fuse into mature multinuclear osteoclasts. Mononuclear osteoclastic cells are also thought to have a bone-resorption activity (Domon and Wakita, 1991). Although a traditional pit assay measures mainly large and visible pits formed by mature osteoclasts, this assay system is also able to measure the initial bone resorption by immature mononuclear osteoclasts.

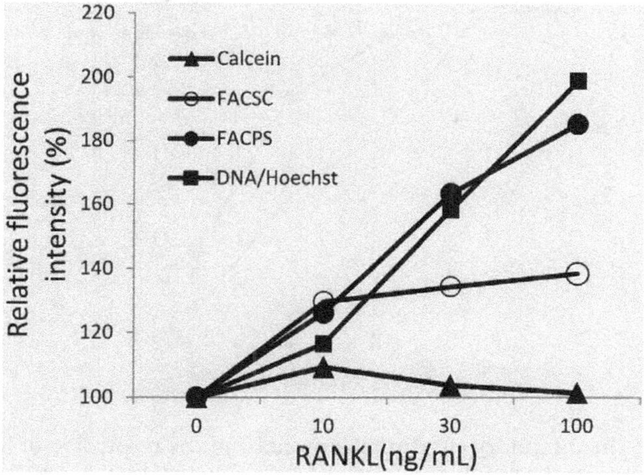

Figure 8.4 Comparison of coating with fluorescent polyanions and calcein. RAW264 cells were cultured on the CaP coating with calcein, FACSC, FACPS or DNA/Hoechst 33258 (DNA/Hoechst). The resorption activity induced with RANKL was evaluated. The data expressed as the percentage to the control (n = 3). Coating of CaP with FACPS or DNA/Hoechst 33258 is suitable for detecting the resorbing activity of osteoclasts. A part of the figure was reproduced with modification from Figures 3A and 4A of Miyazaki *et al.*, 2011. *Analytical Biochemistry*. 410: 7–12 with permission of Elsevier.

8.3.4 Evaluation of Drugs for Osteoporosis

Many drugs for regulating osteoclast differentiation or bone resorption have been found and their function elucidated. These include bisphosphonates, estrogen and its derivatives, vitamin D3, vitamin K and calcitonin. Among them, bisphosphonates, one of the major drugs for osteoporosis, are thought to directly act to osteoclasts and inhibit bone resorption by inducing apoptosis of osteoclasts (Rogers *et al.*, 1996). Figure 8.5 shows the dose-dependent inhibitory effects of alendronate on the CaP resorption induced by RANKL. The inhibitory effect measured by the fluorescence intensity in the medium was well correlated with the result of pit area, and similar to those noted in previous report (Rogers *et al.*, 1996). Other drugs for osteoporosis such as pamidronate and β-estradiol also inhibit resorption activity using this assay system (Miyazaki *et al.*, 2011). Furthermore, as well as the drugs already existing, this assay system is useful for discovering candidates for new drugs (Kim *et al.*, 2012).

In this system, some cautionary points are needed. When using some variants of CS, osteoclastic activity may be controlled by CS or CS-binding protein such as a bone morphogenic protein or the fibroblast growth factor family (Miyazaki *et al.*, 2008; Miyazaki *et al.*, 2010). Another point is that the drug in the medium may affect the fluorescence measurement, for example,

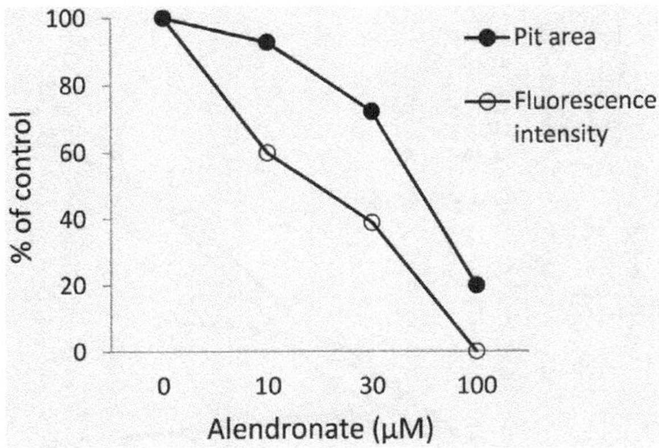

Figure 8.5 The inhibitory effect of alendronate on the resorption of RAW264 cells under the stimulation with RANKL. CaP plates coated with FACPS were used. The decrease of fluorescence intensity of the culture supernatant and the pit area were well correlated. The data expressed as the percentage to the control ($n = 3$).

the autofluorescence of β-estradiol increases the fluorescence intensity when using Hoechst 33258/DNA for CaP labeling.

8.4 Concluding Remarks

This chapter described the new, simple and useful assay method for measuring osteoclastic resorption activity as an alternative method for the traditional pit assay. The point of this method is to use a CaP (carbonate apatite) plate coated with fluorescent-labeled natural polyanionic product such as CS and DNA. The tight binding of fluorescent polyanions to CaP enables to catch the resorption activity of osteoclasts stably. This assay method can evaluate the bone-resorption inhibitory activity of drugs for osteoporosis correlated well with traditional pit assay.

Summary Points

- This chapter focuses on the analysis method to measure the CaP resorption activity by osteoclasts as an alternative method for the traditional pit assay.
- Traditional pit assay measures the area or depth of pits by image-analyzing software after photographing under the microscope, whereas the presented method only measures the fluorescence intensity of the medium.
- A CaP plate labeled with fluorescent polyanion is used in this method. The fluorescent polyanion is released into the medium depending on the CaP resorption of osteoclasts.

- Polyanions such as CS and DNA are available in this assay method. It is important to use long polymer and strong negatively charged molecule for the assay, because these negative charged polyanions can bind tightly with coated CaP and reduce the spontaneous release.
- It is suitable to use carbonate apatite as CaP, because carbonate apatite resembles the native CaP in bone tissue and tends to reveal higher osteoclast formation.
- The drug for osteoporosis such as bisphosphonate is available to evaluate the inhibition activity of bone resorption. It well correlates the fluorescence intensity and the pit area.

Key Facts about Fluorescent Polyanions and CaP Resorption

1. Glycosaminoglycans are large unbranched polysaccharides composed of repeating disaccharide units with the configurations containing an amino sugar and uronic acid. Glycosaminoglycans compose proteoglycan side chains and exist in extracellular matrix. Five primary chains are identified such as chondroitin sulfate, dermatan sulfate, keratan sulfate, heparin/heparan sulfate and hyaluronan. There are some structural isomer variants modified with sulfate groups at different positions. Generally, the molecular weight of glycosaminoglycan is much larger than thousands of daltons.
2. CS exists in bone and cartilage tissue. CS in bone tissue is thought to contribute the nucleation and subsequent growth processes of hydroxyapatite crystals.
3. Calf thymus DNA is a widely used double-stranded DNA. It is used in studies of DNA-binding molecules that modulate DNA structure and function, DNA degradation, and physicochemical studies as a substrate.
4. Calcein is the fluorescent high-affinity calcium ion indicator in the aqueous phase at physiological pH. Calcein and its analog are also used in calcium-ion behavior or assay reagent in the field of cell biology.
5. In this assay, nonsintered CaP is used. Although some kinds of sintered CaP are provided commercially, it is difficult to coat the sintered CaP particle tightly to the culture plate. The dense structure of sintered CaP does not allow the fluorescent-labeled polyanion osmosis to inside the particle.
6. Many commercially available culture media contain phenol red as a pH indicator. Phenol red is fluorescent and affects undesired background fluorescence when measuring resorption assay.
7. Osteoblasts mineralize the matrix by promoting the depositing of CaP (hydroxyapatite) crystals in the sheltered interior of extracellular membrane-limited matrix vesicles and by expanding hydroxyapatite outside of the vesicles, resulting in the propagation of the crystals within the collagenous extracellular matrix.

Definitions of Words and Terms

Bisphosphonates. These are a series of synthetic analogs of carbon-substituted pyrophosphate. Bisphosphonates are used as the drugs to treat osteoporosis, bone metastasis and multiple myeloma by inhibiting the bone resorption of the osteoclasts.

Carbonate apatite. This is a crystalline form of calcium phosphate, varying proportions of the phosphate groups replaced by carbonate ions. Unlike the highly crystalline structure of hydroxyapatite, carbonate apatite in bone has a lower degree of crystallinity that allows turnover and remodeling easily.

Chondroitin polysulfate (CPS). This is a chemically sulfated CS produced from native CS chain. The sulfate groups are introduced to almost of all hydroxyl groups in glucuronic acid and *N*-acetyl-galactosamine.

Fluorescence. This is a type of luminescence created by electromagnetic excitation. It is generated when a substance absorbs light energy at a short wavelength and then emits light energy at a longer wavelength.

Hydroxyapatite. This is one of the calcium phosphates having the formula $Ca_{10}(PO_4)_6(OH)_2$, and the crystal lattice is characteristic of the hexagonal system. Hydroxyapatite is the prototype for the mineral component of bone and teeth.

Image-analyzing software. This is used to capture and analyze the image. When measuring pit area, the image is converted to the appropriate gray-scale image and used for the measurement of all the resorbed regions. Free software, such as NIH Image or Image J, is available.

Osteoporosis. This is a disease in which the density and quality of bone are reduced with an increased risk of fractures. It increases in women after menopause in association with reduced levels of endogenous estrogen.

Osteoclast function. The osteoclast main function is to degrade bone matrix. Bone matrix is consisted from CaP and matrix proteins, such as collagen and noncollagenous protein.

Polyanion. This is a polymer with at least one anionic functional group and has an overall negative charge. There are many kinds of polyanions such as polysaccharides, acidic polyamino acids and polynucleotides.

Sintered CaP. Sintering means that when the powdered material is heated to a temperature below the melting point, the particles is fusing together and one solid piece is created. Sintered CaP is provided for the bone substitute as a medical use.

Slurry. A mixture of a liquid (water) and any of several finely divided insoluble substances, such as clay, cement and plaster.

Tartrate-resistant acid phosphatase (TRAP). This is an acid phosphatase expressed in osteoclasts, and is used widely as a marker to detect osteoclastic phenotype. It is different from the other acid phosphatases regarding resistance to tartrate inhibition.

List of Abbreviations

CaP	Calcium phosphate
CS	Chondroitin sulfate
CSC	Chondroitin sulfate C
CPS	Chondroitin polysulfate
DNA	Deoxyribonucleic acid
Em	Emission wavelength
Ex	Excitation wavelength
FA	Fluoresceinamine
FACS	Fluoresceinamine-labeled chondroitin sulfate
FACSC	Fluoresceinamine-labeled chondroitin sulfate C
FACPS	Fluoresceinamine-labeled chondroitin polysulfate
M-CSF	Macrophage colony-stimulating factor
RANKL	Receptor activator of the NF-kappa B ligand
RAW cells	RAW264 or RAW264.7 cells
TRAP	Tartrate-resistant acid phosphatase

References

Baron, R., Neff, L., Louvard, D. and Courtoy, P. J., 1985. Cell-mediated extracellular acidification and bone resorption: evidence for a low pH in resorbing lacunae and localization of a 100 kD lysosomal membrane protein at the osteoclast ruffled border. *Journal of Cell Biology*. 101: 2210–2222.

Coxon, F. P., Thompson, K., Roelofs, A. J., Ebetino, F. H. and Rogers, M. J., 2008. Visualizing mineral binding and uptake of bisphosphonate by osteoclasts and non-resorbing cells. *Bone*. 42: 848–860.

Delaissé, J. M., Andersen, T. L., Engsig, M. T., Henriksen, K., Troen, T. and Blavier, L., 2003. Matrix metalloproteinases (MMP) and cathepsin K contribute differently to osteoclastic activities. *Microscopy Research and Technique*. 61: 504–513.

Domon, T. and Wakita, M., 1991. Electron microscopic and histochemical studies of the mononuclear osteoclast of the mouse. *American Journal of Anatomy*. 192: 35–44.

Eanes, E. D., Gillessen, I. H. and Posner, A. S., 1965. Intermediate states in the precipitation of hydroxyapatite. *Nature*. 208: 365–367.

Glade, M. J. and Krook, L., 1982. Glucocorticoid-induced inhibition of osteolysis and the development of osteopetrosis, osteonecrosis and osteoporosis. *Cornell veterinarian*. 72: 76–91.

Greenwalt, D., 2008. Methods and tools for detecting collagen degradation. *US Pat.*, 7,405,037.

Hori, M., Takahashi, H., Konno, T., Inoue, J. and Haba, T., 1985. A classification of in vivo bone labels after double labeling in canine bones. *Bone*. 6: 147–154.

Kim, J. L., Kang, S. W., Kang, M. K., Gong, J. H., Lee, E. S., Han, S. J. and Kang, Y. H., 2012. Osteoblastogenesis and osteoprotection enhanced by flavonolignan silibinin in osteoblasts and osteoclasts. *Journal of Biological Chemistry*. 113: 247–259.

Kim, Y. J., Sah, R. L., Doong, J. Y. and Grodzinsky, A. J., 1988. Fluorometric assay of DNA in cartilage explants using Hoechst 33258. *Analytical Biochemistry*. 174: 168–176.

Matsumoto, M., Sudo, T., Saito, T., Osada, H. and Tsujimoto, M., 2000. Involvement of p38 mitogen-activated protein kinase signaling pathway in osteoclastogenesis mediated by receptor activator of NF-kappa B ligand (RANKL). *Journal of Biological Chemistry*. 275: 31155–31161.

Minkin, C., 1982. Bone acid phosphatase: tartrate-resistant acid phosphatase as a marker of osteoclast function. *Calcified Tissue International*. 34: 285–290.

Miyazaki, T., Miyauchi, S., Tawada, A., Anada, T., Matsuzaka, S. and Suzuki, O., 2008. Oversulfated chondroitin sulfate-E binds to BMP-4 and enhances osteoblast differentiation. *Journal of Cellular Physiology*. 217: 769–777.

Miyazaki, T., Miyauchi, S., Tawada, A., Anada, T. and Suzuki, O., 2010. Effect of chondroitin sulfate-E on the osteoclastic differentiation of RAW264 cells. *Dental Materials Journal*. 29: 403–410.

Miyazaki, T., Miyauchi, S., Anada, T., Imaizumi, H. and Suzuki, O., 2011. Evaluation of osteoclastic resorption activity using calcium phosphate coating combined with labeled polyanion. *Analytical Biochemistry*. 410: 7–12.

Ogamo, A., Matsuzaki, K., Uchiyama, H. and Nagasawa, K., 1982. Preparation and properties of fluorescent glycosaminoglycuronans labeled with 5-aminofluorescein. *Carbohydrate Research*. 105: 69–85.

Panda, D. K., Miao, D., Tremblay, M. L., Sirois, J., Farookhi, R., Hendy, G. N. and Goltzman, D., 2001. Targeted ablation of the 25-hydroxyvitamin D 1α-hydroxylase enzyme: evidence for skeletal, reproductive, and immune dysfunction. *Proceedings of the National Academy of Sciences of the United States of America*. 98: 7498–7503.

Quinn, J. M., Elliott, J., Gillespie, M. T. and Martin, T. J., 1998. A combination of osteoclast differentiation factor and macrophage-colony stimulating factor is sufficient for both human and mouse osteoclast formation in vitro. *Endocrinology*. 139: 4424–4427.

Raisz, L. G., 2005. Pathogenesis of osteoporosis: concepts, conflicts, and prospects. *Journal of Clinical Investigation*. 115: 3318–3325.

Rath, N. C., Huff, G. R., Huff, W. E. and Balog, J. M., 2000. Factors regulating bone maturity and strength in poultry. *Poultry Science*. 79: 1024–1032.

Rogers, M. J., Chilton, K. M., Coxon, F. P., Lawry, J., Smith, M. O., Suri, S. and Russell, R. G., 1996. Bisphosphonates induce apoptosis in mouse macrophage-like cells in vitro by a nitric oxide-independent mechanism. *Journal of Bone and Mineral Research*. 11: 1482–1491.

Seldin, D. C., Seno, N., Austen, K. F. and Stevens, R. L., 1984. Analysis of polysulfated chondroitin disaccharides by high-performance liquid chromatography. *Analytical Biochemistry*. 141: 291–300.

Stenbeck, G. and Horton, M. A., 2000. A new specialized cell-matrix interaction in actively resorbing osteoclasts. *Journal of Cell Science*. 113: 1577–1587.

Tamura, T., Takahashi, N., Akatsu, T., Sasaki, T., Udagawa, N., Tanaka, S. and Suda, T., 1993. New resorption assay with mouse osteoclast-like multinucleated cells formed in vitro. *Journal of Bone and Mineral Research*. 8: 953–960.

Tas, A. C., 2003. A review of bone substitutes in bone remodeling: influence of materials chemistry and porosity. In: Sundar, V., Rusin, R. P. and Rutiser, C. A. (ed.) *Bioceramics: Materials and Applications IV*, Wiley, Hoboken, NJ, USA, vol. 147, pp. 15–24.

Yasuda, H., Shima, N., Nakagawa, N., Yamaguchi, K., Kinosaki, M., Mochizuki, S., Tomoyasu, A., Yano, K., Goto, M., Murakami, A., Tsuda, E., Morinaga, T., Higashio, K., Udagawa, N., Takahashi, N. and Suda, T., 1998. Osteoclast differentiation factor is a ligand for osteoprotegerin/osteoclastogenesis-inhibitory factor and is identical to TRANCE/RANKL. *Proceedings of the National Academy of Sciences of the United States of America*. 95: 3597–3602.

Zhang, Z., Welte, T., Troiano, N., Maher, S. E., Fu, X. Y. and Bothwell, A. L., 2005. Osteoporosis with increased osteoclastogenesis in hematopoietic cell-specific STAT3-deficient mice. *Biochemical and Biophysical Research Communications*. 328: 800–807.

Väänänen, H. K. and Laitala-Leinonen, T., 2008. Osteoclast lineage and function. *Archives of Biochemistry and Biophysics*. 473: 132–138.

CHAPTER 9

Methods for the X-Ray Diffraction Patterns of Nanocalcium in Milk

CHING-HSIANG CHEN*[a,b], LIANG-YIH CHEN[a], AND HSIAO-CHIEN CHEN[c]

[a]Department of Chemical Engineering, National Taiwan University of Science and Technology, 43, Section 4, Keelung Road, Taipei, Taiwan; [b]Protrustech Corporation Limited, 3F-1, 293, Sec. 3, Dongmen Road, Tainan, Taiwan; [c]Department of Biochemistry, Taipei Medical University, 250, Wu-Hsing Street, Taipei, Taiwan
*E-mail: bluse.chen@gmail.com

9.1 Overview of Calcium in Food Nanotechnology

9.1.1 Function of Calcium in Human Body

Calcium is an essential and the fifth most abundant element, which is behind carbon, hydrogen, oxygen, and nitrogen, in the human body. Approximately 99% of calcium is existent in bone, which is in the form of hydroxyapatite, to provide strength and flexibility to the skeletal system and to serve as a reservoir of calcium. As calcium concentration is low in blood, osteoclast is responsible for bone disintegration to release calcium. Reversely, osteoblast controls bone formation. The other 1% of calcium is the fluid type appearing both intra- and extracellular. Fifty percent of calcium is chelated

Food and Nutritional Components in Focus No. 10
Calcium: Chemistry, Analysis, Function and Effects
Edited by Victor R. Preedy

Published by the Royal Society of Chemistry, www.rsc.org

by compound or protein. The other fifty percent is the ionized form in soft tissue, blood, and extracellular fluid (Constanzo, 2010). Even though the low concentration of fluid calcium is present, it deeply involves the specific functions in biophysiology, such as blood coagulation, muscle contraction, cardiac functions and nervous transmission (Brass *et al.*, 1978; Fye, 1984; Ohtsuki, 1999; Atlas *et al.*, 2001; Bregestovski and Spitzer, 2005; Sejersted, 2011).

In the human body, the calcium plays an important role in blood clotting. When liquid blood contacts with a foreign surface following damage to a blood vessel, prothrombinase converts a plasma protein, prothrombin. Then calcium is the catalyst in the conversion of prothrombin to thrombin, which converts the soluble fibrinogen to insoluble polymerized fibrin, leading to formation of a meshwork for preventing further loss of blood. In muscle, the calcium concentration could be relative to muscle contraction. At lower calcium concentration, the binding among troponin C (TnC), troponin T (TnT), and troponin I (TnI) is weak. Meanwhile, TnT binds tropomyosin closely that covers the binding site between actin and myosin, blocking their interaction. At the same time, the strong binding between TnI and actin also prevents their interaction and inhibits ATPase activity, and thus keeps muscle in a relaxation state. As intracellular calcium increases, binding of calcium with TnC will cause a change in conformation of TnC to reduce the binding ability of TnI–actin and TnT–tropomyosin, and to allow the actin to interact with myosin for activating the actin–myosin–ATP for muscle contraction (Ohtsuki, 1999). The calcium supports nerve conduction during nerve impulse to muscle fiber. When the nerve impulse reaches the synaptic terminal, the calcium ions will flow into the synaptic terminal through an open voltage-gated calcium channel. The high concentration of calcium activates calcium-sensitive protein and triggers the fusion of the neurotransmitter vesicles with the synaptic membrane for releasing the neurotransmitter (Atlas *et al.*, 2001; Bregestovski and Spitzer, 2005).

Therefore, maintaining the suitable calcium concentration in the human body is important. It has been known that calcium deficit causes osteoporosis for skeletal systems (Cohn *et al.*, 1974). Hypocalcemia results in the symptoms of neuromuscular irritability, cardiovascular complications, depression, and feeling irritable (Rutecki and Whittier, 1998; Miller *et al.*, 2001). On the contrary, hypercalcemia causes symptoms including abnormal heart rhythms, peptic ulcer disease, kidney stones, dementia, and depression (Vella *et al.*, 1999; Schmaldienst *et al.*, 2001; Grundfast *et al.*, 2003; Christensen and White, 2007). Recently, Alzheimer's disease is being focused upon as a key research area in many countries, which has been found to be one of the factors relative to calcium dysregulation in brain (Small, 2009; Woods and Padmanabhan, 2012). In addition, calcium is associated with colorectal cancer, lung cancer, and prostate cancer, which has been reported in several studies (Yang *et al.*, 2010; Schwartz and Skinner, 2012; Kim *et al.*, 2013). Schwarz *et al.* reported the proliferation and apoptosis of cancer cell is dependent on the ionized calcium concentration (Schwarz *et al.*, 2013).

Rising intracellular concentration can enhance cytotoxic T lymphocytes and natural killer cells in killing cancer cells. Therefore, calcium is an extremely important and indispensable element in life. The daily requirement of calcium from foods or supplements must be sufficient. For the daily diet, only a small amount of ingested calcium can be absorbed by the gastrointestinal tract through mass-transport processes. The United States Institute of Medicine has reported the recommendations for the calcium intake from foods and supplements throughout the life cycle of humans in various stages (Office of Dietary Supplements, 2013).

9.1.2 Food Nanotechnology

Nanofood has nowadays become very popular in the scientific world and it is one of the nanotechnologies. The definition of nanotechnology means that one of the dimensions in 3D structure is toward 10^{-9} m. The properties of nanosized foodstuffs require new characterization techniques and result in novel phenomena in foodstuffs because the interaction is between atomic or molecular quantum confinement and bulk (Chau *et al.*, 2007; Bugusu, 2008). The technology of nanoanalysis can not only provide the useful information for the microstructural composition of food, but also correlate among the rheological properties, configuration and functionality of foods. Therefore, the development of nanotechnology between the food and the nanoanalysis becomes more important in food ingredient industries. Nanotechnology now has very high potential to provide health-food ingredients for food industry, such as the product design by applying the natural colloids and functional compositions and the combination of the novel nanofood ideas and traditional manufactures.

Food nanotechnology can be divided into microstructural modifications and equipment applications. The definition of nanofood means food produced by using food nanotechnology on packaging processes, manufactures or any productions. Usually, food nanotechnology can be employed to develop organic and green processes, especially in additional applications of agricultural products and the production of the health food with multifunctional properties, such as health restoration, enhancement of physique, prevention of disease, increase of nutrition, postponement of senility, and immunity adjustment. Lots of studies have focused on the functionality, digestibility, preparations, safety and characterization for cellulose, chitosan, and starch. Recently, it was found that many food factories have added nanoiron or nanocalcium to improve the nutrition of foods. Partial biological technology factories have investigated nanotechnology in the productions of health foods with nanoiron and nanocalcium to increase the solubility for infant, elderly persons and digestive disease patients to absorb the necessary nutrient (Weiss *et al.*, 2006). In addition, some investigations have shown that the intake combining with vitamin D could effectively enhance the calcium absorption in small intestines (Christopher Nordin, 2010). Also, some factors would affect the absorbing quality in our daily life. Excepting physiological diseases, food contains excess oxalic acid or fat would form the insoluble calcium salt (Swartzman *et al.*, 1973; Schroder *et al.*, 2005). Besides, the competition with magnesium that influences the absorption is noted (Balch and Balch, 1997). A main application of nanotechnology is the development of functionality and specialty of foods to achieve high bioavailability, high solubility, and rapid release of flavor (Sanguansri and Augustin, 2006; Weiss *et al.*, 2006). However, nanoparticles in nanofood create thermodynamic instability due to their specific features, such as high surface energies, small sizes, and high surface areas. Moreover, the surfaces of nanoparticles contain many dangling bonds and they tend to aggregate together. If the sizes of nanoparticles increase, the specific physical and chemical properties of the particles of them will be lost. This implies that the different manufacture parameters can change the properties of nanoparticles. For instance, the phase and the crystalline of nanoparticles will change through dissolution and particle regeneration to cause another worse physiological reaction. How to keep their physical and chemical properties would become a basic requirement on the development of the nanotechnology. Nevertheless, the mainly absorbing quality is depended on the dissolution ability of calcium species. The most common types from dietary supplements are calcium carbonate and calcium citrate. Calcium carbonate possesses of highest calcium concentration and the lowest price. The poorly soluble property reduces its efficiency. In contrast, calcium citrate performs with good solubility, improving the calcium absorption. However, it can be inferred that the objective of the long-term development in nanotechnologies would help to understand the properties, the changes of structure and phase of food materials.

9.1.3 Analysis of Calcium

To monitor the daily intake of calcium, food labels should provide sufficient information. The development of accurate biosensors for quantifying calcium in foods has become an important field in food analysis. So far, various techniques for detecting calcium have been proposed. The determination of calcium content in water quality through ethylenediaminetetraacetic acid (EDTA) titrimetric method is according to the International Organization for Standardization 6058 (ISO 6058) (ISO 6058 Water quality, 1984). The applicable range in water is from 2 to 100 mg L^{-1}. ISO 12081 describes how the test portion in milk is precipitated by trichloroacetic acid forming calcium oxalate before titrating with potassium permanganate (ISO 12081, 1998). However, the error value will occur during the titrating process. In addition, the sensitivity and detection limit are not good enough.

Flame atomic absorption and emission spectrometry (FAAES) has been adopted as one of the most suitable methods for routine analysis of milk in the ppm range (Zhou *et al.*, 2002; Liu and Gao, 2004; Noël *et al.*, 2008). This technique is possessed of relative simplicity and high accuracy. Regrettably, in addition to the low sensitivity, the radiation source must be changed to match each element, meaning that only a single element can be analyzed each time. On the contrary, inductively coupled plasma atomic emission spectrometry (ICP-AES) can detect multielements simultaneously (Sze *et al.*, 2007; van Hulzen *et al.*, 2009; Terol *et al.*, 2010). The sample preparation is similar to the FAAES that is divided into two methods, wet and dry digestion. The former method usually mixes with acid to accelerate digestion. For the latter method, dry ash is obtained through hot-plate, furnace or microwave heating. In an ICP-AES system, the high operation temperature prevents the formation of polyatomic species during the measuring process, thus improving the limits of detection. Compared with FAAES, it has high sensitivity and lower detection limits (Zhang *et al.*, 2009). Although the above two techniques are more sensitive and selective than others, the high cost of the instrument and complex operation have been of concern. In addition, the experimental results are obtained from the total calcium concentration only in food, which cannot determinate the initially ionic state of calcium that has presented in the liquid. A calcium sensor based on an electrochemical technique is being widely developed due to the portable, rapid, sensitive, and simple characteristics (Bratov *et al.*, 2000; Gemene and Bakker, 2009; Hernández *et al.*, 2010). It consists of a polymeric membrane for transporting selected ions, resulting in a potential difference corresponding to the ion activity. The ion-selective electrode can maintain the analytic status without destruction.

Summarizing the above methods in determining calcium, the analytic samples are all in an ionic or ionized state. They are not easy to apply in distinguishing the sources of the calcium salt. Therefore, finding the right structure of additive calcium salts would be the one of the important factors to understand the phase transformation of the calcium during any treatments. X-ray diffraction technology can provide useful structural information of nanoclusters and analyze the quantities of additives in the food or identify the phase variation for food through different procedures. For the X-ray diffraction, characterizing the crystallinity of materials ground as tiny powders is convenient. According to diffraction technology, the plane distance can be used to identify the crystal structures of materials (Cullity and Stock, 2001). In addition, the diffraction peak intensity will be changed by varying the plane atomic density. Therefore, the integrated intensity of the diffraction peak and diffraction angle can be applied to characterize the existence of certain crystal materials.

9.2 X-Ray Diffraction Technology

X-rays were discovered in 1985 by the physicist Roentgen. Unlike ordinary light, these rays were invisible, but they traveled in straight lines and affected photographic film in the same way as light. On the other hand, they were much more penetrating than light and could easily pass through the human body, wood, quite thick pieces of metal, and other "opaque" objects. Now, we know that the X-ray is electromagnetic radiation of exactly the same nature as light but of very much shorter wavelength. The unit of measurement in the X-ray region is the angstrom (Å), equal to 10^{-10} m, and X-rays used in diffraction have wavelengths lying approximately in the range of 0.5–25 Å. X-rays therefore occupy the region between gamma and ultraviolet rays in the complete electromagnetic spectrum (Cullity and Stock, 2001).

9.2.1 X-Ray Generation

X-rays are produced when any electrically charged particle of sufficient kinetic energy is rapidly decelerated. Electrons are usually used for this purpose, the radiation being produced in an X-ray tube that contains a source of electrons and two metal electrodes. The high voltage maintained across these electrodes, some tens of thousands of volts, rapidly draws the electrons to the anode, which they strike with very high velocity. When the rays coming from the target are analyzed, they are found to consist of a mixture of different wavelengths, and the variation of intensity with wavelength is found to depend on the tube voltage. When the tube voltage is raised, the intensity of all wavelengths increases. In addition, both the short-wavelength limits (SWL) and the position of the maximum shift to short wavelengths, as shown in Figure 9.1. The radiation represented by such curves is called heterochromatic, continuous, or white radiation. White radiation is also called bremsstrahlung, German for "braking radiation", because it is caused by electron deceleration. The continuous spectrum is due to the rapid deceleration of the electrons hitting the target since any decelerated charge emits energy. However, when the voltage on the X-ray tube is raised above a certain critical value, characteristic of the target metal, sharp intensity maxima appear at certain wavelengths, superimposed on the continuous spectrum. These are very narrow and the wavelength is characteristic of the target metal used. They are called characteristic lines. These lines fall into several sets, referred to as K, L, M, *etc.* (Williams and Carter, 1996).

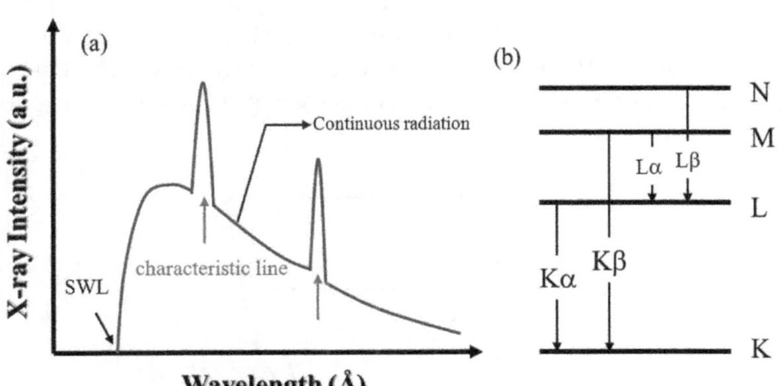

Figure 9.1 (a) A typical X-ray spectrum consisting of continuous radiation (bremesstrahlung) and characteristic lines. (b) The possible electron transition that gives to K, L and M characteristic X-ray sources.

9.2.2 Bravais Lattices

Turning from the properties of the X-ray, we now consider the geometry and structure of crystals in order to discover what there is in general that enables them to diffract the X-rays. A crystal may be defined as a solid composed of atoms arranged in a pattern periodic in three dimensions. For crystals, it is often convenient to ignore the actual atoms composing the crystal and their periodic arrangement in space. In fact, we can consider a set of imaginary points that has a fixed relation in space to the atoms of the crystal and that may be regarded as a sort of framework or skeleton on which the actual crystal is built. A set of points so formed has an important property: it constitutes a point lattice, which is defined as an array of points in space so arranged that each point has identical surroundings. Since all the cells of the lattice are identical, we may choose any one, which is called the unit cell. The size and shape of the unit cell can turn be described by three vectors, $\vec{\mathbf{a}}$, $\vec{\mathbf{b}}$, $\vec{\mathbf{c}}$ drawn from one corner of the cell taken as the origin (Figure 9.2). They may also be described in terms of their lengths (a, b, c) and the angles between them (α, β, γ). These lengths and angles are the lattice parameters of the unit cell. In 1848, French crystallographer Bravais demonstrated that there are fourteen possible point lattices and no more. Now, we call these fourteen point lattices Bravais lattices, as shown in Figure 9.3 (Warren, 1990).

9.2.3 Bragg Spectrometer

In 1912, German physicist von Laue (1879–1960) reasoned that if crystals were composed of regularly spaced atoms they might act as scattering centers for the X-ray. The wavelength of the X-ray is equal to the interatomic distance in the crystals, and it should be possible to diffract the X-ray by means of the crystals. Under this concept, experiments to test this hypothesis were carried out. The X-rays were diffracted by the crystal out of the primary beam to form a pattern of spots on the photographic plate. The account of these experiments was read by the English physicist, W. H. Bragg (1862–1942) and his son, W. L. Bragg (1890–1971). They successfully analyzed the Laue experiments and were able to express the necessary conditions for diffraction in a

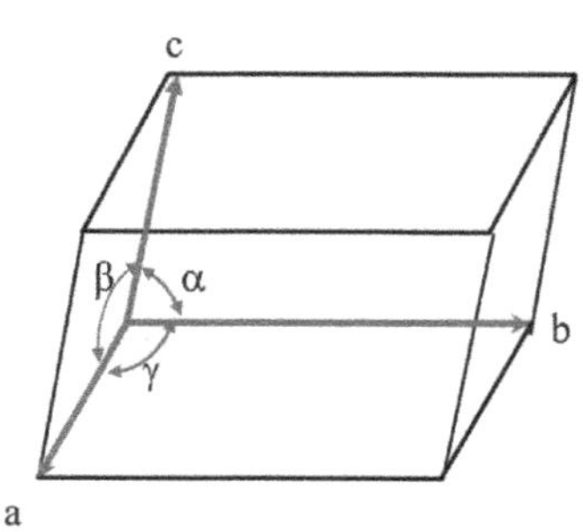

Figure 9.2 Schematic diagram of a unit cell.

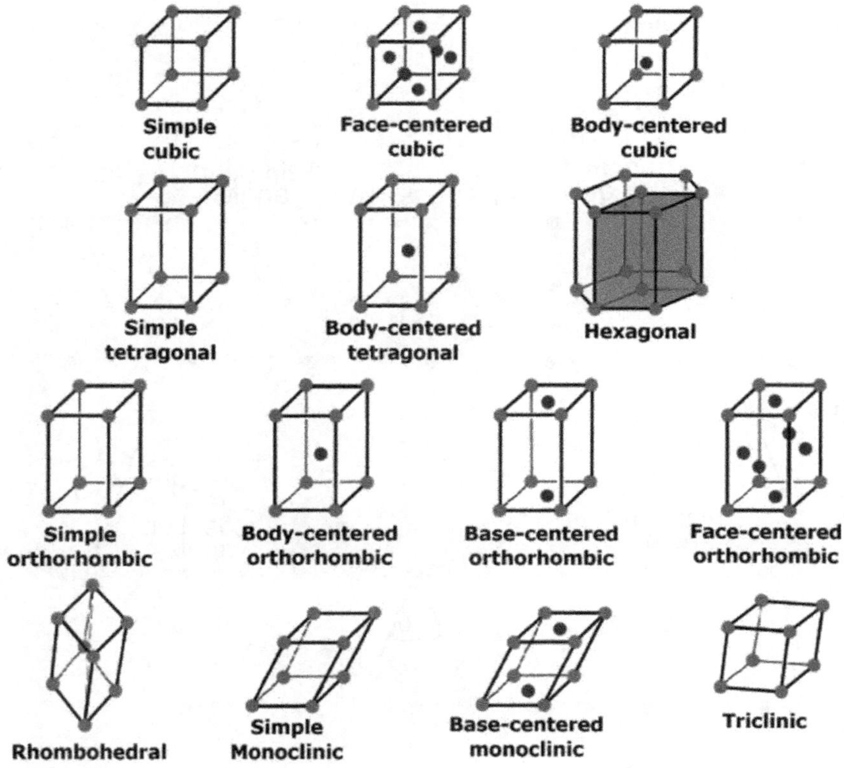

Figure 9.3 Bravais lattices in three dimensions.

considerably simpler mathematical form, as shown in eqn (9.1). Usually, we called it as Bragg's Law (Cullity and Stock, 2001).

$$\lambda = 2d\sin\theta \tag{9.1}$$

From Bragg's Law, diffraction is essentially a scattering phenomenon in which a large number of atoms cooperate. Since the atoms are arranged periodically on a lattice, the rays scattered by them have a definite phase relation between them. These phase relations are such that destructive interference occurs in most directions of scattering, but in a few directions constructive interference take place and diffracted beams are formed. Considering a set of $(h\ k\ l)$ planes is shown in Figure 9.4. The path difference ABC between rays scattered by adjacent $(h\ k\ l)$ planes must be two whole wavelengths. In general, an nth-order reflection from the $(h\ k\ l)$ planes of spacing d' may be considered as a first-order reflection from the $(nh\ nk\ nl)$ planes of spacing $d = d'/n$. This convention is in accord with the definition of Miller indices since $(nh\ nk\ nl)$ are the Miller indices of planes parallel to the $(h\ k\ l)$ planes but with $1/n$th the spacing of the latter (Cullity and Stock, 2001).

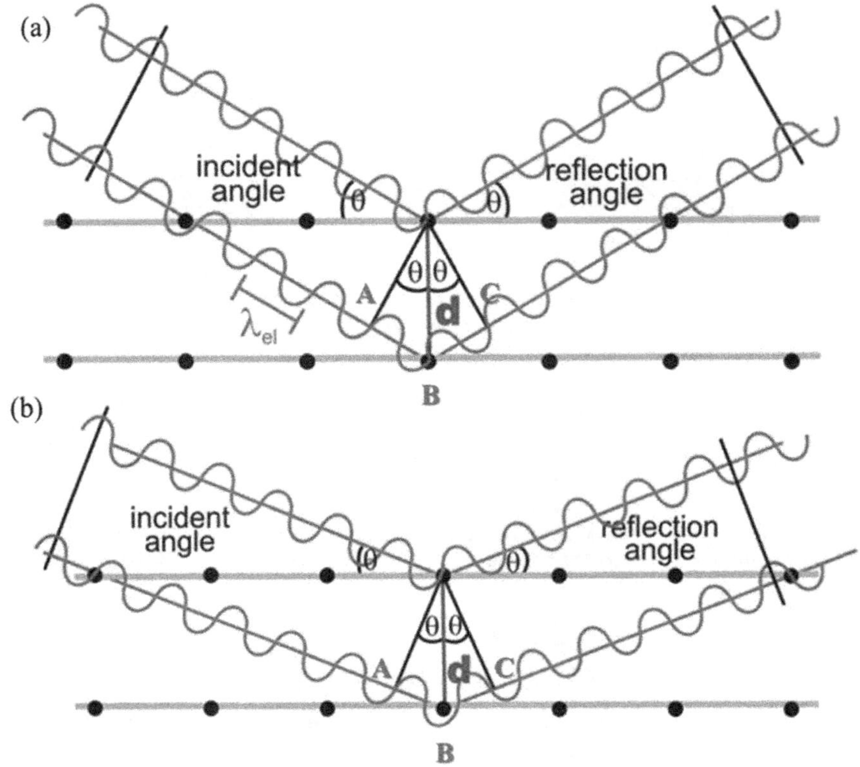

Figure 9.4 (a) Constructive interference of reflected waves (reflected waves in phase, *i.e.*, maxima are superimposed); (b) destructive interference of reflected waves (in the two reflected waves, maximum and minimum of the respective wave amplitudes are superimposed).

Experimentally, we can determine the spacing d of various planes in a crystal by using X-rays of known wavelength as λ and measuring the diffraction angle θ. The essential features of an X-ray spectrometer are shown in Figure 9.5. The X-rays from the tube **T** are incident on a crystal **C** that may be set at any desired angle to the incident beam by rotation about an axis through the center of the spectrometer circle (**O**). **D** is a counter that measures the intensity of the diffracted X-ray. It can also be rotated with respect to **O** and set at any desired angular position. In use, the crystal is positioned so that its reflecting planes make some particular angle θ with the incident beam, and **D** is set at the corresponding angle 2θ (Cullity and Stock, 2001).

In addition to the qualitative analysis, quantitative analysis by diffraction is based on the fact that the intensity of the diffraction pattern of a particular phase in a mixture of phases depends on the concentration of the phase in the mixture. The relation between intensity and concentration is

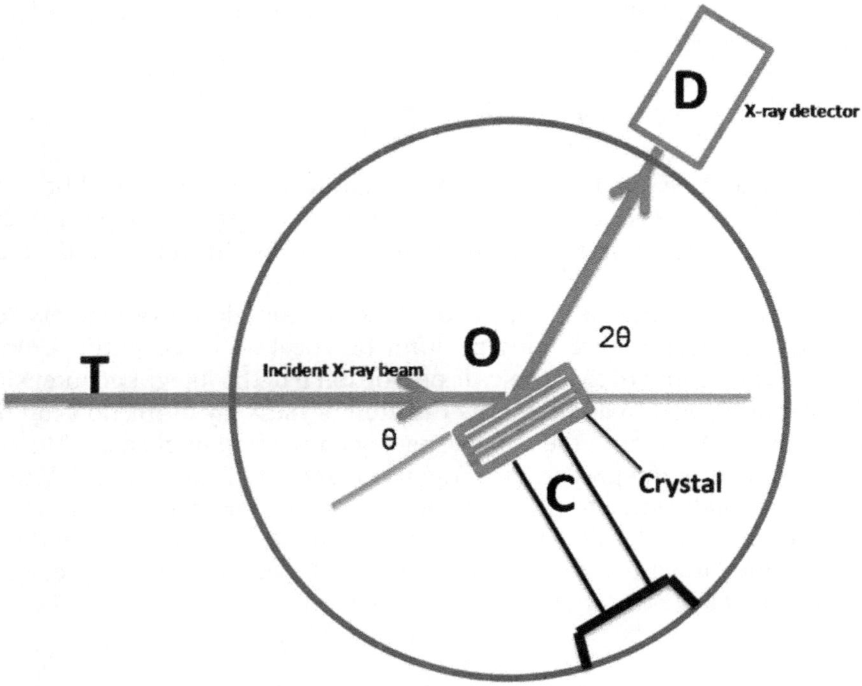

Figure 9.5 Schematic illustration of a Bragg spectrometer.

not generally linear, because the diffraction intensity depends on the absorption coefficient of the mixture and this varies with the concentration. For the intensity diffracted by a single phase powder specimen in a diffractometer is

$$I = I_0\left(\frac{A\lambda^3}{32\pi r}\right)\left[\left(\frac{\mu_0}{4\pi}\right)^2\frac{e^4}{m^2}\right]\left(\frac{1}{V^2}\right)\left[|F|^2 p\left(\frac{1+\cos^2\theta}{\sin^2\theta\cos\theta}\right)\right]\left(\frac{e^{-2M}}{2\alpha}\right) \quad (9.2)$$

I is the integrated intensity per unit length of diffraction line (J s^{-1} m^{-1}), I_0 is the intensity of incident beam (J s^{-1} m^{-2}), A is the cross-sectional area of incident beam (m^2), λ is the wavelength of the incident beam (m), r is the radius of diffractometer circle (m), μ_0 is $4\pi \times 10^{-7}$ m kg C^{-2}, e is the charge of an electron, n is the mass of an electron (kg), V is the volume of the unit cell (m^3), F is the structure factor, p is a multiplicity factor, θ is the Bragg angle, e^{-2M} is the temperature factor and α is the linear absorption coefficient (m^{-1}) (Cullity and Stock, 2001). We can simplify eqn (9.2) considerably for special cases. Considering a mixture of two phases, α and β, we now concentrate on a particular line of α phase and the I of eqn (9.2) becomes I_s, which is the intensity of the selected line of the α phase in the mixture. Then we need to substitute α_m for α, where α_m is the linear absorption coefficient of the mixture. In this modified equation, all factors are constant and independent of the concentration of α except C_α and μ_m. Therefore, we can write

$$I_\alpha = \frac{K \cdot C_\alpha}{\alpha_m} \tag{9.3}$$

where K is a constant. The value of K is unknown. However, it will be canceled out if we measure the ratio to the intensity of some standard reference line. The concentration can then be found from this ratio (Cullity and Stock, 2001).

In summary of this section, for the mixture containing various crystal additives in food, the corresponding diffraction peaks can be identified. Simply, the composition of the species is proportional to the integrated intensity of diffraction peak. In addition, the function of the X-ray diffraction can be divided into: (A) identify the atomic spacing and period of crystal additives in food to realize the properties of the polymer, metal, oxide and the other solid-state condenser, and (B) every species has its characteristic diffraction peak. When the peak value is identical to the standard, we can regard them as the same lattice period. Therefore, X-ray diffraction technology can be applied to evaluate the crystal compound in food.

9.3 Structure of Nanocalcium in Milk

In order to demonstrate the crystal structures of nanocalcium additives in food, two different milk powders are selected from the market to study the structures by using the X-ray diffraction technique. One of the milk powders contains nanocalcium additives. The X-ray diffraction pattern was recorded with a copper anode X-ray tube (Cu Kα1 = 1.54051 Å and Kα2 = 1.54433 Å, the average wavelength is 1.5418 Å) using a MTI Mini X-ray powder diffractometer (MTI, MD-10 precision). The milk powders were packed tightly in sample holders and the sample holder is rotated during the measurement in order to get good signal uniformity. Each sample was exposed to the X-ray beam at 25 kV and 0.4 mA. The collecting region of the diffraction angle (2θ) was from 20° to 60° at 0.02° interval size by a cambered surface CCD and the integration time was set at 100 s.

Figure 9.6a shows the X-ray diffraction patterns of the normal milk and two other milk samples. It was found that the feature of the normal milk was totally different from that of both the two milk samples. It showed only one amorphous peak between 20° and 35°, indicating no crystallized calcium compounds in normal milk. However, the base of the milk powder was amorphous and would have a minimum influence on the identification of the X-ray diffraction. This may imply that the diffraction peaks in both the milk samples are contributed from the crystal structures of nanocalcium additives. In addition, the two milk samples showed that partial peaks are located at different diffraction angles. It can be inferred that the nanocalcium in the milk sample 1 has different crystal structure as compared with that of the milk sample 2. To understand the details of the phase transformation of the nanoscale

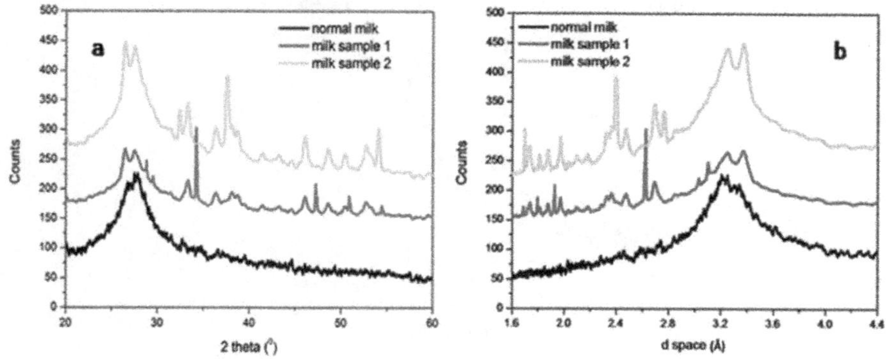

Figure 9.6 The X-ray diffraction patterns as displayed by (a) diffraction angle and (b) *d*-space of the normal milk, the milk sample 1 and the milk sample 2.

calcium additives, the *x*-axis of the X-ray diffraction pattern has been changed to be displayed as *d*-space for data identification, as shown in Figure 9.6b.

To understand the detailed structures of calcium additives in the milk samples, X-ray diffraction patterns of both milk samples are applied to compare with the Joint Committee on Powder Diffraction Standards (JCPDS) database, which is provided by the International Centre for Diffraction Data (ICDD). Figure 9.7 showed the results of the X-ray diffraction pattern of the milk sample 1 compared with JCPDS database. It is found that the structure includes two phases. One is the orthorhombic phase of calcium carbonate and its lattice parameters are $a = 5.74$ Å, $b = 4.961$ Å, and $c = 2.929$ Å (JCPDS 240-0025). The corresponding diffraction *d*-spaces of the orthorhombic phase in the milk sample are 3.396 Å, 3.273 Å, 2.700 Å, 2.484 Å, 2.372 Å, 2.341 Å, 2.329 Å, 2.189 Å, 2.106 Å, 1.977 Å, 1.882 Å, and 1.877 Å. The other is the hexagonal phase of calcium hydroxide and its lattice parameters are $a = b = 3.607$ and $c = 4.979$ Å (JCPDS 84-1275). The corresponding diffraction *d*-spaces of the phase are 3.124 Å, 2.646 Å, 1.947 Å, 1.804 Å, and 1.696 Å.

Figure 9.8 reveals the comparison of the milk sample 2. The results showed that the structure of the calcium additive is distinguished as two kinds of phases. The first structure of the calcium additive is the orthorhombic phase of calcium carbonate and its lattice parameters and the corresponding diffraction *d*-spaces are the same as that of the milk sample 1. However, it can be inferred that the calcium carbonate with orthorhombic phase would be one of the main nanocalcium additives in the milk market. The second structure of the calcium additive in milk sample 2 is the cubic phase of calcium oxide and its lattice parameters are $a = b = c = 4.804$ Å (JCPDS-821690). The corresponding diffraction *d*-spaces are 2.774, 2.402, and 1.688 Å.

Based on our understanding, milk powders contain many kinds of calcium complexes including calcium formate, calcium methylate, calcium acetate, calcium caseinate, *etc.* There are no marked peaks to fit with those structures in the X-ray diffraction patterns due to the fact that the content of other calcium complexes is too low or the structure of them is amorphous, which corresponds to the diffraction pattern of the normal milk powder.

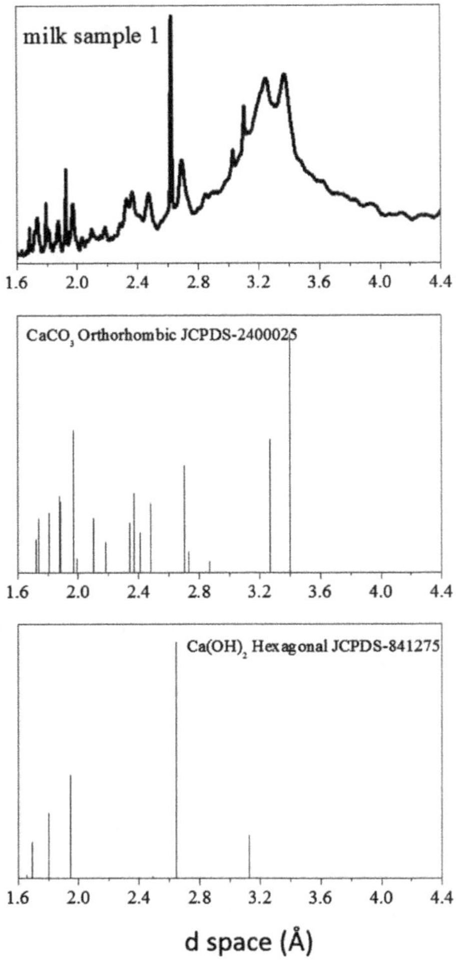

Figure 9.7 The X-ray diffraction patterns of the milk sample 1 compared with the JCPDS database.

The X-ray diffraction technique can only be applied to detect the crystallized compounds. Herein, the technique emphasizes the formation of the main crystal phases of the calcium compounds. However, there are some methods that can be applied to evaluate the amorphous structures and low content species, such as X-ray absorption spectroscopy, small-angle X-ray scattering, surface-enhanced Raman scattering, *etc.*

9.4 Conclusion

The observations of the two milk samples showed that the main structure of the calcium additive is orthorhombic crystal. The milk sample 1 shows the existence of the calcium hydroxide phase and the milk sample 2 shows

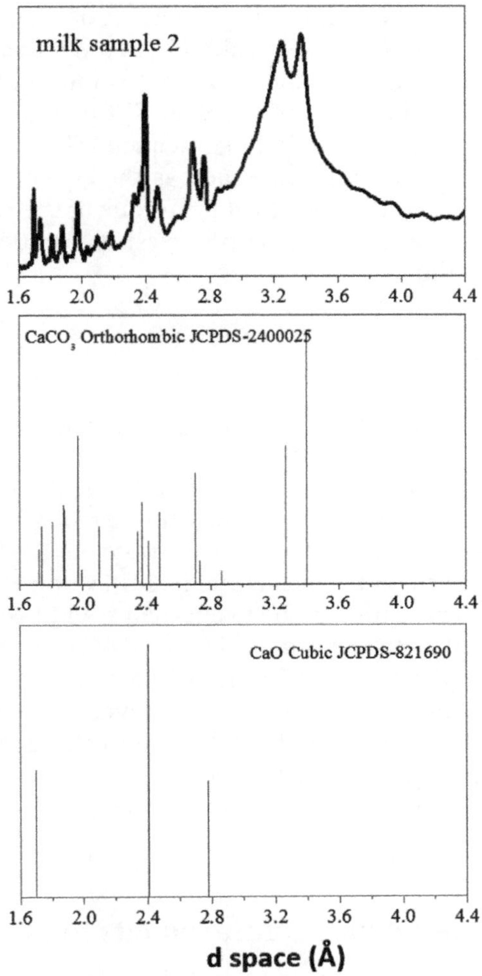

Figure 9.8 The X-ray diffraction patterns of the milk sample 2 compared with the JCPDS database.

the calcium oxide phase. Before mixing milk powders, one of the possibilities to generate the second phase is that the side reaction or poor reaction parameter to generate impurity phase during the calcium carbonate powder formation. After mixing with milk and calcium carbonate additive, the impurity phase may be generated due to the further phase transformation of the calcium carbonate by the variation of a control factor during the milk powder fabrication, such as drying temperature, dehydration rate, *etc.* We believe that the existence of impurity phase will influence the dispersion and dissolution of the nanocalcium additives in milk. It may further influence the digestion and metabolism of the nanocalcium in human body. Herein, we proposed a simple methodology *via* the X-ray diffraction technique. It can effectively analyze the phase transformation of the additives in nanofood after producing, and precisely guide the level of the structural stability. For example, if the level of the structural stability of the additives in nanofood is lower, the intensity of the X-ray diffraction peak of the phase index would be very weak. It could be anticipated that the future development of foodstuff-related programs, such as the conventional technology in foodstuff science, and the analytical technique of the nanofood fields in domestic industries, which could be progressively improved by using the X-ray diffraction method.

Summary Points

- The focus of the first part in this chapter is to explain the function of nanocalcium in the human body.
- Nanocalcium provides functions in blood coagulation, blood clotting, muscle contraction, cardiac functions and nervous transmission.
- The second part of the chapter is to illustrate the X-ray diffraction technology and the function of structural identifications.
- The structural correlation of the nanocalcium additives in milk samples is demonstrated as an orthorhombic crystal of calcium carbonate and it would vary the phase with different treatments.
- The simple methodology is proposed to evaluate the phase transformation and the structural stability of the nanoscale additives *via* the X-ray diffraction technique in food science.

Key Facts

Key Facts of Nanocalcium Additive and its Importance

1. Calcium is an essential element and the suitable calcium concentration in human body is important.
2. Fifty percent of calcium is chelated by compounds or proteins. The other fifty percent is in the ionized form in soft tissue, blood, and extracellular fluid.
3. Calcium is the catalyst in the conversion of prothrombin to thrombin, leading to the formation of a meshwork for preventing further loss of blood.
4. The daily requirement of calcium from foods or supplements must be sufficient.
5. For the daily diet, only a small amount of ingested calcium can be absorbed by the gastrointestinal tract through mass-transport process.
6. Nanocalcium additive can be produced by nanotechnology to improve the nutrition of foods, to increase the solubility for infants, and to help elderly persons and digestive-disease patients to absorb the necessary nutrient.

Key Facts of Structural Identification by Using X-ray Diffraction

1. X-rays are produced when any electrically charged particle of sufficient kinetic energy is rapidly decelerated.
2. A crystal is defined as a solid composed of atoms arranged in a pattern periodic in three dimensions.
3. The wavelength of the X-ray is equal to the interatomic distance in the crystals, and it is possible to diffract the X-ray by means of the crystals.
4. X-rays are diffracted by the crystal out of the primary beam to form a pattern of spots on the photographic plate and Bragg's law determines the positions of the diffraction peaks.
5. Every species has its characteristic X-ray diffraction peaks.
6. The function of the X-ray diffraction peaks can be searched and compared with JCPDS database to identify the atomic spacing and period of crystal additives in food to realize the properties of polymers, metals, oxides and the other solid-state condensers.

Definitions of Terms

Nanotechnology. The definition of nanotechnology means that one of the dimensions in 3D structure is toward 10^{-9} m. The properties of nanosized materials and foodstuffs require new characterization techniques and result in novel phenomena due to the fact that the interaction is between atomic or molecular quantum confinement and bulk.

Nanocalcium additive. Nanocalcium additive can be produced by nanotechnology to improve the nutrient of foods, to increase the solubility of calcium in foods, and to help digestion and absorption of calcium in human body.

Flame atomic absorption and emission spectrometry (FAAES). FAAES is applied to analyze species quantitatively. It can be divided into the FAAS and FAES systems. The two systems use aqueous sample introduction through the nebulizer and into the burner head. This works well for samples that are already in the aqueous phase or can be digested in acid such as soil, atmospheric particles, and tissue samples. All aqueous samples for FAAS and FAES contain some amount of strong acid (1–5%). The acid acts to keep the metal analytes in the dissolved phase and to avoid adsorption of metal ions to sample container and instrument surfaces.

X-ray diffraction. X-ray diffraction is one of the possible methods to identify crystal structures. It follows Bragg's law to refine the diffraction peaks according to the difference in lattices, unit cells and atomic positions.

Bragg's law. The diffraction is essentially a scattering phenomenon in which a large number of atoms cooperate. The details of structural parameters in crystals are able to express the necessary conditions for diffraction in a considerably simpler mathematical form as $\lambda = 2d \sin\theta$, where λ is the wavelength of incident beam (m), θ is the Bragg angle, and d is the space distance.

Bravais lattices. A crystal is made up of a periodic arrangement of one or more atoms repeated at each lattice point. The crystal looks the same

when viewed from any equivalent lattice point. There are 14 possible Bravais lattices in three-dimensional space. If two Bravais lattices have isomorphic symmetry groups, they are often considered equivalent.
Miller indices. A family of lattice planes is determined by three integers h, k, and l. The Miller indices are written as $(h\ k\ l)$ and each index denotes a plane orthogonal to a direction (h, k, l) in the and each index denotes a plane orthogonal to a direction (h, k, l) in the basis of the reciprocal lattice vectors.

List of Abbreviations

TnC	Troponin C
TnT	Troponin T
TnI	Troponin I
ATPase	Adenosine triphosphatease
EDTA	Ethylenediaminetetraacetic acid
FAAES	Flame atomic absorption and emission spectrometry
ICP-AES	Inductively coupled plasma atomic emission spectrometry
SWL	Short-wavelength limits
JCPDS	Joint Committee on Powder Diffraction Standards
ICDD	International Centre for Diffraction Data
Å	Angstrom, unit of length
α, β, γ	The angles of unit cell
λ	Wavelength of incident beam (X-ray)
d	The distance of lattice planes
θ	Bragg's angle
$(h\,k\,l)$	Miller index of lattice plane
I_0	Intensity of incident beam (X-ray)
I	Integrated intensity per unit length of diffraction line
J	Joule, unit of energy
s	Second, unit of time
m	Meter, unit of length
cm	Centimeter, unit of length
A	Cross-sectional area of incident beam (X-ray)
r	Radius of diffractometer circle
μ_0	Permeability of free space, $4\pi \times 10^{-7}$ H m^{-1}, H is Henry, unit of inductance
e	Charge of electron, 1.6×10^{-19} C, C is coulomb, unit of electric charge
n	Mass of electron, 9.11×10^{-31} kg
kg	Kilogram, unit of mass
V	Volume of unit cell
F	Structure factor
p	Multiplicity factor
α	Linear absorption coefficient (cm^{-1})
α_m	Linear absorption coefficient of mixture (cm^{-1})
C_α	Concentration of α phase

References

Atlas, D., Wiser, O. and Trus, M., 2001. The voltage-gated Ca^{2+} channel is the Ca^{2+} sensor of fast neurotransmitter release. *Cellular and Molecular Neurobiology*. 21: 717–731.

Balch, J. F. and Balch, P. A., 1997. *Prescription for Nutritional Healing*. 2nd edn. Avery, Garden City Park, NY.

Brass, E. P., Forman, W. B., Edwards, R. V. and Lindan, O., 1978. Fibrin formation: effect of calcium ions. *Blood*. 52: 654–658.

Bratov, A., Abramova, N., Domínguez, C. and Baldi, A., 2000. Ion-selective field effect transistor (ISFET)-based calcium ion sensor with photocured polyurethane membrane suitable for ionised calcium determination in milk. *Analytica Chimica Acta*. 408: 57–64.

Bregestovski, P. and Spitzer, N., 2005. Calcium in the function of the nervous system: new implications. *Cell Calcium*. 37: 371–374.

Bugusu, B., 2008. Improving food through nanoscience. *Food Technology*. 908: 34–39.

Chau, C. F., Wu, S. H. and Yen, G. C., 2007. The development of regulations for food nanotechnology. *Trends in Food Science & Technology*. 18: 269–280.

Christensen, M. D. and White, H. K., 2007. Dementia assessment and management. *Journal of the American Medical Directors Association*. 8: 89–98.

Christopher Nordin, B. E., 2010. Evolution of the calcium paradigm: the relation between vitamin D, serum calcium and calcium absorption. *Nutrients*. 2: 997–1004.

Cohn, S. H., Ellis, K. J., Wallach, S., Zanzi, I., Atkins, H. L. and Aloia, J. F., 1974. Absolute and relative deficit in total skeletal calcium and radial bone mineral in osteoporosis. *Journal of Nuclear Medicine*. 15: 428–435.

Constanzo, L. S., 2010. *Physiology*. 4th edn. WB Saunders, PA.

Cullity, B. D. and Stock, S. R., 2001. *Elements of X-ray Diffraction*. 3rd edn. Addison-Wesley Publishing Company Inc., New Jersey, U.S.A, pp. 89–184.

Fye, W. B., 1984. Sydney Ringer, calcium, and cardiac function. *Circulation*. 69: 849–853.

Gemene, K. L. and Bakker, E., 2009. Measurement of total calcium by flash chronopotentiometry at polymer membrane ion-selective electrodes. *Analytica Chimica Acta*. 648: 240–245.

Grundfast, M. B., Still, C. D. and Komar, M. J., 2003. Hypercalcemia and peptic ulcer disease-related milk-alkali syndrome. *Nutrition in Clinical Practice*. 18: 250–252.

Hernández, R., Riu, J. and Rius, F. X., 2010. Determination of calcium ion in sap using carbon nanotube-based ion-selective electrodes. *Analyst*. 135: 1979–1985.

van Hulzen, K. J. E., Sprong, R. C., van der Meer, R. and van Arendonk, J. A. M., 2009. Genetic and nongenetic variation in concentration of selenium, calcium, potassium, zinc, magnesium, and phosphorus in milk of Dutch Holstein-Friesian cows. *Journal of Dairy Science*. 92: 5754–5759.

Kim, K.-Z., Shin, A., Kim, J., Park, J. W., Park, S. C., Choi, H. S., Chang, H. J., Kim, D. Y. and Oh, J. H., 2013. Association between CASR polymorphisms, calcium intake, and colorectal cancer risk. *PLoS One*. 8: 59628.

Liu, L. X. and Gao, G. F., 2004. Determination of calcium and lithium in lubricating grease by non-complete digestion-flame atomic emission spectrometry. *Metallurgical Analysis*. 24: 13–15.

Miller, G. D., Jarvis, J. K. and McBean, L. D., 2001. The importance of meeting calcium needs with foods. *Journal of the American College of Nutrition*. 20: 168–185.

Noël, L., Carl, M., Vastel, C. and Guérin, T., 2008. Determination of sodium, potassium, calcium and magnesium content in milk products by flame atomic absorption spectrometry (FAAS): a joint ISO/IDF collaborative study. *International Dairy Journal*. 18: 899–904.

Ohtsuki, I., 1999. Calcium ion regulation of muscle contraction: the regulatory role of troponin T. *Molecular and Cellular Biochemistry*. 190: 33–38.

Rutecki, G. W. and Whittier, F. C., 1998. Decision points in hypocalcemia: Is emergent therapy required? – Complications may include tetany, seizures, and arrhythmias. *Journal of Critical Illness*. 13: 84–90.

Sanguansri, P. and Augustin, M. A., 2006. Nanoscale materials development – a food industry perspective. *Trends in Food Science & Technology*. 17: 547–556.

Schmaldienst, S., Dittrich, E., Pietschmann, P., Niederle, B., Becherer, A. and Watschinger, B., 2001. A patient with evidence of two underlying diseases causing hypercalcaemia. *Nephrology Dialysis Transplantation*. 16: 2423–2425.

Schroder, B. G., Griffin, I. J., Specker, B. L. and Abrams, S. A., 2005. Absorption of calcium from the carbonated dairy soft drink is greater than that from fat-free milk and calcium-fortified orange juice in women. *Nutrition Research*. 25: 737–742.

Schwartz, G. G. and Skinner, H. G., 2012. A prospective study of total and ionized serum calcium and time to fatal prostate cancer. *Cancer Epidemiology Biomarkers and Prevention*. 21: 1768–1773.

Schwarz, E. C., Qu, B. and Hoth, M., 2013. Calcium, cancer and killing: the role of calcium in killing cancer cells by cytotoxic T lymphocytes and natural killer cells. *Biochimica et Biophysica Acta, Molecular Cell Research*. 1833: 1603–1611.

Sejersted, O. M., 2011. Calcium controls cardiac function – by all means. *Journal of Physiology*. 589: 2919–2920.

Small, D. H., 2009. Dysregulation of calcium homeostasis in Alzheimer's disease. *Neurochemical Research*. 34: 1824–1829.

Swartzman, J. A., Hintz, H. F. and Schryver, H. F., 1973. Inhibition of calcium absorption in ponies fed diets containing oxalic acid. *American Journal of Veterinary Research*. 39: 1621–1623.

Sze, K. L., Yeung, W. S. B. and Fung, Y. S., 2007. Separation and determination of metal cations in milk and dairy products by CE. *Electrophoresis*. 28: 4156–4163.

Terol, A., Paredes, E., Maestre, S. E., Prats, S. and Todolí, J. L., 2010. High-temperature liquid chromatography inductively coupled plasma atomic emission spectrometry hyphenation for the combined organic and inorganic analysis of foodstuffs. *Journal of Chromatography A*. 1217: 6195–6202.

Vella, A., Gerber, T. C., Hayes, D. L. and Reeder, G. S., 1999. Digoxin, hypercalcaemia, and cardiac conduction. *Postgraduate Medical Journal*. 75: 554–556.

Warren, B. E., 1990. *X-ray Diffraction*. 2nd edn. Addison-Wesley Publishing Co., New Jersey, USA.

Weiss, J., Takhistov, P. and McClements, D. J., 2006. Functional materials in food nanotechnology. *Journal of Food Science*. 71: R107–R116.

Williams, D. B. and Carter, C. B., 1996. *Transmission Electron Microscopy: Basic Diffraction Imaging Spectroscopy*. Plenum press, New York.

Woods, N. K. and Padmanabhan, J., 2012. Neuronal calcium signaling and Alzheimer's disease. *Advances in Experimental Medicine and Biology*. 740: 1193–1217.

Yang, H., Zhang, Q., He, J. and Lu, W., 2010. Regulation of calcium signaling in lung cancer. *Journal of Thoracic Disease*. 2: 52–56.

Zhang, J. M., Wang, M., Ge, X. P., Wu, J. Z., Ge, Y., Li, S. P. and Chang, J., 2009. Convertibility of the data determined by ICP-AES and FAAS for soil available K and Na. *Spectroscopy and Spectral Analysis*. 29: 1405–1408.

Zhou, F. Q., Huang, Y. A. and Yi, L. H., 2002. Determination of zinc, iron, calcium in powdered milk by FAAS-suspension sample introduction technique. *Natural Science Journal*. 24: 67–68.

Office of Dietary Supplements, 2013. National Institutes of Health, Dietary supplement fact sheet: calcium. Available at: http://ods.od.nih.gov/factsheets/Calcium-QuickFacts.

ISO 6058 Water quality, 1984. Determination of calcium content – EDTA titrimetric method.

ISO 12081 Milk, 1998. Determination of calcium content – titrimetric method.

CHAPTER 10

Using Food-Frequency Questionnaires for Calcium Intakes

B. PAMPALONI[a] AND M. L. BRANDI*[a]

[a]Department of Surgery and Translational Medicine,
University of Florence, Viale Pieraccini, 6, 50139 Firenze, Italy
*E-mail: marialuisa.brandi@unifi.it

10.1 Introduction

Collecting information on the dietary profiles of a population is an important matter for several reasons, mainly to understand and characterize the risk of specific foodborne illnesses. The best strategy in lowering the occurrence of several chronic diseases is prevention, and the adoption of proper dietary choices is a major factor that can have beneficial effects on the relative risk reduction (Maghsoudi *et al.*, 2013).

The method most used to estimate nutrient intake is the food-frequency questionnaire (FFQ), even if an ideal method does not exist for all study populations or nutrient types. In general, the FFQ is used due to its simple format, cost efficiency, possibility to be self-administered, and ability to provide insight regarding long-term and usual dietary intake. Figure 10.1 shows an example of FFQ question.

Food and Nutritional Components in Focus No. 10
Calcium: Chemistry, Analysis, Function and Effects
Edited by Victor R. Preedy

Published by the Royal Society of Chemistry, www.rsc.org

HOW OFTEN DO YOU EAT CHEESE?

☐ Never

☐ Everyday

☐ Twice a day

☐ Number of times per | 1 | 2 | 3 | 4 | 5 | 6 | week

If less, how many times per | 1 | 2 | 3 | month?

- What is the average portion? (See Figures 10.2B and C for portion size)

☐ Less than 13 B	☐ 13 C
☐ 13 B	☐ More than 13 C

- What kind?

☐ Parmesan	☐ Emmental
☐ Pecorino	☐ Other………

Figure 10.1 Example of FFQ question. Example of question in a FFQ studied to evaluate calcium intake. The intake of cheese is an important question that require some specification about what kind of cheese is consumed.

FFQs are frequently incorporated into large research protocols, such as randomized controlled trials, cross-sectional, case-control, and cohort studies, to have an outcome verification of eating habits that can be important to define the characteristics of the study population (Svensson *et al.*, 2013). The most important question is whether with FFQs you can get a correct estimate of the nutrient intake. The utility of FFQs may be limited due to poor design and improper implementation; the questionnaire often overestimates, or underestimates, intakes. To evaluate relative or absolute nutrient consumption, FFQs should incorporate portion size questions, with or without accompanying photographs of foods (Figure 10.2). Moreover, it may also be useful that the FFQ is accompanied by specific instructions for completion (Figure 10.3). To be eligible for use, the questionnaire has to be validated against a more detailed and accurate gold-standard method for intake assessment. Validation studies measure the degree of association between a new FFQ and the reference gold-standard method. Correlation indexes are the most frequently applied statistical procedures, comparing means or medians of food consumption, or nutrient intakes, from the two methods. If the two methods are in agreement, the FFQ may be used for studies in the same population used for validation. The principal gold-standard methods used to validate FFQs are dietary recall or diet recording, relative to 3 to 14 days, and the 24 h dietary recall, both of

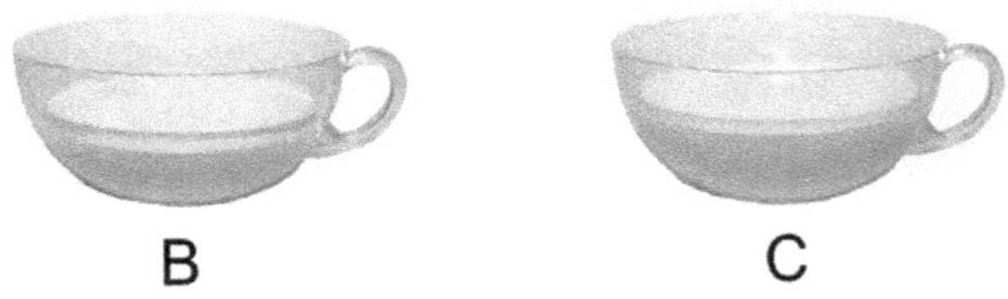

Figure 10.2 Example of photograph of portion size. Each photograph should show almost two portions of different dimensions (B medium portion, C large portion).

which are considered more accurate than the FFQ. In effect, as reported by Serra, "the dietary record does not rely on memory, whereas the FFQ does and also employs cognitive processes for calculating the frequency of intake". With the FFQ many details of dietary intake are not measured, so the quantification is not as accurate as with recalls or records. In evaluating intakes by FFQ, the referred time periods (typically 6 months or 1 year) should be defined, and the season should also be considered (Serra-Majem *et al.*, 2009).

Another important aspect is the food list, which should be completed with care. Even if the great variability of an individual's diet is difficult to fully capture with a finite food list, the pertinence of the food list is crucial to the FFQ method A typical FFQ may contain 100 or more food items, but much depends on the goal we want to achieve with the questionnaire. A comprehensive food list is needed in order to assess energy intake. Similarly, if the nutrient of interest is highly correlated with other nutrients, unless the whole diet is assessed it may not be possible to explore this. If the purpose is very specific, *e.g.* to estimate the intake of a single nutrient or a food group, a comprehensive food list may be unnecessary, and the list can be reduced to a range of 15–30 foods. The brief list should include the principal source of the nutrient of interest (Table 10.1). To study calcium intake, the list will need to include all dairy products and all foods that contain calcium. In general, assessment of calcium intake can be a problem because this mineral occurs in a limited number of foods in high concentrations. Therefore, small daily variations in food consumption can determine a great variability in the estimation of true calcium intake. Supplement use (type and dosage) should also be evaluated (Serra-Majem *et al.*, 2009).

10.2 Using FFQs to Assess Dietary Habits in Epidemiological Studies and Calcium Intakes

Chronic diseases are often complex diseases with multifactorial etiology. An accurate knowledge of people's dietary patterns are more predictive of health outcome and allow development of strategies to counteract bad habits, thereby reducing the risk factors for many chronic degenerative diseases.

Dietary pattern analysis, which examines the effects of overall diet, is recognized as an alternative approach of assessing dietary exposures in

INSTRUCTIONS FOR COMPLETING THE FOOD-FREQUENCY QUESTIONNAIRE

START BY CHECKING THE FREQUENCY OF CONSUMPTION OF EACH FOOD OR DRINK.

REMEMBER! YOU NEED TO THINK ABOUT WHAT YOU EAT THROUGHOUT THE DAY: BREAKFAST, MORNING SNACK, LUNCH, AFTERNOON SNACK, AND DINNER

• When you consume a food or drink more than 6 times a week, which means that the consumption is every day, then you have to check "every day".

☐ Never

☒ Every day

☐ Twice a day

☐ Number of times per

1	2	3	4	5	6

week

• If you consume or drink a food more than 3 times per month, which means that the consumption is at least 1 time per week, then check "Number of times per week"

☐ Never

☐ Everyday

☐ Twice a day

☐ Number of times per

~~1~~ (X)	2	3	4	5	6

week

• If you consume a food or drink a few times a week, indicate the number of times on the table:

☒ Number of times per

1	2	~~3~~ (X)	4	5	6

week

• If you consume a food or drink only a few times per month and not per week, indicate the number of times on the table:

☐ Number of times per month

1	~~2~~ (X)	3

BE CAREFUL!: You must put only one answer. If the response is "every day", it cannot be "Number of times per week" or "Number of times per month"

Figure 10.3 Example of instructions for completing FFQ. As the FFQ may be self-administered, complete instructions for compilation can be needed.

Table 10.1 List of foods for a FFQ to assess calcium intake in children population.

(1)	Milk
(2)	Yogurt
(3)	Cereals for breakfast
(4)	Biscuits
(5)	Packaged snacks
(6)	Bread
(7)	Pasta/rice
(8)	Grated parmesan on pasta
(9)	Pizza
(10)	Meat or sliced meats
(11)	Fish
(12)	Eggs
(13)	Cheese
(14)	Mozzarella
(15)	Legumes (chickpeas, beans, lentils, peas, *etc.*)
(16)	Vegetables
(17)	Fruit
(18)	Ice cream (do not count sorbet)
(19)	Milk chocolate
(20)	Desserts
(21)	Soft drinks

nutritional epidemiology and FFQ has been shown to be the principle measure of foods intake to obtain dietary patterns *via* factor analysis (Loy and Jan Mohamed, 2013). Traditional single nutrient or food analysis is shown to have several limitations. Interactive or synergistic action among nutrients, or foods, which are highly intercorrelated causes difficulty in examining their effects separately. Moreover, the single effect of a nutrient, or food, may be too small to be detected, or statistically significant associations may occur by chance, when several nutrients and foods are tested independently (Jacobs and Steffen, 2003).

Although the causes of chronic diseases are to be found in the diet "*in toto*", and in particular in inappropriate lifestyles, calcium remains an important nutrient that may be related to several pathologies. Thus, the evaluation of dietary calcium intake can often be used to take preventive actions.

10.2.1 Dietary Assessment in Different Ethnic Groups

World globalization and migration processes, which lead to the coexistence in the same country of different ethnic groups, have highlighted how differences in eating habits may correspond to strong variations for the risk factors of several diseases.

Dietary habits vary widely among regions and cultural groups, and FFQs need to be designed for specific populations. The use of FFQs in different populations or subgroups of populations allows the differences among specific dietary habits that can correspond to an enhanced risk factor disease to be evidenced.

FFQ can be used to investigate differences in nutrient intake in different ethnic groups. The study of Newby *et al.*, regarding dietary habits of black and white citizens of different regions of the US, shows that there are lower intakes of potassium, magnesium, and calcium for black men compared to white men. Similar results have been reported, with lower intakes of calcium, magnesium, and potassium among blacks as opposed to whites in the Lower Mississippi Delta, and among those in the Delta region compared to a national sample of the US. It is possible to hypothesize that the observed differences in dietary intakes play an important role in the health disparities and chronic disease risks observed among racial groups and regions in the US (Newby *et al.*, 2011).

10.2.2 Obesity

Dietary intervention is an important aspect in the treatment of excess body weight. Most research on the effects of diet on weight control has focused on the optimal combination of macronutrients, whereas the role of micronutrients has received less attention. Decrease in energy intake is the most important component of any weight-loss strategy. The inverse relationship between calcium intake and adiposity has been observed in several observational studies, and the possibility that dietary calcium and/or dairy products (the main source of calcium) may affect energy balance, with an antiobesity effect, has proved very interesting. In effect, dietary calcium appears to play a central role in the regulation of energy metabolism and obesity risk. The antiobesity mechanism seems to involve dietetic calcium as a modulator of circulating calcitriol, which successively regulates adipocyte intracellular calcium and, finally, this mechanism adjusts lipid accumulation (Zemel and Miller, 2004; Da Silva Ferreira *et al.*, 2013). In a study on 268 Brazilian premenopausal women, the eating patterns were assessed by a semiquantitative FFQ. Dietary analysis considered the usual assumptions of calcium, energy, proteins, carbohydrates and lipids over the previous 6 months. Results show that participants who have high dietetic calcium consumption (>600 mg per day) compared to those who have low calcium (<600 mg per day) consumption, had a lower odd ratio for prevalent overweight, obesity, abdominal obesity, insulin resistance, low HDL-cholesterol, and systolic BP above the recommended levels. All these observations suggest that, in a sample of healthy women, calcium intake could be inversely associated with the development of some cardiovascular risk factors such as excessive total and abdominal adiposity, insulin resistance, and elevated BP, and directly associated with serum levels of adiponectin and HDL-cholesterol (Da Silva Ferreira *et al.*, 2013).

10.2.3 Diabetes Type 2

The main risk factors for type-2 diabetes (T2D) are obesity and failure of physical activity. Nevertheless, there is accumulating evidence that diet composition plays a role in determining the risk of T2D. Several studies have confirmed that lifestyle interventions, including dietary changes, are able to reduce the risk of progression from impaired glucose tolerance to overt T2D.

Dairy products represent an important food group in the diets of all countries, but the relationship between dairy intakes, glucose regulation and T2D risk are still unknown. Thus, it is important to assess their dietary intake and establish the possible correlations with T2D risk (Struijk *et al.*, 2013).

As stated by Struijk *et al.* "there is evidence that dairy products providing high-quality protein are a good source of fat-soluble vitamins (vitamin A, D and K2), water-soluble vitamins (vitamin B12 and riboflavin), and minerals (calcium *in primis*, then magnesium and potassium). Previous studies have shown that these nutrients are beneficially related to T2D risk".

The association between calcium and vitamin D intake and the risk of T2D was prospectively examined in the Nurses' Health Study in 20 years of follow-up. Dietary intakes of 83 779 women aged 30–55 years, living in 11 states of the US, assessed with a semiquantitative validated FFQ, contributed to the follow-up analysis. Women in the group with the highest daily intakes of calcium and vitamin D (*i.e.*, 1200 mg of calcium and 800 IU of vitamin D obtained through diet + supplements) had a 33% lower risk of T2D compared with women in the lowest intake group (600 mg of calcium and 400 IU vitamin D). In the conclusions, the authors stated that "both vitamin D and calcium intakes were inversely associated with development of T2D, and the benefits of the two nutrients appear to be additive. For both vitamin D and calcium, intake from supplements, rather than from diet, were significantly associated with a lower risk of type-2 diabetes" (Pittas *et al.*, 2006).

Despite these observations further studies on pathophysiological mechanisms and large clinical trials are necessary to clarify the role and usefulness of certain minerals for prevention and management of T2D.

10.2.4 Stroke

Nearly 50% of stroke incidence in both genders can be related to hypertension. Higher systolic blood pressure (BP), with or without elevated diastolic BP, is accompanied with greater stroke incidence rate. Therefore, it seems that BP management can be helpful in stroke prevention, and the management of BP is strictly related to nutrition. Several studies have highlighted the relationship between dairy products intake and stroke (Maghsoudi *et al.*, 2013). Moreover an inverse relationship between dietary calcium and/or dairy intake and BP has been reported in various epidemiological studies showing that a reduced intake of calcium was associated with higher levels of BP and/or an increased risk of developing hypertension (Da Silva Ferreira *et al.*, 2013).

A survey conducted in Iran on 129 subjects, male and female, who experienced stroke and nonstroke, demonstrates that mean dairy consumption in subjects with stroke was lower than in the control group. Nutritional habits, including calcium and dairy products intake, were evaluated by FFQ containing 168 food items. The author observed that "In regard to the variety of greatly differing composition of dairy products, we tried to assess nearly all types of Iranian dairy foods. So, it is believable that we were able to evaluate daily dairy intake, adequately" (Maghsoudi *et al.*, 2013).

10.2.5 Cardiometabolic Risk Factors

Cardiovascular disease (CVD) remains the major cause of morbidity and mortality worldwide (WHO, 2011). It is well known that being overweight or obese are just two of the established risk factors for CVD. There are several other factors that relate to lifestyles, such as smoking, physical inactivity, and unhealthy diet, as well as some related medical conditions such as hypertension, elevated blood glucose, and dyslipidemia. Moreover, these risk factors can be combined with each other. Current evidence indicates that the consumption of dairy food can be associated with a reduced risk of CVD and T2D, and lower BP in adults. These effects may be attributable to milk fat, vitamin D, calcium, magnesium, potassium, whey proteins, or the combination of these as part of the unique nutrient package that dairy provides (Rice *et al.*, 2011). Calcium may affect cardiometabolic risk factors by regulating energy metabolism *via* the reduction of lipogenesis and enhancement of lipolysis in adipocyte, increasing both fecal fat excretion and fat oxidation (Abreu *et al.*, 2014). Therefore, to understand the relationship between diet and CVD is a very important issue, and to assess dairy products intakes, with a specific FFQ can be fundamental. However, the conclusions of the studies strictly related to calcium intake and CVD are controversial.

In a prospective cohort study, conducted by Chan *et al.* on 3139 old Chinese men and women, higher dietary calcium intake was associated with a decreased risk of all-cause and CVD mortality. The authors concluded that "High intake of dietary calcium was associated with reduced risk of all-cause mortality and possibly CVD mortality in older people with low habitual calcium intake. The association between calcium intake and mortality is likely dependent on the habitual calcium intake. High calcium intake may have a protective effect on mortality in populations, such as Chinese with comparatively low habitual calcium intake, while it may increase mortality risk in populations with high habitual calcium intake" (Chan *et al.*, 2013).

In a cohort of 48 645 Swedish men aged 45–79 years who were free of cancer, CVD, and diabetes, and did not use dietary supplements at recruitment, dietary calcium intake was assessed by a 96-item FFQ. Calcium intake was associated with a significantly lower rate of all-cause mortality and a non-significant lower risk of CVD death when the highest intake tertile of dietary calcium (mean = 1953 mg per day) was compared with the lowest (990 mg per day). The conclusion was that "the inverse association between dietary calcium intake and CVD mortality was not statistically significant" (Kaluza *et al.*, 2010).

As epidemiological studies are very important for prevention, preventive actions should be planned and started with young people, although it is not often easy to study children and adolescent groups. Communication campaigns that encourage young people to eat low-fat dairy products, with the goal to achieve adequate calcium intake, are taking place in all developed countries.

Special attention is paid to milk consumption and to daily breakfast. Moreover, milk can now be associated to cardiometabolic risk prevention in young people.

10.2.6 Cardiometabolic Risk Factors in Young People

Regarding cardiometabolic risk factors (CMRS), Abreu *et al.* noted that the inverse association between dairy product consumption and cardiometabolic risk, which is sufficiently clear in adults, is relatively unexplored in adolescents. Therefore, a cross-sectional study of about 500 adolescents in Portugal, using an FFQ to evaluate dairy products intake, was planned. The results showed that "Adolescents with appropriate milk intake were less likely to have high CMRS than those with low milk intake. No association was found between CMRS and total dairy, yogurt, and cheese intake. Only milk intake seems to be inversely related to CMRS in adolescents" (Abreu *et al.*, 2014).

10.2.7 Gastroesophageal Reflux Disease

Gastroesophageal reflux disease (GERD) is a chronic disease usually caused by the reflux of acidic gastric and duodenal contents into the distal esophagus. GERD is a very diffuse condition that affects men and women worldwide. Although often it is not a serious disease, it may heavily affect people's daily life and dietary habits. A survey was carried out on 537 Han Chinese recruited from the Endoscopy Center of Tongji Hospital. All subjects were asked to complete a detailed FFQ that included a total of 120 kinds of food, based on the foods with high intake frequency in China. The authors report that "high intake of calcium, meat, oils, and salt is associated with an increased risk for RE, while high intake of protein, carbohydrate, calories from protein (%), vitamin C, grains and potatoes, fruits, and eggs correlates with a reduced risk for RE in Han Chinese" (Wu *et al.*, 2013).

10.2.8 Kidney Stones

Lower dietary calcium is a well-known risk factor for calcium kidney-stone formation. To make a dietary assessment of a population in order to know its calcium intakes, it would be important to determine the population risk factor for kidney stones. Several observational studies have reported the association between lower dietary calcium and an increased risk of incidence of kidney stones. The mechanism by which higher calcium intake may decrease the stone risk is still unclear. In contrast to dietary calcium, higher supplemental calcium intake is not associated or is associated with an increased, rather than a decreased, risk of kidney-stone formation.

In the study of Curhan, *et al.*, 116 671 females aged 25–42 years completed an initial questionnaire, followed up by using biennial mailed questionnaires that inquired about lifestyle practices, and newly diagnosed disease.

Assessment of the questionnaires showed that a higher dietary calcium intake was strongly associated with a reduced risk of kidney stones after adjusting for age. In contrast to this, the intake of supplemental calcium was not significantly associated with risk of stone formation (Curhan *et al.*, 2004).

Taylor and Curhan show results of a large prospective study published in 2013. Fifty-six years of follow-up study was performed in the Health Professionals Follow-up Study (HPFS) in 30762 men, and in the Nurses' Health Study (NHS) I and II in 94164 and 101701 women, respectively. The study investigates the effects of dairy dietary calcium and nondairy dietary calcium, to establish if there are differences in kidney-stone risk. The baseline assessment was done by a semiquantitative FFQ that asked about the annual average consumption of more than 130 food and 22 beverages. Subsequently, a version of this FFQ has been mailed to the participants every 4 years. The important conclusion is that higher dietary calcium is independently associated with a lower risk of symptomatic kidney stones and restriction of dietary calcium should not be recommended as a means of preventing calcium kidney stones (Taylor and Curhan, 2013).

10.2.9 Colorectal Cancer (CRC)

Colorectal cancer is classified as a diet-related type of cancer. High-calorie diet, high red-meat consumption, overcooked red-meat consumption, high saturated-fat consumption, excessive alcohol consumption, cigarette smoking, sedentary lifestyle, and obesity are considered to increase the incidence of CRC, while consumption of fiber, fresh fruit and vegetables, pro- and prebiotics, selenium, vitamin E and folic acid, and a high-calcium diet could have a protective effect (Pampaloni *et al.*, 2013a).

Evidence from epidemiological studies has confirmed unequivocally that calcium is one of the most important factors decreasing the risk of developing colorectal malignancies. Therefore, a low intake of dietary calcium was correlated with colorectal cancer.

Ruder *et al.* in a large prospective study, investigated diet during adolescence and 10 years before baseline (ages 40–61 years) in relation to colorectal cancer. Participants in the NIH-AARP Diet and Health Study ($n = 292797$) completed a 124-item FFQ about diet in the past 12 months and two 37-item FFQs about diet at ages 12–13 years and 10 years previously. The author's conclusion is that "there is evidence that diet in adolescence and midlife may modify the risk of colon and rectal cancer. Specifically, individuals with a high consumption of vitamin A or vegetables during ages 12–13 years had a reduction in colon cancer risk after age 50 years. Furthermore, a high intake of calcium, vitamin A, or vitamin C during ages 40–61 years was inversely associated with colon cancer after age 50 years. Similar protective associations were observed among high consumers of fruit or milk during ages 40–61 years" (Ruder *et al.*, 2011).

Concerning calcium and milk intake and colorectal cancer risk, the literature offers several contrasting conclusions. In their study, Ruder *et al.* show that "after control for recent adult intake, the associations for calcium and milk intakes during adolescence in relation to colon cancer were attenuated". Nevertheless, as highlighted by the same authors: "prospective dietary data collected during adolescence showed a positive association between dairy product intake and colorectal cancer risk in the British-based Boyd-Orr cohort. By contrast, a recent case-control study of participation in government-sponsored school milk programs, indicated a reduced colorectal cancer risk with program participation, including evidence of a dose–response effect with increased milk consumption" (Ruder *et al.*, 2011).

10.2.10 Calcium and Bone

As we have seen, calcium is an important element for health and is implicated in several pathologies. Undoubtedly, the most important associations between dietetic calcium intake and well-being are those related to bone health.

Calcium, vitamin D, and proteins are the three essential nutrients that influence the acquisition and maintenance of bone structure. The relationship between calcium intake and bone-mineral density (BMD) has been widely investigated. A large number of publications collected in this area has made it possible to conclude that BMD in children and adolescents increases as calcium intake increases (Greer and Krebs, 2006; Rizzoli, 2008; Yin *et al.*, 2010). Therefore, high calcium intake influences skeletal calcium retention during bone growth and is associated with achievement of the optimal genetically programmed peak bone mass, as well as reduced risk of osteoporosis and lower fracture risks in adults and the elderly. Osteoporosis is a multifactorial disease characterized by loss of bone mass and microarchitectural deterioration of bone tissue, which leads to a consequent increase in the risk of skeletal fractures.

The development of a preventive osteoporosis strategy has become a priority in European countries due to the aging of the population, and also in the US and Australia osteoporosis is a growing health issue.

In the European Union, 40% of women and 13% of men over 50 years of age will have a fragility fracture. One citizen out of eight, after the age of fifty, has suffered a spinal fracture, and over the next 50 years the number of hip fractures will more than double. The SOF (Study of Osteoporotic Fractures) is a multicenter, observational US study, started in 1986, its primary aim was the description of the risk factors for osteoporotic fractures. Interestingly, SOF prospectively evaluated whether low bone mass was predictive of femoral fracture in the same place, and found that the reduction of a standard deviation for age (z score) increased the risk of fracture by seven times with respect to the control (http://sof.ucsf.edu.).

Osteoporosis prevention starts with a healthy lifestyle and a balanced diet, properly rich in calcium.

If low calcium intake is a widespread problem across countries and age groups, to have an adequate calcium intake is very important throughout all life. Several false dietary habits can determine low calcium intake, for example, to stop drinking milk at breakfast, or completely skip breakfast.

By identifying individuals who are at risk of osteoporosis through inadequate calcium intake, early treatment or prevention of osteoporosis is possible. Nevertheless, identification of inadequate calcium intake is a complex issue.

Several studies have consistently shown that many children and adolescents have poor dietary habits that do not meet recommended dietary guidelines for calcium. In children and adolescents assessing nutrients intake is critical in examining the effects of nutrition on future bone health in order to develop a preventive plan. Low intakes of micronutrients, including calcium, have been reported in European, US, and Australian adolescents.

Dietary intakes of 991 Australian adolescents 17 years old, was assessed using a self-reported semiquantitative FFQ. Data reported that intakes for calcium, folic acid, and vitamins D and E were below recommendations (Gallagher *et al.*, 2014). Similar results were found in Belgian preschool children, and in Malaysian children aged 6 months to 12 years (Vereecken and Maes, 2010; Poh *et al.*, 2013). Low levels of calcium intake are becoming increasingly widespread among children and adolescents, and this may inhibit the attainment of optimal peak bone mass. In order to achieve the goal of recommended nutrients intake in adolescent population nutrient supplements were often used. Studies reported that approximately 24% of Australian adolescents consumed supplements. The prevalence of supplement use among adolescents varies: 20–26% in the US, 11–45% in Europe. However, the intake of some micronutrients, including calcium, folic acid and vitamins D and E, can remain lower than the recommendations (Gallagher *et al.*, 2014).

Several FFQs have been specifically designed for calcium and validated for use in several population subgroups, most commonly in adult and elderly women, and less frequently in children and adolescent, although the eating habits can be very different from those of adults. Thus, it is necessary to build valid and efficient calcium assessment tools for those subjects.

To fill out an FFQ may not be easy for children, even if Livingstone and Robson found that from 8 years of age there is a rapid increase in children's ability to report food intake. Moreover, Baranowski found a strong agreement (82% for all foods) between children's self-reported frequencies and meal observations (Baranowski *et al.*, 1986; Livingstone and Robson, 2000).

Recently, Pampaloni *et al.* validated an FFQ for calcium assessment in Italian children aged 9–10 years. 21 food items were selected for inclusion in the list based on: examination of food composition tables, epidemiologic data on food consumption patterns of the Italian diet, tastes of children, and relative importance as calcium sources. The questionnaire included: milk and dairy products, yogurt, ice cream, cheese, pizza, and chocolate milk. Meat, fish, cereals and bread, vegetables, and fruits were also included. As for milk and yogurt, the amount of calcium does not vary significantly between different types of products, for Italian cheeses, instead, there is a marked difference between aged and fresh products and it was necessary to specify, in the FFQ, the diverse types of cheese. A 7 day dietary record was used as a gold standard for validation process (Pampaloni *et al.*, 2013b).

A lot of data are available for adults. In a meta-analysis done by Kanis *et al.*, results showed that the estimate of calcium intake based on self-reported milk consumption was not useful for the prediction of fractures. Gonnelli *et al.*, reported data from several epidemiological surveys that indicate that calcium intake, in a large proportion of American and European people, is below recommended levels; consequently, numerous public health recommendations focus on encouraging adequate calcium intake (Kanis *et al.*, 2005; Gonnelli *et al.*, 2009).

In the Wlodarek *et al.* study an assessment of calcium intake, was performed using a FFQ containing a list of the forty-nine dairy products most common in Poland (*e.g.* milk, yoghurt, other milk beverages, rennet cheese, cottage cheese) and of meals containing dairy products, as well as portion sizes. The conclusions were that a "lower BMD and previous fractures were associated with lower reported dairy calcium intake. Due to the insufficient consumption of dairy products, a number of women were unable to satisfy their calcium requirement exclusively from their diet" (Wlodarek *et al.*, 2012).

In a large study by Bruyere conducted on 8524 patients from nine European countries, calcium intake assessment was performed with a validated self-questionnaire. In the conclusion of the paper, the authors claim that "the study shows insufficient dietary calcium intakes in postmenopausal osteoporotic women (whatever the age) in nine European countries. A substantial part of our study population does not achieve the general recommendation for calcium intake in such a population". Moreover, as reported in the same paper, other epidemiological studies have described low dietary calcium intake in European subjects. In 131 healthy postmenopausal women, Andon *et al.* found a mean calcium consumption of 606 (±302) mg per day. In late postmenopausal women aged 65 years and older Jensen *et al.* described a mean calcium consumption of 793 (±280) mg per day. In 184 elderly women, Pfister *et al.* found a mean dietary calcium intake of 563.4 mg per day. The combined dietary calcium intake and supplement calcium intake in elderly women participating in NHANES III (Third National Health and Nutrition Examination Survey) was 864 (±16) mg per day. In a recent study, the mean dietary calcium intake, reported by 167 postmenopausal women aged 40–65 years, was 924 (±347) mg per day (Bruyere *et al.*, 2008).

On the other hand, over time, the subjects often suspend the intake of dairy products for various reasons, for example because they are considered too high in calories, or due to episodes of lactose intolerance. In this case, it is important to note that it is now acknowledged that rich calcium mineral waters represent an optimal alternative source of dietary calcium. An FFQ can be useful in supporting proper nutrition, by annually monitoring potential changes in eating habits.

A study conducted by Sanders on approximately 1950 Australian women showed that an annual administration of a short FFQ to verify calcium intake induces the subjects to maintain an adequate consumption. As reported by the authors in the conclusion "This study devised an efficient method to provide feedback on calcium intake in older women. The improvements were modest but significant and most apparent in those with a low intake at baseline. The decreased proportion of these women with an inadequate intake of calcium 12 and 24 months later, suggests this might be a practical, low-cost strategy to maintain an adequate calcium intake among older women" (Sanders *et al.*, 2010).

10.3 Conclusion

In conclusion we can state that the FFQ is the most widespread tool for assessing nutritional habits. The FFQ is recognized by the international scientific community as an essential tool for evaluating nutritional status and dietary habits of a population or parts of it. In this sense, the FFQ is a fundamental tool for developing strategies and plans for the prevention of chronic degenerative disease on a large scale. The use of FFQs to assess calcium intake itself often represents an important preventive action.

Summary Points

- An accurate knowledge of people's eating habits makes it possible to develop strategies to counteract bad habits. Chronic diseases are often complex diseases with multifactorial etiology. The analysis of the overall diet is a good approach of assessing dietary exposures in nutritional epidemiology.
- Obesity. Dietary intervention is an important aspect in the treatment of excess body weight. The inverse relationship between calcium intake and total and/or abdominal adiposity has been observed in several observational studies. The possibility that dietary calcium and/or dairy products may affect energy balance, with an antiobesity effect, has proved very interesting.
- Type-2 diabetes (T2D). Diet composition plays a fundamental role in affecting T2D risk. The role of dairy products in this topic is largely unknown. Because dairy products represent an important food in the diets of all countries, it is important to assess their dietary intake and establish the correlation with T2D that may arise from this.
- Stroke. Several studies have highlighted the relationship between dairy products intake and stroke.
- Cardiometabolic risk factors (CVD). CVD remains the major cause of morbidity and mortality worldwide. Current evidence indicates that the consumption of dairy foods is associated with a reduced risk of CVD. These effects may be attributable to calcium, milk fat, vitamin D, magnesium, potassium, whey proteins, or the combination of these, all contained in dairy products.
- Kidney stones. Several studies have reported the association between lower dietary calcium and an increased risk of incidence of kidney stones. The mechanism by which higher calcium intake may decrease the stone risk is still unclear.

- Colorectal cancer (CRC). Colorectal cancer is classified as a diet-related type of cancer. Evidence from epidemiological studies has confirmed unequivocally that calcium is one of the most important factors decreasing the risk of developing colorectal malignancies.
- Calcium and bone. Osteoporosis is a multifactorial disease characterized by loss of bone mass and microarchitectural deterioration of bone tissue, which leads to a consequent increase in the risk of skeletal fractures. Calcium, vitamin D, and proteins are the three essential nutrients that influence the acquisition and maintenance of bone structure.

Key Facts

Key Facts of Food-Frequency Questionnaires and their Use

1. Collecting information on the dietary profiles of a population is an important matter to understand and characterize the risk of specific foodborne illnesses.
2. By the food-frequency questionnaire (FFQ) method it is possible to record foods eaten in the last week, or month, or year.
3. This method is the most used to estimate nutrient intake, even if an ideal method does not exist for all study populations or nutrient types.
4. The use of FFQs in different populations or subgroups of populations allows the differences among specific dietary habits that can correspond to an enhanced risk factor disease to be evidenced.
5. Although the causes of chronic diseases are to be found in the diet "*in toto*", and in particular in inappropriate lifestyles, calcium remains an important nutrient that can have many relationships with several pathologies.
6. The evaluation of dietary calcium intake can often be used to take preventive actions.

List of Abbreviations

FFQ	Food-Frequency Questionnaire
DM	Diabetes Mellitus
CVD	Cardiovascular disease
HDL	High-Density Lipoprotein
BP	Blood Pressure
T2D	Type-2 diabetes
CMRS	Cardiometabolic risk factors
GERD	Gastroesophageal reflux disease
HPFS	Health Professionals Follow-up Study
NHS	Nurses' Health Study
CRC	Colorectal cancer
BMD	Bone-Mineral Density

References

Abreu, S., Moreira, P., Moreira, C., Mota, J., Moreira-Silva, I., Santos, P. C. and Santos, R., 2014. Intake of milk, but not total dairy, yogurt, or cheese, is negatively associated with the clustering of cardiometabolic risk factors in adolescents. *Nutrition Research*. 34(1): 48–57.

Baranowski, T., Dworkin, R., Henske, J. C., Clearman, D. R., Dunn, J. K., Nader, P. R. and Hooks, P. C., 1986. The accuracy of children's self reports of diet: family health project. *Journal of the American Dietetic Association*. 86: 1381–1385.

Bruyere, O., De Cock, C., Mottet, C., Neuprez, A., Malaise, O. and Reginster, J. Y., 2008. Low dietary calcium in European postmenopausal osteoporotic women. *Public Health Nutrition*. 12(1): 111–114.

Chan, R., Leung, J. and Woo, J. A., 2013. Prospective cohort study examining the associations of dietary calcium intake with all-cause and cardiovascular mortality in older Chinese community-dwelling people. *PLoS One*. 8(11): e80895. DOI: 10.1371/journal.pone.0080895.

Curhan, G. C., Willett, W. C., Knight, E. L. and Stampfer, M. J., 2004. Dietary factors and the risk of incident kidney stones in younger women: Nurses' Health Study II. *Archives of Internal Medicine*. 164(8): 885–891.

Da Silva Ferreira, T., Simas Goncalves Torres, M. R. and Sanjuliani, A. F., 2013. Dietary calcium intake is associated with adiposity, metabolic profile, inflammatory state and blood pressure, but not with erythrocyte intracellular calcium and endothelial function in healthy pre-menopausal women. *British Journal of Nutrition*. 110: 1079–1088.

Gallagher, C. M., Black, L. J. and Oddy, W. H., 2014. Micronutrient intakes from food and supplements in Australian adolescents. *Nutrients*. 6: 342–354.

Gonnelli, S., Rossi, S., Montomoli, M., Caffarelli, C., Cuda, C., Lazzeri, G., Giacchi, M. and Nuti, R., 2009. Accuracy of different reduced versions of a validated food-frequency questionnaire in Italian men and women. *Calcified Tissue International*. 85(3): 221–227.

Greer, F. R. and Krebs, N. F., 2006. American academy of pediatrics committee on nutrition optimizing bone health and calcium intakes of infants, children, and adolescents. *Pediatrics*. 117(2): 578–585.

Jacobs, Jr, D. R. and Steffen, L. M., 2003. Nutrients, foods, and dietary patterns as exposures in research: a framework for food synergy. *American Journal of Clinical Nutrition*. 78: 508S–513S.

Kaluza, J., Orsini, N., Levitan, E. B., Brzozowska, A., Roszkowski, W. and Wolk, A., 2010. Dietary calcium and magnesium intake and mortality: a prospective study of men. *American Journal of Epidemiology*. 171(7): 801–807.

Kanis, J. A., Johansson, H., Oden, A., De Laet, C., Johnell, O., Eisman, J. A., Mc Closkey, E., Mellstrom, D., Pols, H., Reeve, J., Silman, A. and Tenenhouse, A., 2005. A meta-analysis of milk intake and fracture risk: low utility for case finding. *Osteoporosis International*. 16(7): 799–804.

Livingstone, M. B. E. and Robson, P. J., 2000. Measurement of dietary intake in children. *Proceedings of the Nutrition Society*. 59: 279–293.

Loy, S. L. and Jan Mohamed, H. J., 2013. Relative validity of dietary patterns during pregnancy assessed with a food frequency questionnaire. *International Journal of Food Sciences and Nutrition*. 64(6): 668–673.

Maghsoudi, Z., Askari, G., Ghiasvand, R., Khorvash, F., Iraj, B., Shokri, N. and Darvishi, L., 2013. Dairy consumption and stroke risk. *International Journal of Preventive Medicine*. 4(suppl. 2): S294–S299.

Newby, P. K., Noel, S. E., Grant, R., Judd, S., Shikany, J. M. and Ard, J., 2011. Race and region are associated with nutrient intakes among black and white men in the United States. *Journal of Nutrition*. 141: 296–303.

Pampaloni, B., Mavilia, C., Bartolini, E., Tonelli, F., Brandi, M. L. and D'Asta, F., 2013a. Phytoestrogens and colon cancer. In Hany El-Shemy (ed.) *Soybean - Bio-Active Compounds*. InTech. ISBN: 978-953-51-0977-8.

Pampaloni, B., Bartolini, E., Barbieri, M., Piscitelli, P., Di Tanna, G. L., Giolli, L. and Brandi, M. L., 2013b. Validation of a food-frequency questionnaire for the assessment of calcium intake in schoolchildren aged 9–10 years. *Calcified Tissue International*. 93: 23–38.

Pittas, A. G., Dawson-Hughes, B., Li, T., Van Dam, R. M., Willett, W. C., Manson, J. E. and Hu, F. B., 2006. Vitamin D and calcium intake in relation to type 2 diabetes in women. *Diabetes Care*. 29: 650–656.

Poh, B. K., Ng, B. K., Siti Haslinda, M. D., Nik Shanita, S., Wong, J. E., Budin, S. B., Ruzita, A. T., Ng, L. O., Khouw, I. and Norimah, A. K., 2013. Nutritional status and dietary intakes of children aged 6 months to 12 years: findings of the nutrition survey of Malaysian children (SEANUTS Malaysia). *British Journal of Nutrition*. 110: S21–S35.

Rice, B. H., Cifelli, C. J., Pikosky, M. A. and Miller, G. D., 2011. Dairy components and risk factors for cardiometabolic syndrome: recent evidence and opportunities for future research. *Advances in Nutrition*. 2: 396–407.

Rizzoli, R., 2008. Nutrition: its role in bone health best practice & research. *Journal of Clinical Endocrinology and Metabolism*. 22(5): 813–829.

Ruder, E. H., Thiébaut, A. C. M., Thompson, F. E., Potischman, N., Subar, A. F., Park, Y., Graubard, B. I., Hollenbeck, A. R. and Cross, A. J., 2011. Adolescent and mid-life diet: risk of colorectal cancer in the NIH-AARP diet and health study. *American Journal of Clinical Nutrition*. 94: 1607–1619.

Sanders, K. M., Stuart, A. L., Kotowicz, M. A. and Nicholson, G. C., 2010. Annual feedback is an effective tool for a sustained increase in calcium intake among older women. *Nutrients*. 2: 1018–1025.

Serra-Majem, L., Pfrimer, K., Doreste-Alonso, J., Ribas-Barba, L., Sánchez-Villegas, A., Ortiz-Andrellucchi, A. and Henrıquez-Sánchez, P., 2009. Dietary assessment methods for intakes of iron, calcium, selenium, zinc and iodine. *British Journal of Nutrition*. 102: S38.

Struijk, E. A., Heraclides, A., Witte, D. R., Soedamah-Muthu, S. S., Geleijnse, J. M., Toft, U. and Lau, C. J., 2013. Dairy product intake in relation to glucose regulation indices and risk of type 2 diabetes. *Nutrition, Metabolism & Cardiovascular Diseases*. 23: 822–828.

Svensson, A., Renstro, F., Bluck, L., Lissner, L., Franksand, P. W. and Larsson, C., 2013. Dietary intake assessment in women with different weight and pregnancy status using a short questionnaire. *Public Health Nutrition*. 15: 1–10.

Taylor, E. N. and Curhan, G. C., 2013. Dietary calcium from dairy and non-dairy sources, and risk of symptomatic kidney stones. *Journal of Urology*. 190: 1255–1259.

Vereecken, C. M. and Maes, L., 2010. Comparison of a food frequency questionnaire with an online dietary assessment tool for assessing pre-school children's dietary intake. *Journal of Human Nutrition and Dietetics*. 23: 502–510.

WHO, 2011. *Global Status Report on Noncommunicable Diseases 2010*. WHO, Geneva.

Wlodarek, D., Glabska, D., Kolota, A., Adamczyk, P., Czekajlo, A., Grzeszczak, W., Drozdzowska, B. and Pluskiewicz, W., 2012. Calcium intake and osteoporosis: the influence of calcium intake from dairy products on hip bone mineral density and fracture incidence – a population-based study in women over 55 years of age. *Public Health Nutrition*. 17(2): 383–389.

Wu, P., Zhao, X.-H., Ai, Z.-S., Sun, H.-H., Chen, Y., Jiang, Y.-X., Tong, Y.-L. and Xu, S.-C., 2013. Dietary intake and risk for reflux esophagitis: a case-control study. *Gastroenterology Research and Practice*. 691026: 9 pages.

Yin, J., Zhang, Q., Liu, A., Du, W., Wang, X., Hu, X. and Ma, G., 2010. Calcium supplementation for 2 years improves bone mineral accretion and lean body mass in Chinese adolescents. *Asia Pacific Journal of Clinical Nutrition*. 19(2): 152–160.

Zemel, M. B. and Miller, S. L., 2004. Dietary calcium and dairy modulation of adiposity and obesity risk. *Nutrition Reviews*. 62(4): 125–131.

CHAPTER 11

Calcium Digestibility Using In Vivo, In Vitro and Ex Vivo Techniques

F. ROSSI*[a]

[a]Istituto di Scienze Degli Alimenti e della Nutrizione, Facoltà di Scienze Agrarie Alimentari ed Ambientali, Università Cattolica del Sacro Cuore, *via* Emilia Parmense 84, 29100 Piacenza, Italy
*E-mail: filippo.rossi@unicatt.it

11.1 Introduction

Calcium is one of the most important nutrients present in our diets, it is the main component of bones and teeth, and it is required for muscle contraction and for blood coagulation. Furthermore, as second messenger, it is involved in several important metabolic pathways.

For this reason, all database on food composition has data on Ca content, although this information has a limited usefulness if not linked to information on Ca digestibility for each food. As a consequence there is strong interest in the obtainment of value on Ca availability in foods. In addition, the availability of methods for measuring Ca digestibility or bioavailability, is useful in research designed to compare the effectiveness of different foods as a Ca source.

Food and Nutritional Components in Focus No. 10
Calcium: Chemistry, Analysis, Function and Effects
Edited by Victor R. Preedy

Published by the Royal Society of Chemistry, www.rsc.org

At present, researchers involved in research on Ca metabolism are using the following methods:

- input–output Ca balance in animal models (mainly rats and pigs);
- *in vitro* Ca digestibility, determined with multienzymatic methods;
- *ex vivo* studies with human cell cultures;
- *in vivo* experiments on humans volunteers using ^{43}Ca-labeled food, or monitoring bone mineralization.

A complication in the development of methods for the evaluation of Ca digestibility is related to the fact that Ca absorption from the intestine can happen in two ways:

- passive, *via* a paracellular route;
- active, *via* a transcellular route.

In the active transport, Ca is taken up by calbindin, a Ca-binding protein the synthesis of which is induced by vitamin D as a result of poor intake of Ca. If the amount of Ca is not limited, there is not reason to consume ATP to increase the absorption of a nonlimiting nutrient, so the level of calcitriol is low and passive transport is predominant (Guéguen and Pointillart, 2000).

To be absorbed Ca must be in a soluble form, usually Ca^{2+}, or linked to a molecule that can cross the gut wall.

Some factors, such as fiber, oxalate, tannins or phytate content negatively affect Ca digestibility. On the other hand, there are dietary factors like lactose or phosphopeptides that facilitate intestinal absorption of Ca (Guéguen and Pointillart, 2000; Camara-Martos and Amaro-Lopez, 2002). All these factors should be taken into account in order to obtain a reliable estimate of Ca availability.

11.2 Methods for Estimating Ca Digestibility

11.2.1 *In Vitro* Methods

Studies on humans or animal models take these factors into account but are time consuming, expensive and often pose ethical problems. *In vitro* methods are good alternatives, but the complexity of the situation is not easily reproducible with this approach.

The methods more used are based on the procedure proposed by Miller *et al.* (1981) and standardized by Luten *et al.* (1996).

They generally comprise two steps, where sample food is incubated with enzymes for the simulation of gastrointestinal digestion. The first is the simulation of gastric hydrolysis and is performed with pepsin–HCl at pH 2 for 120 min in a shaking water bath at 37 °C. Perales *et al.* (2005), working on baby foods, reduced this incubation time to 30 min to consider the lower digestive efficiency of babies. For the same reason these authors reduced the enzyme concentration.

The second step can be slightly modified depending on whether the solubility or the dialyzability of Ca have to be measured. In the first hypothesis pH is increased to 5.0 with $NaHCO_3$, then a pancreatin–bile extract mixture is added and the whole is incubated for 2 h at 37 °C (Perales *et al.*, 2005). Luten *et al.* (1996) and Jovanì *et al.* (2001) increase pH up to 7.5. The standardized method proposed by Luten *et al.* (1996) is reported in Table 11.1.

For the determination of dialyzable Ca, after the pepsin–HCl attack, an aliquot (20 mL) of pepsin digest is placed in a flask together with a dialysis bag containing 25 mL of water and $NaHCO_3$ equivalent to the titratable acidity previously measured and the incubation is continued for 30 min. The pH is increased to 7.5 and a mixture of pancreatin and bile extracts is added. The flask and its content are incubated at 37 °C for 2 h. During incubation free Ca or Ca included in complexes having a molecular weight less than the cutoff of dialysis bag (10–12 000 Da), move inside the bag. At the end of the incubation period the dialysis tube is removed and the Ca content in the dialyzate is determined (Roig *et al.*, 1999; Jovanì *et al.*, 2001; Perales *et al.*, 2005).

Since about 10% of Ca absorption takes place in the colon (Bronner and Pansu, 1999). Lombardi-Boccia *et al.* (1998) added a further process to the simulation of gastrointestinal digestion. They fermented the intestinal digest with a human fecal inoculum. The fermentation lasted 14 h and at the end, the samples were centrifuged and the supernatant analyzed for Ca content.

Table 11.1 Standardized procedure proposed by Luten *et al.* (1996).

Step	Reagents	Incubation time
1. Gastric digestion	10 g of sample in 80 mL of water[a] 6 mL 6 M HCl 3 g of pepsin solution (16 g in 100 mL 0.1 N HCl) If pH is not 2.0, more 6 M HCl may be added Water until to 100 mL	120 min in a water bath (100–120 strokes per minute)
Stop digestion	Put in ice	90 min
Titratable acidity	20 g of pepsin digest 5 g of pancreatic mixture pH adjusted to 7.5 with 0.5 M NaOH	30 min
2. Intestinal digestion and dialysis	20 g of pepsin digest 250 mm dialysis tube (25 g water + $NaHCO_3$ equivalent in moles to NaOH used for titratable acidity) pH readjusted to 7.5 with 0.5 M NaOH	30 min
	5 g of pancreatic mixture (pancreatin 4 g L^{-1} and bile extracts 25 g L^{-1} of 0.1 M $NaHCO_3$)	120 min
3.	Dialysis bag is removed and rinsed with water. Internal content was analyzed for Ca content	

[a]For the estimation of Ca digestibility, instead of a unique sample size, Perales *et al.* (2005) proposed that the weight of sample be calculated to obtain a Ca amount close to 40 mg.

In Figure 11.1, the several steps of methods proposed for *in vitro* determination of Ca digestibility are summarized.

To avoid exogenous Ca contamination, reagents should be treated with a chelating resin before incubation with food samples.

All these approaches are less expensive and laborious than *in vivo* experiments but estimate only the amount of Ca potentially available for absorption and not the fraction *in vivo* really absorbed.

Magnano San Lio *et al.* (2011) determined Ca digestibility of different foods in a balance studies, using rats as animal model and regressed these data against Ca-solubility values obtained with the enzymatic digestion followed by centrifugation (Perales *et al.*, 2005). The correlation between the two sets of data was low ($r^2 = 0.0613$) (Figure 11.2), suggesting that this simple *in vitro* methods is not very suitable to predict Ca digestibility.

In the paper of Perales *et al.* (2005) Ca solubility was positively affected by the Ca content of tested foods and resulted inversely correlated with Ca dialyzability ($r = -0.826$) and Ca uptake by Caco-2 cells ($r = -0.932$), while dialyzed Ca showed a linear relationship with Ca uptake by Caco-2 cells ($r = 0.705$). A possible explanation for these results could be the fact that solubilized Ca is composed not only of Ca ions, but also of Ca-containing

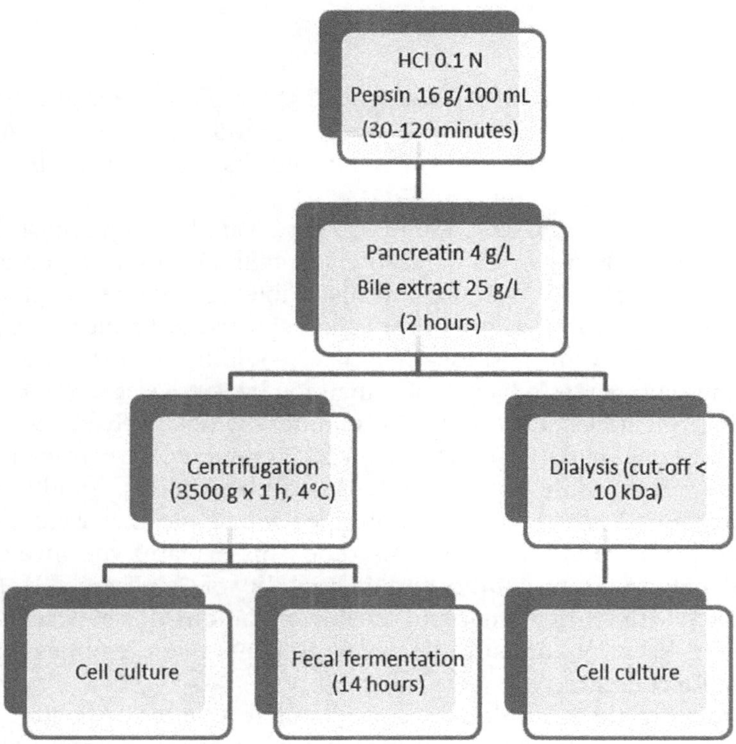

Figure 11.1 Brief summary of the *in vitro* digestion methods proposed by several authors.

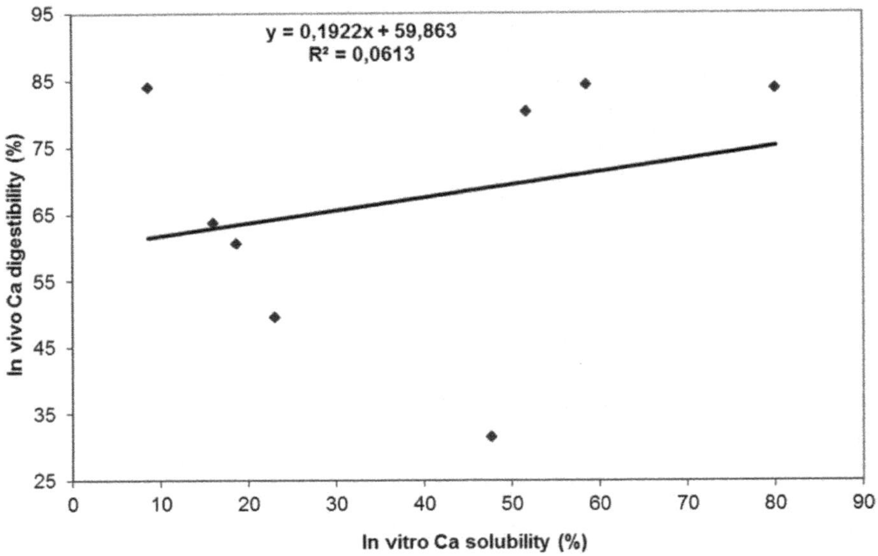

Figure 11.2 Correlation between *in vitro* and *in vivo* calcium digestibility. (Reprinted with permission from Magnano San Lio *et al.* Copyright by Springer-Verlag).

molecules, of various molecular weight. If the molecular weight is higher than the cutoff value of the dialysis tube, Ca cannot pass into the bag and a similar limitation, but with a lower cutoff value, is applicable to the absorption by cell culture.

Unlike Perales *et al.* (2005), Lombardi-Boccia *et al.* (1998) reported a linear relationship between Ca solubility and dialyzability in legumes. Furthermore, the values of soluble and dialyzable Ca obtained in this latter research were sharply lower than the value reported by Perales *et al.* (2005). A possible explanation of these differences could be the lower Ca content of legumes compared to baby-foods analyzed by Perales *et al.* (2005). Since according to Perales *et al.* (2005) Ca solubility is positively affected by Ca content we could hypothesize that a low Ca content of legumes leads to low solubility values, while high Ca content, as in baby-foods, could be due to the inclusion of this mineral in molecules with molecular weight higher than the cutoff value of the dialysis tube This explains the inverse relationship between the solubility and dialyzability of Ca observed by Perales *et al.* (2005) with Ca-rich foods and the linear relationships between the two parameters seen by Lombardi-Boccia *et al.* (1998) with legumes having a medium Ca content.

According to these results, Caco-2 cells offer a better estimation of the complexity of the intestinal absorption process and the inclusion of a step with cultured human cells (Caco-2 or HT-29) after *in vitro* digestion, will provide results more close to *in vivo* digestibility.

11.2.2 *In Vivo* Studies

11.2.2.1 Animal Models

11.2.2.1.1 What is the Best Animal Model? The main aim of an animal model is to provide data that reflect human metabolism, the main species used in animal experiments are: rats, guinea pigs and, to a lesser extent, pigs. All three species have peculiar characteristics that affect the reliability of obtained data. Like humans, pigs are omnivores and eat two or three times a day, while rats and guinea pigs basically eat grains and have no well-defined feeding times. Anyway, rats can eat almost all foods, but they have to be mixed to form a pelleted feed. The gut physiology of pigs resembles that of a human one, particularly as concerns the large intestine, while guinea pigs are coprophagous; a characteristic that humans, fortunately, do not possess. In pigs, the endogenous fecal loss of Ca is higher than in rats or guinea pigs and this is a confounding factor that must be considered when planning balance experiments with swine.

One of the main advantages of using small-size animals, like rats or guinea pigs, is the easier handling in comparison to pigs. Furthermore, the quantities of feces or urine to collect and store are obviously less in experiments involving rats than in trials where pigs are the models.

A rat skeleton continuously grows and this makes it a good model for studying Ca bioavailability but not osteoporosis since this pathology can be investigated only in bones where growth has stopped. A similar bone metabolism can also be found in pigs.

11.2.2.1.2 Animal Balance Studies. Animal studies are usually classical balance experiments that provide data for Ca digestibility (apparent or true) and net retention.

Apparent Ca absorption is calculated according to the formula given below:

$$\frac{\text{Ca intake} - \text{Ca in feces}}{\text{Ca intake}} \times 100$$

If the fecal loss of endogenous Ca is estimated, true digestibility can be calculated by subtracting endogenous loss to fecal excretion. The endogenous loss of Ca can be estimated in two ways: estimating fecal Ca in animal fed Ca-free diet, or injecting a stable or radioactive isotope of Ca at the beginning of the trial and dosing its feces.

Ca retention is calculated as:

$$\frac{\text{Ca intake} - (\text{Ca in feces} + \text{Ca in urine})}{\text{Ca intake}}$$

True Ca retention can be obtained after the calculation of endogenous urine loss, using the same approach already described for true digestibility.

Another parameter that can be determined in studies on animal models, is the Ca content in specific bones, usually the femur.

Growing rats are the model most used (Larsen *et al.*, 2000; Nestares *et al.*, 2003) but sometimes mice are used (Singh *et al.*, 2007). It is important to use young animals because they have the highest Ca requirements and bone metabolism is shifted towards an anabolic state. This is the best situation to highlight differences in Ca bioavailability in foods.

Balance studies are carried out in metabolic cages where, after an adaptation period ranging from 3 to 14 days, food intake, feces and urine are recorded and collected for 7 days and successively analyzed for Ca content (Chang *et al.*, 1994; Larsen *et al.*, 2000; Nestares *et al.*, 2003). In these papers, experiments were not longer than 28 days and Larsen suggested that studies where the main monitored parameter is the bone mineralization, should last for at least 28 days.

Sarria and Vaquero (2001), used suckling rats to study Fe and Zn digestibility in infant formula, an approach that could also be suitable for Ca. In this experiment no adaptation period was adopted and animals, housed in metabolic cages, suckled different types of milk-replacer for 7 days. During this period, food intake, feces and urine were collected, recorded and analyzed. Ca digestibility was calculated using the formulas previously described.

In order to highlight significant differences in Ca availability by means of the determination of Ca content in femur, the experiment should last for at least 28 days.

As discussed by Weaver *et al.* (1987) and Heaney *et al.* (1988), the rat model can be useful for ranking Ca availability from different foods, but these data must not necessarily be considered as absolute values.

Ca bioavailability is determined also in farm animals (pigs and poultry) or in companion animals (cats and dogs) but usually these animals are not used as models, but as final targets. Only pigs could be interesting models, particularly growing piglets.

11.2.2.2 Humans Studies

11.2.2.2.1 Balance Studies. Ca digestibility can be measured by ingest–excreta studies also in humans, and some experiments have been carried out (Rosado *et al.*, 1992; Vermorel *et al.*, 2004). In these experiments different amounts of Ca were given to volunteers, and Ca digestibility resulted affected by mineral intake.

As depicted in Figure 11.3 the relationship between Ca intake and Ca digestibility is better described by a polynomial equation ($r^2 = 0.5725$) than logarithmic ($r^2 = 0.4059$) or linear ($r^2 = 0.2529$) ones. This pattern is consistent with a transport system where active sites are progressively saturated, with a decreasing efficiency for very high Ca intake, probably the prevalence of passive diffusion transport. However, this equation is based only on 6 data from 3 studies, so it must be treated with caution.

A second method to measure Ca digestibility is based on the administration of a single Ca load and the collection of feces for 12 h. The intestine must be cleaned with isotonic solutions before the test meal is ingested. After 12 h,

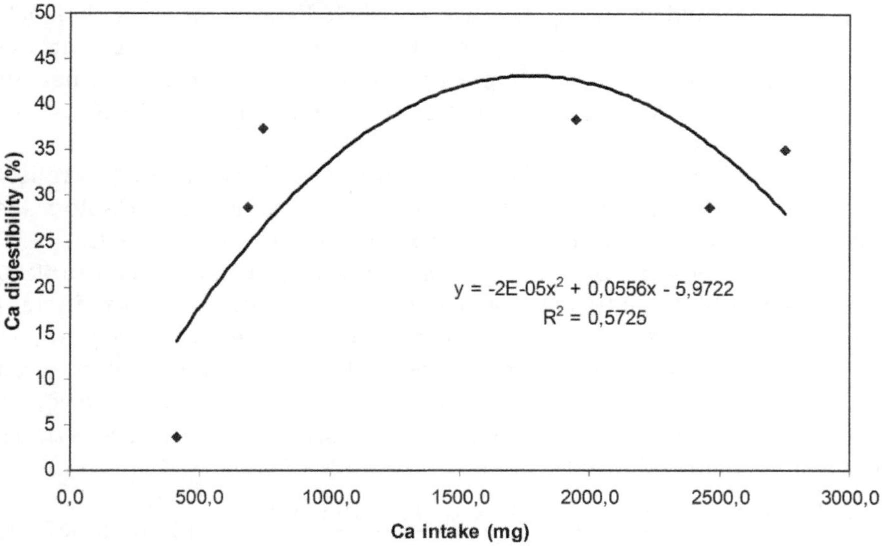

Figure 11.3 Relationship between Ca digestibility in humans and the intake of the mineral.

a second intestinal lavage is done and unexcreted fecal residue are collected (Guéguen and Pointillart, 2000). Although not really time consuming, this method is quite invasive and unphysiological.

Ca sources can be ranked determining the urinary excretion of a Ca load ingested *per os*. Since humans excrete large amounts of Ca in urine, it is to possible estimate the absorbability of Ca by measuring the amount of Ca excreted after the administration of a Ca load in fasting subjects. This method does not provide the real digestibility of the mineral, but can be useful to compare different Ca sources. In spite of this limitation, the method is simple and fast, since the first urine sample can be taken 3 or 4 h after Ca intake. In order to facilitate the comparison between different subjects, data of Ca excretion should be expressed as a function of urinary creatinine (Pak *et al.*, 1975).

11.2.2.2.2 Methods using Isotopes. A less complicated approach for carry out a balance study, is use a radioactive (^{45}Ca, ^{47}Ca) or stable (^{42}Ca, ^{44}Ca, ^{46}Ca or ^{48}Ca) Ca isotope. This marker can be intrinsic or extrinsic to food. Examples of the first group are milk labeled by injecting cows with Ca isotope or plants grown with isotope-supplemented fertilizer. The use of an extrinsic Ca isotope is easier than the use of labeled foods, a time and money-consuming procedure. This approach is based on the assumption of a rapid equilibrium between Ca in the food and the isotope. This is true for dairy foods (Buchowski *et al.*, 1989) but not for some legumes and vegetables where Ca can be bound to phytate or oxalate (Weaver *et al.*, 1987), and as a consequence the bioavailability of Ca can be overestimated.

Another method involving the use of labeled Ca is the "double-label" method (Chanard *et al.*, 1974). Here, a Ca isotope is administered intravenously, while a second and different Ca isotope is ingested as test meal or food. Blood samples are taken at different times (0.5 – 1 – 1.5 – 2 – 3 – 4 – 5 – 6 – 8 – 12 – 20 – 48 h) and a curve of Ca levels is drawn.

The area under the curve of the injected isotope is considered to reflect a 100% digestibility, and the ratio between the curve of the two isotopes is considered to be equal to the fractional absorption (from 0 to 1) of Ca.

Parra *et al.* (2007) adopted a single Ca isotope to determine Ca absorption (not Ca digestibility) of fresh or pasteurized yogurt in subjects with different levels of lactose digestion. They used $^{43}CaCO_3$ extrinsically incorporated to the yogurt and homogenized. After an overnight fast, the volunteers received the test meal. Blood samples were taken at fasting and 1 h after the yogurt was ingested. Also in this case, the total Ca absorbed is expressed as a trapezoidal area calculated with the following formula:

$$\frac{\text{Blood Ca at fasting state} - \text{blood Ca 1 h after the meal}}{2} \times \text{sampling time (1 h)}$$

If urine is collected for 24 h before the consumption of labeled yogurt, it is possible to calculate the percentage of excreted and retained Ca of tested foods.

These methods are suitable for comparing different sources of Ca, but they do not provide data on Ca digestibility.

11.2.2.2.3 Bone Mineralization as an Index of Ca Bioavailability. While previous methods were quite fast, although with some differences, methods based on the evaluation of bone mineralization required prolonged experimental periods, although they represent the most reliable technique of estimating the bioavailability of Ca from different sources. In animal models the items that can be measured are: mineral content, mineral density, breaking strength, or morphometric parameters obviously these analysis can be performed only on killed animals.

In humans, the evaluation of bone mineralization is carried out using double X-ray absorptiometry (DEXA) or quantitative computed tomography (QCT). The first method can be used to evaluate the whole skeleton, while the second one focuses on lumbar vertebrae.

A DEXA scan sends out two X-ray beams with different energy levels (high and low), that are directed on volunteer's bone. The bone-mineral density can be determined, after subtraction of soft-tissue absorption, from the difference in the absorption of each beam by bone.

While DEXA uses two X-ray beams, the QCT use a single X-ray beam and provides three-dimensional volumetric BMD analysis (DEXA gives two-dimensional images). QCT is also more sensitive to change in metabolic bone, see Figure 11.4.

In order to increase the sensitivity of the method and use it to rank foods according to the bioavailability of Ca, the experiments must be carried out on young subjects, which having great Ca requirements can amplify the difference in the Ca bioavailability of different foods. In addition, the Ca content

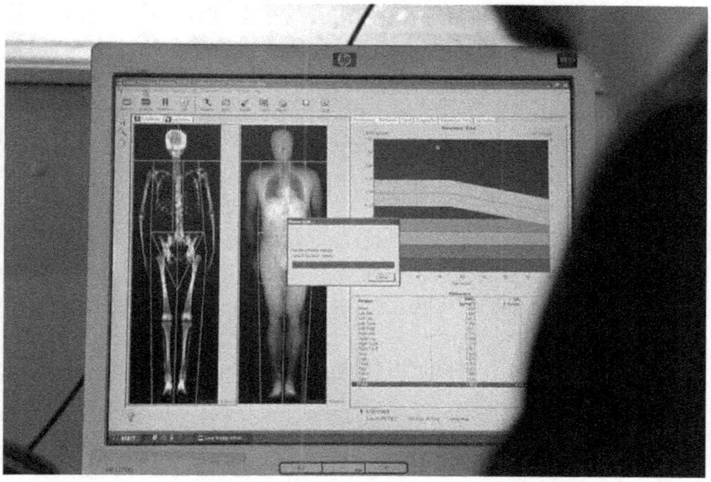

Figure 11.4 A full-body scan performed with a DEXA scanner. http://www.bris.ac.uk/alspac/resources/visualidentity/powerpoint-template-and-image-bank.html. Author photo: Nick Smith: nick@smithphotography.com.

of the basal diet must be as low as possible, otherwise it will be very difficult to extrapolate the bioavailability of Ca from the tested food if it represents only a small fraction of the ingested mineral. To compare data from different sets of experiments, the other characteristics of basal diet affecting Ca absorption of utilization, such as: protein content, fiber, P, Na, oxalate and phytate, must be kept constant.

11.2.3 Cell Culture

At present there is no availability of culturable intestinal cells expressing structural and functional properties of gut epithelium, so human intestinal tumor cells are used. Cells from adenocarcinoma cancer colon have been shown to be suitable for research purposes. In fact, these cells can differentiate and show many of the functional and morphological properties of mature enterocyte, including the epithelial layer (Ferraretto *et al.*, 2007). Two tumor cell lines have been used in experiments to study the Ca uptake on cell culture: Caco-2 and HT-29.

Caco-2 can be driven to differentiate in enterocyte after cell confluence (Rousset *et al.*, 1985) or by means of specific culture media (Jumarie and Malo, 1991; Lentz *et al.*, 2000; Ferruzza *et al.*, 2012). In line HT-29, differentiation may be induced by reducing the glucose level (<2 g L^{-1}) of medium or supplementing it with Na-butyrate, or substituting glucose with galactose or using drugs such as: forskolin, colchicine, nocodazole or taxol (Rousset *et al.*, 1985; Ophir *et al.*, 1995).

After differentiation, Caco-2 cells develop a monolayer of absorbent cells, expressing tight junction, microvillus of brush-border with typical enzymes, as alkaline phosphatase and sucrose-isomaltase, and transporter for sugars, amino acids, oligopeptides, vitamins and minerals.

Differentiated HT-29 cells express microvillus brush border with typical enzymes in the apical side Na^{+}–K^{+}/ATPase pump in the basolateral side.

In HT-29 cells, the type of culture media affects cell morphology and the differentiation requested for experiments on Ca absorption can be obtained by growing the cells in glucose-limited conditions. The RPMI (Roswell Park Memorial Institute medium) with a low glucose concentration (13.9 mM L^{-1}) or the DMEM modified medium (Dulbecco's modified eagle medium) where glucose is substituted for galactose give the best results in terms of cell differentiation. RPMI-grown cells show a polarized monolayer, with a well-developed brush border. Tight junctions, adherence junctions and desmosome are also present. In galactose-supplemented DMEM, the brush border shows a lower number of microvillus compared to mediums containing low concentrations of glucose.

An interesting improvement of the *in vitro* methods is the addition of a further step, represented by a human colon carcinoma cell line Caco-2 or HT-29 that can absorb solubilized Ca. The combination of *in vitro* model with Caco-2 cells has been proposed by Jovanì *et al.* (2001) to estimate the bioavailability of Ca from infant formulas and used also by Perales *et al.* (2005) who measured the uptake of dialyzed Ca by a human colon carcinoma cell line (Caco-2).

Ferraretto *et al.* (2001) measured Ca uptake in HT-29 tumor cells, since they had not tested a food but pure casein phosphopeptides, a very rich Ca source, it was not necessary to perform an enzymatic digestion of food matrix.

Summary Points

- This chapter focus on the methods for estimating Ca digestibility.
- Ca can be absorbed in two ways: a passive route (when Ca intake is high) and an ATP-consuming active transport (when Ca intake is low). This make determination of Ca digestibility difficult.
- *In vitro* methods based on enzymatic attack, are time and money saving but cannot describe the complexity of intestinal Ca absorption.
- To improve the reliability of *in vitro* methods, they can be followed by a step based on Ca absorption by differentiated human tumor cells (Caco-2 or HT-29).
- Animal studies can use rats, guinea pigs or pigs. The gut physiology of pigs is the most similar to human one, but for handling reasons, rats are preferred as animal models.
- Studies performed on humans comprise: balance studies, experiment with isotope and the use of bone mineralization as index of Ca bioavailability.

Key Facts

Listed here are the key facts of intestinal Ca absorption, including the definition of active and passive transport and the role of vitamin D_3.

1. Ca absorption takes place mainly in the small intestine and to a lesser extent in the large intestine. In the small intestine we have different situations – active transport across cells is predominant in the duodenum and upper jejunum, while in the ileum, passive diffusion is predominant. In the small intestine a limited but non-negligible contribution to Ca absorption comes from endocytosis. In the large intestine, passive diffusion is the only way by which dietary Ca is absorbed.
2. Active transport systems allow a substance to cross a cell membrane against its concentration gradient. This kind of transport uses energy coming from ATP hydrolysis or from the electrochemical gradient. In the human gut this process can be divided into two stages: entry across the brush border of enterocyte, diffusion through the cytoplasm into the extracellular liquid through the basolateral membrane. The first step is driven by a positive electrochemical gradient generated by the maintenance of a very low Ca concentration in the cytoplasm. This low level of Ca is mainly obtained thanks to the extrusion of Ca by means of a Ca-ATPase pump, regulated by Ca binding protein (CaBP) and calmodulin. A less important route of Ca excretion is a Na—Ca pump (3 Na^+ for 2 Ca^{2+}).
3. Vitamin D_3 or calcitriol, regulates the transcription of DNA codifying for CaBP and activates the Ca-ATPase pump. This increases Ca extrusion and maintaining a low intracellular Ca concentration promotes Ca uptake. Calcitriol synthesis is indirectly regulated by plasma Ca concentration; when mineral blood levels fall, parathormone (PTH) levels are increased and, in turn, PTH promotes conversion of calcidiol to calcitriol.
4. This active transport is saturable and negatively correlated with blood Ca. This could be considered a quite logical evolutionary approach. If the plasma Ca level is elevated (high Ca intake) there is no need to consume ATP for the absorption of nonlimiting nutrients, while when Ca intake is low there is also a low blood Ca content and consequently it is conceivable to spend ATP for the intestinal uptake of Ca.
5. Passive transport is based on the existence of an electrochemical gradient between two compartments. The mineral moves from major (intestinal lumen) to minor concentration (enterocyte) determining the movement of Ca from major to minor concentration. This type of transport is not saturable, is not dependent on vitamin D_3 and increases with the gradient of Ca concentration. All dietary factors that improve Ca solubility within the ileum increase the concentration gradient of this mineral and consequently the passive diffusion.

Listed here are the key facts of human tumor cell lines used as a model of the human absorptive intestine. The main characteristics of Caco-2 and HT-29 cell lines are given.

1. The Caco-2 cell line is a cell culture obtained from human epithelial colorectal cancer. Although this cell line is derived from a colon carcinoma, it is possible, by culturing them under specific conditions, to differentiate them in cells resembling enterocytes. Several clones of differentiated Caco-2 cell lines are available. They do not always express in the same way all the characteristics of the native line, but within a single clone, they are homogeneous. However, this heterogeneity between different clones, makes it difficult to compare results obtained with different clones of Caco-2 cells.

   Differentiated Caco-2 cells are not individual cells, but a confluent monolayer of polarized epithelial cells, grown on an appropriate holder. This polarized epithelium provides a physical–chemical barrier to the passage of small molecules, like Ca ions or Ca-containing peptides. For this reason the Caco-2 differentiated cells are an *in vitro* model of the small intestine that can be used to predict the absorption of nutrients and drugs.

2. HT-29 cells. Like Caco-2, HT-29 cells, derive from a human colorectal adenocarcinoma diagnosed in a 44 year old woman. If undifferentiated HT-29 cells are grown under glucose-limiting conditions or culture media is supplemented with inducers (forskolin) a cell differentiation can be obtained. A unique characteristic of HT-29 cells is their property to became differentiated and polarized, after galactose supplementation of culture medium.

   Differentiated HT-29 cells, form a polarized monolayer with apical brush-border and tight junctions. As for Caco-2 cells, the HT-29 monolayer represents a good *in vitro* model to carry out experiments on intestinal absorption.

List of Abbreviations

DEXA	Double X-ray absorptiometry
QCT	Quantitative computed tomography
RPMI	Roswell Park Memorial Institute medium
DMEM	Dulbecco's modified eagle medium
CaBP	Calcium-binding protein
PTH	Parathormone
BMD	Bone-mineral density

References

Bronner, F. and Pansu, D., 1999. Nutritional aspects of Ca absorption. *Journal of Nutrition*. 129: 9–12.

Buchowski, M. S., Swizral, K. C., Lengemann, F. W., van Campen, D. and Miller, D. D., 1989. A comparison of intrinsic and extrinsic tracer methods for estimating calcium bioavailability to rats from dairy foods. *Journal of Nutrition*. 119: 228–234.

Chanard, J. A., Assailly, J., Bader, C. and Funck-Brentano, J. L., 1974. A rapid method for measurement of fractional intestinal absorption of Ca. *Journal of Nuclear Medicine*. 15: 588–592.

Chang, M. J., Bailey, J. W. and Collins, J. L., 1994. Dietary tannins from cowpeas and tea transiently alter apparent calcium absorption but not absorption and utilization of protein in rats. *Journal of Nutrition*. 124: 283–288.

Camara-Martos, F. and Amaro-Lopez, M. A., 2002. Influence of dietary factors on calcium bioavailability. *Biological Trace Element Research*. 89: 43–52.

Ferraretto, A., Signorile, A., Gravaghi, C., Fiorilli, A. and Tettamanti, G., 2001. Casein phosphopeptides influence calcium uptake by cultured human intestinal HT-29 tumor cells. *Journal of Nutrition*. 131: 1655–1661.

Ferraretto, A., Gravaghi, C., Donetti, C., Cosentino, S., Donida, B. M., Bedoni, M., Lombardi, G., Fiorilli, A. and Tettamanti, G., 2007. New methodological approach to induce a differentiation phenotype in Caco-2 cells prior to post-confluence stage. *Anticancer Research*. 27: 3919–3925.

Ferruzza, S., Rossi, C., Scarino, M. L. and Sambuy, Y., 2012. A protocol for differentiation of human intestinal Caco-2 cells in asymmetric serum-containing medium. *Toxicology In Vitro*. 26: 1252–1255.

Guéguen, L. and Pointillart, A., 2000. The bioavailability of dietary calcium. *Journal of the American College of Nutrition*. 19: 119–136.

Heaney, R. P., Weaver, C. M. and Recker, R. R., 1988. Calcium absorbability from spinach. *American Journal of Clinical Nutrition*. 47: 707–709.

Jovanì, M., Barberà, R., Farrè, R. and de Aguilera, E. M., 2001. Calcium, iron and zinc uptake from digests of infant formulas by Caco-2 cells. *Journal of Agricultural and Food Chemistry*. 49: 3480–3485.

Jumarie, C. and Malo, C., 1991. Caco-2 cells cultured in serum-free medium as a model for the study of enterocytic differentiation *in vitro*. *Journal of Cell Physiology*. 149: 24–33.

Larsen, T., Thilsted, S. H., Kongsbak, K. and Hansen, M., 2000. Whole small fish as a rich calcium source. *British Journal of Nutrition*. 83: 191–196.

Lentz, K. A., Hayashi, J., Lucisano, L. J. and Polli, J. E., 2000. Development of a more rapid, reduced serum culture system for Caco-2 monolayers and application to the biopharmaceutics classification system. *International Journal of Pharmaceutics*. 200: 41–51.

Lombardi-Boccia, G., Lucarini, M., Di Lullo, G., Del Puppo, E., Ferrari, A. and Carnovale, E., 1998. Dialysable, soluble and fermentable calcium from beans (*Phaseolus vulgaris* L.) as a model for *in vitro* assessment of the potential calcium availability. *Food Chemistry*. 61: 167–172.

Luten, J., Crew, H., Flynn, A., Van Dael, P., Kastenmayer, P., Hurrell, R., Deelstra, H., Shen, L.-H., Fainveather-Tait, S., Hickson, K., Farrè, R., Schlemmer, U. and Frarhlichj, W., 1996. Interlaboratory trial on the determination of the *in vitro* iron dialyzability from food. *Journal of the Science of Food and Agriculture*. 72: 415–424.

Magnano San Lio, E., Bruschi, S., Mulazzi, A., Calabrese, G., Piva, G. and Rossi, F., 2011. *In vivo* and *in vitro* digestibility of the calcium contained in foods of animal and plant origin. *Mediterranean Journal of Nutrition and Metabolism*. 4: 105–110.

Miller, D. D., Schricker, B. and Cederblad, A., 1981. An *in vitro* method for estimation of iron availability from meals. *American Journal of Clinical Nutrition*. 34: 2248–2256.

Nestares, T., Barrionuevo, M., Lopez-Frias, M., Vidal, C. and Urbano, G., 2003. Effect of different soaking solutions on nutritive utilization of minerals (calcium, phosphorus and magnesium) from cooked beans (*Phaseolus vulgaris* L.) in growing rats. *Journal of Agricultural and Food Chemistry*. 51: 515–520.

Ophir, I., Cohen, E. and Ben-Shaul, Y., 1995. Apical polarity in human colon carcinoma cell lines. *Tissue & Cell*. 27: 659–666.

Pak, C. Y. C., Kaplan, R., Bone, H., Towsen, J. and Waters, O., 1975. A simple test for the diagnosis of absorptive, resorptive and renal hypercalciuria. *New England Journal of Medicine*. 292: 497–500.

Parra, M. D., Martinéz de Morentin, B. E., Cobo, J. M., Lenoir-Wijnkoop, I. and Martinez, J. A., 2007. Acute calcium assimilation from fresh or pasteurized yoghurt depending on the lactose digestibility status. *Journal of the American College of Nutrition*. 26: 288–294.

Perales, S., Barberà, R., Lagarda, M. J. and Farrè, R., 2005. Bioavailability of calcium from milk-based formulas and fruit juices containing milk and cereals estimated by *in vitro* methods (solubility, dialyzability, and uptake and transport by Caco-2 cells). *Journal of Agricultural and Food Chemistry*. 53: 3721–3726.

Roig, M. J., Alegrìa, A., Barberà, R., Farrè, R. and Lagarda, M. J., 1999. Calcium bioavailability in human milk, cow milk and infant formulas comparison between dialysis and solubility methods. *Food Chemistry*. 65: 353–357.

Rosado, J. L., Lopez, P., Morales, M., Munoz, E. and Allen, L. H., 1992. Bioavailability of energy, nitrogen, fat, zinc, iron and calcium from rural and urban Mexican diets. *British Journal of Nutrition*. 68: 45–48.

Rousset, M., Laburthe, M., Pinto, M., Chevalier, G., Rouyer-Fessard, C., Dussaulx, E., Trugnan, G., Boige, N., Brun, J. L. and Zweibaum, A., 1985. Enterocytic differentiation and glucose utilization in the human colon tumor cell line Caco-2: modulation by forskolin. *Journal of Cell Physiology*. 123: 377–385.

Sarria, B. and Vaquero, M. P., 2001. Zinc and iron bioavailability in a powder or in-bottle-sterilized infant formula estimated by *in vitro* and in suckling rats. *Journal of Nutritional Biochemistry*. 12: 266–273.

Singh, G., Arora, S., Sharma, G. S., Sindhu, J. S., Kansal, V. K. and Sangwan, R. B., 2007. Heat stability and calcium bioavailability of calcium-fortified milk. *LWT-Food Science and Technology*. 40: 625–631.

Vermorel, M., Coudray, C., Wils, D., Sinaud, S., Tressol, J. C., Montaurier, C., Vernet, J., Brandolini, M., Bouteloup-Demange, C. and Rayssiguier, Y., 2004. Energy value of a low-digestible carbohydrate, NUTRIOSE® FB, and its impact on magnesium, calcium and zinc apparent absorption and retention in healthy young men. *European Journal of Nutrition*. 43: 344–352.

Weaver, C. M., Martin, B. R., Ebner, J. S. and Krueger, C. A., 1987. Oxalic acid decreases calcium absorption in rats. *Journal of Nutrition*. 117: 1903–1906.

CHAPTER 12

Determining Calcium Bioavailability Using Caco-2 Cells

AMPARO ALEGRÍA[a], GUADALUPE GARCIA-LLATAS[a], AND REYES BARBERÁ*[a]

[a]Nutrition and Food Science Area, Faculty of Pharmacy, University of Valencia, Avda Vicente Andrés Estellés s/n 46100 Burjassot, Valencia, Spain
*E-mail: reyes.barbera@uv.es

12.1 Introduction

Calcium is an essential macromineral required for biological functions such as nerve conduction, muscle contraction, mitosis, blood coagulation, and structural support of the skeleton (Miller *et al.*, 2001). The main source of calcium in the diet is milk and dairy products, which supply about 70% of the global intake – only a few green vegetables and dried fruits being good sources of calcium, while drinking water provides 6–7% (Guéguen and Pointillart, 2000).

From a nutritional point of view, it is necessary to precisely define the amount of the element available for absorption and utilization. Among the dietary factors that influence bioavailability, dietary fiber (phytates) found in bran, most cereals and seeds can form insoluble complexes with calcium, thereby reducing its absorbability. On the other hand, the inhibitory effect of phytates upon calcium absorption depends on the phytate to calcium

Food and Nutritional Components in Focus No. 10
Calcium: Chemistry, Analysis, Function and Effects
Edited by Victor R. Preedy

Published by the Royal Society of Chemistry, www.rsc.org

molar ratio (Etcheverry *et al.*, 2012). The inhibiting effect of phytic acid upon calcium absorption can be countered by fructo-oligosaccharides. The fructo-oligosaccharides are fermented in the colon, where the organic acids formed lower the pH value and make calcium more soluble. In addition, they stimulate bacterial proliferation in the colon, resulting in an increased secretion of phytase preventing phytate binding to calcium (Cámara-Martos and Amaro-López, 2002).

Fibers seem to have no direct effect upon calcium absorbability. Soluble fiber with a low content of sulfate and carboxyl groups does not have a significant effect upon calcium bioavailability (Cámara-Martos and Amaro-López, 2002).

Other inhibitors of calcium bioavailability are oxalates and tannins. Lipids, especially milk fats, form insoluble soaps with calcium; however, the amount of fat in the diet does not affect calcium absorption (Guéguen and Pointillart, 2000).

Other milk components such as proteins, casein phosphopeptides (CPPs), lactose and organic acids (*i.e.* citric, tartaric, malic and ascorbic acids) appear to favor calcium bioavailability. CPPs can bind calcium and inhibit the formation of insoluble calcium phosphates. Nevertheless, it has also been reported that CPPs lack any effect upon calcium bioavailability, perhaps due to incomplete casein proteolysis (Etcheverry *et al.*, 2012). Whey proteins bind calcium and protect it from potential precipitating anions such as phosphates in the intestine. It has been described that lactose competes with inhibitors such as phosphates, keeping calcium soluble within the bowel lumen, but it does not act by increasing soluble calcium in the lumen. Other studies support the view that lactose acts upon the intestinal mucosa to increase its permeability, facilitating the passive diffusion of calcium (Guéguen and Pointillart, 2000).

Most governments and health organizations (Guéguen and Pointillart, 2000; WHO/FAO, 2006) underscore the importance of calcium enrichment of food products. Calcium carbonate is the least costly calcium source, and is often used in products where clarity or a dry and chalky mouth feel would likely not be noticeable. However, it functions best when stabilizers and ingredients to reduce interaction with proteins are employed to keep this insoluble calcium form suspended, and to minimize protein–calcium complexation and sedimentation. Calcium phosphate in turn is a low-solubility calcium source used in foods and nonclear beverages. Calcium citrate, lactate, lactate gluconate, and gluconate are calcium salts that prove to be more soluble over a broad pH and concentration range, while other highly soluble forms of calcium, such as calcium chloride, have undesirable taste acceptance. Lastly, calcium fumarate would appear to be useful for clear beverages, due to its taste and solubility characteristics (Rafferty *et al.*, 2007).

Most of the *in vitro* techniques consist of a simulation of gastrointestinal digestion followed by determination of how much of the element is soluble or dialyzes through a membrane of a certain pore size. The element solubility or dialysability can be used to establish trends in the bioavailability or relative bioavailability values of minerals in foods. These *in vitro* mineral bioavailability assays have been improved by incorporating cell cultures (Caco-2 cell line).

12.2 Caco-2 Cells

12.2.1 Generalities

The cell line Caco-2 under culture conditions, and after spontaneous differentiation, exhibits typical structural and functional characteristics of human small-intestinal enterocytes, and have been used as a validated *in vitro* model in studies of calcium bioavailability (Pinto *et al.*, 1983).

In culture, these cells exhibit three different stages: undifferentiated (subconfluence); polarized and nonhomogeneously differentiated (confluence around day 15); and polarized and homogeneously differentiated (postconfluence) (Figure 12.1). The differentiation process is rapid (15–20 days) exhibiting morphological characteristics of human enterocytes, such as cell organization in a polarized monolayer with the presence of intercellular tight junctions and dome formation on an impermeable support (indicating the presence of transepithelial ion transport). Brush-border microvilli are located only at the apical surface and are associated with the activity of enzymes such as alkaline phosphatase, sucrase-isomaltase, lactase, *etc.* In turn, the basolateral membrane is associated with Na/K-dependent ATPase activity and hormonal receptors (Pinto *et al.*, 1983; Hidalgo *et al.*, 1989). Enzymatic activities and transepithelial electrical resistance (TEER) measurements have been used as a marker of cultured cell differentiation, and monolayer integrity (Figure 12.1).

Standardization of Caco-2 culture conditions provides simple, reproducible, more economic and ethical alternative to animal testing (Langerholc *et al.*, 2011).

Caco-2 cells grown as monolayers on solid supports or porous supports (with apical and basolateral compartments) allow us to evaluate uptake or uptake and transport across the cell monolayer, respectively. These models have also been used to study the kinetics of transport of nutrients such as calcium, to elucidate intestinal absorption mechanisms, or explore the influence of food components upon calcium bioavailability.

12.2.2 Methodological Aspects of Caco-2 Cell Assays

Different methodological aspects of bicameral *in vitro* systems with cell cultures have been reviewed by Ekmekcioglu (2002). Prior to the addition of solutions or soluble or dialyzable fractions to the cell culture, it is important to adjust the osmolarity in the test solutions to isotonic values before assays are performed. In the human gastrointestinal tract, the luminal osmolarity is about 290–300 mosmol per kg H_2O. This is important, since components are transported against a gradient, and it is well known that they can compete at the absorption site or can mutually modulate their transport rates.

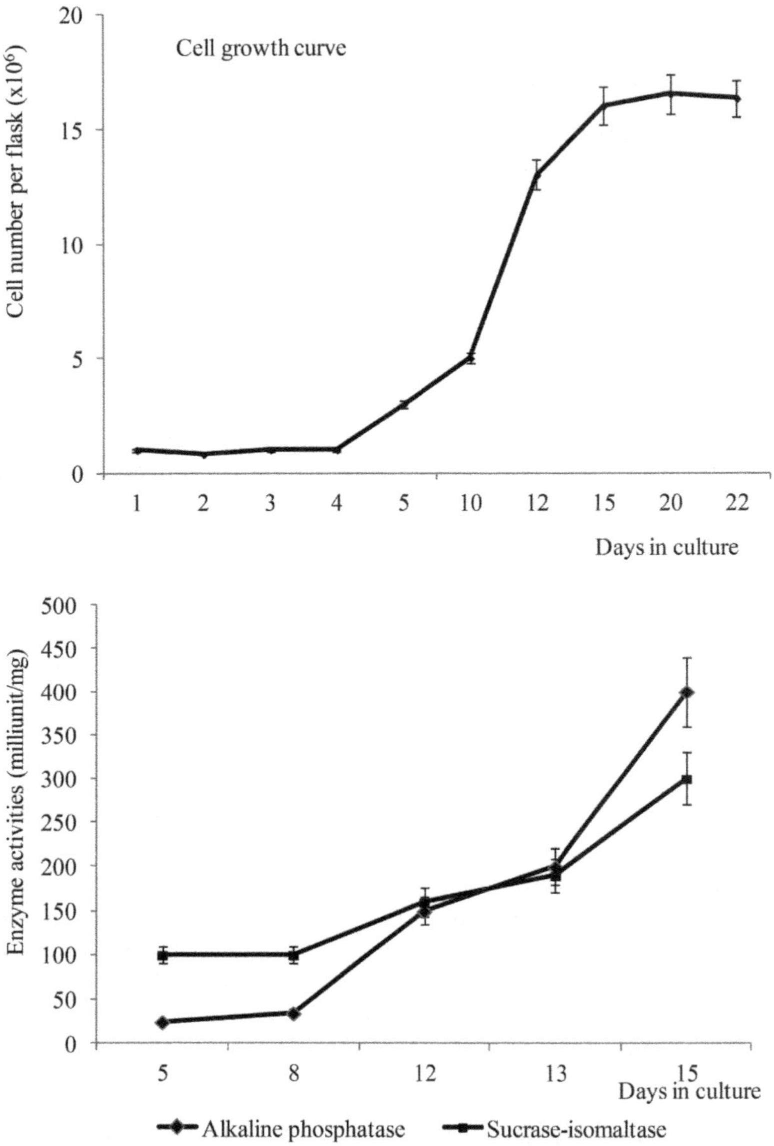

Figure 12.1 Cell growth curve and brush-border membrane associated enzyme activities during Caco-2 cell growth. The process of differentiation of Caco-2 cells is completed within 15–20 days. Differentiated Caco-2 cells exhibit development of brush-border microvilli associated with hydrolase enzyme activities. Caco-2 cells have been used as a validated model of intestinal epithelium to evaluate mineral bioavailability (own data, expressed as mean and standard deviation).

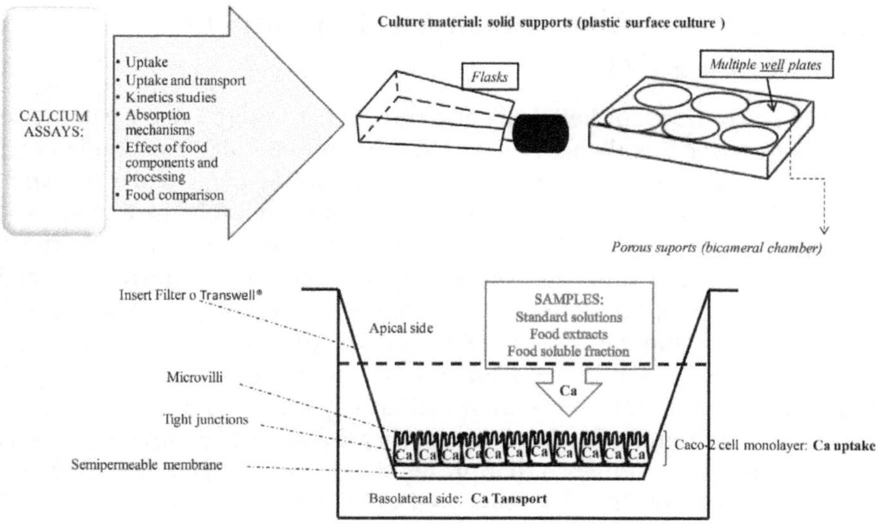

Figure 12.2 Scheme of different types of Caco-2 cell assays. The Caco-2 cells in culture conditions on solid supports present some structural characteristics as the human small intestinal enterocytes. Different calcium assays can be carried out with Caco-2 cells as shown in the figure. To evaluate both uptake and transport a bicameral chamber with a semipermeable membrane must be used, however, uptake estimation needs a nonporous solid support.

On the other hand, the transport area is different in the human small intestine (absorptive area of about 100 000 cm^2, without considering microvilli, *versus* an area of 4.72 cm^2 in filter-grown Caco-2 cells).

When the soluble fraction obtained with the *in vitro* digestion method is added to the Caco-2 cell monolayer, detachment of the cell monolayer from the flask (making it impossible to separate absorbed from unabsorbed mineral) has been observed. This is caused by the proteolytic activity of the digest, which needs to be inactivated. This is possible by adding fetal bovine serum (FBS) to the digest (the amount of FBS needed would be greater than 20% v/v, which is too much and unfeasible from a physiological and economic point of view) or heating to 100 °C for 4 min before addition to the cell monolayer. This is not necessary in the case of dialyzates, because proteolytic enzymes do not cross the dialysis membrane used, and are retained outside the dialysis bag (Jovaní *et al.*, 2001; Viadel *et al.*, 2006a). Figure 12.2 provides a schematic representation of assays that can performed with the Caco-2 cell model in relation to calcium – including sample types and treatments (standard solutions, extract or digest of foods), culture conditions, and study objectives (addressed in greater detail below).

12.3 Calcium Absorption Mechanisms in Caco-2 Cells

The kinetics of transepithelial calcium transport through Caco-2 cell monolayers is described as the sum of a saturable transcellular mechanism controlled by the action of vitamin D and a nonsaturable paracellular transport process. Saturable calcium transport is increased 4.3-fold in Caco-2 cells treated with 10–8 M of 1,25-$(OH)_2D_3$, though in the same way as in the *in vivo* setting, nonsaturable calcium transport is not significantly increased. On the other hand, 1,25-$(OH)_2D_3$-mediated calcium transport is apparently dose-dependent and can be seen with physiological levels (10^{-10} M) of the hormone. These observations show that calcium transport in Caco-2 cells is similar to that found in the mammalian small intestine (Giuliano *et al.*, 1991a). Furthermore, it has been shown that the Caco-2 cell line possesses a vitamin D receptor similar to that present in the mammalian intestine, and the concentration of this receptor is highest in confluent differentiated cells (Giuliano and Wood, 1991b). Thus, the Caco-2 cell line is a potentially useful model for studying vitamin D-induced calcium transport.

Fleet and Wood (1999) demonstrated that the regulation of calcium transport in Caco-2 cells involves a transcellular pathway and is due to the specific regulation of cellular events by 1,25-$(OH)_2D_3$.

In contrast to the above, Blais *et al.* (1997), found most of the calcium crossing the cell monolayer to use the paracellular pathway, considering that mannitol transport was not significantly different from calcium transepithelial transport – at least for a calcium concentration of 1 mM in the medium. The authors indicated that Caco-2 cells cannot be used to study calcium transcellular transport because this calcium transport pathway represents only a small percentage of total calcium transepithelial transport.

12.4 Bioavailability of Calcium-Enriched Foods

Milk and dairy products are among the foods most widely used for calcium enrichment. Perales *et al.* (2006) evaluated calcium bioavailability by using Caco-2 cells in calcium-enriched milks compared to the equivalent nonenriched samples. Results showed that calcium fortification exerted a positive effect on calcium transport and uptake. However, calcium-retention percentages were similar in enriched and nonenriched samples. A significant linear correlation ($p < 0.05$) was obtained between calcium uptake and the amount of soluble calcium added to the cells, which could be explained because milk was fortified with calcium and vitamin D, and the latter increases calcium uptake by Caco-2 cells. The percentage calcium uptake and calcium content in milks also showed a significant linear correlation ($p < 0.05$).

Recently, calcium-enriched meat products (hamburger, bologna and dry fermented sausages) have been developed (Soto *et al.*, 2014). The absolute transport was not significantly different ($p < 0.05$) among the samples (independently of food matrix or type of calcium salt- lactate or citrate malate). However, the transport percentage was slightly higher in bologna sausages, though no significant differences were observed in relation to the fresh meat products and control milk, independently of the type of calcium salt added ($p < 0.05$). In contrast, in the dry fermented sausages, the percentage transport decreased significantly ($p < 0.05$) with both types of calcium salts. The type of calcium salt was not the limiting factor in calcium transport; nevertheless, the greater fat content of the dry fermented sausages could have interfered with calcium crossing of the cell membrane by forming indigestible calcium soaps with fatty acids. These products would not be well absorbed in the gastrointestinal tract, and calcium absorption therefore would be decreased (Soto *et al.*, 2014).

12.5 Influence of Food Components and Processing upon Calcium Bioavailability

12.5.1 Fiber and Phytic Acid

There is strong evidence of the adverse effects of dietary fiber upon mineral bioavailability. In differentiated Caco-2 cells, cellulose or barley fiber extract did not affect ^{45}Ca transport and uptake across the cell monolayers, but wheat fiber extract reduced both transport and uptake by about 20%. This effect has been attributed to the phytate in wheat fiber extract, since dephytinization eliminates the inhibitory action (Kennefick and Cashman, 2000).

Inositol hexaphosphate (IP6, phytate) significantly ($p < 0.05$) inhibited fractional apical-basolateral $^{45}CaCl_2$ transport in Caco-2 cells at levels of 100 $\mu mol\ L^{-1}$ and above. The fractional inhibition of calcium transport increased with the concentration of InsP6 (0–1 $mmol\ L^{-1}$), but the amount of calcium transported still increased with rising concentrations of calcium (0.5–10 $mmol\ L^{-1}$) (Phillippy, 2006).

Dephytinization of infant cereals reconstituted with water improved the percentage calcium uptake by about four-fold, and to a greater degree than the same infant cereals reconstituted with a follow-on formula. This shows the positive effect of eliminating phytic acid, and the influence of the matrix used for reconstituting dephytinized infant cereals upon calcium bioavailability (Frontela *et al.*, 2009a). In gluten-free infant cereals, Frontela *et al.* (2009b) found dephytinization to improve calcium uptake (%) efficiency (2.48) *versus* nondephytinized infant cereal (0.39). The lack of significant effects of dephytinization upon calcium bioavailability in most of the infant cereals analyzed could be explained by the binding of calcium phytate to the membrane of the Caco-2 cells and by the low calcium to phytate ratio (≤ 0.18) in the infant cereals. Only in gluten-free cereals (with the highest phytate to calcium ratio) was an increase in transport and uptake efficiency observed after dephytinization.

Caco-2 cell calcium uptake from biscuits was found to be affected by enrichment with mixtures of pseudocereals and legumes with inulin. Significant improvement in calcium uptake (21.8%) was achieved only by using inulin plus amaranth flour. The addition of soy flour, carob flour or oat fiber did not result in statistically significant differences in relation to reference biscuits (without inulin and pseudocereals and legumes). Apple fiber had a markedly strong negative impact upon calcium uptake (Vitali *et al.*, 2011).

12.5.2 Calcium–Iron Interaction

Calcium has been reported to influence iron transport across both the apical and basolateral membranes of enterocytes. In Caco-2 cells, Thompson *et al.* (2010) have demonstrated the negative effect of calcium as an inhibitor of nonheme iron absorption. Calcium significantly decreased iron-induced ferritin formation at calcium concentrations of 1.25 and 2.5 mM, and the calcium effect proved dose dependent, though only after 16–24 h of incubation. This effect could be attributed to calcium inhibitory action at the apical cell membrane, decreasing expression of the iron transporter DMT1 (divalent metal transporter 1) at the cell membrane. However, there was no effect on the levels of ferroportin (FPN) (another iron transporter) in the same cell membranes.

In the apical membrane of Caco-2 cells, calcium (100 μM) after 1.5 h did not affect iron uptake or DMT1 expression, though FPN serosal exporter abundance at the basolateral membrane was found to decrease. However, after 4 h of exposure to calcium, DMT1 and FPN expression increased and there was increased FPN at the membrane, suggesting a rebound effect. Thus, the effect of calcium on iron absorption may be of short duration, and adaptation may occur with time (Lönnerdal, 2010).

12.5.3 Maillard Reaction Products

Several studies have explored the influence of Maillard reaction products (MRPs) upon calcium availability and the intestinal absorption of calcium in Caco-2 cells (Delgado-Andrade *et al.*, 2005, 2006). There have been descriptions of the formation of calcium complexes with different melanoidins, which presented different calcium binding abilities that may be related to structural differences of the MRP. In this sense, the addition of a solution containing the gastrointestinal digest from a heated casein–glucose–fructose mixture (2.5 mM calcium concentration) to Caco-2 cells significantly increased calcium transport with respect to the same sample without heating and a solution containing calcium but without digestion products. However, the transport efficiency decreased significantly due to lower calcium solubility and greater calcium precipitation after *in vitro* digestion of the samples. Both findings have been related to the formation of CPPs during digestion and which may function as carriers of minerals (Seiquer *et al.*, 2001). Similar calcium transport across Caco-2 cells was obtained when cultures were exposed to a bioaccessible fraction of MRP-rich or MRP-low diets, with similar calcium availability. These findings are in line with the *in vivo* results, indicating that high MRP intake has no apparent effect upon dietary calcium bioavailability (Mesías *et al.*, 2009).

12.5.4 Peptides and Proteins

Casein phosphopeptides are known to improve the absorption of calcium by intestinal cells, and although the underlying mechanism remains unclear, they can form soluble complexes with calcium, enhancing paracellular absorption (Seiquer *et al.*, 2001). However, Consentino *et al.* (2010) suggested a second possible mechanism for the action of CPPs. Caco-2 cells treated with CPPs obtained from casein hydrolyzate and vitamin D_3, in the presence of an extracellular calcium concentration of 2.5 mM, showed an increased responsiveness to CPPs and uptake of calcium ions *versus* undifferentiated Caco-2 cells or Caco-2 cells differentiated without the addition of vitamin D_3. The authors suggested that a receptor such as calcium-sensing receptor (CaSR), or calcium channels such as TRPV6 and L-type channels, which regulate apical calcium absorption and are modulated by vitamin D_3 could explain the intervention of CPPs in calcium absorption.

Soybean phosphopeptides display a high level of calcium-binding activity compared with such peptides in the absence of phosphorylation. They prevent calcium phosphate precipitation and induce an increase in cytosolic calcium concentration in Caco-2 cells similar to free calcium ions due to an influx of calcium. These facts suggest a possible application for these peptides as a calcium carrier capable of preventing poor absorption of dietary calcium (Lee *et al.*, 2005).

The effect upon calcium uptake of different bovine milk proteins, as fortifiers of human milk, *versus* a commercial human milk fortifier has been evaluated using an *in vitro* digestion and Caco-2 cell culture model. The results indicated that calcium uptake was not different with any of them, except for unfortified human milk and human milk added with colostrums, which showed significantly lower cell uptakes. In human milk this can simply be explained by lower calcium levels in human milk. In human milk with colostrum, calcium formed a complex that was soluble and small enough to dialyze through the 15 kDa membrane. The high affinity of the complex for calcium could impair transfer of the mineral to the cells and to aggregations of different complexes at the brush border – thus entrapping the calcium and impairing the action of proteolytic enzymes with subsequent calcium release (Etcheverry *et al.*, 2004).

Recent studies have examined the possibility of the value-added utilization of eggshell proteins as a potential food ingredient or nutraceutical to increase calcium absorption. In this sense, whole eggshell proteins and one of the proteins isolated (molecular mass 21 kDa and N-terminal sequence

Met-Ala-Val-Pro-Gln-Thr-Met-Val-Gln), have shown an increase in calcium transport across Caco-2 cells monolayers (Daengprok *et al.*, 2003).

12.5.5 Processing

Cooking improves calcium uptake by Caco-2 cells from foods such as legumes. This may be due to decreases in the antinutritional factor contents responsible for a decrease in calcium bioavailability, secondary to losses in soaking and cooking water, which were discarded as is usual in cooking practices. The traditional cooking of legumes (soaked in tap water for 16 h (overnight) at room temperature, followed by draining of the soaking water, the addition of fresh tap water and cooking) significantly ($p < 0.05$) increased the uptake of calcium from white beans and lentils, and decreased the uptake from chick peas. Soaking has been documented to be an effective treatment for removing antinutritional factors that are eliminated on draining the soaking liquid. Traditional cooking methods reduced tannins, phytates, soluble oxalate and total phenol contents in white beans, chickpeas and lentils. Furthermore, traditional cooking produced a decrease in the inositol hexaphosphate (IP6) contents of over 30% in white beans and lentils, but of only 3.8% in chickpeas. The lesser reduction in oxalate and IP6 contents in chickpeas might explain the even negative effect of traditional cooking upon calcium uptake by Caco-2 cells *versus* white beans or lentils (Viadel *et al.*, 2006a).

In lentils, calcium retention, transport and uptake (retention plus transport) in Caco-2 cells increased significantly ($p < 0.05$) between two- and 6-fold when the legumes were cooked by an industrial processing *versus* traditional cooking. Soluble oxalate contents were higher in traditionally cooked lentils *versus* ready-to-eat lentils, which could at least partially explain the higher calcium retention, transport, and cell uptake values from ready-to-eat lentils, along with other possible effects of processing (Viadel *et al.*, 2006b).

In milk-based fruit beverages, high-pressure processing (HPP) (400 MPa s^{-1}, 36 °C for 5 min) *versus* conventional thermal treatment (TT) (90 °C for 30 s) did not modify calcium retention and uptake efficiency, and slightly decreased transport efficiency in Caco-2 cells – thus indicating that HPP can be used as an alternative to TT in the manufacture of functional foods (Cilla *et al.*, 2011).

In bakery products, dough fermentation resulted in a significant increase in calcium transport and uptake efficiency with respect to the baked samples. The effect of processing (fermentation and baking) upon the lowering of phytate content is similar, however, it did not show a clear relationship between phytate content and calcium availability of bakery products (Frontela *et al.*, 2011).

Table 12.1 summarizes several studies of calcium bioavailability from standard solution or resembling food systems. The effect of food components and processing upon calcium uptake and/or transport in Caco-2 cells are shown in Tables 12.2 and 12.3, respectively. Finally, Table 12.4 lists studies of Ca bioavailability from different food groups.

Table 12.1 Caco-2 cell assays with calcium standard solutions or resembling food systems. Different studies with calcium standard solution or simulated food systems evaluating the influence of components upon calcium uptake and transport (%), correlation to cell differentiation state, and the influence of vitamin D. Different Caco-2 cell models and culture and assays conditions are described.

Sample	Aim of study	*In vitro* method	Results	References
Inositol hexaphosphate (IP6)	To evaluate Ca transport in presence de IP6	Caco-2 cells: American type culture collection	Ca transport (%) depending on Ca concentration (0.5–10 mmol L^{-1}):	Phillippy, (2006)
		Cells: 25–50 passage number; 1.6×10^5 cells per well; bicameral chambers	Without IP6 (0.34–0.19)	
		Assay: 22–23 days after seeding	With IP6 (0.08–0.25)	
Semisynthetic diets with heated (C) or unheated (HC) casein–glucose–fructose mixture	To assess the influence of heating casein with glucose and fructose in Ca transport	Caco-2 cells: European collection of cell culture		Seiquer *et al.* (2001)
		Cells: 22–30 passage number; 100 000 cells per cm^2; bicameral chambers (Transwell®, 24 mm diameter, 4.7 cm^2 area, 3 µm pore size). TEER 500–650 Ω cm^{-2}	Ca transport (%): Ca standard solution (24.75), soluble fraction of C (39.33), soluble fraction of HC (45.59), soluble fraction of C : HC low Ca (44.13)	
		Assay: 21 days after seeding	Ca transport efficiency[a] (%): Soluble fraction of C (36.64), soluble fraction of HC (9.96), soluble fraction of C : HC low Ca (9.64)	

(*continued*)

Table 12.1 (*continued*)

Sample	Aim of study	*In vitro* method	Results	References
Commercial CPP derived of casein hydrolysate (CE90 CPP III)	To evaluate CPP effect on Ca uptake and if state of cell differentiation and vitamin D_3 is relevant	Caco-2 cells: Istituto Zooprofilattico Sperimentale di Brescia (BS TCL 87) *Cells*: 7–12 passage number (undifferentiated Caco-2) and 40–44 days (differentiated Caco-2). 7.5 × 10^4 cells per cm^2; glass coverslip (24 mm diameter, thickness 0.13–0.17 mm) in Petri dishes (35 mm) *Assay:* 48 h after seeding Addition of 1280 μM CPP and 2 mM Ca solution to undifferentiated cells with and without vitamin D_3 and differentiated cells Cytoplasmic calcium measurement by Furan-2 fluorescence assays and external standards of calcium	Responsive cells (% cells with intracellular Ca increments ≥ 20 nM): undifferentiated (1.5), undifferentiated with vitamin D_3 (37.3), differentiated (55.9) Mean single maximum intracellular Ca concentration rise[b] (nM): undifferentiated (86.7), undifferentiated with vitamin D_3 (124.7), differentiated (10 811.5) Mean total intracellular Ca concentration[c] (nM): undifferentiated (23.5), undifferentiated with vitamin D_3 (52.5), differentiated (1233.7)	Cosentino *et al.* (2010)
Soybean phosphopeptides	To demonstrate that soybean phosphopeptides improve Ca binding and uptake	Caco-2 cells: Korea line bank *Cells*: 20–40 passage number. 2.6 × 10^4 cells per cm^2; glass coverslip in Petri dishes (35 mm) *Assay*: cytoplasmic calcium measurement by Furan-2 fluorescence assays	Intracellular Ca concentration (nM): Blanc (~60), addition $CaCl_2$ (~120), addition soybean phosphopeptides (~120)	Lee *et al.* (2005)

[a](% soluble Ca – % transported Ca)/100.
[b]For each single cell by subtracting the baseline from the peak value after CPP administration and averaging all the analyzed cells.
[c]Percentage of responsive cells × mean single maximum intracellular Ca rise of cellular field constituted by 100 cells.

Table 12.2 Influence of different food components on calcium bioavailability in Caco-2 cells. Different studies on the influence of food components and processing corresponding to different foods upon calcium uptake or/and transport in Caco-2 cells. Different Caco-2 cell models and culture and assays conditions are described.

Sample	Aim of study	*In vitro* method	Results	References
Wheat bran and barley hull fiber extracts	Effect on Ca absorption of two cereal-based fiber extracts and role of associated phytate	Caco-2 cells: European collection *Cells*: 25–50 passage number; 2.5×10^5 cells per cm^2; bicameral chambers Transwell®; TEER > 200 Ω cm^{-2} *Assay*: 18 days after seeding (cells fully differentiated)	Ca uptake (nmol per well): cellulose (3.5), barley (3.8), wheat (2.5) Dephytinized: cellulose (3.4), barley (3.3), wheat (3.5) Ca transport rate (nmol per well per min): Cellulose (0.68), barley (0.62), wheat (0.55) Dephytinized: cellulose (0.65), barley (0.65), wheat (0.67)	Kennefick and Cashman (2000)
Infant cereals	Effect of dephytinization	Caco-2 cells: European collection *Cells*: passage number; 50 000 cells per cm^2; plastic flask 75 cm^2 *Assay*: 19–21 days after seeding (cells fully differentiated)	Ca uptake (%): Ca solution not detected Infant cereals: without phytase (3.32–6.50), with phytase (3.24–19.5)	Frontela *et al.* (2009a)
		Caco-2 cells: European collection *Cells*: passage number; 50 000 cells per cm^2; bicameral chambers Transwell® *Assay*: 21 days after seeding (cells fully differentiated)	Ca (%): Infant cereals without phytase: retention (0.51–2.80); transport (0.22–4.9); retention + transport (0.23–5.99) Infant cereals with phytase: retention (0.34–2.80); transport (0.25–3.24); retention + transport (0.43–3.31)	Frontela *et al.* (2009b)
Biscuits	Effect of addition pseudocereals/inulin mixtures	Caco-2 cells: American type culture collection *Cells*: 50–60 passage number; 35×10^4 cells per well; bicameral chambers Transwell®; TEER > 300 Ω cm^{-2}	Ca uptake (%): biscuits (36.57) Biscuits with inulin and: soy flour (30.71), amaranth flour (35.20), carob flour (26.39), oat fiber (38.95) or apple fiber (28.91)	Vitali *et al.* (2011)

(*continued*)

Table 12.2 (*continued*)

Sample	Aim of study	*In vitro* method	Results	References
Diets	Effect of high intake of Maillard reaction products (MRPs)	Caco-2 cells: European collection *Cells*: 30–35 passage number; 100 000 cells per cm^2; bicameral chambers (Transwell®, 24 mm diameter, 4.7 cm^2 area, 3 µm pore size). TEER 500–650 $\Omega\ cm^{-2}$	Ca transport (µg per well): Poor MRP diet (1.40), rich MRP diet (1.52)	Mesías *et al.* (2009)
		Assay: 21 days after seeding	Ca transport (%): Poor MRP diet (12.58), rich MRP diet (12.65) Ca transport efficiency[a] (%): Poor MRP diet (1.26), rich MRP diet (1.90)	
Human milk (HM)	Effect of different bovine protein *versus* a commercial fortifier	Caco-2 cells: American type culture collection *Cells*: 30–35 passage number; 50 000 cells per cm^2, collagen-treated 6-well plates with plastic inserts (dialysis membrane molecular weight cut off of 15 000 Da) *Assay*: 16 days after seeding	Ca uptake (% relative to HM with commercial fortifier): HM with commercial fortifier (100), HM (~10), HM fortifier with: colostrums (~65), α-lactalbumin (~90), casein phosphopeptides (~80–110), caseinate (~100), whey protein concentrate (~125)	Etcheverry *et al.* (2004)
Chicken eggshell	Effect of protein of chicken eggshell	Caco-2 cells: American type culture collection *Cells*: 25–60 passage number; 2×10^5 cells per mL, bicameral chambers (Transwell®, 12 mm diameter, 0.4 µm pore size). TEER 350–400 $\Omega\ cm^{-2}$	Ca uptake (µg per well): Control (3.75), soluble eggshell proteins (3.60), HPLC-isolated fractions (2.59–3.88)	Daengprok *et al.* (2003)
		Assay: 15–20 days after seeding	Ca transport (µg per well): Control (49.6), soluble eggshell proteins (81.5), HPLC-isolated fractions (51.2–83.1), identified protein from HPLC-fractions (molecular mass of 21 kDa) (91.2)	

[a](% soluble Ca × % transported Ca)/100.

Table 12.3 Effect of food processing in calcium bioavailability. Summary of the influence of processing and food matrixes upon calcium uptake or/and transport in Caco-2 cells. Different Caco-2 cell models and culture and assays conditions are described. The main results, expressed as calcium uptake and transport are summarized.

Sample	Aim of study	*In vitro* method	Results	Reference
Legumes	Effect of cooking and legume species	Caco-2 cells: European collection *Cells*: 54–60 passage number; 50 000 cells per cm^2; plastic flask 75 cm^2 *Assay*: 14–16 days after seeding (cells fully differentiated)	Ca uptake (%): raw legume, traditional cooking, ready-to-eat: white beans (3.57, 18.77, not detected), chickpeas (14.58, 2.76, not detected), lentils (2.45, 6.54, 7.58)	Viadel *et al.* (2006a)
		Caco-2 cells: European collection *Cells*: 50–60 passage number; 35 × 10^4 cells per well; bicameral chambers Transwell®; TEER > 250 Ω cm^{-2} *Assay*: 19–21 days after seeding (cells fully differentiated)	Ca (retention, transport, retention + transport nmol cm^{-2} h): Raw legumes: white beans (21.96, 11.53, 33.49); chickpeas (23.72, 18.02, 41.74) Lentils: raw (16.19, 6.95, 23.14), traditional cooking (10.97, 6.05, 17.02); ready-to-eat (24.53, 37.39, 61.92)	Viadel *et al.* (2006b)
Milk-based fruit beverages	Effect of processing and food matrix	Caco-2 cells: American type culture collection *Cells*: 36–46 passage number; 25 000 cells per cm^2; bicameral chambers Transwell®; TEER > 250 Ω cm^{-2} *Assay*: 10–12 days after seeding	Ca (%): Beverages: retention (5.86–9.74), transport (6.56–15.63), retention + transport (15.54–21.59) Thermally treated beverages: retention (1.44–3.91), transport (11.93–19.30), retention + transport (12.83–19.83) High-pressure processing: retention (3.91–6.27), transport (6.46–13.71), retention + transport (10.96–19.97)	Cilla *et al.* (2011)

Table 12.4 Calcium bioavailability in Caco-2 cells from different foods. Several studies have used Caco-2 cells to compare calcium bioavailability from similar food groups, or to evaluate the impact of calcium enrichment with different compounds upon calcium bioavailability.

Sample	Aim of study	*In vitro* method	Results	Reference
Mineral waters	To estimate Ca bioavailability	Caco-2 cells: American type culture collection *Cells*: 23–40 passage number; 3×10^5 cells per filter; bicameral chambers Transwell®; TEER > 250 Ω cm^{-2} *Assay*: 10–12 days after seeding	Ca fractional transport (%) between 1.65 and 6.72 Presence de anions such (HCO_3^- or SO_4^{2-}), interaction with Mg ions or modulation of the paracellular Ca-transport pathway could explain the differences in the Ca transport among the mineral waters	Ekmekcioglu *et al.* (1999)
Infant formulas (milk and soy based)	To compare Ca bioavailability	Caco-2 cells: European collection *Cells*: 53–55 passage number; 50×10^3 cells per cm^2; 75 cm^2 plastic flask *Assay*: 15–18 days after seeding (cells fully differentiated checked by sucrose-isomaltase and alkaline phosphatase activities)	Ca uptake (%): Milk-based: 5.9–7.2 Soy-based: 7–13.5	Jovaní *et al.* (2001)
Infant foods	To evaluate Ca bioavailability	Caco-2 cells: European collection *Cells*: 70–80 passage number; 35×10^4 cells per filter; bicameral chambers Transwell® *Assay*: 19–21 days after seeding (cells fully differentiated)	Ca (%) infant formulas: retention (0.1–1.2), transport (1.6–17.4), retention + transport (1.9–18.6) Fruit juices: retention (1.8–1.9), transport (16.9–20.7), retention + transport (18.7–22.5)	Perales *et al.* (2005)

Calcium-enriched and nonenriched milks	Effect of Ca enrichment on bioavailability	Caco-2 cells: European collection *Cells*: 70–90 passage number; 35×10^4 cells per filter; bicameral chambers Transwell® (24 mm diameter, 0.4 µm pore size). TEER > 250 Ω cm^{-2} *Assay*: 19–21 days after seeding (cells fully differentiated)	Ca (%): Nonenriched samples: retention (0.71–0.79), transport (6.06–6.76), retention + transport (6.82–7.44) Enriched samples: retention (0.72–0.89), transport (7.09–7.53), retention + transport (7.87–8.24)	Perales *et al.* (2006)
School meals (eight dishes)	To evaluate Ca retention and transport	Caco-2 cells: European collection *Cells*: 50–60 passage number; 35×10^4 cells per filter; bicameral chambers Transwell® (24 mm diameter, 0.4 µm pore size). TEER > 250 Ω cm^{-2} *Assay*: 21 days after seeding	Ca (%): Retention: Macaroni with tuna (40.6) – fried hake (1.1) Transport: lentils (36.9)-chicken in breadcrumbs (not detectable) Retention + transport: macaroni with tuna (56.3) – fried hake (3.3)	Cámara *et al.* (2007)
Calcium-enriched meat products (hamburgers, bologna and dry fermented sausages)	To study Ca bioavailability	Caco-2 cells: CacoReady™ kit *Cells*: 21 day cell barrier integrated in 24 HTS Transwell plates. TEER > 1000 Ω cm^{-2}	Ca transport (calcium citrate malate and lactate, %): bologna sausage (12.71 and 11.35) > hamburger (9.85 and 9.98) > dry fermented sausage (7 and 6.75)	Soto *et al.* (2014)

Summary Points

- This chapter focuses on calcium bioavailability and Caco-2 cells.
- Calcium bioavailability is the amount of calcium available for absorption and utilization by the organism.
- Caco-2 cells constitute a cell culture validated as a human intestinal model.
- Intestinal mechanisms for calcium absorption comprise a saturable transcellular mechanism controlled by the action of vitamin D, and a nonsaturable paracellular pathway.
- Dietary fiber exerts an inhibitory effect upon calcium bioavailability.
- Dephytinization of cereal products improves calcium bioavailability.
- Calcium bioavailability is not affected by the presence of Maillard reaction products.
- Casein phosphopeptides improve calcium absorption, though the underlying mechanisms remain unclear.
- Soaking and cooking of legumes improves calcium uptake by Caco-2 cells by facilitating the removal of antinutritional factors.

Key facts

Caco-2 Cells as a Model to Study Calcium Bioavailability

1. Cell cultures are a rapid, easy and economical alternative and complementary method for studying calcium bioavailability compared with *in vivo* studies in animals or humans.
2. Caco-2 cells are the most widely used cell culture, because they exhibit morphological and functional characteristics of human enterocytes.
3. Caco-2 cells are grown in monolayers and therefore can be grown in flasks to estimate calcium uptake, or in filters or dialysis membranes for studying calcium transport between two compartments (apical and basal).
4. In studies of calcium bioavailability with Caco-2 cells, absorption mechanisms from calcium standard solutions or food resembling systems, effects of food components, and processing have been evaluated.
5. This *in vitro* model has also been used to evaluate calcium enrichment with different compounds and to compare calcium bioavailability from similar or different food groups.

Definitions of Words and Terms

Bioavailability. Proportion of a nutrient that is released from the food matrix and can be absorbed and utilized by the organism.
Casein phosphopeptides (CPPs). Phosphorylated peptides obtained from milk casein through tryptic digestion or gastrointestinal digestion.

Cell culture. Growth of cells in a favorable artificial environment composed of nutrient solutions, a suitable surface to support their growth, and ideal conditions of temperature, humidity, and gaseous atmosphere.

Dietary fiber. Structural residues from vegetable cells that resist human enzymatic hydrolysis in the digestive tract.

Enriched or fortified foods. Foods with one or more nutrients added, in order to prevent or correct a deficiency in the population or in specific population groups.

Phytates. Mineral complexes of phytic acid that are the major storage form of phosphorus and are known as one of the principal agents limiting mineral bioavailability.

Maillard reaction products (MRPs). Compounds resulting from the condensation of free amino groups with carbonyl groups during thermal processing or storage conditions.

Retention or uptake. Mineral content in the cell monolayer.

Soluble fraction or digest. Fraction obtained by centrifugation after simulated gastrointestinal digestion.

Transport. Fraction of mineral (calcium) determined in the basal chamber of a bicameral cell culture system.

List of Abbreviations

CaSR	Calcium-sensing receptor
CPPs	Casein phosphopeptides
DMT	Divalent metal transporter
FBS	Fetal bovine serum
FPN	Ferroportin
HPP	High-pressure processing
IP6	Inositol hexaphosphate
MRPs	Maillard reaction products
TEER	Transepithelial electrical resistance
TT	Thermal treatment
WPC	Whey protein concentrate

References

Blais, A., Aymard, P. and Lacour, B., 1997. Paracellular calcium transport across Caco-2 and HT-29 cell monolayers. *Pflügers Archiv-European Journal of Physiology*. 434: 300–305.

Cámara-Martos, F. and Amaro-López, M. A., 2002. Influence of dietary factors on calcium bioavailability. A brief review. *Biological Trace Element Research*. 89: 43–52.

Cámara, F., Barberá, R., Amaro, M. A. and Farré, R., 2007. Calcium, iron, zinc and copper transport and uptake by Caco-2 cells in school meals: influence of protein and mineral interactions. *Food Chemistry*. 100: 1085–1092.

Cilla, A., Lagarda, M. J., Alegría, A., de Ancos, B., Cano, M. P., Sánchez-Moreno, C., Plaza, L. and Barberá, R., 2011. Effect of processing and food matrix on calcium and phosphorus bioavailability from milk-based fruit beverages in Caco-2 cells. *Food Research International*. 44: 3030–3038.

Cosentino, S., Gravaghia, C., Donettic, E., Donidaa, B. M., Lombardi, G., Bedonic, M., Fiorillia, A., Tettamantia, G. and Ferraretto, A., 2010. Casein phosphopeptide-induced calcium uptake in human intestinal cell lines HT-29 and Caco2 is correlated to cellular differentiation. *Journal of Nutritional Biochemistry*. 21: 247–254.

Daengprok, W., Garnjanagoonchorn, W., Naivkul, O., Pornsinlpati, P., Issigonis, K. and Mine, Y., 2003. Chicken eggshell matrix proteins enhance calcium transport in the human intestinal epithelial cells, Caco-2. *Journal of Agricultural and Food Chemistry*. 51: 6056–6061.

Delgado-Andrade, C., Seiquer, I. and Navarro, M. P., 2005. Comparative effects of glucose-lysine *versus* glucose-methionine Maillard reaction products consumption: *in vitro* and *in vivo* calcium availability. *Molecular Nutrition & Food Research*. 49: 679–684.

Delgado-Andrade, C., Seiquer, I. and Navarro, M. P., 2006. Changes in calcium absorption and subsequent tissue distribution induced by Maillard reaction products: *in vitro* and *in vivo* assays. *Journal of the Science of Food and Agriculture*. 86: 271–278.

Ekmekcioglu, C., 2002. A physiological approach for preparing and conducting intestinal bioavailability studies using experimental systems. *Food Chemistry*. 76: 225–230.

Ekmekcioglu, C., Pomazal, K., Steffan, I., Schweiger, B. and Marktl, W., 1999. Calcium transport from mineral waters across Caco-2 cells. *Journal of Agricultural and Food Chemistry*. 47: 2594–2599.

Etcheverry, P., Grusak, M. A. and Fleige, L. E., 2012. Application of in vitro bioaccessibility and bioavailability methods for calcium, carotenoids, folate, iron, magnesium, polyphenols, zinc, and vitamins B_6, B_{12}, D and E. *Frontiers in Physiology*. 3: 1–22. DOI: 10.3389/fphys.2012.00317.

Etcheverry, P., Wallingford, J. C., Miller, D. D. and Glahn, R. P., 2004. Calcium, zinc, and iron bioavailability from a commercial human milk fortifier: a comparison study. *Journal of Dairy Science*. 87: 3629–3637.

Fleet, J. C. and Wood, R. J., 1999. Specific $1,25(OH)_2D_3$-mediated regulation of transcellular calcium transport in Caco-2 cells. *American Journal of Physiology*. 276: G958–G964.

Frontela, C., Ros, G. and Martínez, C., 2009a. Iron and calcium availability from digestion of infant cereals by Caco-2 cells. *European Food research and Technology*. 228: 789–797.

Frontela, C., Ros, G. and Martínez, C., 2011. Phytic acid content and "*in vitro*" iron, calcium and zinc bioavailability in bakery products: the effect of processing. *Journal of Cereal Science*. 54: 173–179.

Frontela, C., Scarino, M. L., Ferruzza, S., Ros, G. and Martínez, C., 2009b. Effect of dephytinization on bioavailability of iron, calcium and zinc from infant cereals assessed in the Caco-2 cell model. *World Journal of Gastroenterology*. 15: 1977–1984.

Giuliano, A. R., Franceschi, R. T. and Wood, R. J., 1991a. Characterization of the vitamin D receptor from the Caco-2 human colon carcinoma cell line. Effect of cellular differentiation. *Archives of Biochemistry and Biophysics*. 285: 261–269.

Giuliano, A. R. and Wood, R. J., 1991b. Vitamin-D regulated calcium transport in Caco-2 cells: unique in vitro model. *American Journal of Physiology*. 260: G207–G212.

Guéguen, L. and Pointillart, A., 2000. The bioavailability of dietary calcium. *Journal of the American College of Nutrition*. 19: 119S–136S.

Hidalgo, I. J., Raub, T. and Borchardt, R. T., 1989. Characterization of the human colon carcinoma cell line (Caco-2) as a model system for intestinal epithelial permeability. *Gastroenterology*. 96: 736–749.

Jovaní, M., Barberá, R., Farré, R. and Martin de Aguilera, E., 2001. Calcium, iron and zinc uptake from digests of infant formulas by Caco-2 cells. *Journal of Agricultural and Food Chemistry*. 49: 3480–3485.

Kennefick, S. and Cashman, K. D., 2000. Inhibitory effect of wheat fibre extract on calcium absorption in Caco-2 cells: evidence for a role of associated phytate rather than fibre *per se*. *European Journal of Nutrition*. 39: 12–17.

Langerholc, T., Maragkoudakis, P. A., Wollgast, J., Gradisnik, L. and Cencic, A., 2011. Novel and established intestinal cell line models – an indispensable tool in food science and nutrition. *Trends in Food Sciences and Technology*. 22: S11–S20.

Lee, S. H., Uang, J. I., Hong, S. M., Hahm, D. H., Lee, S. Y., Kim, I. H. and Choi, S. Y., 2005. Phosphorylation of peptides derived from isolated soybean protein: effects on calcium binding, solubility and influx into Caco-2 cells. *Biofactors*. 23: 121–128.

Lönnerdal, B., 2010. Calcium and iron absorption-mechanisms and public health relevance. *International Journal for Vitamin and Nutrition Research*. 80: 293–299.

Mesías, M., Siquer, I. and Navarro, M. P., 2009. Influence of diets rich in Maillard reaction products on calcium bioavailability. Assays in male adolescents and in Caco-2 cells. *Journal of Agricultural and Food Chemistry*. 57: 9532–9538.

Miller, G. D., Jarvis, J. K. and McBean, L. D., 2001. Review: the importance of meeting calcium needs with foods. *Journal of the American College of Nutrition*. 20: 168–185.

Perales, S., Barberá, R., Lagarda, M. J. and Farré, R., 2005. Bioavailability of calcium from milk-based formulas and fruit juices containing milk and cereals estimated buy in vitro methods (solubility, dialyzability and uptake and transport by Caco-2 cells). *Journal of Agricultural and Food Chemistry*. 53: 3721–3726.

Perales, S., Barberá, R., Lagarda, M. J. and Farré, R., 2006. Fortification of milk with calcium: effect on calcium bioavailability and interactions with iron and zinc. *Journal of Agricultural and Food Chemistry*. 54: 4901–4906.

Phillippy, B. Q., 2006. Transport of calcium across caco-2 cells in the presence of inositol hexakisphosphate. *Nutrition Research*. 26: 146–149.

Pinto, M., Robine-Leon, S., Appay, M. D., Kedinger, M., Triadou, N., Dussaulx, E., Lacroix, B., Simon-Assmann, P., Haffen, K., Fogh, J. and Zweibaum, A., 1983. Enterocyte-like differentiation and polarization of the human colon carcinoma cell line Caco-2 in culture. *Biology of the Cell*. 47: 323–330.

Rafferty, K., Walters, G. and Heaney, R. P., 2007. Calcium fortificants: overview and strategies for improving calcium nutriture of the US population. *Journal of Food Science*. 72: 152–158.

Seiquer, I., Aspe, T., Vaquero, P. and Navarro, P., 2001. Effects of heat treatment of casein in the presence of reducing sugars on calcium bioavailability: *in vitro* and *in vivo* assays. *Journal of Agricultural and Food Chemistry*. 49: 1049–1055.

Soto, A. M., Morales, P., Haza, A. I., García, M. L. and Selgas, M. D., 2014. Bioavailability of calcium from enriched meat products using Caco-2 cells. *Food Research International*. 55: 263–270.

Thompson, B. A. V., Sharp, P. A., Elliot, R. and Fairweather-Tait, S. J., 2010. Inhibitory effect of calcium on non-heme iron absorption may be related to translocation of DMT-1 at the apical membrane of enterocytes. *Journal of Agricultural and Food Chemistry*. 58: 8414–8417.

Viadel, B., Barberá, R. and Farré, R., 2006a. Effect of cooking and legumes species upon calcium, iron and zinc uptake by Caco-2 cells. *Journal of Trace Elements in Medicine and Biology*. 20: 115–120.

Viadel, B., Barberá, R. and Farré, R., 2006b. Uptake and retention of calcium, iron and zinc from raw legumes and the effect of cooking on lentils in Caco-2 cells. *Nutrition Research*. 26: 591–596.

Vitali, D., Radic, M., Cetina-Cizmek, B. and Dragojevic, I. V., 2011. Caco-2 cell uptake of Ca, Mg and Fe from biscuits as affected by enrichment with pseudocereal/inulin mixtures. *Acta Alimentaria*. 40: 480–489.

WHO/FAO, 2006. Guidelines on food fortification with micronutrients. Available at: http://www.who.int/nutrition/publications/micronutrients/guide_food_fortification_micronutrients.pdf. Accessed 3 December 2013.

Section IV
Function and Effects

CHAPTER 13

Adolescents and Dietary Calcium

MARTA MESÍAS*[a], ISABEL SEIQUER[a], AND
M. PILAR NAVARRO[a]

[a]Instituto de Nutrición Animal, Estación Experimental del Zaidín, Consejo Superior de Investigaciones Científicas (CSIC), Camino del Jueves, 18100 Armilla, Granada, Spain
*E-mail: mmesias@ictan.csic.es

13.1 The Role of Calcium in Adolescence

Calcium is an essential nutrient for skeletal mineralization since about 99% of its body reserves are deposited in bone and teeth. In addition to the structural function, calcium is involved in many metabolic and cellular functions and, consequently, this element is essential for optimal growth and development. Adolescence is characterized by an accelerated growth rate associated with a rapid muscular, skeletal, and sexual development. Adolescents acquire 15–25% of their adult size during this period (Bailey *et al.*, 2000), which involves huge demands of nutrients, particularly calcium, so that an adequate intake of this mineral is fundamental during this period.

In the body, two phenomena are produced simultaneously, the synthesis of new bone and the modeling–remodeling of previously synthesized bone. The bone acts as a reservoir of calcium maintaining extracellular homeostasis and transferring the mineral when blood concentration decreases from the normal values (9.0–10.2 mg dL^{-1}). As a consequence, a dynamic balance

Food and Nutritional Components in Focus No. 10
Calcium: Chemistry, Analysis, Function and Effects
Edited by Victor R. Preedy

Published by the Royal Society of Chemistry, www.rsc.org

exists between calcium in the extracellular medium and that found in bone, so that about 500 mg of this mineral enter and leave daily from the bones. In chronic calcium-deficiency situations derived from a continual inadequate intake or poor intestinal absorption, circulating calcium concentration is maintained at the expense of skeletal mass. Therefore, mineral deficiency leads to inadequate mineralization of bone matrix, resulting in rickets in children and adolescents and, together with other risk factors, contributing to possible osteoporosis in adulthood.

During the adolescence, approximately 45–50% of total adult skeletal mass is completed, reaching around 90% at the age of 17 years. Therefore, adequate calcium intake during growth is extremely important to reach the optimum peak bone mass, which protects against osteoporosis in the adult age. Moreover, the adolescent growth spurt, which corresponds to the peak accrual rate of bone mass, is related with the risk of fractures, and high calcium intakes have been associated with a protective effect. Peak height-gain velocity for girls starts at the age of 10 years (6 cm per year) and reaches a maximum by the age of 12 years (9 cm per year). For boys, by the age of 12 the mean height-gain velocity is 5 cm per year and increases to an average peak of 10 cm per year by the age of 14. Mean height gain velocity is close to zero at the age of 15 and 17 in girls and boys, respectively (Matkovic *et al.*, 2004).

Due to the high proportion of body calcium present in bone and the importance of bone as the major calcium reservoir, the development and maintenance of bone are the major determinants of this mineral need. In the body, both bone formation and resorption are continuous processes and, specifically in adolescents, bone formation predominates over bone resorption. For that reason, adolescents' diets should be balanced and adjusted to their requirements in order to provide suitable amounts of energy and nutrients to meet calcium recommendations and to promote mineral utilization. In this sense it has been reported that bone health is highly conditioned by the diet, both concerning the different food groups and the whole diet (Seiquer *et al.*, 2008).

13.2 Recommended Calcium Intakes

Calcium intake during adolescence has a direct relationship with bone mineralization and a positive calcium balance during this stage is mandatory for maximum peak bone mass to be achieved. Usually calcium intake and balance are positively correlated until a maximum value considered as threshold. In this respect, balance studies performed in adolescents have shown an increase in calcium retention in parallel with increased calcium intake, but only up to an intake of 1300 mg per day (Jackman *et al.*, 1997). This fact suggests that growth in humans may be limited by calcium intake until a plateau is reached, so that a suboptimal intake limits the body's ability to store calcium as bone tissue and, on the contrary, calcium intake higher than the requirements does not increases calcium storage in bones. Due to the high calcium needs as a consequence of the growth, low intake during puberty may limit bone mineralization, thus, if calcium intake is below 500 mg per day in childhood, more than 50% of ingested calcium must be retained to obtain adequate mineral accretion (Mølgaard *et al.*, 1999).

During childhood and adolescence an adequate calcium intake is mandatory, but a high percentage of American and European adolescents fail to meet the recommendations. Table 13.1 shows the average calcium intake (mg per day) in adolescents from different countries, according

Table 13.1 This table shows a revision of the average calcium intake in adolescents from different countries, as revised by Mesías *et al.* (2011).[a]

	Ca (mg per day)				
Country	Male	Female	Dietary methodology	Sample size	Age (years)
Austria	819	667	7 days WR (10–14 years)	n.a. (10–14 years)	10–12
	733	688	3 days WR (15–18 years)	102 (15–18 years)	13–14
	675	718			15–18
Bolivia	862	750	24 h recall	45	16–19
Brazil	635	576	3 days record + FFQ	121	10–18
Canada	1873/1159	—	FFQ/24 h recall	55	14–16
Costa Rica	619	523	3 days record	275	13–18
Denmark	1239	1144	7 days record	105	10
England	786	753	2 × 3 days record	379	11–12
Estonia	855	648	48 h recall + FFQ	341	12
	825	592			15
Finland	—	1267	FFQ + 3 days record	198	11–15
France	1108	972	DH	94	10
	1278	1079			14
	1322	1065			16
Germany	726	667	HBS	3565	10–12
	835	746			13–14
	920	813			15–18
Greece	905	771	24 h recall + 3 days WR	471	14–16
Hungary	612	596	3 × 24 h record	n.a.	12–13
	777	667			14–15
Ireland	1030	830	DH	1015	12
	1180	790			15
Italy	773	638	3 × 4 days record + FFQ	233	15–19
Netherlands	—	1134	FFQ + 3 days record	184	11–15
Norway	933	785	4 days record (13 years)	1009 (13 years)	13
	1400	1000	FFQ (16–19 years)	n.a. (16–19 years)	16–19
Panama	—	440	24 h recall + FFQ	180	12–17
Poland	—	831	FFQ + 3 days record	200	11–15
Portugal	—	1224	FFQ	254	15–17
Scotland	1010	723	4 days record	218	16–17
Spain	959	835	2 × 24 h recall	203	10–17
USA	1247	1014	FFQ	1521	15–16

[a]WR: weighed food record; n.a.: not available; FFQ: food-frequency questionnaire; DH: diet history; HBS: Household budget survey.

Table 13.2 This table shows the recommended calcium intakes for European and American adolescents according to the European and American recommendations.

Calcium intake recommendations (mg per day)					
Europe (Scientific Committee for Food, 1993)			USA (National Research Council, 2011)		
Age	Boys	Girls	Age	Boys	Girls
11–14	1000	800	9–13	1300	1300
15–17	1000	800	14–18	1300	1300
18+	700	700	19+	1000	1000

to age and sex, according to bibliography revised by Mesías *et al.* (2011). Among boys, countries such as Costa Rica, Brazil or Hungary showed the lowest levels of calcium intakes, whereas male adolescents from Canada, Norway or USA had the highest consumption of this mineral. In girls, the lowest intakes were found in countries like Brazil, Costa Rica or Panama, while the highest ones were recorded in Denmark, Finland or Portugal. In general, the daily calcium intake of boys was often higher than that of the girls. As can be observed, some European and American adolescents are failing to meet the current recommendations (Table 13.2); some of them meet European (Scientific Committee for Food, 1993) but not American recommendations (National Research Council, 2011) and only a few groups reach 1300 mg per day. Therefore, these data confirm that calcium consumption in the adolescent diet frequently fails to meet the body's needs during the growth spurt.

13.3 Main Food Sources of the Daily Calcium Intake

Milk and dairy products are the major source of dietary calcium for children and adolescents and contribute from 42% to 70% of the total calcium consumed by adolescents from different countries (Mesías, 2007, Figure 13.1). Therefore, incorporation of these products into the diet of adolescents may be the best strategy to meet calcium recommendations and to achieve optimal bone mineralization. In this regard, Kim *et al.* (2013) have recently reported that the intake of milk and milk products in adolescents, particularly in girls, can improve the bone-mineral density. Among dairy products, fluid milk is the most significant calcium contributor, followed by cheese, ice cream, and yoghurt. Sometimes, cheese may contribute to calcium intake in a similar or even higher proportion than cereals (Mesías, 2007). In addition, other foods such as chocolate or pizzas, which have cheese or milk incorporated, may also be important contributors to dietary calcium among adolescents.

Over the last years milk intake has been decreased as a consequence of the increased consumption of carbonated beverages among boys and girls. Soft

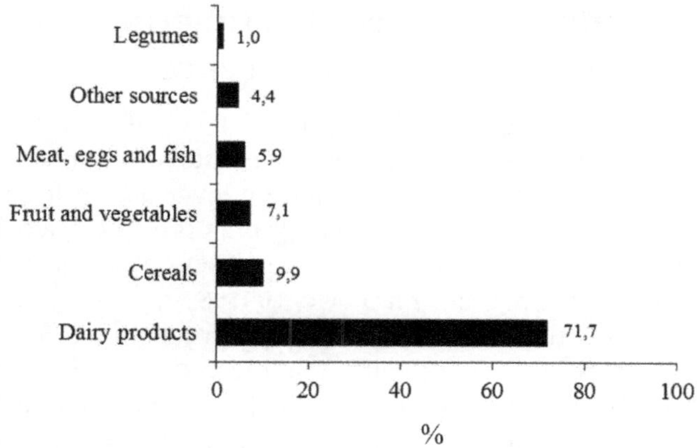

Figure 13.1 This represents the percentage of contribution of the different sources of food to the daily calcium intake among Spanish male adolescents aged 11–14 years consuming their habitual diets. Data are based on results reported by Mesías (2007).

drinks may negatively affect bone mineralization because, on the one hand, they can replace milk intake thus decreasing milk consumption and, on the other hand, because of their phosphorus and caffeine content. It is known that the phosphoric acid content of soft drinks may limit calcium absorption and contribute to bone loss, and caffeine has been associated with reduced bone-mineral density and increased fracture risk. Moreover, overall calcium intake has decreased because of the current dietary habits of adolescents. These habits are linked with a high consumption of snacks and fast foods, as well as with a great preference for eating more meals and snacks at restaurants and fast-food establishments, where they consume fewer fruit and dairy products in comparison with their home consumption.

An important part of calcium in the diet may also be provided by cereals; legumes, in spite of their relative high calcium content, contribute little to dietary calcium due to their low consumption among young people (Mesías, 2007). On the other hand, meat and eggs are minority calcium sources (Seiquer *et al.*, 2008). In the case of fish, although its calcium content is higher than in meat and eggs, the low consumption among adolescents usually leads to a scarce contribution to total dietary calcium. It has been reported that the initial bone status adjusted by height and weight in 10 year-old children is positively associated with the intake of small fish and dairy products, and negatively with the preference by boys for meat (Hirota *et al.*, 2005). Therefore, in order to favor calcium utilization, more fish and less meat consumption should be recommended during adolescence. Nutrients found in abundance in fruit and vegetables may be protective for bone health, as discussed below. Moreover, drinking water, including mineral water, may provide 6–7% of daily calcium intake (Guéguen and Pointillart, 2000).

13.4 Determinant Factors Affecting the Calcium Absorption during Adolescence

Puberty is associated with a high rate of dietary calcium absorption in order to satisfy the increased calcium requirements for intensive adolescent growth. To reach this aim, adolescents have a low calcium fecal excretion, which allows an increase in mineral absorption. Studies carried out in adolescents with calcium intakes of 1000–1100 mg per day report a fecal excretion rate of 600–700 mg per day (Matkovic, 1991; Seiquer *et al.*, 2008), values which are lower than those showed for an adult population (760 mg per day) (Guéguen and Pointillart, 2000). During adolescence there is not only an increase in the absolute value of absorbed calcium, but the effectiveness of the process is also enhanced, thus increasing the fractional absorption rate or digestibility (Peacock, 1991).

13.4.1 Hormonal Factors

Modifications in calcium absorption and metabolism in adolescents are associated with the maturation of the hypothalamo–pituitary–gonadal axis, which includes the gonadotropin-releasing hormone (GnRH). At a preprogrammed time in a child's life, there is an increase in the amplitude of GnRH pulses that triggers a cascade of events including increases in the amplitude of follicle-stimulating hormone (FSH) and luteinizing hormone (LH) pulses, followed by a marked elevation in gonadal sex steroidal output, which in turn increases growth hormone (GH) in the hypophysis and insulin-like growth factor-1 (IGF-1) production in different tissues (Mauras *et al.*, 1996). Moreover, sex steroids can act centrally by regulating GH secretion and peripherally modulating GH responsiveness (Figure 13.2). Testosterone stimulates GH secretion, but it seems to do so after being aromatized to estradiol. Many of these hormones have an influence on calcium absorption: GH enhances intestinal calcium absorption, increasing 1,25-dihydroxyvitamin D production by stimulating renal 1-α-hydroxylase (Saggese *et al.*, 2002). In addition, GH and IGF-1 stimulate sex steroids secretion. Both estrogens and androgens participate in the regulation of calcium fluxes and bone-calcium deposition, increasing calcium absorption and retention. Many of these hormones may favor calcium absorption across 1,25-dihydroxyvitamin D, the principal enhancer of this mineral absorption at any stage of life but especially in the pediatric period. According to several authors, during puberty the efficacy of conversion of 25-hydroxyvitamin D to 1,25-dihydroxyvitamin D increases to meet the needs for skeletal growth (Saggese *et al.*, 2002).

As a consequence of the hormonal changes produced in this period, an increase in total calcium absorption is found among pubertal (Tanner stages 2–4) but not prepubertal (Tanner stage 1) adolescents (Abrams *et al.*, 1997).

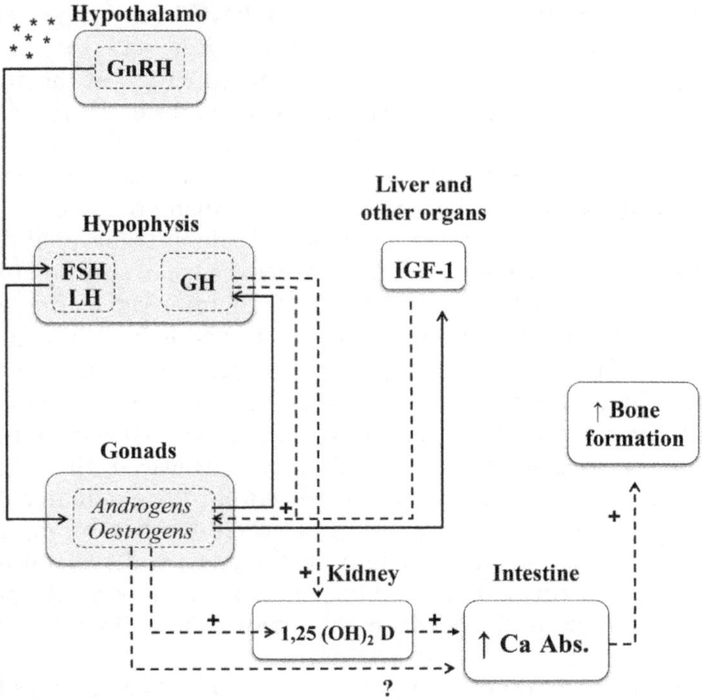

Figure 13.2 This represents the different hormonal changes during adolescence which are determinants of calcium absorption. GnRH: Gonadotropin-releasing hormone. FSH: follicle-stimulating hormone. LH: luteinizing hormone. GH: growth hormone. IGF-1: insulin-like growth factor-1. Abs.: absorption. ***: Pulses; ——: Hormonal changes; ------: Effects of calcium absorption.

13.4.2 Dietary Factors

Since the calcium absorption efficacy increases during periods of high physiological requirements, it is important that diets provide enough mineral that allows the maximum bone density to be achieved. Calcium absorption values of 324 mg per day (Seiquer *et al.*, 2008), 355 mg per day (Abrams *et al.*, 1997) and 425 mg per day (Matkovic, 1991) have been reported in several studies with adolescents. In addition, a positive relationship between calcium intake and absorption has been observed when mineral consumption of the subjects ranged from 750 to 1400 mg Ca per day (Abrams *et al.*, 1997). It is important to highlight that the main factors affecting absorption efficiency are body needs, which are determined at this stage of development by the growth speed (Guéguen and Pointillart, 2000). Nevertheless, diet may content positive factors that improve calcium availability.

The availability of calcium for bone development and maintenance may be conditioned by the amount of calcium in the diet but also by food ingredients. As some components of diet have been suggested to be enhancers or inhibitors of calcium absorption, it is necessary to identify food components and/or functional food ingredients that maximize calcium bioavailability from foods. Thus, phytates found in bran and most cereals and seeds, oxalates in spinach or walnuts, and tannins in tea, can form insoluble complexes with calcium, thereby reducing its absorption. On the contrary, milk and dairy products are considered good sources of calcium due to their high mineral content and calcium bioavailability. Milk nutrients may promote bone mineralization because in addition of being a major source of calcium, milk provides phosphates, magnesium, proteins, and as yet unidentified nutrients likely to favor bone health. Esterle *et al.* (2009) observed that milk consumption increased the calcium balance and promoted mineral accretion on the lumbar spine in female adolescents. These authors suggested that IGF-1 is likely to be involved, as IGF-1 levels were positively associated with milk intake. However, the mechanism by which milk may increase serum IGF-1 is not fully elucidated. Vitamin D is also an essential factor for intestinal calcium absorption and plays a central role in maintaining calcium homeostasis and skeletal integrity. Adequate levels of this vitamin are obtained by suitable intake and sufficient exposure to sunlight, which is the major source of vitamin D in the organism.

Fatty acid composition of the dietary fat may also affect calcium bioavailability; it has been shown that a high ratio unsaturated/saturated fatty acids has beneficial effects on calcium absorption. As commented, fish may be a good source of calcium because its protein can promote calcium absorption, but also because omega-3 fatty acid might promote calcium transport. Moreover, the positive effect of fish fat in calcium utilization is promoted when it is consumed together with olive oil (Pérez-Granados *et al.*, 2000), as it commonly occurs in the Mediterranean diet. Olive oil may be another contributor to enhance calcium utilization, since oleuropein, an olive oil phenolic compound, reduces bone loss (Puel *et al.*, 2004). On the other hand, studies in humans (Van den Heuvel *et al.*, 1999) have revealed a positive effect of apparent calcium absorption after consumption of oligofructose, which may also diminish the negative effects of phytic acid. A diet rich in cereals, fruits, and vegetables can increase the presence of these prebiotic products in the digestive system, thus improving calcium absorption when demands for the mineral are high.

Therefore, a balanced diet, adjusted to the requirements of adolescents, would provide a stimulus for growth associated with greater absorption of calcium. In this respect, it has been demonstrated that the consumption of a varied and balanced diet by male adolescents aged 11–14 years may increase calcium absorption, from 325 g per day when subjects are on their usual diet to 463 g per day if they consume a diet based on Mediterranean patterns, without significantly varying the total calcium intake (Seiquer *et al.*, 2008).

13.5 Determinant Factors Affecting Calcium Retention during Adolescence

It has been reported that the most important determinants of calcium retention in the body are dietary calcium intake and urinary calcium excretion. In adults urinary calcium depends on the intake, however, in adolescents, calcium excretion is independent of the consumption, representing obligatory renal losses since its use is a matter of priority for the organism. During childhood, urinary calcium excretion increases from ~40 mg per day in young children to ~80 mg per day just before puberty. Nevertheless, during the peak of maximum growth, this value decreases, especially among males. Calcium excretion rises to reach the values of adulthood (100–250 mg per day) by the end of adolescence, which probably reflects the decreasing needs of the skeleton for calcium (Peacock, 1991). When calciuria is related to body weight (mg kg^{-1} per day) the reduced levels of urinary calcium observed during puberty are more pronounced, showing minimum levels at the age of the maximum growth speed (13–15 years in boys and 11–14 years in girls) (Manz *et al.*, 1999) (Figure 13.3). Several studies carried out with adolescents show urinary calcium values of around 90 and 67 mg per day for 9–14 year-old girls and boys, respectively (Abrams *et al.*, 1997), 56 mg per day in 11 year-old girls (Manz *et al.*, 1999), and 100 mg per day for 11–14 year-old boys (Seiquer *et al.*, 2008). In relation with body weight, values around 2 mg kg^{-1} per day have been observed in male adolescents (Manz *et al.*, 1999; Mesías, 2007), while excretion rates of up to 4 mg kg^{-1} per day are associated with hypercalciuria (Manz *et al.*, 1999).

Urinary losses of calcium may be affected not only by the physiological status but also by certain dietary components. Thus, excessive protein intake, particularly animal protein, generally leads to an increase in calcium urinary losses, although the opposite occurs when this fact has been studied in

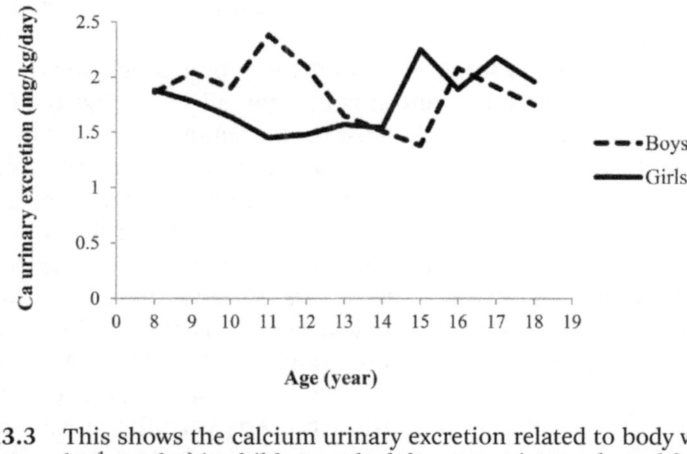

Figure 13.3 This shows the calcium urinary excretion related to body weight (mg kg^{-1} per day) in children and adolescents. Figure adapted from Mesías *et al.* (2011).

adolescents. Independent factors can be related to urinary mineral excretion, but total urinary excretion is determined by the metabolic effect of the overall diet. Nutritional intervention studies have shown that a high intake of fruits and vegetables, which provide organic salts of potassium and magnesium with a buffering effect, decreases urinary calcium intake (Jones *et al.*, 2001). On the contrary, low phosphorus and high sodium and caffeine intake, for instance from salty snacks and soft drinks, are associated with increased urinary calcium.

It is known that bone mass and markers of bone turnover change during adolescence. Skeletal growth is a complex process that includes growth in length, growth in width, and bone maintenance. These three mechanisms involve the formation and degradation of collagen matrix; thus, during puberty an increase in bone formation and resorption markers is produced (Rauch *et al.*, 2002). Several studies have shown that certain markers of bone formation such as bone alkaline phosphatase and osteocalcin, increase significantly during puberty (Mora *et al.*, 1999). Similarly, bone resorption markers such as pyridinoline (Pyr) and deoxypyridinoline (DPyr) present higher values in early puberty (Tanner stage II) and lower ones in late puberty (Tanner stage V) (Mora *et al.*, 1999; Rauch *et al.*, 2002).

The increase in urinary bone markers is an indicator of higher bone mass as well as of more active bone turnover. A positive correlation has been described between the urinary excretion of these markers and the growth velocity in healthy children. The increase in DPyr excretion is more marked among male than female adolescents, which suggests a more intense metabolic activation in the former (Rauch *et al.*, 2002). In healthy boys, levels of Pyr and DPyr begin to increase when they are around 10 years old, reaching a maximum at 12 years and decreasing later. In girls, however, such a marked peak is not observed and the levels begin to diminish from 11 years (Fujimoto *et al.*, 1995). The large pubertal increase in the levels of these markers suggests that they may be relatively more sensitive as indicators of skeletal health during puberty.

During adolescence, mineral retention must be enough for the requirements of a strong growth rate and it is promoted by the physiological state and the diet consumed. From the absorbed calcium, adolescents retain in the skeleton the necessary mineral to saturate this compartment and the rest is excreted by the urine. The average retention during this stage is close to 210 mg Ca per day and may be as high as 400–500 mg per day in the pubertal period (NIH Consensus Conference, 1994). Maximum values of calcium retention have been observed midway through puberty (Tanner stage III), and so increased calcium intake is advised during this period in order to ensure adequate bone mineralization (Mølgaard *et al.*, 1999). In this regard, maximum values of retention have been estimated at the age of 10–12 years in girls and 12–13 years in boys. Regarding bone density, the maximum values would be reached between 10 and 14 years in girls and 13–18 years in boys, because, as mentioned previously, bone accretion continues when the longitudinal growth has stopped, being able to extend during the third decade of life. During each of the 2 years surrounding this age, boys gain on average of 407 g of bone mineral, while girls gain around 322 g (Whiting *et al.*, 2004). A summary of the amount of calcium absorbed and retained for adolescents according to the data reported by different authors is shown in Figure 13.4.

13.5.1 Hormonal Factors

Hormonal changes affecting calcium absorption and retention begin 2–3 years before puberty, when an acceleration of growth is observed. Thus, higher levels of GH and of sexual steroids during the prepubertal period have

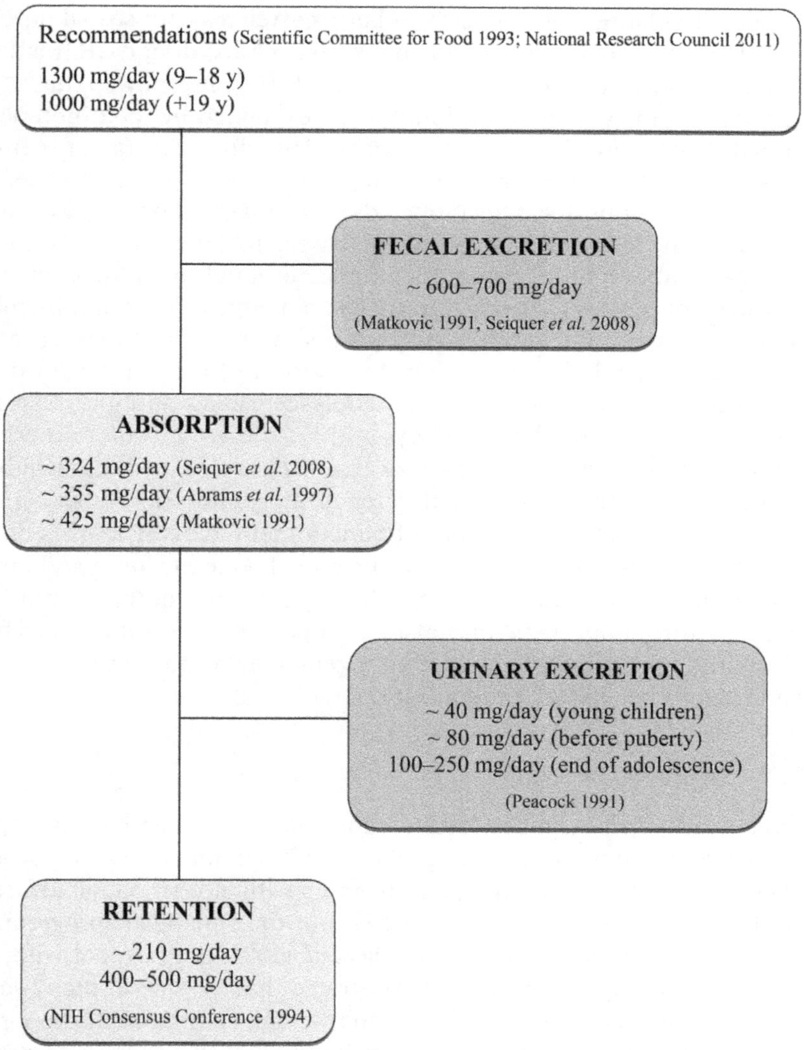

Figure 13.4 This summarizes the calcium absorption and retention in adolescents according to data reported by various authors.

a positive influence on bone-mineral density and the accumulation of calcium in the skeleton. GH deficiency may decrease bone turnover, and the balance between bone formation and bone resorption might be uncontrolled. Fujimoto *et al.* (1995) showed that urinary Pyr and DPyr levels in children with GH deficiency were lower than those in healthy children. This finding suggests that bone turnover might be slightly decreased in GH-deficient children. Urinary marker levels in these GH-deficient children increased significantly during GH therapy. The effects of GH on bone turnover may be partly mediated by locally produced IGF-1, which is a beneficial factor on skeletal development and bone formation. IGF-1 serum levels increase and reach a maximum during puberty, at 14.5 years in girls and 15.5 years in boys. The increase in circulating IGF-1 at early puberty correlates with sexual development and results from the interaction between sex steroids and GH (Matkovic *et al.*, 2004). A study of adolescent boys aged 13–15 years showed that 11.5% of the variation in calcium retention may be explained by the difference in serum IGF-1 concentration (Hill *et al.*, 2008). Therefore, this factor seems to be a major regulator of bone growth during childhood and adolescence.

With the secretion of sex hormones during puberty, bone growth accelerates and bone-mass accumulation increases. In females, the accretion rate increases about 4-fold before menarche, although bone mass changes little or even decreases thereafter. In males, bone-mass accretion increases approximately 6-fold during puberty with a slower but still marked accretion at many skeletal sites thereafter. In addition, there are gender differences in the porosity of bone between adolescent boys and girls that may reflect greater bone remodeling in boys at this time. As a result of these differences, males have a larger bone size and greater thickness after puberty than females, but there is little difference in volumetric density. Estrogens are an important determinant of bone-mineral density in girls during puberty. These hormones can decrease the rate of bone turnover, and inhibit the osteoclastic resorption of bone by affecting bone cell differentiation and function. In turn, androgens may also be important determinants of bone density, although it is thought that estrogens play a more important role than androgens in skeletal mineralization.

13.5.2 Dietary Factors

Age, body weight, height, pubertal status, and body-mass index have an important influence on bone-mass development and bone-mineral density in children and adolescents. In addition, diet, together with other modifiable factors, also influences the attainment or otherwise of the genetically programmed peak bone mass. In this regard, Slemenda *et al.* (1991) suggest that, although studies of twin young adults have demonstrated that 60–80% of the variation in peak bone mass is determined by genetic factors, an important proportion depends on the lifestyle during infancy, including diet and physical activity.

Several studies have shown the beneficial effect of high calcium intakes on pubertal children. Hill *et al.* (2008) reported that calcium intake is the main predictive factor of calcium retention in adolescent males, which is maximum when the threshold intake is satisfied. In a study of Spanish adolescents, it was observed that calcium intake and retention were positively correlated (Mesías, 2007). It was deduced that calcium consumption in this study was below the threshold value and therefore that an increase in calcium intake would have improved the acquisition of bone mineral. Bone metabolism depends intensely on the nutritional status and bone turnover can be reduced in malnourished children. With an adequate diet, calcium bioavailability (% retention/intake) is favored, reaching values around 36.5% for boys and 29.6% for girls, or even higher when diets provide suitable amounts of mineral (Bailey *et al.*, 2000; Seiquer *et al.*, 2008). Thus, as mentioned above, the dietary habits of adolescents are an important factor to meet calcium requirements and, consequently, the needs for pubertal growth.

It has been estimated that a relatively high protein intake, including those from animal sources, is associated with increased bone mineral mass and a reduced incidence of osteoporotic fractures, and therefore such intake is necessary for optimal bone metabolism during growth (Conigrave *et al.*, 2008). Proteins can stimulate intestinal calcium absorption and enhance IGF-1, exerting a positive activity on skeletal development and bone formation. However, proteins, particularly those from animal sources, might also be deleterious for bone health, by inducing chronic metabolic acidosis, which in turn would be responsible for increased calciuria and accelerated mineral dissolution. Animal proteins provide sulfur-containing amino acids, the metabolism of which produces an acid load responsible for increasing urinary calcium. Therefore, a lower contribution of this kind of protein protects bone resorption. On the contrary, it has been suggested that although dietary protein intake typically lowers the pH of urine, the pH of the extracellular fluid is undisturbed due to efficient regulatory control by the kidneys. Therefore, it may be concluded that a varied protein intake, from different sources of high biological value, is likely to have a complementary effect in improving bone accretion.

Fruit and vegetables are another food source that may affect bone mass, due to the above-mentioned buffering effect. Several studies have shown the link between fruits/vegetables consumption and peak bone mass acquisition in boys and girls. Whiting *et al.* (2004) suggested that girls consuming adequate amounts of this food group showed a greater bone mineral trajectory than those consuming fewer than 5 servings per day. In the same way, subjects with an intake of 10 servings per day of fruits and vegetables presented a higher total body bone mineral content than did those consuming 1 serving per day (Vatanparast *et al.*, 2005). Moreover, fruit and vegetables provide vitamin K, which is an essential cofactor for osteoblastic activity and natural antioxidants like phytestrogens, which seem to play a role in bone metabolism. Phytestrogens, like estrogens, stimulate human osteoblasts and modulate osteoclast activity, thus preventing bone resorption (Chiechi and Micheli, 2005).

A healthy dietary model that incorporates practically all these positive factors is the Mediterranean diet, characterized by moderate levels of animal protein, abundant fresh fruits and vegetables, cereals and fish, and little saturated fat. This diet consumption, together with a healthy lifestyle, is a wise habit to maintain lifelong in order to fight osteoporosis and osteoporotic fractures. It has been observed that dietary calcium utilization during adolescence may be greatly improved by a diet based on the Mediterranean patterns (Seiquer *et al.*, 2008).

13.6 Other Factors Affecting Bone-Mass Acquisition

Together with genetic and ethnic factors, body weight and dietary habits, the accretion of bone mass during adolescence may be influenced by lifestyle factors such as physical activity and smoking. It has been reported that both exercise and calcium intervention have an overall beneficial impact on bone acquisition during childhood and adolescence. Physical activity during the critical growing years has a significant impact on the accrual of bone mass, showing an osteogenic effect. Moreover, exercise may indirectly increase bone mass by increasing muscle mass, which is a determinant of bone development (Vicente-Rodriguez *et al.*, 2004). In this sense, a relatively low calcium intake may be compensated by regular physical activities in the accrual of peak bone mass.

The current lifestyle habits of many adolescents are related to inadequate dietary intake, insufficient physical exercise and high sedentary activities, spending too much time watching TV or using the computer. These habits result in an increased level of overweight and obesity, which are a serious public health problem in both developed and developing countries. It is known that body weight gain interferes with both the acquisition and loss of bone mass. In this regard, it has been reported that overweight children have a lower bone mass and bone area relative to their body weight than children with a healthy body weight, which may predispose them to fractures. This fact may be probably explained due to calcium intake being negatively correlated with body fat percentage and the body-mass index (BMI), which has been demonstrated in adolescents (Lederer Goldberg *et al.*, 2009).

On the other hand, factors such as smoking and excessive alcohol consumption are associated with low bone mass and fracture risk and they have been identified as risk factors for osteoporosis (Keen, 2007). A prevalence of concurrent alcohol and tobacco use among European and American adolescents of 20–25% has been reported; an inverse relationship between bone-mineral density and smoking has been found, due to its negative effect on calcium absorption and estrogen metabolism, and moreover, heavy alcohol consumption hinders calcium absorption and damages bone cells (Seiquer *et al.*, 2012).

13.7 Conclusions

Adolescents are at nutritional risk due to their special alimentary habits and their high nutrient requirements. In this stage of life, calcium requirements are increased due to the accelerated growth of this period and the need to attain the optimal peak bone mass, which protects against future resorption and osteoporosis. Hormonal changes associated with puberty have a great influence on calcium absorption and retention, thus promoting calcium metabolism and enhancing bone mineralization.

Calcium intake during adolescence is decisive to achieve the maximum peak bone mass; however, not only calcium level but the whole diet affects dietary calcium utilization. As attainment of optimum bone mass seems to be the best protection against osteoporosis, a disease whose incidence is increasing progressively in Western countries, the lifestyle and dietary habits of adolescents should be aimed to promote calcium absorption and the formation of the bone mass. Mediterranean patterns, together with physical activity, should be recommended among adolescent population as a useful dietary model to facilitate the utilization of dietary calcium and prevent degenerative diseases such as osteoporosis.

Summary Points

- This chapter focuses on calcium importance during adolescence.
- Calcium requirements are increased during adolescence due to the accelerated growth and development.
- An adequate intake of calcium is fundamental during adolescence to attain the optimal peak bone mass.
- A high percentage of American and European adolescents fail to meet the calcium recommendations.
- Milk and dairy products are the major source of dietary calcium for adolescents.
- The current lifestyle habits of many adolescents are related to inadequate dietary intake, insufficient physical exercise and high sedentary activities.
- The lifestyle and dietary habits of adolescents should be aimed to promote calcium absorption and thus the formation of the bone mass.
- Calcium absorption and retention are influenced by hormonal changes associated with puberty.
- To reach the maximum peak bone mass protects against future resorption and osteoporosis.
- Mediterranean patterns are a useful dietary model that facilitates the utilization of dietary calcium.

Key Facts

Key Facts of Calcium

1. Calcium is an essential nutrient for skeletal mineralization.
2. About 99% of the corporal reserves are deposited in bone and teeth.
3. Calcium needs are increased during adolescence.
4. An adequate calcium intake during growth is decisive to reach the maximum peak bone mass.

5. Hormonal changes associated with puberty have a great influence on calcium absorption and retention.
6. Dietary factors and the whole composition of diet may affect calcium availability.

Key Facts of Adolescence

1. Adolescence is the period of time from puberty to adulthood.
2. It is characterized by an accelerated growth rate associated with a rapid muscular, skeletal, and sexual development.
3. During adolescence nutritional requirements are increased.
4. Many adolescents do not reach nutritional requirements.
5. Currently lifestyle habits of many adolescents are related to inadequate dietary intake, insufficient physical exercise and high sedentary activities.

Key Facts of Osteoporosis

1. Osteoporosis is a worldwide health problem.
2. It is characterized by low bone-mineral density and structural deterioration of bone tissue
3. Osteoporosis leads to bone fragility and an increased susceptibility to fractures.
4. To obtain the maximum peak bone mass protects against osteoporosis in the adulthood.
5. Prevention of osteoporosis in adulthood begins in childhood.

Definitions of Words and Terms

Adolescence. Period from puberty to adulthood characterized by an accelerated growth rate associated with a rapid muscular, skeletal, and sexual development.

Calcium. Essential nutrient for skeletal mineralization since about 99% of the corporal reserves are deposited in bone and teeth.

Osteoporosis. Worldwide health problem characterized by low bone-mineral density and structural deterioration of bone tissue, leading to bone fragility and an increased susceptibility to fractures.

Peak bone mass. Amount of bone tissue present at the end of the skeletal maturation.

Absorption. Intake – fecal excretion.

Retention. Absorption – urinary excretion.

Digestibility. (Absorption/intake) × 100.

Bioavailability. (Retention/intake) × 100.

Growth hormone. Hormone that stimulated growth and cell reproduction and regeneration.

Mediterranean diet. Nutritional model considered as one of the healthiest dietary models, characterized by high consumption of olive oil, legumes, unrefined cereals, fruits and vegetables, moderate intake of animal protein, dairy products and wine, and low consumption of meat and meat products and saturated fat.

List of Abbreviations

ABS	Absorption
DH	Diet history
DPyr	Deoxipyridinoline
FFQ	Food-frequency questionnaire
FSH	Follicle-stimulating hormone
GH	Growth hormone
GnRH	Gonadotropin-releasing hormone
HBS	Household budget survey
IGF-1	Insulin-like growth factor-1
LH	Luteinizing hormone
n.a.	Not available
Pyr	Pyridinoline
WR	Weighed food record

Acknowledgements

This research was supported by the Spanish Ministry of Education and Science.

References

Abrams, S. A., Grusak, M. A., Staff, J. and O'Brien, K. O., 1997. Calcium and magnesium balance in 9–14-y-old children. *American Journal of Clinical Nutrition*. 66: 1172–1177.

Bailey, D. A., Martin, A. D., Mckay, H. A., Whiting, S. and Mirwald, R., 2000. Calcium accretion in girls and boys during puberty: a longitudinal analysis. *Journal of Bone and Mineral Research*. 15: 2245–2250.

Chiechi, L. M. and Micheli, L., 2005. Utility of dietary phytestrogens in preventing postmenopausal osteoporosis. *Current Topics in Nutraceutical Research*. 3: 15–28.

Conigrave, A. D., Brown, E. M. and Rizzoli, R., 2008. Dietary protein and bone health: roles of amino acid-sensing receptors in the control of calcium metabolism and bone homeostasis. *Annual Review of Nutrition*. 28: 131–155.

Esterle, L., Sabatier, J. P., Guillon-Metz, F., Walrant-Debray, O., Guaydier-Souquieres, G., Jehan, F. and Garabédian, M., 2009. Milk, rather than other foods, is associated with vertebral bone mass and circulating IGF-1 in female adolescents. *Osteoporosis International*. 20: 567–575.

Fujimoto, S., Kubo, T., Tanaka, H., Miura, M. and Seino, Y., 1995. Urinary pyridinoline and deoxypyridinoline in healthy children and in children with growth hormone deficiency. *International Journal of Food Sciences and Nutrition*. 80: 1922–1928.

Guéguen, L. and Pointillart, A., 2000. The bioavailability of dietary calcium. *Journal of the American College of Nutrition*. 19: 199–136.

Hill, K., Braun, M., Kern, M., Martin, B. R., Navalta, J. W., Sedlock, D. A., McCabe, L., McCabe, G. P., Peacock, M. and Weaver, C. M., 2008. Predictors of calcium retention in adolescent boys. *Journal of Clinical Endocrinology and Metabolism*. 93: 4743–4348.

Hirota, T., Kusu, T. and Hirota, K., 2005. Improvement of nutrition stimulates bone mineral gain in Japanese school children and adolescents. *Osteoporosis International*. 16: 1057–1064.

Jackman, L. A., Millaje, S. S., Martin, B. R., Word, O. B., McCabe, G. P. P., Peacock, M. and Weaver, C. M., 1997. Calcium retention in relation to calcium intake and postmenarcheal age in adolescent females. *American Journal of Clinical Nutrition*. 66: 327–333.

Jones, G., Riley, M. D. and Whiting, S., 2001. Association between urinary potassium, urinary sodium, current diet, and bone density in prepubertal children. *American Journal of Clinical Nutrition*. 73: 839–844.

Keen, R., 2007. Osteoporosis: strategies for prevention and management. *Best Practice & Research, Clinical Rheumatology*. 21: 109–122.

Kim, S. H., Kim, W. K. and Kang, M. H., 2013. Effect of milk and milk products consumption on physical growth and bone mineral density in Korean adolescents. *Nutrition Research and Practice*. 7: 309–314.

Lederer Goldberg, T. B., da Silva, C. C., Lopes Peres, L. N., Nogueira Berbel, M., Braz Heigasi, M., Cabral Ribeiro, J. M., Suzuki, K., Aparecida Josué, L. M. and Dalmas, J. C., 2009. Calcium intake and its relationship with risk of overweight and obesity in adolescents. *Archivos Latinoamericanos de Nutricion*. 59: 14–21.

Manz, F., Kehrt, R., Lausen, B. and Merkel, A., 1999. Urinary calcium excretion in healthy children and adolescents. *Pediatric Nephrology*. 13: 894–899.

Matkovic, V., 1991. Calcium metabolism and calcium requirements during skeletal modeling and consolidation of bone mass. *American Journal of Clinical Nutrition*. 54: 245–260.

Matkovic, V., Badenhop-Stevens, N., Ha, E. J., Crncevic-Orlic, Z. and Clairmont, A., 2004. Nutrition and bone health in children and adolescents. In: Holick, M. F. and Dawson-Hughes, B. (ed.) *Nutrition and Bone Health*. Humana Press, Totowa, New Jersey, pp. 173–195.

Mauras, N., Rogol, A. D., Haymond, M. W. and Veldhuis, J. D., 1996. Sex steroids, growth hormone, insulin-like growth factor-1: neuroendocrine and metabolic regulation in puberty. *Hormone Research*. 45: 1–2.

Mesías, M., 2007. Importancia de la dieta en la digestibilidad y metabolismo de hierro y calcio en la adolescencia. Influencia del consumo de productos de la reacción de Maillard. Doctoral Dissertation. University of Granada, Spain.

Mesías, M., Seiquer, I. and Navarro, M. P., 2011. Calcium nutrition in adolescence. *Critical Reviews in Food Science and Nutrition*. 51: 195–209.

Mølgaard, C., Thomsen, B. L. and Michaelsen, K. F., 1999. Whole body bone mineral accretion in healthy children and adolescents. *Archives of Disease in Childhood*. 81: 10–15.

Mora, S., Pitukcheewaonont, P., Kaufman, F. R., Nelson, J. C. and Gilsanz, V., 1999. Biochemical markers of bone turnover and the volume and the density of bone in children at different stages of sexual development. *Journal of Bone and Mineral Research*. 14: 1664–1671.

National Research Council, 2011. In: Ross, C., Taylor, C. L., Yaktine, A. L. and del Valle, H. L. (ed.) *Dietary Reference Intakes for Calcium and Vitamin D*. The National Academies Press, Washington, DC.

NIH Consensus Conference. NIH Consensus Development Panel on Optimal Calcium Intake, 1994. Optimal calcium intake. *JAMA, the Journal of the American Medical Association*. 272: 1942–1948.

Peacock, M., 1991. Calcium absorption efficiency and calcium requirements in children and adolescents. *American Journal of Clinical Nutrition*. 54: 261–265.

Pérez-Granados, A. M., Vaquero, M. P. and Navarro, M. P., 2000. Biodisponibilidad de calcio y fósforo en ratas alimentadas con grasa de sardina cruda o frita con aceite de oliva. *Food Science and Technology International*. 6: 387–397.

Puel, C., Quintin, A., Agalias, A., Mathey, J., Obled, C., Mazur, A., Davicco, M., Lebecque, P., Skaltsounis, A. L. and Coxam, V., 2004. Olive oil and its main phenolic micronutrient (oleuropein) prevent inflammation-induced bone loss in the ovariectomised rat. *British Journal of Nutrition*. 92: 119–127.

Rauch, F., Georg, M., Stabrey, A., Neu, C., Blum, W. F., Remer, T., Manz, F. and Schoenau, E., 2002. Collagen markers deoxypyridinoline and hydroxylysine glycosides: pediatric reference data and use for growth prediction in growth hormone-deficient children. *Clinical Chemistry*. 48: 315–322.

Saggese, G., Baroncelli, G. I. and Bertelloni, S., 2002. Puberty and bone development. *Best Practice & Research, Clinical Endocrinology & Metabolism*. 16: 53–64.

Scientific Committee for Food, 1993. Nutrient and Energy Intakes for the European Community. Opinion Adopted by the Scientific Committee on Food on 11 December 1992. Reports of the Scientific Committee for Food, Thirty-First Series. European Commission, Luxembourg. Available at: http://ec.europa.eu/food/fs/sc/scf/reportsen.html. Accessed 10 Nov 2013.

Seiquer, I., Mesías, M. and Navarro, M. P., 2012. Osteoporosis, nutrition and adolescence. In: Dionyssiotis, Y. (ed.) *Osteoporosis*. In Teach, Croatia, pp. 411–434.

Seiquer, I., Mesías, M., Muñoz Hoyos, A., Galdó, G. and Navarro, M. P., 2008. A Mediterranean dietary style improves calcium absorption in healthy male adolescents. *Journal of the American College of Nutrition*. 27: 454–462.

Slemenda, C. W., Christian, J., Williams, C. J., Norton, J. A. and Johnston, C. C., 1991. Genetic determinants of bone mass in adult women: a reevaluation of the twin model and the potential importance of gene interaction on heritability estimates. *Journal of Bone and Mineral Research*. 6: 561–567.

Van den Heuvel, E., Muys, T., van Dokkum, W. and Schaafsma, G., 1999. Oligofructose stimulates calcium absorption in adolescents. *American Journal of Clinical Nutrition*. 69: 544–548.

Vatanparast, H., Baxter-Jones, A., Faulkner, R. A., Bailey, D. A. and Whiting, S. J., 2005. Positive effects of vegetable and fruit consumption and calcium intake on bone mineral accrual in boys during growth from childhood to adolescence: The University of Saskatchewan Pediatric Bone Mineral Accrual Study. *American Journal of Clinical Nutrition*. 82: 700–706.

Vicente-Rodriguez, G., Ara, I., Perez-Gomez, J., Serrano-Sanchez, J. A., Dorado, C. and Calbet, J. A., 2004. High femoral bone mineral density accretion in prepubertal soccer players. *Medicine and Science in Sports and Exercise*. 36: 1789–1795.

Whiting, S. J., Vatanparast, H., Baxter-Jones, A., Faulkner, R. A., Mirwald, R. and Bailey, D. A., 2004. Factors that affect bone mineral accrual in the adolescent growth spurt. *Journal of Nutrition*. 134: 696–700.

CHAPTER 14

The Influence of Protein Intake on Calcium Balance

E. ROUY[a] AND D. TOME*[a]

[a]AgroParisTech, UMR0914 Nutrition Physiology and Ingestive Behavior, 16 rue Claude Bernard, 75005 Paris, France
*E-mail: daniel.tome@agroparistech.fr

14.1 Introduction

In western countries, proteins represent 10 to 20% of dietary energy intake. The main contributors to protein intake are usually animal products such as meat, milk and eggs. Because of their wide usage in food products, wheat and other cereal proteins are also a significant contributor to protein in the diet. The quality of dietary protein is linked to its capacity to provide indispensable amino acids (phenylalanine, valine, threonine, tryptophan, isoleucine, methionine, leucine, lysine, and histidine). The main function of amino acids is to support protein synthesis in the body. Protein and amino acids have also been shown to interact with calcium metabolism. These interactions between protein intake and calcium metabolism are very complex and remain debated.

Food and Nutritional Components in Focus No. 10
Calcium: Chemistry, Analysis, Function and Effects
Edited by Victor R. Preedy

Published by the Royal Society of Chemistry, www.rsc.org

14.2 Protein Intake and Calcium Metabolism: Interacting at Multiple Levels

14.2.1 Protein Intake Increases Urinary Calcium Excretion

Dietary protein increases urinary calcium excretion. Urinary calcium is one of the two major ways of losing calcium for the body, the main one being fecal calcium. Calcium can also be lost by sweat, but in much smaller amounts. The findings that an increase in protein intake induces hypercalciuria have been consistent over the last 20 years, with an increase of urinary calcium of about 0.5 to 2 mg for each additional gram of dietary protein ingested (Calvez *et al.*, 2012).

Evidence that protein quality is likely to influence the calcium excretion in urine remains inconsistent. On the one hand, some studies show that urinary calcium is more important in people consuming an omnivorous diet than in vegans; on the other hand, some studies report no difference between a high and a low meat diet. These discrepancies in the results could be explained by the fact that the interaction with other nutrients should be considered in a whole-food approach. For example, the phosphorus contained in some foods decreases urinary calcium, potentially counteracting the hypercalciuric effect of protein intake. For standardization purposes most studies are performed with purified protein such as casein, wheat gluten or egg proteins and do not use a whole-food approach. All studies using purified proteins show an increase of urinary calcium with protein intake.

The increased amount of calcium in the urine with high protein intake is related to the renal handling of calcium. More precisely, increased protein consumption raises the glomerular filtration rate and reduces the reabsorption of calcium by peritubular capillaries. Those two mechanisms result in a higher calcium concentration in the urine. The change in calcium renal handling seems to be independent of parathyroid hormone (PTH), the main regulator of calcemia (Jajoo *et al.*, 2006; Ceglia *et al.*, 2009; Hunt *et al.*, 2009).

The increased urinary calcium with protein intake raises numerous questions about calcium–protein interaction. The main one is the question of the origin of the calcium (Figure 14.1). As calcium is critical for cell signaling, blood calcium is very tightly regulated to avoid the dramatic consequences of a small calcium change. Hence, increased calcium loss in the urine can only have two origins – increased intestinal absorption of calcium or solicitation of calcium storage in bone. It is possible that both of these compensatory pathways can be active at the same time.

14.2.2 Does Protein Intake Increase Calcium Absorption?

One of the key elements of calcium metabolism is the intestinal calcium absorption that controls the amount of calcium from ingested food that enters the body. Calcium absorption occurs by passive transport in two parts of the intestine, the jejunum and the ileum, and by a vitamin D dependent transport in the duodenum.

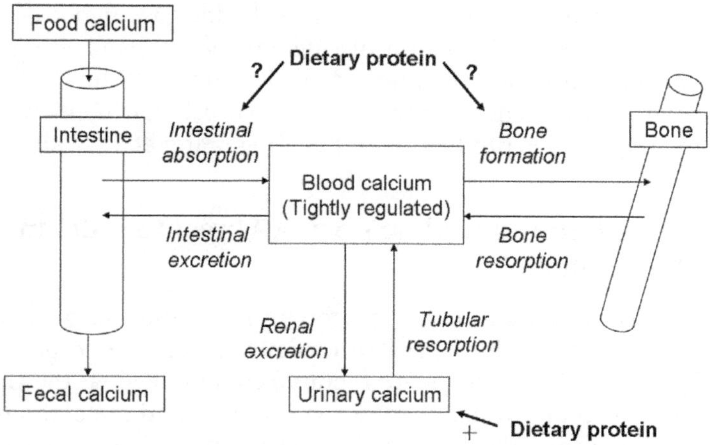

Figure 14.1 Main calcium compartments and fluxes in the body. Dietary protein increases urinary calcium. The origin of the calcium is discussed as it could come from an increase of intestinal absorption or a decrease of bone formation.

For a long time, calcium absorption was calculated as the difference between ingested calcium and fecal calcium. Using this method for the measurement of the effect of protein on calcium absorption gave no clear results. This could be due to the fact that fecal calcium is hard to measure with any accuracy, likely leading to errors in the evaluation of calcium absorption. However, the new, more reliable methods using stable isotopes do not yield consistent results either. Some clinical studies show a positive effect of protein on calcium absorption, whereas others show no effect (Ceglia *et al.*, 2009; Hunt *et al.*, 2009).

The amount of calcium in the diet is one of the origins of the variability in the results for the effect of dietary protein on calcium absorption. Indeed, it seems that protein increases calcium absorption when dietary calcium intake is low, but that this effect fades away when dietary calcium is high (Hunt *et al.*, 2009).

Calcium availability is also a determinant of calcium absorption as some molecules binding to calcium impede its absorption by reducing the amount of calcium available in the intestinal tract. Calcium is supposed to be less bioavailable in plant-derived food, because of the presence of phytates and fibers, both nondigestible molecules. Conversely, some proteins like milk casein can bind to calcium after acid treatment, improving its solubility and absorption.

Calcium is absorbed in the duodenum only if the pH is lower than 6, otherwise calcium will not dissolve into ions. Hence, management of gastric acid (responsible for pH lowering in the intestinal tract) is critical to calcium absorption. It has been shown that some amino acids could contribute to increased intestinal calcium absorption by stimulating gastric acid secretion. However, this has only been shown *in vitro* and *ex vivo* (Busque *et al.*, 2005).

The effect of a low-protein diet on calcium metabolism has also been studied. The consequence of a reduction of protein in the diet is a secondary hyperparathyroidism. The rise of PTH is likely due to a hypocalcemic situation caused by the perturbation of calcium metabolism. From the few existing clinical studies on low protein diets, it seems that the hyperparathyroidism is caused by a reduction of calcium absorption, which lowers blood calcium and triggers the production of PTH (Kerstetter *et al.*, 2003).

14.2.3 Is Protein-Induced Acidosis a Threat to Calcium Metabolism?

It has been shown that an increased protein intake induces a small acidosis, *i.e.* a small drop of blood pH. For obvious reasons blood pH is tightly regulated, hence buffer systems are required to fight against the variations of blood pH. The main ways of dealing with excess acid are carbon-dioxide excretion through the lungs and acid excretion in the urine. However, calcium could also be mobilized as a buffer system if the acid level is not managed by the other organs. Calcium would be mobilized from bone storage and then excreted in the urine (Lemann *et al.*, 2003).

Hence, protein-induced acidosis could be one of the explanations for the increased urinary calcium with protein consumption. This hypothesis is reinforced by the fact that supplementation with alkalinizing molecules such as bicarbonate lowers calciuria (Barzel and Massey, 1998). The potential renal acid load (PRAL) represents the potential acidifying effect of a food in the diet. PRAL is usually high for protein-rich food (Table 14.1). The net acid excretion (NAE), which is the amount of acid excreted by the body, is also positively linked to protein intake. Finally, urinary pH is negatively related to protein intake. However, the acid theory has been disputed as, in a healthy subject, the lung and kidney handling of acids is largely sufficient to deal with the small pH reduction due to protein intake (in the usual range of protein intake), and because of this buffer system, PRAL is not a good indicator of acidosis.

Table 14.1 PRAL (potential renal acid load) of some common foods. PRAL is given per 100 g of food and per gram of protein in the food. PRAL of animal protein is not necessarily higher than that of plant proteins (After Massey, 2003).

Food	PRAL/100 g	PRAL/g of protein
Hard cheese	69.3	2.77
Brown rice	4.2	1.79
Peanuts	27.7	1.19
Beef	26.0	1.11
Spaghetti	11.2	0.93
Lentils	4.5	0.50
Peas	1.2	0.17
Egg white	1.7	0.16
Milk	0.3	0.10
Broccoli	−0.8	−0.17
Potato	−1.2	−0.57

The effects of protein on the acid–base balance are supposed to be due to their sulfur amino acid content. The degradation of such amino acids leads to the release of sulfuric acid in the blood. Not all proteins contain the same amount of sulfur amino acids. Nondairy animal proteins are the richest in sulfur amino acids and are linked to an increased NAE. Conversely, there is no relationship between dairy or plant proteins and NAE. Sulfuric acid could explain up to 44% of the increased urinary calcium linked to protein intake, which means that other mechanisms are active at the same time (Zemel *et al.*, 1981). The acid-ash hypothesis was initially based on the fact that when pH is low, the solubility of the calcium salts increases. Although this is true from a physicochemical point of view, it has been criticized because bone matrix is not in direct contact with the extracellular fluid. There is a cellular barrier that separates bone matrix from the extracellular fluid. Among those cells covering bone are the bone-forming osteoblasts and the bone-resorbing osteoclasts. They are both sensitive to pH variation. However, the pH fall necessary to inhibit osteoblasts and to stimulate osteoclasts is much greater than that induced by protein intake (Bonjour, 2005).

14.2.4 The Calcium-Sensing Receptor also Senses Amino Acids

The calcium-sensing receptor (CaSR) is a membrane receptor signaling the presence of extracellular calcium. It is found at different key locations in the body such as the gut, the thyroid and parathyroid glands. There is some evidence from cell-culture studies that the CaSR could also be capable of sensing some amino acids (the aromatic, aliphatic and polar ones). Amino acid sensing by the CaSR provides a link between the protein and calcium metabolisms at the cellular level (Figure 14.2). But CaSR response to amino acids is not the same as its response to calcium – different signaling pathways are triggered. Moreover, it seems that a minimal extracellular calcium level is required for activation of the CaSR by amino acids (Conigrave and Brown, 2006).

The activation of the CaSR by amino acids seems to have numerous effects on calcium metabolism. Amino acid-activated CaSR could reduce PTH secretion and increase calcitonin secretion in C-cells. Calcitonin is a hormone reducing calcemia by reducing bone resorption, calcium absorption in the intestine and calcium reabsorption in the renal tubule. In the intestinal tract, CaSR could also activate the secretion of gastric acid, and in renal tubules the CaSR could increase calcium excretion. In summary, the activation of CaSR by amino acids would decrease calcium release by bone (low PTH, high calcitonin) and increase calcium urinary excretion. At the same time, the secretion of gastric acid would increase calcium absorption and avoid a drop in blood calcium (Conigrave *et al.*, 2008).

While this hypothesis provides an explanation for all the consequences of increased protein intake (increased calcium absorption and excretion, no effect on bone), more evidence is still needed to decide on the importance of CaSR activation by amino acids.

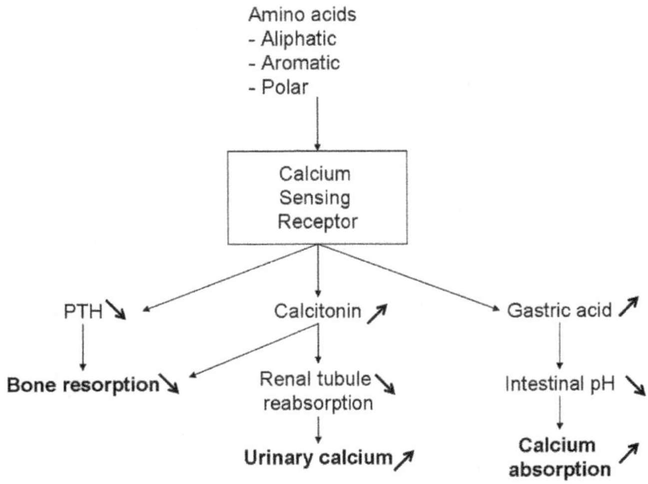

Figure 14.2 Activation of the calcium-sensing receptor by amino acids. Hypothesis of consequences on calcium metabolism of the activation of the calcium-sensing receptor by amino acids. The overall effect is positive for bone health.

14.2.5 Overall Effect of Protein Intake on Calcium Balance

Calculation of the calcium balance is one of the best ways to understand the effects of protein intake on calcium metabolism. Indeed, calcium balance is calculated as the calcium intake by the diet minus the sum of all calcium losses (urinary and fecal; sweat is often considered negligible). Calcium balance takes into account all the stages, from calcium absorption to renal handling of calcium. If the losses are higher than the intake, it means that the body is losing calcium, the lost calcium coming from bone resorption. Conversely, if the intake is higher than the losses, the body is gaining calcium, which will strengthen the skeleton. However, and despite the use of stable isotopes, there is no clear consensus on the issue. Some studies observed negative calcium balance due to the increased loss of calcium in the urine when protein intake was increased (Reddy *et al.*, 2002). Other studies found no effect of protein ingestion on calcium balance (Kerstetter *et al.*, 2005).

It seems that the nature of the protein is important. When the protein source was meat, there was no modification of calcium absorption or urinary calcium. But when the protein was given in a purified form, there was a reduction of fecal calcium loss that offset the urinary loss. An explanation could be the phosphorus content of the complex food. Most protein-rich foods are also rich in phosphorus, which has a hypocalciuric effect and contributes to a better calcium balance. Such findings underline the fact that despite their usefulness for standardization, purified proteins are not fully representative of the mechanisms really occurring when protein intake is increased. Indeed, the associated nutrients and the digestion kinetics are very different between purified proteins and actual food.

Finally, it should be noted that fecal calcium loss is about ten times greater than urinary calcium loss. This means that even a small imprecision in the measurement of fecal calcium can lead to a false calcium balance measurement. Hence, interpretation of calcium balance calculations must be cautious (Calvez *et al.*, 2012).

14.3 Effects of Protein Intake on Calcium Metabolism and Implications for Health

14.3.1 Protein Intake and Bone: Anabolic or Catabolic?

Bone is the major calcium store in the body but also has numerous functions such as support for the body and protection of the main organs. The balance between bone formation and bone resorption determines bone dynamics. If bone formation is higher than resorption, blood calcium is used to strengthen the bone. Conversely, if bone resorption is higher than bone formation, calcium is released in the blood but bone weakens. Given the health burden of osteoporosis, prevention of bone loss is of great importance in the aging population.

There are two conflicting views on the effect of protein intake on bone. On the one hand, the increased calciuria could be due to bone resorption. On the other hand, there is probably an anabolic effect of protein on bone by IGF-1 stimulation. The majority of epidemiological studies find a positive association between protein intake and bone-mineral density (BMD) in a variety of populations including men, children and pre- or postmenopausal women (Whiting *et al.*, 2002; Chevalley *et al.*, 2008; Thorpe *et al.*, 2008a). There is only one study that links protein intake with a BMD reduction (Metz *et al.*, 1993). Measurement of BMD is the most widely used measure to assess bone quality but it is not able to evaluate some parameters that influence the occurrence of fracture, such as the risk of a fall.

It is harder to evaluate fracture risk than to measure BMD, but fracture risk provides more complete information. Data obtained on the relationship between protein intake and fracture risk are conflicting. Some studies conclude that a high protein intake is associated with a reduced risk of fracture and others find an increased risk (Dargent-Molina *et al.*, 2008; Thorpe *et al.*, 2008b). In the debate to determine whether proteins have a positive effect or a negative effect on bone, comments were raised on the studies finding a positive association of protein intake and fracture risk. These observations are related to the fact that, in one study, despite a positive association between the level of protein intake and fracture of the wrist, there was a weak inverse association between protein intake and hip fracture ($p = 0.018$). In the other study, the main finding is that there is no association between protein intake and risk of fracture. There was an increased risk of fracture related to protein intake only for the lowest quartile of calcium intake (Bonjour, 2005). This could indicate that the consequences for bone of increased protein consumption depend on the calcium status of the subject.

Finally, a meta-analysis of all studies in the field showed that there was no negative relationship between protein and bone. So it seems that the amount of protein in the diet is associated with an increased BMD, although the impact on fracture risk is still questionable (Darling *et al.*, 2009). However, being derived from observational studies, epidemiological data cannot establish a causal link between two observations.

Changes in BMD during nutritional interventions are difficult to measure due to their small amplitude. The few existing studies focus on very specific populations and the results are conflicting. One study showed that proteins reduce the bone mineral content in bedridden patients (Zwart *et al.*, 2005). Contrary to this, an increase in protein intake resulted in an increase in BMD in elderly hospitalized patients (Tkatch *et al.*, 1992). Finally, a study of protein supplementation for one year in postmenopausal women reported no significant results (Arjmandi *et al.*, 2005). Studies measuring the effects of protein on bone turnover markers are more numerous, but again the results are inconclusive. Some studies report a negative effect of proteins with increased bone resorption. Other studies find a positive effect, although the reasons vary: increased formation or decreased resorption. The only meta-analysis on intervention studies concluded a positive effect of total protein in lumbar spine BMD studies, which suggests that the effect of protein on bone is positive (Darling *et al.*, 2009). In general, intervention studies do not last long enough to measure a significant effect. Indeed, because of the very slow rate of change in bone, it takes a very long time and is costly to study and detect the possible consequences of nutritional interventions. It is therefore difficult to reach conclusions about the impact on bone of increased protein consumption.

Epidemiological and clinical studies seem to rule out a negative effect of proteins on the skeleton. They rather seem to indicate that there is a positive effect of dietary protein on bone. Proteins slightly increase BMD, but it is possible that this effect is too weak to influence the fracture risk in any measurable way. In addition, we cannot rule out publication bias that would tend to overrepresent studies with a positive effect on bone at the expense of studies showing no effect.

14.3.2 Protein Intake, Urinary Calcium and Kidney-Stone Formation

One of the functions of the kidney is to get rid of the nitrogen produced during protein catabolism and considered as waste. A greater dietary protein intake means more nitrogen for the kidney to handle. As previously mentioned, the consequences are an increased glomerular filtration rate. In some animal models, a high protein diet also caused renal hypertrophy. It is not certain if these observations are due to an overload of renal function or a physiological adaptation of the kidney. Several studies on the effect of a high-protein diet on the kidney concluded that it increased chronic kidney disease only in people already at risk. Healthy people can consume a relatively high-protein diet with no consequence on their renal function (Martin *et al.*, 2005).

There is a concern that a high protein diet could promote the formation of kidney stones. Kidney stones are formed by the concretion of crystals from minerals in the urine. Kidney stones can be from a variety of origins, but the more current are formed of calcium oxalate. Kidney stones lead to pain and health problems when their sizes are too great for them to be evacuated with the urine. Dietary proteins cause an increase in urinary calcium and a decrease in urinary pH, two factors likely to promote kidney-stone formation. Indeed, from a physicochemical point of view, an increased calcium concentration and an acidic environment promote the appearance of crystals. The literature on the question is conclusive only for men, in whom increased animal protein consumption is linked to an increase in kidney-stone formation. However, it seems that this negative effect of protein intake only occurs in subjects with pre-existing metabolic dysfunctions (Hess, 2002).

In conclusion, a healthy kidney with no chronic disease of metabolic dysfunction is able to adapt to relatively high protein consumption without negative health consequences. Conversely, protein consumption is deleterious for subjects with an already injured kidney or a stone forming disease.

14.4 Conclusions

Given the discrepancies of the results in most of the areas where protein and calcium metabolisms interact, it is hard to end with a clear conclusion. However, it seems that an increased protein intake is not harmful to calcium metabolism despite the increased urinary calcium. Although the results are not straightforward, recent studies and meta-analyses concluded that there is no evidence that protein intake stimulates bone resorption. On the contrary, it is more likely that bone anabolism is stimulated by the high IGF-1 level due to the protein intake. The calcium urinary losses would be offset by increased intestinal absorption. The increased calcium intestinal absorption and bone accrual should lead to a positive calcium balance. However, in calcium-balance studies, there is little evidence of this positive effect. Given the complexity of both calcium and protein metabolisms, there is every reason to believe that there is more than one mechanism underlying their interactions. In fact, the acid-ash hypothesis and the anabolic effect by IGF-1 are not incompatible and could be active at the same time. This would mean that, from a whole-diet point of view, a high protein intake is beneficial only if alkalinizing foods are also part of the diet to counteract the mild acidosis.

Summary Points

- This chapter focuses on the effect of protein consumption on calcium metabolism.
- Increased consumption is known to increase urinary calcium (hypercalciuria).

- Protein-induced hypercalciuria can be explained by increased intestinal calcium absorption or by increased bone resorption.
- The impact of protein intake on intestinal calcium absorption is debated, but there are some indications that absorption could be improved.
- There is increasing evidence that proteins have a small positive effect on bone, especially on lumbar spine bone-mineral density.
- This anabolic effect of proteins on bone is probably due to increased IGF-1.
- The increased calcium and acid urinary excretions caused by protein intake could cause kidney stones.
- It seems that protein intake only increases kidney-stone formation in men already subject to a stone-forming disease.

Key Facts

Key Facts on Protein

1. Protein is one of the three major nutrients in the diet, with fat and carbohydrates.
2. In western countries, protein consumption is high in most of the population.
3. A high protein intake increases urinary calcium loss, which could be a sign of negative calcium balance.
4. The question of the effect of protein on the calcium economy is important as it can have consequences on both bone and kidney health.

Key Facts on Osteoporosis

1. Osteoporosis is a disease characterized as a reduction of bone-mineral density (BMD), which leads to a weakening of bone and an increased risk of fracture.
2. More than 8.9 million fractures occur every year due to osteoporosis. The lifetime risk for an osteoporotic fracture has been estimated to be 30–40% in developed countries.
3. Osteoporotic fractures are very incapacitating and usually cause a loss of independence for the patient.

Key Facts on Kidney Health

1. Kidneys are two bean-shaped organs located in the abdominal cavity. They are mainly known for their activity of blood filtration, but they have numerous other roles. They ensure homeostasis by regulating fluid balance, electrolytes losses and acid–base balance. They also produce hormones such as vitamin D. The human kidney is also able to produce glucose from lactate, glycerol and glutamine.

2. Kidney function is closely linked to dietary protein consumption as the kidneys are in charge of the disposal of wastes produced by protein metabolism. Renal filtration occurs in two phases, a first phase of excretion by ultrafiltration and then a phase of reabsorption. The reabsorption phase is vital as it reduces the loss of some important molecules in the urine.
3. Glomerular filtration rate and tubular reabsorption are used to characterize renal function. The glomerular filtration rate is the measurement of the amount of liquid filtered by the kidneys per unit of time. Tubular reabsorption is the amount of liquid reabsorbed by the kidney.

Definitions of Words and Terms

Protein. Large molecules made of chains of amino acids. They have multiple biological functions (structural, enzymatic, transport, *etc.*). From a dietary point of view, proteins are one of the three macronutrients. They provide energy and amino acids to the body.

Protein quality. The quality of a protein is its ability to cover the needs in essential amino acids of the body. The FAO and WHO have determined that the most reliable method is the Protein Digestibility Corrected Amino Acid Score (PDCAAS). This method compares the first limiting amino acid in one gram of the considered protein to a score corresponding to the essential amino acid requirements of a 2- to 5-year-old child.

Calciuria. Calcium concentration in the urine.

Calcemia. Calcium concentration in the blood.

Intestinal absorption. After digestion in the stomach, most nutrients, including amino acids and calcium, are absorbed in the intestine. The intestinal absorption determines what amount of a given nutrient will enter the blood.

Stable isotopes method. Method of measurement of intestinal calcium absorption. A calcium isotope is given with a meal and another one is given by intravenous infusion. The absorption is determined by the calculation of the oral isotope on the intravenous isotope.

Acidosis. Disturbance of the acid–base balance in the body implying an acidification of the blood. Acidosis starts when pH is below the normal 7.35 level.

Amino acid. Chemical compounds able to form protein chains. Some amino acids cannot be synthesized by the human body. These essential amino acids have to be supplied in the diet in order to ensure protein synthesis.

Osteoblast. Cell type present in the bone. Osteoblasts ensure the formation and mineralization of new bone matrix. The mineralization process reduces blood calcium.

Osteoclast. Cell type present in the bone. Osteoclasts are responsible for the resorption of bone matrix. During resorption, calcium is released in the blood.

Fracture risk. Measurement of the probability of fracture in a group. Although requiring large groups and long studies, it is the best measurement of bone health. Indeed, as fracture is the very event that needs to be avoided, it includes all direct and indirect parameters.

Bone-mineral density. Amount of mineral in grams in a bone area or a bone volume. As it is a noninvasive measurement, bone-mineral density is used for prevention and in many scientific studies.

List of Abbreviations

PTH	Parathyroid hormone
PRAL	Potential Renal Acid Load
NAE	Net Acid Excretion
CaSR	Calcium-Sensing Receptor
BMD	Bone-Mineral Density
IGF-1	Insulin-like Growth Factor-1

References

Arjmandi, B. H., Lucas, E. A., Khalil, D. A., Devareddy, L., Smith, B. J., McDonald, J., Arquitt, A. B., Payton, M. E. and Mason, C., 2005. One year soy protein supplementation has positive effects on bone formation markers but not bone density in postmenopausal women. *Nutrition Journal*. 4: 8.

Barzel, U. S. and Massey, L. K., 1998. Excess dietary protein can adversely affect bone. *Journal of Nutrition*. 128: 1051–1053.

Bonjour, J. P., 2005. Dietary protein: an essential nutrient for bone health. *Journal of the American College of Nutrition*. 24: 526S–536S.

Busque, S. M., Kerstetter, J. E., Geibel, J. P. and Insogna, K., 2005. L-Type amino acids stimulate gastric acid secretion by activation of the calcium-sensing receptor in parietal cells. *American Journal of Physiology: Gastrointestinal and Liver Physiology*. 289: G664–G669.

Calvez, J., Poupin, N., Chesneau, C., Lassale, C. and Tome, D., 2012. Protein intake, calcium balance and health consequences. *European Journal of Clinical Nutrition*. 66: 281–295.

Ceglia, L., Harris, S. S., Abrams, S. A., Rasmussen, H. M., Dallal, G. E. and Dawson-Hughes, B., 2009. Potassium bicarbonate attenuates the urinary nitrogen excretion that accompanies an increase in dietary protein and may promote calcium absorption. *Journal of Clinical Endocrinology and Metabolism*. 94: 645–653.

Chevalley, T., Bonjour, J. P., Ferrari, S. and Rizzoli, R., 2008. High-protein intake enhances the positive impact of physical activity on BMC in prepubertal boys. *Journal of Bone and Mineral Research*. 23: 131–142.

Conigrave, A. D. and Brown, E. M., 2006. Taste receptors in the gastrointestinal tract. II. L-Amino acid sensing by calcium-sensing receptors: implications for GI physiology. *American Journal of Physiology: Gastrointestinal and Liver Physiology*. 291: G753–G761.

Conigrave, A. D., Brown, E. M. and Rizzoli, R., 2008. Dietary protein and bone health: roles of amino acid-sensing receptors in the control of calcium metabolism and bone homeostasis. *Annual Review of Nutrition*. 28: 131–155.

Dargent-Molina, P., Sabia, S., Touvier, M., Kesse, E., Breart, G., Clavel-Chapelon, F. and Boutron-Ruault, M. C., 2008. Proteins, dietary acid load, and calcium and risk of postmenopausal fractures in the E3N French women prospective study. *Journal of Bone and Mineral Research*. 23: 1915–1922.

Darling, A. L., Millward, D. J., Torgerson, D. J., Hewitt, C. E. and Lanham-New, S. A., 2009. Dietary protein and bone health: a systematic review and meta-analysis. *Journal of the American College of Nutrition*. 90: 1674–1692.

Hess, B., 2002. Nutritional aspects of stone disease. *Endocrinology and Metabolism Clinics of North America*. 31: 1017–1030.

Hunt, J. R., Johnson, L. K. and Fariba Roughead, Z. K., 2009. Dietary protein and calcium interact to influence calcium retention: a controlled feeding study. *American Journal of Clinical Nutrition*. 89: 1357–1365.

Jajoo, R., Song, L., Rasmussen, H., Harris, S. S. and Dawson-Hughes, B., 2006. Dietary acid-base balance, bone resorption, and calcium excretion. *Journal of the American College of Nutrition*. 25: 224–230.

Kerstetter, J. E., O'Brien, K. O. and Insogna, K. L., 2003. Dietary protein, calcium metabolism, and skeletal homeostasis revisited. *American Journal of Clinical Nutrition*. 78: 584S–592S.

Kerstetter, J. E., O'Brien, K. O., Caseria, D. M., Wall, D. E. and Insogna, K. L., 2005. The impact of dietary protein on calcium absorption and kinetic measures of bone turnover in women. *Journal of Clinical Endocrinology and Metabolism*. 90: 26–31.

Lemann, Jr, J., Bushinsky, D. A. and Hamm, L. L., 2003. Bone buffering of acid and base in humans. *American Journal of Physiology: Renal, Fluid and Electrolyte Physiology*. 285: F811–F832.

Martin, W. F., Armstrong, L. E. and Rodriguez, N. R., 2005. Dietary protein intake and renal function. *Nutrition and Metabolism*. 2: 25.

Metz, J. A., Anderson, J. J. and Gallagher, Jr, P. N., 1993. Intakes of calcium, phosphorus, and protein, and physical-activity level are related to radial bone mass in young adult women. *American Journal of Clinical Nutrition*. 58: 537–542.

Reddy, S. T., Wang, C. Y., Sakhaee, K., Brinkley, L. and Pak, C. Y., 2002. Effect of low-carbohydrate high-protein diets on acid-base balance, stone-forming propensity, and calcium metabolism. *American Journal of Kidney Diseases*. 40: 265–274.

Thorpe, M., Mojtahedi, M. C., Chapman-Novakofski, K., McAuley, E. and Evans, E. M., 2008a. A positive association of lumbar spine bone mineral density with dietary protein is suppressed by a negative association with protein sulfur. *Journal of Nutrition*. 138: 80–85.

Thorpe, D. L., Knutsen, S. F., Beeson, W. L., Rajaram, S. and Fraser, G. E., 2008b. Effects of meat consumption and vegetarian diet on risk of wrist fracture over 25 years in a cohort of peri- and postmenopausal women. *Public Health Nutrition*. 11: 564–572.

Tkatch, L., Rapin, C. H., Rizzoli, R., Slosman, D., Nydegger, V., Vasey, H. and Bonjour, J. P., 1992. Benefits of oral protein supplementation in elderly patients with fracture of the proximal femur. *Journal of the American College of Nutrition*. 11: 519–525.

Whiting, S. J., Boyle, J. L., Thompson, A., Mirwald, R. L. and Faulkner, R. A., 2002. Dietary protein, phosphorus and potassium are beneficial to bone mineral density in adult men consuming adequate dietary calcium. *Journal of the American College of Nutrition*. 21: 402–409.

Zemel, M. B., Schuette, S. A., Hegsted, M. and Linkswiler, H. M., 1981. Role of the sulfur-containing amino acids in protein-induced hypercalciuria in men. *Journal of Nutrition*. 111: 545–552.

Zwart, S. R., Davis-Street, J. E., Paddon-Jones, D., Ferrando, A. A., Wolfe, R. R. and Smith, S. M., 2005. Amino acid supplementation alters bone metabolism during simulated weightlessness. *Journal of Applied Physiology*. 99: 134–140.

CHAPTER 15

Bioaccessibility of Calcium in Legumes

Mª JESÚS LAGARDA[a], ANTONIO CILLA[a], AND REYES BARBERÁ*[a]

[a]Nutrition and Food Chemistry, Faculty of Pharmacy, University of Valencia, Avda Vicente Andrés Estellés s/n 46100 Burjassot, Valencia, Spain
*E-mail: reyes.barbera@uv.es

15.1 Introduction

Calcium is an essential mineral required for the diverse physiological and biochemical functions in the human body: bone health, muscle contraction, blood clotting, nerve conduction and enzyme regulation (Guéguen and Pointillart, 2000). An adequate intake of calcium is important in all the stages of life, but particularly in infants and young children, in order to ensure optimum development, growth and health (Untoro *et al.*, 2005).

Although animal foods, especially milk and dairy products, are rich dietetic sources of calcium, in developing countries this nutrient is obtained mainly through food grains by a majority of the population. Staple diets in developing countries include cereals and legumes.

Legumes are a low-cost source of proteins and also of minerals, including calcium. Legume grains are processed before consumption. The most commonly used domestic methods for processing legumes include soaking for different time periods, dehulling of soaked seeds, ordinary and pressure cooking, and germination, and these processes have been reported to be

Food and Nutritional Components in Focus No. 10
Calcium: Chemistry, Analysis, Function and Effects
Edited by Victor R. Preedy

Published by the Royal Society of Chemistry, www.rsc.org

beneficial for enhancing the nutritive value of some food legumes – including chickpeas, peas, moth beans, mung beans and rice (Duhan *et al.*, 2002).

From a nutritional point of view, it is interesting to know not only the calcium content in legumes but also its bioavailability, *i.e.*, the fraction of ingested calcium that is used for normal physiological functions or storage. The first step in the bioavailability process is bioaccessibility (BA), which has been defined as the fraction of a compound that is released from its food matrix in the gastrointestinal tract and thus becomes available for intestinal absorption (Fernández-García *et al.*, 2009). Dietetic factors (element supply, chemical form, solubility and interaction with other dietary components (enhancers and inhibitors)) are the main determinants of BA, due to their influence upon absorption. The present study therefore addresses studies on the bioaccessibility of calcium in legumes and the factors that can influence BA.

15.2 Calcium Content in Legumes

Table 15.1 summarizes the calcium content of the principal legumes used for human consumption, both raw and processed. The highest content of minerals always corresponds to the raw material product. These contents, when considering the raw product, are between 42 (*Lens culinaris*) and 377 (*Phaseolus vulgaris*) mg/100 g of dry substance. Although content can differ between varieties, it seems that lentils afford the lowest concentration of calcium. The mineral distribution in the grain is not homogeneous. Lombardi-Boccia *et al.* (1998) indicated that the hull in *Phaseolus vulgaris* is particularly rich in calcium, accounting for about 70% of the total seed calcium content. Legume grains are generally processed before consumption, depending upon cultural and taste preferences. It is a general practice to soak legume seeds overnight prior to processing and cooking. In general, a decrease in calcium content is observed in processed legumes. Duhan *et al.* (2002) observed a loss in total calcium content with soaking, and the dehulling of soaked seeds was seen to produce a further significant reduction in calcium content. These authors observed a decrease in calcium content with the prolongation of soaking (with a 5% and 11% loss after 6 and 18 h of soaking, respectively). This loss in calcium content may be attributed to leaching of the mineral into the soaking medium. The loss of calcium reaches 25% with soaking and dehulling. Cooking of unsoaked seeds was found to cause no loss, whereas significant losses were noticed when soaked dehulled seeds were cooked. It seems that no loss of minerals occurred during cooking, but some of the overall loss was attributable to the fact that the seeds were soaked and dehulled prior to cooking.

In the same sense, ElMaki *et al.* (2007) reported a major decrease in calcium content of three different cultivars (Serege, Giza and RO21) of *Phaseolus vulgaris* as the days of soaking were increased. After one day of soaking, the decrease in calcium content was 42%, 22% and 33%, respectively; whereas after three days of soaking the respective decreases were 49%, 31% and 41%

Table 15.1 Calcium content (mg/100 g dry matter) in legumes. Contents of calcium in the principal legumes used in the diets of the population. There are detailed the contents of the raw product, after being submitted to different processes, such as soaking, dehulling, germination, ordinary cooked (domestic, or to atmospheric pressure), high-pressure cooked (or manufacturer), microwave cooked, *etc*. All the values are express as dry weight.[a]

Legume	Raw	Processed	References
Beans (*Phaseolus vulgaris* L.)		171^{RF}	Wyatt and Triana-Tejas (1994)
	154	104° 137^{P}	Sebastiá *et al.* (2001)
	197		Sandberg (2002)
	102.8	83.2^{O} 68.4^{P}	Quinteros *et al.* (2002)
	102.4		Sahuquillo *et al.* (2003)
	321–377	$161–218^{O}$ $188–240^{S}$ $148–201^{S+O}$	ElMaki *et al.* (2007)
	126	111°	Abebe *et al.* (2007)
	321–377	$367–470^{G}$	Al-Numair *et al.* (2009)
Butter bean (*Phaseolus lunatus* L.)	106		Elhardallou and Walker (1995)
Chickpeas (*Cicer arietinum* L.)	103	91° 93^{M} 118^{P}	Sebastiá *et al.* (2001)
	124		Sandberg (2002)
	122.4	95.2^{O} 96.4^{M} 109.2^{P}	Quinteros *et al.* (2002)
	80.1		Sahuquillo *et al.* (2003)
	222	176^{G} 63^{G+D}	Ghavidel and Prakash (2007)
	213		Abebe *et al.* (2007)
Cowpea (*Vigna catjang*)	87	75^{G} 54^{G+D}	Ghavidel and Prakash (2007)
Green gram (*Phaseolus aureus*)	136	114^{G} 71^{G+D}	Ghavidel and Prakash (2007)
Lentils (*Lens culinaris* L.)	151		Elhardallou and Walker (1995)
	77	49° 54^{M} 141^{P}	Sebastiá *et al.* (2001)
	71		Sandberg (2002)
	58.1	50.6^{O} 49^{M} 49.6^{P}	Quinteros *et al.* (2002)
	53.4		Sahuquillo *et al.* (2003)
	77	63^{G} 50^{G+D}	Ghavidel and Prakash (2007)
	42		Abebe *et al.* (2007)
	97	50° 55^{P} 57^{M}	Hefnawy (2011)
Pigeon pea (*Cajanus cajan*)		$216^{O,P}$ $193–205^{S}$ $199^{S+O,\ S+P,\ G}$ $162^{S+D+(O\ or\ P)}$	Duhan *et al.* (2002)
Soybean sees	300		Frossard *et al.* (2000)
	201		Sandberg (2002)
		139°	Kamchan *et al.* (2004)

[a]RF = refried; D = dehulled; G = germinated; M = cooked by microwave; O = ordinary cooked; P = pressure cooked; S = soaked.

with respect to the initial calcium content. The authors indicated that soaking of the seeds, with and without cooking, significantly ($p \leq 0.01$) reduced the calcium content of white bean cultivars. The loss of calcium during the treatment may be attributed to its leaching out into the discarded water.

After the processes of germination, in many cases the observed loss of the element was not due to germination as such but to the prior soaking stage. Duhan *et al.* (2002) did not observe changes in calcium content during germination at 30 °C during 24, 36 or 48 h. Ghavidel and Prakash (2007) compared the effect of germination (during 24 h) with dehulling, and found a further decline in calcium levels after dehulling (following germination), which may be attributed to the presence of these minerals in the hull portion.

On the other hand, Al-Numair *et al.* (2009) found major contents in calcium in *Phaseolus vulgaris* to major time of germination (2 to 6 days) increases being reached of up to 25% *vs.* nongerminated legumes.

On considering the processed product, calcium contents are between 49 (*Lens culinaris*, ordinary or pressure cooked, or microwaved) and 470 mg/100 g (*Phaseolus vulgaris*, after germination). Although it does not always happen, it seems that legumes cooked at high pressure present greater concentrations of calcium compared with the same legumes cooked by traditional methods (Sebastiá *et al.*, 2001).

The differences found in the processed legumes are related not only to the cultivars, but also to differences in processing, such as the conditions of temperature, time, proportions between weight of legumes and cooking water volume, soaking, *etc.*

15.3 Calcium Bioaccessibility Studies in Legumes

A high mineral content is not always related to high mineral BA, since calcium must be in soluble form within the gastrointestinal tract in order to be absorbed. In this context, *in vitro* methods designed to assess calcium BA in legumes are useful for comparing food matrixes, or processing or cooking conditions. In comparison with human or animal studies, these methods are less expensive, faster, and offer better control of the experimental variables (Sandberg, 2002). However, they cannot replace *in vivo* studies, but can serve as useful preliminary screening techniques that help us identify the most promising food matrices, processing conditions, staple crops, cultivars, growing conditions, *etc.*, and their relative capacity to influence nutrient bioavailability (Etcheverry *et al.*, 2012).

Increased bioavailability of minerals in food legumes has been attributed to various processing and cooking methods, including domestic cooking, autoclaving or industrial cooking, and also soaking.

Studies on the solubility or BA of calcium in legumes are scarce – iron being the most extensively investigated mineral in this regard. Different estimators have been used to evaluate calcium BA, such as hydrochloric acid extractability, solubility or dialysis after simulated gastrointestinal digestion, obtaining a soluble fraction through centrifugation (solubility) or a fraction that dialyses through a membrane (dialysis). Cooked sprouts (soaked for 12 h at 37 °C, with germination for 68 and 48 h for chickpeas or black gram beans, respectively, followed by cooking until soft) exhibit the greatest hydrochloric acid extractability (%) of calcium (bioaccessibility estimator), followed by sprouting, autoclaving (soaking and cooked at 1.05 kg cm^{-2} 15 min) and ordinary cooking (soaked and cooked until soft). Calcium extractability does not differ in soaked seeds (for 12 h at 37 °C) *versus* unprocessed (raw) seeds (Jood and Kapoor, 1997). However, in pigeon peas, soaking increases calcium extractability up to 6% *versus* untreated sample, though the greatest percentage increment in calcium extractability corresponds to germination for 48 h (27%), followed by pressure cooking of soaked-dehulled peas (21%), and ordinary cooking of soaked pigeon peas (13%) (Duhan *et al.*, 2002).

Hydrochloric acid extractability of calcium increases after the soaking of white beans for three days, and a further increase has been observed after cooking the soaked seeds. This indicates that a successive increase in the calcium extractability of white beans occurred with prolongation of the soaking period and cooking of the soaked seeds. Reduction in phytic acid as a result of soaking and cooking may explain the higher calcium extractability (ElMaki *et al.*, 2007).

The beneficial effect of germination upon calcium BA is also observed in soybeans, where calcium dialyzability is two-fold greater in soybean sprouts *versus* mature soybean seeds. The amount of oxalate seems to have the strongest negative correlation to calcium dialyzability ($r = -0.79$), compared with dietary fiber ($r = -0.61$) and phytate ($r = -0.42$). On considering the correlation of calcium dialyzability and dietary fiber, phytate and oxalate content, the phytate and oxalate contents were found to be a good predictor of calcium BA in legumes ($r = -1$) (Kamchan *et al.*, 2004).

Dehulling germinating legumes is associated with a significant ($p < 0.05$) increase (two to three times with respect to raw legume) in calcium BA. This could be attributed to a simultaneous reduction of phytic acid, tannins and dietary fiber. High negative correlation coefficient values of bioavailable calcium *versus* phytic acid, tannin and total dietary fiber in legumes have been obtained (Ghavidel and Prakash, 2007).

Dehulling greatly improves calcium dialyzability or solubility when compared to whole-seed values, with increases of up to 97% being found in cotyledon of white and mottled beans. The similar dialyzability in white and colored beans indicates that tannins (which are present only in the colored beans) does not bind calcium in insoluble complexes. Cooking was not found to affect dialyzable or soluble calcium – a fact that could be explained by a small reduction in total phytate content after cooking, since the method employed for phytic acid determination does not discriminate between phytic acid and inositol phosphates with fewer phosphate groups that might originate upon cooking, and this could positively influence calcium BA. Cotyledon of the beans contained almost all the phytate content of the legumes; however, it exhibits greater calcium dialyzability or solubility than whole seeds. This observation suggests that other inhibitors such as oxalic and dietary fiber present in the hull fraction are the main factors contributing to the lesser calcium dialyzability or solubility in whole seeds (Lombardi-Boccia *et al.*, 1998).

Whole grains have higher fiber contents than their respective decorticated forms (in the case of chickpeas, 27.3% *versus* 8.76%, and in the case of green gram beans 22.3% *versus* 8.37%) (Hemalatha *et al.*, 2007).

In different cultivars of sprouting white beans, calcium hydrochloric acid extractability was seen to increase (2- to 4-fold) *versus* nonsprouting legume, being more accentuated after 6 days of sprouting. This observation is consistent with the decrease in phytic acid and polyphenol contents with sprouting (Al-Numair *et al.*, 2009).

Although soaking and/or cooking was seen to result in a decrease in the calcium content of white beans, calcium hydrochloric acid extractability (an index of calcium BA) increased significantly (2- to 3-fold) *versus* untreated legumes. This improvement may be explained by the soaking and cooking, which release the calcium in free form, thereby increasing its hydrochloric acid extractability and the reduction of phytic acid by soaking and cooking (ElMaki *et al.*, 2007).

15.3.1 Dietetic Factors Affecting Calcium Bioaccessibility. Influence of Processing

It has been reported that binding to several legume components, such as oxalate, phytate, and polyphenols can affect mineral solubility and bioavailability. Table 15.2 shows the principal contents of these calcium absorption inhibitors. Phytic acid (myoinositol hexaphosphate) is a legume component with well-known effects on mineral solubility. Phytate is negatively charged under physiological conditions and thus strongly chelates metal ions, including calcium, forming insoluble complexes in the gastrointestinal tract that cannot be digested or absorbed in humans because of the absence of intestinal phytase enzymes. In whole legumes the phytate content ranges from 0.17–9.15% and, as has been mentioned previously, it is uniformly distributed throughout the cotyledons, where it is associated with protein. Hence, when the hull or seed coat of the legumes is removed, their phytate concentration increases. The critical molar ratio above which calcium absorption is compromised by phytate is uncertain. To ensure absorption of calcium from a meal, it is important to consider the phytate: calcium molar ratios. It has been suggested that the desirable molar ratios are less than 0.17 (Umeta *et al.*, 2005).

Legume grains are generally processed before consumption. Common domestic or industrial methods include soaking for different time periods, dehulling of soaked seeds, cooking and germination, and these have been reported to lower the levels of antinutrients and thus influence mineral BA.

Soaking has been documented to be an effective treatment for removing (*via* an internal leaching process) antinutritional factors such as phytates,

Table 15.2 Inhibitors compounds upon calcium bioaccessibility from legumes. Contents of main inhibitors components (tannins, phytic acid, polyphenols, dietary fiber and oxalates) upon calcium bioaccessibility corresponding to different legumes are described.[a]

Legumes	Tannins (mg/100 g)	Phytic acid (mg/100 g)	Polyphenols (mg/100 g)	Dietary fiber (g/100 g)	Oxalates (mg/100 g)	References
Soybean (*Glycine max* L.)			—		—	El-Adawy *et al.* (2000)
Bean (*Phaseolus vulgaris*)						
Lupin seed (*Lupinus albus*) flours						
Raw	252, 55, 227	711, 670, 899		3.6–4.5		
Soaked 0.5% HCO_3/9,8,7 h respectively	79, 31, 16	561, 564, 707		4.3–6.1		
Soybean (*Glycine max* L.)		1000–2220				Frossard *et al.* (2000)
Pigeon pea (*Cajanus cajan*)	—		—	—	—	Duhan *et al.* (2002)
Untreated		857				
Soaking (6–18 h)		804–620				
Soaking and dehulling		600				
Cooking (unsoaked, soaked, soaked and dehulling)		822, 725, 580				
Pressure cooking (unsoaked, soaked, soaked and dehulling)		820, 720, 550				
Germination (24–48 h)		560–470				
Beans (*Phaseolus vulgaris* L.)			—	—		
Chickpeas (*Cicer arietinum* L.)	—					Quinteros *et al.* (2003)
Lentils (*Lens culinaris*)						
Raw					86.5, 64.4, 144.2	
Traditional cooked					11.5, 18.5, 26.4	
Microwave cooked					nd, 3.4, 9.6	
Ready-to-eat					3.6, 1.1, 6.3	
Beans (*Phaseolus vulgaris* L.),						Mañez *et al.* (2002)
Chickpeas (*Cicer arietinum* L.)						
Lentils (*Lens culinaris*						
Raw						
Traditional cooked				1870, 1600, 1510		
Microwave cooked				1220, 1540, 960		
Ready-to-eat				nd, 1150, 830		
				1040, 650, 630		

(*continued*)

Table 15.2 (*continued*)

Legumes	Tannins (mg/100 g)	Phytic acid (mg/100 g)	Polyphenols (mg/100 g)	Dietary fiber (g/100 g)	Oxalates (mg/100 g)	References
Soybean young seeds (*Canjasus indicus/Canjanus canjans*)		292.1		8.3	4.6	Kamchan *et al.* (2004)
Soybean seeds (*Glycine max* L.) cooked		397.7		9.4	5.5	
Mottled beans (*Phaseolus vulgaris*)						Lombardi-Boccia *et al.* (1998)
Raw (whole seed, cotyledon, hull)		1100, 1180, 60				
Cooked (whole seed, cotyledon)		860, 1100, nd				
White beans (*Phaseolus vulgaris*)						
Raw (whole seed, cotyledon, hull)		810, 830, 40				
Cooked (whole seed, cotyledon)		690, 810, nd				
Long beans (*Vigna sinensis*) boiled		164				Chan *et al.* (2007)
Mug beans (*Phaseolus Radiates*) boiled, flour raw, flour porridge		766, 433, 581				
Soybean snack (*Glycine Max*) fried		667				
Three white beans (*Phaseolus vulgaris*)cultivars	—				—	ElMaki *et al.* (2007)
Untreated		353–458	219–676			
Cooked		297–371	166–552			
Soaked and cooked		256–361	165–489			
Green gram (*Phaseolus aureus*)	660, 590, 360	610, 500, 290	—	20, 20.5, 15.2	—	
Cowpea (*Vigna catjana*)	470, 340, 250	600, 480, 290		27.4, 28.1, 23		Ghavidel and Prakash (2007)
lentil (*Lens culinaris*)	750, 610, 360	190, 150, 100		16.5, 17, 12		
Chickpea (*Cicer arietinum*)	530, 440, 300	480, 380, 240		27.2, 28, 20.2		
Raw, germinated, germinated and dehulled						
White beans (*Phaseolus vulgaris*)	—			—		Al-Numair *et al.* (2009)
Sprouting (0–6 days)		458–219	676–112			
Lentils (*Lens culinaris*)						Hefnawy (2011)
Raw	128	411				
Boiling	91	260				
Autoclaving	82	240				
Microwave cooking	84	250				

[a]nd = not determined.

tannins, trypsin inhibitors and hemagglutinin activity, which are eliminated with the discarded soaking solutions, but the effects vary according to the legume cultivars and soaking conditions, such as the type of soaking solution, soaking period and temperature. Soya beans, lupin and bean seeds soaking in 0.5% sodium bicarbonate for 9, 7 or 8 h, respectively, reduce antinutritional factors (tannins (43–93%) and phytic acid (16–21%) (El-Adawy *et al.*, 2000). In this same sense, water soaking of pigeon peas diminishes the phytic content between 6 and 28%, depending on the soaking time (6–18 h). There is an accumulative effect of soaking and dehulling, with phytic acid losses of 30% (Duhan *et al.*, 2002).

In white beans, soaking causes losses in phytic acid (9–16%) and polyphenols (0.8–23%) dependent upon the type of culture and soaking time (1–3 days). Phytate loss during soaking may be due to leaching of phytate ions into the soaking water under the influence of a concentration gradient, which regulates the diffusion rate (ElMaki *et al.*, 2007).

Soaking of legume flours under optimal conditions can markedly reduce their inositol hexaphosphate (IP6) and pentaphosphate (IP5) contents through a combination of passive diffusion of water-soluble sodium or potassium phytate and hydrolysis of phytate by endogenous phytases. Only modest reductions in IP5 + IP6 are achieved after soaking whole seeds or legumes (Gibson *et al.*, 2010).

A reduction of up to 36% in phytic acid content has been observed in germinated pigeon peas, possibly due to the hydrolytic activity of phytase (Duhan *et al.*, 2002). Phytic acid and polyphenol contents decreased significantly ($p < 0.01$) with an increase in sprouting time (0–6 days). Reductions in phytic acid content of up to 38% and in polyphenol content (53%) have been observed after a sprouting time of 6 days (Al-Numair *et al.*, 2009).

Hurrell (2003) reported that there are three ways to decrease the inhibitory effect of phytic acid upon mineral absorption. The first method is to remove phytic acid, the second involves enzymatic phytic acid degradation, and the third technique involves adding compounds to the food that prevents phytate–mineral binding. Because phytic acid in legume seeds is contained in the protein bodies of the cotyledon and not in the aleurone layer, the manufacture of protein isolates or concentrates tends to increase the phytic acid level. The most effective way to decrease phytic acid is through enzymatic degradation. Commercial phytases can completely degrade phytic acid in two hours when added to an aqueous slurry of the complementary food held at optimum pH for phytase activity. Traditional legume preparation processes such as soaking, germination and fermentation can also activate native phytases and substantially degrade phytic acid.

Germination of different legumes has been found to decrease the phytic acid content (18–21%). In turn, germination increased the soluble dietary fiber fraction and decreased the insoluble dietary fiber fraction, with a significant reduction ($p < 0.05$) in tannin content. On dehulling the germinated legumes, phytic acid and tannins were decreased 47–52% and 43–52%, respectively. In cotyledons there was little detectable phytic acid and tannin, indicating that most of the phytic acid and tannin are present in the seed coat (Ghavidel and Prakash, 2007).

Germination and fermentation induce the enzymatic hydrolysis of IP6 and IP5, lowering inositol phosphates through the action of endogenous or microbial phytase enzymes, respectively. For example, germination leads to an increase in phytase activity in most legumes, through *de novo* synthesis, activation of intrinsic phytase, or both. Germination also decreases the content of certain tannins and other polyphenols in legumes (*e.g.*, *Vicia faba*) and red sorghum, as a result of the formation of polyphenol complexes with proteins. The organic acids produced during fermentation lower the pH value, thus generating a pH that is optimum for allowing the intrinsic phytases in legume flours to further degrade the phytic acid (Gibson *et al.*, 2010).

Pressure cooking (of soaked and dehulled) pigeon peas diminishes phytic content by 36% *versus* 32% in the case of ordinary cooking. The loss of phytic acid during cooking could probably be explained on the basis that phytase activity at a temperature of 40–55 °C may degrade IP6 to IP5 or to lower molecular weight forms (Duhan *et al.*, 2002).

A lesser reduction in phytic acid content after the ordinary cooking (until soft) of unsoaked white beans (13–25%) *versus* beans cooked after soaking (21–27%) has been reported. Reductions in polyphenol contents were similar after cooking (17–26%) and in soaked and cooked seeds (19–28%) (ElMaki *et al.*, 2007).

Tannins (46.04–49.10%) and phytic acid (30.93–41.32%) in lentils were significantly reduced ($p < 0.05$) by different cooking processes (boiling, autoclaving or microwave cooking) *versus* raw lentils, though there were no differences between the different cooking methods (Hefnawy, 2011).

Cooking (domestic or industrially) was responsible for a loss of soluble oxalate contents in legumes. Losses are higher in industrial cooking processes (85.7–92.9%) than in domestic cooking, and of these latter processes, microwave cooking is more effective (losses of 79.6–85.9%) than usual domestic cooking (losses of 19.5–59.6%) (Quinteros *et al.*, 2003).

Cooking (ordinary and microwave) produced a decrease (*versus* raw product) in IP6 contents of over 30–40% in white beans, lentils and chickpeas. The greatest losses (40–60%) corresponded to ready-to-eat legumes (industrial processing). These decreases were accompanied by an increase in IP5 contents, with the exception of lentils, where IP6 was probably degraded to less phosphorylated compounds (Mañez *et al.*, 2002). Whether the adverse effect of inositol phosphates upon calcium absorption is restricted to the higher inositol phosphates such as inositol hexaphosphate and inositol pentaphosphate is currently unknown (Gibson *et al.*, 2010).

Tables 15.3 and 15.4 summarize the studies on calcium solubility or dialysis, respectively, in legumes. The objective of most of these studies is to evaluate the effect of dietetic factors or/and processing upon calcium bioaccessibility.

Table 15.3 Calcium bioaccessibility (solubility method) from legumes. Summary of the studies that have applied *in vitro* solubility methods to compare calcium bioaccessibility between legumes and/or or study influence of foods components or processing.

Sample matrix	Aim of study	*In vitro* method	Results	References
Soy protein concentrate + Ca/Zn/phytate	Study interactions and effect on solubility of protein, phytate, Zn and Ca	*Gastric digestion*: 1.0 N HCl with magnetic stir bar 20 min + 30 min. pH adjusted to 2.0 ± 0.05 and stirred 30 min *Intestinal digestion*: adjust pH 5.5 ± 0.05 and stirred 30 min	Increasing pH from 2.0 to 5.5 significantly ($p < 0.01$) decreased soluble calcium over 40% due to the formation of insoluble complexes between Ca and phytic acid. Soluble Ca decreased as the concentration of Ca was reduced at pH 2.0 and 5.5. Soluble Ca is not altered by increasing Zn concentration	Gifford-Steffen and Clydesdale (1993)
Pinto beans (*Phaseolus vulgaris*) cooked and prepared	Evaluate phytate content and mineral solubility in Mexico's food	*Gastric digestion*: pepsin–HCl solution pH 1.35/metabolic shaker waterbath/90 min/37 °C. Centrifugation 3000 rpm/45 min. Supernatant filtration (Whatman no 44 filter paper) *Intestinal digestion*: an aliquot of gastric digest pH adjusted to 7.5 with NaOH//metabolic shaker waterbath/90 min/37 °C. Centrifugation 3000 rpm/45 min. Supernatant filtration (Whatman no 44 filter paper)	61.29% of the total Ca is in the soluble fraction	Wyatt and Triana-Tejas (1994)
Butter beans (*Phaseolus lunatus* L.), and lentils (*Lens culinaris* L.) raw and cooked With/without Ca added or with fe, Zn, Mg and Cu	Investigate Ca-binding capacity of legumes and influence of other minerals	*Simulated small intestine*: pancreatin-buffer solution pH 7/moderate shaking 37 °C/24 h Filtration (Whatman no. 1 filter paper)	Solubility percentages: raw *vs.* cooked legumes: butter beans (77, 90), lentils (36, 95) Raw legumes showed a significant ($p < 0.01$) increase in Ca-binding after cooking with addition of Ca alone or with the other minerals Ca binding can be classified as follows: butter beans > lentils. These results can be attributed partly to the content of phytic acid content and may contribute to reduce Ca solubility	Elhardallou and Walker (1995)

(*continued*)

Table 15.3 (*continued*)

Sample matrix	Aim of study	*In vitro* method	Results	References
Chick pea (*Cicer arietinum*) Blackgram (*Vigna mungo*)	Influence of processing on mineral extractibility	*Gastric digestion*: extraction with HCl 0.03 N/shaking/4 h/37 °C Filtration (Whatman no. 42 filter paper)	Ca extractability (%): chick pea *vs.* blackgram: unprocessed seeds (70–77, 48–59), soaked seeds (70–75, 47–58), cooked seed (74–84, 53–63), autoclaved seeds (79–89, 58–69), sprouted seeds (82–90, 60–71), cooked sprouts of seeds (84–95, 63–76).	Jood and Kapoor (1997)
Pigeon pea (*Cajanus cajan*)	Influence of domestic processing and cooking methods on mineral extractibility	*Gastric digestion*: extraction with HCl 0.03 N/shaking/3 h/37 °C Filtration (Whatman no. 42 filter paper)	Ca extractability (%): soaking (51.3–54.2), soaking and dehulling (61.3), ordinary cooking (55.2–60.2), pressure cooking (56.2–62.0), germination (61.3–64.3). Negative correlation between phytic acid content and Ca HCl–extractibility	Duhan *et al.* (2002)
White beans (*Phaseolus vulgaris*)	Influence of processing on mineral extractability	*Gastric digestion*: extraction with HCl 0.03 N/shaking/3 h/37 °C Filtration (Whatman no. 42 filter paper)	Ca extractability (%): control samples (24.4–44%), soaking (49.5–55.4), soaking and cooking (62.2–67.1)	ElMaki *et al.* (2007)
White beans (*Phaseolus vulgaris* L.), chick pea (*Cicer arietinum* L.) lentils (*lens culinaris* L.)	Estimate mineral bioaccessibility	*Gastric digestion*: adjust pH 2/pepsin solution shaking water bath/2 h/37 °C. Ice for 10 min to stop the pepsin digestion *Intestinal digestion*: an aliquot of gastric digest pH adjusted to 5/pancreatin–bile extract mixture/shaking water bath/2 h/37 °C. Ice for 10 min to stop the intestinal digestion. Adjust pH 7.2. Centrifugation 3200 rpm/25 min/20 °C	Ca solubility (%): white beans (30.8, 33.7), chick peas (22.3, 29.1), lentils (28.8, 46.6)	Sahuquillo *et al.* (2003)

School menus with leguminous base	Estimate mineral bioaccessibility	*Gastric digestion*: adjust pH 2/pepsin solution shaking water bath/2 h/37 °C. Ice for 10 min to stop the pepsin digestion *Intestinal digestion*: an aliquot of gastric digest pH adjusted to 5/pancreatin–bile extract mixture/shaking water bath/2 h/37 °C. Ice for 10 min to stop the intestinal digestion. Adjust pH 7.2. Centrifugation 3500*g*/25 min/20 °C	Ca solubility (%): lentils with sausage (35.3), stew (a dish that contain chickpeas) (91.7).	Cámara *et al.* (2005)
Faba bean (*Vicia Faba* L), azuki bean (*Vigna angularis* L) and mug bean (*Phaseolus radiates* L) seeds	Influence of germination and cooking on *in vitro* Fe, Ca and Zn bioaccessibility	*Salivar and Gastric digestion*: α-amylase solution/30 min/37 °C. Adjust pH 4/stomach medium (lipase, pepsin)/1 h/37 °C *Intestinal digestion*: adjust pH to 6. Pancreatin–bile extract mixture//30 min/37 °C Centrifugation 3000*g*/15 min/4 °C	Ca bioaccessibility (mg): soaking causes significant reduction *vs.* raw legume in faba bean (20.5 *vs.* 24.2) and azuki bean (32.1 *vs.* 37.8) No differences were observed on Ca bioaccessibility after different germination periods or after pressure cooking or microwaving	Luo *et al.* (2013)

Table 15.4 Calcium bioaccessibility (dialyzability methods) from legumes. Several studies that have used *in vitro* dialyzability methods to compare calcium bioaccessibility between legumes and/or evaluate influence of processing.

Sample matrix	Aim of study	*In vitro* method	Results	References
Mottled and white bean (*Phaseolus vulgaris*) Raw and cooked	Determinate *in vitro* availability of Ca Influence of dehulling in Ca dialysis or solubility	**Dialyzability** *Gastric digestion*: adjust pH to 2.0 ± 0.05. Pepsin solution/shaking water bath/2 h/37 °C *Intestinal digestion*: dialysis tubing (cutoff 6–8000 Da) containing $NaHCO_3$ is introduced in digest 30 min/37 °C. When the pH reached 5, addition of pancreatin–bile extract mixture 2 h/37 °C **Soluble Ca** After gastrointestinal digestion, the soluble Ca was determined in retentates by centrifugation 3500*g*/20 min	Dialyzable and soluble Ca (%) in both beans was about 14% Dehulling improved Ca dialyzable with increases of up to 97% Cooking did not affect dialyzable or soluble Ca	Lombardi-Boccia *et al.* (1998)
White beans (*Phaseolus vulgaris* L.), chick pea (*Cicer arietinum* L.) lentils (*lens culinaris L*)	Influence of processing on mineral dialyzability	*Gastric digestion*: adjust pH to 2.0. Pepsin solution/shaking water bath/2 h/37 °C *Intestinal digestion*: dialysis tubing (—cutoff 12 000–1400 Da) containing $NaHCO_3$ is introduced in digest 90 min/37 °C. Pancreatin–bile extract mixture 2 h/37 °C	Cooking (traditional and microwave treatments reduced the dialyzability of Ca (11–14%) *vs.* raw legumes (14–21%) The dialyzability of Ca in the three legumes was higher in industrial processing (15–24%) than in traditional or microwave cooking (12–13%)	Sebastiá *et al.* (2001)

Soybean young seeds (*Canjasus indicus*/*Canjasus canjans*), soybean seeds (*Glycine max*) cooked	Study the bioavailability of Ca from plant foods and the presence of Ca inhibitors factors	*Gastric digestion*: adjust pH to 2. Pepsin solution/shaking water bath/2 h/37 °C *Intestinal digestion*: dialysis tubing (—cutoff 6000–8000 Da) containing $NaHCO_3$ is introduced in digest 30 min/37 °C. Pancreatin–bile mixture 2 h/37 °C	Ca dialyzability (%): young soybean seeds (15.4), soybean mature seeds (11.1), soybean sprouts (20.3) Phytate and dietary fiber were the main inhibitors factors of Ca bioavailability	Kamchan *et al.* (2004)
School menus with leguminous base	Estimate mineral bioaccessibility	*Gastric digestion*: adjust pH to 2.0. Pepsin solution/shaking water bath/2 h/37 °C *Intestinal digestion*: dialysis tubing (—cutoff 10 000–12 000 Da) containing $NaHCO_3$ is introduced in digest 45 min/37 °C. Pancreatin–bile extract mixture 2 h/37 °C	Ca dialyzability (%): lentils with sausage (10.3), stew (a dish that contain chickpeas) (30.5)	Cámara *et al.* (2005)
Green gram (*Phaseolus aureus*) cowpea (*Vigna catjang)* lentil (*Lens culinaris*) chickpea (*Cicer arietinum)*	Influence of germinated and dehulled legumes on Ca and Fe bioavailability	*Gastric digestion*: adjust pH to 2. After 15 min readjusted pH to 2 if is necessary. Pepsin solution/shaking water bath (100–120 strokes per min)/2 h/37 °C. Ice for 90 min to stop the pepsin digestion *Intestinal digestion*: dialysis tubing (cutoff 10 000 Da) containing $NaHCO_3$ is introduced in digest 30 min/37 °C. Pancreatin mixture 2 h/37 °C	Germination and dehulling increased bioavailable Ca (%) *vs.* raw legumes: green gram (24.7, 40.5 *vs.* 15.7); cowpea (38.2, 54.8 *vs.* 22.6); lentil (46.5, 59.6 *vs.* 29.3); chickpea (32.9, 48.1 *vs.* 19.3) Negative correlation coefficients were found between bioavailable Ca and phytic acid, tannin or total dietary fiber	Ghavidel and Prakash (2007)

Summary Points

- This chapter focuses on calcium bioaccessibility in legumes.
- The highest content of minerals always corresponds to the raw material product, with the exception of some germinated products.
- Legume grains are generally processed before consumption, depending upon cultural and taste preferences. It is a general practice to soak legume seeds overnight prior to processing and cooking.
- A major decrease in calcium content is found in relation to the duration of soaking.
- Calcium BA has been estimated by hydrochloric acid or extractability or solubility/dialysis after simulated gastrointestinal digestion.
- Oxalate, phytate, tannins, dietetic fiber and polyphenols are the main antinutritional components that can negatively affect calcium bioaccessibility in legumes.
- Traditional legume preparation processes (soaking, germination and fermentation) can activate native phytases and substantially degrade phytic acid.
- Cooking (domestic or industrially) is responsible for a loss of soluble oxalate contents in legumes.
- Dehulling the germinating legumes is associated with significant enhancement of calcium bioaccessibility.

Key Facts

In Vitro Bioavailability of Ca in Legumes

1. Legumes are a low-cost source of minerals, including calcium.
2. Calcium distribution in legumes is not homogeneous. In general, the hull is particularly rich in calcium, with lesser amounts in the cotyledon.
3. Although the calcium content in legumes is generally higher in dairy products, its bioaccessibility is lower.
4. To evaluate calcium bioaccessibility from legumes, solubility methods comprising hydrochloric acid extractability (gastric stage) or complete simulation of gastrointestinal digestion are used.
5. Solubility and dialysis methods are useful *in vitro* techniques for comparing calcium bioaccessibility among different legumes, different processing methods, or for assessing the influence of dietary components.

Definitions of Words and Terms

Bioavailability. Refers to the fraction of the ingested mineral that is used for normal physiological functions or storage.
Bioaccessibility. Refers to the fraction of a mineral that is released from its food matrix in the gastrointestinal tract and thus becomes available for intestinal absorption.

Cotyledons. The mature legume has three major components, hull, cotyledon and embryonic axis, which on average represent 89%, 10% and 1%, respectively, of the total seed weight. The cotyledon contains the main reserve of substances, basically proteins and carbohydrates.
Dehulling. Removing the hull. Decortication of seeds. Produces refined cotyledons.
Dialysis. Refers to the soluble mineral fraction that dialyses through a membrane of a certain pore size after simulated gastrointestinal digestion.
Dietary fiber. Components of plants that are not digested.
Essential minerals. Those minerals that must be consumed with the diet and are necessary for the human organism.
Germination. A natural process occurring during the growth period of seeds in which they meet the minimum conditions for growth and development. The process starts with the uptake of water by the quiescent dry seed and terminates with the emergence of the embryonic axis, usually the radical axis.
HCl extractability. Refers to soluble mineral extracted with HCl 0.03 N.
Pressure cooking. Involves processing foods at very high pressure for a short period of time. Such cooking is faster than the traditional method.
Soaking. A process in which the legume is submerged in water for a period of time before cooking. It can represent a simple prolongation of washing of the seeds, and can also facilitate dehulling of the seeds.
Solubility. Refers to the soluble mineral fraction obtained by centrifugation after simulated gastrointestinal digestion.

List of Abbreviations

BA Bioaccessibility
IP6 inositol hexaphosphate
IP5 inositol pentaphosphate

References

Abebe, Y., Bogale, A., Hambidge, K. M., Stoecker, B. J., Bailey, K. and Gibson, R. S., 2007. Phytate, zinc, iron and calcium content of selected raw and prepared foods consumed in rural Sidama, Southern Ethiopia, and implications for bioavailability. *Journal of Food Composition and Analysis.* 20: 161–168.

Al-Numair, K. S., Ahmed, S. B., Al-Assaf, A. H. and Alamri, M. S., 2009. Hydrochloric acid extractable minerals and phytate and polyphenols contents of sprouted faba and white beans cultivars. *Food Chemistry.* 113: 997–1002.

Cámara, F., Amaro, M. A., Barberá, R. and Clemente, G., 2005. Bioaccessibility of minerals in school meals: comparison between dialysis and solubility methods. *Food Chemistry.* 92: 481–489.

Chan, S. S. L., Ferguson, E., Bailey, K., Fahmida, U., Harper, T. B. and Gibson, R. S., 2007. The concentrations of iron, calcium, zinc and phytate in cereals and legumes habitually consumed by infants living in East Lombok, Indonesia. *Journal of Food Composition and Analysis*. 20: 609–617.

Duhan, A., Khetarpaul, N. and Bishnoi, S., 2002. Content of phytic acid and HCl-extractability of calcium, phosphorus and iron as affected by various domestic processing and cooking methods. *Food Chemistry*. 78: 9–14.

El-Adawy, T. A., Rahma, E. H., El-Bedawy, A. A. and Sobihah, T. Y., 2000. Effect of soaking process on nutritional quality and protein solubility of some legume seeds. *Nahrung*. 44: 339–343.

Elhardallou, S. B. and Walker, A. F., 1995. Binding of Ca by three starchy legumes in the presence of ca alone or with Fe, Zn, Mg and Cu. *Food Chemistry*. 52: 379–384.

ElMaki, H. B., AbdelRahaman, S. M., Idris, W. H., Hassan, A. B., Babiker, E. E. and El Tinay, A. H., 2007. Content of antinutritional factors and HCl-extractability of minerals from white bean (*Phaseolus vulgaris*) cultivars: influence of soaking and/or cooking. *Food Chemistry*. 100: 362–368.

Etcheverry, P., Grusak, M. A. and Fleige, L. E., 2012. Application of in vitro bioaccessibility and bioavailability methods for calcium, carotenoids, folate, iron, magnesium, polyphenols, zinc, and vitamins B_6, B_{12}, D, and E. *Frontiers in Physiology*. 3: 317. DOI: 10.3389/fphys.2012.00317.

Fernández-García, E., Carvajal-Lérida, I. and Pérez-Gálvez, A., 2009. In vitro bioaccessibility assessment as a prediction tool of nutrient efficiency. *Nutrition Research*. 29: 751–760.

Frossard, E., Bucher, M., Mächler, F., Mozafar, A. and Hurrell, R., 2000. Potential for increasing the content and bioavailability of Fe, Zn and Ca in plants for human nutrition. *Journal of the Science of Food and Agriculture*. 80: 861–879.

Ghavidel, R. A. and Prakash, J., 2007. The impact of germination and dehulling on nutrients, antinutrients, in vitro iron and calcium bioavailability and in vitro starch and protein digestibility of some legume seeds. *LWT – Food Science and Technology*. 40: 1292–1299.

Gibson, R. S., Bailey, K. B., Gibbs, M. and Ferguson, E. L., 2010. A review of phytate, iron, zinc, and calcium concentrations in plant-based complementary foods used in low-income countries and implications for bioavailability. *Food and Nutrition Bulletin*. 31(suppl. 2): S134–S146.

Gifford-Steffeni, S. R. and Clydesdale, F. M., 1993. Effect of varying concentrations of phytate, calcium, and zinc on the solubility of protein, calcium, zinc, and phytate in soy protein concentrate. *Journal of Food Protection*. 56: 42–46.

Guéguen, L. and Pointillart, A., 2000. The bioavailability of dietary calcium. *Journal of the American College of Nutrition*. 19(suppl. 2): 119S–136S.

Hefnawy, T. H., 2011. Effect of processing methods on nutritional composition and anti-nutritional factors in lentils (*Lens culinaris*). *Annals of Agricultural Science*. 56: 57–61.

Hemalatha, S., Platel, K. and Srinivasan, K., 2007. Zinc and iron contents and their bioaccessibility in cereals and pulses consumed in India. *Food Chemistry*. 102: 1328–1336.

Hurrell, R. F., 2003. Influence of vegetable protein sources on trace element and mineral bioavailability. *Journal of Nutrition*. 133: 2973S–2977S.

Jood, S. and Kapoor, A. C., 1997. Improvement in bioavailability of minerals of chickpea and blackgram cultivars through processing and cooking methods. *International Journal of Food Sciences and Nutrition*. 48: 307–312.

Kamchan, A., Puwastien, P., Sirichakwal, P. P. and Kongkachuichai, R., 2004. In vitro calcium bioavailability of vegetables, legumes and seeds. *Journal of Food Composition and Analysis*. 17: 311–320.

Lombardi-Boccia, G., Lucarini, M., Di Lullo, G., Del Puppo, E., Ferrari, A. and Carnovale, E., 1998. Dialysable, soluble and fermentable calcium from beans (*Phaseolus vulgaris* L.) as model for in vitro assessment of the potential calcium availability. *Food Chemistry*. 61: 167–172.

Luo, Y.-W., Xie, W.-H., Jin, X.-X., Wang, Q. and Zai, X.-M., 2013. Effects of germination and cooking for enhanced in vitro iron, calcium and zinc bioaccessibility from faba bean, azuki bean and mung bean sprouts. *CyTA – Journal of Food*. 11: 318–323.

Mañez, G., Alegría, A., Farré, R. and Frígola, A., 2002. Effect of traditional, microwave and industrial cooking on inositol phosphate content in beans, chickpeas and lentils. *International Journal of Food Sciences and Nutrition*. 53: 503–508.

Quinteros, A., Farré, R. and Lagarda, M. J., 2002. Contenidos de calcio, magnesio, hierro, cinc y Fósforo en legumbres crudas y sometidas a distintos procesos de cocción. *Revista Amazónica de Investigación Alimentaria*. 2: 97–102.

Quinteros, A., Farré, R. and Lagarda, M. J., 2003. Effect of cooking on oxalate content of pulses using an enzymatic procedure. *International Journal of Food Sciences and Nutrition*. 54: 373–377.

Sahuquillo, A., Barberá, R. and Farré, R., 2003. Bioaccessibility of calcium, iron and zinc from three legume samples. *Nahrung*. 47: 438–441.

Sandberg, A. S., 2002. Bioavailability of minerals in legumes. *British Journal of Nutrition*. 88(suppl. 3): S281–S285.

Sebastiá, V., Barberá, R., Farré, R. and Lagarda, M. J., 2001. Effects of legume processing on calcium, iron and zinc contents and dialyzabilities. *Journal of the Science of Food and Agriculture*. 81: 1180–1185.

Umeta, M., West, C. E. and Fufa, H., 2005. Content of zinc, iron, calcium and their absorption inhibitors in food commonly consumed in Ethiopia. *Journal of Food Composition and Analysis*. 18: 803–817.

Untoro, J., Karyadi, E., Wibowo, L., Erhardt, J. and Gross, R., 2005. Multiple micronutrient supplements improve micronutrient status and anemia but not growth and morbidity of Indonesian infants: a randomized, double-blind, placebo-controlled trial. *Journal of Nutrition*. 135: 639S–645S.

Wyatt, C. J. and Triana-Tejas, A., 1994. Soluble and insoluble Fe, Zn, Ca, and phytates in foods commonly consumed in northern Mexico. *Journal of Agricultural and Food Chemistry*. 42: 2204–2209.

CHAPTER 16

Calcium – Function and Effects: Rice Calcium and Phytic Acid Levels

JIANFEN LIANG*[a], YIFAN HE[a], QIAN GAO[a], XUAN WANG[a], AND M. J. ROBERT NOUT[b]

[a]College of Food Science and Nutritional Engineering, China Agricultural University, P.O. Box 294, Tsinghua East Road, Beijing 100083, P. R. China; [b]Laboratory of Food Microbiology, Wageningen University, Bornse Weilanden 9, 6700 AA Wageningen, The Netherlands
*E-mail: liangjf@cau.edu.cn

16.1 Introduction

Rice is the seed of the monocot plant *Oryza sativa* or *Oryza glaberrima*. Basically, rice is a starchy crop, which acts as the primary food source for more than half of the world population. According to data of 2010, rice is the second-highest produced grain worldwide, after maize. In Asia, it serves as the major source of energy, protein, thiamine, riboflavin, niacin, iron and calcium in the diet (Juliano, 1997). Rice is also the most important staple food in China. In 2012, the national production was estimated at 20.6 million tons (FAO, 2014). Previous reports indicated a rice consumption in China of 251 g per capita per day, accounting for 30.4% of the supply of dietary energy, 19.5% of dietary protein, and 2.5% of dietary fat (Kennedy *et al.*, 2002).

Food and Nutritional Components in Focus No. 10
Calcium: Chemistry, Analysis, Function and Effects
Edited by Victor R. Preedy

Published by the Royal Society of Chemistry, www.rsc.org

Rice does not fulfill all nutritional requirements. It does not supply minerals adequately because of its low content of minerals, and moreover a loss of minerals takes place during processing. In addition, rice contains phytic acid. Phytic acid is the most important antinutritional factor (ANF) impeding the availability of divalent minerals. It forms complexes with mineral ions, such as Fe, Zn and Ca, thereby reducing their bioavailability.

The concentrations of minerals and mineral inhibitors vary due to genetic diversity, soil and climate of cultivation, and agricultural practices (such as irrigation and fertilization), postharvest conditions and processing prior to consumption (Phuong *et al.*, 1999; Liang *et al.*, 2007). Genetic factors determine the seed morphology and to some extent the levels of minerals, which mostly occur as phytic acid chelates. Cultivation practice affects the uptake of minerals from soil and their distribution throughout the plant (Slingerland *et al.*, 2009).

Rice has the highest food yield and food energy yield among all the important cereals, and contrary to maize that is predominantly used for feed or industrial purposes, 95% of rice is for food use, so even a small change in its nutritive value is highly significant. Total mineral content (ash content) of brown rice and white rice (milled rice, polished rice) is about 1–1.5% and 0.5–0.8%, respectively. Calcium, iron, zinc and cadmium are the most relevant mineral elements in rice. Rice also contains phytic acid, approximately 1% in brown (cargo) rice, and much less in white (polished) rice. Phytic acid is of interest because of its antinutritional effect, although at low concentrations it also has positive effects on human health. In this chapter we focus on calcium, in relation to the antinutritional effect of phytic acid.

16.2 Calcium in Rice

Calcium plays an important role in human health since it participates in building stronger, denser bones early in life and keeps bones strong and healthy later. Approximately 99% of the body's calcium is stored in the bones and teeth. In addition to the obvious structural role of the skeleton, it also functions in muscle and nerve activity, coagulation of blood, and other physiological phenomena. Notably, calcium deficiencies constitute predisposing conditions for some common chronic diseases. They increase the risk of cancer, particularly of breast, colon, and prostate gland, as well as of metabolic disorders, *e.g.* metabolic syndrome, and hypertension (Peterlik and Cross, 2005).

It is vital, therefore, that adequate dietary calcium is consumed at all stages of life. Dairy products are excellent sources of calcium, both qualitatively and quantitatively. Both the amount in diet, and the absorption of dietary calcium in foods are critical factors. Food made of diverse ingredients will contain different levels of calcium, whereas it may contain antinutritional factors such as oxalic acid and phytic acid, which bind calcium and reduce its absorption. Green leafy vegetables, seeds and legumes are good sources of calcium. Some vegetables such as kale, celery, collard, pak-chee-lao, Chinese cabbage and soybean sprouts, containing low levels of dietary fiber, phytic acid and oxalate, have a higher calcium dialyzability (20–39%) than Indian mulberry and sesbania leaves (11–18%); and even 1.3–1.6 times that of milk powder. In contrast, amaranth, wild betal and white and black sesame seeds that contain high levels of oxalate (680–2620 mg/100 g) had the lowest dialyzable calcium (Achiraya *et al.*, 2004).

16.2.1 Calcium Contents in Rice Products

Calcium contents of rice products reported in different studies are summarized in Table 16.1.

Rice and rice products are poor sources of calcium, and only a few studies have been published. Calcium contents in brown rice depend on variety and cultivation location. Generally, white polished rice contains lower levels of calcium than the original brown cargo rice, and this loss is mainly caused by polishing (Doesthale *et al.*, 1979). Levels of calcium in other rice products are mainly due to processing and/or added ingredients. For example, higher levels of calcium in infant foods result from ingredients formulation and fortification (Liang *et al.*, 2010).

From Table 16.1, we can also see significant differences in calcium contents published by different researchers. This may stem from both the rice products sampled as well as the methods of analysis used. Atomic absorbance spectrophotometry (AAS), energy-dispersion X-ray, and inductively coupled plasma (ICP)-AAS or ICP-OES (optical emission spectrophotometry)

Table 16.1 Calcium contents of rice products (mg/100 g, dry basis). Calcium contents of rice products published by different researchers are summarized. There are not many studies on calcium nutrition of rice products. The data showed a big variation of different products, and indicated the effects of processing and materials used.

Products	Average	Lowest value	Highest value	Number of samples	References
Brown rice					
Worldwide	12.1	3.7	26.4	1259	Gregorio *et al.* (2009)
Bangladesh	22.2	17	32	10	Tamanna *et al.* (2013)
China	28.6	24.0	33.0	3	Liang *et al.* (2010)
Germinated rice	38.7	—	—	1	Liang *et al.* (2010)
Parboiled rice	1.9	1.3	2.3	9	Tamanna *et al.* (2013)
Milled rice	3.4	2.52	4.74	4	Ma *et al.* (2005)
Milled rice	6.8	5.0	11.0	8	Liang *et al.* (2010)
Cooked rice (polished)	8.23	7.08	9.07	4	Ma *et al.* (2005)
Rice noodle	14.01	—	—	1	Ma *et al.* (2005)
Rice noodle	27.8	3.0	75	7	Liang *et al.* (2010)
Rice cracker	26.7	6	59	7	Liang *et al.* (2010)
Infant food	455.0	408	532	4	Liang *et al.* (2010)

are presently the main methods for mineral analysis. Procedures for pretreatment of samples, their digestion and detection limits of equipment also affect the absolute levels of minerals reported in rice (Liang *et al.*, 2007).

16.2.2 Availability of Calcium from Rice Products

Low bioavailability of dietary calcium is one of the main reasons for deficiencies. Factors such as the physical state of calcium (soluble or insoluble) in food, the presence of inhibitors of mineral uptake (*e.g.* oxalic acid, phytic acid), the individual calcium status of the consumer, will influence the absorption of calcium by the human body. It was reported that the *in vitro* solubility of calcium ranged between 0 and 87%, with lowest average (12%) in brown rice and highest average (50%) in infant foods (Liang *et al.*, 2010). Mineral solubility could be improved by various strategies. For calcium in rice products, phytic acid is the main factor that inhibits its availability, since only traces of oxalic acid are present. Enzymatic pretreatment of white rice with phytase was shown to significantly improve mineral solubility (Liang *et al.*, 2009).

16.3 Phytic Acid in Rice

16.3.1 Phytic Acid and its Chemical Determination

16.3.1.1 Phytic Acid

Phytic acid, or myo-inositol 1,2,3,4,5,6 hexakis dihydrogen phosphate (also known as inositol hexaphosphoric acid or myo-inositol hexakisphosphate-IP6), has the molecular formula $C_6H_{18}O_{24}P_6$ and molecular weight 660.04. Phytic acid is a saturated cyclic acid and a natural plant compound found in most cereal grains, legumes, roots, tubers, nuts and oilseeds. Phytic acid has a unique structure that explains its characteristic properties. Phytic acid has 12 replaceable protons and at physiological pH, the phosphate groups of phytic acid are negatively charged, allowing them to form complexes (chelates) with multivalent cations, positively charged proteins, and starch (either directly or indirectly) (Figure 16.1).

Phytic acid is the principal storage form of phosphorus in many plant tissues, especially in bran and seed. Phytate occurs in cereals as a mixed potassium, magnesium, and calcium salt of phytic acid (De Stefano *et al.*, 2002). Phytate is heat resistant and cannot be digested by humans or nonruminant animals. Phytate can be degraded by phytase, either as endogenous enzyme in seeds and accumulated during germination, or by exogenous microbial enzymes, to myo-inositol or lower inositol phosphates (IP1–IP5), while releasing phosphorus and minerals at the same time.

Endogenous phytase, which is mainly located in the outer layer of cereals kernels, is activated and accumulated during seed germination, and acts on phytic acid to release inorganic phosphate, which is then utilized for plant

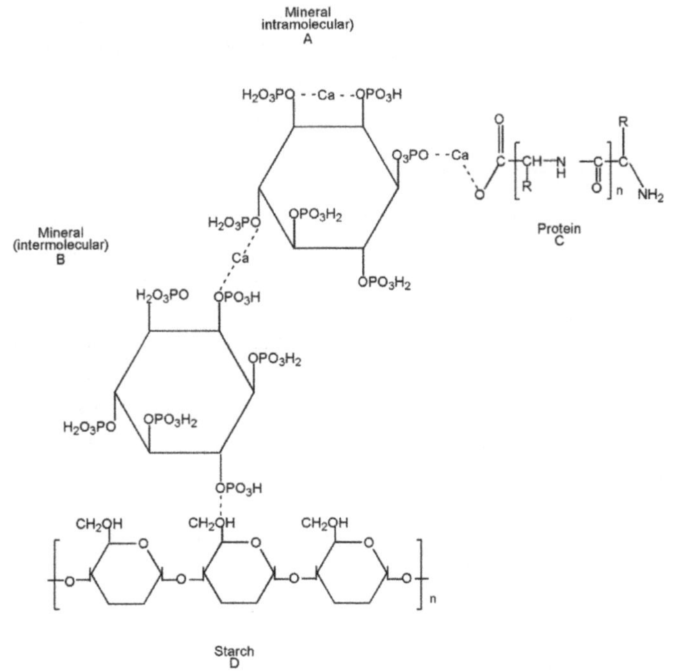

Figure 16.1 Chelates of phytic acid with different compounds in food. Chelates of phytic acid with minerals, protein and starch are illustrated. Calcium is used as model mineral. (A) Mineral chelated in one molecular of phytic acid; (B) mineral chelated in two molecules of phytic acid; (C) chelate with protein; (D) chelate with starch. The figure is adapted and modified from Oatway *et al.* (2001).

growth, and serves as a natural buffer in grains as well. With degradation and the release of phosphate by phytase, lower inositol phosphate esters, such as pentakis-, tetrakis-, tris-, bis- and monophosphate, and even inositol are generated (Lönnerdal *et al.*, 1989). Steeping in demineralized water with 0.5 mM calcium chloride and 0.1 M hydrogen peroxide at 20 °C for 24 h led to the lowest phytase activity in brown rice, which was only 1/4 that of raw rice. Phytase activity increased with prolonged sprouting time and gradually reached stable levels in the sprouting stage. After 5 d sprouting, the phytase activity in brown rice reached 320–340 U kg^{-1}(unless stated otherwise, all are expressed on dry matter basis), which was about 1.5 times higher than in the raw material. Compared to extreme conditions (either strong acidic or alkaline), steeping in demineralized water at neutral pH (6.8) provided the optimum pretreatment prior to sprouting at 25 °C, to activate maximum levels of phytase (Ou *et al.*, 2011).

It was reported that application of microbial phytase in rice fractions led to a decrease of phytic acid content and an increase of free phosphate. In white rice, in kernel as well as ground flour, phytic acid contents decreased to levels below the detection limit after treatment with 500 U kg^{-1} dry matter phytase, at 55 °C, pH 2.5–4 for 30 min. An increase of soluble phosphorous indicated stronger effects on flour. Under the same conditions, added phytase had no significant effect on phytic acid content in brown-rice kernels, which suggested that testa and aleurone layers acted as a barrier to influx of, *e.g.*, phytase and diffusive loss of matter.

16.3.1.2 Chemical Determination of Phytic Acid

Among the different techniques proposed to determine phytic acid, the most widely accepted are the AOAC anion-exchange method, and the high-performance liquid chromatography (HPLC) method. The anion-exchange method was adopted as the China National Standard Analysis Method. The method consists of subsequent extraction, separation on an anion-exchange resin, and spectrophotometric detection of the reaction product with ferric chloride ($FeCl_3$) and sulfosalicylic acid. Phytic acid in samples is extracted by a solution of sodium sulfate (100 g L^{-1})–hydrogen chloride (1.2%) at room temperature for 2 h. The extract is then filtered or centrifuged for clarification. The clarified extract is mixed with sodium hydroxide (30 g L^{-1}) and distilled water followed by separation on an anion resin column (resin, AG1-X4, 100–200 mesh). The column is then washed with sodium hydroxide (0.05 mol L^{-1}). Phytic acid is eluted with sodium chloride (0.7 mol L^{-1}). Next, the absorbance of a mixture of eluate and $FeCl_3$ (0.03%) and sulfosalicylic acid (0.3%) is measured at 500 nm using a spectrophotometer (Ma *et al.*, 2005). The limitation of this method is that it cannot distinguish IP6 from other inositol phosphates (IP1–IP5), so the detected result is higher than the real value of IP6.

The HPLC method can identify different forms of inositol phosphates. This method is sensitive and suitable for detection of low concentrations.

Phytic acid is normally recognized as an ANF in cereals and legumes. It combines minerals and proteins, and influences their solubility and bioavailability. The intake of much phytic acid might be one important factor causing deficiency of essential nutrients.

In recent years, some beneficial health effects of phytic acid have also been reported. Phytic acid is an effective antioxidant (Cornforth, 2002). It has also been reported that phytic acid can reduce the toxicity of some heavy metals due to its strong chelating capacity (Persson *et al.*, 1998).

16.3.2 Phytic Acid Contents of Rice and Rice Products

16.3.2.1 Phytic Acid in Brown Rice

Rice cultivation is ideally suited to countries and regions with low labor costs and high rainfall; however, rice can be grown practically anywhere, even on a steep hill or mountain area with the use of water-controlling terrace systems. It was reported that rice consumption in sub-Saharan Africa is increasing yearly and will rise by 50% from 2005 to 2015 (Gregorio *et al.*, 2009). Currently, it is estimated that more than fifty thousand varieties of rice exist in China, of which about two hundred and thirty are grown at a commercial scale.

Phytic acid contents can fluctuate because phytates are the main form of storage phosphorous in cereal grains, representing about 70% of total phosphorous in the seed. During the development of the rice seed, phosphorous is deposited in the form of mixed phytates, together with some mineral cations, such as potassium, magnesium, calcium, iron, and zinc. This process is influenced by genetic characteristics, environmental conditions and agricultural practice. Table 16.2 presents published data of phytic acid in brown rice.

Different phytic acid contents were found in brown rice from different sources. The highest phytic acid content was 19.4 mg g^{-1}, in commercial brown rice from China, and the lowest was 4.1 mg g^{-1}, from Bangladesh. Phytic acid ranged from 4.05–6.35, 6.9–19.4 and 8.6–17.6 mg g^{-1}, respectively, in brown rice from Bangladesh, China and Korea. Combined effects of rice

Table 16.2 Phytic acid content (mg g^{-1}) of brown rice from Bangladesh, China and Korea. Phytic acid contents in brown rice from different countries were summarized. Rice from Bangladesh had the smallest variation, while rice from China had the biggest phytic acid contents range. Different sources of samples, either by cultivation practices of sample collections support the effects of genetic characteristics, agricultural practice and cultivation location on phytic acid contents in rice.

Country	Average	Lowest	Highest	Number of samples	Origin of samples	References
Bangladesh	5.06	4.05	6.35	10	Collected from Bangladesh Rice Research Institute	Tamanna *et al.* (2013)
	9.6	7.2	11.9	56	Collected from regional Academy of Agricultural Science	Liang *et al.* (2007)
	17.5	14.9	19.4	3	Purchased at supermarkets in Beijing	Liang *et al.* (2010)
	8.73	6.9	10.3	72	Grown at the experimental farm	Liu *et al.* (2005)
	8.15	6.75	9.42	24	Grown at four ecologically different locations	Liu *et al.* (2005)
China	8.68	6.99	10.34	29	Grown at the experimental farm	Wu *et al.* (2007)
Korea	12.6	8.6	17.6	68	—	Lee *et al.* (1997)

genotype, environment (*e.g.* soil, temperature, rainfall), and cultivation practice underlie such significant differences. The contribution of genetic characteristics to phytic acid contents was demonstrated by growing different varieties under the same environmental conditions, with the same cultivation practice (Liu *et al.*, 2005; Wu *et al.*, 2007). A comparison of 24 rice varieties, grown in four ecologically different locations of China, *i.e.* Hangzhou, Jiaxing, Changzhou and Xi'an, showed that location represented a greater contribution to phytate levels than cultivars (Liu *et al.*, 2005). The environmental effect appears to be predominant in determining phytic acid.

In addition, sensitivity and accuracy of determination method may also influence results. Tamanna *et al.* (2013) selected ten high-yield rice varieties cultivated in Bangladesh, which differed in their morphology, 5 long-, 3 medium- and 2 short-grain. The lowest contents and the smallest variation of phytic acid contents were found for those genotypes may be the result of low sensitivity of the determination method. They determined phytic acid contents with the following method. Phytic acid reacts with $FeCl_3$ and form ferric phytate. The available ferric ion after reaction is determined by developing a blood-red color with potassium thiocyanate (Tamanna *et al.*, 2013). The AOAC method, which gives higher values, was applied in the other studies.

16.3.2.2 Phytic Acid in Rice Products

Variation of phytic acid contents gives the possibility for production of rice with low phytic acid contents in grains. Using suitable cultivars for a given location is an effective approach to achieve high yield and control the phytic acid contents in rice.

The rough rice, or paddy, contains inedible and edible parts. The grains are first "milled" using a huller to remove the outer inedible part, *i.e.* the husk (or hull). The product of "milling" is the edible part, which is called brown rice. Brown rice is not widely consumed because its sensory properties are not appreciated by most consumers, especially in Asia. Brown rice is polished to remove the bran, thereby creating polished rice (also called white rice). White rice is commonly consumed as cooked rice or used as raw material for food. An overview of commercial rice products (Liang *et al.*, 2010) is listed in Table 16.3.

White rice is the most popular rice product all over the world, whereas the other rice products have their more specific consumer markets. For example, parboiled rice is mainly consumed in South Asia, Europe, Africa and the Middle East; germinated rice is appreciated by Japanese. Rice noodle is frequently consumed as breakfast in the South of China.

Phytic acid contents in processed rice products are summarized in Table 16.4.

Germinated rice contained the highest concentration of phytic acid of all processed rice products, followed by parboiled rice, both of them had a similar phytic acid concentration as brown rice (Liang *et al.*, 2007; Tamanna *et al.*, 2013). Rice noodles had the lowest level.

Table 16.3 Commercial rice products.

Category	Product	Processing
Primarily processed products[a]	Brown rice	Hulled only
	Germinated rice	Brown rice, germinated until a sprout length of 0.5–1.0 mm is reached
	Parboiled rice	Paddy is parboiled, followed by hulling, milling and polishing
	White rice	Milling and polishing to remove the outer layer of brown rice and obtain nice appearance and edible quality
Intensively processed products[b]	Rice noodle[c]	White rice soaked in water, ground with or without water, steamed, extruded, cooled and dried
	Rice noodle[d]	White rice together with other ingredients, such as starch or soy protein soaked or not soaked in water, ground, steamed, extruded, cooled and dried
	Rice crackers	Rice mixed with water, pulped, molded, puffed, and baked
	Infant foods	White rice ground to powder, roasted, enzyme-treated, drum dried and formulated with other ingredients

[a]Primarily processed products: still can recognize the shape of rice kernel, include brown, white and germinated rice.
[b]Mainly originate from further processing of white rice, include rice noodles, rice crackers and rice-based infant foods.
[c]Ingredients of rice noodles are white rice and water, no other cereal materials added.
[d]Other cereal materials were used together with rice.

Table 16.4 Phytic acid contents (mg g^{-1}) of rice products. Phytic acid contents of different types of intensively processed rice products published by different researchers are summarized. Germinated rice and rice noodle by Ma *et al.* only had one sample so there were no lowest and highest values for those products. Germinated and parboiled rice had the same levels of phytic acid contents as brown rice. Affected by processing procedures, rice noodle had the lowest level.

Products	Average	Lowest	Highest	Number of samples	References
Germinated rice	13.1	—	—	1	Liang *et al.* (2010)
Parboiled rice	5.34	4.25	6.65	9	Tamanna *et al.* (2013)
Milled rice	1.15	0.55	1.83	4	Ma *et al.* (2005)
Milled rice	1.6	0.6	2.4	8	Liang *et al.* (2010)
Cooked rice	0.31	0.14	0.38	4	Ma *et al.* (2005)
Rice noodle	0.14	—	—	1	Ma *et al.* (2005)
Rice noodle	1.2	0	4.1	7	Liang *et al.* (2010)
Rice cracker	1.4	0.8	2.6	7	Liang *et al.* (2010)
Infant food	2.3	0.9	5.9	4	Liang *et al.* (2010)

Phytic acid contents of most rice products are much lower than that of brown rice. High-pressure steaming for parboiled rice production had no significant effect on phytic acid contents, and parboiled rice (without milling and polishing) had even higher phytic acid contents than raw brown rice (Tamanna *et al.*, 2013). Commercial-scale germination duration to produce germinated rice is apparently insufficient to activate and accumulate endogenous phytase, which could have reduced phytic acid and thus lowered phytic acid contents (Liang *et al.*, 2008a). The phytic acid content in milled rice is much lower than in brown rice, due to milling and polishing (Liang *et al.*, 2008b). The combined effects of loss during soaking, degradation by endogenous phytase and micro-organisms in rice-noodle processing brought out the lowest phytic acid content in rice noodles (Liang *et al.*, 2009).

16.4 Location of Calcium, Phosphorous and Phytic Acid in Rice Kernel

Brown rice (hulled rice) is composed of surface bran (6–7% by weight), endosperm (~90%) and embryo (2–3%). As in most cereal grains, the kernel does not reveal a homogeneous structure from its outer (surface) to its inner (central) part (Itani *et al.*, 2002). Milling by abrasion is an important mechanical procedure for the production of white rice, which is referred to as milled, polished or whitened rice when 8–10% of the outer layer (mainly bran) of brown rice is removed (Kennedy *et al.*, 2002). As a consequence, information on the location of components will greatly help to improve sensory properties and retain essential nutrients in white rice. Early studies described the effect of milling on levels of minerals and their spatial distribution in relation to approximate milling degrees, such as lightly milled, reasonably milled and well milled, or as fractions I, II and III, respectively (Kennedy and Schelstraete, 1975; Tabekhia and Luh, 1979). A study on the effect of milling on mineral and trace element composition of raw and parboiled rice indicated that milling rates of brown rice of 5% and 10% led to phosphorous decrease from 349 mg/100 g to, respectively, 167 and 101 mg/100 g, while for parboiled rice they were reduced from 350 mg/100 g to 219 and 138 mg/100 g (Doesthale *et al.*, 1979).

Development of precisely controlled milling machines, and X-ray imaging enabled the study of the spatial distribution of compounds and elements in rice kernels. Studies on Indian rice indicated that milling extent significantly influences losses of magnesium and calcium (Bajaj *et al.*, 1989). The distribution of minerals such as magnesium, potassium, phosphorus, calcium and sulfur in quinoa seeds was mapped by X-ray fluorescent microscopy techniques (Emoto *et al.*, 2004).

Using two varieties of short-grain brown rice as samples, precise abrasive milling was used to obtain a range of milling degrees, and X-ray imaging methods were applied to map the distribution of different minerals.

16.4.1 Location of Elements with X-Ray Microscope

X-ray images indicated that a high calcium density layer is located in the outer layer of the brown-rice kernel. Most calcium is found in rice bran, in contrast to the location of zinc (Liang *et al.*, 2008b). The distribution of calcium in the kernel differs with variety (Figure 16.2). In Ji 307, a distinct region with a high density of calcium occurs around the kernel. The embryo is the part that contains the highest density of calcium. Very little calcium is located in the inner bran layer and core endosperm. In Nanjing No. 1, a high calcium density in the bran, but not in the embryo is observed. On the other hand, some calcium is located in the endosperm. Thus, considering the human nutrition of calcium, it would be easier to maintain more calcium in white rice of Nanjing No. 1 than in Ji 307.

Images of spatial distribution of phytic acid (as phosphorous) in the whole brown-rice kernel and in the embryo obtained by X-ray fluorescent scanning are shown in Figure 16.3. Phosphorous is distributed unevenly in the whole brown rice kernel. The highest density of phosphorous occurred at the boundary of embryo and endosperm (Figure 16.3a). In contrast with the high density of zinc in embryo, there is not much phosphorous located in the embryo, which is similar to the case of calcium. The whole kernel is practically surrounded by a high-density phosphorous layer that tends to decrease from the surface region inward. When compared with other varieties, we observed that Ji 307 has a similar phosphorous distribution compared with Bijing 37, both being short-grain japonica rice. The distribution of phosphorous in the rice kernels suggested that at least the outer layer should be removed if we want to significantly decrease phytic acid in milled rice, since 70–85% of phosphorous occurs as phytic acid in rice.

16.4.2 Location of Phytic Acid by Abrasive Method

From Figure 16.4 we conclude that retention of phytic acid decreases and mass loss increases with prolonged milling. Although it was observed earlier that all minerals (including phosphorus) decrease from the outermost fraction (Itani *et al.*, 2002) it appears here that more than 70% of phytic acid was removed by milling off 8% of the outmost layer, which was accomplished by 60 s milling time. In practice, this 8% of kernel represents mainly the bran. With progressive milling, the rate of loss of phytic acid and mass decreased. This may be caused by the harder structure of the inner parts of the rice kernel. The results also tell us that about 80% of kernel weight only contains less than 5% total phytic acid.

X-ray images showed the location of elements in rice kernels, while abrasive milling approaches enabled a quantification of the distribution of phytic acid. In conclusion, the location of the minerals (phosphorous, calcium and zinc) differs per element and rice variety. From the fact that calcium occurs in the same parts where phosphorous predominates, we could expect that most of the calcium is present in the form of phytate. Consequently, the bioavailability of calcium could be improved when phytic acid is degraded.

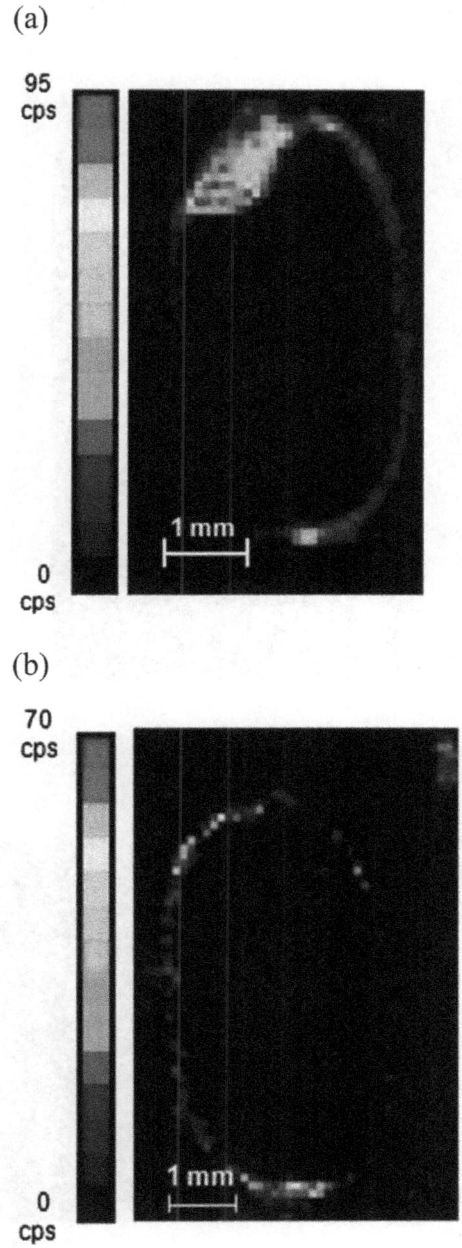

Figure 16.2 Location of calcium in short-grain rice ((a) variety Ji 307; (b) variety Nanjing No. 1). Both varieties are japonica rice (short grain). The images illustrate the calcium location in brown-rice kernels. Most of the calcium in Ji 307 is located in the outer layer of the kernel and the calcium in the endosperm part is below the detection limits, while a low density of calcium in the endosperm of the kernel of Nanjing No. 1 was measured. The highest density of calcium in Ji 307 is at the embryo part.

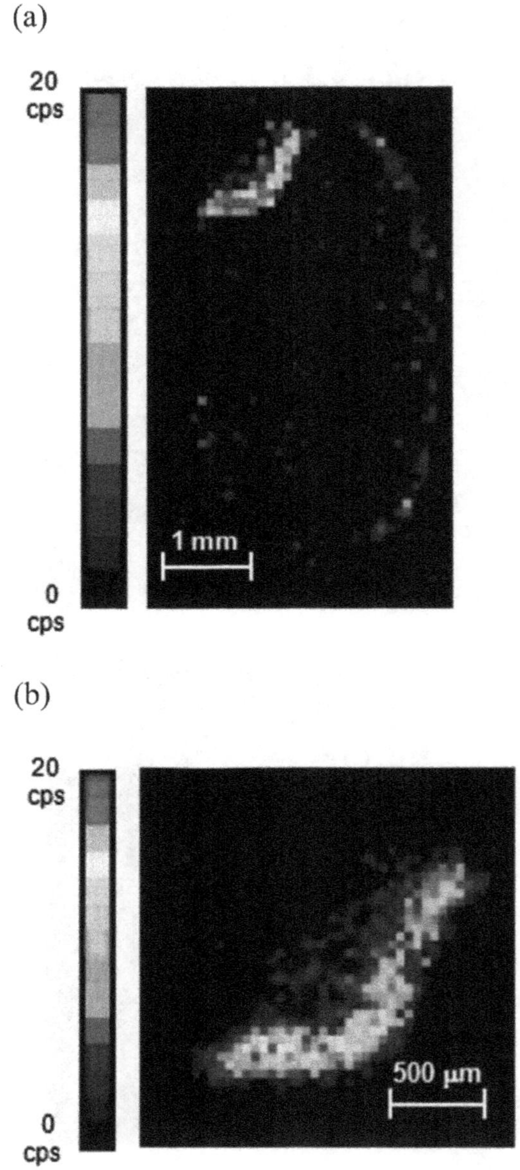

Figure 16.3 Location of phosphorous in rice Ji 307. (a) Whole kernel; (b) embryo. Images of phosphorous location (as an index of phytic acid) indicate that phosphorous is distributed unevenly in the whole brown rice kernel. The highest density occurred at the boundary of the embryo and endosperm (a and b).

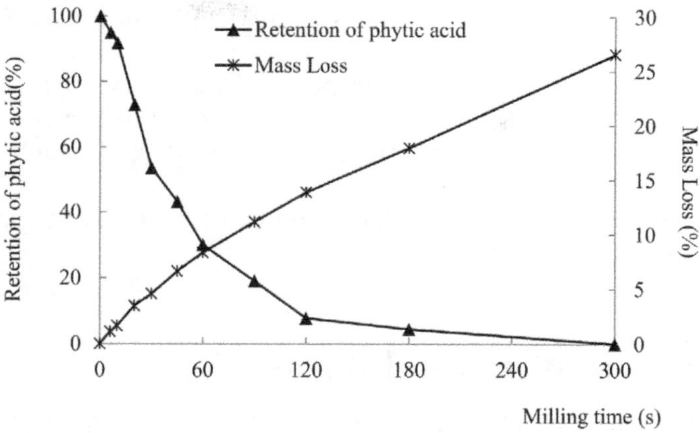

Figure 16.4 Retention of phytic acid and mass loss during milling (rice variety Ji 307). Retention of phytic acid in milled rice and mass loss caused by milling are shown for comparison. Retention of phytic acid in milled rice is the result of uneven distribution of phytic acid in the kernel, while mass loss is mainly influenced by the hardness of different layers of the kernel. For example, for Ji 307, when we remove 10% of the outer layer of brown rice, about 75% phytic acid would be removed at the same time.

16.5 Conclusion

With the new paradigm shift from green revolution to nutritional revolution, one must realize that any small increase of nutritional value of rice could improve the health of the rice-eating population. Promising approaches for enhancing the bioavailability of minerals in rice products include: (i) increase of mineral levels by controlled loss during milling, supplementation or fortification, (ii) increase bioavailability through added enhancers or by removal of inhibitors or (iii) combinations thereof. The genetic variation and significant effect of location provided hope for plant scientist to develop high minerals and low phytic acid contents rice by traditional breeding methods for human health. The uneven distribution of minerals and inhibitors in the rice kernel inspired food scientists to develop and apply new technologies for improved rice nutrition. Fortification with minerals should take into account their interactions, palatability especially for iron compounds, and opportunities for enhancement of bioavailability by enhancing components from food ingredients.

Summary Points

- The chapter focuses on calcium and phytic acid in rice.
- Rice contains low levels of calcium and relatively high phytic acid.
- Contents of calcium and phytic acid varied with rice genotypes and cultivation locations.

- Calcium and phytic acid are distributed unhomogeneously in rice kernel.
- Most calcium locates in the bran part and more than 70% phytic acid locates in less than 8% of the outermost layer of rice kernel.
- Soaking, germination and treatment with phytase could degrade phytic acid.
- Calcium nutrition of rice products could be improved by supplementation, fortification and enhancers.

Definitions of Words and Terms

Antinutritional factor (ANF). Factor that negatively affects digestion and/or absorption of nutrients, such as protein and minerals. For example, oxalic acid is a ANF for calcium. This is because when oxalic acid present, calcium in food will be formed into an insoluble state and can no longer be absorbed by humans.

Brown rice. The name comes from the color, also known as hulled rice. It is the edible part of paddy obtained after removing of husk with a huller. Brown rice is normally milled and polished for white color and nice taste.

Germinated rice. Brown rice is soaked for water absorption and then followed with germination under certain conditions till the sprout length is 0.5–1.0 mm.

Rice milling. Processing procedure of rice production, the technology is based on an abrasive mechanism. Milling removes the outer layer of brown rice (bran) and a white color product is obtained.

Rice noodle. White rice soaked in water or not soaked, ground with or without water, a mixture of rice flour and water steamed, extruded for noodles, heated, cooled to obtain final fresh noodle or dehydrated to dried rice noodle.

Parboiled rice. Paddy (rough rice) is subjected to a steaming or parboiling process followed with drying to remove water. The same processing as normal paddy is undertaken. Parboiling causes nutrients from the outer husk, especially thiamine, to move into the grain itself.

Phytase. General name of enzymes that can degraded phytic acid (phytate). Phytase can be intrinsic components in seeds and activated by germination, or substances isolated from micro-organisms.

Phytate. Salt form of phytic acid. Mineral cations include sodium, potassium, calcium, zinc and so on.

Phytic acid. Myo-inositol 1,2,3,4,5,6 hexakis dihydrogen phosphate (also known as inositol hexaphosphoric acid or myo-inositol hexakisphosphate-IP6). The molecular formula is $C_6H_{18}O_{24}P_6$ and molecular weight 660.04.

White rice. Also known as milled rice, polished rice, it is the most popular and normal rice product on market.

List of Abbreviations

AAS Atomic absorption spectrophotometry
ICP Inductively coupled plasma
OES Optical emission spectrophotometry
ANF Antinutritional factor
HPLC High-performance liquid chromatography
$FeCl_3$ Ferric chloride

Acknowledgements

We gratefully acknowledge Kouichi Tsuji and Kazuhiko Nakano from the Department of Applied Chemistry, Osaka City University (Osaka, Japan) who made the X-ray images for brown rice.

References

Achiraya, K., Prapasri, P., Prapaisri, P. S. and Ratchanee, K., 2004. In vitro calcium bioavailability of vegetables, legumes and seeds. *Journal of Food Composition and Analysis*. 17: 311–320.

Bajaj, M., Arora, C. L., Chhibba, I. M. and Sidhu, J. S., 1989. Extended milling of Indian rice. III. Effect on mineral composition. *Chemie, Mikrobiologie, Technologie der Lebensmittel*. 12: 58–60.

Cornforth, D. P., 2002. Potential use of phytate as an antioxidant in cooked meats. In: Reddy, N. R. and Sathe, S. K. (ed.) *Food Phytates*. CRC Press, Boca Raton, US, pp. 190–205.

De Stefano, C., Giuffrè, O., Milea, D., Rigano, C. and Sammartano, S., 2002. Speciation of phytate ion in aqueous solution. Noncovalent interaction with biogenic polyamines. *Chemical Speciation and Bioavailability*. 15: 29–36.

Doesthale, Y. G., Devara, S., Rao, S. and Belavady, B., 1979. Effect of milling on mineral and trace element composition of raw and parboiled rice. *Journal of the Science of Food and Agriculture*. 30: 40–46.

Emoto, T., Sato, Y., Konishi, Y., Ding, X. and Tsuji, K., 2004. Development and applications of grazing exit micro X-ray fluorescence instrument using a polycapillary X-ray lens. *Spectrochimica Acta, Part B: Atomic Spectroscopy*. 59: 1291–1294.

FAO statistic data, Crop National Production China year, 2012. Available at: http://data.fao.org/. Accessed 14 February 2014.

Gregorio, G. B., Htut, T. and Cabuslay, G. S., 2009. Breeding for micronutrient enriched rice. In: Bañuelos, G. S. and Lin, Z. Q. (ed.) *Development and Use of Biofortified Agricultural Products*. CRC Press, Boca Raton, US, pp. 181–203.

Itani, T., Tamaki, M., Arai, E. and Horino, T., 2002. Distribution of amylose, nitrogen, and minerals in rice kernels with various characters. *Journal of Agricultural and Food Chemistry*. 50: 5326–5332.

Juliano, B. O., 1997. *Rice products in Asia.* Regional Office for Asia and the Pacific, Laguna, Philippines. vol. 38, pp. 1–42.

Kennedy, G., Burlingame, B. and Nguyen, V. N., 2002. Nutritional contribution of rice and impact of biotechnology and biodiversity in rice-consuming countries. In: Dat, V. T. (ed.) *Proceeding of the 20th session of the international rice commission.* Food and Agriculture Organization of the United Nations, Bangkok, Thailand, pp. 59–69.

Kennedy, B. M. and Schelstraete, M., 1975. Chemical, physical and nutritional properties of high-protein flours and residual kernel from the over-milling of uncoated milled rice. *Cereal Chemistry*. 52: 173–182.

Lee, H. H., Rhee, H. I., Lee, S. Y., Kim, C. H. and Choi, Y. S., 1997. Contents of phytic acid and minerals of rice cultivars from Korea. *Journal of Food Science and Nutrition*. 2: 301–303.

Liang, J., Han, B. Z., Han, L., Nout, M. J. R. and Hamer, R. J., 2007. Iron, zinc and phytic acid content of selected rice varieties from China. *Journal of the Science of Food and Agriculture*. 87: 504–510.

Liang, J., Han, B. Z., Han, L., Nout, M. J. R. and Hamer, R. J., 2008a. Effect of soaking, germination and fermentation on phytic acid, and total and *in vitro* soluble zinc in brown rice. *Food Chemistry*. 110: 821–828.

Liang, J., Li, Z., Tsuji, K., Nakano, K., Nout, M. J. R. and Hamer, R. J., 2008b. Milling characteristics and distribution of phytic acid and zinc in long-, medium- and short-grain rice. *Journal of Cereal Science*. 48: 83–91.

Liang, J., Han, B. Z., Han, L., Nout, M. J. R. and Hamer, R. J., 2009. Effect of soaking and phytase treatment on phytic acid calcium, iron and zinc in rice fractions. *Food Chemistry*. 115: 789–794.

Liang, J., Han, B. Z., Han, L., Nout, M. J. R. and Hamer, R. J., 2010. *In vitro* solubility of calcium, iron and zinc in relation to phytic acid levels in common rice-based consumer products in China. *International Journal of Food Science and Nutrition*. 61: 40–51.

Liu, Z., Cheng, F. and Zhang, G., 2005. Grain phytic acid content in japonica rice as affected by cultivar and environment and its relation to protein content. *Food Chemistry*. 89: 49–52.

Lönnerdal, B., Sandberg, A. S., Sandström, B. and Kunz, C., 1989. Inhibitory effects of phytic acid and other inositol phosphates on zinc and calcium absorption in suckling rats. *Journal of Nutrition*. 119: 211–214.

Ma, G., Jin, Y., Piao, J., Kok, F., Bonnema, G. and Jacobsen, E., 2005. Phytate, calcium, iron, and zinc contents and their molar ratios in foods commonly consumed in China. *Journal of Agricultural Food Chemistry*. 53: 10285–10290.

Oatway, L., Vasanthan, T. and Helm, J. H., 2001. Phytic acid. *Food Reviews International*. 17: 419–431.

Ou, K., Cheng, Y., Xing, Y., Lin, L., Nout, R. and Liang, J., 2011. Phytase activity in brown rice during steeping and sprouting. *Journal of Food Science and Technology*. 48: 598–603.

Persson, H., Türk, M. and Nyman, M., 1998. Binding of Cu^{2+}, Zn^{2+}, and Cd^{2+} to inositol tri-, tetra-, penta-, and hexaphosphate. *Journal of Agricultural and Food Chemistry*. 46: 3194–3200.

Phuong, T. D., Chuong, P. V., Tong Khiemc, D. and Kokot, S., 1999. Elemental content of Vietnamese rice. Part 1. Sampling, analysis and comparison with previous studies. *Analyst*. 124: 553–560.

Peterlik, M. and Cross, H. S., 2005. Vitamin D and calcium deficits predispose for multiple chronic diseases. *European Journal of Clinical Investigation*. 35: 290–304.

Slingerland, M., Zhang, F., Stomph, T. J., Gao, X., Liang, J. and Jiang, W., 2009. Biofortification in a food chain approach for rice in China. In: Bañuelos, G. S. and Lin, Z. Q. (ed.) *Development and Use of Biofortified Agricultural Products*. CRC Press, Boca Raton, US, pp. 181–203.

Tabekhia, M. M. and Luh, B. S., 1979. Effect of milling on macro and micro minerals and phytate of rice. *Deutsche Lebensmittel Rundschau*. 75: 57–62.

Tamanna, S., Parvin, S., Kumar, S., Dutta, A. K., Ferdoushi, A., Siddiquee, M. A., Biswas, S. K. and Howlader, M. Z. H., 2013. Content of some minerals and their bioavailability in selected popular rice varieties from Bangladesh. *International Journal of Current Microbiology and Applied Sciences*. 2: 35–43.

Wu, W., Cheng, F., Liu, Z. and Wei, K., 2007. Difference of phytic acid content and its relation to four protein composition contents in grains of twenty-nine japonica rice varieties from Jiangsu and Zhejiang province, China. *Rice Science*. 14: 311–314.

CHAPTER 17

Adding Calcium to Foods and Effect on Acrylamide

NESLIHAN GÖNCÜOĞLU TAŞ[a], AYTÜL HAMZALIOĞLU[a], TOLGAHAN KOCADAĞLI[a], AND VURAL GÖKMEN*[a]

[a]Food Engineering Department, Hacettepe University, Beytepe 06800, Turkey
*E-mail: vgokmen@hacettepe.edu.tr

17.1 Introduction

Calcium is an important mineral in human metabolism, thus calcium-rich foods play a role in many aspects of human health. Although calcium is provided by a wide variety of foods, fortification may be needed for special groups of people. In the European Union, the following forms of calcium are authorized for addition to foods and for use in food supplements: carbonate, chloride, citrate, malate, gluconate, glycerophosphate, lactate, hydroxide, oxide, acetate, L-ascorbate, bisglycinate, pyruvate, salts of orthophosphoric acid, succinate, L-lysinate, malate, L-pidolate, L-threonate, sulfate (European Food Safety Authority, 2012a).

In addition to its nutritional contributions, calcium salts are especially added to foods in cheese making, for firming texture of fresh-cut fruits and vegetables, as a leavening agent, preservative and for many other technological purposes as summarized in Table 17.1. Moreover, calcium salts play an important role in mitigation of acrylamide formation during heating.

Food and Nutritional Components in Focus No. 10
Calcium: Chemistry, Analysis, Function and Effects
Edited by Victor R. Preedy

Published by the Royal Society of Chemistry, www.rsc.org

Table 17.1 Some calcium salts added to foods and their technological functions. The table content is adapted from the scientific report of European Food Safety Authority (2012a). It shows the types of calcium salts, their purpose of usage and food categories that calcium salts are added to.

	Functional classes	Food category
Calcium carbonate	Acidity regulator, anti-caking agent, color, firming agent, flour treatment agent, stabilizer	Coffee, tea, dried pasta and noodles, dried whey and whey products, frozen battered fish, fish products, pasteurized cream, salt, UHT creams
Calcium chloride	Firming agent, stabilizer, thickener	Dried whey and whey products, fermented vegetable, pasteurized cream, sterilized and UHT creams
Calcium lactate	Acidity regulator, firming agent, flour treatment agent	Coffee, tea, fermented milks, fermented vegetable, pasteurized cream, salt substitutes, UHT creams, reduced fat creams
Calcium acid pyrophosphate	Raising agent	Bakery products

17.2 Acrylamide Formation Mechanism

Formation of acrylamide in heat-treated foods was first reported in 2002 (Mottram *et al.*, 2002; Stadler *et al.*, 2002; Tareke *et al.*, 2002). Acrylamide has been detected in various heat-treated foods. The presence of acrylamide has been considered as an important food-related crisis by international authorities as it has been classified as probably carcinogenic to humans (Group 2A) by the International Agency for Research on Cancer (IARC, 1994). Bakery products, potato chips, breakfast cereals and roasted coffee are important sources of acrylamide. Acrylamide concentrations of certain heat-treated foods are given in Table 17.2.

Several researchers have established that the main pathway of acrylamide formation in foods is linked to the Maillard reaction and, in particular, the amino acid asparagine (Yaylayan *et al.*, 2003; Zyzak *et al.*, 2003). Acrylamide is formed through Maillard reaction in the presence of asparagine and carbonyl compounds at temperatures higher than 100 °C (Figure 17.1). The first step in the reaction is formation of a Schiff base upon condensation of a carbonyl group with the amine group of asparagine. Then, the Schiff base formed in the Maillard reaction rearranges to the more stable Amadori compound. However, the Amadori compound does not yield acrylamide by decarboxylation (Stadler *et al.*, 2004). Decarboxylation of the Schiff base can occur directly through Schiff betaine or Schiff betaine rearranges to oxazolidine-5-one intermediate that is known to decarboxylate easily to form azomethine ylide (decarboxylated Schiff base) (Yaylayan *et al.*, 2003; Zyzak *et al.*, 2003). Azomethine ylide rearranges by tautomerization to form decarboxylated Amadori compound,

Table 17.2 Acrylamide concentration of certain heat-treated foods. Data is adapted from the scientific report of European Food Safety Authority (2012b). Data were acquired in European countries between 2007 and 2010. Note that bakery, potato and coffee products are the main dietary source of exposure due to the high consumption.

Foods	Mean (μg kg^{-1})	Maximum (μg kg^{-1})
Breads and rolls	136	2430
Pastry and biscuits	288	6798
Breakfast cereals	149	1649
Nuts and oilseeds	111	1658
Potato, baked	147	1027
Potato, chips	956	5500
French fries	410	5269
Coffee, brewed	17	245
Coffee, roasted	314	3025
Coffee, decaffeinated	331	590
Dried fruits	47	258
Alcoholic beverages	17	84
Baby foods, canned	58	399
Baby foods, dry powder	237	470
Baby foods, biscuits	121	1100

which then forms acrylamide *via* β-elimination (Stadler *et al.*, 2004). This rearrangement occurs if there is hydroxyl functionality on the β position to the nitrogen atom as is the case with α-hydroxycarbonyls like glucose. On the other hand, azomethine ylide might decompose directly to acrylamide or 3-aminopropionamide (3-APA) (Zyzak *et al.*, 2003). 3-APA easily deaminates to form acrylamide (Granvogl and Schieberle, 2006). It was revealed that almost 1% of asparagine converted to acrylamide in the asparagine–hexose sugar model systems. Most of the Schiff base formed rearranges to an Amadori compound, which contributes to color and flavor formation. The rate-limiting step of acrylamide formation is decarboxylation of the Schiff base (Blank, 2005).

Several factors, such as the initial concentration of reactants and their ratio, temperature and time of processing, pH and water activity, have been shown to influence the levels of acrylamide in heat-processed foods (Gökmen and Şenyuva, 2006; Gökmen *et al.*, 2007; Gökmen and Palazoğlu, 2008; Acar and Gökmen, 2009). Acrylamide formation depends on the time–temperature history of the processed foods. Its formation generally increases with increase in thermal load as in bakery and fried potato products. Contrary to this, as is the case with coffee roasting, after a given period acrylamide concentration starts to decrease due to the limited amount of asparagine found in green coffee beans and due to prolonged roasting to obtain dark roasted coffee (Şenyuva and Gökmen, 2005).

Acrylamide formation is directly related with asparagine concentration (Gökmen and Şenyuva, 2006). In the presence of high amounts of asparagine, carbonyl compounds might be limiting in the acrylamide formation

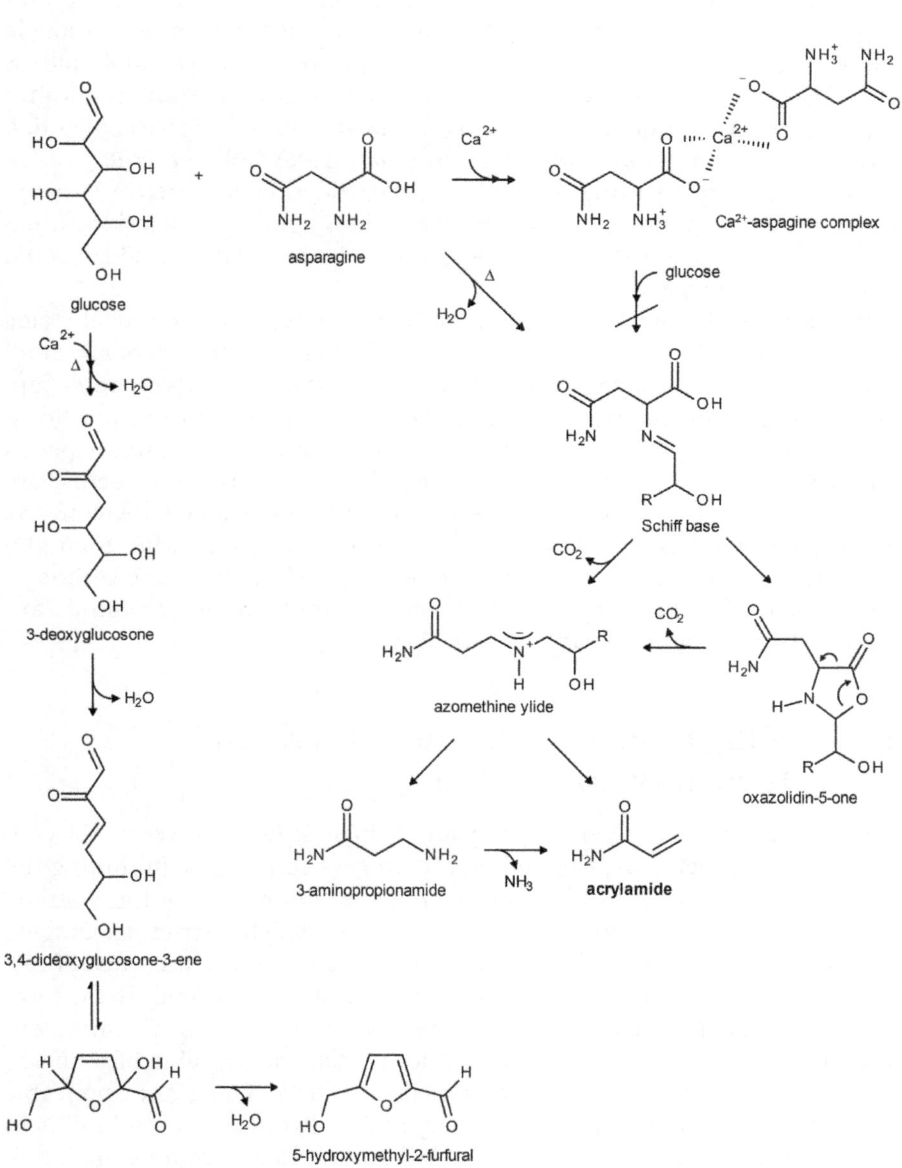

Figure 17.1 Mechanisms of acrylamide and 5-hydroxymethyl-2-furfural formation *via* a Maillard reaction and mitigation of acrylamide formation in the presence of calcium ions. Acrylamide is formed from asparagine in the presence of carbonyl compounds *via* a Maillard reaction. Ca^{2+} ions prevent formation of the Schiff base and thus, of acrylamide during heating. Ca^{2+} promotes sugar dehydration causing formation of 5-hydroxymethyl-2-furfural. Adapted from Yaylayan *et al.* (2003), Zyzak *et al.* (2003), Stadler *et al.* (2004), Gökmen and Şenyuva (2007b).

such as in potatoes. Hydroxy carbonyl compounds such as fructose and glucose are found in foods in high amounts compared to other carbonyls and contribute to acrylamide formation significantly (Yaylayan and Stadler, 2005). Glyoxal, methylglyoxal, 5-hydroxymethyl-2-furfural (HMF) and other sugar degradation products containing carbonyl groups promote acrylamide formation (Amrein *et al.*, 2004; Amrein *et al.*, 2006; Gökmen *et al.*, 2012). Additionally, carbonyls originating from lipid oxidation or bioactive carbonyl compounds also participate in Maillard reaction, yielding acrylamide formation (Hidalgo *et al.*, 2010; Hamzalıoğlu and Gökmen, 2012; Kocadağlı *et al.*, 2012; Hamzalıoğlu *et al.*, 2013).

Besides the chemical reactivity of carbonyl sources their physical states have a crucial role under low moisture conditions. Aldose sugars are more reactive than ketose sugars in liquid reaction systems. However, water rapidly evaporates on the surface of foods during heating. In dry conditions, molecular mobility of the reactants is restricted and thus reaction depends on melting of the reactants. The melting points of glucose and fructose are 157 °C and 127 °C, respectively. Hence, fructose more efficiently converts asparagine to acrylamide since reaction starts to proceed earlier than glucose (Robert *et al.*, 2004). Similarly, the reactivity of the bioactive carbonyls found to be related directly with their melting points in the acrylamide formation (Hamzalıoğlu and Gökmen, 2012).

17.3 Mitigation Mechanism of Acrylamide Formation by Calcium

Basic strategies for mitigation of acrylamide include either lowering the concentration of precursors, controlling the process conditions, inhibiting the Maillard reaction or providing competitive inhibition. One of these strategies includes usage of mono- or divalent cations, which restricts asparagine to form the Schiff base during the Maillard reaction in the presence of carbonyl compounds. These cations are found or added to the foods in the form of salts. Sodium and calcium salts are the most studied forms of monovalent and divalent cations. It is known that not all salts are capable of reducing acrylamide formation. Calcium chloride is the most effective and applicable form of calcium in the mitigation of acrylamide due to its high solubility in water. Calcium lactate is another soluble form found as a commercial product to mitigate acrylamide formation. Usage of calcium carbonate is limited as it is not soluble in water.

The effect of mono- and divalent cations on acrylamide formation was investigated both in model systems and foods (Gökmen and Şenyuva, 2007a). At amounts equivalent to those of asparagine and fructose, added Ca^{2+} was found to prevent acrylamide formation completely, whereas Na^{+} almost halved the acrylamide formed in the model system at 150 and 180 °C (Figure 17.2).

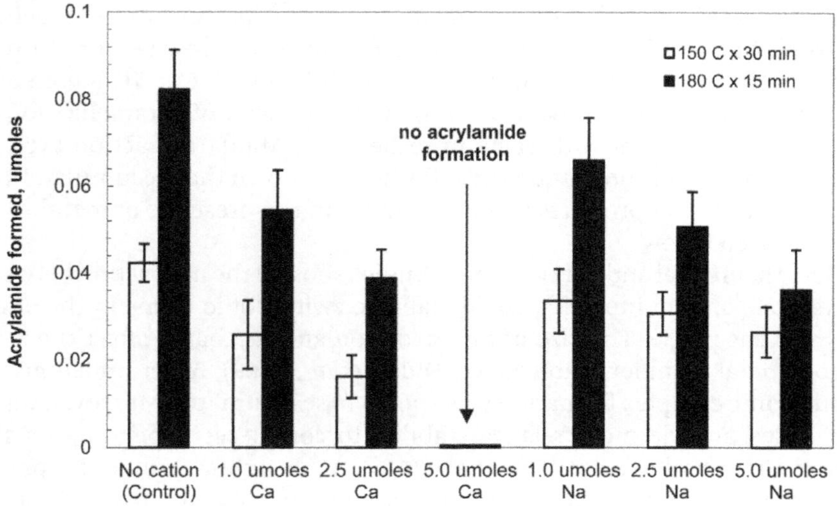

Figure 17.2 Amounts of acrylamide formed (μmol) in the asparagine–fructose model system as influenced by the presence of Ca^{2+} and Na^{+} in the reaction mixture. Model systems were prepared in 100 μL water containing asparagine (5 μmol) and fructose (5 μmol). Reactions were held in sealed glass tubes at 150 and 180 °C. Note that Ca^{2+}, at amounts equivalent to those of asparagine and fructose, was found to prevent acrylamide formation completely, whereas Na^{+} almost halved the acrylamide formed in the reaction mixture. Data are from Gökmen and Şenyuva (2007a), with permission from the publisher.

Lindsay and Jang (2005) have hypothesized that ionic associations involving the ions and charged groups of asparagine and related intermediates were likely to be involved in acrylamide mitigation with Ca^{2+}. Zhou *et al.* (2003) have also reported that Ca^{2+} may form a complex with poly(acrylamide), but such a complex was not confirmed between the Ca^{2+} ion and acrylamide. The Ca^{2+} ion is known to form complexes preferably with carboxyl groups found in proteins, pectic substances and organic acids. Ionic interaction between Ca^{2+} ions and the carboxyl group of asparagine becomes limited in the presence of these compounds because of competitive behavior. Thus, calcium concentration applied or added to foods is an important factor in the mitigation of acrylamide formation. Additionally, variations of the amount of these competitive groups might yield different results even in the same product category.

The first step in the acrylamide formation mechanism is the formation of a Schiff base between the carbonyl and α-amino group of asparagine by means of the dehydration of the *N*-glycosyl compound. The proposed mechanism of mitigation of acrylamide formation with Ca^{2+} ions is based on the prevention of formation of the Schiff base and thus of acrylamide during heating. It was proved that both the Schiff bases of asparagine and acrylamide were not observed when Ca^{2+} was present in the reaction mixture. Meanwhile, the reaction proceeds to form brown-colored products even if the Schiff base formation is inhibited with Ca^{2+} (Gökmen and Şenyuva, 2007a). The brown color might be the result of caramelization of sugars in high extent with calcium rather than Maillard reaction even in the presence of amino compounds. It was also shown that accumulation of sugar degradation products was much more in the presence of metal ions, as discussed below.

Both theoretical and spectroscopic studies showed the interaction between metal ions and amino acids could stabilize zwitterionic forms in the aqueous and gas phase. The size of the metal ion and its charge affect the zwitterion–metal complex stabilization (Bush *et al.*, 2008). A net preference of zwitterionic complex formation was shown for calcium ion. Moreover, water molecules provide more relative stability to complexes depending on the side chain of amino acids (Qin *et al.*, 2013). As the amino group of asparagine gets protonated in the zwitterionic complex of calcium, its nucleophilic strength becomes diminished to carbonyl groups. These findings also reveal why asparagine does not form Schiff base during Maillard reaction in the presence of calcium (Figure 17.1).

It has also been suggested that reduction of acrylamide formation with cations may be based on pH drop. Metal cations are known to cause pH reduction when added to certain foods. It was observed that plant tissues contacting with mono- and divalent cation salts become more acidic depending on the increased salt concentration. The mechanism of pH drop is mainly based on ionization of carboxyl groups during interaction with cations (Levine and Ryan, 2009). Mestdagh *et al.* (2008) investigated the importance of pH drop by addition of calcium and magnesium cations in a potato model system. They indicated that addition of $CaCl_2$ to potato cuts was known to provoke a pH drop since HCl reduces when calcium crosslinks the pectin. The data given by the authors revealed that cations provided a major mitigation by a direct effect, rather than the pH drop caused by interaction with pectin. On the other hand, Levine and Ryan (2009) asserted that pH drop was the main mechanism to mitigate acrylamide formation by addition of calcium ions to model wheat products. They showed that calcium ions caused additional pH reduction during heat treatment, even adjusting the initial pH to control by NaOH. Therefore, they proposed that acrylamide reduction was no longer attributed mostly to the direct effects of calcium on the reaction mechanisms since the larger pH drop likely became a more important factor as an indirect effect. The pH drop during heating is caused by the formation of carboxylic acids by the cleavage of dicarbonyl compounds originating from sugar degradation. Reduction of pH is the result of caramelization since the Maillard reaction is prevented at the beginning by the direct effect of calcium not by the pH reduction occurring during thermal treatment. Therefore, the effect of calcium on acrylamide formation could not be only attributed to pH drop, ignoring the direct effect of calcium in the mechanism.

17.4 Mitigation of Acrylamide Formation in Foods by Calcium Salts

Among the food items, potato products such as French fries and crisps contain highest amounts of acrylamide. Acrylamide formation takes place mainly at the surface and in near-surface regions, because during the process of frying, the conditions in this part of the potato strip become favorable for acrylamide formation, as a result of simultaneous drying (Gökmen *et al.*, 2006). Therefore, any treatment that inhibits the reaction responsible for the formation of acrylamide on the potato surface would be a viable approach to limit the amount of acrylamide formed during frying. The amount of acrylamide was reported to be decreasing by dipping potato strips into the solutions of sodium chloride and calcium chloride (Table 17.3). Besides, when water was used alone for the pretreatments it has only a limited effect on the amount of acrylamide formed during frying. Hence, the inhibition of acrylamide formation during frying was mainly attributed to the presence of mono- or divalent cations in the potato strips after the dipping treatment, rather than the reduction of acrylamide precursors by dipping (Gökmen and Şenyuva, 2007a).

Mestdagh *et al.* (2008) investigated the impact of chemical pretreatments on the acrylamide formation and sensorial quality of potato crisps. Addition of 0.1 M $CaCl_2$ to the blanching water of potatoes resulted with 93% of acrylamide reduction, while 0.5 M $CaCl_2$ addition provided 64% reduction. Sensorial analysis showed that higher concentrations of $CaCl_2$ cause bitter aftertaste and a more crispy texture. Besides, $CaCl_2$ added to the blanching water did not change product color compared to water. Puracal Ca100 and

Table 17.3 Effect of dipping into calcium chloride and sodium chloride solution on the inhibition of acrylamide formation during frying. Glucose, fructose and asparagine concentrations of potato strips were 0.45, 1.82 and 1.86 g kg^{-1} referring to fresh weight. The concentration of acrylamide formed upon frying at 170 °C for 5 min in potato strips without pretreatment (control) was 711 ± 33 µg kg^{-1} ($n = 3$). Note that dipping in calcium solution effectively mitigates acrylamide formation. Data are from Gökmen and Şenyuva (2007a), with permission from the publisher.

Dipping		Loss by dipping treatment (%)				
Solution	Time (min)	Glucose	Fructose	Asparagine	Acrylamide (µg kg^{-1})	Inhibition (%)
Water	15	3	4	3	655 ± 18	8
	30	7	9	8	628 ± 59	12
	60	17	21	19	589 ± 41	17
0.1 M NaCl	15	6	5	7	430 ± 58	40
	30	11	9	12	328 ± 39	54
	60	17	18	20	302 ± 41	58
0.1 M $CaCl_2$	15	5	6	6	151 ± 14	79
	30	9	8	11	60 ± 8	92
	60	19	17	21	40 ± 3	95

Ca200 provided a similar reduction in acrylamide formation. Addition of calcium salts also lowered the absorption of oil upon frying. However, it was also indicated that these salts might lead to a more rapid oil turnover rate during industrial frying. A further study on an industrial scale was also performed by Vinci *et al.* (2011) in French fries. It was shown that lower acrylamide levels were obtained in samples treated with calcium lactate solution at a concentration of 0.08 and 0.16 M. However, the sensorial properties were found to be unacceptable by the panelists at concentrations above 0.3 M. Similarly, a crunchier/harder texture was also reported.

Application of $CaCl_2$ to the potatoes is possible on an industrial scale. Although there are some other efficient inhibitors of acrylamide, $CaCl_2$ seems to be a suitable choice because of its low price and acceptable taste. Immersion of potato slices into 5 g L^{-1} $CaCl_2$ solution before frying resulted in 85% reduction in acrylamide in the commercially produced fried crisps (Ou *et al.*, 2008).

Mulla *et al.* (2011) tested acrylamide reduction in extruded snacks prepared from potato flour:semolina blend by using different additives such as citric acid, calcium salts, amino acids, vitamins and their combinations. Addition of 50 μmol g^{-1} $CaCl_2$ to the blend decreased acrylamide 65% in the extruded snack, while 10 μmol g^{-1} citric acid together with 50 μmol g^{-1} $CaCl_2$ showed no reduction beyond 65%. This must be due to the chelation of calcium by citrate, thus avoiding any synergetic effect. However, no such conclusions were drawn in the study. Addition of 50 μmol g^{-1} $CaCl_2$ was the most effective way to reduce acrylamide in extruded snacks with high sensory scores.

The effect of calcium salts on acrylamide formation in bread, savory biscuits and sweet biscuits was studied by Sadd *et al.* (2008). Commercial fortified flour with 0.3% calcium carbonate provided 30% acrylamide reduction in bread crust compared to base flour. 1% calcium chloride application to the surface of bread dough caused reduction of acrylamide by more than 60%. Among the tested calcium salts in savory and sweet biscuits, calcium chloride application was found to be the most effective one with 60% reduction in acrylamide. On the other hand, calcium propionate application promoted acrylamide formation in both biscuits (Sadd *et al.*, 2008).

Acrylamide is formed on the surface and nearby layers of bakery products because of the suitable temperature and moisture conditions. For this reason, it is important to test mitigation strategies for bakery products in crust-like model. Cookies without calcium chloride having thicknesses of 10 and 1 mm had 128 and 552 ng g^{-1} of acrylamide. After increased amounts of $CaCl_2$ addition, a significant reduction in acrylamide content was observed in both cookies (Acar and Gökmen, 2009).

The effect of $CaCl_2$, NaCl and KCl on acrylamide reduction was investigated with a cereal-based model system. The acrylamide content in heated cereal model without salts was 603 μg kg^{-1} whereas it was 345 μg kg^{-1} in NaCl or KCl added formula and 70 μg kg^{-1} in $CaCl_2$ added formula. $CaCl_2$ was found to be the most efficient salt in mitigation of acrylamide in cereal-based models (Kukurova *et al.*, 2009).

$Ca(OH)_2$ is another source of Ca^{2+} used for nixtamalization process in tortilla chips in order to husk corn grain easily. Addition of $Ca(OH)_2$ to the corn flour not only eases the process, but also helps to decrease the acrylamide formation during frying of tortilla chips. Increasing the $Ca(OH)_2$ concentration from 1.0 g/100 g to 1.5 and 2.0 g/100 g in the corn flours caused an additional reduction of 52 and 36%, respectively. Controlling $Ca(OH)_2$ concentration in the corn flour has been offered as an effective way to control acrylamide in the fried products made from nixtamalized corn flours (Salazar *et al.*, 2013).

There are some commercial products applied in the food industry like PURACAL ACT from Purac containing soluble calcium salts used to mitigate acrylamide. These products reduce the acrylamide content besides improving texture, reducing fat absorption, providing firmness and crispiness to the snack foods. Addition of these calcium salts is not restricted by regulations except for cereal flours and related products. The Food and Drug Administration (FDA) have specified requirements for total calcium contents of standardized cereal flours and related products. It was remarked that enriched flours might contain 2.116 g kg^{-1} of total calcium. It is also mandatory in Canadian Legislations that all white flours should be enriched with B vitamins, iron and folic acid and may contain 1.4 g kg^{-1} of calcium besides some other additives. Similarly, 0.3% standard calcium fortification of flours is also obligatory in the UK for nutritional reasons.

17.5 Effect of Calcium Addition on Sugar Degradation in Foods and Model Systems

5-Hydroxymethyl-2-furfural is formed through hexose sugar dehydration and Maillard reaction during thermal processes or storage. It is formed by removal of three molecules of water from hexose sugars, as shown in Figure 17.1. It is indicated that HMF is not a serious concern for human health. However, its metabolites are receiving attention because of their genotoxic effects. As it is found in widely consumed foods such as processed fruits, coffee, dairy foods and bakery products, its daily intake might reach 150 mg. This amount is several-fold higher than the total intake of other food processing contaminants such as acrylamide. Furfural is another process contaminant formed from pentose sugar degradation. It may also be formed *via* decarboxylation of HMF.

Gökmen and Şenyuva (2007b) reported that heating a glucose–asparagine mixture without any cation at 150 °C for 20 min resulted in HMF formation, whereas no formation was noted for furfural. The presence of cations in the reaction mixture influences the rate of decomposition of glucose. Accordingly, formation of HMF and furfural in the glucose–asparagine model system was promoted due to the rapid decomposition of glucose in the presence of cations. Moreover, with the increasing amounts of Ca^{2+}, Mg^{2+}, Na^{+}, Fe^{3+}, K^{+} and Zn^{2+}, formation of both HMF and furfural increased (Figure 17.3). 80% of asparagine in glucose-asparagine model system remained unreacted,

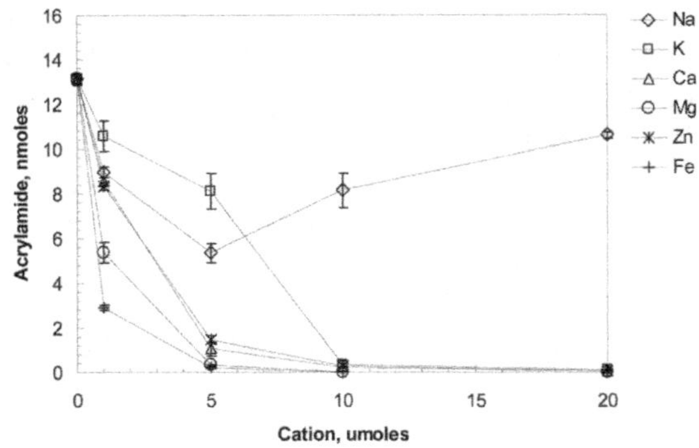

(a)

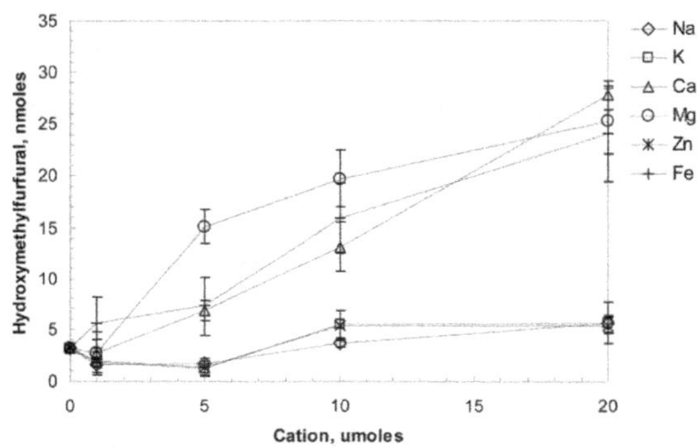

(b)

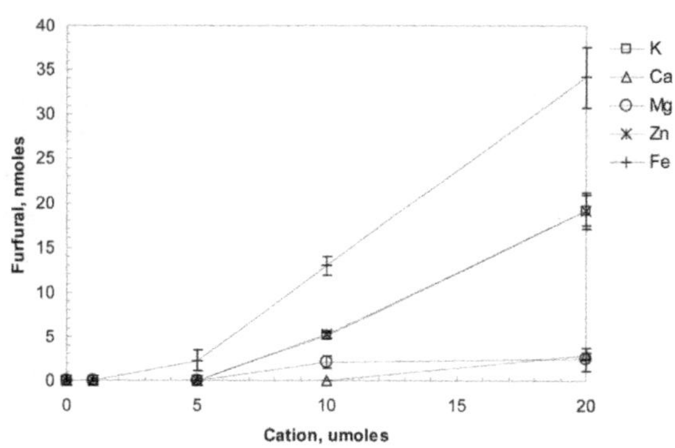

(c)

Figure 17.3 Influences of some monovalent, divalent, and trivalent cations on the amounts of (a) acrylamide, (b) 5-hydroxymethyl-2-furfural, and (c) furfural formed after heating the mixture of glucose and asparagine Reaction mixtures were containing 10 μmol glucose and 10 μmol asparagine in 100 μL water. Reactions were performed at 150 °C for 20 min in sealed glass tubes. Note that rapid decomposition of glucose yielded 5-hydroxymethyl-2-furfural (b) and furfural (c) in the presence of cations while a Maillard reaction is restricted. Thus, acrylamide formation (a) was decreased. Data are from Gökmen and Şenyuva (2007b), with permission from the publisher.

while most of the glucose decomposed in the presence of cations. Thus, the reaction pathway proceeded mainly toward the dehydration of glucose to HMF and furfural (Figure 17.1).

The effect of the calcium salts of lactic acid (Puracal Act 100 and Puracal Act 200) and $CaCl_2$ on both acrylamide and HMF formation was tested in cookies by Acar *et al.* (2012). The ratio of calcium and lactate were 23% and 35% by weight for Puracal Act 100 and 20% and 44% by weight for Puracal Act 200, respectively.

The HMF concentration of control cookies baked at 205 °C for 11 min was found to be 1.97 mg kg^{-1}. Addition of 0.5% of $CaCl_2$ to the cookie formulation increased HMF to 4.2 mg kg^{-1}. The HMF concentrations of cookies formulated with 0.5% and 1.0% of Puracal Act 100 were determined to be 2.8 mg kg^{-1} and 3.3 mg kg^{-1}, respectively, which were lower than those of cookies formulated with Puracal Act 200 and $CaCl_2$. When 0.5 and 1.0% of calcium derivatives were added to the formulas, there was no significant effect even if the types of calcium derivatives were different. On the other hand, addition of 0.5% calcium derivatives reduced acrylamide 70% without any sensorial change. In this manner, the calcium concentration should be taken into consideration while modifying recipes. According to this study, HMF levels of cookies were still too low (<5.0 mg kg^{-1}) when cookies were formulated with calcium derivatives at amounts ranging between 0.1% and 1.0% (Acar *et al.*, 2012).

The adverse effects of mitigation strategies on sensory properties make their applications limited in the food industry. Calcium salts that are already added to foods for fortification could be considered a good solution in the minimization of acrylamide at moderate concentrations. Sensory properties are adversely affected at high concentrations together with formation of undesirable process contaminants like HMF and furfural. For this reason, the concentration of calcium salts used in mitigation applications should be taken into consideration. Besides the concentration of calcium salts, the type of salt is also important. Among the calcium salts, calcium chloride and calcium lactate were found to be the most effective and applicable ones due to their high solubility in water. This is a successful approach for the foods having high acrylamide risk such as bakery products, potato products and tortilla chips. It is also possible to apply these soluble calcium salts during processing of some commercial products in the food industry.

Summary Points

- This chapter focuses on how addition of calcium affects the acrylamide formation in foods.
- Acrylamide is a probable carcinogen to humans.
- Acrylamide is formed from asparagine *via* a Maillard reaction during heating.
- Calcium prevents reaction of asparagine to form acrylamide.
- Pretreatments of potato products with calcium solutions are an effective way of mitigating acrylamide formation.
- These pretreatments result in more crispy and firm potato products.

- Fortification of bakery products with calcium reduces the amount of acrylamide formed during baking.
- Calcium lactate and calcium chloride have no adverse effect on the sensory properties of foods at low concentrations.
- Calcium promotes sugar dehydration causing formation of 5-hydroxymethyl-2-furfural.
- The strategy is industrially applicable.

Key Facts of Mitigation of Acrylamide Formation with Calcium

This list explains the key facts of mitigation of acrylamide formation with calcium including acrylamide formation and its mechanism, the role of calcium in mitigation mechanism, the types of effective calcium salts and other cations.

1. Acrylamide is a process contaminant formed in foods *via* a Maillard reaction during heating.
2. Maillard reaction occurs between amino group of an amino acid or protein, and a carbonyl compound.
3. Asparagine is the responsible amino acid leading to acrylamide in a Maillard reaction.
4. Amino group of asparagine reacts with a carbonyl compound and condenses to form a Schiff base in the first step of a Maillard reaction.
5. Calcium interacts ionically with carboxyl group of asparagine and block the formation of a Schiff base due to stabilization of calcium-asparagine complex. Therefore, protonated amino groups of asparagine do not form a Schiff base.
6. Calcium is known to form complexes preferably with carboxyl groups found in proteins, pectic substances and organic acids. Therefore, the amount of freely available calcium is important to interact with asparagine.
7. The reaction proceeds to form brown-colored products even if the Schiff base formation is inhibited with calcium.
8. Calcium chloride and lactate are effective calcium salts that could be used in food formulations due to their solubility in water.
9. Some other cations such Mg^{2+}, Na^{+}, Fe^{3+}, K^{+} and Zn^{2+} are also effective in mitigation of acrylamide formation.

Definition of Words and Terms

Acrylamide. Carcinogenic compound formed *via* Maillard reaction in the presence of asparagine and a carbonyl compound during heating process.

3-Aminopropionamide (3-APA). A biogenic amine formed from azomethine ylide of asparagine and easily deaminates to form acrylamide.

Azomethine ylide. An iminium ion that occurs *via* decarboxylation of a Schiff base during Maillard reaction.
Chelation. Complexation of metal ions with a chelating agent containing at least two functional groups capable of giving electrons easily.
5-Hydroxymethyl-2-furfural (HMF). A furanic compound that is formed *via* loss of three molecules of water from hexose sugars during caramelization or Maillard reaction.
Maillard reaction. The reaction takes place between the amino group of an amino acid and a carbonyl compound. It may occur in foods both during heating at elevated temperatures or even at room temperature.
Schiff base. An *N*-substituted imine (having a carbon–nitrogen double bond) that is formed *via* condensation of the amino group of asparagine and carbonyl compound with removal of one molecule of water during Maillard reaction.
Sugar dehydration. Drying of sugar molecule by loss of water from carbon backbone during heating.
Zwitterion. A molecule containing both negative and positive charges on it with a net zero charge. Amino acids may form zwitterions by protonation of amino group and deprotonation of carboxyl group depending on the pH.

List of Abbreviations

3-APA	3-Aminopropionamide
FDA	Food and Drug Administration
HMF	5-Hydroxymethyl-2-furfural
IARC	International Agency for Research on Cancer

References

Acar, O. C. and Gökmen, V., 2009. Investigation of acrylamide formation on bakery products using a crust-like model. *Molecular Nutrition and Food Research*. 53(12): 1521–1525.

Acar, O. C., Pollio, M., Di Monaco, R., Fogliano, V. and Gökmen, V., 2012. Effect of calcium on acrylamide level and sensory properties of cookies. *Food and Bioprocess Technology*. 5(2): 519–526.

Amrein, T. M., Andres, L., Manzardo, G. G. G. and Amado, R., 2006. Investigations on the promoting effect of ammonium hydrogencarbonate on the formation of acrylamide in model systems. *Journal of Agricultural and Food Chemistry*. 54(26): 10253–10261.

Amrein, T. M., Schonbachler, B., Escher, F. and Amado, R., 2004. Acrylamide in gingerbread: critical factors for formation and possible ways for reduction. *Journal of Agricultural and Food Chemistry*. 52(13): 4282–4288.

Blank, I., 2005. Current status of acrylamide research in food: measurement, safety assessment, and formation. *Maillard Reaction: Chemistry at the Interface of Nutrition, Aging, and Disease*. 1043: 30–40.

Bush, M. F., Oomens, J., Saykally, R. J. and Williams, E. R., 2008. Effects of alkaline earth metal ion complexation on amino acid zwitterion stability: results from infrared action spectroscopy. *Journal of the American Chemical Society*. 130(20): 6463–6471.

European Food Safety Authority, 2012a. Scientific opinion on the tolerable upper intake level of calcium. In, vol. 10(7), p. 2814.

European Food Safety Authority, 2012b. Update on acrylamide levels in food from monitoring years 2007–2010. In, vol. 10(10), p. 2938.

Gökmen, V., Acar, O. C., Köksel, H. and Acar, J., 2007. Effects of dough formula and baking conditions on acrylamide and hydroxymethylfurfural formation in cookies. *Food Chemistry*. 104(3): 1136–1142.

Gökmen, V., Kocadaglı, T., Göncüoğlu, N. and Mogol, B. A., 2012. Model studies on the role of 5-hydroxymethyl-2-furfural in acrylamide formation from asparagine. *Food Chemistry*. 132(1): 168–174.

Gökmen, V. and Palazoğlu, T. K., 2008. Acrylamide formation in foods during thermal processing with a focus on frying. *Food and Bioprocess Technology*. 1(1): 35–42.

Gökmen, V., Palazoğlu, T. K. and Şenyuva, H. Z., 2006. Relation between the acrylamide formation and time-temperature history of surface and core regions of French fries. *Journal of Food Engineering*. 77(4): 972–976.

Gökmen, V. and Şenyuva, H. Z., 2006. A simplified approach for the kinetic characterization of acrylamide formation in fructose-asparagine model system. *Food Additives and Contaminants*. 23(4): 348–354.

Gökmen, V. and Şenyuva, H. Z., 2007a. Acrylamide formation is prevented by divalent cations during the Maillard reaction. *Food Chemistry*. 103(1): 196–203.

Gökmen, V. and Şenyuva, H. Z., 2007b. Effects of some cations on the formation of acrylamide and furfurals in glucose–asparagine model system. *European Food Research and Technology*. 225(5–6): 815–820.

Granvogl, M. and Schieberle, P., 2006. Thermally generated 3-aminopropionamide as a transient intermediate in the formation of acrylamide. *Journal of Agricultural and Food Chemistry*. 54(16): 5933–5938.

Hamzaloğlu, A. and Gökmen, V., 2012. Role of bioactive carbonyl compounds on the conversion of asparagine into acrylamide during heating. *European Food Research and Technology*. 235(6): 1093–1099.

Hamzalıoğlu, A., Mogol, B. A., Lumaga, R. B., Fogliano, V. and Gökmen, V., 2013. Role of curcumin in the conversion of asparagine into acrylamide during heating. *Amino Acids*. 44(6): 1419–1426.

Hidalgo, F. J., Delgado, R. M., Navarro, J. L. and Zamora, R., 2010. Asparagine decarboxylation by lipid oxidation products in model systems. *Journal of Agricultural and Food Chemistry*. 58(19): 10512–10517.

IARC, 1994. Acrylamide. In: *Monographs on the Evaluation of Carcinogenic Risks to Humans*. International Agency for Research on Cancer, Lyon, vol. 60. pp. 389–433.

Kocadağlı, T., Goncuoglu, N., Hamzalıoğlu, A. and Gökmen, V., 2012. In depth study of acrylamide formation in coffee during roasting: role of sucrose decomposition and lipid oxidation. *Food and Function*. 3(9): 970–975.

Kukurova, K., Ciesarova, Z., Bednarikova, A. and Markova, L., 2009. Effect of inorganic salts on acrylamide formation in cereal matrices. *Czech Journal of Food Sciences*. 27: S425–S428.

Levine, R. A. and Ryan, S. M., 2009. Determining the effect of calcium cations on acrylamide formation in cooked wheat products using a model system. *Journal of Agricultural and Food Chemistry*. 57(15): 6823–6829.

Lindsay, R. C. and Jang, S., 2005. Model systems for evaluating factors affecting acrylamide formation in deep fried foods. *Chemistry and Safety of Acrylamide in Food*. 561: 329–341.

Mestdagh, F., De Wilde, T., Delporte, K., Van Peteghem, C. and De Meulenaer, B., 2008. Impact of chemical pre-treatments on the acrylamide formation and sensorial quality of potato crisps. *Food Chemistry*. 106(3): 914–922.

Mottram, D. S., Wedzicha, B. L. and Dodson, A. T., 2002. Acrylamide is formed in the Maillard reaction. *Nature*. 419(6906): 448–449.

Mulla, M. Z., Bharadwaj, V. R., Annapure, U. S. and Singhal, R. S., 2011. Effect of formulation and processing parameters on acrylamide formation: A case study on extrusion of blends of potato flour and semolina. *LWT - Food Science and Technology*. 44: 1643–1648.

Ou, S. Y., Lin, Q., Zhang, Y. P., Huang, C. H., Sun, X. and Fu, L., 2008. Reduction of acrylamide formation by selected agents in fried potato crisps on industrial scale. *Innovative Food Science and Emerging Technologies*. 9(1): 116–121.

Qin, P.-H., Zhang, W. and Lu, W.-C., 2013. Theoretical study of hydrated Ca^{2+}-amino acids (glycine, threonine and phenylalanine) clusters. *Computational & Theoretical Chemistry*. 1021: 164–170.

Robert, F., Vuataz, G., Pollien, P., Saucy, F., Alonso, M. I., Bauwens, I. and Blank, I., 2004. Acrylamide formation from asparagine under low-moisture Maillard reaction conditions. 1. Physical and chemical aspects in crystalline model systems. *Journal of Agricultural and Food Chemistry*. 52(22): 6837–6842.

Sadd, P. A., Hamlet, C. G. and Liang, L., 2008. Effectiveness of methods for reducing acrylamide in bakery products. *Journal of Agricultural and Food Chemistry*. 56(15): 6154–6161.

Salazar, R., Arámbula-Villa, G., Luna-Bárcenas, G., Figueroa-Cárdenas, J. D., Azuara, E. and Vázquez-Landaverde, P. A., 2013. Effect of added calcium hydroxide during corn nixtamalization on acrylamide content in tortilla chips. *LWT-Food Science and Technology*. 56(1): 87–92.

Şenyuva, H. Z. S. and Gökmen, V., 2005. Study of acrylamide in coffee using an improved liquid chromatography mass spectrometry method: investigation of colour changes and acrylamide formation in coffee during roasting. *Food Additives and Contaminants*. 22(3): 214–220.

Stadler, R. H., Blank, I., Varga, N., Robert, F., Hau, J., Guy, P. A., Robert, M. C. and Riediker, S., 2002. Acrylamide from Maillard reaction products. *Nature*. 419(6906): 449–450.

Stadler, R. H., Robert, F., Riediker, S., Varga, N., Davidek, T., Devaud, S., Goldmann, T., Hau, J. and Blank, I., 2004. In-depth mechanistic study on the formation of acrylamide and other vinylogous compounds by the Maillard reaction. *Journal of Agricultural and Food Chemistry*. 52(17): 5550–5558.

Tareke, E., Rydberg, P., Karlsson, P., Eriksson, S. and Tornqvist, M., 2002. Analysis of acrylamide, a carcinogen formed in heated foodstuffs. *Journal of Agricultural and Food Chemistry*. 50(17): 4998–5006.

Vinci, R. M., Mestdagh, F., Van Poucke, C., Kerkaert, B., de Muer, N., Denon, Q., Van Peteghem, C. and De Meulenaer, B., 2011. Implementation of acrylamide mitigation strategies on industrial production of French fries: challenges and pitfalls. *Journal of Agricultural and Food Chemistry*. 59(3): 898–906.

Yaylayan, V. A. and Stadler, R. H., 2005. Acrylamide formation in food: a mechanistic perspective. *Journal of AOAC International*. 88(1): 262–267.

Yaylayan, V. A., Wnorowski, A. and Locas, C. P., 2003. Why asparagine needs carbohydrates to generate acrylamide. *Journal of Agricultural and Food Chemistry*. 51(6): 1753–1757.

Zhou, J. S., Sun, G., Han, Z. R., Liao, G. Z., Wu, R. G., Zhang, Q. L. and Liu, F. Q., 2003. Static light scattering study of the formation of partially hydrolyzed poly(acrylamide)/calcium(II) complexes. *Abstracts of Papers of the American Chemical Society*. 226: U377–U378.

Zyzak, D. V., Sanders, R. A., Stojanovic, M., Tallmadge, D. H., Eberhart, B. L., Ewald, D. K., Gruber, D. C., Morsch, T. R., Strothers, M. A., Rizzi, G. P. and Villagran, M. D., 2003. Acrylamide formation mechanism in heated foods. *Journal of Agricultural and Food Chemistry*. 51(16): 4782–4787.

CHAPTER 18

Addition of Calcium to Gluten and Nongluten Flours for Breadmaking

URSZULA KRUPA-KOZAK*[a]

[a]Department of Chemistry and Biodynamics of Food, Institute of Animal Reproduction and Food Research of the Polish Academy of Sciences, Tuwima St. 10, 10-748 Olsztyn, Poland
*E-mail: u.krupa-kozak@pan.olsztyn.pl

18.1 Introduction

The history of breadmaking probably began in the Neolithic era when the hunter-gatherer way of life was replaced by the domestication of crops and animals. Our ancestors learned how to cultivate plants and the agricultural revolution began. A new diet appeared, predominantly based on cereals, marking an important turning point in human history. Cereals provide an important amount of well-balanced macro- and micronutrients. At present, the contribution of cereal products to the estimated dietary intake of several minerals and trace elements is a crucial part of the total intake in Western countries. However, this change in dietary habits, produced unexpected problems and generated a variety of previously unknown food antigens. The introduction of gluten-containing grains probably represented an evolutionary challenge that created the conditions for human diseases related to gluten exposure. Nowadays, it is clear that gluten intolerance is common, since

Food and Nutritional Components in Focus No. 10
Calcium: Chemistry, Analysis, Function and Effects
Edited by Victor R. Preedy

Published by the Royal Society of Chemistry, www.rsc.org

the human body is still vulnerable to the toxic effects of these proteins. Insufficient adaptation of gastrointestinal and immunological responses could be seen as the main reason for the "gluten-sensitivity epidemic".

Calcium is an essential ion in the human body, necessary to maintain its proper functionality, while at the same time being the most abundant element. Additionally, calcium could be considered to be a functional ingredient because an adequate life-long intake of this mineral can reduce the risk of osteoporosis in the elderly (Cashman and Flynn, 1999). Although calcium supplementation together with vitamin D may reduce bone loss and fracture incidence, some foods (such as white bread) are deficient in calcium. The processing of wheat grain into flour impoverishes the content of this nutrient as it passes to the bran fraction during milling, leading to the low nutritional value of white flour. Therefore, calcium supplementation of white flour is recommended. Currently, food innovation is based on the concept of functional food based on the use of nutrition knowledge in the food industry to improve consumer health. Bread has been a staple food for ages. Its ubiquitous consumption places it in a position of global importance in international nutrition. However, the rising prevalence of coeliac disease and other gluten allergies parallels the increasing demands for gluten-free products, including bread.

This chapter reviews the existing literature concerning the technological and nutritional aspects of calcium in both traditional and gluten-free breadmaking. Additionally, calcium recommendations and its role in the human body, gluten-related disorders and the nutritional value of cereals are described.

18.2 Calcium – Role and Demands

18.2.1 Calcium in the Human Body

In the human body, calcium constitutes above one kilogram of body mass. More than 99% of calcium is deposited in the bones and teeth. In bones, calcium provides structure and mechanical strength and serves as a reservoir that can be tapped to maintain extracellular calcium concentration, regardless of intake. Moreover, calcium is necessary to maintain the proper functionality of the circulatory (regulating blood pressure, activating blood coagulation and haemoglobin formation) and neuromuscular systems (regulating the release of neurotransmitters, maintaining tissue excitability, impulse conduction); is a cofactor of several hormones and enzymes, and influences the immunological system, see Table 18.1.

Serum calcium is tightly regulated and does not fluctuate with changes in dietary intake. To provide calcium homeostasis, the coordinated actions of three processes: absorption in the intestine, reabsorption from the kidney and exchange from bones are critical. Calcium balance is regulated through the calciotropic hormones: parathyroid hormone (PTH), 1,25-dihydroxyvitamin D and calcitonin.

Nevertheless, even an adequate calcium intake with the diet may not ensure a proper calcium balance, as calcium absorption in the intestine is regulated

Table 18.1 The main physiological functions of calcium and the effects of deficiency of this element. Reproduced with permission from Brzóska and Moniuszko-Jakoniuk (1997).

Physiological functions of calcium	Effects of calcium deficiency
Structuring and homeostasis of the skeleton	Impairment of bone mineralization (rachitis, osteomalacia, osteoporosis)
Regulation of all cell functions including secretion, excitation, locomotion, differentiation and proliferation	Increased risk for osteoporosis, particularly in postmenopausal women (increased rate of bone loss)
Cofactor for a whole series of important enzymatic reactions	Secondary hyperparathyroidism
Preserving and stabilizing of cell structures	Increased blood pressure, hypertension (including hypertension during pregnancy)
Stabilization and regulation of membranes permeability	Increased risk for the development of cardiovascular disease
Regulation of hormone secretion and neurotransmitter release	Tetany
Regulation of heart, smooth and striated muscle contraction	Disturbance of extra- and intracellular Ca^{2+} concentration gradient and thus impairment of all cell functions and Ca^{2+}-regulated processes
Regulation of vascular tone and blood pressure	
Regulation of blood clotting	
Regulation of neuromuscular excitability	

by several factors. An adverse calcium balance, beyond an inadequate calcium supply, may result from poor intestine absorption caused by infection, digestion disease or pathology in the intestine morphology. Even serious disturbances in the blood–calcium balance are not manifested by obvious clinical symptoms, as both hypocalcaemia and hypercalcaemia may be symptoms of other diseases. The group of calcium deficiency risk includes, in particular children and adolescents, in which a calcium deficiency may cause the growth problems and difficulties in the peak bone mass achievement, as well as persons with chronic intestinal diseases (coeliac disease, Crohn's disease). In the elderly, a nutritional calcium deficiency causes osteoporosis and increases fracture risk as a consequence of decreased bone-mineral density (BMD). The relationship between gluten-related diseases and bone disorders is well documented. Newly diagnosed or inadequately treated CD patients often suffer from skeletal disorders, have reduced BMD and increased bone fragility, leading to a high prevalence of bone fractures (Krupa-Kozak, 2014).

18.2.2 Recommendations and Guidelines

Calcium is an essential nutrient required in substantial amounts (FAO/WHO, 1998), see Figure 18.1.

Among the general population, calcium intake does not meet those recommendations. The United States Food and Drug Administration (2008)

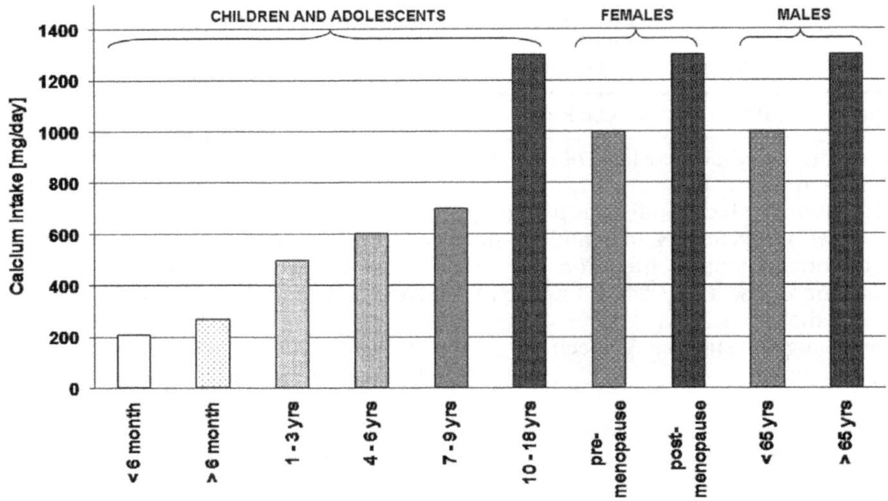

Figure 18.1 The adequate daily calcium intake requirements.

has accepted health claims related to calcium and osteoporosis for foods and supplements, that recently was expanded to include vitamin D and it reads as follows: "Adequate calcium and vitamin D as part of a healthful diet, along with physical activity, may reduce the risk of osteoporosis in later life". Diet based on gluten-free products is also low in calcium (Thompson *et al.*, 2005). Calcium supply in the diet of CD patients is additionally reduced due to a decreased intake of milk and dairy products as secondary lactose intolerance is common in CD. The calcium status of individuals suffering from gluten intolerance depends on the length of time they have lived with an active-yet-undiagnosed disease, the extent of damage to the gastrointestinal (GI) tract and the degree of malabsorption. A daily calcium intake for people with CD should be higher than the recommended daily allowance (RDA) because of latent malabsorption in many patients, thus the increased calcium intake could potentially compensate for the reduced calcium absorption (Pazianas *et al.*, 2005).

18.3 Gluten Sensitivity

Wheat, barley, rye and probably oat proteins are the major triggers of gluten sensitivity. A wide spectrum of gluten-related disorders exist, the best known of which are mediated by the adaptive immune system: coeliac disease (CD) and wheat allergy (WA). CD is a chronic small intestinal, immune-mediated enteropathy occurring in genetically predisposed individuals precipitated by exposure to dietary gluten and related prolamines, characterized by specific autoantibodies against tissue transglutaminase (anti-tTG) and endomysium (anti-EMA). Screening studies have shown that the general prevalence of CD is about 1%, although large unexplained differences have been observed across different European countries (0.3–2.4%) (Mustalahti *et al.*, 2010). WA is an adverse reaction to the ingestion of wheat proteins, caused by an immunological mechanism. Although it is less common than CD, wheat allergy affects a high proportion of food-allergy sufferers in Northern Europe. Besides CD and WA, there are cases of gluten reactions in which neither allergic nor autoimmune mechanisms are involved, see Table 18.2.

Gluten sensitivity (GS), known also as noncoeliac gluten sensitivity (NCGS) is triggered by gluten ingestion in the absence of coeliac-specific antibodies and of classical coeliac villous atrophy, with variable human leukocyte antigen (HLA) status and a variable presence of first generation antigliadin

Table 18.2 Characteristic of the main gluten-related disorders.[a]

	Coeliac disease	Wheat allergy	Gluten sensitivity
Definition	Immune-mediated enteropathy triggered by the ingestion of gluten in susceptible individuals	Adverse immunologic reaction to wheat proteins	Gluten sensitivity with negative immunoallergy tests to wheat or negative CD serology with normal duodenal histopathology
Epidemiology	Approx. 1%	Children 0.5–9%; adults 0.4–3%	Unknown
Pathogenesis	Autoimmune reaction to gluten; genetic predisposition; environmental factors	Allergic reaction to wheat proteins	Neither allergic nor autoimmune, rather immune-mediated reaction to gluten and other wheat proteins
Clinical presentation	Gastrointestinal symptoms (chronic diarrhoea, malabsorption symptoms, weight loss) and/or extraintestinal symptoms (anaemia, osteoporosis, neurological disturbances)	Baker's asthma, allergic rhinitis, atopic dermatitis, urticaria, anaphylaxis	Extraintestinal symptoms ("foggy mind", headache, fatigue, joint and muscle pain, dermatitis, depression, anaemia) combined with gastrointestinal symptoms
Diagnosis	CD symptoms, positive CD serology, determination of HLA-DQ2/DQ8, small-intestine biopsy	Specific skin prick tests, wheat specific serum IgE assays, gluten challenge	Exclusion criteria for CD and WA, gluten challenge
Treatment	Gluten-free diet	Gluten-free diet	Gluten-free diet

[a]CD – coeliac disease; WA – wheat allergy; HLA – human leukocyte antigen; IgE – immunoglobulin E.

antibodies (anti-AGA) (Catassi *et al.*, 2013). The prevalence of NCGS is still unknown. However, NCGS and wheat sensitivity occur very often in adults with IBS and IBS-like symptoms (Biesiekierski *et al.*, 2011). Taking into account the very high prevalence of IBS in the general population (16–25%) (Krosgaard *et al.*, 2013), the prevalence of NCGS could well be higher than CD.

18.4 Calcium in Traditional Breadmaking

18.4.1 Nutritional Value of Cereals

Cereals provide important amounts of most nutrients. Nutritional recommendations and guidelines emphasize the role of cereal products in the everyday diet as a source of well-balanced macro- and micronutrients, see Figure 18.2.

18.4.1.1 Macronutrients

Among carbohydrates, starch is the most abundant polysaccharide of grains, providing a bulk nutrient and energy source in the human diet. Endosperm, the main part of the grain (80–85%), contains mostly starch and is low in proteins and lipids (Rosell, 2011). Cereals are rich in dietary fiber, including resistant starch, cellulose and other complex polysaccharides, such as arabinoxylans, β-glucans, pectins and arabinogalactans, together with lignin. Cereal proteins are mainly located in the germ, the small inner core that

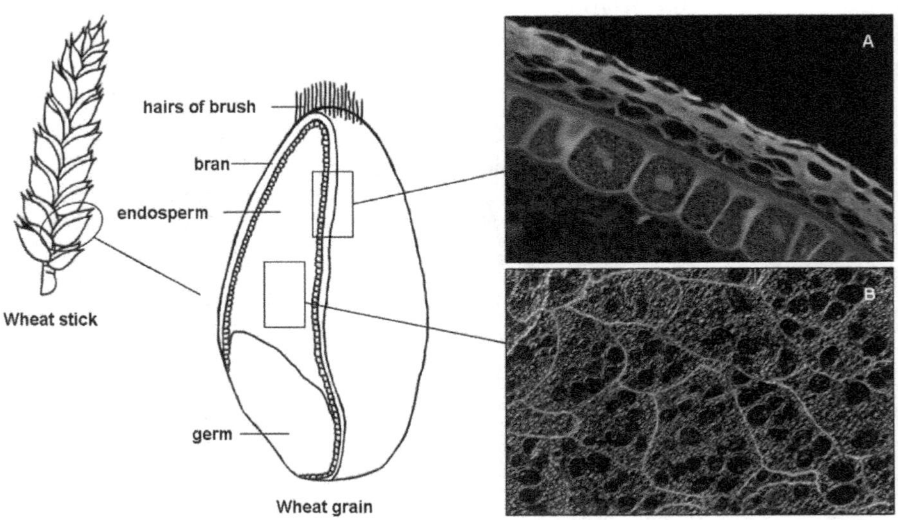

Figure 18.2 The anatomical structure and microstructure of wheat kernel. (A) Aleurone layer; (B) endosperm cells with starch granules. Scanning electron microscope pictures reproduced with permission from Błaszczak and Fornal (2008).

represents approximately 21% of the grain. They comprise up to 80% gluten proteins (gliadins and glutenins) that have high contents of proline and glutamine, at the expense of essential amino acids, particularly lysine and, to a lesser extent, threonine. Although lipids comprise only about 1.5–7.0% of cereal grains, they include a range of components such as essential fatty acids, fat-soluble vitamins (Bock, 2000) and phytosterols.

18.4.1.2 Main Micronutrients

Cereals provide some vitamins, especially of group B (thiamine, riboflavin, niacin and pyridoxine), but also a moderate amount of biotin and folic acid, while being poor in ascorbic acid (vitamin C) and vitamins A, D, E and K (Bock, 2000). Additionally, grains contain about 1.5–2.5% minerals. Bran, which represents 7% of the grain, contains the majority of the minerals (40–70%) (Rosell, 2011). Wheat, rye, barley and oats are classified as a moderate source of calcium (100–200 mg/10 g) and magnesium (100–200 mg/100 g) (Dewettinck *et al.*, 2008). Studies on the distribution of minerals within the kernel of common wheat have shown that the endosperm contains about 50% calcium, whereas about 25% is found in the aleurone layer (Pomeranz, 1988). Other elements in cereals are presented in small (iron, zinc, copper) or trace quantities (Bock, 2000).

18.4.2 The Breadmaking Process

The first bread, while not in the same form as we know it today, was made around 10 000 years ago, after which it became a basic dietary item. Egyptians made the "art of breadmaking" popular through the world. Since that time, the bread "story" has been reiterated for generations, and bread is still today one of the most frequently consumed food products. There is general agreement concerning the great importance of bread for many aspects of humanity and civilization. However, in the last two decades the consumption of traditional bread has declined due to change of eating patterns and increasing number of bread substitutes, such as breakfast cereals and fast foods, whereas a growing interest in ethnic and specialty breads has been observed. Sliced and packaged bread is also becoming increasingly popular in some countries.

In traditional breadmaking, cereals, milled into flour, are used to make different types of breads with many shapes, structure and sensory characteristics. Wheat (*Triticum aestivum*) is the most important crop for breadmaking because of its superior baking performance compared to all other cereals. Apart from wheat, rye (*Secale cereale*) is also suitable for the preparation of leavened breads. The nutritive value of bread could be increased if, for example, barley (*Hordeum vulgare*) and oats (*Avena sativa*) are incorporated into the bread formula. A basic bread formula is composed mainly of cereal flour (always 100%) and water (approximately 50% of flour amount by weight), affecting the texture of the crumb, whereas other ingredients, like leavening agent (yeast or leavening agent), salt, sugar as well as shortening agent, are present in smaller amounts. However, large-scale production and consumer demand for higher bread quality, convenience, increased nutritional value and longer shelf-life, have generated a need for functional additives (emulsifiers, antistaling agents) and supplements in breadmaking.

To convert the above-mentioned ingredients into a dough, and then into a light, aerated, well-expanded and palatable loaf of bread, several subsequent processes, defined as breadmaking, need to be performed (Scanlon and Zghal, 2001). To briefly define the breadmaking process, it is divided into three major stages: mixing, proofing and baking. Initially, mixing disperses the ingredients homogeneously, hydrates the flour components, supplies the energy for protein network development and incorporates air into the dough. The majority of the water added to make up the dough is absorbed by hydrophilic groups of protein molecules. Under conditions providing sufficient dough hydration and when the dough is optimally developed by the mixing process, the disruption of the initially spherical protein particles takes place, together with the stretching and alignment of proteins, which leads to the formation of a three-dimensional viscoelastic structure with gas-retaining properties. Proteins of wheat flour (gliadin and glutenin) form gluten, which is vital to the proper development of the crumb structure.

During dough proofing or fermentation, the leavening agent (yeast) generates gas (CO_2), resulting in the growth of air bubbles previously incorporated during mixing. Each yeast cell forms a center around which carbon dioxide bubbles are released. As the fermentation progresses, the expansion of the dough proceeds as air bubbles are inflated to large volumes with thin cell walls. Kneading or remixing of the dough favors the release of large gas bubbles, resulting in a more even distribution of the bubbles within the dough. Baking is the last step in breadmaking process. In the oven, the final bread crumb structure is achieved by conversion of the liquid phase to a solid phase. Temperature is the dominating factor in various physicochemical and biological changes taking place during bread baking. The initial rise in temperature intensifies the yeast fermentation and production of gas inside the dough, thus causing the expansion of the dough volume (Scanlon and Zghal, 2001). CO_2 is also released from the bread when the bubble walls start to break under pressure, making the porous structure more continuous and open to the outside of the bread. With baking time, both protein denaturation and starch gelatinization (60–85 °C) occur, affecting the diffusion of water. Starch gelatinization results in an amorphous structure in the final bread. Gluten changes its characteristics from elastic to semirigid, contributing to the change from dough to crumb. The crust is formed during baking. Sugars and proteins break down products participating in Maillard or nonenzymatic browning reactions, which both contribute to the formation of bread flavor and color.

Cauvain and Young (2007) identified four major breadmaking methods: straight dough bulk fermentation, sponge and dough, rapid processing, and mechanical dough development. The first method is the straight dough method, where the mixing of ingredients is performed in one step, while in the sponge and dough method the mixing of ingredients is performed in two steps. During the first step, the leavening agent is prepared. The yeast is mixed with a certain quantity of water and flour and the mixture is then left to develop for a few hours and is then mixed with the rest of the ingredients. The Chorleywood bread process is an example of the mechanical method of breadmaking in which all the ingredients are mixed in an ultrafast mixer for several minutes. In the past few decades, sub-zero and low temperature have been introduced for the production of bread in order to adapt the breadmaking process to consumer demands. The frozen dough method, retarded or interrupted proofing method, and the partial baking method are techniques facilitating the launch of a great number of new baking products to the market without compromising quality.

18.4.3 Cereals Processing *vs.* Calcium Content

Refined cereals do not have the same nutritional and health benefits as grain or wholemeal. Processing affects the levels of several nutrients, including calcium, and modifies the bioavailability of these compounds. White cereal flours are generally deficient in calcium. There are several factors responsible for calcium shortage in traditional bread, including two of decisive significance: milling and breadmaking.

Milling is the critical process affecting the concentrations of inorganic elements in wheat-derived food products. As the outer parts of the kernel, especially the aleurone layer and the germ, are richer in minerals than the starchy endosperm, conventional milling reduces their content in flour and concentrates them in the milling residues (Slavin *et al.*, 2001), see Figure 18.3.

Flours differ in their extraction rate, which is defined as the proportion of flour by weight, derived by milling from a known quantity of grains. White flour is generated when the extraction rate is 75% or less. Early study by Pederson *et al.* (1989) indicated that since wholemeal wheat flour (extraction rate 100%) is twice richer in calcium (0.44 mg g^{-1}) than white wheat flour

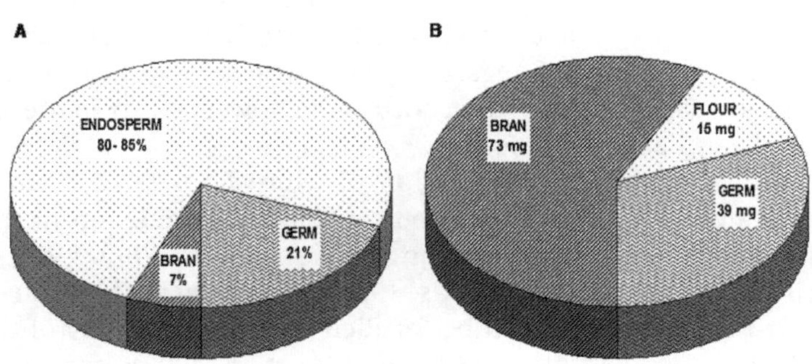

Figure 18.3 Composition of wheat grain and the calcium content in wheat fractions separated during milling process. (A) Proximate composition of wheat grain in %, (B) calcium content in wheat fractions in mg/100 g.

(extraction rate 66–0.23 mg g^{-1}), the extraction rate corresponds with calcium concentration. However, even if grains are a moderate source of calcium, its bioavailability is decreased due to the formation of an insoluble complex with phytic acid (PA). Cereal grains contain 1–4% of PA or its salts (phytates, *myo*-inositol hexakis phosphates or InsP6) concentrated in the grain aleurone layers. PA has a strong chelating capacity and is considered to be an antinutritional factor. Inositol phosphate analysis showed that whole wheat flour contained significantly higher amounts than wheat flour (Frontela *et al.*, 2011).

Although calcium content is especially influenced by milling, breadmaking can also alter the bioavailability of calcium in cereal products. The breadmaking process consists of three major stages: mixing, proofing and baking. As mentioned previously, PA complexes calcium that impairs its absorption in humans. During dough fermentation, the content of PA decreases due to the action of endogenous grain phytase, an enzyme hydrolyzing phytic acid into lower *myo*-inositol phosphate esters with a lower capacity to bind minerals (Leenhardt *et al.*, 2005). Phytate-degrading enzymes also exist in yeast and lactic acid bacteria isolated from sourdough (Lopez *et al.*, 2000; Turk *et al.*, 2000). Flour extraction rate, dough pH, temperature, water content and fermentation time all contribute to phytate reduction. The phytate content decreases in the fermentation stage, but usually not to such an extent as to greatly improve the mineral bioavailability in whole-wheat products.

18.4.4 Calcium Supplements and Additives in Breadmaking

The selection of the appropriate calcium supplement should be based on the functional and technological properties associated with the respective product, such as solubility, calcium content, taste and bioavailability, whereas in the case of a specific food application, medical conditions (such as lactose intolerance, impaired gastric acid secretion and high-risk of kidney-stone formation) should also be taken into consideration before selection of an appropriate supplement. The dominant calcium supplement in bakery production worldwide is calcium carbonate, containing the largest calcium percentage (40%), while at the same time being the most cost effective. Calcium citrate provides 21% of calcium and is better absorbed than calcium carbonate. Moreover, it is absorbed regardless of gastric acidity; thus, individuals producing less gastric acids or taking drugs that lower acidity in the stomach may optimally utilize this salt form.

In the United Kingdom, fortification of white flour (except wholemeal flour) with calcium is mandatory; flour is fortified in calcium in the form of calcium carbonate in quantities ranging from 235 to 390 mg of Ca/100 g of flour (Bread and Flour Regulations, 1998). According to the United States enrichment standards, flour can be fortified to contain 211 mg of Ca/100 g (Code of Federal Regulations, 1996). Ranhotra *et al.* (2000) have shown that flour can be fortified to contain calcium at much higher levels than the US standard without adversely affecting white bread quality and that this calcium is well absorbed and retained.

Several studies analyzed the influence of different calcium salts on dough and bread quality. Wang and Tang (1995) indicated the improvement of the dough expansion without adverse effects on the quality of calcium lactate fortified bread. Sudha and Leelavathi (2008) analyzed the effect of different calcium salts on farinograph (water absorption, dough development time, and dough stability) and alveograph (tenacity, extensibility, and deformation energy of dough) properties and demonstrated that calcium salts affected dough stability and extensibility-to-tenacity ratio. Bread has a short shelf-life and a number of chemical and physical changes occur, commonly known as staling. Therefore, to improve the shelf-life of bakery products, a variety of preservatives are used. Calcium propionate is commonly applied in bread as an antifungal agent. The addition of 0.3% calcium propionate prolongs the shelf-life of low-salt bread to 10–12 days (Belz *et al.*, 2012). However, the authors indicate that the addition of sourdough fermented using the specifically selected antifungal *Lactobacillus amylovorus* DSM 19280 can replace a calcium propionate addition and guarantee the product safety of low-salt bread.

18.5 Calcium in Gluten-Free Breadmaking

18.5.1 Technological and Nutritional Quality of Gluten-Free Baked Products

The term "gluten free", by definition, refers only to foods containing less than 20 ppm of gluten (Codex Alimentarius Commission, 2007). In addition, the label "very low gluten" may be used for foods such as bread produced using cereals that have been specially processed to remove most of the gluten and contain less than 30 mg of gluten per daily portion.

It is generally accepted that cereals must be avoided in gluten-sensitivity disorders. However, a cereal-free and gluten-free diet is extremely difficult to adhere to due to their ubiquity in human foods. In gluten-free breadmaking, the use of wheat, rye, barley and hybrids (Kamut®, spelt and triticale) is not allowed. Instead, wheat starch, as well as other starches from naturally gluten-free sources such as corn, potato, rice or tapioca have been used in gluten-free formulations (Moore *et al.*, 2004; Krupa-Kozak *et al.*, 2011b). However, the lack of gluten proteins in the GF formulation presents a great challenge to both cereal technologists and bakers. Gluten, due to its unique properties, contributes to batter emulsification and viscoelasticity, provides cohesiveness to dough during processing, retains leavening gases, sets the crumb structure and imparts elasticity to the bread texture. Many studies on gluten-free baking products have been focused on the design of a gluten-free matrix to overcome the negative impact on the absence of gluten network. A wide range of hydrocolloids have been incorporated to the GF baked products to provide a proper network necessary to hold the carbon dioxide produced during proofing and mimic the viscoelastic properties of gluten. These resulted in the improvement of the technological properties, acceptability and shelf-life of gluten-free baked products (Lazaridou *et al.*, 2007). However, microstructure analysis suggested that hydrocolloids alone are not sufficient to fully replace gluten. Therefore, proteins (dairy proteins, egg proteins) and proteins crosslinking enzymes (transglutaminase) were used to improve the textural characteristics of GF bread (Moore *et al.*, 2006).

However, a poor structure and texture are not the only problems of gluten-free products. They are characterized by a low content of nutritional components, mineral compounds (iron, calcium), B vitamins (thiamine, riboflavin and niacin), folate and dietary fiber (Thompson *et al.*, 2005). Dairy ingredients are used in the GF bread, resulting in nutritional benefits (increase in proteins), improved volume, appearance and sensory aspects (Nunes *et al.*, 2009; Krupa-Kozak *et al.*, 2013). Additionally, nutrient-dense gluten-free grains, such as amaranth, buckwheat, quinoa and teff incorporated in the GF bakery product, add variety and improve its nutritional quality (Alvarez-Jubete *et al.*, 2009; Krupa-Kozak *et al.*, 2011a).

18.5.2 Calcium in Gluten-Free Baked Products

Very few gluten-free products are fortified in calcium as their wheat-containing counterparts, making calcium supplementation desirable or necessary. However, studies concerning the influence of calcium supplements on the rheological properties of gluten-free dough, as well as reports addressing the relations between those compounds and the technological properties and sensory-quality attributes of the end products are limited.

Krupa-Kozak *et al.* (2012) analyzed the influence of different organic and nonorganic calcium salts (calcium lactate, calcium citrate, calcium chloride and calcium carbonate) on gluten-free dough behavior during mixing–heating–cooling assessed by Mixolab® (Chopin, France). Gluten-free dough lacks a natural viscoelastic network, which makes it necessary to completely design polymeric matrices to meet breadmaking requirements. The absence of gluten determines the rheological properties of gluten-free dough, which is similar to cake batter rather than wheat dough. The rheological behavior of gluten-free dough, mainly composed of corn and potato starches (data unpublished), is presented in Figure 18.4.

In the initial stage of mixing and the beginning of heating, a very low consistency of the GF dough is observed due to the high hydration and very low protein content. Regarding the heating phase, two different slopes were detected, corresponding to the gelatinization of potato starch and corn starch. The addition of calcium salts influenced the dough consistency. Krupa-Kozak *et al.* (2012) indicated that calcium citrate and carbonate promoted an increase in the dough consistency during heating. Conversely, lactate and chloride induced the opposite effect, resulting in delayed corn starch swelling, and calcium chloride lowered the maximum consistency reached. A high increase in the consistency during cooling suggests that dough presents a high retrogradation tendency and, consequently, the baked product prepared from this dough would undergo a high staling rate during storage (Matos and Rosell, 2013). Therefore, it must be emphasized that the addition

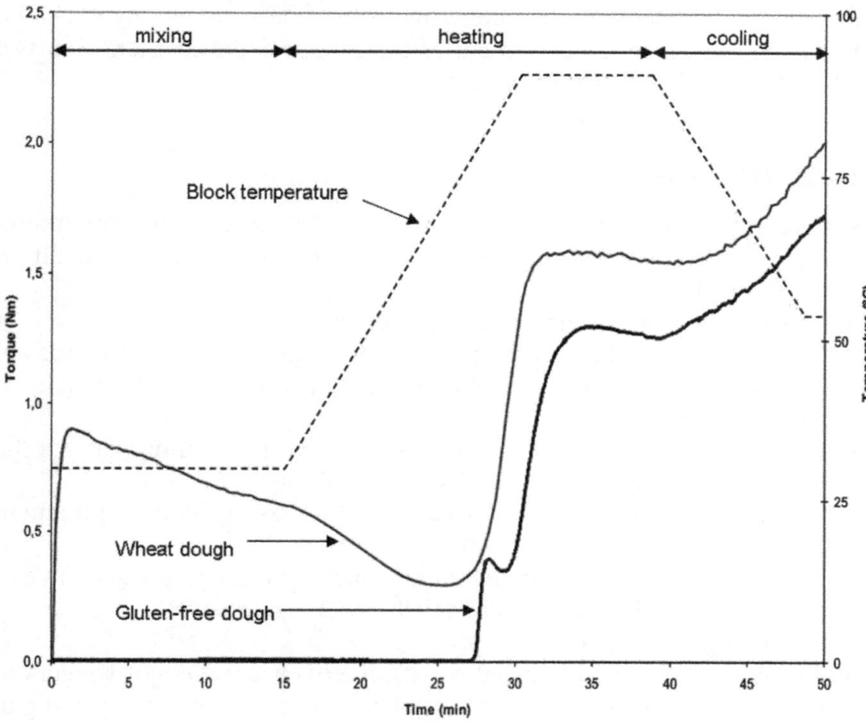

Figure 18.4 Comparison of gluten-free (data unpublished) and wheat dough (Rosell *et al.*, 2007) consistency. Dough consistency was studied using Mixolab® (Chopin, France), which measured in real time the torque (expressed in N m) produced by passage of dough between two kneading arms, thus allowing the study of its physicochemical behavior.

of calcium salts to a gluten-free formulation modifies the dough performance during mixing–heating–cooling, which could have a major impact on the resulting gluten-free bread.

In general, the technological quality of baked gluten-free products is low; they exhibit crumbling texture, poor mouth feel, color and flavor. Additionally, the evaluation of the mechanical properties of gluten-free bread crumb is difficult, because the distribution of cells within the bread crumb is not homogeneous and differences in texture properties between the upper and the bottom part of bread slices occur, influenced by the differences in the thickness of the cell walls and the moisture content (Krupa-Kozak *et al.*, 2011b). Calcium supplements may favorably influence the baking characteristics and consumer opinions concerning the appearance and sensory quality of gluten-free breads. Study by Krupa-Kozak *et al.* (2011b) demonstrated that the application of a mixture of two organic calcium supplements: calcium caseinate (low lactose dairy products of the beneficial influence on the sensory quality of gluten-free bread) and calcium citrate (rich in elemental calcium) results in GF bread with the increased specific volume, fine-colored crust and pleasant sensory characteristics. Additionally, this study indicated that it is possible to obtain GF bread with a high-calcium content that is a good source of calcium (FAO/WHO, 1998).

Summary Points

- Cereals provide important amounts of well-balanced macro- and micronutrients and in traditional breadmaking, they are used to make different types of bread.
- Wheat produces supreme baking performance.
- Wheat proteins (gliadin and glutenin) form gluten that contributes to the viscoelastic properties of dough, and the structure and elasticity of bread crumb.
- The technological processing of cereals, especially milling, affects the levels of calcium and modifies its bioavailability.
- White wheat flour is deficient in calcium, therefore calcium supplementation is mandatory or recommended.
- The dominant calcium supplement in bakery production is calcium carbonate, containing the largest calcium percentage, while at the same time being the most cost effective.
- Wheat, barley, rye and probably oat proteins are the major triggers of gluten-related disorders, such as coeliac disease, wheat allergy and gluten sensitivity.
- In the case of any gluten sensitivity, the consumption of cereals and cereal-based products containing gluten is forbidden, however, the elimination of gluten presents a great challenge to both the cereal technologist and the bakers.
- The quality of baked gluten-free products is low; they exhibit crumbling texture, poor mouth feel and short shelf-life.
- Absence of gluten determines the rheological properties of gluten-free dough that resemble liquid batter.
- Calcium supplements added to a gluten-free formulation modify the dough performance and have a great impact on the technological properties of the resulting gluten-free bread.
- While gluten-free products are characterized by a low content of several nutrients, including calcium, calcium supplements may influence favorably the nutritional quality of the obtained gluten-free product resulting in calcium-rich fine-colored gluten-free bread of pleasant sensory characteristics.

Key Facts of Breadmaking

1. The first bread was made probably 10 000 years ago. The domestication of cereals was an important turning point in human history, starting the agricultural revolution.

2. Bread is still a staple food, although its consumption has declined by about 14% during the period 1999–2003. The growing interest in ethnic and specialty breads is observed.
3. Nowadays, bread could be considered as a functional food. Bread is a "carrier" of several bioactive compounds and supplements that could reduce the risk of several diseases. New varieties of bread such as a low-caloric bread, high-fiber bread, or a low-salt bread are of consumers demand.

Definitions and Explanations

Bran. The external part of the cereal grain covering the kernel, discarded during grain milling into white flour. It is mainly composed of carbohydrates, especially dietary fiber, but contain also proteins, B vitamins and minerals.

Breadmaking. A course of subsequent processing operations leading to conversion of basic ingredients such as flour, water, leavening agent and salt into dough, then to bread.

Gluten. A mixture of wheat proteins (gliadin and glutenin) possessing the unique viscoelastic properties, playing an important role in the production of cereals baked products. Gluten provides cohesiveness to dough during processing, retains leavening gases, sets the crumb structure and imparts elasticity to the bread texture.

Gluten sensitivity. A spectrum of disorders triggered by gluten consumption manifested by a variety of immunological, symptomatic and/or morphological symptoms.

Gluten-free diet. A special diet, and at the same time the only method of gluten-related disorders treatment, that strictly exclude consumption of products containing gluten.

Proofing. The process when the leavening agent (yeast or chemicals) in the dough is allowed to expand resulting in dough-volume enlargement.

Food supplementation. A process of food enrichment in one or several nutrients (vitamins, minerals, *etc.*). The main assumption of supplementation is to increase the intake of deficient nutrient with food when it is consumed in insufficient quantities.

Osteoporosis. A bone disease, characterized by low bone-mineral density (T-score ≤ -2.5), leading to bone weakening and increased fragility to fractures.

Calciotropic hormones. A group of hormones that has a specific regulatory effect on the regulation of calcium metabolism and bone homeostasis.

Hydrocolloids. Polymers mainly derived from natural sources, that disperse in water and depending on the quantity of water available form a gel or sol. They are widely applied in food production as compounds influencing viscosity and texture.

Maillard reaction. A chemical reaction between an amino acid and reducing sugar induced by heating responsible for the formation of flavor and color that occurs on crusts during bread baking.

List of Abbreviations

Anti-AGA	Antigliadin antibodies
Anti-EMA	Antiendomysium antibodies
Anti-tTG	Antitransglutaminase antibodies
BMD	Bone-mineral density
CD	Coeliac disease
GF	Gluten free
GI	Gastrointestinal
GS	Gluten sensitivity
HLA	Human leukocyte antigen
IBD	Inflammatory bowel disease
IBS	Irritable bowel syndrome
IgE	Immunoglobulin E
NCGS	Noncoeliac gluten sensitivity
PA	Phytic acid
PTH	Parathyroid hormone
RDA	Recommended daily allowance
WA	Wheat allergy

References

Alvarez-Jubete, L., Arendt, E. K. and Gallagher, E., 2009. Nutritive value and chemical composition of pseudocereals as gluten-free ingredients. *International Journal of Food Science and Nutrition*. 60: 240–257.

Belz, M. C. E., Mairinger, R., Zannini, E., Ryan, L. A. M., Cashman, K. D. and Arendt, E. K., 2012. The effect of sourdough and calcium propionate on the microbial shelf-life of salt reduced bread. *Applied Microbiology and Biotechnology*. 96: 493–501.

Biesiekierski, J. R., Newnham, E. D., Irving, P. M., Barrett, J. S., Haines, M., Doecke, J. D., Shepherd, S. J., Muir, J. G. and Gibson, P. R., 2011. Gluten causes gastrointestinal symptoms in subjects without celiac disease: a double-blind randomized placebo-controlled trial. *American Journal of Gastroenterology*. 106: 508–514.

Błaszczak, W. and Fornal, J., 2008. Application of microscopy methods in food analysis. *Polish Journal of Food and Nutrition Sciences*. 58: 183–198.

Bock, M. A., 2000. Minor constituents of cereals. In: Kulp, K. and Ponte, Jr., J. G. (ed.) *Handbook of Cereal Science and Technology*. Marcel Dekker Inc., New York, pp. 479–504.

Bread and Flour Regulations, 1998. No. 141. Available at: http://www.legislation.gov.uk/uksi/1998/141. Accessed 23 December 2013.

Brzóska, M. M. and Moniuszko-Jakoniuk, J., 1997. The influence of calcium content in diet on cumulation and toxicity of cadmium in the organism. *Archives in Toxicology*. 72: 63–73.

Cashman, K. D. and Flynn, A., 1999. Optimal nutrition: calcium, magnesium and phosphorus. *Proceedings of the Nutrition Society*. 58: 477–487.

Catassi, C., Bai, J. C., Bonaz, B., Bouma, G., Calabrò, A., Carroccio, A., Castillejo, G., Ciacci, C., Cristofori, F., Dolinsek, J., Francavilla, R., Elli, L., Green, P., Holtmeier, W., Koehler, P., Koletzko, S., Meinhold, C., Sanders, D., Schumann, M., Schuppan, D., Ullrich, R., Vécsei, A., Volta, U., Zevallos, V., Sapone, A. and Fasano, A., 2013. Non-celiac gluten sensitivity: the new frontier of gluten related disorders. *Nutrients*. 5: 3839–3853.

Cauvain, S. P. and Young, L. S., 2007. *Technology of Breadmaking*. Springer Science + Business Media, LLC, New York, USA, pp. 26–46.

Code of Federal Regulations (CFR), 1996. Food and drugs. Available at: http://www.gpo.gov/fdsys/pkg/CFR-2012-title21-vol2/pdf/CFR-2012-title21-vol2-sec137-165.pdf. Accessed 23 December 2013.

Codex Alimentarius Commission, 2007. *Draft Revised Codex Standard for Foods for Special Dietary Use for Persons Intolerant to Gluten Joint FAO/WHO Food Standards Programme*, WHO, Geneva, Switzerland.

Dewettinck, K., Van Bockstaele, F., KüHne, B., Van de Walle, D., Courtens, T. M. and Gellynck, X., 2008. Nutrition value of bread: influence of processing, food interactions and consumer perception. *Journal of Cereal Sciences*. 48: 243–457.

FAO/WHO, 1998. Vitamin and Mineral Requirements in Human Nutrition: Report of a Joint FAO/WHO Expert Consultation. Bangkok. Thailand. Geneva: World Health Organization.

Frontela, C., Ros, G. and Martinez, C., 2011. Phytic acid content and "*in vitro*" iron, calcium and zinc bioavailability in bakery products: the effect of processing. *Journal of Cereal Science*. 54: 173–179.

Krosgaard, L. R., Engsbro, A. L. and Bytzer, P., 2013. The epidemiology of irritable bowel syndrome in Denmark. A population-based survey in adults <50 years of age. *Scandinavian Journal of Gastroenterology*. 48: 523–299.

Krupa-Kozak, U., 2014. Pathologic bone alterations in celiac disease: etiology, epidemiology, and treatment. *Nutrition*. 30: 16–24.

Krupa-Kozak, U., Wronkowska, M. and Soral-Śmietana, M., 2011a. The effect of buckwheat flour on microelements and proteins content in gluten-free bread. *Czech Journal of Food Sciences*. 2011(2): 103–108.

Krupa-Kozak, U., Bączek, N. and Rosell, C. M., 2013. Application of dairy proteins as technological and nutritional improvers of calcium-supplemented gluten-free bread. *Nutrients*. 5: 4503–4520.

Krupa-Kozak, U., Troszyńska, A., Bączek, N. and Soral-Śmietana, M., 2011b. Effect of organic calcium supplements on the technological characteristic and sensory properties of gluten-free bread. *European Food Research and Technology*. 232: 497–508.

Krupa-Kozak, U., Wronkowska, M., Bączek, N. and Rosell, M. C., 2012. Breadmaking performance and technological characteristic of gluten-free bread with inulin supplemented with calcium salts. *European Food Research and Technology*. 235: 545–554.

Lazaridou, A., Duta, D., Papageorgiou, M., Belc, N. and Biliaderis, C. G., 2007. Effects of hydrocolloids on dough rheology and bread quality parameters in gluten-free formulations. *Journal of Food Engineering*. 79: 1033–1047.

Leenhardt, F., Levrat-Verny, M. A., Chanliaud, E. and Rémésy, C., 2005. Moderate decrease of pH by sourdough fermentation is sufficient to reduce phytate content of whole wheat flour through endogenous phytase activity. *Journal of Agricultural and Food Chemistry*. 53: 98–102.

Lopez, H. W., Ouvry, A., Bervas, E., Guy, C., Messager, A., Demigne, C. and Remesy, C., 2000. Strains of lactic acid bacteria isolated from sour doughs degrade phytic acid and improve calcium and magnesium solubility from whole wheat flour. *Journal of Agricultural and Food Chemistry*. 48: 2281–2285.

Matos, M. E. and Rosell, C. M., 2013. Quality indicators of rice based gluten-free bread-like products: relationships between dough rheology and quality characteristics. *Food and Bioprocess Technology*. 6: 2331–2341.

Moore, M. M., Heinbockel, M., Dockery, P., Ulmer, H. M. and Arendt, E. K., 2006. Network formation in gluten-free bread with application of transglutaminase. *Cereal Chemistry*. 83: 28–36.

Moore, M. M., Schober, T. J., Dockery, P. and Arendt, E. K., 2004. Textural comparison of gluten-free and wheat based doughs, batters and breads. *Cereal Chemistry*. 81: 567–575.

Mustalahti, K., Catassi, C., Reunanen, A., Fabiani, E., Heier, M., McMillan, S. Murray, L., *et al.*, 2010. The prevalence of celiac disease in Europe: results of a centralized, international mass screening project. *Annals of Medicine*. 42: 587–595.

Nunes, M., Ryan, L. and Arendt, E. K., 2009. Effect of low lactose dairy powder addition on the properties of gluten-free batters and bread quality. *European Food Research and Technology*. 229: 31–41.

Pazianas, M., Butcher, G. P., Subhani, J. M., Finch, P. J., Ang, L., Collins, C., *et al.*, 2005. Calcium absorption and bone mineral density in celiacs after long term treatment with gluten-free diet and adequate calcium intake. *Osteoporosis International*. 16: 56–63.

Pederson, B., Knudsen, K. E. B. and Eggum, B. O., 1989. Nutritive value of cereal products with emphasis on the effect of milling. *World Review of Nutrition and Dietetics*. 60: 1–5.

Pomeranz, Y., 1988. Chemical composition of kernel structures. In: Pomeranz, Y. (ed.) *Wheat Chemistry and Technology*. American Association of Cereal Chemists, St. Paul, MN, USA, vol. 1, pp. 97–158.

Ranhotra, G. S., Gelroth, J. A. and Leinen, S. D., 2000. Utilization of calcium in breads highly fortified with calcium as calcium carbonate or as dairy calcium. *Cereal Chemistry*. 77: 293–296.

Rosell, C. M., 2011. The Science of doughs and bread quality. In: Preedy, V. R., Watson, R. R. and Patel, V. B., (ed.) *Flour and Breads and their Fortification in Health and Disease Prevention*. Academic Press, Elsevire Inc., London, UK, pp. 3–14.

Rosell, M. C., Collar, C. and Haros, M., 2007. Assessment of hydrocolloid effects on the thermo-mechanical properties of wheat using the Mixolab. *Food Hydrocolloids*. 21: 452–462.

Scanlon, M. G. and Zghal, M. C., 2001. Bread properties and crumb structure. *Food Research International*. 34: 841–864.

Slavin, J. L., Jacobs, D. and Marquart, L., 2001. Grain processing and nutrition. *Critical Reviews in Biotechnology*. 21: 49–66.

Sudha, M. L. and Leelavathi, K., 2008. Influence of micronutrients on rheological characteristics and bread-making quality of flour. *International Journal of Food Sciences and Nutrition*. 59: 105–115.

Thompson, T., Dennis, M., Higgins, L. A., Lee, A. R. and Sharrett, M. K., 2005. Gluten-free diet survey: are Americans with coeliac disease consuming recommended amounts of fibre, iron, calcium and grain foods?. *Journal of Human Nutrition and Dietetics*. 18: 163–169.

Turk, M., Sandberg, A. S., Carlsson, N. G. and Andlid, T., 2000. Inositol hexaphosphate hydrolysis by baker's yeast. Capacity, kinetics, and degradation products. *Journal of Agricultural and Food Chemistry*. 48: 100–104.

United States Food and Drug Administration, 2008. Food labeling: health claims; calcium, vitamin D, and osteoporosis. Available at: http://www.ecfr.gov/cgi-bin/text-idx?c=ecfr&sid=b39601aa6d95670faf7f0c388a191312&rgn=div8&view=text&node=21:2.0.1.1.2.5.1.3&idno=21. Accessed 23 December 2013.

Wang, S. and Tang, Y. R., 1995. Fortification of Ca and Zn in wheat flour. *Journal of Zhengzhou Grain College*. 16: 26–34.

CHAPTER 19

Calcium-Fortified Soymilk: Function and Health Benefits

LILY STOJANOVSKA*[a], MUTAMED AYYASH[a,b], AND VASSO APOSTOLOPOULOS[a]

[a]College of Health and Biomedicine, Centre for Chronic Disease, Victoria University, PO Box 14428, St Albans, Vic 8001, Australia; [b]Department of Food Science, United Arab Emirates University, Al-Ain 15551, UAE
*E-mail: lily.stojanovska@vu.edu.au

19.1 Introduction

Soymilk is extracted from soybeans by soaking soybeans in water and grinding, resulting in a product that resembles milk. The process dates back to around 25–225 AD with limestone slabs found in China with carvings of soymilk preparation. The proportion of carbohydrates, protein and fat of soymilk is similar to that of cow's milk, however, the type of each differs significantly (Ferrie, 2013) (Table 19.1). Cow's milk contains the carbohydrate lactose, whereas soymilk does not contain lactose but complex carbohydrates. Soy milk protein is rich is phytoestrogens, with many health benefits to postmenopausal women, breast cancer, heart disease and osteoporosis sufferers, while cow's milk protein consists of casein and whey, popular amongst body builders as effective muscle-building proteins. The fat contained in full cream cow's milk is butterfat, high in saturated fat, whilst, soymilk is high in polyunsaturated fats of soy oil, with high omega-3 and omega-6 content. Due

Food and Nutritional Components in Focus No. 10
Calcium: Chemistry, Analysis, Function and Effects
Edited by Victor R. Preedy

Published by the Royal Society of Chemistry, www.rsc.org

Table 19.1 Nutritional facts of nonfortified, full-fat soymilk compared to full-fat cow's milk.[a]

Nutritional fact	Soymilk per 100 g	Cow's milk per 100 g
Calories	54	60
Protein	3.3 g	3.2 g
Total fat	1.8 g	3.25 g
Saturated fat	0.2 g	1.9 g
Polyunsaturated fat	1.0 g	0.2 g
Monounsaturated fat	0.4 g	0.8 g
Cholesterol	0.0 mg	10 mg
Carbohydrate	6 g	4.5 g
Sugar	4 g	5.3 g
Fiber	0.6 g	0.0 g
Calcium	25 mg	117 mg
Potassium	118 mg	143 mg
Sodium	51 mg	40 mg
Vitamin A	0.0 g	0.0 g
Vitamin B	0.0 g	0.0 g
Vitamin C	0.0 g	0.0 g
Vitamin D	0.0 g	0.0 g

[a]Source includes: USDA, United States Department of Agriculture.

Table 19.2 Calcium content of selected milk products.[a]

Milk product	Calcium per serve	Serving
Regular milk, full fat	304 mg	250 mL
Regular milk, full fat, from powder	875 mg	100 mL
Reduced-fat milk	367 mg	250 mL
Reduced-fat milk, calcium fortified	520 mg	250 mL
Reduced-fat evaporated milk	713 mg	250 mL
Skim milk	341 mg	250 mL
Regular soymilk	309 mg	250 mL
Reduced fat soymilk	367 mg	250 mL
Low fat soymilk	367 mg	250 mL
Rice milk, calcium fortified	182 mg	250 mL

[a]Source includes: Food standards Australia and New Zealand, NUTTAB 2010.

to increasing health concerns, such as, inflammatory bowel disease, lactose intolerance, and cholesterol control, and, a protein alternative for vegetarians, soymilk has received considerable attention in the last 20 years in the western diet. Soymilk however, contains very little calcium, as calcium is bound to the insoluble beans pulp (Tables 19.2 and 19.3). Hence, manufacturers add calcium (calcium fortification) to soymilk products. Calcium intake positively affects bone mass and is crucial for development and maintenance of a healthy skeleton (Gueldner *et al.*, 2008). Indeed, soy intake has been shown to protect against bone fracture in postmenopausal women (Zhang *et al.*, 2005) and in a Singapore Chinese Health Study (Koh *et al.*, 2009).

Table 19.3 Calcium content of selected foods.[a]

Food	Calcium per serve	Serving
Tofu	832 mg	250 mL
Lecithin, soy granules	120 mg	100 g
Yoghurt, natural, full fat	386 mg	Small tub
Yoghurt, natural, low fat	488 mg	Small tub
Cheddar cheese	160 mg	1 slice
Parmesan cheese	204 mg	1 slice
Mozzarella cheese	200 mg	1 slice
Camembert cheese	121 mg	1 wedge
Sardines, canned	486 mg	Small can
Pink salmon, canned	279 mg	Small can
Red salmon, canned	203 mg	Small can
Chickpeas, canned	90 mg	1 cup
Boiled egg	21 mg	Medium
Licorice	34 mg	1 stick
Cheesecake	163 mg	1 slice
Tahini	66 mg	1 tablespoon
Almonds	30 mg	10 almonds
Dried figs	160 mg	6 figs
Cinnamon, ground	1228 mg	100 g
Cocoa powder	149 mg	100 g
Parsley	12 mg	1 tablespoon
Cucumber	68 mg	1 cup
Celery	31 mg	1 cup
Wine, red	7 mg	100 mL
Water, tap	2 mg	100 mL
Bread, white flour	62 mg	100 g
Bread, white flour, calcium fortified	228 mg	100 g
Bread, mixed grain	101 mg	100 g
Bread, wholemeal flour	92 mg	100 g

[a]Source includes: Food standards Australia and New Zealand, NUTTAB 2010.

19.2 Calcium Fortification and Bioavailability

Dietary calcium is fundamental to bone health. Soymilk contains 80% less calcium than cow's milk, hence, manufacturers of soymilk, fortify soymilk with calcium, which appears to be an effective method to increase calcium intake and the total amount of absorbed calcium (Lopez-Huertas *et al.*, 2006) (Tables 19.1 and 19.2). Calcium in cow's milk is in the form of caseinate-phosphate complex that is released during digestion leading to its high bioavailability, however, calcium bioavailability in soymilk is affected by the type of fortificant used (Lopez-Huertas *et al.*, 2006). Tricalcium phosphate and calcium carbonate are the most common calcium fortificants, with, calcium carbonate being absorbed more readily by humans, compared to tricalcium, and is similar to that of cow's milk (Heaney *et al.*, 2000; Zhao *et al.*, 2005). Other commercial salts used include, calcium chloride, calcium phosphate, calcium citrate malate, calcium lactate and calcium gluconate, calcium lactate gluconate. The bioavailability of these salts depends not only on their nature, but also on the nature of proteins present in soymilk, which act as carriers during absorption. With some fortificants and some types of soymilk, the calcium may interact with the soymilk proteins particularly after heat processing, resulting in sedimentation and gelation. Hence, stabilizers and emulsifiers are added to maintain calcium in suspension as well as to improve mouth-feel and appearance (Singh *et al.*, 2007). A study evaluating calcium absorption between fortified soymilk (calcium, as calcium phosphate) and cow's milk in a randomized single-blind acute crossover study, demonstrated comparable calcium absorption amongst the two forms (Tang *et al.*, 2010a). Similarly, calcium sulfate and calcium malate are absorbed comparably to cow's milk (Smith *et al.*, 1987; Martin *et al.*, 2002). The presence of phytate in calcium-fortified soymilk may interfere with mineral absorption, and lactic acid bacteria, such as *Lactobacillus acidophilus*, *Lactobacillus casei*, *Lactobacillus plantarum* and *Lactobacillus fermentum*, produce the enzyme phytase that degrades phytates, leading to improved calcium bioavailability and absorption. The addition of *Lactobacillus* probiotics to calcium-fortified soymilk, results in an increase in calcium solubility that is related to lowered pH associated with production of lactic and acetic acids. In addition, fermentation increases the level of conversion of isoflavones into biologically active compounds, hence, the addition of probiotics to calcium-fortified soymilk increases the calcium bioavailability due to increased calcium solubility (Tang *et al.*, 2007). A comparison study was conducted amongst *Lactobacillus acidophilus* (ATCC4962, ATCC33200, ATCC4356, ATCC4161), *Lactobacillus casei* ASCC290, *Lactobacillus plantarum* ASCC276 and *Lactobacillus fermentum* VRI-003, on phytase activity and phytate degradation in calcium-fortified soymilk. *Lactobacillus acidophilus* ATCC4161 produced the highest phytase activity, although all strains showed phytase activity, with no phytate degradation noted (Tang *et al.*, 2010b).

19.3 Soy Isoflavones

Isoflavones are plant compounds, belonging to phytoestrogens or plant estrogens. Soy foods are a rich source of isoflavones and plant estrogens are similar in structure and function to that of human estrogen. The concentration and composition of isoflavone in soymilk differs markedly, in fact, in 40 different brands of Australian commercial soymilks, the total isoflavone concentration varied up to 5-fold and was not consistent between repeat purchases (Setchell and Cole, 2003). The total isoflavones in soybeans are, 37% daidzein, 57% genistein and 6% glycitein (Database, 2014). The bioavailability of isoflavones in plasma and urine as analyzed in 12 young adult women after receiving soymilk as part of their liquid diet, demonstrated that daidzein was a more bioavailable soymilk isoflavone than genistein (Xu *et al.*, 1994). Hence, the bioavailability of daidzein may be sufficient to exert some health-protective effects.

Soy isoflavones have an effect in preventing estrogen-related bone loss by inhibiting bone resorption. The Chinese are amongst the highest population of soy consumers, and, a double-blind, placebo-controlled, randomized trial was conducted to assess the effects of soy isoflavones on bone loss in postmenopausal Chinese women. Three groups of subjects, 230 subjects in total, were either assigned to daily doses of placebo (no isoflavones), mid-dose (40 mg isoflavones), or high dose (80 mg isoflavones), and, women in the high-dose group demonstrated higher bone mineral content at the hip and trochanter compared to placebo and mid-dose groups, especially in postmenopausal women with low initial bone mass (Chen *et al.*, 2003). In addition, postmenopausal women with osteoporosis, or, with at least 3 risk-factors of osteoporosis, were assigned to 4 treatment groups: soymilk containing isoflavones, transdermal progesterone, a combination of both, or placebo (Lydeking-Olsen *et al.*, 2004). Significant bone loss (bone mineral content and bone-mineral density) was noted in the placebo and combined treatment groups, with no change in the group that took soymilk containing isoflavones (Lydeking-Olsen *et al.*, 2004). Hence, daily intake of soymilk containing isoflavones prevents bone loss in postmenopausal women. Intake of transdermal progesterone had bone-sparing effects and when combined with isoflavones resulted in a negative interaction with greater bone loss compared to transdermal progesterone alone. 454 healthy postmenopausal Chinese women were analyzed for dietary intake of soy protein and isoflavones on the bone-mineral density and content at the spine, hip and total body (Ho *et al.*, 2003). No differences were noted in the first 4 years of menopause, however, amongst the later postmenopausal women, a dose–response relationship was noted amongst increasing soy intake and increasing bone-mineral density values at the hip and total body (Ho *et al.*, 2003). Furthermore, the bioavailability of soy isoflavones depends greatly upon the relative ability of gut microflora to degrade these compounds, as demonstrated in 7 women on a soymilk liquid diet (Xu *et al.*, 1995). As such, the addition of probiotics of bacteria *Lactobacillus* with or without yeast *Saccharomyces boulardii* to ferment soymilk, increases isoflavone content in fermented soymilk and improves the biological functionality of soymilk (Tang *et al.*, 2007; Rekha and Vijayalakshmi, 2010).

19.4 Soymilk and the Immune Response

Inflammation plays a pivotal role in the development of chronic diseases, and tumor necrosis factor alpha (TNF-alpha), interleukin (IL)-1 and IL-6 are key mediators of inflammatory responses. *In vitro* studies using a macrophage cell line, the addition of hydrolyzates of soymilk inhibits the expression of nitric oxide, IL-1, TNF-alpha and proinflammatory enzymes (nitric oxide synthase and cyclooxygenase) (Dia *et al.*, 2014). Likewise, in a study involving 12 women who ingested soymilk containing isoflavones for 16 weeks, a significant decrease in serum TNF-alpha and IL-1 levels were noted, but not for IL-6 (Huang *et al.*, 2005). In addition, cultures of monocytes from soymilk ingested subjects, inhibited lipopolysaccharide induced TNF-alpha production by up to 55% (Huang *et al.*, 2005). Conversely, in a randomized study, in 31 postmenopausal women, 3 servings of soymilk a day, had no effect on serum TNF-alpha, IL-1, IL-6 and oxidative stress (Beavers *et al.*, 2009). In diabetic patients with nephropathy, soymilk consumption for 4 weeks had no effects on TNF-alpha, IL-6, C-reactive protein and malondialdehyde levels. Furthermore, soymilk comprising high levels of isoflavones results in higher B cell populations, lower plasma concentrations of 8-hydroxy-2-deoxy-guanosine, an oxidative marker of DNA damage, and, inhibits reactive nitrogen species, with no effects in interferon gamma, IL-2, or C-reactive protein (Ryan-Borchers *et al.*, 2006). The anti-inflammatory effect of isoflavones in soymilk leads to their importance in immune modulation and prevention against DNA damage. Moreover, soymilk could be used as nutritional rehabilitation, as experimental malnutrition models in rats demonstrated that soymilk quickly recovers muscle proteins, phagocytic activity and gut mucosa IgA-secreting cells (Fontenla de Petrino *et al.*, 2007). Interestingly, soymilk inhibits protease-activated receptor 2 cleavage, affects cytoskeletal and cell-surface organization and reduces keratinocyte phagocytosis, leading to depigmentation of the skin (Paine *et al.*, 2001). These findings imply that soymilk may be used as an alternative approach to depigmentation.

19.5 Soymilk and Bone Health

Soy protein aids the promotion of bone health. Factors that affect the excretion of calcium play an important role in determining calcium balance and bone-mineral density. High levels of calcium excretion have been proposed as one factor contributing to the high rates of osteoporosis. In comparison to animal proteins, soy protein causes much less calcium to be excreted in the urine, possibly due to the lower levels of sulfur amino acid content in soy protein. In addition, isoflavones directly inhibit bone resorption, hence, the importance of soy in the diet for bone health. In a Chinese population of developing adolescent girls, where the diet is low in calcium, 104 girls were followed for 1 year after calcium-fortified soymilk supplementation, and were compared to 95 girls in a control group (Ho *et al.*, 2005). No differences were noted amongst both groups for bone-mineral density and content at the spine, height and weight, however, significant association was noted with supplementation with a percentage increase in bone-mineral density and content at the hip (Ho *et al.*, 2005). Thus, 2 glasses a day of calcium-fortified soymilk supplementation is effective for bone-mass development in adolescent girls. In postmenopausal women, there is rapid bone loss and the risk of osteoporosis increases substantially. A total of 141 Chinese women without osteoporosis and postmenopausal for more than 2 years were assigned to consume either, cow's milk or calcium-fortified soymilk (both containing

250 mg calcium), or, placebo for 18 months. The women in the placebo group experienced reduction in bone-mineral density at the hip and femoral neck compared to both calcium groups (Gui *et al.*, 2012). Hence, consumption of at least 250 mg of calcium per day over 18 months prevents bone-mineral density loss in postmenopausal women.

19.6 Soymilk Properties in Health and Disease

Soymilk consumption in the western culture, is particularly popular due to the recognition of its health benefits (Table 19.5), with evidence demonstrating that there are positive outcomes to postmenopausal women, breast and colon cancer, cholesterol, heart disease, ischemic stroke, and osteoporosis (McCarron and Heaney, 2004).

19.7 Effects of Soymilk in Cholesterol

Soymilk has antioxidant properties and if incorporated into a low-fat diet, can reduce serum cholesterol and low-density lipoprotein (LDL) concentrations. Due to decreased saturated fats in soymilk and increase in polyunsaturated fats in soymilk compared to cow's milk, soymilk consumption should enhance its benefits on lipoprotein concentrations, reducing the risks of atherosclerosis development. In 42 healthy males either on a tofu diet or a lean meat diet for 1 month, resulted in lower total cholesterol and triglyceride levels in the tofu diet as compared to the lean meat diet. In a randomized controlled trial on 60 premenopausal healthy females who consumed soymilk for 2 menstrual cycles, had significant reduced total serum cholesterol concentrations (Takatsuka *et al.*, 2000). Likewise, consumption of soymilk by either 42 healthy Nigerian Africans for 3 weeks significantly lowered total plasma cholesterol, triglyceride and LDL concentrations and increased high-density lipoprotein (HDL) cholesterol concentrations (Onuegbu *et al.*, 2011). Hence in normolipidemic subjects, soy products (Table 19.4) including soymilk has hypocholesterolemic effects. Similar effects of soymilk intake with plasma and liver lipid profiles have also been noted in numerous high-cholesterol fed animal models. In hypercholesterolemic subjects soymilk enhanced with plant sterols, decreased plasma total cholesterol, LDL cholesterol, triglyceride concentrations, as well as reduced intestinal cholesterol absorption and cholesterol synthesis, as compared to cow's milk (Rideout *et al.*, 2009). Similarly, in 19 hyperlipidemic patients, undergoing hemodialysis, consuming soy protein for 12 weeks led to major decreases in total cholesterol levels (18.6%), triglyceride levels (43%), LDL cholesterol levels (25.8%), apolipoprotein levels (15.4%) and insulin levels (49.8%), with significant increases to HDL cholesterol levels (17%), compared to milk proteins (Chen *et al.*, 2005). Conversely, in a 4 week, single-blind randomized controlled trial in 32 postmenopausal women, soymilk showed no hypocholesterolemic effects (Beavers *et al.*, 2010).

Table 19.4 List of soy products and their origin.[a]

Soy products	Origin
Cheonggukjang	Korea
Doubanjiang	China
Douchi	China
Edamame	Japan
Miso	Japan
Natto	Japan
Soy nuts	China
Soy sauce	China
Soymilk	China
Tauco	Indonesia
Tempeh	Indonesia
Tofu	China

[a]Sources includes: The soy info center; http://www.soyinfocenter.com.

Table 19.5 Summary of health benefits of soymilk consumption.

Health or disease	Summary of effects
Bone health	Prevents estrogen-related bone loss
	Higher bone mineral content
	Higher bone-mineral density
	Inhibits bone resorption
Cholesterol	Reduces total serum cholesterol
	Reduces triglyceride levels
	Reduces LDL-cholesterol concentrations
	Increases HDL-cholesterol concentrations
Hormones	Decreases circulating levels of hormones
	Decreases circulating levels of gonadotropins
	Decreases beta-estradiol
	Decreases serum luteal estradiol levels
	Decreases progesterone
	Decreases adrenal androgen levels
	Increases menstrual cycle length
Immune system	Inhibits protein kinase
	Inhibits DNA topoisomerases
	Inhibits signal transduction enzymes
Metabolic syndrome disorders	Decreases diabetic phenotype
	Decreases adipose specific proteins
	Prolongs glycemic control indices
	Mild changes to weight, body-mass index
	Mild changes to body-fat composition
Cancer	*In vitro* inhibits proliferation of cancer cells
	In animal models, protection against cancer growth
	Anticarcinogenic effects
	Benefits to breast, prostate and colon cancer

19.8 Effects of Soymilk in Metabolic Syndrome Disorders

An increase in sedentary lifestyles with decreasing physical activity has led to obesity, type-2 diabetes, hypertension, hypercholesteremia and increased proinflammatory state. These constitute the metabolic syndrome, leading to atherosclerosis and other cardiovascular related diseases. Soy proteins have been shown to ameliorate the diabetic phenotype, adipose-specific proteins, total liver adiposity and prolong glycemic control indices (Davis *et al.*, 2007). However, in a randomized double-blind placebo-controlled trial in 180 healthy Chinese women, soy proteins had no effects on insulin resistance, glycemic control, blood pressure and endothelial molecules, even though reduction of systolic blood pressure and soluble intercellular adhesion molecule-1 were noted in subjects with initial elevated blood pressure, relative to milk-protein consumption. In addition, the effects of soy proteins on weight and body-fat loss was investigated in 180 Chinese women with mild hyperglycemia for 6 months. Mild but significant changes in body weight, body-mass index and body-fat percentage were noted. Furthermore, renal dysfunction is significantly reduced in high soy protein diets with lower levels of proteinuria and dilated tubules, in several animal models of obesity and metabolic syndrome (Davis *et al.*, 2007). In diabetic patients with nephropathy, soymilk consumption for 4 weeks is associated with reduction in systolic blood pressure compared with consumption of cow's milk (Miraghajani *et al.*, 2013).

19.9 Effects of Soymilk in Cancer

In the Asian population the incidence of breast, prostate and colon cancer is significantly lower compared with western populations, and it has been suggested that diets low in animal products and high in soy foods, may reduce risk of disease. Soybeans contain, isoflavones (genistein), which have anticarcinogenic effects and are weakly estrogenic, that decrease circulating levels of hormones and gonadotropins. In 10 healthy women who consumed a soya containing diet, had 25% reduced beta-estradiol and 45% reduced progesterone with no effects on luteinizing hormone and follicle stimulating hormone, compared to nonsoy diet (Lu *et al.*, 2001). In addition, soymilk feeding decreases adrenal androgen levels and increases menstrual cycle length by 2 days, and such effects may account at least in part in a decreased incidence of breast cancer (Lu *et al.*, 1996; Nagata *et al.*, 1998). A 7 month soy intervention study in 20 healthy, cycling premenopausal women, demonstrated a significant reduction in serum luteal estradiol levels, with no changes in the follicular phase estradiol and follicular or luteal phase progesterone (Wu *et al.*, 2000). The, reduction in the ovarian hormone levels is related to decreased risk of breast cancer, and, soy diets may have implications for reducing breast-cancer risk by dietary intervention. Conversely, in 37 643 British women taking part in the European Prospective Investigation into Cancer and Nutrition, the consumption of isoflavone intake was calculated from food-frequency questionnaires on consumption of soy foods and soymilk, and no significant associations were noted in the relative risk of breast cancer between vegetarian diets and those rich in soy products (Travis *et al.*, 2008) (Table 19.4).

In addition to soymilk having estrogenic activity, the isoflavone, genistein, is a protein kinase inhibitor, inhibits DNA topoisomerases and enzymes involved in signal transduction (Messina *et al.*, 1994). *In vitro*, genistein inhibits the proliferation of cancer cells and in animal models genistein exhibits protective effects against the growth of cancer (Messina *et al.*, 1994). Using gene-expression profiles of phytoestrogens isolated from soymilk, it was noted that phytoestrogens mitigate estrogenic signals in estrogen receptor positive breast cancer cells, however, in late-stage cells where estrogen receptor is lacking the phytochemicals contribute to tumor-promoting transcriptions (Dip *et al.*, 2008). Importantly, the addition of *Lactobacillus casei* Shirota (Kaga *et al.*, 2013) to soymilk further suppresses tumor growth in a chemically induced cancer model, and the addition of *Streptococcus thermophiles* and *Bifidobacterium infantis* (Lai *et al.*, 2013) inhibits proliferation of colon-cancer cell lines *in vitro*. The addition of 6 different probiotics to soymilk enhances the *in vitro* antihypertensive effects (angiotensin I-converting enzyme inhibitory activity) and production of bioactive aglycones, demonstrating its potential effects on hypertension, cancer (breast and prostate) and osteoporosis (Yeo and Liong, 2010). In males, soymilk consumption has been associated with a reduced risk of prostate cancer by significantly decreasing the concentration of estrone, and not, estradiol, testosterone or sex hormone binding globulin (Nagata *et al.*, 2001). Interestingly, topical application of soymilk to ultraviolet irradiated skin, eliminates thymine dimers and apoptotic cells in the epidermis, thus, preventing sunlight-induced skin damage and reduces the risk of tumor formation and progression (Huang *et al.*, 2004).

19.10 Adverse Effects of Soymilk

It is evident that soymilk consumption has positive effects in cancer, osteoporosis and cholesterol levels (Table 19.5). However, soymilk is not an ideal protein on its own, as it is deficient in the essential amino acid, methionine, and contains very low levels of lysine, as compared to milk protein, casein. In addition, raw soybean meal contains endogenous inhibitors of digestive enzymes and lectins leading to poor digestion. Hence, their elimination during processing by fractionation is required or are inactivated by heat treatment. Although most commercially available soy products that have been heat treated, still contain up to 20% of inhibitor enzymes. An increase in consumption of soymilk, has in part been due to its lack of lactose content, as opposed to cow's milk. However, allergies to soy proteins are becoming as issue especially with its increase intake in the last decade. In fact, at least 15 soybean proteins have been shown to be recognized by sera from soy-sensitive patients, causing severe allergic reactions in humans. Moreover, the high isoflavone content of soymilk has been in dispute regarding its potential harm to breast-cancer patients or to high-risk breast-cancer patients, in enhancing the development of breast cancer. Although some *in vitro* studies suggest this, epidemiological studies report that high soymilk consumption does not have an effect on or improves prognosis of breast cancer. In addition, patients receiving thyroid drug therapy have negative interactions with soymilk and isoflavones that directly affect the absorption of the thyroid drug rather than a direct effect on thyroid function. The Cochrane Review found no effects on endometrial function, however, in a 5 year placebo-controlled study of isoflavone consumption through soymilk and other soy products in 376 women, demonstrated that the incidence of endometrial hyperplasia was increased, with no cases of endometrial cancer.

19.11 Conclusion

Soymilk has numerous nutritional attributes. It is lactose free, cholesterol free, contains high-quality protein including isoflavones, is low in saturated fats and high in polyunsaturated fats and contains omega-3 fatty acids. Soymilk is also beneficial to the prevention and treatment of heart disease including lowering of cholesterol levels, cancer and osteoporosis. It is clear that soy products, including soymilk should be incorporated into the diet.

Summary Points

- This chapter focuses on soymilk's benefits to health and disease.
- Soy products contains a low calcium content compared with dairy products.
- Fortification of soy products with calcium is an alternative method to compensate the deficiency in calcium.
- Calcium absorption and bioavailability from calcium-fortified soymilk is comparable to that of cow's milk.
- Soymilk fortification might be an effective way to increase calcium intake although calcium bioavailability in soymilk is highly determined by the type of fortificant used.
- Fermentation of soymilk with probiotics has the potential benefit of increasing calcium bioavailability and absorption.
- Reduction in phytic acid content of soymilk increases the absorption of calcium and isoflavones.
- Soy isoflavones have an effect in preventing estrogen-related bone loss by inhibiting bone resorption.
- Soymilk has anti-inflammatory effects.
- Soymilk aids the promotion of bone health.
- Soymilk reduces cholesterol levels in healthy and hypercholesterolemic subjects, has beneficial effects to cancer and to metabolic syndrome disorders.

Key Facts

Key Facts of Soymilk History

The key facts of the history of soymilk, its name terminology over the years, methods of production, the first countries to commercialize soymilk products and other soymilk products that are available.

1. Soymilk has been popular in the Asian culture for over 2000 years.
2. Soymilk is called Doujiang in the Chinese language, where Dou means bean, and Jiang derived from the word Chiang, means liquid, beverage or drink.
3. Soymilk is a liquid extraction of soybeans, resembling that of milk.
4. The nutritional composition, texture and flavor are similar to that of cow's milk.
5. Traditional soymilk is filtered and soybean pulp is removed.
6. Modern soymilk production is in suspended form comprising all the original soybean.
7. Many soymilk products are manufactured using soy protein isolates.
8. Over the years soymilk has had many terminologies, originally is was the "solution of legumin" (1889), "this filtered stuff" (1889), "the fresh milky liquid" (1895), "the milky white liquid" or "the filtrate of the cooked soybeans resembles milk" (1899), "soy-bean milk" (1897), "soya or soya milk" (1911), "vegetable milk, synthetic milk, or artificial milk made from soy bean" (1915), "soymilk" and "soy bean milk" (1916), "bean milk" (1916), "soy-bean milk or vegetable milk" (1916), and "soy-bean milk" (1923). In 1940 the term "soymilk" spelt as one word, was devised and is the term widely used today. Although in the 1970s it was also called "soybean milk and soy beverage".
9. Plain soymilk contains only soybeans and water, however, most soymilks contain sweeteners, oil, salt and vanilla in order to have similar texture and taste to that of cow's milk.
10. Most soymilks undergo lactic acid fermentation during manufacture.
11. 1950–1969 saw the success of Hong Kong's Vitasoy, the first "soymilk soft drink" to be commercialized. In the 1970s soymilk soft drink production became increasingly popular in countries outside Hong Kong, although some countries in Asia started production earlier.
12. Countries that produce soymilk locally for consumption, are, Vietnam (1911), India (1933), Philippines (1938), Hong Kong (1950), Japan (1950), Malaysia (1952), Indonesia (1957), Thailand (1958), Taiwan (1962), Korea (1973), China (1975), and SriLanka (1981).
13. European manufacturers of soymilk began with Tetra Pak International and Alfa Laval in Sweden in 1952. As early as 1916 soymilk production was known to be in the US.
14. Many soymilk products are available today that include, soymilk ice cream, soymilk whipped cream, soymilk custards and puddings, soymilk mayonnaise, soymilk shakes, soy nog and soymilk cheeses.

Key Facts of Calcium-Fortification Method

The key facts of the method for calcium-fortified soymilk production according to the US 4,906,482A patent.

1. Soymilk does not have lactose or cholesterol and contains low levels of calcium.
2. Fortification of soymilk with calcium results in coagulation and precipitation of soy proteins.
3. Chemicals are added to chelate calcium ions to avoid soy protein precipitation.
4. Polyphosphate is added as a chelating agent for calcium and provides microbial stability of soymilk. This aids in the appropriate amounts of calcium to be added, to be similar to that present in cow's milk.
5. Types of calcium fortifiers used are, calcium carbonate, calcium chloride, calcium citrate, calcium phosphate, calcium gluconate, calcium lactate, with calcium chloride being preferred.
6. However, a large portion of the added calcium remains nutritionally unavailable.
7. The, "Method of making calcium fortified soymilk", according to the US 4,906,482A patent, filed in 1988 is:

- Wash soybeans.
- Soak soybeans in water, 3 times its volume, store at 4 °C for 12–18 h to loosen the hulls.
- Wash and rinse soybeans to remove and soften hulls.
- Chop the dehulled soybeans and add water to form a slurry. The content can also be added to a blender for 5 min to aid in slurry formation.
- Cook the soybean slurry mixture for 0.5–2.0 min at 220 °F.
- Immediately cool the slurry.
- Extract the soymilk by pressing and sieving (to extract milk and remove the solids).
- The soymilk is now ready for calcium fortification that is usually performed at room temperatures and constant pH of 6.7–6.85.
- Add polyphosphate to soymilk, 0.5–1.0 g per 100 mL of soymilk, mix.
- Add calcium to soymilk, 0.2–0.75 g per 100 mL of soymilk, mix.
- The calcium-fortified soymilk homogenized, pasteurized or sterilized according to methods conventionally available.
- The resultant soymilk contains 1.7 mg of calcium per 1.0 mL of soymilk, and has a shelf-life in the fridge of 3 weeks.
- Flavors or sweeteners could be added, usually, 7.0 g per 100 mL soymilk.

Key Facts of Osteoporosis

This table lists the key facts of osteoporosis, including its meaning, what it is, the risks factors and the importance of calcium in the prevention of osteoporosis.

1. In 1820, French physician Jean Lobstein named the disease "osteoporosis" because patients' bones had larger than normal holes or pores. The term "osteoporosis" is from the Greek *osteon*, meaning "bone," and *poros*, meaning "passage" or "space".
2. Osteoporosis weakens bones and increases the risk of unexpected fractures. Serious consequences can occur with some fractures.
3. It can be prevented with a healthy diet and by staying physically active.
4. Many risk factors can lead to osteoporosis; age, family history, sexual hormones, drinking alcohol, low-calcium diet, and low vitamin D.
5. Women are more likely to develop osteoporosis because they typically have less bone mass to start with than men do. Women undergoing the menopause also experience a sudden loss of estrogen, a sex hormone that helps build and maintain healthy bones.
6. Osteoporosis is treated by prescription medications, stopping the use of alcohol and cigarettes, and adequate exercise.
7. A healthy diet with enough calcium and vitamin D decreases the risk of osteoporosis.
8. Calcium-fortified soymilk has been shown to have beneficial effects on bone-mineral density and bone mineral content, contributing to overall bone health.

Definitions of Words and Terms

Bioavailability. The rate at which the administered drug or food product is available in the systemic circulation.

Consumption or ingestion. Eating something through the mouth.

Daidzein and genistein. Biologically active isoflavones present in high amounts in unprocessed plant foods and are hydrolyzed into aglycones.

Estrogens. Female sex hormones with a steroid structure, and their activity is mediated by binding to the estrogen receptor.

Fermentation. The conversion of sugars into ethanol. With food products it is usually the conversion of carbohydrates to alcohols, carbon dioxide or acids, by using yeast or bacteria, including probiotics.

Fortification. Food fortification is referred to as the enrichment or addition of micronutrients, such as trance elements and vitamins. Calcium fortification of soymilk, refers to the addition of calcium to soymilk.

High-density lipoprotein. Also known as HDL, or "good cholesterol", and is associated with removing fats and cholesterol from cells, within artery wall atheroma and transports it back to the liver for reutilization. Higher levels of HDL correlate with fewer problems with cardiovascular disease.

Low-density lipoprotein. Also known as LDL, or "bad cholesterol", and is associated with health problems including cardiovascular disease.

Phytoestrogens. Plant-derived estrogens that are similar to human estrogen.

Probiotics. Good bacteria that are required for healthy gut microflora. Consumption of probiotics is beneficial by improving the balance of flora in the digestive tract.

Soymilk. Also known as soymilk beverage, it is a drink made from soybeans. It consists of oil, water, salt and protein, and is produced by soaking soybeans in water.

List of Abbreviations

HDL	High-density lipoprotein
IL	Interleukin
LDL	Low-density lipoprotein
TNF	Tumor necrosis factor

References

Beavers, K. M., Serra, M. C., Beavers, D. P., Cooke, M. B. and Willoughby, D. S., 2009. Soymilk supplementation does not alter plasma markers of inflammation and oxidative stress in postmenopausal women. *Nutrition Research*. 29: 616–622.

Beavers, K. M., Serra, M. C., Beavers, D. P., Hudson, G. M. and Willoughby, D. S., 2010. The lipid-lowering effects of 4 weeks of daily soymilk or dairy milk ingestion in a postmenopausal female population. *Journal of Medicinal Food*. 13: 650–656.

Chen, S. T., Ferng, S. H., Yang, C. S., Peng, S. J., Lee, H. R. and Chen, J. R., 2005. Variable effects of soy protein on plasma lipids in hyperlipidemic and normolipidemic hemodialysis patients. *American Journal of Kidney Diseases*. 46: 1099–1106.

Chen, Y. M., Ho, S. C., Lam, S. S., Ho, S. S. and Woo, J. L., 2003. Soy isoflavones have a favorable effect on bone loss in Chinese postmenopausal women with lower bone mass: a double-blind, randomized, controlled trial. *Journal of Clinical Endocrinology and Metabolism*. 88: 4740–4747.

Database U. S. D. A., 2014. Isoflavones Contents of Food. http://www.isoflavones.info.

Davis, J., Higginbotham, A., O'Connor, T., Moustaid-Moussa, N., Tebbe, A., Kim, Y. C., Cho, K. W., Shay, N., Adler, S., Peterson, R. and Banz, W., 2007. Soy protein and isoflavones influence adiposity and development of metabolic syndrome in the obese male ZDF rat. *Annals of Nutrition & Metabolism*. 51: 42–52.

Dia, V. P., Bringe, N. A. and de Mejia, E. G., 2014. Peptides in pepsin–pancreatin hydrolysates from commercially available soy products that inhibit lipopolysaccharide-induced inflammation in macrophages. *Food Chemistry*. 152: 423–431.

Dip, R., Lenz, S., Antignac, J. P., Le Bizec, B., Gmuender, H. and Naegeli, H., 2008. Global gene expression profiles induced by phytoestrogens in human breast cancer cells. *Endocrine-Related Cancer*. 15: 161–173.

Ferrie, S., 2013. *Soy versus Dairy: Which Milk is Better for You?* The Conversation 5 March.

Fontenla de Petrino, S., Prchal, A., Fontenla, M., Cena, A. M., Gomez, J., Pintos, S. and Peral, M. C., 2007. Recovery from experimental malnutrition with soymilk: immunological and genetic aspects. *Nutricion Hospitalaria.* 22: 244–251.

Gueldner, S. H., Grabo, T. N. and Newman, E. D., 2008. *Osteoporosis: Clinical Guidelines for Prevention, Diagnosis and Management.* Springer Publishing Group, New York.

Gui, J. C., Brasic, J. R., Liu, X. D., Gong, G. Y., Zhang, G. M., Liu, C. J. and Gao, G. Q., 2012. Bone mineral density in postmenopausal Chinese women treated with calcium fortification in soymilk and cow's milk. *Osteoporosis International.* 23: 1563–1570.

Heaney, R. P., Dowell, M. S., Rafferty, K. and Bierman, J., 2000. Bioavailability of the calcium in fortified soy imitation milk, with some observations on method. *American Journal of Clinical Nutrition.* 71: 1166–1169.

Ho, S. C., Guldan, G. S., Woo, J., Yu, R., Tse, M. M., Sham, A. and Cheng, J., 2005. A prospective study of the effects of 1-year calcium-fortified soy milk supplementation on dietary calcium intake and bone health in Chinese adolescent girls aged 14 to 16. *Osteoporosis International.* 16: 1907–1916.

Ho, S. C., Woo, J., Lam, S., Chen, Y., Sham, A. and Lau, J., 2003. Soy protein consumption and bone mass in early postmenopausal Chinese women. *Osteoporosis International.* 14: 835–842.

Huang, M. T., Xie, J. G., Lin, C. B., Kizoulis, M., Seiberg, M., Shapiro, S. and Conney, A. H., 2004. Inhibitory effect of topical applications of nondenatured soymilk on the formation and growth of UVB-induced skin tumors. *Oncology Research.* 14(7–8): 387–397.

Huang, Y., Cao, S., Nagamani, M., Anderson, K. E., Grady, J. J. and Lu, L. J., 2005. Decreased circulating levels of tumor necrosis factor-alpha in postmenopausal women during consumption of soy-containing isoflavones. *Journal of Clinical Endocrinology and Metabolism.* 90: 3956–3962.

Kaga, C., Takagi, A., Kano, M., Kado, S., Kato, I., Sakai, M., Miyazaki, K., Nanno, M., Ishikawa, F., Ohashi, Y. and Toi, M., 2013. *Lactobacillus casei* Shirota enhances the preventive efficacy of soymilk in chemically induced breast cancer. *Cancer Science.* 104: 1508–1514.

Koh, W. P., Wu, A. H., Wang, R., Ang, L. W., Heng, D., Yuan, J. M. and Yu, M. C., 2009. Gender-specific associations between soy and risk of hip fracture in the Singapore Chinese Health Study. *American Journal of Epidemiology.* 170: 901–909.

Lai, L. R., Hsieh, S. C., Huang, H. Y. and Chou, C. C., 2013. Effect of lactic fermentation on the total phenolic, saponin and phytic acid contents as well as anti-colon cancer cell proliferation activity of soymilk. *Journal of Bioscience and Bioengineering.* 115: 552–556.

Lopez-Huertas, E., Teucher, B., Boza, J. J., Martinez-Ferez, A., Majsak-Newman, G., Baro, L., Carrero, J. J., Gonzalez-Santiago, M., Fonolla, J. and Fairweather-Tait, S., 2006. Absorption of calcium from milks enriched with fructo-oligosaccharides, caseinophosphopeptides, tricalcium phosphate, and milk solids. *American Journal of Clinical Nutrition*. 83: 310–316.

Lu, L. J., Anderson, K. E., Grady, J. J. and Nagamani, M., 1996. Effects of soya consumption for one month on steroid hormones in premenopausal women: implications for breast cancer risk reduction. *Cancer Epidemiology, Biomarkers & Prevention*. 5: 63–70.

Lu, L. J., Anderson, K. E., Grady, J. J. and Nagamani, M., 2001. Effects of an isoflavone-free soy diet on ovarian hormones in premenopausal women. *Journal of Clinical Endocrinology and Metabolism*. 86: 3045–3052.

Lydeking-Olsen, E., Beck-Jensen, J. E., Setchell, K. D. and Holm-Jensen, T., 2004. Soymilk or progesterone for prevention of bone loss – a 2 year randomized, placebo-controlled trial. *European Journal of Nutrition*. 43: 246–257.

Martin, B. R., Weaver, C. M., Heaney, R. P., Packard, P. T. and Smith, D. L., 2002. Calcium absorption from three salts and CaSO(4)-fortified bread in premenopausal women. *Journal of Agricultural and Food Chemistry*. 50: 3874–3876.

McCarron, D. A. and Heaney, R. P., 2004. Estimated healthcare savings associated with adequate dairy food intake. *American Journal of Hypertension*. 17: 88–97.

Messina, M. J., Persky, V., Setchell, K. D. and Barnes, S., 1994. Soy intake and cancer risk: a review of the in vitro and in vivo data. *Nutrition and Cancer*. 21: 113–131.

Miraghajani, M. S., Najafabadi, M. M., Surkan, P. J., Esmaillzadeh, A., Mirlohi, M. and Azadbakht, L., 2013. Soy milk consumption and blood pressure among type 2 diabetic patients with nephropathy. *Journal of Renal Nutrition*. 23: 277–282 e271.

Nagata, C., Takatsuka, N., Inaba, S., Kawakami, N. and Shimizu, H., 1998. Effect of soymilk consumption on serum estrogen concentrations in premenopausal Japanese women. *Journal of the National Cancer Institute*. 90: 1830–1835.

Nagata, C., Takatsuka, N., Shimizu, H., Hayashi, H., Akamatsu, T. and Murase, K., 2001. Effect of soymilk consumption on serum estrogen and androgen concentrations in Japanese men. *Cancer Epidemiology, Biomarkers & Prevention*. 10: 179–184.

Onuegbu, A. J., Olisekodiaka, J. M., Onibon, M. O., Adesiyan, A. A. and Igbeneghu, C. A., 2011. Consumption of soymilk lowers atherogenic lipid fraction in healthy individuals. *Journal of Medicinal Food*. 14: 257–260.

Paine, C., Sharlow, E., Liebel, F., Eisinger, M., Shapiro, S. and Seiberg, M., 2001. An alternative approach to depigmentation by soybean extracts *via* inhibition of the PAR-2 pathway. *Journal of Investigative Dermatology*. 116: 587–595.

Rekha, C. R. and Vijayalakshmi, G., 2010. Bioconversion of isoflavone glycosides to aglycones, mineral bioavailability and vitamin B complex in fermented soymilk by probiotic bacteria and yeast. *Journal of Applied Microbiology*. 109: 1198–1208.

Rideout, T. C., Chan, Y. M., Harding, S. V. and Jones, P. J., 2009. Low and moderate-fat plant sterol fortified soymilk in modulation of plasma lipids and cholesterol kinetics in subjects with normal to high cholesterol concentrations: report on two randomized crossover studies. *Lipids in Health and Disease*. 8: 45.

Ryan-Borchers, T. A., Park, J. S., Chew, B. P., McGuire, M. K., Fournier, L. R. and Beerman, K. A., 2006. Soy isoflavones modulate immune function in healthy postmenopausal women. *American Journal of Clinical Nutrition*. 83: 1118–1125.

Setchell, K. D. and Cole, S. J., 2003. Variations in isoflavone levels in soy foods and soy protein isolates and issues related to isoflavone databases and food labeling. *Journal of Agricultural and Food Chemistry*. 51: 4146–4155.

Singh, G., Arora, S., Sharma, G. S., Sindhu, J. S., Kansal, V. K. and Sangwan, R. B., 2007. Heat stability and calcium bioavailability of calcium-fortified milk. *LWT-Food Science and Technology*. 40: 625–631.

Smith, K. T., Heaney, R. P., Flora, L. and Hinders, S. M., 1987. Calcium absorption from a new calcium delivery system (CCM). *Calcified Tissue International*. 41: 351–352.

Takatsuka, N., Nagata, C., Kurisu, Y., Inaba, S., Kawakami, N. and Shimizu, H., 2000. Hypocholesterolemic effect of soymilk supplementation with usual diet in premenopausal normolipidemic Japanese women. *Preventive Medicine*. 31: 308–314.

Tang, A. L., Shah, N. P., Wilcox, G., Walker, K. Z. and Stojanovska, L., 2007. Fermentation of calcium-fortified soymilk with Lactobacillus: effects on calcium solubility, isoflavone conversion, and production of organic acids. *Journal of Food Science*. 72: M431–M436.

Tang, A. L., Walker, K. Z., Wilcox, G., Strauss, B. J., Ashton, J. F. and Stojanovska, L., 2010a. Calcium absorption in Australian osteopenic post-menopausal women: an acute comparative study of fortified soymilk to cow's milk. *Asia Pacific Journal of Clinical Nutrition*. 19(2): 243–249.

Tang, A. L., Wilcox, G., Walker, K. Z., Shah, N. P., Ashton, J. F. and Stojanovska, L., 2010b. Phytase activity from *Lactobacillus* spp. in calcium-fortified soymilk. *Journal of Food Science*. 75: M373–M376.

Travis, R. C., Allen, N. E., Appleby, P. N., Spencer, E. A., Roddam, A. W. and Key, T. J., 2008. A prospective study of vegetarianism and isoflavone intake in relation to breast cancer risk in British women. *International Journal of Cancer*. 122: 705–710.

Wu, A. H., Stanczyk, F. Z., Hendrich, S., Murphy, P. A., Zhang, C., Wan, P. and Pike, M. C., 2000. Effects of soy foods on ovarian function in premenopausal women. *British Journal of Cancer*. 82: 1879–1886.

Xu, X., Harris, K. S., Wang, H. J., Murphy, P. A. and Hendrich, S., 1995. Bioavailability of soybean isoflavones depends upon gut microflora in women. *Journal of Nutrition*. 125: 2307–2315.

Xu, X., Wang, H. J., Murphy, P. A., Cook, L. and Hendrich, S., 1994. Daidzein is a more bioavailable soymilk isoflavone than is genistein in adult women. *Journal of Nutrition*. 124: 825–832.

Yeo, S. K. and Liong, M. T., 2010. Angiotensin I-converting enzyme inhibitory activity and bioconversion of isoflavones by probiotics in soymilk supplemented with prebiotics. *International Journal of Food Sciences and Nutrition*. 6: 161–181.

Zhang, X., Shu, X. O., Li, H., Yang, G., Li, Q., Gao, Y. T. and Zheng, W., 2005. Prospective cohort study of soy food consumption and risk of bone fracture among postmenopausal women. *Archives of Internal Medicine*. 165: 1890–1895.

Zhao, Y., Martin, B. R. and Weaver, C. M., 2005. Calcium bioavailability of calcium carbonate fortified soymilk is equivalent to cow's milk in young women. *Journal of Nutrition*. 135: 2379–2382.

CHAPTER 20

The Effect of Prebiotics on Calcium Absorption and Utilization

STEVEN JAKEMAN[a] AND CONNIE WEAVER*[b]

[a]Department of Food Science, Purdue University, 745 Agriculture Mall Drive, West Lafayette, IN 47907-2009, USA; [b]Department of Nutrition Science, Purdue University, 700 W State Street, West Lafayette, IN 47907-2059, USA
*E-mail: weavercm@purdue.edu

20.1 Introduction

Widely known as the primary mineral constituent of bone, calcium likely encourages other health-promoting effects such as reduced risk for high blood pressure, colon cancer, and cardiovascular disease (NIH, 2013). However, in an evaluation of adults and adolescents in 20 countries, individuals in all groups, except for young men, had a substantial risk of inadequate calcium intake (Looker, 2006). Furthermore, calcium absorption by postmenopausal women, a segment of the population with great risk for low calcium status, is only approximately 30% or less from one serving of dairy or 300 g supplements. One strategy to enhance calcium status is to increase the absorption efficiency of this mineral.

Historically, calcium-absorption enhancement strategies have primarily focused on identifying the most soluble form of calcium (Reinwald *et al.*, 2008). Recently, efforts in several fields, including nutrition, have begun to

Food and Nutritional Components in Focus No. 10
Calcium: Chemistry, Analysis, Function and Effects
Edited by Victor R. Preedy

Published by the Royal Society of Chemistry, www.rsc.org

focus on leveraging microbial cell populations in and on the human body. Such cells outnumber host cells 10 to 1, and some postulate that over the course of millennia, humans have evolved to rely on microbial colonies as a kind of symbiote-dependent superorganism. Indeed, human health parameters modulated by microbes include brain function, cancer development, and nutrient absorption (Aziz *et al.*, 2013). Due to their ability to modulate health through gut bacteria, prebiotics, which are thought to aid the absorption of certain minerals, have become the focus of much calcium-related research.

20.2 What are Prebiotics?

Prebiotics are dietary components, primarily carbohydrates, that are selectively fermented by beneficial microbes in the large intestine, resulting in positive health outcomes for the human host. These compounds come from a variety of sources, including plants, animals, and synthetic production.

Disaccharides, oligosaccharides, and polysaccharides, carbohydrate members of the prebiotic family vary in both chain length and bond character. Where digestible saccharides primarily contain $\alpha(1 \rightarrow 4)$, $\alpha(1 \rightarrow 6)$ bonds, and the occasional $\beta(1 \rightarrow 4)$ bond (in the case of lactose) between sugar molecules, prebiotics additionally contain an array of $\alpha(1 \rightarrow 2)$, $\beta(1 \rightarrow 2)$, $\beta(1 \rightarrow 3)$, $\beta(1 \rightarrow 6)$, and other bonds that are resistant to human enzymes. It is possible that variation in chain length and bond type could be responsible for differences in prebiotic effectiveness.

The health gains of prebiotics are primarily a result of their consumption and fermentation by beneficial gut microbiota, which result in two distinct benefits: the feeding of good intestinal bacteria and the release of short-chain fatty acids. First, by providing targeted nutrition to good bacteria, such as bifidobacteria, prebiotics encourage these colonies to grow and become metabolically active, providing the human host several favorable outcomes. These include crowding out pathogenic species, relieving symptoms of lactose intolerance, producing B-vitamins, controlling blood-cholesterol levels, relieving gastrointestinal distress, and modulating nutrient absorption (Manning and Gibson, 2004).

The second distinct benefit is the production and release of short-chain fatty acids (SCFAs) through fermentation. Short-chain fatty acids include acetate, butyrate, and propionate, and are associated with several health gains (Hijova and Chmelarova, 2007). As the preferred energy source of colonic mucosal cells, butyrate is important to colon health and function. When SCFAs are eaten as part of the diet the majority are absorbed by the small intestine; prebiotics, however, are not broken down in the small intestine, but are instead fermented into SCFAs in the colon. Eating foods that are subsequently fermented into SCFAs is one of the only ways to consistently provide these important nutrients to the large intestine. Thus, prebiotics not only serve as nutrients for gut microbiota, but also for the large intestine itself.

20.3 Calcium Absorption

The body absorbs calcium both actively and passively, potentially at multiple points through the digestive tract. Passive absorption occurs through paracellular diffusion along the length of the intestine. Active absorption occurs through calcium transporters primarily in the small intestine, though recent evidence suggests that the large intestine houses active transport also. Zhao *et al.* (2005) showed through kinetic modeling that when rats are fed a small bolus dose of calcium, the calcium is absorbed monophasically. However, when rats are fed a large bolus dose of calcium they absorb it biphasically. This suggests two relatively efficient sites of calcium absorption, the first, and potentially the second, being saturable. The timing and biphasic nature of calcium absorption when the dose is high is consistent with absorption in the small and large intestines. This suggests that either passive diffusion is more effective in the large intestine than in the small intestine, or that the large intestine is also capable of active calcium transport.

20.4 Prebiotic Mediated Calcium Absorption

Prebiotics have been linked with increased calcium absorption in the large intestine, research suggests. For example, ovariectomized (OVX) rats have diminished calcium absorption and are used to model the postmenopausal condition. Mitamura *et al.* (2004) found that a raffinose intervention restored calcium absorption in OVX rats. This benefit was not seen in rats that were cecocolonectomized, suggesting that extracting the cecum and colon removed the site of calcium absorption. Furthermore, prebiotics remain essentially unaffected throughout digestion until they arrive in the large intestine. The exact process by which prebiotics enhance calcium absorption is unclear. However, potential mechanisms that prebiotics may work through include changing the luminal environment of the large intestine, modifying the composition of the large intestine, or by mediating the gut-microbial population of the large intestine.

20.4.1 Changes in the Lumen Environment

One popular hypothesis is that the short-chain fatty acids generated from carbohydrate fermentation are the driving force behind increased calcium absorption. The theory is that as SCFA concentration increases, the luminal pH of the colon decreases. This causes greater ionization and solubility of dietary calcium, leading to greater active transport and passive diffusion across the membrane of the large intestine.

While the argument is compelling, a good amount of evidence suggests that decreased luminal pH is not the only player, or even a dominant player, in prebiotic-mediated enhanced calcium absorption. Mineo *et al.* (2001), using the Ussing chamber method to examine rat colon segments, found

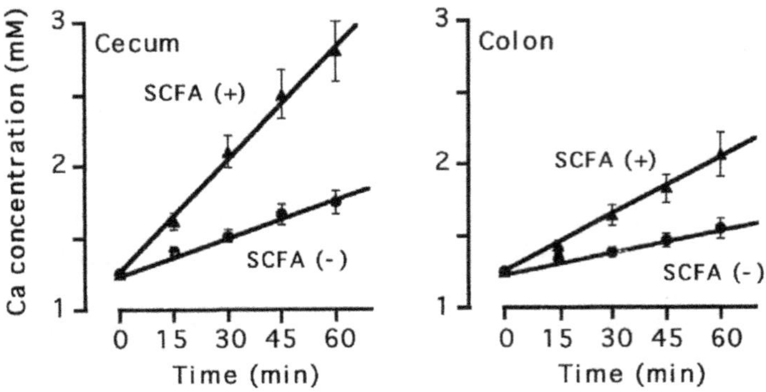

Figure 20.1 Short-chain fatty acids increase calcium absorption in rat cecal and colon segments *ex vivo*. In this experiment, a 130 mM SCFA mixture was added to the mucosal side of cecal and colon segments engaged in a Ussing chamber. Calcium concentration was measured on the serosal side of the segments taken in 15 min intervals. This shows that segments treated with SCFA were more efficient at absorbing calcium over time than segments not treated with SCFA ($p < 0.001$). Reprinted from Mineo *et al.*, 2001. Short-chain fatty acids enhance diffusional Ca transport in the epithelium of the rat cecum and colon. *Life Sciences*. 69: 521, with permission from Elsevier.

that treating the cells with SCFAs increased calcium absorption, while treating them with HCl did not, indicating that solely decreasing luminal pH does not drive calcium absorption in the large intestine (Figure 20.1).

Interestingly, Yonezawa *et al.* (2007) found that treating cells from the breast cancer cell line MCF-7 with SCFAs butyrate, propionate, or acetate increased cellular concentrations of calcium. When G-protein GPR43 was silenced, the effect observed due to propionate was inhibited. Thus, it was hypothesized that SCFAs had a signaling effect on the cell, which encouraged calcium uptake.

Other mechanisms may play a role. For example, in a randomized trial examining the effect of several prebiotic fibers on calcium absorption in rats, Weaver *et al.* (2010) showed that cecal content weight was much more correlated with enhanced calcium absorption than cecal SCFA concentration. Differences in cecal content weight between interventions are likely a function of the water-holding capacity of individual prebiotics. It has been proposed that prebiotics leading to increased cecal content weight may improve calcium absorption due to increased molecular mobility and greater mineral solubilization. Other postulated mechanisms involve changes to the large intestine.

20.4.2 Changes to the Large Intestine

Some investigators have proposed that prebiotics enhance calcium absorption by changing the large intestine itself. Fukushima *et al.* (2012) found that Caco-2 cells treated with SCFAs developed significantly more calbindin-D9k, a vitamin-D-mediated calcium transporter, than untreated cells. Interestingly, neither vitamin D receptor, nor the cell-proliferation marker CDX2 increased in Caco-2 cells treated with SCFAs. Similarly, Nzeusseu *et al.* (2006) showed that in rats, inulin and oligofructose consumption increased the expression of calbindin-D9k, though the effect seen by inulin was much stronger.

Prebiotics may also increase calcium absorption by increasing colon cell count and surface area. Nzeusseu *et al.* (2006) saw an increase in cecal wall weight in rats fed inulin or oligofructose. Perez-Conesa *et al.* (2007) observed a similar effect when they provided an intervention to four-week-old Sprague-Dawley rats consisting of diets containing 1.2%, 5%, and 10% GOS for 30 days. At the end of the trial, they analyzed cecal and colon cell density and crypt depth, as well as femur and tibia mineral contents. They found that under all interventions, the proximal colon experienced a significant increase in crypt depth and cell density compared to the control. Similarly, in the distal colon all prebiotic interventions significantly increased crypt depth, while cell density was only significantly increased in rats consuming 1.2% and 5% diets. As expected, all GOS interventions significantly increased calcium content in both the femur and tibia compared to control groups, implying a connection between increased colon cell count and surface area, and calcium uptake. Interestingly, researchers saw similar results when feeding the rats with probiotics—live exogenous beneficial microbes. Perhaps calcium absorption is increased when prebiotics are consumed because of changes in gut-microbial populations, rather than changes to the human cells of the large intestine.

20.4.3 Changes to the Gut-Microbial Population

Changing gut-microbial populations may also affect calcium absorption. These changes may occur through consuming prebiotics or probiotics. Few studies have been performed to determine the effect of consuming probiotics, live exogenous beneficial microbes, on mineral absorption. Narva *et al.* (2004) showed that postmenopausal women who consumed milk fermented by *Lactobacillus helveticus* had increased calcium absorption above those who consumed milk fermented by other lactic acid bacteria. Kruger *et al.* (2009) showed that true calcium absorption in male rats consuming feed containing *Lactobacillus rhamnosus* for 3 weeks was significantly increased over those consuming a control diet. These rats also experienced a positive, nonsignificant increase in fractional calcium retention. Importantly, in this study, the intervention group did consume more total calcium than the control group did. Kruger *et al.* (2009) also showed that the bones of OVX rats that consumed *Lactobacillus rhamnosus* were more like the bones of sham-operated rats than the bones of other OVX rats that did not consume the probiotic. Perez-Conesa *et al.* (2006) showed that consuming the probiotics *Bifidobacterium bifidum* and *Bifidobacterium longum* either alone, or in conjunction with prebiotics, increased apparent calcium absorption. In this study, the increase in calcium absorption was also correlated with an increase in crypt depth and cell density in the proximal and distal colon. It is unclear how probiotics increase calcium absorption, but it may occur due to a change in luminal environment, large-intestine structure, or through some other metabolic activity unaccounted for by specific bacterial species.

Like probiotics, prebiotics can change the makeup of gut-microbial populations. Weaver *et al.* (2011) showed that intestinal microbial populations in rats fed 2%, 4%, 6%, or 8% GOS diets were similar within a dose of GOS, but different between GOS levels. Similarly, Clavijo-Gutierrez (2013) found that gut-microbial populations of adolescent girls became similar when they moved from a self-selected diet to a controlled diet. When soluble corn fiber (SCF), a prebiotic, was added to the controlled diet, another microbial population shift occurred. Gut microbiota between individuals on the SCF-containing diet were similar; however, gut microbiota between individuals on SCF-containing and non-SCF diets varied. Dietary changes may transform the composition of gut-microbial populations by giving an advantage to those organisms that can best utilize new nutrients. This selection for specific microbes may lead to increased calcium absorption.

20.5 Types of Prebiotics

20.5.1 Disaccharides

As one category of the prebiotic family, disaccharides are carbohydrates consisting of two sugar units. Of these prebiotics, perhaps the most studied is lactulose (4-*O*-β-D-galactopyranosyl-D-fructose). This man-made prebiotic is created during heat-sterilization of milk, and was one of the first nondigestible carbohydrates to be linked with enhancing gut-microbial populations (Seki *et al.*, 2007).

Seki *et al.* (2007) investigated the effects of 2 g lactulose per day and 4 g lactulose per day on 24 young adult men in a double-blind, placebo-controlled, dual-stable isotope study. They found that subjects who consumed 4 g lactulose in a test meal experienced significantly higher calcium absorption and urinary excretion than the control group. They did not, however, experience a change in bone-resorption markers. It is unclear if calcium balance improved, or if the additionally absorbed calcium was merely excreted.

20.5.2 Oligosaccharides

While experts sometimes disagree on the exact number of polymerized sugar units required to make an oligosaccharide, definitions usually range from 3 to 10 sugar units. Many oligosaccharides exist, but the most common classes of prebiotic oligosaccharides are fructo-oligosaccharides and galacto-oligosaccharides.

Fructo-oligosaccharides (FOS) include a variety of compounds varying in chain length, primarily composed of fructose molecules connected by β(1 → 2) linkages and capped with a terminal glucose. While some FOS are synthetically made, many are found in such plants as bananas, Jerusalem artichoke, chicory, asparagus, and onions.

Fructo-oligosaccharides have frequently been tied to enhanced calcium absorption. For instance, Morohashi *et al.* (1998) conducted a 3 day metabolic study on rats that revealed that calcium absorption, excretion, and balance were all increased when the rats were fed a diet with 5% FOS. Similarly, van den Heuvel *et al.* (1999) found that adolescent males who consumed 15 g FOS per day for 9 days experienced significantly higher fractional calcium absorption. However, Martin *et al.* (2010) saw no effect on adolescent girls who consumed 9 g per day of a FOS/inulin blend for 3 weeks.

Another common prebiotic type, galacto-oligosaccharides (GOS) are found as a mixture of oligosaccharides of varying length. Similar to FOS, GOS is comprised of galactose units bound by various β(1 → 2), β(1 → 3), β(1 → 4), and β(1 → 6) linkages with a terminal glucose. Galacto-oligosaccharides are found naturally in milk and can be created synthetically by the enzymatic action of glycosidic hydrolases on lactose (Figure 20.2).

Like fructo-oligosaccharides, galacto-oligosaccharides (GOS) have been shown to increase calcium absorption. van den Heuvel *et al.* (2000) saw that postmenopausal women who consumed 20 g GOS per day experienced increased calcium absorption. The intervention did not affect calcium excretion. Because the storage site for calcium is bone, it is inferred that this excess calcium was being used to create new bone. In another study, dos Santos *et al.* (2011) showed that both gastrectomized and sham-operated rats experienced increased calcium absorption when they were fed 5% GOS diets. Similarly, Weaver *et al.* (2011) showed that rats consuming diets enriched with 2%, 4%, 6%, and 8% GOS had a net increase in calcium balance in a dose-dependent manner. Whisner *et al.* (2013) also documented a positive effect from GOS intervention. In this study, adolescent girls were fed 0 g GOS,

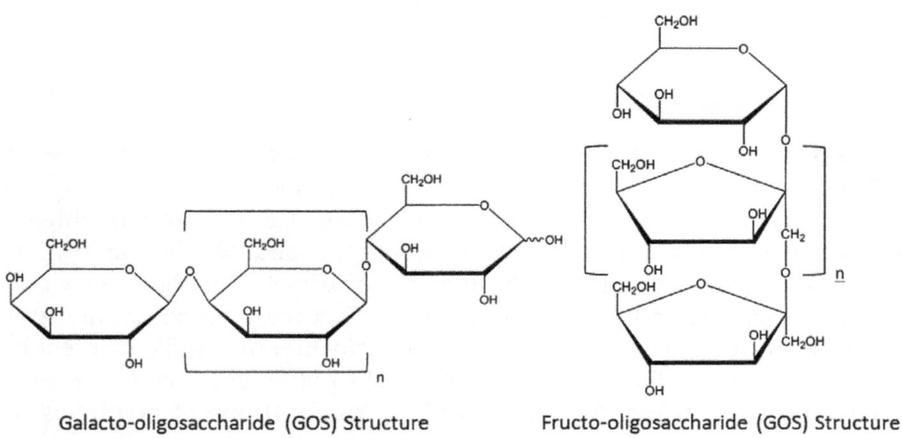

Figure 20.2 The basic structure of galacto-oligosaccharides (GOS) and fructo-oligosaccharides (FOS).

2.5 g GOS and 5 g GOS twice daily for 3 weeks. Both interventions, low and high dose, were shown to significantly increase calcium absorption, though the benefit was not dose-dependent.

20.5.3 Polysaccharides

Polysaccharides are composed of many sugar subunits and include cellulose, hemicellulose, gums, polyglycans, and polyfructans. Some, like cellulose and hemicellulose, are not well fermented and thus cannot be prebiotics. Gums, polyglycans, and polyfructans, on the other hand, are well fermented, which may mean they have prebiotic potential. Thus, calcium absorption research on these compounds may be warranted. For example, Kawase *et al.* (2007) showed that, under *ex vivo* conditions, intestinal segments from rats fed a diet with 7.5% gum arabic for 10 days had more efficient calcium absorption than intestinal segments from control rats. The most common polyfructan used as a prebiotic is inulin, the parent compound to FOS. Coudray *et al.* (2005a) found that rats on a diet supplemented with 10% inulin had increased calcium absorption over animals on nonsupplemented diets.

20.5.4 Synbiotics

Occasionally, prebiotics are eaten in tandem with probiotics, an external source of beneficial bacteria. When these two products, prebiotics and probiotics, are consumed at the same time it is deemed a synbiotic. Examples of synbiotics products include yogurt and kefir. Synbiotics, like prebiotics and probiotics, have been shown to increase calcium absorption. For instance, Perez-Conesa *et al.* (2006) fed weanling rats a diet containing probiotics but no GOS, as well as diets containing 1.2%, 5%, or 10% GOS with and without probiotics. At the end of the study, all interventions showed significantly higher apparent calcium retention than the control group.

20.5.5 Mixed Prebiotics

It has been hypothesized that the various types of prebiotics are effective at different locations in the large intestine. Some prebiotics are small and ferment quickly. Others take longer to ferment because they are long, highly branched, and include many kinds of bonds. It is possible that rapidly fermenting prebiotics may primarily affect the proximal large intestine, while slowly fermenting prebiotics may primarily affect the transverse and distal colon. However, it is still unknown if the fermentation profile of prebiotics is an important factor in their mineral-absorption-enhancing qualities. Therefore, some researchers have studied combinations of prebiotics to gain an insight into the fermentation profile and its effect on maximization of calcium absorption. Holloway *et al.* (2007) performed a randomized, double-blind, crossover study using a FOS/inulin mixture intervention with postmenopausal women. Participants undergoing the intervention consumed 10 g of the prebiotic mixture each day for 6 weeks. The study showed a significant increase in calcium absorption after the intervention. This effect was primarily seen in subjects who had low lower-spine bone-mineral density.

In a balanced, randomized, crossover design, Griffin *et al.* (2002) fed girls either 8 g FOS per day or 8 g FOS + inulin mixture per day for 3 weeks. In the mixed prebiotic intervention, calcium absorption was significantly higher than placebo, whereas the intervention with only 8 g FOS did not increase calcium absorption. Importantly, there was no inulin-only intervention, and so it is not possible to separate an effect from inulin itself compared to the mixed prebiotic.

Interestingly, Hicks *et al.* (2012) found no difference in absorption between infants consuming formula enriched a with 4 g L^{-1} GOS + polydextrin mixture and those consuming nonenriched formula. Abrams *et al.* (2005) showed that adolescents who received an 8 g per day dose of short- and long-chain inulin type fructans experienced an 8% increase in calcium absorption after 8 weeks, and a 6% increase in calcium absorption after 1 year, over the control group.

20.6 Factors that Influence Prebiotic Mediated Calcium Absorption

As described above many, but not all, studies suggest prebiotics can have a promoting effect on calcium absorption. The following sections review distinct factors that may influence the efficacy of prebiotics on calcium absorption.

20.6.1 Age

One of the many variables that may impact calcium absorption mediated by prebiotics is age. Young people are generally more active than older individuals, inducing more load-bearing events on their skeletons, which encourages bone growth. Hormone levels between older and younger individuals also differs drastically. The most critical age stages for bone health are postmenopause, when changes in hormone levels leads to increased bone loss, and adolescence, when the majority of bone is accrued.

Several studies indicate that loss of calcium absorption due to the postmenopausal condition could be alleviated by prebiotics. In one crossover study (van den Heuvel *et al.*, 2000), a daily dose of 20 g GOS for 9 days increased true calcium absorption in postmenopausal women compared to a placebo. Similarly, another study (Holloway *et al.*, 2007) showed that fractional calcium absorption in postmenopausal women increased significantly when they consumed 5 g FOS per day for 6 weeks. Interestingly, one study (Tahiri *et al.*, 2003) showed no effect of 10 g FOS per day for 5 weeks on women

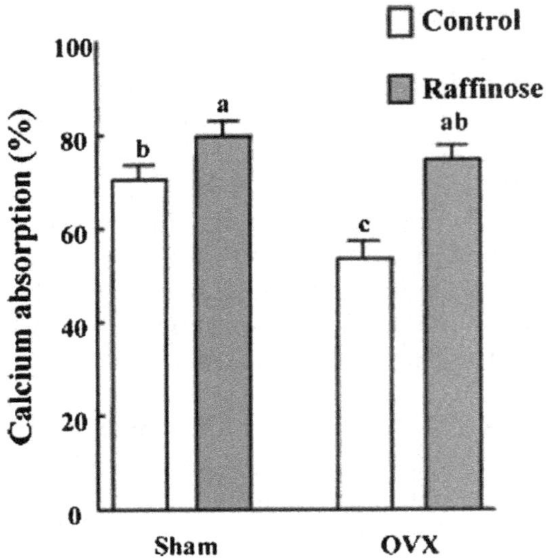

Figure 20.3 The effect of ovariectomy and dietary raffinose on calcium absorption in female rats. Feeding adult female rats raffinose, a prebiotic, increases their calcium absorption. When female rats undergo ovariectomy their calcium absorption decreases. Female rats that undergo ovariectomy but that are also fed raffinose maintain normal calcium absorption. Values not sharing a common letter are significantly different ($p < 0.05$). Reprinted from Mitamura *et al.* Ingestion of raffinose promotes calcium absorption in the large intestine of rats. *Bioscience Biotechnology and Biochemistry*. 68(2): 386. Reproduced with permission.

2–6 years postmenopause. However, they did see a positive trend in women 6 or more years postmenopause. Unfortunately, only six participants fell into this age range. In OVX rats, a model intended to mimic the postmenopausal condition, loss of calcium absorption was restored when the rats were fed a diet with 3% raffinose (Mitamura *et al.*, 2004) (Figures 20.3 and 20.4).

Data for adolescents are less consistent. Perhaps the best example, one study (Griffin *et al.*, 2002) showed that girls who consumed 8 g FOS + inulin mixture per day for 3 weeks experienced significantly higher calcium absorption. Yet a different study (Martin *et al.*, 2010) showed that adolescent girls who consumed 9 g of the same FOS + inulin mixture per day for 3 weeks did not have increased calcium absorption. The primary difference between these two studies was that the effective intervention was provided with orange juice, while the ineffective intervention was provided in ready-to-eat cereal.

In agreement with these studies, Coudray *et al.* (2005b) showed that aged rats, animals suffering from diminished calcium absorption, experienced a greater increase in calcium absorption from inulin intervention than young

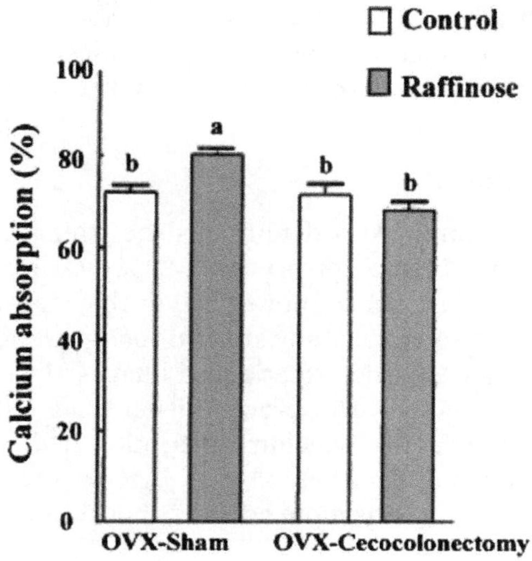

Figure 20.4 The effect of cecocolonectomy and dietary raffinose on calcium absorption in OVX rats. OVX rats have increased calcium absorption when fed raffinose, a prebiotic. Removal of the large intestine does not decrease the calcium absorption in rats that have undergone ovariectomy. However, removal of the large intestine eliminates the benefit of dietary raffinose on calcium absorption in OVX rats, indicating that dietary raffinose relies on the large intestine to increase calcium absorption. Values not sharing a common letter are significantly different ($p < 0.05$). Reprinted from Mitamura *et al.* Ingestion of raffinose promotes calcium absorption in the large intestine of rats. *Bioscience Biotechnology and Biochemistry*. 68(2): 388. Reproduced with permission.

rats did. However, this study used male rats, and was not comparing OVX rats to young female rats.

20.6.2 Intervention Length and Calcium Status

Both intervention length and calcium status of subjects are thought to be important factors in how well prebiotics enhance calcium absorption. In one study (Coudray *et al.*, 2005a), rats were fed a 10% inulin diet for 13 days. Animals on either low- or high-calcium diets experienced increased calcium absorption over animals on normal calcium diets. When the trial was extended to 37 days, the animals on the low-calcium diet retained the enhanced calcium absorption, while the animals on a high-calcium diet did not. This may be because vitamin-D-mediated calcium absorption is downregulated when calcium absorption is high. Therefore, because the inulin enhanced calcium absorption, active calcium absorption may have been hormonally downregulated over time in the rats on high-calcium diets, leading to the loss of the benefit seen in the short term. Supporting this theory, several long-term studies using subjects with adequate calcium intake have shown significant increases in calcium absorption when they were fed prebiotics (Abrams *et al.*, 2005; Holloway *et al.*, 2007; dos Santos *et al.*, 2011; Weaver *et al.*, 2011) (Figure 20.5).

20.6.3 Dose Amount

Intervention dose amount likely determines the strength of the prebiotic response. In one study (Perez-Conesa *et al.*, 2006), rat weanlings that consumed a 10% GOS diet absorbed more calcium than those on 1.2% or 5% GOS diets. Similarly, a separate study showed that growing rats consuming 2%, 4%, 6%, and 8% GOS diets experienced increased calcium absorption in a dose-dependent manner. However, Whisner *et al.* (2013) showed that adolescent girls with adequate calcium absorption had increased calcium absorption when consuming either 2.5 g or 5 g GOS per day. Curiously, in this case, 2.5 g GOS per day was more effective than 5 g GOS per day.

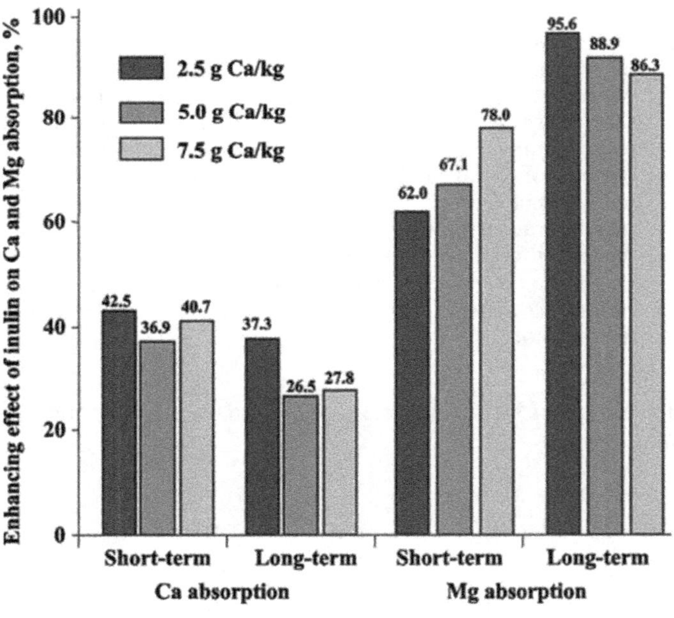

Figure 20.5 Short- and long-term calcium absorption may be influenced by prebiotics based on calcium consumption. In this experiment, rats on low-, normal-, and high-calcium diets experienced increased calcium absorption when fed 10% inulin diets. A portion of this benefit was lost on normal- and high-calcium intake rats, but was maintained in low-calcium intake rats. This may be due to hormonal responses to increased total calcium absorption. With kind permission from Springer Science + Business Media: "Coudray *et al.*, 2005. Stimulatory effect of inulin on intestinal absorption of calcium and magnesium in rats is modulated by dietary calcium intakes – short- and long-term balance studies. *European Journal of Nutrition*. 44: 298, Figure 1, and any original copyright notice displayed with material".

20.6.4 Genetics and Pre-Existing Conditions

Subject genetics may play a role in the effectiveness of prebiotic intervention. Many studies have been performed in rat populations, which while practical are not the ideal representation of human physiology. For instance, the cecum plays a much more substantial role in fermentation in rats than in humans. Additionally, variation across individuals in a species can greatly change calcium absorption. Replogle *et al.* (2014) showed that, in rats, hormonal response to calcium status varied significantly based on the rat phenotype. Weaver *et al.* (2009) has observed similar effects in human populations with different genetic histories. A similar effect could be seen with prebiotic interventions. In fact, some studies have used the terms *responders* and *non-responders* to differentiate individuals with different sensitivity to prebiotics. One study showed that postmenopausal women with lower lumbar BMD were more likely to respond to prebiotic intervention (Holloway *et al.* 2007). Another (Tahiri *et al.*, 2003) found that women more than 6 years postmenopause were more likely to respond than those closer to menopause.

20.7 Prebiotics on Bone

While increasing calcium absorption is a first step towards improved health outcomes, the primary end outcome of concern is bone strength. This is usually quantified by a proxy measure, such as bone-mineral density, or trabecular and cortical thickness. There is evidence that suggests consuming

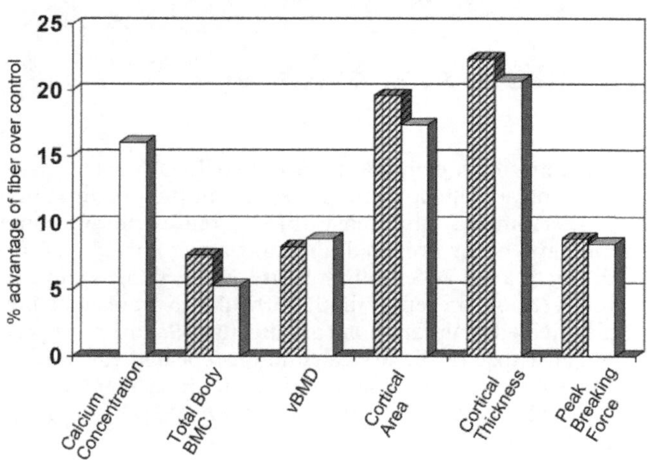

Figure 20.6 Prebiotics can increase bone strength and enhance bone architecture. Soluble corn fiber (SCF; slashed bars) and soluble fiber dextrin (SFD; open bars) fed rats had improved femoral bone quality over control fed rats ($p < 0.05$). Reprinted with permission from Weaver *et al.*, 2010. Novel fibers increase bone calcium content and strength beyond efficiency of large intestine fermentation. *Journal of Agricultural and Food Chemistry*. 58(16): 8955. Copyright 2010 American Chemical Society.

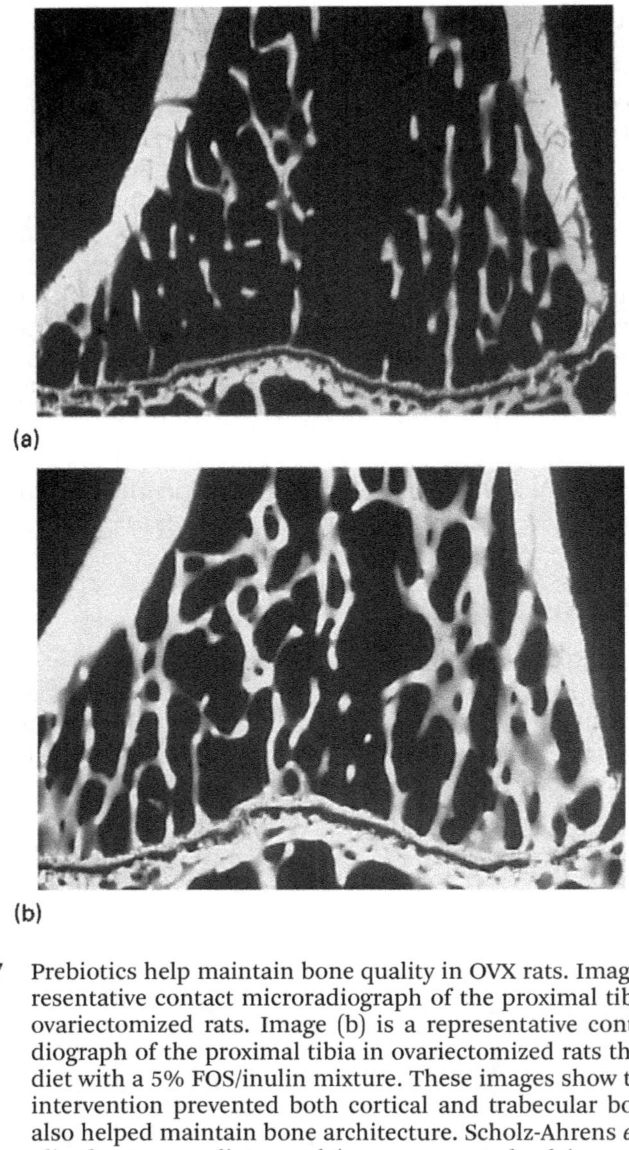

Figure 20.7 Prebiotics help maintain bone quality in OVX rats. Image (a) is a representative contact microradiograph of the proximal tibia in control ovariectomized rats. Image (b) is a representative contact microradiograph of the proximal tibia in ovariectomized rats that were fed a diet with a 5% FOS/inulin mixture. These images show that prebiotic intervention prevented both cortical and trabecular bone loss, and also helped maintain bone architecture. Scholz-Ahrens *et al.* Effect of oligofructose or dietary calcium on repeated calcium and phosphorus balances, bone mineralization and trabecular structure in ovariectomized rats. *British Journal of Nutrition.* 88(4): 368, reproduced with permission.

prebiotics improves these parameters. Abrams *et al.* (2005) showed that after 1 year of consuming a FOS + inulin mixture, adolescents had significantly greater whole-body bone-mineral content and bone-mineral density than the control group. Similarly, Roberfroid *et al.* (2002) found that a diet consisting of 0.5 and 1.0% inulin increased whole-body bone-mineral density and whole-body bone-mineral content in rats. Another study

(Nzeusseu *et al.*, 2006) showed that, in rats, a 5% GOS intervention for 3 months significantly increased trabecular bone area over control. Additionally, a 5% inulin intervention for 3 months increased trabecular bone area more than the GOS intervention. It also increased trabecular bone area in the lower lumbar area, which the GOS intervention did not.

The ultimate test of bone quality is bone strength. Weaver *et al.* (2011) showed that young growing rats had greater volumetric bone-mineral density and peak breaking strength when fed a diet with 2%, 4%, 6%, and 8% GOS for 8 weeks. The same research group (Weaver *et al.*, 2010) compared the effect of eight different prebiotic fibers on rat calcium absorption, bone quality measurements, and peak breaking force. These fibers included two resistant starches, soluble corn fiber (SCF), soluble fiber dextrin (SFD), pullulan, polydextrose, inulin, and a FOS + inulin mixture. Except SCF and pullulan, the prebiotics were effective at increasing calcium content of the rats' bones. While the only fiber to significantly increase femur uptake of calcium was the FOS + inulin mixture, only SFD and SCF significantly increased femoral peak breaking strength. The parameters that both SCF and SFD significantly increased, and the inulin/FOS mixture did not, were total femoral BMD and cortical thickness. These data indicate that some prebiotics are more effective at enhancing bone formation, even when others may increase calcium uptake more efficiently (Figure 20.6).

Scholz-Ahrens *et al.* (2002) found that in OVX rats, tibial bone was spared with an intervention of high calcium, FOS, or a combination of the two. Interestingly, the spared bone had a different trabecular architecture depending on whether the intervention included high calcium, FOS, or both. This indicates that some of the benefit of prebiotics to bone structure and strength may be in addition to the benefit of greater calcium absorption (Figure 20.7).

20.8 Conclusion and Future Work

The gastrointestinal tract hosts many trillions of microbes, and through prebiotics humans might be able to unlock some of their health-promoting potential. The majority of reports evaluating the effect of prebiotics on calcium absorption indicate that various types of prebiotics enhance calcium absorption and bone deposition. Further research is necessary to determine how calcium absorption is enhanced, and whether bone strength is truly enhanced beyond the effect of increased calcium absorption.

Summary Points

- Millions of people worldwide do not consume enough calcium.
- Calcium absorption occurs in both the small and large intestine.
- Prebiotics can help increase calcium absorption in the large intestine.
- Prebiotics are carbohydrates that humans cannot digest, but that can be fermented by bacteria found in the large intestine.

- The fermentative byproducts of prebiotic fermentation are short-chain fatty acids.
- Several theories exist that may explain the increase in calcium absorption associated with prebiotics.
- The lumen environment in the large intestine is changed by prebiotics.
- Increased cell number and density are correlated with consuming prebiotics.
- The microbial population of the gut is modified by dietary prebiotics.
- Many factors, including calcium status, subject age, genetics, and load of prebiotic intervention may mediate the absorption-enhancing prebiotic effect.
- Increased calcium absorption due to prebiotic intervention has been shown to lead to increased bone strength.

Key Facts

Key Features of Prebiotics

1. The term *prebiotic* to describe a dietary component that could modulate gut-microbial populations for host health was first used in 1995.
2. In normal animals, prebiotics have been shown to increase colonic crypt depth and the number of colonic epithelial cells. Germ-free animals, those with no gut microbiota, do not experience the same changes.
3. Prebiotics are found in many foods including energy bars, soups, candy, dairy products, and more.
4. Prebiotics exist both as soluble and insoluble fiber, but are not digested by human enzymes.
5. Dietary prebiotics are broken down through fermentation by gut microbiota.
6. Dietary prebiotics can alter gut microbiota to enhance the proportion of bacteria that can ferment fibers.
7. ISAPP is a professional society for those interested in pre- and probiotics.

Definitions of Words and Terms

Active transport. The transport of particles into a cell through the use of protein complexes called transporters.

Cecocolonectomy. Removal of the cecum and colon.

Disaccharide. A carbohydrate with two sugar units.

Fructo-oligosaccharide. Oligosaccharides consisting of fructose chains and branches with a terminal glucose.

Galacto-oligosaccharide. Oligosaccharides consisting of galactose chains and branches with a terminal glucose.

Inulin. Polysaccharides consisting of fructose chains and branches with a terminal glucose.

Lactose. A milk sugar consisting of one glucose and one fructose.
Oligosaccharides. Carbohydrates between three and ten sugars long.
Ovariectomized (OVX). Removal of the ovaries. This procedure is generally performed on rodents in order to model the postmenopausal condition of older human women.
Passive diffusion. The movement of particles in a solution across a membrane down a concentration gradient.
Polysaccharides. Carbohydrates more than ten sugar units long.
Prebiotic. A dietary component, primarily carbohydrates, that is not digested by human enzymes, but that is fermented by bacteria in the large intestine and benefits host well-being.
Probiotics. Dietary bacteria intended to colonize the lower gut and confer health benefits on the host.
Short-chain fatty acid (SCFA). A nonesterified fatty acid containing between 2 and 6 carbons. The SCFAs produced by the fermentation of prebiotics are between 2 and 4 carbons long. These include acetic acid, propionic acid, and butyric acid, respectively.
Synbiotics. The combination of prebiotics and probiotics.

List of Abbreviations

SCFA	Short-chain fatty acid
FOS	Fructo-oligosaccharide
GOS	Galacto-oligosaccharide
OVX	Ovariectomized
SCF	Soluble corn fiber
SFD	Soluble fiber dextrin

References

Abrams, S. A., Griffin, I. J., Hawthorne, K. M., Liang, L., Gunn, S. K., Darlington, G. and Ellis, K. J., 2005. A combination of prebiotic short- and long-chain inulin-type fructans enhances calcium absorption and bone mineralization in young adolescents. *American Journal of Clinical Nutrition*. 82(2): 471–476.

Aziz, Q., Dore, J., Emmanuel, A., Guarner, F. and Quigley, E. M., 2013. Gut microbiota and gastrointestinal health: current concepts and future directions. *Neurogastroenterology & Motility*. 25(1): 4–15.

Clavijo-Gutierrez, A. P., 2013. Response of human gut microbiota to diet supplementation with soy or soluble corn fiber. Ph.D. Dissertation, Purdue University.

Coudray, C., Feillet-Coudray, C., Tressol, J. C., Gueux, E., Thien, S., Jaffrelo, L., Mazur, A. and Rayssiguier, Y., 2005a. Stimulatory effect of inulin on intestinal absorption of calcium and magnesium in rats is modulated by dietary calcium intakes – short- and long-term balance studies. *European Journal of Nutrition*. 44(5): 293–302.

Coudray, C., Rambeau, M., Feillet-Coudray, C., Tressol, J. C., Demigne, C., Gueux, E., Mazur, A. and Rayssiguier, Y., 2005b. Dietary inulin intake and age can significantly affect intestinal absorption of calcium and magnesium in rats: a stable isotope approach. *Nutrition Journal*. 4: 29.

Fukushima, A., Aizaki, Y. and Sakuma, K., 2012. Short-chain fatty acids increase the level of calbindin-D9k messenger RNA in Caco-2 Cells. *Journal of Nutritional Science and Vitaminology*. 58(4): 287–291.

Griffin, I. J., Davila, P. M. and Abrams, S. A., 2002. Non-digestible oligosaccharides and calcium absorption in girls with adequate calcium intakes. *British Journal of Nutrition*. 87(suppl. 2): S187–S191.

van den Heuvel, E. G. H. M., Muys, T., van Dokkum, W. and Schaafsma, G., 1999. Oligofructose stimulates calcium absorption in adolescents. *American Journal of Clinical Nutrition*. 69(3): 544–548.

van den Heuvel, E. G. H. M., Schoterman, M. H. C. and Muijs, T., 2000. Transgalactooligosaccharides stimulate calcium absorption in postmenopausal women. *Journal of Nutrition*. 130(12): 2938–2942.

Hicks, P. D., Hawthorne, K. M., Berseth, C. L., Marunycz, J. D., Heubi, J. E. and Abrams, S. A., 2012. Total calcium absorption is similar from infant formulas with and without prebiotics and exceeds that in human milk-fed infants. *BMC Pediatrics*. 12: 118.

Hijova, E. and Chmelarova, A., 2007. Short chain fatty acids and colonic health. *Bratislavske Lekarske Listy*. 108(8): 354–358.

Holloway, L., Moynihan, S., Abrams, S. A., Kent, K., Hsu, A. R. and Friedlander, A. L., 2007. Effects of oligofructose-enriched inulin on intestinal absorption of calcium and magnesium and bone turnover markers in postmenopausal women. *British Journal of Nutrition*. 97(2): 365–372.

Kawase, A., Hirata, N., Tokunaga, M., Matsuda, H. and Iwaki, M., 2007. Gum arabic enhances intestinal calcium absorption in rats. *Journal of Health Science*. 53(5): 622–624.

Kruger, M. C., Fear, A., Chua, W. H., Plimmer, G. G. and Schollum, L. M., 2009. The effect of *Lactobacillus rhamnosus* HN001 on mineral absorption and bone health in growing male and ovariectomised female rats. *Dairy Science & Technology*. 89(3–4): 219–231.

Looker, A. C., 2006. Dietary Calcium. *Calcium in Human Health*. Humana Press, Totowa, New Jersey, pp. 105–122.

Manning, T. S. and Gibson, G. R., 2004. Prebiotics. *Best Practice & Research in Clinical Gastroenterology*. 18(2): 287–298.

Martin, B. R., Braun, M. M., Wigertz, K., Bryant, R., Zhao, Y. D., Lee, W., Kempa-Steczko, A. and Weaver, C. M., 2010. Fructo-oligosaccharides and calcium absorption and retention in adolescent girls. *Journal of the American College of Nutrition*. 29(4): 382–386.

Mineo, H., Hara, H. and Tomita, F., 2001. Short-chain fatty acids enhance diffusional Ca transport in the epithelium of the rat cecum and colon. *Life Sciences*. 69(5): 517–526.

Mitamura, R., Hara, H. and Aoyama, Y., 2004. Ingestion of raffinose promotes calcium absorption in the large intestine of rats. *Bioscience Biotechnology and Biochemistry*. 68(2): 384–389.

Morohashi, T., Sano, T., Ohta, A. and Yamada, S., 1998. True calcium absorption in the intestine is enhanced by fructooligosaccharide feeding in rats. *Journal of Nutrition*. 128(10): 1815–1818.

Narva, M., Nevala, R., Poussa, T. and Korpela, R., 2004. The effect of *Lactobacillus helveticus* fermented milk on acute changes in calcium metabolism in postmenopausal women. *European Journal of Nutrition*. 43(2): 61–68.

National Institute of Health, 2013. Calcium Fact Sheet for Consumers. Available at: http://ods.od.nih.gov/factsheets/Calcium-QuickFacts/. Accessed December 15, 2013.

Nzeusseu, A., Dienst, D., Haufroid, V., Depresseux, G., Devogelaer, J. P. and Manicourt, D. H., 2006. Inulin and fructo-oligosaccharides differ in their ability to enhance the density of cancellous and cortical bone in the axial and peripheral skeleton of growing rats. *Bone*. 38(3): 394–399.

Perez-Conesa, D., Lopez, G., Abellan, P. and Ros, G., 2006. Bioavailability of calcium, magnesium and phosphorus in rats fed probiotic, prebiotic and synbiotic powder follow-up infant formulas and their effect on physiological and nutritional parameters. *Journal of the Science of Food and Agriculture*. 86(14): 2327–2336.

Perez-Conesa, D., Lopez, G. and Ros, G., 2007. Effects of probiotic, prebiotic and synbiotic follow-up infant formulas on large intestine morphology and bone mineralisation in rats. *Journal of the Science of Food and Agriculture*. 87(6): 1059–1068.

Reinwald, S., Weaver, C. M. and Kester, J. J., 2008. The health benefits of calcium citrate malate: a review of the supporting science. *Advances in Food and Nutrition Research*. 54: 219–346.

Replogle, R. A., Li, Q., Wang, L., Zhang, M. and Fleet, J. C., 2014. Gene-by-diet interactions influence calcium absorption and bone density in mice. *Journal of Bone and Mineral Research*. 29(3): 657–665.

Roberfroid, M. B., Cumps, J. and Devogelaer, J. P., 2002. Dietary chicory inulin increases whole-body bone mineral density in growing male rats. *Journal of Nutrition*. 132(12): 3599–3602.

dos Santos, E. F., Tsuboi, K. H., Araujo, M. R., Andreollo, N. A. and Miyasaka, C. K., 2011. Dietary galactooligosaccharides increase calcium absorption in normal and gastrectomized rats. *Revista do Colégio Brasileiro de Cirurgiões*. 38(3): 186–191.

Scholz-Ahrens, K. E., Acil, Y. and Schrezenmeir, J., 2002. Effect of oligofructose or dietary calcium on repeated calcium and phosphorus balances, bone mineralization and trabecular structure in ovariectomized rats. *British Journal of Nutrition*. 88(4): 365–377.

Seki, N., Hamano, H., Iiyama, Y., Asano, Y., Kokubo, S., Yamauchi, K., Tamura, Y., Uenishi, K. and Kudou, H., 2007. Effect of lactulose on calcium and magnesium absorption: a study using stable isotopes in adult men. *Journal of Nutritional Science and Vitaminology*. 53(1): 5–12.

Tahiri, M., Tressol, J. C., Arnaud, Y., Bornet, F. R. J., Bouteloup-Demange, C., Feillet-Coudray, C., Brandolini, M., Ducros, V., Pepin, D., Brouns, F., Roussel, A. M., Rayssiguier, Y. and Coudray, C., 2003. Effect of short-chain fructooligosaccharides on intestinal calcium absorption and calcium status in postmenopausal women: a stable-isotope study. *American Journal of Clinical Nutrition*. 77(2): 449–457.

Weaver, C. M., Martin, B. R., Nakatsu, C. H., Armstrong, A. P., Clavijo, A., McCabe, L. D., McCabe, G. P., Duignan, S., Schoterman, M. H. C. and van den Heuvel, E. G. H. M., 2011. Galactooligosaccharides improve mineral absorption and bone properties in growing rats through gut fermentation. *Journal of Agricultural and Food Chemistry*. 59(12): 6501–6510.

Weaver, C. M., Martin, B. R., Story, J. A., Hutchinson, I. and Sanders, L., 2010. Novel fibers increase bone calcium content and strength beyond efficiency of large intestine fermentation. *Journal of Agricultural and Food Chemistry*. 58(16): 8952–8957.

Weaver, C. M., McCabe, L. D., McCabe, G. P., Braun, M., Martin, B. R., DiMeglio, L. A. and Peacock, M., 2009. Vitamin D status and calcium metabolism in adolescent black and white girls on a range of controlled calcium intakes. *Journal of Clinical Endocrinology & Metabolism*. 93: 3907.

Whisner, C. M., Martin, B. R., Schoterman, M. H. C., Nakatsu, C. H., McCabe, L. D., McCabe, G. P., Wastney, M. E., van den Heuvel, E. G. H. M. and Weaver, C. M., 2013. Galacto-oligosaccharides increase calcium absorption and gut bifidobacteria in young girls: a double-blind cross-over trial. *British Journal of Nutrition*. 110(7): 1292–1303.

Yonezawa, T., Kobayashi, Y. and Obara, Y., 2007. Short-chain fatty acids induce acute phosphorylation of the p38 mitogen-activated protein kinase/heat shock protein 27 pathway *via* GPR43 in the MCF-7 human breast cancer cell line. *Cellular Signalling*. 19(1): 185–193.

Zhao, Y. D., Martin, B. R., Wastney, M. E., Schollum, L. and Weaver, C. M., 2005. Acute *versus* chronic effects of whey proteins on calcium absorption in growing rats. *Experimental Biology and Medicine*. 230(8): 536–542.

CHAPTER 21

Health Aspects of Calcium in Drinking Water

A. ATA ALTURFAN*[a] AND EBRU EMEKLI ALTURFAN[b]

[a]Institute of Forensic Sciences, Cerrahpasa Medical Faculty, Istanbul University, Kocamustafapasa, 34098 Istanbul, Turkey; [b]Department of Biochemistry, Faculty of Dentistry, Marmara University, Guzelbahce Buyukciftlik Sok, No: 6 Nisantası, Istanbul, Turkey
*E-mail: ataalturfan@gmail.com

21.1 Drinking Water

Drinking water has traditionally been extracted from groundwater sources like wells or drawn from lakes and streams. Collecting from rainwater directly has also been a way of providing drinking water. As the technology advanced new methods of providing drinking water were improved, such as ocean water desalination, melting of ice and used water reclamation by membranes (Ong *et al.*, 2009). As different water resources were introduced issues such as quality started to arise. For instance, low mineral intakes from foods and water are common problem in many parts of the world. These inadequate intakes may lead to subclinical deficiencies of some important elements such as iron, magnesium, zinc and calcium in the world (Durlach *et al.*, 1985; Couzy *et al.*, 1995; Garzon and Eisenberg, 1998). Accordingly, many studies have been conducted on the mineral content in drinking water and their relationships to various diseases. This sensitive issue and the related health consequences are still inadequately understood.

Food and Nutritional Components in Focus No. 10
Calcium: Chemistry, Analysis, Function and Effects
Edited by Victor R. Preedy

Published by the Royal Society of Chemistry, www.rsc.org

21.2 Calcium and Water Hardness

Generally, hard water is defined as water containing a high concentration of calcium and magnesium ions. On the other hand, hardness can be also caused by several other dissolved metals; cations, such as barium, aluminum, iron, strontium, zinc, and manganese. Monovalent ions including sodium and potassium do not lead to hardness normally. On the other hand they have the capacity to come together with anions in the water in order to produce stable salts. The type of anion present in the salts helps to distinguish between the two kinds of hardness as carbonate hardness and noncarbonate hardness. Alkalinity is defined as the capacity of water to neutralize acids and is dependent on compounds like carbonate, bicarbonate, hydroxide, and also borate, silicate, and phosphate. Two types of hardness can be defined. Carbonate hardness is named as temporary hardness since it can be removed by boiling water. On the other hand noncarbonate hardness cannot be removed by boiling the water, therefore it is called permanent hardness (Figure 21.1). It is important to separate the two types of hardness because the method of removing differs for the two types. A crucial factor indicating the hardness is the ratio of calcium and magnesium in water. Limestone (the precursor of calcium in the water) and dolomite (precursor of magnesium) are the most common sources of hardness in water (Prepas *et al.*, 2001; Sengupta, 2013).

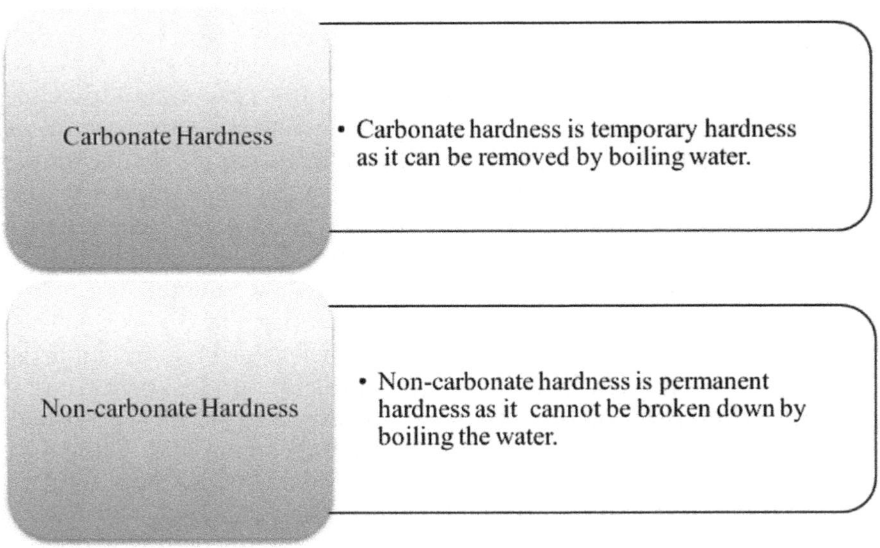

Figure 21.1 Types of water hardness. Two types of hardness are defined. Carbonate hardness is also called as temporary hardness because it can be removed by boiling water. However, noncarbonate hardness cannot be broken down by boiling the water, therefore it is known as permanent hardness.

Calcium is a very important component of water. It occurs in water naturally. Calcium functions as an important determinant of water hardness, a pH stabilizer as well as giving water a better taste. On the other hand, drinking water is essential for life. The concentration of calcium in drinking water varies from one supply to another, mineral-rich drinking waters generally provide substantial contributions to total intakes of calcium in some populations. On the other hand, water-treatment processes can influence concentration of minerals and therefore the total calcium intake for some people (Ong *et al.*, 2009). Water hardness was initially measured as the sum of all polyvalent cations concentration in water such as Ca, Mg, Sr, Ba, Fe, Al, Mn, but since the ions apart from Ca and Mg play only a minor role, later it was generally explained as the sum of the Ca and Mg concentrations that is evaluated by the EDTA titrimetric method (Kozisek, 2003). Water hardness can be defined as very soft, soft, medium hard, hard and very hard. As will be estimated being extreme degrees, very soft and very hard are not the desired levels of Ca and Mg. Drinking waters are generally originated from low and medium mineralized underground and surface waters and calcium and magnesium are found as simple ions Ca^{2+} and Mg^{2+}. The Ca level varies from tens to hundreds of mg L^{-1}, whereas the Mg level varies from units to tens of mg L^{-1}. Since magnesium is present in lower amounts compared with calcium, usually it is also less abundant in waters than calcium. The ratio of Ca to Mg reaches up to 10 in common underground and surface waters, however, a common Ca to Mg ratio is about 4, corresponding to a substance ratio of 2.4 (Kozisek, 2003). Decomposition of calcium and magnesium from aluminosilicates and limestone leads to the liberation of calcium and magnesium in water. Also, these elements may be added to drinking water intentionally during treatment. This may occur during underground waters deacidification using calcium hydroxide or filtration through some compounds such as $CaCO_3$, $MgCO_3$ and MgO counteracting acidity, and also during the low-mineralized water stabilization by adding CaO and CO_2 (Amavis *et al.*, 1976; Kozisek, 2003).

21.3 Water as a Source of Calcium and Bioavailability of Calcium in Water

For most people worldwide dairy products especially milk, are most important sources of calcium in the diet. Although the concentration of calcium in milk is consistent, the calcium concentration in water is very variable. Many factors affect calcium solubility such as whether the ingestion of calcium or the precipitation in the bottom of the container, atmospheric carbon dioxide, pH or some anions (such as phosphates, nitrates, chlorides and sulfates) (Weaver and Nieves, 2009).

The absorbtion and utilization of calcium determine the bioavailability of calcium. The factors that affect calcium bioavailability from food are similar to the factors that affect calcium bioavailability from water (Weaver and

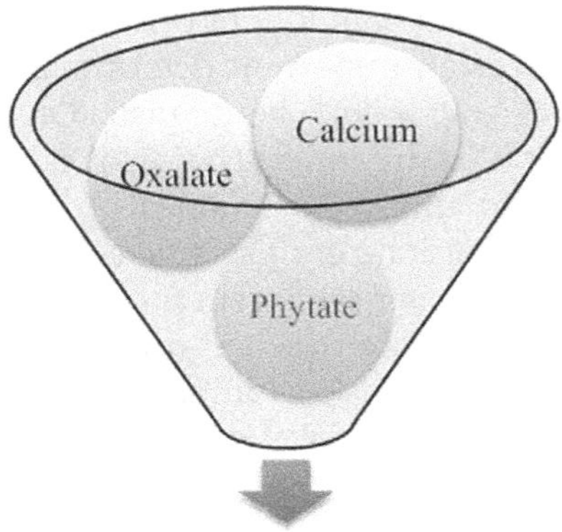

Figure 21.2 Factors that affect the bioavailability of calcium. Bioavailability of calcium from water or other sources in the diet is influenced by the presence of some ions. In particular, calcium absorption has been shown to be improved by the consumption of calcium and inhibited by oxalate and phytate.

Heaney, 2006). In particular, the presence of some ions influence the calcium bioavailability from water or other dietary sources. Calcium absorption is improved by the consumption of calcium by about 5% (Recker, 1985), and inhibited by oxalate and phytate. Oxalate forming insoluble salts with calcium reduces calcium absorption (Weaver and Heaney, 2006) (Figure 21.2).

21.4 Calcium Content of Drinking Water in Different Regions

The mineral content of drinking water varies in different regions of the world. In most of the Asian drinking-water supplies calcium is in the range of 2–80 mg L^{-1} and magnesium is below 20 mg L^{-1} (Ong *et al.*, 2009). The possible relation between gastrointestinal-tract cancer risk and the hardness of drinking water have been evaluated by researchers. Accordingly, Yang *et al.* reported an inverse relationship between the hardness of water and different diseases, like coronary mortality, cerebrovascular disease and gastrointestinal tract cancers in people living in Taiwan and China (Yang *et al.*, 1996, 1997, 1999; Yang, 1998; Yang and Hung, 1998). Kumar *et al.* reported ranges of 8–343 mg L^{-1} and 5–235.6 mg L^{-1} for calcium and magnesium, respectively, in mineral contents in the well water (Kumar *et al.*, 2006).

A limited number of studies have been conducted in African and Eastern Mediterranean countries. Mineral contents of well waters have been reported to be generally higher than those of surface waters (Ong *et al.*, 2009). Olajire and Imeokparia reported that the calcium concentration in the Osun River in Nigeria was between 2 and 60.8 mg L^{-1}, however, the levels in groundwater sources ranged from 3 to 52 mg L^{-1} (Olajire and Imeokparia, 2001).

The tap water mineral content among North American cities varied significantly. Azoulay *et al.* evaluated water sources in San Jose in California and reported that calcium levels varied from 9 to 60 mg L^{-1} (Azoulay *et al.*, 2001). In England, drinking water is generally considered as "very hard" exhibiting calcium carbonate equivalent more than 200 mg L^{-1} however Wales, Cornwall and north-west England are defined as the softer-water areas having hardness ranging from 0 to 200 mg L^{-1} (Ong *et al.*, 2009).

21.5 Potential Effects of Drinking-Water Hardness on Health

21.5.1 Effects of the Hardness of Drinking Water on Cardiovascular Diseases

The effects of hardness of the drinking water have been investigated for more than 40 years. Most of these studies focused on the magnesium and calcium content of drinking water. Studies were mainly about the effects on cardiovascular mortality (Kobayashi, 1957; Durlach *et al.*, 1998). This association was first suggested in Japan, a significant correlation between drinking-water quality and mortality from cerebrovascular disease has been reported (Schroeder, 1960). The correlation between mortality from cardiovascular disease (CVD) in males aged 45–64 years and water hardness in the USA have been shown and suggested "soft water, hard arteries" (Hewitt and Neri, 1980). Many studies were performed thereafter and many papers were published reporting on water hardness in relation with CVD (Kozisek, 2003; WHO, 2005). Beginning from 1978 the WHO pointed out the importance of drinking-water mineral composition and also warned the public against cation-exchange sodium cycle softening in water treatment (WHO, 1978; Eisenberg, 1992; WHO, 2005).

In the next few years some researchers suggested magnesium as the major agent that has a beneficially effective against CVD progression, whereas calcium had only a supportive effect (Durlach, 1988). Studies investigating the role of magnesium in cardiovascular mortality were designed and an inverse correlation was shown (Rockett, 1994).

The role of magnesium as the major factor of water hardness and its protective effects were confirmed in the 1980s rather than new perspectives on the issue (Kozisek, 2003). However later in the 1990s criticisms began against the studies conducted about water hardness and cardiovascular mortality.

The progression of a disease is not simple but often complex and diseases generally have multiple risk factors. Accordingly, the relationship between drinking-water consumption and other risk factors for cardiovascular disease is complex. Rockett developed a model referred as the web of causation emphasizing other factors that may be important in the onset of cardiovascular disease (Punsar and Karvonen, 1979). Using current innovations in methods the previous studies were revised, noting that in some studies even the water was not analysed for the magnesium and calcium content but focused on water hardness only. Moreover, in some studies factors affecting CVD morbidity such as eating habits, alcohol consumption, age, socioeconomic factors were not taken into account (Luoma *et al.*, 1983; Kozisek, 2003; WHO, 2005). After this kind of criticism based on these weak association between water hardness and health the WHO concluded that "the available data were not enough to come to a conclusion that association was causal and no health-based guideline value for water hardness was proposed" (Sauvant and Pepin, 2002).

In the 1990s most epidemiological studies pointed out and specified the impact of calcium/magnesium and also focused on morbidity instead of CVD. In these studies the protective effect of magnesium and calcium in drinking water against CVD was confirmed. Sauvant and Pepin reviewed the epidemiological studies about the relationship between drinking-water composition and CVD between 1990 and 2000 (Sauvant and Pepin, 2002).

In Sweden three remarkable studies were conducted on magnesium and calcium content in drinking water and CVD. One of them was performed in 1991 by Rylander *et al.* They reported a significant inverse correlation between water hardness and mortality from CVD. This study was conducted on both males and females and a significant inverse correlation was also reported between the magnesium content in drinking water and mortality from CVD in males (Rylander *et al.*, 1991). The other study was a case-control study that reported significantly lower mortality rate in females living in calcium supplied water areas (>70 mg L^{-1}) compared with areas where drinking water calcium level was <31 mg L^{-1} (Rubenowitz *et al.*, 1999). In another report of a study performed in Sweden the correlation between drinking-water Mg levels and morbidity and mortality from acute myocardial infarction (AMI) was investigated in females by Rubenowitz *et al.* (2000). Based on their results it was concluded that magnesium prevented primarily sudden death from AMI. On the other hand, again in Sweden, Gyllerup *et al.* reported that the climate or the cold index had a greater effect on mortality from CVD than water hardness (Gyllerup *et al.*, 1991).

Hard-water supplementation has been shown to decrease the incidence rate of AMI 20.6 per 10 000 population whereas in areas supplied with soft water the rate was high in Germany (Teitge, 1990) (Table 21.1). Mortality from CVD was found to be lower (19%) in areas with hard water (161 mg $CaCO_3$ L^{-1}) when compared with areas with soft water (39 mg $CaCO_3$ L^{-1}) in USA (Erb, 1997).

In general, the beneficial effects of hard drinking-water supplementation on CVD and AMI have been attributed mainly to the magnesium content rather than the calcium content (Bloom and Peric-Golia, 1989; Rylander, 1996). According to the report of the World Health Organisation (WHO) on the "Calcium and Magnesium in Drinking-water" the total magnesium intake has been indicated to be at least 450–500 mg per day, and drinking water should contain a minimum of 25–50 mg L^{-1} (Altura and Altura, 2009).

Catling *et al.* reviewed epidemiological studies evaluating a meta-analysis of case-control studies (Catling *et al.*, 2008). One case-control study was

Table 21.1 Results of some studies investigating the effects of hardness of the drinking water on cardiovascular disease. In general, the beneficial effects of hard drinking water supplementation on CVD and AMI have been mainly attributed to the magnesium content.

Authors	Reports
Teitge (1990)	The incidence rate of AMI has been shown to decrease due to hard water supplementation compared with the areas supplied with soft water in Germany
Rylander *et al.* (1991)	A significant inverse correlation has been reported between water hardness and mortality from CVD in both males and females with a significant inverse correlation between the magnesium content in drinking water and mortality from CVD in males
Gyllerup *et al.* (1991)	The climate or the cold index has been reported to have a higher effect on mortality from CVD than water hardness
Rubenowitz *et al.* (1999)	Significantly lower mortality rate has been reported in females living in calcium-supplied water areas
Rubenowitz *et al.* (2000)	Magnesium has been reported to prevent primarily sudden death from AMI in the study performed in Sweden investigating the correlation between drinking-water Mg levels and morbidity and mortality from AMI

identified that linked water hardness with deaths from CVD that found no significant association. However, there have been studies that linked water magnesium and/or calcium with deaths from CVD. The study of Rubenowitz *et al.* was the only one reporting a protective effect of calcium in drinking water on mortality of females due to acute myocardial infarction (Rubenowitz *et al.*, 1996). On the other hand, other studies suggested a significant protective effect of magnesium in drinking water against mortality from acute myocardial infarction (Rubenowitz *et al.*, 1996, 2000), stroke (Yang, 1998) and hypertension (Yang and Chiu, 1999). No protective effect has also been reported from water hardness, magnesium or calcium (Rosenlund *et al.*, 2005). However, there are questions about the magnesium content of drinking water and the number of people living in the area to see an effect if it existed (Calderon and Hunter, 2009).

Based on the results of the ecological studies reports, a protective association between CVD mortality and water hardness, calcium or magnesium levels exists; however, it should be noted that results are not consistent. Although different studies report a decrease in CVD mortality risk with increasing magnesium levels in drinking water, there is not enough evidence showing an association with water hardness or calcium levels. Most of the studies focus on magnesium rather than calcium. Accordingly, Calderon and Hunter suggested in their review that the epidemiological evidence for the water hardness and cardiovascular disease hypothesis is still not proven (Calderon and Hunter, 2009). Therefore, more studies are necessary to illuminate the effect of calcium in drinking water on CVD.

21.5.2 Effects of Calcium in Water on Bone Metabolism

It has been reported that high calcium levels in water suppresses parathyroid hormone (PTH) compared with low-calcium water (Guillemant *et al.*, 2000, 2002; Meunier *et al.*, 2005). Meunier *et al.* performed a study in 180 postmenopausal women aged 70.1 ± 4.0 years. The women were randomized as; (1) high-calcium water, (2) low-calcium water, (3) high-calcium water plus supplementation with 400 IU of vitamin D_2 per day or (4) calcium plus vitamin D_3 supplement for 6 months (Meunier *et al.*, 2005). Based on their results biochemical markers of bone turnover (osteocalcin, alkaline phosphatase, serum type collagen C-telopeptide, serum PTH) significantly decreased in the group on high-calcium mineral water compared with low-calcium mineral-water-consuming group. The vitamin D_2 supplement did not effect beneficially. The high-calcium water group was affected similarly with the calcium plus vitamin D_3 supplemented group. In another study designed in 15 young in a crossover design men groups were: (1) 600 mg calcium treatment as high-calcium mineral water as three divided doses (2) a single dose of tricalcium phosphate and (3) a low-calcium water. The inhibition of serum PTH and a bone resorption marker, serum C-terminal telopeptide (CTX) were compared in these three experimental periods. Initially the biochemical markers were suppressed, after the second dose of the high-calcium water, the biochemical indices remained suppressed throughout the test period and returned to the control levels after ingestion of the low-calcium water. The authors concluded that one oral intake of water containing a very moderate dose of calcium acutely inhibited intact parathyroid hormone (iPTH) secretion and bone resorption (Guillemant *et al.*, 2002).

21.5.3 Effects of Calcium in Water on Kidney Stones

Most kidney stones are composed of calcium salts that occur as calcium oxalate and to a lesser extent as calcium phosphate. The rest of the calcium salts are uric acid, struvite and cystine stone. Stone formation in urine occurs when the urine is supersaturated depending on the chemical free ion activity. When the urine is undersaturated the stone will not grow and it may even dissolve. When the urinary ion excretion increases and urine volume decreases, increased free ion activity will lead to the stone formation. Nephrolithiasis is the name given to the kidney stones and they are dependent on genetic, metabolic, nutritional and environmental factors. Hypercalciuria,

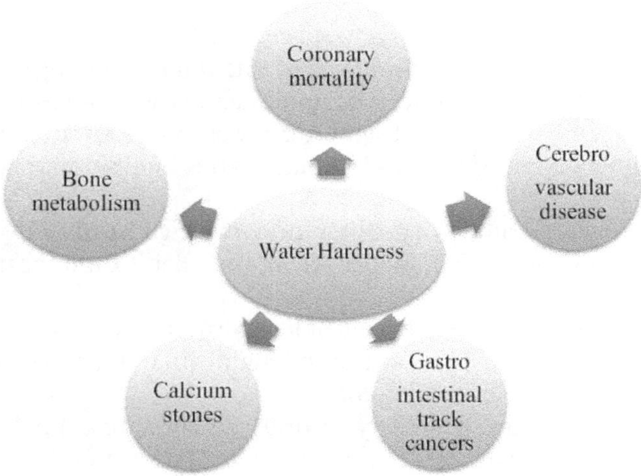

Figure 21.3 Potential effects of water hardness on the organism. Water hardness has been related with bone metabolism, calcium stones, gastrointestinal track cancers, cerebrovascular disease and coronary mortality.

hypocitraturia, hyperuricosuria, hyperoxaluria, cystinuria and infections are among the metabolic factors involved in stone formation. It is still not clear if the water hardness, particularly calcium in water, affects the urinary-stone formation. No association has been shown between water hardness and the incidence of urinary stone formation. The significance of the correlation between water hardness and urinary calcium, citrate, and magnesium levels is not clear. Nevertheless, in order to prevent calcium nephrolithiasis, intake of soft water has been suggested instead of hard water as it is associated with a lower risk for recurrence of calcium stones (Bellizzi *et al.*, 1999; Cotruvo and Bartram, 2009) (Figure 21.3).

21.6 Conclusion

Calcium and magnesium are the most prevalent and abundant metals in our body. Therefore, sufficient intake of these metals are very important. In particular, general public need to be informed by governments about the contents of drinking water that they utilize. In the literature many studies have reported the protective effect of water hardness on CVD mortality. However most of these studies focused on the magnesium content rather than calcium. Investigators need to focus on the calcium and magnesium levels in addition to other minerals and trace elements in soft and hard waters and in softened waters mostly. Accordingly further studies are necessary in order to come to a final conclusion about the "health aspects of calcium in drinking water".

Summary Points

- This chapter focuses on the effects of calcium in drinking water.
- Occurring naturally in water calcium functions as an important determinant of water hardness, a pH stabilizer as well as giving water a better taste.
- The mineral content of drinking water varies in different regions of the world.
- There are many studies in the literature investigating the effects of water hardness and most of them focused on the calcium and the magnesium content.
- In general, the beneficial effects of hard drinking water on cardiovascular disease risk have been attributed mainly to the magnesium content rather than the calcium content.
- Further studies are necessary in order to come to a final conclusion about the "health aspects of calcium in drinking water".

Key Facts

Key Facts of Potential Health Effects of Drinking-Water Hardness

1. An inverse relationship between water hardness and various diseases, including coronary mortality, cerebrovascular disease and gastrointestinal tract cancers have been reported.
2. Some researchers suggested magnesium as the major agent that is beneficially effective against cardiovascular disease progression, whereas calcium had only a supportive effect.
3. High calcium levels in water have been reported to suppress parathyroid hormone (PTH) compared with low-calcium water.
4. Significantly decreased biochemical markers of bone turnover (osteocalcin, alkaline phosphatase, serum-type collagen C-telopeptide, serum PTH) have been reported due to high-calcium mineral water consumption compared with a low-calcium mineral-water-consuming group.
5. In order to prevent calcium nephrolithiasis, intake of soft water has been suggested instead of hard water as it is associated with a lower risk for recurrence of calcium stones.

Key Facts of Calcium and Water Hardness

1. Hard water is defined as water containing a high concentration of calcium and magnesium ions.
2. Water hardness can also be caused by several other dissolved metals; such as divalent or multivalent cations including aluminum, barium, strontium, iron, zinc, and manganese.
3. Sodium and potassium do not cause hardness normally but they may come together with anions in the water to form stable salts.

4. Alkalinity is the capacity of water to neutralize acids and is dependent on compounds such as carbonate, bicarbonate, hydroxide, and also borate, silicate, and phosphate.
5. Two types of hardness can be defined as the carbonate hardness that is called temporary hardness and the noncarbonate hardness called permanent hardness.

Definitions of Words and Terms

Acute myocardial infarction (AMI). AMI is the medical term for a heart attack. It defines the situation when blood stops flowing properly to a part of the heart and the heart muscle is injured due to not receiving oxygen.

Alkalinity. The capacity of water to neutralize acids and it depends on compounds such as carbonate, bicarbonate, hydroxide, and also borate, silicate, and phosphate.

Bioavailability. This is the proportion of the administered substance available for use or storage and capable of being absorbed.

C-terminal telopeptide (CTX). This is a biomarker in serum that is used to measure the rate of bone turnover.

Calcium. This is an essential element for living organisms, with the symbol Ca and with the atomic number 20.

Cardiovascular disease (CVD). This is the one of the leading causes of death worldwide referring to cardiac disease, vascular diseases of the brain and kidney, and peripheral arterial disease.

Cystinuria. This is an inherited autosomal recessive disease. Patients are characterized by the formation of cystine stones in the kidneys, ureter, and bladder.

Hard water. Water containing a high concentration of calcium and magnesium ions and several other dissolved metals.

Hypercalciuria. This is the term referring to elevated urinary calcium excretion.

Hyperoxaluria. This is the term referring to the excessive excretion of oxalate in urine.

Hyperuricosuria. This is the term referring to the presence of excessive amounts of uric acid in the urine.

Hypocitraturia. This is the term referring to a low amount of citrate in the urine that is an important risk factor for the formation of kidney stones.

Intact parathyroid hormone (iPTH). Intact PTH is measured commonly for the differential diagnosis of calcium homeostasis problems. In general, less than 5–25% of total immunoreactive is the intact hormone.

Nephrolithiasis. This is the kidney stone, a solid and crystal aggregation formed in kidneys from dietary minerals also known as a renal calculus.

Parathyroid hormone (PTH). The hormone secreted by parathyroid glands as a polypeptide containing 84 amino acids and acting to increase the concentration of calcium in the blood.

List of Abbreviations

AMI Acute myocardial infarction
CVD Cardiovascular disease
CTX C-terminal telopeptide
iPTH Intact parathyroid hormone
PTH Parathyroid hormone
WHO World Health Organisation

References

Altura, B. M. and Altura, B. T., 2009. Atherosclerosis and magnesium. In: Cotruvo, J. and Bartram, J. (ed.) *Calcium and Magnesium in Drinking-Water: Public Health Significance*. World Health Organization, Geneva, Switzerland, p. 80.

Amavis, R., Hunter, W. J. and Smeets, G. P. M., 1976. Hardness of drinking water and public health. In: *Proceedings of the European Scientific Colloquium Luxembourg 1975*. Oxford, Pergamon Press, p. 440.

Azoulay, A., Garzon, P. and Eisenberg, M. J., 2001. Comparison of the mineral content of tap water and bottled waters. *Journal of General Internal Medicine*. 16: 168–175.

Bellizzi, V., De Nicola, L., Minutolo, R., Russo, D., Cianciaruso, B., Andreucci, M., Conte, G. and Andreucci, V. E., 1999. Effects of water hardness on urinary risk factors for kidney stones in patients with idiopathic nephrolithiasis. *Nephron*. 81: 66–70.

Bloom, S. and Peric-Golia, L., 1989. Geographic variation in the incidence of myocardial calcification associated with acute myocardial infarction. *Human Pathology*. 20: 726–731.

Calderon, R. and Hunter, P., 2009. Epidemiological studies and the association of cardiovascular disease risks with water hardness. In: Cotruvo, J. and Bartram, J. (ed.) *Calcium and Magnesium in Drinking-Water: Public Health Significance*. World Health Organization, Geneva, Switzerland, vol. 10, p. 138.

Catling, L. A., Abubakar, I., Lake, I. R., Swift, L. and Hunter, P. R., 2008. A systematic review of analytical observational studies investigating the association between cardiovascular disease and drinking water hardness. *Journal of Water and Health*. 6(4): 433–442.

Cotruvo, J. and Bartram, J., 2009. Expert consensus. In: Cotruvo, J. and Bartram, J. (ed.) *Calcium and Magnesium in Drinking-Water: Public Health Significance*. World Health Organization, Geneva, Switzerland, p. 3.

Couzy, F., Kastenmayer, P., Vigo, M., Clough, J., Munoz-Box, R. and Barclay, D. V., 1995. Calcium bioavailability from a calcium- and sulfate-rich mineral water, compared with milk, in young adult women. *American Journal of Clinical Nutrition*. 62: 1239–1244.

Durlach, J., Bac, P., Durlach, V., Rayssiguier, Y., Bara, M. and Guiet-Bara, A., 1998. Magnesium status and ageing: an update. *Magnesium Research*. 11: 25–42.

Durlach, J., 1988. The importance of magnesium in water. In: Durlach, J. (ed.) *Magnesium in Clinical Practice*. London, John Libbey & Co Ltd, pp. 221–222.

Durlach, J., Bara, M. and Guiet-Bara, A., 1985. Magnesium level in drinking water and cardiovascular risk factor: a hypothesis. *Magnesium*. 4: 5–15.

Eisenberg, M. J., 1992. Magnesium deficiency and sudden death. *American Heart Journal*. 124: 544–549.

Erb, B. D., 1997. Water hardness and cardiovascular death rates in Tennessee. *Tennessee Medicine*. 90: 314–316.

Garzon, P. and Eisenberg, M. J., 1998. Variation in the mineral content of commercially available bottled waters: implications for health and disease. *American Journal of Medicine*. 105: 125–130.

Guillemant, J., Accarie, C., de la Gueronniere, V. and Guillemant, S., 2002. Calcium in mineral water can effectively suppress parathyroid function and bone resorption. *Nutrition Research*. 22: 901–910.

Guillemant, J., Le, H.-T., Accarie, C., Tezenas du Montecel, S., Delabroise, A. M., Arnaud, M. J. and Guillemant, S., 2000. Mineral water as a source of dietary calcium: acute effects on parathyroid function and bone resorption in young men. *American Journal of Clinical Nutrition*. 71: 999–1002.

Gyllerup, S., Lanke, J., Lindholm, L. H. and Schersten, B., 1991. Water hardness does not contribute substantially to the high coronary mortality in cold regions of Sweden. *Journal of Internal Medicine*. 230: 487–492.

Hewitt, D. and Neri, L. C., 1980. Development of the "water story": some recent Canadian studies. In: Calabrese, E. J., Moore, G. S., Tuthill, R. W. and Sieger, T. L. (ed.) *Drinking Water and Cardiovascular Disease*. Pathotox Publishers, Park Forest South, Illinois, USA, pp. 51–63.

Kobayashi, J., 1957. On geographical relations between the chemical nature of river water and death rate from apoplexy. *Berichte des Ohara Institut fur Landwirtschftliche Biologie*. 11: 12.

Kozisek, F. 2003. Health significance of drinking water calcium and magnesium, National Institute of Public Health, 2003. Available at http://midasspring.com/typed%20documents/HealthSignificance.pdf. Accessed on 31 December 2013.

Kumar, M., Singhi, S. and Mahajan, R. K., 2006. Trace level determination of U, Zn, Cd, Pb and Cu in drinking water samples. *Environmental Monitoring and Assessment*. 112: 283–292.

Luoma, H., Aroma, A., Helminen, S., Murtoma, H., Kiviluoto, L., Punsar, S. and Knekt, P., 1983. Risk of myocardial infarction in Finnish men in relation to fluoride, magnesium and calcium concentration in drinking water. *Acta medica Scandinavica*. 213: 171–176.

Meunier, P. J., Jenvrin, C., Munoz, F., de la Gueronniere, V., Garnero, P. and Menz, M., 2005. Consumption of a high calcium mineral water lowers biochemical indices of bone remodeling in postmenopausal women with low calcium intake. *Osteoporosis International*. 16(10): 1203–1209.

Olajire, A. A. and Imeokparia, F. E., 2001. Water quality assessment of Osun River: studies on inorganic nutrients. *Environmental Monitoring and Assessment*. 69(1): 17–28.

Ong, J. N., Grandjean, A. C. and Heaney, R. P., 2009. The mineral composition of water and its contribution to calcium and magnesium intake. In: Cotruvo, J. and Bartram, J. (ed.) *Calcium and Magnesium in Drinking-Water: Public Health Significance*. World Health Organization, Geneva, Switzerland, p. 37.

Prepas, E. E., Pinel-Alloul, B., Chambers, P. A., Murphy, T. P., Reedyk, S., Sandland, G. and Serediak, M., 2001. Lime treatment and its effects on the chemistry and biota of hardwater eutrophic lakes. *Freshwater Biology*. 46: 1049–1060.

Punsar, S. and Karvonen, M. J., 1979. Drinking water quality and sudden death: observations from West and East Finland. *Cardiology*. 64: 24–34.

Recker, R. R., 1985. Calcium absorption and achlorhydria. *New England Journal of Medicine*. 313: 70–73.

Rockett, I., 1994. Population and health: an introduction to epidemiology. *Population Bulletin*. 49(3): 11.

Rosenlund, M., Berglind, N., Hallqvist, J., Bellander, T. and Bluhm, G., 2005. Daily intake of magnesium and calcium from drinking water in relation to myocardial infarction. *Epidemiology*. 16(4): 570–576.

Rubenowitz, E., Axelsson, G. and Rylander, R., 1996. Magnesium in drinking water and death from acute myocardial infarction. *American Journal of Epidemiology*. 143(5): 456–462.

Rubenowitz, E., Axelsson, G. and Rylander, R., 1999. Magnesium and calcium in drinking water and death from acute myocardial infarction in women. *Epidemiology*. 10: 31–36.

Rubenowitz, E., Molin, I., Axelsson, G. and Rylander, R., 2000. Magnesium in drinking water in relation to morbidity and mortality from acute myocardial infarction. *Epidemiology*. 11: 416–421.

Rylander, R., 1996. Environmental magnesium deficiency as a cardiovascular risk factor. *Journal of Cardiovascular Risk*. 3: 4–10.

Rylander, R., Bonevik, H. and Rubenowitz, E., 1991. Magnesium and calcium in drinking water and cardiovascular mortality. *Scandinavian Journal of Work, Environment & Health*. 17: 91–94.

Sauvant, M.-P. and Pepin, D., 2002. Drinking water and cardiovascular disease. *Food and Chemical Toxicology*. 40: 1311–1325.

Schroeder, H. A., 1960. Relation between mortality from cardiovascular disease and treated water supplies. *Journal of the American Medical Association*. 172: 1902–1908.

Sengupta, P., 2013. Potential Health Impacts of Hard Water. *International Journal of Preventive Medicine*. 4(8): 866–875.

Teitge, J. E., 1990. Incidence of myocardial infarct and the mineral content of drinking water. *Zeitschrift für die gesamte innere Medizin und ihre Grenzgebiete*. 45: 478–485.

Weaver, C. M. and Heaney, R. P. (ed.), 2006. *Calcium in Human Health*. Humana Press, Totowa, NJ, p. 129.

Weaver, C. M. and Nieves, J. W., 2009. Calcium and magnesium: role of drinking water in relation to bone metabolism. In: Cotruvo, J. and Bartram, J. (ed.)

Calcium and Magnesium in Drinking-Water: Public Health Significance. World Health Organization, Geneva, 2009, p. 96.

WHO, 1979. Health effects of the removal of substances occurring naturally in drinking water, with special reference to demineralized and desalinated water. Report on a Working Group, Brussels, 20–23 March 1978. Euro Reports and Studies 16. World Health Organization, Copenhagen.

WHO, 2005. *Nutrients in Drinking Water. Water, Sanitation and Health Protection and the Human Environment*. World Health Organization, Geneva, Switzerland, pp. 92–95.

Yang, C. Y., 1998. Calcium and magnesium in drinking water and risk of death from cerebrovascular disease. *Stroke*. 29: 411–414.

Yang, C. Y. and Chiu, H. F., 1999. Calcium and magnesium in drinking water and the risk of death from hypertension. *American Journal of Hypertension*. 12: 894–899.

Yang, C. Y., Chiu, H. F., Chiu, J. F., Cheng, M. F. and Kao, W. Y., 1997. Gastric cancer mortality and drinking water qualities in Taiwan. *Archives of Environmental Contamination and Toxicology*. 33(3): 336–340.

Yang, C. Y., Chiu, J. F., Chiu, H. F., Wang, T. N., Lee, C. H. and Ko, Y. C., 1996. Relationship between water hardness and coronary mortality in Taiwan. *Journal of Toxicology and Environmental Health*. 49: 1–9.

Yang, C. Y. and Hung, C. F., 1998. Colon cancer mortality and total hardness levels in Taiwan's drinking water. *Archives of Environmental Contamination and Toxicology*. 35: 148–151.

Yang, C. Y., Tsai, S. S., Lai, T. C., Hung, C. F. and Chiu, H. F., 1999. Rectal cancer mortality and total hardness levels in Taiwan's drinking water. *Environmental Research*. 80: 311–316.

CHAPTER 22

Calcium in Saliva and Impact on Health

AYSEN YARAT*[a], EBRU EMEKLI ALTURFAN[a], AND SERAP AKYUZ[b]

[a]Department of Biochemistry, Faculty of Dentistry, Marmara University, Guzelbahce, Buyukciftlik Sok, No: 6 Nisantası, Istanbul, Turkey;
[b]Department of Pediatric Dentistry, Faculty of Dentistry, Marmara University, Guzelbahce Buyukciftlik Sok, No: 6 Nisantası, Istanbul, Turkey
*E-mail: ayarat@marmara.edu.tr

22.1 Importance of Calcium in Biological Systems

Soon after the discovery of calcium (Ca) as an element by Davy in 1808, its presence and main role in mammalian bones and other mineralized tissues were recognized. However, it was much later revealed that calcium ions could play an important role in other tissues as well (Forsén and Kordel, 1994; Peacock, 2010). Our organism usually has a total calcium concentration of the order of a few milimolar (mM) in the body fluids (Forsén and Kordel, 1994). The concentration is observed to be within the narrow limits, 2.45 mM in adult serum. In human saliva it is reported to be between 0.2 and 5 mM (Table 22.1) (Forsén and Kordel, 1994; Chiappin *et al.*, 2007; Yarat, 2008). A number of important physiological and biochemical processes are regulated by calcium ions. These are muscle contraction, blood coagulation, secretory processes, membrane transport, enzyme reactions, cell division and growth. The calcium ion has been conclusively regarded as the primary

Food and Nutritional Components in Focus No. 10
Calcium: Chemistry, Analysis, Function and Effects
Edited by Victor R. Preedy

Published by the Royal Society of Chemistry, www.rsc.org

Table 22.1 Plasma and saliva calcium concentrations (mM) (Chiappin *et al.*, 2007; Yarat, 2008). This table shows calcium concentrations in the plasma and in the different types of saliva samples. It provides a comparison of calcium concentrations from different sources.

	Calcium (mM)
Plasma	2.4
Whole unstimulated saliva	0.5–2.8
Whole stimulated saliva	0.2–4.7
Parotid unstimulated saliva	0.5–2.1
Parotid stimulated saliva	0.2–2.7
Submandibular unstimulated saliva	0.5–5
Submandibular stimulated saliva	0.7–3.7
Sublingual stimulated saliva	1.7–2.95

intracellular factor regulating the secretion in exocrine gland cells, including fluid secretion in salivary glands. Furthermore, the proper extracellular fluid, periostal concentrations of calcium and phosphate ions are necessary for bone mineralization. In order to ensure the normal operation of these processes, the plasma calcium ion concentration is maintained within very narrow limits. Deviation of ionized calcium from the normal range may result in many disorders and it may even be life threatening (Granner, 1990; Baker and Worthley, 2002; Theobald, 2005; Ambudkar, 2011; Bhattacharya, 2012).

22.2 Saliva Production, Composition and Salivary Calcium Concentration

Saliva is a variable fluid that originates mainly from major salivary glands (parotid, sublingual and submandibular) and minor salivary glands. The contribution of each gland to the total unstimulated saliva secretion is 65% for submandibular, 23% for parotid, 8% for Von Ebner and 4% for sublingual glands (Chiappin *et al.*, 2007). Some salivary components have a nonglandular origin such as fluids originating from oropharyngeal mucosa (oral cells, bacteria, fungi, virus, gastrointestinal reflux, upper airways secretions) and crevicular fluid. Crevicular fluid can be regarded as a transudate of plasma. Saliva may also contain food debris and blood-derived compounds that are transferred actively or passively like plasma proteins, plasma cells in the case of inflammation or mucosal lesions (Chiappin *et al.*, 2007).

The salivary glands are composed of acini, in which the initial or primary saliva is produced, being isotonic compared to plasma. Intercalated ducts connect the several acini and the secreted saliva is drained to the oral cavity through striated and excretory ducts. During this passage, the levels of some electrolytes alter due to active ionic transport and this gives the oral fluid its hypotonic character. In general during ageing, the number of acini as well as the protein biosynthesis activity decrease, however, this can be very different among individuals. There is a typical secretion of each type of salivary glands.

The parotis gland produces a serous fluid, whereas the submandibular gland produces a seromucous secretion, while the sublingual gland secretes only mucous saliva. Situated on the buccal mucosa of the lips and on the palate, the minor glands, produce a viscous secretion (Aps and Martens, 2005).

In salivary glands, saliva is stored in secretion granules and in the acini. These granules are filled with water, and electrolytes and proteins are dissolved therein. Without stimulation a fluid is produced *via* vesicles and secreted by salivary glands. The saturation of the glands with blood requires energy and has a great importance. There is also a paracellular source of fluid, coming from the interstitium, especially when the saliva is stimulated and the latter obviously depends on the water supply of the body. When less water is available such as the cases of fever or diarrhea a low volume of saliva may lead to dry mouth (Aps and Martens, 2005).

Salivary glands' secretion is normally a watery fluid with electrolytes and a complex mixture of proteins, even though mucous gland secretions are often viscous owing to the discharge of large molecular weight mucins. Depending on the origin of the stimulation (sympathetic or parasympathetic) and the type of salivary gland, the final composition of saliva varies. Initially saliva is isotonic, as it is formed in the acini and it becomes hypotonic later as it passes through the duct network (Melvin *et al.*, 2005).

Healthy adults normally produce 500–1500 mL of saliva per day, at a rate of approximately 0.5 mL min^{-1} but several physiological and pathological conditions can alter saliva production in terms of quantity and quality, *e.g.*, smell and taste stimulation, chewing, psychological and hormonal status, drugs, age, hereditary influences, oral hygiene and physical exercise. Sodium, potassium, chloride and bicarbonate are the most abundant electrolytes in saliva whereby calcium, magnesium and phosphate are present in less concentrations. Only potassium and phosphate concentrations in saliva are higher than plasma concentrations. Saliva and plasma calcium concentrations are positively correlated (Aps and Martens, 2005; Chiappin *et al.*, 2007; Yarat, 2008; Teixeira *et al.*, 2012).

Salivary calcium concentration is usually measured by ion-selective electrodes, atomic absorption spectroscopy and by the traditional spectrophotometric methods similar to other biological fluids (Forsén and Kordel, 1994).

Salivary calcium and other electrolytes are actively transported into the acini and striated ducts from serum (Figure 22.1). However mucous glands can only be stimulated cholinergically that causes ion fluxes of both Na^+/K^+ and Ca^{2+} over the cell membranes. A fluid results that is rich in electrolytes and in high volume. Medications effective in the central nervous or the peripheral nervous systems, as well as their side effects, will alter salivary composition and salivary volume. Moreover, certain systemic diseases such as cystic fibrosis, multiple sclerosis, graft-*versus*-host disease, diabetes mellitus, alcoholic liver cirrhosis, acquired human immunodeficiency syndrome, epilepsy, burning-mouth syndrome and kidney dysfunction may cause similar salivary alterations. The circadian rhythm regulates both the saliva volume and salivary electrolyte concentrations (Aps and Martens, 2005).

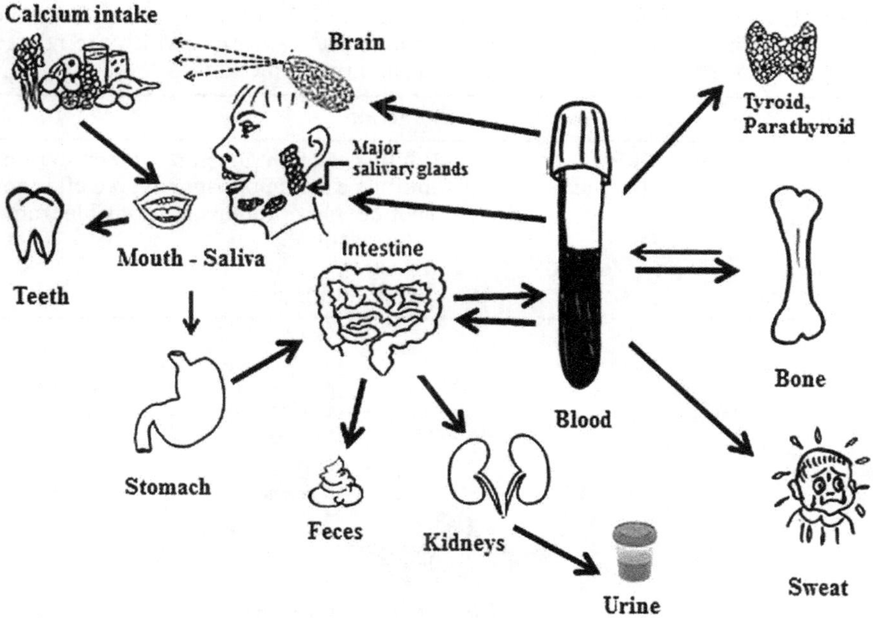

Figure 22.1 Schematic representation of calcium route. This figure shows the calcium route in the organism. The calcium level is regulated by many factors, which may prevent the significant increase in blood calcium level therefore the excretion of elevated calcium in saliva, sweat or urine is also possible.

Although there are some conflicting reports about the relation of salivary flow rate and calcium concentration, the calcium concentration in saliva is mainly dependent on pH and salivary flow rate. Teixeira *et al.* reported no significant correlation between salivary calcium ion concentrations, salivary flow rate and buffering capacity (Teixeira *et al.*, 2012).

Calcium ions are secreted by salivary proteins. Calcium is found in the saliva as free (ionized) calcium or bound (total) calcium depending on the pH of saliva. Since ionized calcium provides a balance between dental enamel and surrounding fluids, it gains a major role in the protection against caries attack. There is a considerable correlation between total and ionized calcium in saliva. Calcium secretion increases parallel to the increase in salivary flow rate (De Almeida *et al.*, 2008). Nonionized calcium can be linked to inorganic ions such as inorganic phosphate, bicarbonate, fluoride, to small organic ions like citrate, and to macromolecules including statherin, histidine-rich peptides, and proline-rich proteins (PRPs) (Table 22.2). A particular aspect of the combination of calcium is its strong link with α-amylase, where it acts as a cofactor required for the enzyme function (Morris, 2011).

At pH 6–7.5 stimulated parotid saliva, 45% of total calcium is in a complexed form. Some calcium ions are complexed with CO_2. When CO_2 escape by air bubbles, pH and calcium concentration also increase, whereas

Table 22.2 Salivary proteins related with salivary calcium and teeth. This table includes the list of the common salivary proteins, which are related with salivary calcium and teeth, and their functions.

Protein	Function
Proline-rich proteins (PRPs)	Inhibits calcium phosphate precipitation
Tyrosine-rich proteins (statherin)	Inhibits calcium phosphate precipitation
Mucins	They are glycoproteins, responsible from lubrication
Amylase	Digestion
Glycoproteins	Pellicle formation

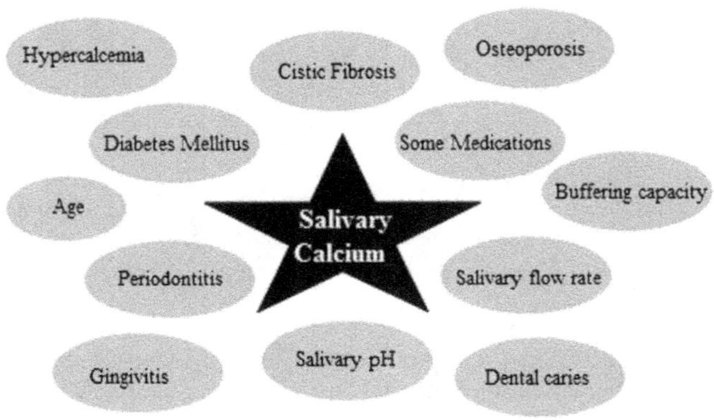

Figure 22.2 Conditions affecting salivary calcium concentration. This figure includes the conditions associated with the changes in salivary calcium concentrations.

phosphate concentration decreases. Calcium concentration is higher in the submandibular glands compared to the parotid glands. However, high calcium concentrations in saliva may be related mostly to the parotid glands (Almstahl and Wikstrom, 2003; Yarat, 2008; Moreira *et al.*, 2009).

Calcium concentration in human parotid saliva has been reported to be increased in patients with hypercalcemia and in some other diseases (Dilena *et al.*, 1983).

Salvolini *et al.* studied 100 healthy subjects from both genders, aged between 10 and 80 years, and suggested that the concentrations of calcium and phosphorus were not related to age (Salvolini *et al.*, 1999; Wu *et al.*, 2008). However, in contrast, several studies reveal that calcium and phosphate concentrations in saliva show a clear increase with increasing age (Sevon *et al.*, 2008) (Figure 22.2).

The characteristics of saliva of normal-weight, overweight and obese children have been determined and the results of the biochemical analysis of saliva showed that total protein, amylase, urea, phosphate, triglycerides and calcium concentrations did not differ among groups. However, urea, phosphate and calcium concentrations differed significantly between unstimulated and stimulated saliva in the normal-weight and overweight groups, they have been reported to be lower in stimulated saliva compared to unstimulated saliva. In the overweight group, total protein levels also differed between saliva samples, whereas the obese group showed no considerable difference in the studied biochemical analysis (De Campos *et al.*, 2013).

Salivary calcium levels have been reported to be different in osteoporotic and osteopenic women compared with the controls. In postmenopausal osteoporosis, calcium concentrations have been shown to be significantly above the cut-off point. Salivary calcium concentrations have been found under the cut-off point in 67.5% of patients with osteoporosis, while 60% of women without osteoporosis had salivary calcium concentration under the cut-off point (Rabiei *et al.*, 2013).

Singh *et al.* showed that salivary calcium concentration was significantly higher in postmenopausal women with oral dryness (OD) compared to those without OD (Singh *et al.*, 2013). The low level of estrogen in postmenopausal has been suggested as the possible mechanism reducing calcium absorption in intestine that leads to decreased serum calcium level resulting in increased serum parathyroid hormone (PTH) level. The increased serum PTH level removes calcium from bones by resorption and the calcium level increases in the serum. However, the calcium level is regulated by many factors, which may prevent the significant increase in calcium level, therefore the excretion of elevated calcium in saliva or urine is also possible (Agha-Hosseini *et al.*, 2009; Singh *et al.*, 2013).

22.3 The Role of Salivary Calcium in Maintenance of Tooth Integrity

Saliva functions can be classified under 5 major categories that serve to maintain oral health and create an appropriate ecologic balance: (1) lubrication and protection, (2) buffering action and clearance, (3) maintenance of tooth integrity, (4) antibacterial activity, (5) taste and digestion (Humphrey and Williamson, 2001).

Similar to all other mineralized tissues, teeth must be resistant to chemical as well as physical traumas. Dissolution is the major chemical threat for teeth and can be prevented by surrounding teeth with a solution that is supersaturated with minerals. Saliva protects the parts above the gum margin (Bited-sized Tutorials, 2014). The saliva composition is influenced by mainly Ca^{2+} and other bivalent cations such as Mg^{2+}, Zn^{2+}, *etc.*, that play an essential role within the human organism (Elena, 2012). Saliva plays a major role in maintaining the physical and chemical integrity of dental enamel by balancing remineralization and demineralization (De Almeida *et al.*, 2008) (Table 22.3).

The active concentrations of salivary free calcium, phosphate, and fluoride and the salivary pH are the main factors controlling the stability of dental enamel (De Almeida *et al.*, 2008). Calcium and phosphate, with an alkaline

Table 22.3 Key facts of salivary calcium in maintenance of tooth integrity. This table includes key facts of salivary calcium and other salivary factors that take place in de/remineralization processes.

1. Salivary calcium has critical importance in the balancing de/remineralization process
2. Dissolution is the major chemical threat for enamel integrity and can be prevented by surrounding teeth with a solution that is supersaturated with minerals
3. The proline-rich proteins (PRPs) and a tyrosine-rich proteins (statherin) inhibiting calcium and phosphate precipitation, help the supersaturation of saliva
4. Under physiological conditions pellicle facilitates mineralization of enamel
5. When salivary pH decreases below the "critical pH" mineral dissolution from enamel begins
6. The concentration of calcium in saliva varies depending on the salivary flow. Diseases like cystic fibrosis and some medications may lead to an increase in salivary calcium levels
7. Excess salivary calcium creates a tendency to develop supra- or subgingival calculus
8. Studies on the oral conditions leading to enamel caries still continue

or neutral environment are necessary for remineralization. In saliva, calcium and phosphate act as common ions in enamel and dentin, leading to a slower dissolution rate of mineral (Buzalaf *et al.*, 2012). Teeth are constantly being replenished with saliva. The calcium and phosphate ion concentrations in saliva are variable. Hydroxyapatite $Ca_{10}(PO_4)_6(OH)_2$ is the mineral component of the enamel, dentin and cementum. Saturated with Ca^{2+} ions and PO_4^{3-} at neutral pH values, hydroxyapatite is in equilibrium with the oral environment. This is due to the presence of the PRPs and a tyrosine-rich proteins (statherin) that are able to inhibit calcium and phosphate precipitation. High concentrations (supersaturated levels) of calcium and phosphate in saliva are maintained with respect to apatite in this way. Salivary proteins produced by acinar cells have been shown to bind to dissolution sites and to cover crystallites by specific absorption (Buzalaf *et al.*, 2012). They are present in the enamel pellicle formed initially and in "mature" pellicles that are a 0.1–1.0 μm thick layer of macromolecular material on the dental mineral surface. Exposed solid surfaces in the oral cavity are covered by a proteinaceous layer referred to as the acquired pellicle that is an organic film free of bacteria covering oral hard and soft tissues. It is composed of glycoproteins and proteins, including several enzymes (Buzalaf *et al.*, 2012). Pellicle covers teeth surface from saliva and prevents fresh calcium phosphate from being continuously laid down (Bited-sized Tutorials, 2014). Pellicle acts as a diffusion barrier, slowing both attacks of bacterial acids and loss of dissolved calcium and phosphate ions. Calcium has been suggested to diffuse in and out of the pellicle easily, allowing free calcium trade-off between the saliva and the absorbed pellicle under physiological conditions, which may potentially facilitate the mineralization of enamel (Ash *et al.*, 2013).

Mineral is not deposited on erupted teeth and calcium phosphate solution requires a nucleating center for hydroxyapatite deposition and the surfaces of teeth are coated in an acquired pellicle of proteinaceous material derived from saliva masking crystals underlying. As the fluid surrounding the tooth becomes increasingly acidic, the supersaturation point is reached and any further decrease in pH leads to mineral dissolution. This is called the "critical pH" which is normally between pH 5.2–5.5 depending on the particular saliva composition of the individual. Caries and tooth erosion are the clinical impacts of tooth mineral dissolution (Bited-sized Tutorials, 2014).

The reverse happens when conditions become more alkaline. The salivary calcium phosphate is destabilized and calculus is caused by precipitation of a calcium phosphate salt. Accordingly salivary calcium and phosphorus have been implicated in the deposition of calculus over teeth. A high level of salivary calcium is needed for the resistance to dental decay. Dental calculus represents mineralized bacterial plaque (Bited-sized Tutorials, 2014). Despite inhibitory action of statherin and PRPs calculus is formed in plaque in saliva and it may be due to failure in diffusing into calcifying plaque. Proteolytic enzymes of oral bacteria or lysed leukocytes may suppress inhibitory proteins. Bacteria in plaque may produce their own inhibitors. There are four different crystals of calcium phosphate: brushite, octacalcium phosphate, hydroxyapatite and whitlockite that are present both in the recent and old calculus. When relatively low plaque pH is present and concomitant high Ca/P ratio in saliva, brushite is formed that may later develop into hydroxyapatite and whitlockite. When supragingival plaque mineralizes, octacalcium phosphate is formed and is gradually changed into hydroxyapatite. In the presence of alkaline pH and anaerobic condition and concomitant presence of magnesium or zinc and carbonate, a large amount of whitlockite is formed and this is the stable form of mineralization. In patients who have a tendency to develop supra- or subgingival calculus excess calcium concentration was observed. Therefore, recommending the use of alkaline mouthwashes is important in those individuals for the maintenance of good oral hygiene (Hassan and Al-Sandook, 2005; Duckworth and Huntington, 2006).

The high concentrations of calcium and phosphate in saliva provide ionic exchanges as directed towards the tooth surfaces and this process begins with tooth eruption until posteruptive maturation. Remineralization of a demineralized enamel before cavitation is possible, mostly due to the availability of calcium and phosphate ions in saliva. The concentration of calcium in saliva varies depending on the salivary flow and is not affected by diet. However, diseases like cystic fibrosis and some medications such as pilocarpine may lead to an increase in calcium levels (De Almeida *et al.*, 2008).

When the pH is neutral and there are enough ions of Ca^{2+} and PO_4^{3-} in the oral flora the process of demineralization is reversible. The remineralization mechanism includes precipitation of calcium and phosphate ions in the saliva in a form of insoluble calcium phosphate that is incorporated into demineralized enamel (Elena, 2012) (Figure 22.3).

One of the considerable key salivary parameter in terms of remineralization is the extent of variations in calcium concentration between resting saliva with a low calcium concentration and stimulated saliva with higher

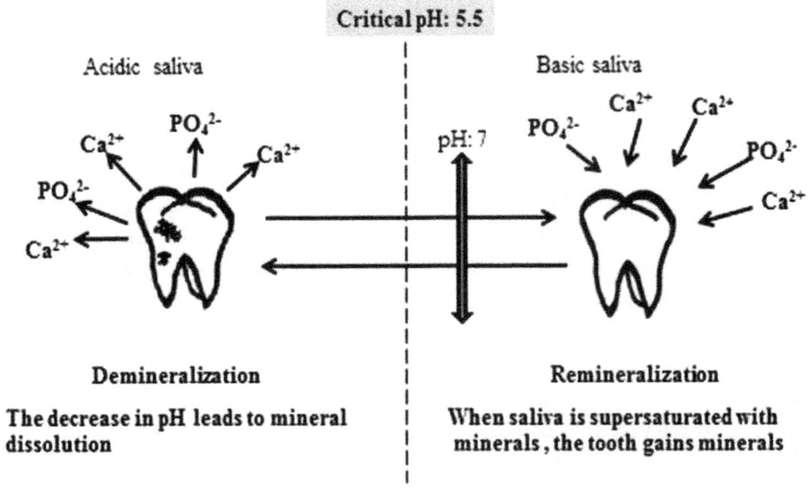

Figure 22.3 This figure explains main conditions of tooth de/remineralization.

calcium concentration. In resting saliva phosphate levels do not vary markedly, however, large fluctuations in calcium concentrations occur in an individual. Since remineralization would not occur when the degree of saliva saturation regarding tooth mineral is low, variations in calcium concentration have important implications for the critical pH and for the possibility of remineralization. In other words, low levels of bioavailable calcium and phosphate ions together with minimal amounts of fluoride (<1 ppm) would enhance remineralization. In contrast, remineralization is not chemically possible under low calcium concentrations (Walsh, 2009).

Buffering capacity, pH, salivary concentrations of calcium and phosphate vary significantly in individuals related with time. These changes affect the possibility of mineral loss and gain directly, in terms of both dental erosion and dental caries. There are closely related phosphoproteins in saliva, enamel, dentine, bone, cementum, and milk that bind and stabilize calcium and phosphate, harmonizing the behavior of these ions in a pH-dependent way. Actually statherins in saliva, casein phosphoproteins in products, and phosphoproteins in tooth structure are remarkably similar. After the demineralization of the hard tissues the remaining phosphoproteins influence the ability of these tissues to remineralize (Walsh, 2009).

Teixeira *et al.* evaluated whether orthodontic fixed appliances induced qualitative and quantitative changes in the saliva of individuals and accordingly flow rate, pH, buffering capacity, amylase activity, total proteins, calcium and glucose of saliva were determined. They found that the saliva of individuals with fixed orthodontic appliances had lower pH, buffering capacity and calcium concentration compared to individuals without any type of orthodontic appliance. They concluded that these oral changes were enough to cause tooth demineralization and patients with orthodontic appliances should consider additional oral hygiene measures (Teixeira *et al.*, 2012).

Remineralization has been suggested to occur at low Ca/Pi molar ratio and high Pi levels within the simulated saliva formulations in a recent *in vitro* study (Hara *et al.*, 2008). Koc-Ozturk *et al.* have found that the salivary free Ca levels in a pregnant group were significantly higher and Pi levels were lower in comparison to a nonpregnant group. They suggested that higher Ca/Pi molar ratio, lower pH and buffering capacity may indicate demineralization in the pregnant group and nonsignificant increase in DMFT (decayed missing filled permanent teeth) values in lactation group compared to pregnancy may also support demineralization. In the same study the possible increase in salivary free calcium level with caries, may be overshadowed due to the lactation physiology (Koc-Ozturk *et al.*, 2013). In another study, it has been reported that salivary free calcium level in the lactation was decreased in lactation period as a consequence of production of maternal milk (Al-Nuaimy and Al-Sandook, 2006).

22.4 Salivary Calcium and Periondontitis

Periodontitis is a chronic inflammatory disease characterized by tissue destruction. The changes in the chemical composition of saliva could be utilized to observe gingivo-periodontal alterations (Koss *et al.*, 2009). Due to its affinity for being readily taken up by plaque, it seems that salivary calcium is an important factor for the onset of periodontitis. There is a close association between high salivary calcium levels and gingivitis, implying an increased risk of periodontitis as high salivary calcium was associated with better dental health compared to subjects having low salivary calcium levels (Sewón *et al.*, 1998). The proposed mechanism could be that more teeth may remain intact and/or become remineralized after initial demineralization in the presence of high salivary mineral content. Salivary calcium increases the level of calcium in plaque making it readily available for remineralization (Sewón *et al.*, 1998).

Acharya *et al.* investigated the association between calcium levels in saliva with periodontal status in a case-control study. In this study, periodontal disease was found to be associated with higher salivary calcium levels, which indicates that the calcium level in saliva could possibly be a risk factor for development of periodontal diseases (Acharya *et al.*, 2011).

Kiss *et al.* investigated the calcium concentrations of saliva between smoker and nonsmoker patients with or without periodontitis. Their findings indicated that patients with periodontitis who smoke exhibit higher salivary calcium levels than nonsmokers. However, they suggested that large-scale controlled studies are necessary (Kiss *et al.*, 2010).

Sutej *et al.* evaluated the effects of tobacco smoking and salivary calcium on the periodontal status of young adults. They suggested an association between an increased level of salivary calcium and development of periodontal disease in moderate cigarette-smoking young healthy subjects but no significant impact on salivary flow or calcium concentrations in unstimulated saliva have been reported (Sutej *et al.*, 2012).

Fiyaz *et al.* conducted a study to evaluate inorganic salivary calcium, phosphate, flow rate and pH of unstimulated saliva and oral hygiene of healthy subjects and patients with periodontitis and patients with dental caries in order to correlate salivary calcium levels with the number of intact teeth. Results showed significantly increased salivary inorganic calcium and phosphate levels, poor oral hygiene status, pH and salivary flow rate in patients with periodontitis when compared with dental caries group and controls. Individuals with higher salivary inorganic calcium, phosphate, pH, flow rate and poor oral hygiene could be at a higher risk for developing periodontitis and may have less dental caries and more intact teeth (Fiyaz *et al.*, 2013).

22.5 Salivary Calcium and Dental Caries

Salivary buffering capacity, pH and calcium concentration are among the risk factors related with dental caries.

Flow rate, Ca, P, and Mg concentrations were measured in resting parotid saliva samples collected from 207 children (mean age 8.96 years) having dental caries. DMFS (decayed missing filled permanent tooth surface) values were one-third higher in boys than in girls, and this increase was along with significantly lowered Ca and P concentrations in saliva. Both P and Ca levels decreased in the high DMFS group and the 9% decrease in Ca level was not found significant in subjects divided into two DMFS groups (Shannon and Feller, 1979).

In healthy children salivary electrolytes were evaluated according to their dentition, sex, age and caries incidence. Dentition groups showed no significant differences in the levels of any of the parameters between the children with and without caries (Kargül *et al.*, 1998).

Black tooth stain in children has been correlated commonly with a low caries experience. Garan *et al.* compared salivary factors and caries indices in children with and without black tooth stain. They found significantly higher salivary buffering capacity and calcium levels, and lower flow rate in children with black tooth stain compared with children without black tooth stain. The authors suggested that the low caries tendency observed in children with black tooth stain may be associated with high salivary calcium and buffering capacity (Garan *et al.*, 2012).

Santosh Kumar *et al.* assessed the effects of acute alcohol consumption on tooth surfaces dissolution and evaluated the difference in the dissolution potential of whisky, beer and wine. They reported decreased salivary pH levels in subjects of all three groups. There was a rise in salivary inorganic phosphate concentration in both whisky and wine groups, while only whisky was able to dissolve calcium from the tooth surfaces (Santosh Kumar *et al.*, 2013).

Laine *et al.* evaluated the effect of diet on the paraffin-stimulated saliva, and on dental caries and reported increased paraffin-stimulated salivary flow rate in children with their current long-term dietary intervention. Salivary calcium concentration was correlated with dental health directly. Higher salivary flow rate has been suggested to be due to higher fiber intake (Laine *et al.*, 2014).

Bardow *et al.* compared the major physicochemical characteristics of stimulated whole saliva in 85 subjects divided into three groups, each with normal saliva flow rates and at least 24 remaining teeth. On a group level, subjects having little dental caries were likely to have a more favorable physicochemical saliva composition in terms of higher calcium, phosphate, bicarbonate, pH, degree of saturation with respect to hydroxyapatite and a lower critical pH. They concluded that stimulated whole saliva, including major physicochemical characteristics, will probably have little predictive value for prospective dental caries (Bardow *et al.*, 2013).

Areias *et al.* examined the association between dental caries and saliva components including phosphate, calcium, potassium, chloride as well as α-amylase in children with Down syndrome. In this study no correlation was found between dental caries and salivary ionic composition as well as with α-amylase secretion rate (Areias *et al.*, 2013).

Ahmadi-Motamayel *et al.* found increased salivary calcium concentration in group with caries than those caries free, but the difference was not statistically significant. Saliva has a crucial role in the prevention of dental caries. Therefore, physical and chemical alterations in the composition of saliva play an important role on the development and progression of caries (Ahmadi-Motamayel *et al.*, 2013).

22.6 Salivary Calcium and Diabetes

Although there is evidence showing that diabetic patients have saliva secretion and composition different from nondiabetic subjects (Alemzadeh and Wyatt, 2003), there are contradictory findings in the literature. Mata *et al.* demonstrated higher salivary calcium concentrations in diabetic individuals compared to healthy subjects (Mata *et al.*, 2004). Similar results were reported in children with type-1 diabetes mellitus (Moreira *et al.*, 2009). On the contrary, Lopez *et al.* observed lower salivary calcium concentrations in diabetic patients (Lopez *et al.*, 2003). Reuterving *et al.* and Edblad *et al.* did not find any difference between healthy and diabetic children regarding calcium concentrations in saliva (Reuterving *et al.*, 1987; Edblad *et al.*, 2001). Moreover, no significant difference in calcium electrolyte concentrations between healthy, controlled diabetic patients and uncontrolled diabetic patients have further been reported (Reznick *et al.*, 2006).

The low blood calcium concentrations in diabetic patients compared to healthy subjects could be related to renal problems such as impaired absorption and altered calcium excretion into urine. This situation may lead to the alterations in salivary calcium levels. Therefore, elevated urine and salivary calcium concentrations are expected in patients with endocrine complications that lead to trace-element disturbances (Abou-Seif and Youssef, 2004; Moreira *et al.*, 2009).

Lasisi and Fasanmade investigated salivary flow and saliva composition in patients with type-2 diabetes mellitus and compared with healthy sex- and age-matched controls. They reported that the levels of total protein, Na^+, Ca^{2+}, Cl^- and HCO_3^- did not change between the two groups. These results suggest that some oral diseases associated with diabetes mellitus may be due to the altered salivary flow rate and glucose and potassium levels (Lasisi and Fasanmade, 2012).

22.7 Salivary Calcium as a Diagnostic Tool

The qualitative and quantitative evaluation of saliva serves as an important tool for diagnosis of various disorders and for monitoring the progression of certain pathologies even during the normal physiological functioning of the body (Kaufman and Lamster, 2002; Teixeira *et al.*, 2012; Rabiei *et al.*, 2013; Singh *et al.*, 2013).

Plasma makes an important contribution to the electrolyte concentrations in whole saliva and therefore variable concentrations of blood and serum products can be found in whole saliva (De Campos *et al.*, 2013).

Gargate *et al.* concluded that salivary calcium levels can be used as a diagnostic marker for osteoporosis in postmenopausal women (Gargate, 2011) (Figure 22.4).

Rabiei *et al.* conducted a study to evaluate whether measurements of salivary calcium level may be a useful screening tool for osteoporosis in postmenopausal women and concluded that salivary calcium concentration differs between women with and without osteoporosis and that saliva may

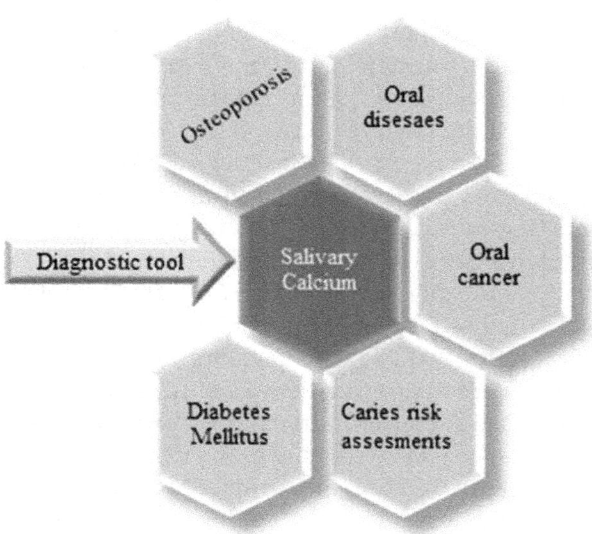

Figure 22.4 Diagnostic importance of salivary calcium. This figure includes some pathologies in which salivary calcium may be used as a diagnostic tool.

constitute a useful tool for screening osteoporosis. Salivary calcium determination may be at least as efficient as a prescreening method for targeting cases where DXA (dual-energy X-ray) for testing high-risk postmenopausal women could be useful (Rabiei *et al.*, 2013).

In addition, salivary calcium levels have been suggested as a possible indicator for oral dryness observed in postmenopausal women (Singh *et al.*, 2013).

In order to identify possible markers that might aid the diagnosis of the illness, Dziewulska *et al.* analyzed the mineral content of saliva in patients with oral cancer. They reported that salivary mineral content in patients with oral squamous cell carcinoma is indicative for oral dehydration; however, no evidence of any salivary mineral markers that would be useful for the diagnosis of oral cancer was found (Dziewulska *et al.*, 2013).

On the other hand, increased calcium concentration in saliva has been introduced as a valuable indicator for the oral health in diabetic patients (Moreira *et al.*, 2009).

Kaur *et al.* evaluated nonmicrobial salivary caries activity parameters and salivary biochemical indicators in predicting dental caries and they suggested that salivary biochemical indicators like calcium, phosphorus and alkaline phosphatase determine caries susceptibility of an individual. Accordingly, these salivary parameters can be used as a diagnostic tool for caries risk assessment (Kaur *et al.*, 2012).

22.8 Conclusion

Despite the difficulty of salivary research, both dental and medical researchers have been investigating saliva for many years. Salivary calcium has great importance with regards to dental and oral health. However, future studies in large samples are needed to fully understand salivary calcium and its impact on health.

Summary Points

- Calcium as an element was first discovered by Davy in 1808. It plays a key role in many important biological processes in our body. Calcium is also one of the inorganic components of saliva. This chapter focuses on calcium in saliva.
- Calcium found in mM concentrations in saliva is originated from blood. It is present as free (ionized) or bound forms in saliva.
- Salivary calcium concentration is mainly dependent on pH and salivary flow and can change due to various factors such as periodontitis, dental caries, pregnancy and diabetes mellitus.
- Salivary calcium provides a balance between dental hard tissues and surrounding fluids. It helps to maintain enamel integrity.
- Salivary calcium may have diagnostic importance.

Key Facts of Salivary Calcium in Maintenance of Tooth Integrity

1. Salivary calcium has critical importance in balancing de/remineralization process.
2. Dissolution is the major chemical threat for enamel integrity and can be prevented by surrounding teeth with a solution that is supersaturated with minerals.
3. The proline-rich proteins (PRPs) and a tyrosine-rich proteins (statherin) inhibiting calcium and phosphate precipitation, help the supersaturation of saliva.
4. Under physiological conditions pellicle facilitates mineralization of enamel.
5. When salivary pH decreases lower than the "critical pH" mineral dissolution from enamel begins.
6. The concentration of calcium in saliva varies depending on the salivary flow. Diseases like cystic fibrosis and some medications may lead to an increase in salivary calcium levels.
7. Excess salivary calcium creates a tendency to develop supra- or subgingival calculus.
8. Researches on the oral conditions leading to enamel caries still continue.

Definitions of Words and Terms

Biofilm. A complex structure of micro-organisms, formed on the teeth following routine oral hygiene procedures.

Biomarker. A substance that is used to determine and evaluate as an indicator of normal biologic or pathologic processes.

Calcium. This is an essential inorganic element and ionizable constituent of a living cell, blood, or other organic matter.

Critical pH. This signifies the pH level (5.5) from which the enamel crystalline structure begins to dissolve.

Dental caries. Multifactorial disease characterized with the dissolution of enamel, dentin and cementum caused by the micro-organisms (*Streptococcus mutans* and other acid-producing bacteria) in the biofilm.

Dental pellicle. This is a protein film that forms on the surface enamel by selective binding of glycoproteins from saliva preventing deposition of salivary calcium phosphate. It forms in seconds after tooth brushing. It also protects teeth from the acids produced by oral micro-organisms after carbohydrate consumption.

Demineralization. The dissolution of enamel crystalline structure caused by the decrease of pH surrounding enamel.

Diabetes mellitus. This is a condition caused by a deficiency of the pancreatic hormone insulin, which results in a failure to metabolize sugar and starch.

Electrolytes. These are substances that ionize when dissolved in suitable solvents such as water.

Enamel. Outermost calcified structure of teeth visible in the mouth.
Hydroxylapatite. The basic component of structure of tooth enamel and bone in the human body, consisting of calcium, phosphate and hydroxyl ions.
Parotid saliva. This is a serous fluid produced by the parotis salivary gland.
Periodontitis. This is the inflammation of the tissue around the teeth, often causing shrinkage of the gums and loosening of the teeth.
Remineralization. Precipitation of calcium and phosphate ions in non-acidic conditions into the enamel.
Saliva. A physically and chemically variable fluid and it originates mainly from three pairs of major salivary glands (parotid, sublingual and sub-mandibular) as well as from a large number of minor salivary glands.
Salivary flow rate. This is the volume of saliva collected per minute and expressed in mL min^{-1}.
Salivary glands. They are exocrine glands that secrete saliva.
Sublingual saliva. This is mucous saliva secreted by sublingual salivary glands.
Submandibular saliva. This is a seromucous secretion produced by sub-mandibular salivary gland.
Stimulated saliva. Saliva that is collected by masticatory action or by gustatory stimulation.
Unstimulated saliva. Saliva that is collected without exogenous gustatory, masticatory, or mechanical stimulation.
Whole saliva. Whole saliva (mixed saliva) is a mixture of oral fluids and includes secretions from both the major and minor salivary glands, in addition to several constituents of nonsalivary origin.

List of Abbreviations

mM	millimolar
Ca	Calcium
OD	Oral dryness
DMFS	Decayed missing filled permanent tooth surface
DMFT	Decayed missing filled permanent teeth
PRPs	Proline-rich proteins
DXA	Dual-energy X-ray

References

Abou-Seif, M. A. and Youssef, A. A., 2004. Evaluation of some biochemical changes in diabetic patients. *Clinica Chimica Acta*. 346: 161–170.

Acharya, A., Kharadi, M. D., Dhavale, R., Deshmukh, V. L. and Sontakke, A. N., 2011. High salivary calcium level associated with periodontal disease in Indian subjects – a pilot study. *Oral Health & Preventive Dentistry*. 9(2): 195–200.

Agha-Hosseini, F., Mirzali-Dizgah, I., Mansourian, A. and Zabihi-Akhtechi, G., 2009. Serum and stimulated whole saliva parathyroid hormone in menopausal women with oral dry feeling. *Oral Surgery, Oral Medicine, Oral Pathology, Oral Radiology, and Endodontics*. 107: 806–810.

Ahmadi-Motamayel, F., Goodarzi, M. T., Hendi, S. S., Abdolsamadi, H. and Rafieian, N., 2013. Evaluation of salivary flow rate, pH, buffering capacity, calcium and total protein levels in caries free and caries active adolescence. *Journal of Dental Oral Hygiene*. 5(4): 35–39.

Alemzadeh, R. and Wyatt, D. T., 2003. Diabetes mellitus. In: Kliegman, R. M., Behrman, R. E., Jensen, H. B. and Stanton, B. F. (ed.) *Nelson Textbook of Pediatrics*. W. B. Saunders, Philadelphia. 17th edn, ch. 583, pp. 1947–1972.

Almstahl, A. and Wikstrom, M., 2003. Electrolytes in stimulated whole saliva in individuals with hyposalivation of different origins. *Archives of Oral Biology*. 48: 337–344.

Al-Nuaimy, K. M. T. and Al-Sandook, T. A., 2006. Salivary calcium level during lactation. *Al-Rafidain Dental Journal*. 6: 12–14.

Ambudkar, I. S., 2011. Dissection of calcium signaling events in exocrine secretion. *Neurochemistry Research*. 36: 1212–1221.

Aps, J. K. M. and Martens, L. C., 2005. Review: the physiology of saliva and transfer of drugs into saliva. *Forensic Science International*. 150: 119–131.

Areias, C., Sampaio-Maia, B., Macho, V., Leal, I., Melo, P. and De Andrade, C., 2013. Does the chemistry in the saliva of down syndrome children explain their low caries prevalence? *European Journal of Paediatric Dentistry*. 14(1): 23–26.

Ash, A., Ridout, M. J., Parker, R., Mackie, A. R., Burnett, G. R. and Wilde, P. J., 2013. Effect of calcium ions on in vitro pellicle formation from parotid and whole saliva. *Colloids and Surfaces, B: Biointerfaces*. 102: 546–553.

Baker, S. B. and Worthley, L. I. G., 2002. The essentials of calcium, magnesium and phosphate metabolism: part I. Physiology. *Critical Care and Resuscitation*. 4: 301–306.

Bardow, A., Lykkeaa, J., Qvis, V., Ekstrand, K., Twetman, S. and Fiehn, N. E., 2013. Saliva composition in three selected groups with normal stimulated salivary flow rates, but yet major differences in caries experience and dental erosion. *Acta Odontology Scandinavian*. 1–8. DOI: 10.3109/00016357.2013.860621. Published online. Ahead of print.

Bhattacharya, S., 2012. Contribution of purinergic receptors to calcium signaling in salivary gland. Ph.D. Thesis. The University of Toledo, ProQuest, UMI Dissertations Publishing, 3549000.

Bite-sized tutorials: saliva and tooth dissolution. University of Newcastle, School of Dental Sciences. http://ncl.ac.uk/dental/oralbiol/oralenv/tutorials/calciumphosphate.html, 2014.

Buzalaf, M. A. R., Hannas, A. R. and Kato, M. T., 2012. Saliva and dental erosion. *Journal of Applied Oral Science*. 20(5): 493–502.

Chiappin, S., Antonelli, G., Gatti, R. and De Palo, E. F., 2007. Saliva specimen: a new laboratory tool for diagnostic and basic investigation. *Clinica Chimica Acta*. 383: 30–40.

De Almeida, P. D. V., Grégio, A. M. T., Machado, M. A. N., de Lima, A. A. S. and Azevedo, L. R., 2008. Saliva composition and functions: a comprehensive review. *Journal of Contemporary Dental Practice*. 9(3): 72–80.

De Campos, M. M., Kobayashi, F. Y., Barbosa, T. D., Costa, S. D., Lucas, B. D. and Castelo, P. M., 2013. Characteristics of salivary secretion in normal-weight, overweight and obese children: a preliminary study: salivary composition and excessive fat tissue. *Odontology*. 1–7. DOI: 10.1007/s10266-013-0103-8. Published online. Epub ahead of print.

Dilena, B. A., Walmsley, R. N. and Fraser, C. G., 1983. Total and ionized calcium in plasma and saliva. *Clinical Chemistry*. 29(10): 1856–1857.

Duckworth, R. M. and Huntington, E., 2006. On the relationship between calculus and caries. In: Duckworth, R. M. (ed.) *The Teeth and Their Environment. Monograph in Oral Sciences*. Karger, Basel, vol. 19, pp. 1–28.

Dziewulska, A., Janiszewska-Olszowska, J., Bachanek, T. and Grocholewicz, K., 2013. Salivary mineral composition in patients with oral cancer. *Magnesium Research*. 26(3): 120–124.

Edblad, E., Lundin, S. A., Sjodin, B. and Aman, J., 2001. Caries and salivary status in young adults with type 1 diabetes. *Swedish Dental Journal*. 25(2): 53–60.

Elena, P. O., 2012. Correlations of the salivary calcium with the etiopathogeny and dental caries therapy in children. Lucian Blaga University of Sibiu, Victor Papilian Faculty of Medicine, Ph.D. Thesis, Sibiu. (Scientific Advisor: Prof. Ph.D. ML. Neamtu).

Fiyaz, M., Ramesh, A., Ramalingam, K., Thomas, B., Shetty, S. and Prakash, P., 2013. Association of salivary calcium, phosphate, pH and flow rate on oral health: a study on 90 subjects. *Journal of Indian Society of Periodontology*. 17(4): 454–460.

Forsén, S. and Kordel, J., 1994. Calcium in biological systems. In: Bertini, I., Gray, H. B., Lippard, S. and Valentine, J. (ed.) *Bioinorganic Chemistry*. University Science Books Mill Valley, California, pp. 106–166.

Garan, A., Akyüz, S., Ozturk, L. K. and Yarat, A., 2012. Salivary parameters and caries indices in children with black tooth stains. *Journal of Clinical Pediatric Dentistry*. 36(3): 285–288.

Gargate, S. B., 2011. A comparative study of salivary calcium level with that of serum estrogen and bone mineral density level to establish the role of salivary calcium as a diagnostic marker for osteoporosis in non-pregnant, Pregnant and Post-Menopausal Women. KLE University, Belgaum, Karnataka. http://hdl.handle.net/123456789/1336.

Granner, D. K., 1990. Hormones that regulate calcium metabolism. In: Murray, R. K., Granner, D. K., Mayes, P. A. and Rodwell, V. W. (ed.) *Harper's Biochemistry*. Appleton and Lange Publishing, USA. 22nd edn, p. 492.

Hara, A. T., Karlinsey, R. L. and Zero, D. T., 2008. Dentine remineralization by simulated saliva formulations with different Ca and Pi contents. *Caries Research*. 42: 51–56.

Hassan, S. A. and Al-Sandook, T. A., 2005. Salivary calcium concentration in patients with high incidence of calculus formation. *Al-Rafidain Dental Journal*. 5(1): 88–90.

Humphrey, S. P. and Williamson, R. T., 2001. A review of saliva: normal composition, flow, and function. *Journal of Prosthetic Dentistry*. 85: 162–169.

Kargül, B., Yarat, A. and Tanboğa, I., 1998. A study of some salivary elements with respect to sex, age and caries in children. *Saudi Dental Journal*. 10(3): 100–106.

Kaufman, E. and Lamster, I. B., 2002. The diagnostic applications of saliva – a review. *Critical Reviews in Oral Biology & Medicine*. 13(2): 197–212.

Kaur, A., Kwatra, K. S. and Kamboj, P., 2012. Evaluation of non-microbial salivary caries activity parameters and salivary biochemical indicators in predicting dental caries. *Journal of Indian Society of Pedodontics and Preventive Dentistry*. 30(3): 212–217.

Kiss, E., Sewon, L., Gorzo, I. and Nagy, K., 2010. Salivary calcium concentration in relation to periodontal health of female tobacco smokers: a pilot study. *Quintessence International*. 41(9): 779–785.

Koc-Ozturk, L., Akyuz, S., Garan, A. and Yarat, A., 2013. Salivary and dental-oral hygiene parameters in 3rd trimester of pregnancy and early lactation: the effect of education. *Marmara Dental Journal*. 1: 1–8.

Koss, M. A., Castro, C. E., Salúm, K. M. and López, M. E., 2009. Changes in saliva protein composition in patients with periodontal disease. *Acta Odontologica Latinoamericana*. 22(2): 105–112.

Laine, M. A., Tolvanen, M., Pienihäkkinen, K., Söderling, E., Niinikoski, H., Simell, O. and Karjalainen, S., 2014. The effect of dietary intervention on paraffin-stimulated saliva and dental health of children participating in a randomized controlled trial. *Archives of Oral Biology*. 59(2): 217–225.

Lasisi, T. J. and Fasanmade, A. A., 2012. Salivary flow and composition in diabetic and non-diabetic subjects. *Nigerian Journal of Physiological Sciences*. 27(1): 79–82.

Lopez, M. E., Colloca, M. E., Paez, R. G., Schallmach, J. N., Koss, M. A. and Chervonagura, A., 2003. Salivary characteristics of diabetic children. *Brazilian Dental Journal*. 14: 26–31.

Mata, A. D., Marques, D., Rocha, S., Francisco, H., Santos, C., Mesquita, M. F. and Singh, J., 2004. Effects of diabetes mellitus on salivary secretion and its composition in the human. *Molecular and Cellular Biochemistry*. 261: 137–142.

Melvin, J. E., Yule, D., Shuttleworth, T. and Begenisich, T., 2005. Regulation of fluid and electrolytes secretion in salivary gland acinar cells. *Annual Review of Physiology*. 67: 445–469.

Moreira, A. R., Passos, I. A., Sampaio, F. C., Soares, M. S. M. and Oliveira, R. J., 2009. Flow rate, pH and calcium concentration of saliva of children and adolescents with type 1 diabetes mellitus. *Brazillian Journal of Medical and Biological Research*. 42(8): 707–711.

Morris, C., 2011. Impact of calcium on salivary α-amylase activity, starch paste apparent viscosity and thickness perception. *Chemosensory Perception*. 3: 112–116.

Peacock, M., 2010. Calcium metabolism in health and disease. *Clinical Journal of the American Society of Nephrology*. 5: S23–S30.

Rabiei, M., Masooleh, I. S., Leyli, E. K. and Nikoukar, L. R., 2013. Salivary calcium concentration as a screening tool for postmenopausal osteoporosis. *International Journal of Rheumatic Diseases*. 16(2): 198–202.

Reuterving, C. O., Reuterving, G., Hagg, E. and Ericson, T., 1987. Salivary flow rate and salivary glucose concentration in patients with diabetes mellitus influence of severity of diabetes. *Diabetes & Metabolism*. 13: 457–462.

Reznick, A. Z., Shehadeh, N., Shafir, Y. and Nagler, R. M., 2006. Free radicals related effects and antioxidants in saliva and serum of adolescents with type 1 diabetes mellitus. *Archives of Oral Biology*. 51: 640–648.

Salvolini, E., Mazzanti, L., Martarelli, D., Di Giorgio, R., Fratto, G. and Curatola, G., 1999. Changes in the composition of human unstimulated whole saliva with age. *Aging*. 11: 119–122.

Santosh Kumar, T., Jyothi, T., Harish, T., Prabu, D. and Suhas, K., 2013. A pilot study into the effect of whisky, wine and beer consumption on tooth surface dissolution. *Journal of Oral Health and Dental Management*. 12(3): 151–154.

Sevon, I., Laine, M. A., Karjalainen, S., Doroguinskia, A., Helenius, H., Kiss, E. and Lehtonen-Veromaa, M., 2008. Effect of age on flow rate, protein and electrolyte composition of stimulated whole saliva in healthy, non-smoking women. *Open Dental Journal*. 2: 89–92.

Sewón, L. A., Karjalainen, S. M., Söderling, E., Lapinleimu, H. and Simell, O., 1998. Associations between salivary calcium and oral health. *Journal of Clinical Periodontology*. 25: 915–919.

Shannon, I. L. and Feller, R. P., 1979. Parotid saliva flow rate, calcium, phosphorous and magnesium concentrations in relations to dental caries experience in children. *Pediatric Dentistry*. 1(1): 16–20.

Singh, B., Sheikh, S., Pallagatti, S., Kaur, K. and Sohi, R., 2013. Evaluation of salivary calcium and salivary parathyroid levels in postmenopausal women with and without oral dryness. *Contemporary Clinical Dentistry*. 4: 488–492.

Sutej, I., Peros, K., Benutic, A., Capak, K., Basic, K. and Rosin-Grget, K., 2012. Salivary calcium concentration and periodontal health of young adults in relation to tobacco smoking. *Oral Health & Preventive Dentistry*. 10(4): 397–403.

Teixeira, H. S., Kaulfuss, S. M. O., Ribeiro, J. S., Pereira, B. R., Brancher, J. A. and Camargo, E. S., 2012. *Calcium, Amylase, Glucose, Total Protein Concentrations, Flow Rate, pH and Buffering Capacity of Saliva in Patients Undergoing Orthodontic Treatment*.

Theobald, H. E., 2005. Dietary calcium and health. *British Nutrition Foundation Nutrition Bulletin*. 30: 237–277.

Yarat, A., 2008. Overview of saliva composition and types. In: Emekli, N. and Yarat, A. (ed.) *Saliva Histology, Physiology, Microbiology and Biochemistry*. Nobel Tip Kitabevleri Ltd. Şti., İstanbul, pp. 211–225.

Walsh, L. J., 2009. Contemporary technologies for remineralization therapies: a review. *International Dentistry SA*. 11(6): 6–16.

Wu, K. P., Ke, J. Y., Chung, C. Y., Chen, C. L., Hwang, T. L., Chou, M. Y., Wong, A. M. K., Hu, C. F. and Lee, Y. C., 2008. Relationship between unstimulated salivary flow rate and saliva composition of healthy children in Taiwan. *Chang Gung Medical Journal*. 31(3): 281–286.

CHAPTER 23

Intestinal Absorption of Calcium

BRIAN R. STEPHENS*[a] AND JAMES S. JOLLIFF[b]

[a]Louisiana State University – Eunice, Division of Math and Science, P.O. Box 1129, Eunice, LA 70535, USA; [b]Akey Inc., 10 Nutrition Way, Brookville, OH 45309, USA
*E-mail: bstephen@lsue.edu

23.1 Introduction

Calcium (Ca^{2+}) is a divalent alkaline metal and ranks as the fifth most abundant element in the human body. The diverse functions of calcium include its role as a cofactor for enzymes such as calmodulin and troponin, the formation of short- and long-term memories in the hippocampus, and participation in signal transduction at the cellular level. In cells, free Ca^{2+} ions can be stored in the endoplasmic reticulum and lysosome, sequestered in the mitochondria, and/or bound to cytosolic calmodulin or other calcium-binding proteins. In the extracellular milieu, Ca^{2+} is necessary for vasodilation and vasoconstriction as well as the mineralization of bone and blood clotting.

The vast majority (>99%) of total-body calcium is stored in bone and teeth as the calcium–phosphate complex hydroxyapatite [$Ca_{10}(PO_4)_6(OH)_2$]; the remaining 11 g (<1%) of calcium participates in a wide variety of physiological processes described above. The differential amount of calcium found in extracellular fluid (1 μmol) compared to the cytosol (0.1 μmol) leads to the generation of a concentration and electrochemical gradient; these gradients

Food and Nutritional Components in Focus No. 10
Calcium: Chemistry, Analysis, Function and Effects
Edited by Victor R. Preedy

Published by the Royal Society of Chemistry, www.rsc.org

Table 23.1 This table shows the recommended daily allowance of calcium. The asterisk denote adequate intake (AI). The table is adapted from Dietary Reference Intakes: The Essential Guide to Nutrient Requirements.

Age	Male	Female	Lactating	Pregnant
0–6 months*	200 mg	200 mg		
7–12 months*	260 mg	260 mg		
1–3 years	700 mg	700 mg		
4–8 years	1000 mg	1000 mg		
9–13 years	1300 mg	1300 mg		
14–18 years	1300 mg	1300 mg	1300 mg	1300 mg
19–50 years	1000 mg	1000 mg	1000 mg	1000 mg
51–70 years	1000 mg	1200 mg		
71+ years	1200 mg	1200 mg		

favor the entrance of calcium from extracellular environment or intestinal lumen to inside of the cell.

The recommended daily allowance (RDA) for calcium ranges from 1000–1300 mg for adults with higher requirements for pregnant women and the elderly (see Table 23.1) (Council, 2006). Dairy products, fortified foods, and some vegetables provide adequate sources of dietary calcium (see Table 23.2) (Council, 2006). During digestion of calcium-containing foods, dietary Ca^{2+} is released from the food matrix into the intestinal lumen where it can be absorbed into the bloodstream through the intestinal epithelium (Bronner and Pansu, 1999; Bronner, 2009).

Calcium absorption occurs *via* two distinct mechanisms: paracellular and transcellular (Bronner, 2003). In active or transcellular transport, intestinal uptake of Ca^{2+} is passive but Ca^{2+} efflux is ATP dependent (Bronner, 2003). This pathway predominates when calcium intake is low and is dependent on vitamin D and parathyroid hormone. In contrast, the passive or paracellular transport describes the movement of calcium ions down the concentration gradient through the cell junctions between enterocytes (Bronner, 2003).

This chapter will discuss and review the mechanism of calcium absorption from the intestine with insights from the literature and animal models.

23.2 Calcium Absorption

Calcium absorption from the intestinal lumen occurs *via* the paracellular and transcellular route (see Figure 23.1) (Wasserman and Fullmer, 1995; Christakos, 2012). In terms of calcium absorption, the paracellular and transcellular pathways can be described as a slower, high-capacity system and a fast, low-capacity system, respectively (Christakos, 2012). Paracellular absorption of calcium describes the passive movement of Ca^{2+} from the intestinal lumen through tight junctions and into the bloodstream (Christakos, 2012). In contrast, the transcellular pathway describes the active uptake of calcium from the intestinal lumen, the apical-to-basolateral trafficking of calcium within the enterocyte, and ATP-dependent extrusion

Table 23.2 This table shows common sources of dietary calcium. This table is adapted from Dietary Reference Intakes: The Essential Guide to Nutrient Requirements.

Food	Milligrams	Percent DV
Yogurt, plain, low fat, 8 ounces	415	42
Mozzarella, part skim, 1.5 ounces	333	33
Sardines, canned in oil, with bones, 3 ounces	325	33
Yogurt, fruit, low fat, 8 ounces	313–384	31–38
Cheddar cheese, 1.5 ounces	307	31
Milk, nonfat, 8 ounces	299	30
Soymilk, calcium-fortified, 8 ounces	299	30
Milk, reduced-fat (2% milk fat), 8 ounces	293	29
Milk, buttermilk, lowfat, 8 ounces	284	28
Milk, whole (3.25% milk fat), 8 ounces	276	28
Orange juice, calcium-fortified, 6 ounces	261	26
Tofu, firm, made with calcium sulfate, 1/2 cup	253	25
Salmon, pink, canned, solids with bone, 3 ounces	181	18
Cottage cheese, 1% milk fat, 1 cup	138	14
Tofu, soft, made with calcium sulfate, 1/2 cup	138	14
Ready-to-eat cereal, calcium-fortified, 1 cup	100–1000	10–100
Frozen yogurt, vanilla, soft serve, 1/2 cup	103	10
Turnip greens, fresh, boiled, 1/2 cup	99	10
Kale, raw, chopped, 1 cup	100	10
Kale, fresh, cooked, 1 cup	94	9
Ice cream, vanilla, 1/2 cup	84	8
Chinese cabbage, bok choi, raw, shredded, 1 cup	74	7
Bread, white, 1 slice	73	7
Pudding, chocolate, ready to eat, refrigerated, 4 ounces	55	6
Tortilla, corn, ready-to-bake/fry, one 6″ diameter	46	5
Tortilla, flour, ready-to-bake/fry, one 6″ diameter	32	3
Sour cream, reduced fat, cultured, 2 tablespoons	31	3
Bread, whole-wheat, 1 slice	30	3
Broccoli, raw, 1/2 cup	21	2
Cheese, cream, regular, 1 tablespoon	14	1

of Ca^{2+} from the basolateral side of the enterocyte into the bloodstream (Wasserman *et al.*, 1992). These complementary pathways work to ensure optimal calcium absorption from the intestine independent of lumenal calcium concentration.

23.3 The Paracellular Pathway

Absorptive enterocytes form a cellular barrier comprising tight junctions that prevent diffusion of the digesta into the blood. The paracellular pathway describes the movement of free Ca^{2+} ions down their concentration gradient from the intestinal lumen to the blood *via* cell junctions. Unlike the transcellular pathway, the paracellular pathway is a passive, slow, high-capacity system for calcium absorption (Wasserman and Fullmer, 1995). Thus, this pathway is nonsaturable, energy independent, and predominates

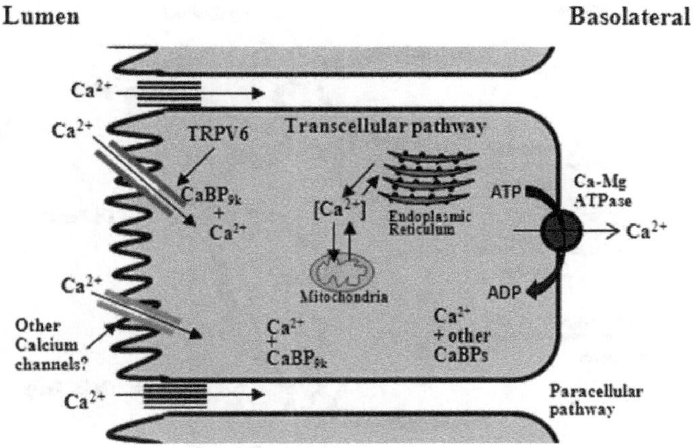

Figure 23.1 Model of vitamin D mediated intestinal calcium absorption. The traditional model of transcellular calcium transport consists of influx through an apical calcium channel (TRPV6), diffusion through the cytosol *via* calbindin-D9k (CaBP9k) and active extrusion at the basolateral membrane through PMCA1 (Ca^{2+}/Mg^{2+}-ATPase) into the bloodstream. Note paracellular absorption of calcium cations through gap junctions. This figure is reproduced from Christakos (2012) with permission from the publisher.

when lumenal concentrations of calcium are high, though it accounts for a small fraction of absorbed calcium. Normal blood calcium concentration ranges between 1.2 and 1.5 mM, thus, when the intestinal luminal Ca^{2+} concentration exceeds >1.5 mM, the paracellular pathway predominates (Wasserman and Fullmer, 1983). While paracellular transport is thought to be independent of vitamin D, there is increasing evidence that the steroid hormone increases gap junction permeability in enterocytes by altering the expression of tight junctions (Fujita *et al.*, 2008; Tudpor *et al.*, 2008).

23.4 The Transcellular Pathway

Enterocytes express a highly selective calcium transporter called TRVP6 along with some other, less-specific channels (*e.g.* DMT-1), which facilitates the uptake of calcium ions across the apical brush-border membrane, through the cytoplasm, and extrusion of Ca^{2+} ions into the blood (Bronner, 1987). Because this route of absorption is mediated by transporters, the transcellular pathway is saturable and nonlinear; this is a rapid, low-capacity system. The transcellular pathway is upregulated and predominates when blood and luminal concentrations of Ca^{2+} are low (*i.e.* the paracellular route cannot keep up with body needs). Three models for the transcellular pathway have been described in the literature: (1) facilitated diffusion, (2) ER-mediated transport, and (3) vesicular transport (Wasserman and Fullmer, 1995) (Figure 23.2).

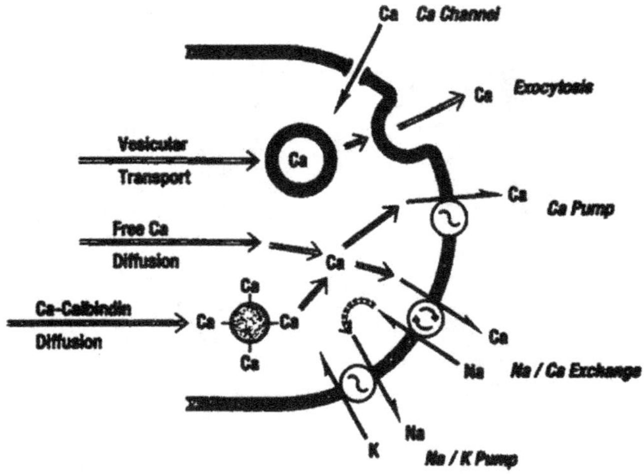

Figure 23.2 Calcium-related events at the basolateral membrane of the enterocyte. Calcium at the basolateral membrane can be exported into bloodstream *via* vesicular fusion or energy-dependent extrusion. The sodium–calcium exchanger uses energy to exchange intracellular calcium for extracellular sodium. The transmembrane gradient for sodium is maintained by the sodium–potassium pump that exchanges intracellular sodium for extracellular potassium. This figure is obtained from Wasserman and Fullmer (1995) and reprinted with permission from the publisher.

23.5 Facilitated Diffusion

In the facilitated diffusion or "cytoplasmic ferry" model, cytoplasmic Ca^{2+} is bound *via* the Ca^{2+}-binding protein calbindin and is transported from the apical to the basolateral side of the enterocyte (Wasserman and Fullmer, 1995). The basolateral side of the enterocyte expresses two distinct pumps, PMCA1 and the NCX, which participate in the pumping of calcium into the blood (Christakos, 2012). PMCA1 is a high-affinity, low-capacity ATP-dependent transporter. The NCX is a low-affinity, high-capacity antiporter and exchanges one Ca^{2+} ion for three Na^{+} ions. NCX is able to transport Ca^{2+} because the cells keep intracellular Na^{2+} concentrations low.

23.6 Endoplasmic Reticulum-Mediated Transport of Calcium

The endoplasmic reticulum spans the enterocyte and expresses the calcium transporters SERCA and IP_3-gated Ca^{2+} channel on the apical and basolateral side, respectively. It has been hypothesized that unbound calcium is actively pumped from the cytosol to the lumen of the endoplasmic reticulum *via* the apically located SERCA pump and released to the basolateral side through the calcium-release transporter, IP3, followed by extrusion into the blood *via* PMCA1 or NCX transporter (Christakos, 2012). Calcium also accumulates in the mitochondria and lysosomes that facilitate ER-mediated trafficking of calcium. Both the mitochondria and lysosome act as tertiary intracellular reservoirs for calcium and contribute to the intracellular homeostasis of Ca^{2+} (Patergnani *et al.*, 2011; Penny *et al.*, 2014).

23.7 Vesicular Transport

Calcium is internalized from the brush-border membrane as calcium-rich endocytic vesicles fuse with lysosomes in the cytoplasm (Wasserman and Fullmer, 1995). This endosomal–lysosomal transport facilitates the movement of Ca^{2+} from the apical side of the enterocyte to the basolateral side of the membrane. The endosome–lysosome fuses with the basolateral membrane, exocytosing calcium into the bloodstream (Larsson and Nemere, 2002). There is also evidence that calcium-rich lysosomes participate in the trafficking of calcium by "tunneling" through the endoplasmic reticulum (Larsson and Nemere, 2002). The apical and basolateral face of the endoplasmic reticulum are in close proximity to TRPV5 pumps and PMCA1 and NCX pumps on the apical and basolateral, respectively. Tunneling contributes to transcellular movement of calcium and refers to calcium uptake into the endoplasmic reticulum, lumenal movement of calcium ions through the endoplasmic reticulum followed by basolateral extrusion into the cytoplasm (Mogami *et al.*, 1997).

23.8 Calcium Homeostasis

Like all minerals, the body needs ways to deal with shortages and excesses of body Ca^{2+}. The primary source of Ca^{2+} is food and the body stores most Ca^{2+} in bone during excesses, though some can be excreted in urine, sweat, and feces. The body keeps the blood Ca^{2+} concentration in a narrow range (1.2–1.5 mM). When blood Ca^{2+} is elevated, the body stores Ca^{2+} in bone. When blood Ca^{2+} is low, the body resorbs Ca^{2+} back from bone or absorbs more from the diet. These changes in Ca^{2+} flux are mediated through hormones and the hormone-like nutrient vitamin D.

The parathyroid glands sense dropping concentrations of Ca^{2+} in the blood and release parathyroid hormone (PTH). PTH travels to bone and kidneys. In bone, PTH signals for the resorption of Ca^{2+} and P. In the kidney, PTH signals the reabsorption of Ca^{2+}, increased excretion of P, and the conversion of 25-hydroxycholecalciferol to 1,25-dihydroxycholecalciferol (25-D to 1,25-D) to increase intestinal absorption of Ca and P. When Ca^{2+} concentrations in the blood are high, calcitonin reverses the effects of PTH.

23.9 Vitamin D

Active vitamin D (1,25-dihydroxycholecalciferol) is a steroid hormone synthesized from a cholesterol and undergoes photoisomerization in the skin and sequential hydroxylation reactions in the liver and kidney. Vitamin D

	Calcium absorption			
	Relative absorption of calcium by segment	Untreated group	Treated group	1,25D-dependent
	%	mg		
Duodenum	7	3	12	9
Jejunum	13	6	23	17
Ileum	80	34	140	106
Total		43	175	132

Figure 23.3 The effect of vitamin D on calcium absorption in renal dialysis patients. Human volunteers were given an initial meal with 301 mg of calcium with or without 1 μg of oral vitamin D. The net absorption of calcium absorption for the untreated group was 43 mg compared with 175 mg, an increase of over 300%. Note the high absorption of calcium in the ileum relative to other segments of the small intestine. The duodenum has a greater rate of absorption, but, due to its short length, the duodenum does not absorb the majority of a mineral. This distinction belongs to the jejunum or ileum. This figure was modified from Wasserman (2004) and reprinted with permission from the publisher.

can be of endogenous or exogenous origins. When 1,25-D is released from the kidney, it targets intracellular vitamin D receptors (VDRs) in enterocytes. These ligand-bound VDRs homodimerize or heterodimerize with RXR and translocate from the cytoplasm into the nucleus where it directs the synthesis of genes involved in calcium absorption. Vitamin D induces calbindin, TRPV6, and PMCA1 expression in enterocytes and nongenomically regulates NCX and alters gap junction permeability to calcium (Nemere, 1995; Walters *et al.*, 2007). Overall, both the genomic and nongenomic actions of vitamin D increase Ca^{2+} flux across the intestinal barrier and absorption into the blood (Figure 23.3).

23.10 Other Factors Affecting Calcium Absorption

In the gastrointestinal tract, calcium absorption principally occurs in the distal portion of the small intestine with negligible absorption occurring in the colon. Diets rich in protein, sodium, and vitamin D increase calcium absorption from the intestine. Furthermore, prebiotics such as inulin and oligofructose has been shown to stimulate intestinal calcium absorption (Abrams *et al.*, 2007). Probiotic organisms such as *Lactobacillus casei* and *Lactobacillus acidophilus* increase the solubility and bioavailability of calcium from calcium-enriched beverages (Tang *et al.*, 2007). In contrast, consumption of oxalate- or phytate-containing foods such as legumes and cereal grains has an inhibitory effect on calcium absorption (Viadel *et al.*, 2006; Morris *et al.*, 2007). Likewise, ethanol in alcoholic beverages as well as tannins and caffeine from coffee or tea decrease calcium absorption from the intestine (Krawitt, 1975; Chang *et al.*, 1994).

23.11 Studies from Other Animal Models

There has been considerable research in the absorption of calcium in pigs (Jolliff and Mahan, 2013; González-Vega and Stein, 2014). Like humans, pigs possess active and passive mechanisms for calcium absorption and predominately absorb calcium from the proximal intestine. Most diet formulations for swine contain inorganic calcium (*e.g.* calcium carbonate) because it is relatively inexpensive and absorption of calcium from organic sources such as plants or cereal grains is poorer in comparison (Jolliff and Mahan, 2013; González-Vega and Stein, 2014).

Much work in swine has focused on the dietary ratio of calcium to phosphorous. Increasing the Ca : P ratio can hinder P digestibility (Fan and Archbold, 2012), but reducing the Ca : P ratio too much can hinder bone mineralization (Létourneau-Montminy *et al.*, 2012). However, the focus of Ca and P research has changed over time with the recognition that phytate not only contains unavailable P (*i.e.* unavailable to be digested and absorbed), but can bind Ca and other cations in available complexes. Thus, the use of synthetic phytase as a feed ingredient and the advent of formulating feed on an available or digestible P basis over the past two decades have changed how people view Ca nutrition in addition to P. In fact, the next step in mineral nutrition in swine may be to use of digestible Ca instead of total Ca in formulating diets.

Summary Points

- Ca^{2+} is a divalent cation involved in a wide array of biological processes including muscle contraction, neurotransmitter release, and signal transduction.
- Dietary sources of calcium are diverse and include dairy, meats, nuts as well as other foods. The RDA for adults (ages 19–50) calcium is about 1 g daily with higher amounts required for pregnant and nursing women as well as the elderly.
- Calcium absorption from the intestinal lumen occurs *via* paracellular and transcellular routes (see Figure 23.1). If one were to plot calcium uptake over time, the paracellular route would be nonsaturable and linear, while the transcellular pathway is saturable and nonlinear. High levels of calcium in the intestinal lumen favor the paracellular pathway; in contrast, the transcellular pathway predominates when luminal concentrations of Ca^{2+} are low.
- Calcium is absorbed from the intestinal lumen into the bloodstream through *passive* or *active* mechanisms. The intestinal lumen is comprised of individual enterocytes that are laterally held together by a network of tight and gap junctions. These junctions provide a barrier between the digested food and the bloodstream. Enterocytes have an apical, brush-border membrane that faces the lumen and a basolateral membrane that faces the bloodstream.
- The *passive* absorption of calcium describes the paracellular movement of Ca^{2+} ions from the intestinal lumen into the bloodstream. This route

is both saturable and energy independent and predominates when calcium content in the intestinal lumen is high.

- The *active* absorption of calcium describes the transcellular movement of Ca^{2+} ion from the apical brush border membrane, intracellular trafficking of calcium to the basolateral membrane by calbindin, followed by extrusion of Ca^{2+} into the bloodstream. This route is both nonsaturable and energy dependent and predominates when calcium content in the intestinal lumen is low.
- Vitamin D, PTH, pregnancy, prebiotics, and probiotics potentiate intestinal calcium absorption, while tannins, the hormone calcitonin, and phytate have an inhibitory effect on intestinal calcium absorption.

Key Facts of Calbindin

The key facts of calcium absorption that include paracellular and transcellular mechanisms as well as other factors that influence calcium absorption from the intestine.

1. Intestinal calbindin (also referred to as calcium-binding protein calbindin-D9K or CaBP9k) is a cytosolic protein that is involved in the transcellular movement of calcium. Calbindin-D9k was first cloned and characterized in 1992 by two teams in Canada and the United Kingdom (Howard *et al.*, 1992; Jeung *et al.*, 1992).
2. The protein is encoded by the *S100G* gene and is constitutively expressed at low levels in the absence of vitamin D. In contrast, the presence vitamin D enhances *S100G* transcription leading to increased protein levels of calbindin-D9k.
3. Calbindin-D9k is 9 kDa protein has two helix-loop-helix motifs called EF hands. These EF hands have strong affinity for calcium and undergo cooperative binding in the presence of Ca^{2+} ions.
4. Recent studies have shown that knockdown of calbindin-D9k reduces but does not abrogate intestinal absorption casting some controversy over the "cytoplasmic ferry" model.

Definition of Words and Terms

Calbindin refers to a group of calcium-binding proteins. During intestinal absorption, these proteins are involved in shuttling calcium from the apical side of the enterocyte to the basolateral side.

Calcitonin a hormone secreted by the thyroid gland. This hormone targets bone, kidney and intestine to reduce the amount of Ca entering plasma.

Calmodulin is a multifunctional calcium-binding protein that is involved in memory formation in the hippocampus.

Enterocytes are individual cells of the small intestine and involved in the uptake, metabolism, and transport of nutrients from the intestine into the bloodstream.

Large intestine refers to the colon structures. These portions are involved in the uptake of water and ions.

Lysosome is an organelle within the cell that is involved in breaking down worn-out organelles as well as serving as a intracellular reservoir and transport for calcium.

Lumen the space or void – often tubular – within intestine, esophagus, or organelle.

Photoisomerization refers to light-induced shape changes in a molecule. In the skin, 7-dehydrocholesterol is converted into previtamin D_3 after exposure to sunlight or UVB radiation.

Parathyroid hormone is a hormone secreted by the parathyroid glands. This hormone targets the intestine and decreases the absorption of calcium from the intestine.

Small intestine refers to the duodenum, jujenum, and ileum. These portions are involved in the uptake of nutrients such as amino acids, lipids, and sugars.

Vitamin D is a steroid hormone synthesized from a cholesterol and undergoes photoisomerization in the skin and sequential hydroxylation reactions in the liver and kidney.

List of Abbreviations

AI	Adequate intake
ATP	Adenosine triphosphate
DMT-1	Divalent metal transporter 1
ER	Endoplasmic reticulum
IP3	Inositol triphosphate
NCX	Na^+/Ca^{2+} exchanger
PMCA1b	Plasma membrane Ca^{2+}-ATPase
PTH	Parathyroid hormone
RDA	Recommended daily allowance
RXR	Retinoid X receptor
SERCA	Sarco/endoplasmic reticulum Ca^{2+}-ATPase
TRPV6	Transient receptor potential cation channel, vanilloid subfamily, subtype 6
VDR	Vitamin D receptor

References

Abrams, S. A., Griffin, I. J., *et al.*, 2007. Effect of prebiotic supplementation and calcium intake on body mass index. *Journal of Pediatrics*. 151(3): 293–298.

Bronner, F., 1987. Intestinal calcium absorption: mechanisms and applications. *Journal of Nutrition*. 117(8): 1347–1352.

Bronner, F., 2003a. Mechanisms and functional aspects of intestinal calcium absorption. *Journal of Experimental Zoology, Part A: Comparative Experimental Biology*. 300(1): 47–52; Bronner, F., 2003b. Mechanisms of intestinal calcium absorption. *Journal of Cellular Biochemistry*. 88(2): 387–393.

Bronner, F., 2009. Recent developments in intestinal calcium absorption. *Nutrition Reviews*. 67(2): 109–113.

Bronner, F. and Pansu, D., 1999. Nutritional aspects of calcium absorption. *Journal of Nutrition*. 129(1): 9–12.

Chang, M. C., Bailey, J. W., *et al.*, 1994. Dietary tannins from cowpeas and tea transiently alter apparent calcium absorption but not absorption and utilization of protein in rats. *Journal of Nutrition*. 124(2): 283–288.

Christakos, S., 2012a. Mechanism of action of 1,25-dihydroxyvitamin D_3 on intestinal calcium absorption. *Reviews in Endocrine & Metabolic Disorders*. 13(1): 39–44; Christakos, S., 2012b. Recent advances in our understanding of 1,25-dihydroxyvitamin D(3) regulation of intestinal calcium absorption. *Archives of Biochemistry and Biophysics*. 523(1): 73–76.

Council, N. R., 2006. *Dietary Reference Intakes: The Essential Guide to Nutrient Requirements*. The National Academies Press, Washington, DC.

Fan, M. Z. and Archbold, T., 2012. Effects of dietary true digestible calcium to phosphorus ratio on growth performance and efficiency of calcium and phosphorus use in growing pigs fed corn and soybean meal-based diets. *Journal of Animal Science*. 90(suppl. 4): 254–256.

Fujita, H., Sugimoto, K., *et al.*, 2008. Tight junction proteins claudin-2 and -12 are critical for vitamin D-dependent Ca^{2+} absorption between enterocytes. *Molecular Biology of the Cell*. 19(5): 1912–1921.

González-Vega, J. and Stein, H. H., 2014. Calcium digestibility and metabolism in pigs. *Asian-Australasian Journal of Animal Sciences*. 27: 1–9.

Howard, A., Legon, S., Spurr, N. K. and Walters, J. R., 1992. Molecular cloning and chromosomal assignment of human calbindin-D9k. *Biochemical and Biophysical Research Communications*. 185(2): 663–669.

Jeung, E. B., Krisinger, J., Dann, J. L. and Leung, P. C., 1992. Cloning of the porcine Calbindin-D9k complementary deoxyribonucleic acid by anchored polymerase chain reaction technique. *Biology of Reproduction*. 47(4): 503–508.

Jolliff, J. S. and Mahan, D. C., 2013. Effect of dietary calcium and phosphorus levels on the total tract digestibility of innate and supplemental organic and inorganic microminerals in a corn-soybean meal based diet of grower pigs. *Journal of Animal Science*. 91(6): 2775–2783.

Krawitt, E. L., 1975. Effect of ethanol ingestion on duodenal calcium transport. *Journal of Laboratory and Clinical Medicine*. 85(4): 665–671.

Larsson, D. and Nemere, I., 2002. Vectorial transcellular calcium transport in intestine: integration of current models. *Journal of Biomedicine and Biotechnology*. 2(3): 117–119.

Létourneau-Montminy, M. P., Jondreville, C., Sauvant, D. and Narcy, A., 2012. Meta-analysis of phosphorus utilization by growing pigs: effect of dietary phosphorus, calcium and exogenous phytase. *Animal*. 6(10): 1590–1600.

Mogami, H., Nakano, K., Tepikin, A. V. and Petersen, O. H., 1997. Ca^{2+} flow via tunnels in polarized cells: recharging of apical Ca^{2+} stores by focal Ca^{2+} entry through basal membrane patch. *Cell.* 88(1): 49–55.

Morris, J., Nakata, P. A., *et al.*, 2007. Increased calcium bioavailability in mice fed genetically engineered plants lacking calcium oxalate. *Plant Molecular Biology.* 64(5): 613–618.

Nemere, I., 1995. Nongenomic effects of 1,25-dihydroxyvitamin D_3: potential relation of a plasmalemmal receptor to the acute enhancement of intestinal calcium transport in chick. *Journal of Nutrition.* 125(suppl. 6): 1695S–1698S.

Patergnani, S., Suski, J. M., Agnoletto, C., Bononi, A., Bonora, M., De Marchi, E., Giorgi, C., Marchi, S., Missiroli, S., Poletti, F., Rimessi, A., Duszynski, J., Wieckowski, M. R. and Pinton, P., 2011. Calcium signaling around Mitochondria Associated Membranes (MAMs). *Cell Communication and Signaling.* 9: 19.

Penny, C. J., Kilpatrick, B. S., Min Han, J., Sneyd, J. and Patel, S., 2014. A "mix-and-match" approach to designing Ca^{2+} microdomains at membrane-contact sites. *Communicative & Integrative Biology.* 7: e29586.

Tang, A. L., Shah, N. P., *et al.*, 2007. Fermentation of calcium-fortified soymilk with Lactobacillus: effects on calcium solubility, isoflavone conversion, and production of organic acids. *Journal of Food Science.* 72(9): M431–M436.

Tudpor, K., Teerapornpuntakit, J., *et al.*, 2008. 1,25-Dihydroxyvitamin D(3) rapidly stimulates the solvent drag-induced paracellular calcium transport in the duodenum of female rats. *Journal of Physiological Sciences.* 58(5): 297–307.

Viadel, B., Barbera, R., *et al.*, 2006. Calcium, iron and zinc uptakes by Caco-2 cells from white beans and effect of cooking. *International Journal of Food Sciences and Nutrition.* 57(3–4): 190–197.

Walters, J. R., Balesaria, S., *et al.*, 2007. The effects of vitamin D metabolites on expression of genes for calcium transporters in human duodenum. *Journal of Steroid Biochemistry and Molecular Biology.* 103(3–5): 509–512.

Wasserman, R. H., 2004. Vitamin D and the dual processes of intestinal calcium absorption. *Journal of Nutrition.* 134(11): 3137–3139.

Wasserman, R. H., Chandler, J. S., *et al.*, 1992. Intestinal calcium transport and calcium extrusion processes at the basolateral membrane. *Journal of Nutrition.* 122(suppl. 3): 662–671.

Wasserman, R. H. and Fullmer, C. S., 1983. Calcium transport proteins, calcium absorption, and vitamin D. *Annual Review of Physiology.* 45: 375–390.

Wasserman, R. H. and Fullmer, C. S., 1995. Vitamin D and intestinal calcium transport: facts, speculations and hypotheses. *Journal of Nutrition.* 125(suppl. 7): 1971S–1979S.

CHAPTER 24

The Calcium-Sensing Receptor in Intestinal Cells

TOHRU HIRA*[a], SHINGO NAKAJIMA[b], AND HIROSHI HARA[a]

[a]Division of Fundamental AgriScience Research, Research Faculty of Agriculture, Hokkaido University, Kita-9, Nishi-9, Kita-ku, Sapooro, 060-8589, Japan; [b]Faculty of Health Science, Hokkaido University, Kita-12, Nishi-5, Kita-ku, Sapporo 060-0812, Japan
*E-mail: hira@chem.agr.hokudai.ac.jp

24.1 Introduction

The calcium-sensing receptor (CaSR) was first identified in 1993 (Brown *et al.*, 1993) in the bovine parathyroid gland as a cell-surface receptor for extracellular calcium ions. CaSR is a G-protein-coupled receptor (GPCR) characterized by seven transmembrane domains. CaSR is a member of class C GPCRs, among which are metabolic glutamate receptors (mGluRs), gamma amino butyrate B ($GABA_B$) receptors, and taste (sweet and umami) receptors. Class C GPCRs have a large extracellular domain called the Venus flytrap module, and they function as homo- or heterodimers. In the case of CaSR, homodimerization is essential for its calcium-sensing activity.

The primary function of CaSR is to sense increased extracellular calcium concentrations in order to regulate calcium homeostasis in the body. In the parathyroid gland, where CaSR is highly expressed, CaSR activation

Food and Nutritional Components in Focus No. 10
Calcium: Chemistry, Analysis, Function and Effects
Edited by Victor R. Preedy

Published by the Royal Society of Chemistry, www.rsc.org

suppresses the release of parathyroid hormone (PTH), which leads to attenuation of bone resorption, reduced calcium reabsorption in the distal tubule of the kidney, and reduced production of $1{,}25(OH)_2D_3$ in the proximal tubule of the kidney. Together, these effects enhance urinary calcium excretion and reduce intestinal calcium absorption. Activation of parathyroid CaSR also suppresses PTH synthesis and parathyroid cellular proliferation. Similarly, CaSR expressed in the kidney contributes to calcium excretion (reduced calcium reabsorption) through inactivation of a renal outer medullary potassium channel in the thick ascending limb of Henle's loop (Wang *et al.*, 1997). Activation of renal CaSR also regulates transport of $NaCl/Mg^{2+}/K^{+}/Cl^{-}/H_2O$ as well as urine acidification and concentration. CaSR expression has been reported in both osteoclasts and osteoblasts of the bone. It has been suggested that CaSR suppresses osteoclast function (calcium resorption and osteoclastogenesis) under conditions of high blood calcium either directly or indirectly *via* osteoblast activation. Thus, the role of CaSR in calcium homeostasis has been well characterized.

24.2 Multiple Agonists for CaSR

Although extracellular calcium is the primary ligand for CaSR, di- and trivalent cations (*e.g.*, Mg^{2+} and Gd^{3+}) also activate the receptor. In addition, organic polycations, including polyamines (spermine), protamine, polylysine, polyarginine, and neomycin are known to be CaSR ligands (Riccardi and Brown, 2010). Importantly, CaSR was found to be activated by L-amino acids, such as Phe, Trp, and His, in CaSR-transfected HEK293 cells *in vitro* (Conigrave *et al.*, 2000). Amino acid-induced CaSR activation requires the presence of extracellular calcium at levels above approximately 1.0 mM. The activation of CaSR by L-amino acids but not D-amino acids indicates the receptor's stereoselectivity. This study suggested a putative role for CaSR in integrating signals arising from distinct classes of nutrients, such as mineral ions and amino acids. As such, CaSR was speculated to modulate the actions of dietary protein on calcium metabolism, as in the case of high-protein diet-induced hypercalciuria. Furthermore, it has been hypothesized that CaSR functions as a nutrient-sensing receptor in the gastrointestinal tract.

A subsequent *in silico* study suggested that some Group C GPCRs, including CaSR, accommodate larger ligands than amino acids, such as peptides (Wang *et al.*, 2006). Indeed, an *in vitro* binding assay and a fluorescence-based functional assay identified a small γ-glutamyl peptide, glutathione (GSH), as a potential ligand for the rat CaSR. GSH is a taste enhancer, termed kokumi taste. Other γ-glutamyl peptides, such as γ-Glu-Val-Gly and γ-Glu-Cys, have been identified as CaSR agonists in *in vitro* screens; these peptides elicit the kokumi taste in humans (Ohsu *et al.*, 2010). These characteristics of the receptor raise the possibility that CaSR binds small peptides, such as dietary protein hydrolyzate (dietary peptides), generated in the intestinal lumen after ingestion of a meal.

24.3 CaSR in the Gastrointestinal Tract

The expression and function of CaSR have been elucidated in various tissues, including the lung, breast, central nervous system, taste buds, and gastrointestinal tract in addition to tissues related to calcium homeostasis (*e.g.*, parathyroid gland, kidney, and bone). This chapter reviews CaSR functions in the intestine, especially as a nutrient-sensing receptor in enteroendocrine cells.

The expression of CaSR in the gastrointestinal tract has been studied since the late 1990s (Butters *et al.*, 1997; Gama *et al.*, 1997; Chattopadhyay *et al.*, 1998). Since 2000, multiple roles for intestinal CaSR have been identified in the stomach, small intestine, large intestine, and enteric nervous system. Although the role of apical CaSR in luminal calcium absorption was not surprising, there remain limited studies on this role to date (Jongwattanapisan *et al.*, 2012). In isolated colonic tissue, apical application of CaSR agonists (neomycin and spermine) stimulated active calcium transport. The effect was observed in proximal but not distal colonic tissue. CaSR agonists failed to affect colonic electrical parameters, such as transepithelial potential difference, short-circuit current, and transepithelial resistance, suggesting that the enhanced calcium transport elicited by CaSR agonists did not involve increased paracellular permeability. Further studies are required to better characterize the role of epithelial CaSR in luminal calcium absorption.

24.3.1 Gut-Hormone Secretion

Serotonin (5-HT) and peptide hormones, such as ghrelin, gastrin, secretin, cholecystokinin (CCK), glucose-dependent insulinotropic polypeptide (GIP), glucagon-like peptide-1 (GLP-1), glucagon-like peptide-2 (GLP-2), and peptide-YY (PYY), are well-known gut hormones. Gut hormones are released from specific hormone-producing cells called enteroendocrine cells. Enteroendocrine cells are distributed throughout the intestinal epithelium and represent approximately 1% of the epithelial cell population. Most gut hormones, except for ghrelin, are immediately released after nutrient ingestion. Upon release, gut hormones regulate various postprandial responses in the gastrointestinal tract, including secretion of gastric acid, pancreatic bicarbonate, pancreatic enzymes, and insulin, gastric and intestinal motility, and satiety. Thus, enteroendocrine cells are major nutrient sensors in the gastrointestinal tract (Dockray, 2003; Gribble, 2012; Tolhurst *et al.*, 2012). The molecular mechanisms underlying luminal nutrient sensing have been elucidated in enteroendocrine cell lines and enteroendocrine cell-specific transgenic animals. In particular, GPCRs, including taste receptors, free-fatty acid receptors, and bile-acid receptors have been implicated in luminal nutrient sensing.

24.3.2 Gastrin and Gastric-Acid Secretion

Gastric-acid secretion from acid-producing parietal cells is induced by histamine, acetylcholine, and gastrin. Dietary peptides, amino acids, and calcium stimulate gastrin and gastric-acid secretion, and CaSR is expressed in gastric cells (Cheng *et al.*, 1999); therefore, it was suggested that CaSR is involved in postprandial gastric-acid secretion. Indeed, Geibel *et al.* reported that CaSR modulated gastric-acid secretion (Geibel *et al.*, 2001). Using isolated gastric glands from rats, they demonstrated that cation ligands of CaSR (Ca^{2+}, Mg^{2+}, and Gd^{3+}) induced acid secretion through the apical H^+–K^+-ATPase. Gadolinium-induced acid secretion was dependent on the omeprazole-sensitive H^+–K^+-ATPase and independent of histamine. A subsequent study demonstrated functional expression of CaSR in freshly isolated human gastric glands and the involvement of CaSR in histamine-induced H^+–K^+-ATPase activation (Dufner *et al.*, 2005).

In isolated rat gastric glands and *ex vivo* whole-stomach preparations, L-amino acids activated CaSR and stimulated acid secretion from gastric parietal cells (Busque *et al.*, 2005). In that study, the ligated stomach preparation was exposed to buffer containing L- or D-Phe for 1 h, and the luminal pH was subsequently measured. Although the test concentration of Phe at 50 mM was relatively high, L- but not D-Phe decreased luminal pH. This finding demonstrated stereoselective acid secretion elicited by Phe. To model gastric-acid secretion, the rate of H^+ release was measured by intracellular pH monitoring of individual parietal cells in isolated gastric glands. Extracellular calcium dose dependently increased H^+ release, which was further amplified by the addition of L-Phe and L-Trp but not L-Leu, D-Phe, or D-Trp. These results were consistent with the allosteric amino acid-sensing property of CaSR (Conigrave *et al.*, 2000). Furthermore, they supported a role for CaSR in amino acid sensing in gastric parietal cells.

CaSR expression was demonstrated in human gastrin-producing endocrine cells (G cells), and gastrin release was stimulated by extracellular calcium and spermine in gastrin cell-enriched cell cultures (Ray *et al.*, 1997; Buchan *et al.*, 2001). It was suggested that dietary calcium activated apical CaSR based on the presence of CaSR immunoreactivity at both the apical and basolateral sides of G cells (Buchan *et al.*, 2001; Conigrave *et al.*, 2002). By using CaSR-null mice, the involvement of CaSR in luminal nutrient-induced gastrin secretion was revealed (Feng *et al.*, 2010). Oral gavage of various CaSR agonists, including calcium, Phe, and the pharmacological agonist cinacalcet, induced gastrin secretion in wild-type mice but not in CaSR-null mice. Interestingly, peptone-induced gastrin secretion was abolished in CaSR-null mice and wild-type mice treated with the CaSR antagonist. These results indicated an essential role for CaSR in nutrient sensing in gastrin-producing enteroendocrine cells. Since there were fewer G cells in CaSR-null mice than in wild-type mice, CaSR was also implicated in G-cell proliferation.

24.3.3 CCK Secretion

CaSR expression was demonstrated in enteroendocrine cells of human colonic tissue using immunohistochemistry (Sheinin *et al.*, 2000). Although a specific gut hormone was not identified, CaSR immunoreactivity was detected in cells expressing chromogranin-A, an enteroendocrine cell marker, in the colonic epithelium.

CCK is produced in enteroendocrine cells called I cells, which are mainly located in the upper small intestine. CCK secretion is stimulated by luminal nutrients, such as fatty acids, dietary peptides, and amino acids. CCK stimulates pancreatic bicarbonate and exocrine enzyme secretions to promote digestion of meal components in the intestinal lumen. CCK induces satiety and suppresses appetite *via* vagal-nerve activation. Another major role for released CCK is inhibition of gastric emptying.

Functional involvement of CaSR in amino acid-induced CCK secretion was first demonstrated in 2008, by using a murine enteroendocrine cell line, STC-1 (Figure 24.1, Hira *et al.*, 2008). STC-1 is a well-known CCK-producing cell line established from a duodenal endocrine tumor in a transgenic mouse. Therefore, it is a useful model for investigating nutrient-induced CCK secretion. Some amino acids, such as Phe and Trp, were known to stimulate CCK secretion, but the sensing mechanism had remained unclear. In STC-1 cells, the CCK secretory response to Phe was enhanced by increasing extracellular calcium levels from the physiological concentration (1.2 mM) to 3.0 mM. Intracellular calcium mobilization in response to Phe was also enhanced by increasing the extracellular calcium concentration. In addition, intracellular calcium mobilization in response to increased extracellular calcium concentration was enhanced by pretreatment with Phe in STC-1 cells. These findings were consistent with the known properties of CaSR. The expression of CaSR mRNA was

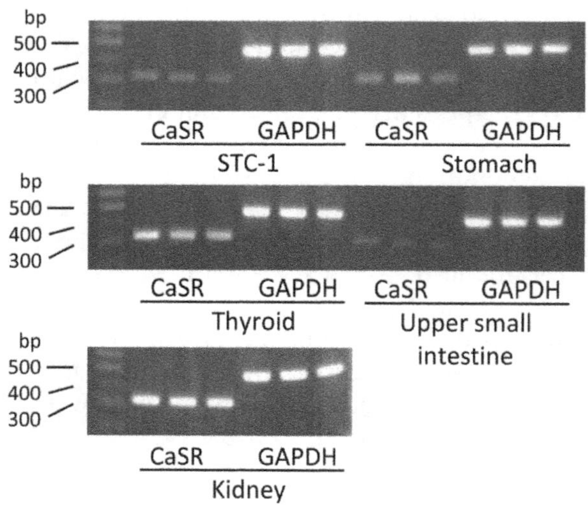

Figure 24.1 CaR expression in enteroendocrine STC-1 cells and mouse tissues (Hira *et al.*, 2008). RNA was extracted from STC-1 cells and mouse tissues (gastric mucosa, upper small intestinal mucosa, thyroid, kidney), and then subjected to RT-PCR with CaR or GAPDH primers.

demonstrated in STC-1 cells by reverse-transcription PCR, and a CaSR antagonist (NPS2143) attenuated Phe-induced CCK secretion and intracellular calcium mobilization. These results demonstrated the involvement of CaSR in amino acid-induced CCK secretion in an enteroendocrine cell line.

Subsequent studies investigated the expression and function of CaSR in CCK-producing cells isolated from the murine intestine. CCK-producing cells were isolated from CCK promoter-driven enhanced GFP transgenic mice (CCK-eGFP mice) (Liou *et al.*, 2011; Wang *et al.*, 2011). Coexpression of CaSR and CCK was demonstrated by immunohistochemistry in frozen sections of intestine prepared from CCK-eGFP and wild-type mice. CaSR mRNA expression was also demonstrated in isolated CCK-producing cells. Furthermore, Phe and Trp induced intracellular calcium mobilization and CCK secretion in isolated CCK-producing cells. These responses were abolished by treatment with a CaSR inhibitor (Wang *et al.*, 2011). CCK-producing cells isolated from CaSR/PTH double-homozygous knockout mice did not respond to these amino acids, further supporting a role for CaSR as an amino acid sensor in CCK-producing enteroendocrine cells (Liou *et al.*, 2011).

Interestingly, CCK secretion induced by tryptone (a casein hydrolyzate) was partially inhibited in the absence of CaSR (Liou *et al.*, 2011). Since CaSR is activated by some peptides, such as protamine, polylysine, polyarginine, and short gamma glutamyl peptides, it was suggested that CaSR regulates dietary peptide-induced CCK secretion.

In STC-1 cells, the CaSR antagonist NPS2143 attenuated CCK secretion induced by various dietary peptides (Figure 24.2, Nakajima *et al.*, 2010, 2012). CaSR-transfected HEK293 cells but not control HEK293 cells displayed intracellular calcium mobilization in response to dietary peptides. Because protein hydrolyzates generally contain some degree of free amino acids, it was suggested that free amino acids contributed to this effect. However, peptide fractions (>500 Da) from which free amino acids were removed still induced CCK secretion, and the secretory response was partially inhibited by a CaSR antagonist. These results demonstrated that CaSR plays a significant role in sensing luminal dietary peptides to activate CCK secretion in enteroendocrine cells.

24.3.4 GLP-1 Secretion

Recently, the involvement of CaSR in GLP-1 secretion was demonstrated (Diakogiannaki *et al.*, 2013). GLP-1 is an incretin hormone that is released from enteroendocrine cells and enhances glucose-induced insulin secretion from pancreatic beta cells. GLP-1-producing enteroendocrine cells are called L cells and are mainly located in the ileum, caecum, and colon. Typically, GLP-1 secretion is thought to be stimulated by luminal glucose and fatty acids. However, several studies have demonstrated that dietary protein,

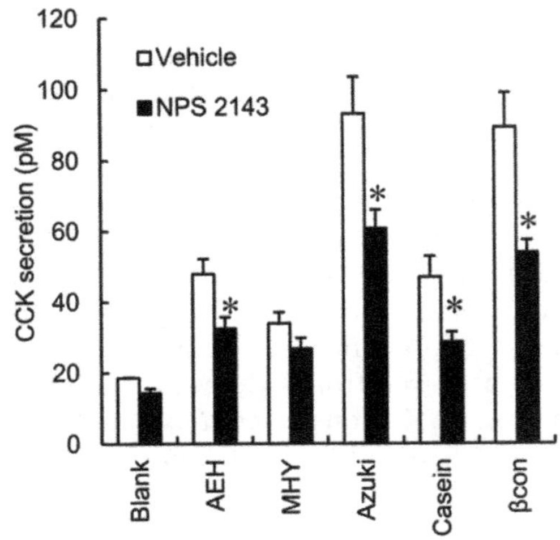

Figure 24.2 CaSR mediates dietary peptide-induced CCK secretion in enteroendocrine STC-1 cells (Nakajima *et al.*, 2012). STC-1 cells were exposed to various protein hydrolyzates at 5 mg mL^{-1} for 60 min after a 10 min pretreatment with 25 μM NPS 2143 or its vehicle (0.1% DMSO). Albumin egg hydrolyzate (AEH); azuki protein hydrolyzate (azuki); casein hydrolyzate (casein); soybean β-conglycinin hydrolyzate (βcon). Significant differences ($P < 0.05$, Student's *t*-test) between vehicle and NPS 2143 treatments are indicated with asterisk signs (*).

peptides, and amino acids induce GLP-1 secretion in cell cultures, animal models, and humans.

GLP-1 secretion was triggered by peptones (enzymatic meat hydrolyzates) and di- and tripeptides in murine mixed primary colonic cultures (Diakogiannaki *et al.*, 2013). Dipeptide-induced GLP-1 secretion involved peptide transporter-1 (PEPT1). In contrast, peptone-induced GLP-1 secretion was abrogated by CaSR antagonists. GLP-1-producing enteroendocrine cells (L cells) were identified and purified based on their cell-specific expression of the fluorescent protein Venus from GLU-Venus transgenic mice. CaSR mRNA was highly expressed in purified L cells as compared to non-L cells. Another previous study used isolated loops of rat small intestine to examine gut-hormone secretion in response to luminal nutrient perfusion (Mace *et al.*, 2012). Luminal amino acids, including L-Phe, L-Trp, L-Asn, and L-Arg, stimulated GLP-1, GIP, and PYY secretion, and these secretory responses were blocked by a CaSR antagonist. This finding suggested the involvement of CaSR in amino acid-induced GLP-1/PYY secretion in L cells and GIP secretion in K cells.

Thus, targeting CaSR in enteroendocrine cells could be a novel and safe strategy for enhancing or reducing secretion of endogenous gut hormones, such as GLP-1 and CCK, in order to control postprandial glycaemia, gastric motility, and appetite.

24.4 Colonic-Fluid Secretion

Secretion of colonic fluid is stimulated by elevated osmotic pressure in the colonic lumen, inflammation, cholera toxin, enterotoxins, abnormal gut motility, and impaired chloride absorption. Increased fluid secretion results in diarrhoea and electrolyte losses.

In rat colonic tissue, CaSR expression was observed both at the apical and basolateral sides of epithelial cells (Cheng *et al.*, 2002). Forskolin-induced net fluid secretion in isolated, perfused crypts was abolished by increasing the extracellular calcium concentration. A subsequent study (Geibel *et al.*, 2006) demonstrated that an allosteric CaSR agonist abrogated fluid and electrolyte secretion induced by cAMP- and cGMP-generating secretagogs, including cholera toxin and heat-stable *Escherichia coli* enterotoxin STa. These effects disappeared in isolated crypts prepared from the CaSR-null mice. Recently, it was suggested that CaSR expressed in tetrodotoxin (TTX)-sensitive cells of the enteric nervous system regulates the suppressive effect of CaSR agonists on secretagog-induced chloride secretion (Cheng, 2012). These findings point to colonic CaSR as a therapeutic target in the prevention of secretory diarrhoea.

24.5 Epithelial Proliferation and Differentiation

CaSR is involved in colonic epithelial proliferation and differentiation through modulating Wnt/β-catenin signaling. Activation of CaSR in colonic myofibroblasts stimulated secretion of bone morphogenetic protein-2 (BMP-2), which enhanced the epithelial barrier function, as evaluated by transepithelial resistance in T-84 cells (Peiris *et al.*, 2007). In addition, CaSR activation in the epithelium downregulated the expression of the BMP antagonist protein Noggin. This suggests that CaSR regulates both secretion of BMP-2 and its downstream effects in the intestine, which enhances the barrier function. In myofibroblasts, CaSR-induced secretion of Wnt5a stimulated sucrose/isomaltase promoter activity through increased expression of caudal homeodomain factor CDX2 (Pacheco and Macleod, 2008).

It has been suggested that dietary calcium protects against colon cancer, and CaSR may mediate this effect. Specifically, colonic crypt proliferation is stimulated by β-catenin signaling, and CaSR activation decreases β-catenin phosphorylation (MacLeod *et al.*, 2007). Moreover, deletion of intestinal CaSR increased crypt proliferation (Rey *et al.*, 2012). Thus, the stimulatory effect of CaSR on colonic differentiation and its suppressive effect on colonic epithelial proliferation may contribute to the preventive effect of dietary calcium on colon cancer.

24.6 Future Perspectives

As described above, CaSR is expressed in the gastrointestinal tract and has diverse roles in the regulating gastrointestinal functions beyond calcium homeostasis. Although CaSR was recently recognized as a nutrient-sensing receptor in the gastrointestinal tract, its role in nutrient absorption remains unclear. The involvement of epithelial CaSR in absorption of

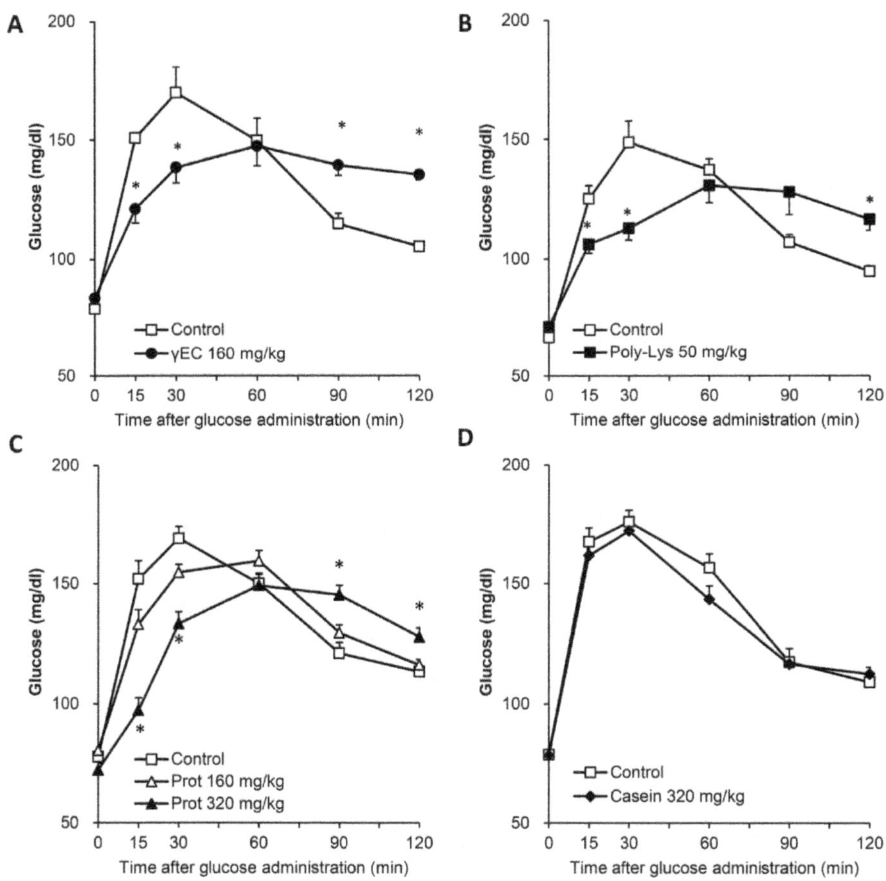

Figure 24.3 Oral administration of CaSR agonist peptides reduces glycemic responses to oral glucose load in rats (Muramatsu *et al.*, 2014). Glucose (2 g kg^{-1}, 8 mL kg^{-1}) solution containing or not containing (control) various peptides/proteins was orally administered in the stomach of conscious rats. (A) γ-Glu-Cys (γEC); (B) poly-D-lysine hydrobromide (poly-Lys); (C) protamine (Prot); (D) casein. Blood samples were collected from the tail vein before and after the oral administration of glucose solution. Plots with asterisks (*) indicate significant differences compared with control treatment at the same time point (Student's *t*-test or Dunnett's test, $P < 0.05$).

calcium, amino acids and peptides could be speculated because of ligand preference of CaSR for these nutrients. Future studies may clarify the regulation or modulation of absorption of these nutrients by intestinal CaSR. Moreover, CaSR may affect the absorption of other nutrients, including glucose and fatty acids. It has been recently demonstrated that activation of gastrointestinal CaSR by peptide agonists lowered gastric emptying and glycemic responses in rats (Figures 24.3 and 24.4, Muramatsu *et al.*, 2014). Further studies are required to understand the physiological relevance of such effect.

The roles of intestinal CaSR described above (Figure 24.5) have been mainly demonstrated in animal and cellular experiments. Therefore, it is essential that future studies investigate the effect of CaSR activation or inactivation in the human intestine. Furthermore, in order to affect intestinal CaSR, specific CaSR agonists or antagonist must be delivered to target cells, including those in the distal intestinal epithelium, myofibroblasts, and enteric neurons. Therefore, such compounds would require several properties, including gastric-acid resistance, low absorbability in the upper intestine, and low toxicity. Based on their effects in enteroendocrine cells, pharmacological or food-derived CaSR agonists may be used clinically to control glycaemia *via* enhancing GLP-1 secretion or to control appetite *via* CCK secretion.

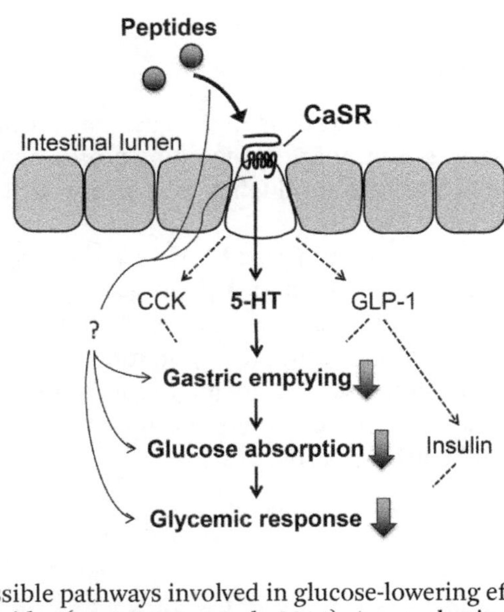

Figure 24.4 Possible pathways involved in glucose-lowering effect of CaSR agonist peptides (Muramatsu *et al.*, 2014). Arrows having solid lines present possible pathways suggested in the present study. Thin solid arrows present unknown pathway possibly involved. Dashed arrows and dashed lines present pathways that do not appear to be involved in the effect of CaSR agonist peptides.

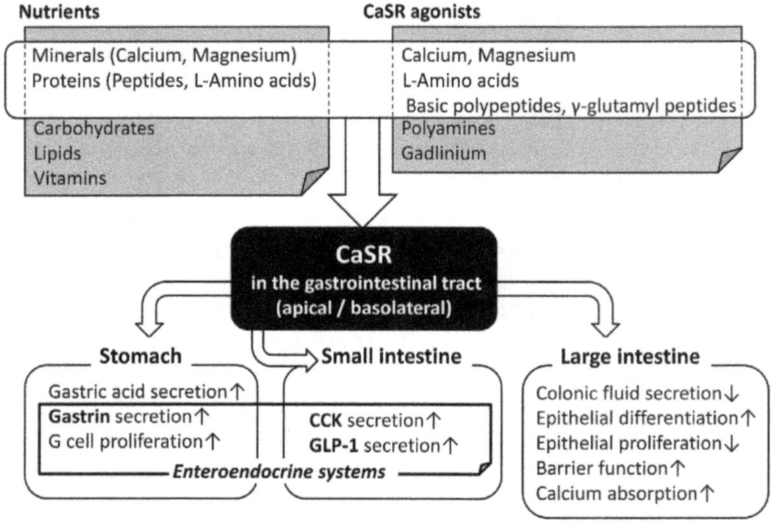

Figure 24.5 Multiple functions of CaSR in the gastrointestinal tract. Functions of CaSR in the gastrointestinal tract are summarized. Some of nutrients such as calcium, magnesium, amino acids and peptides are recognized as agonists for CaSR. Activation of gastric and intestinal CaSRs leads to various physiological responses, respectively. With regard to enteroendocrine systems, CaSR is involved in secretions of gastrin, CCK, GLP-1 induced by luminal amino acids.

Summary Points

- CaSR is expressed in various cells in the gastrointestinal tract.
- CaSR senses various compounds, including amino acids and peptides.
- CaSR functions in gastrin-producing cells and gastric parietal cells to regulate gastric-acid secretion.
- CaSR functions in enteroendocrine I cells to stimulate CCK secretion in response to amino acids and dietary peptides.
- CaSR functions in enteroendocrine L cells to stimulate GLP-1 secretion in response to amino acids and dietary peptides.
- Activation of CaSR supresses secretagog-induced colonic fluid secretion.
- Activation of CaSR stimulates differentiation and suppresses proliferation in intestinal epithelial cells.

Key Facts

CaSR

1. CaSR is a member of class C GPCRs and was originally identified in parathyroid cells.
2. CaSR is activated by increased extracellular calcium concentration.
3. Activation of CaSR in the parathyroid gland suppresses PTH secretion and calcium reabsorption in the kidney.

4. L-Amino acids, basic polypeptides, and gamma glutamyl peptides activate CaSR as allosteric modulators.
5. CaSR functions as a nutrient sensor in enteroendocrine cells.

Enteroendocrine System

1. Enteroendocrine cells produce and release gut hormones in response to nutrient ingestion.
2. Enteroendocrine cells are scattered throughout the intestinal epithelium, represent approximately 1% of the cell population, and are flask shaped.
3. A variety of gut hormones are produced and released from enteroendocrine cells and regulate gastrointestinal functions as well as appetite.
4. Enteroendocrine cells are classified based on the hormones they produce. For example, gastrin-producing cells are called G cells; CCK-producing cells are called I cells; GLP-1-producing cells are called L cells.
5. The gastrointestinal tract is thought to be the largest endocrine organ of the body due to the presence of the enteroendocrine systems.

Nutrient-Sensing Receptors

1. Nutrient-sensing mechanisms in the gastrointestinal tract were recently elucidated beginning in 2000.
2. Taste-sensing receptors are also expressed in the gastrointestinal tract and function as nutrient-sensing receptors.
3. Sweet, bitter, and umami taste receptors are functionally expressed in enteroendocrine cells.
4. GPR41 and GPR43 are functionally expressed in enteroendocrine cells as the short/long-chain fatty-acid receptors.
5. GPR40 and GPR120 are functionally expressed in enteroendocrine cells as the medium/long-chain fatty-acid receptors.

Definitions of Words and Terms

G-Protein coupled receptor (GPCR). GPCR is a family of transmembrane proteins that detect various extracellular signals and molecules such as light, smell, taste, hormones, neurotransmitters, ions. The signal is transduced to intracellular signaling molecules *via* tri-heteromeric G-proteins.
Gut hormones. Gut hormones are produced in specific endocrine cells in the gastrointestinal tract. Most gut hormones are released after a meal ingestion, and regulate various postprandial responses in the gastrointestinal tract and the body.
Gastrin. Gastrin is a gut hormone produced and secreted by gastric G cells. Gastrin release is stimulated by luminal calcium, amino acids and peptides. Gastrin stimulates gastric acid secretion from gastric parietal cells.

Cholecystokinin (CCK). CCK is a gut hormone produced and secreted by enteroendocrine I cells. I cells are mainly located at the upper small intestine. CCK release is stimulated by luminal fatty acids, peptides, and amino acids. CCK stimulates pancreatic enzyme secretion, inhibits gastric emptying and induces satiety through the vagus nerve.
Glucagon-like peptide-1 (GLP-1). GLP-1 is a gut hormone produced and secreted by enteroendocrine L cells. L cells are mainly located at lower small intestine and the large intestine. L cells also produce glucagon-like peptide-2 (GLP-2) and peptide-YY (PYY). GLP-1 enhances glucose-stimulated insulin secretion from pancreatic β cells.
Incretin. Incretins are gut hormones that enhance glucose-stimulated insulin secretion from pancreatic β-cells. GLP-1 and glucose-dependent insulinotropic polypeptide (GIP) are recognized as incretin.
Secretory diarrhoea. Secretory diarrhoea occurs as the result of increased fluid secretion from the intestinal epithelium. Enterotoxins, gut hormones (Vasoactive intestinal polypeptide (VIP) and Substance P) and fatty acids induce excessive fluid secretion.
Wnt. The Wnt signal pathway is involved in tissue development, and widely conserved in vertebrates and invertebrates. Abnormal Wnt signaling is related to colon-cancer development.
β-Catenin. β-Catenin is an intracellular protein that transduces extracellular The Wnt signal. β-Catenin is stabilized by the Wnt signal and forms a complex with other intracellular proteins (TCF/LEF). The complex activates the translation of target genes.
Parathyroid hormone (PTH). PTH is produced in parathyroid gland. PTH enhances bone resorption (calcium release from the bone), calcium reabsorption in the kidney, and intestinal calcium absorption to increase plasma calcium concentration.

List of Abbreviations

BMP	Bone morphogenetic protein
cAMP	Cyclic adenosine monophosphate
CaSR, CaR	Calcium-sensing receptor
CCK	Cholecystokinin
CDX	Caudal homeodomain factor
eGFP	Enhanced green fluorescent protein
GABA	Gamma aminobutyric acid
GIP	Glucose-dependent insulinotropic polypeptide
GLP-1	Glucagon-like peptide-1
GLP-2	Glucagon-like peptide-1
GPCR	G-Protein coupled receptor
GSH	Glutathione
Pept1	Peptide transporter 1
PTH	Parathyroid hormone
PYY	Peptide-YY

STa	Heat-stable *Escherichia coli* enterotoxin
TTX	Tetrodotoxin
5-HT	5-Hydroxytryptophan

References

Brown, E. M., Gamba, G., Riccardi, D., Lombardi, M., Butters, R., Kifor, O., Sun, A., Hediger, M. A., Lytton, J. and Hebert, S. C., 1993. Cloning and characterization of an extracellular Ca(2+)-sensing receptor from bovine parathyroid. *Nature*. 366(6455): 575–580.

Buchan, A. M., Squires, P. E., Ring, M. and Meloche, R. M., 2001. Mechanism of action of the calcium-sensing receptor in human antral gastrin cells. *Gastroenterology*. 120: 1128–1139.

Busque, S. M., Kerstetter, J. E., Geibel, J. P. and Insogna, K., 2005. L-Type amino acids stimulate gastric acid secretion by activation of the calcium-sensing receptor in parietal cells. *American Journal of Physiology: Gastrointestinal and Liver Physiology*. 289(4): G664–G669.

Butters, Jr, R. R., Chattopadhyay, N., Nielsen, P., Smith, C. P., Mithal, A., Kifor, O., Bai, M., Quinn, S., Goldsmith, P., Hurwitz, S., Krapcho, K., Busby, J. and Brown, E. M., 1997. Cloning and characterization of a calcium-sensing receptor from the hypercalcemic New Zealand white rabbit reveals unaltered responsiveness to extracellular calcium. *Journal of Bone and Mineral Research*. 12(4): 568–579.

Chattopadhyay, N., Cheng, I., Rogers, K., Riccardi, D., Hall, A., Diaz, R., Hebert, S. C., Soybel, D. I. and Brown, E. M., 1998. Identification and localization of extracellular Ca(2+)-sensing receptor in rat intestine. *American Journal of Physiology*. 274(1 Pt 1): G122–G130.

Cheng, I., Qureshi, I., Chattopadhyay, N., Qureshi, A., Butters, R. R., Hall, A. E., Cima, R. R., Rogers, K. V., Hebert, S. C., Geibel, J. P., Brown, E. M. and Soybel, D. I., 1999. Expression of an extracellular calcium-sensing receptor in rat stomach. *Gastroenterology*. 116: 118–126.

Cheng, S. X., Okuda, M., Hall, A. E., Geibel, J. P. and Hebert, S. C., 2002. Expression of calcium-sensing receptor in rat colonic epithelium: evidence for modulation of fluid secretion. *American Journal of Physiology: Gastrointestinal and Liver Physiology*. 283(1): G240–G250.

Cheng, S. X., 2012. Calcium-sensing receptor inhibits secretagogue-induced electrolyte secretion by intestine *via* the enteric nervous system. *American Journal of Physiology: Gastrointestinal and Liver Physiology*. 303(1): G60–G70.

Conigrave, A. D., Quinn, S. J. and Brown, E. M., 2000. L-Amino acid sensing by the extracellular Ca^{2+}-sensing receptor. *Proceedings of the National Academy of Sciences of the United States of America*. 97(9): 4814–4819.

Conigrave, A. D., Franks, A. H., Brown, E. M. and Quinn, S. J., 2002. L-Amino acid sensing by the calcium-sensing receptor: a general mechanism for coupling protein and calcium metabolism? *European Journal of Clinical Nutrition*. 56(11): 1072–1080.

Diakogiannaki, E., Pais, R., Tolhurst, G., Parker, H. E., Horscroft, J., Rauscher, B., Zietek, T., Daniel, H., Gribble, F. M. and Reimann, F., 2013. Oligopeptides stimulate glucagon-like peptide-1 secretion in mice through proton-coupled uptake and the calcium-sensing receptor. *Diabetologia*. 56(12): 2688–2696.

Dockray, G. J., 2003. Luminal sensing in the gut: an overview. *Journal of Physiology and Pharmacology*. 54(suppl. 4): 9–17.

Dufner, M. M., Kirchhoff, P., Remy, C., Hafner, P., Müller, M. K., Cheng, S. X., Tang, L. Q., Hebert, S. C., Geibel, J. P. and Wagner, C. A., 2005. The calcium-sensing receptor acts as a modulator of gastric acid secretion in freshly isolated human gastric glands. *American Journal of Physiology: Gastrointestinal and Liver Physiology*. 289(6): G1084–G1090.

Feng, J., Petersen, C. D., Coy, D. H., Jiang, J. K., Thomas, C. J., Pollak, M. R. and Wank, S. A., 2010. Calcium-sensing receptor is a physiologic multimodal chemosensor regulating gastric G-cell growth and gastrin secretion. *Proceedings of the National Academy of Sciences of the United States of America*. 107(41): 17791–17796.

Gama, L., Baxendale-Cox, L. M. and Breitwieser, G. E., 1997. Ca^{2+}-sensing receptors in intestinal epithelium. *American Journal of Physiology*. 273(4 Pt 1): C1168–C1175.

Geibel, J. P., Wagner, C. A., Caroppo, R., Qureshi, I., Gloeckner, J., Manuelidis, L., Kirchhoff, P. and Radebold, K., 2001. The stomach divalent ion-sensing receptor scar is a modulator of gastric acid secretion. *Journal of Biological Chemistry*. 276: 39549–39552.

Geibel, J., Sritharan, K., Geibel, R., Geibel, P., Persing, J. S., Seeger, A., Roepke, T. K., Deichstetter, M., Prinz, C., Cheng, S. X., Martin, D. and Hebert, S. C., 2006. Calcium-sensing receptor abrogates secretagogue-induced increases in intestinal net fluid secretion by enhancing cyclic nucleotide destruction. *Proceedings of the National Academy of Sciences of the United States of America*. 103(25): 9390–9397.

Gribble, F. M., 2012. The gut endocrine system as a coordinator of postprandial nutrient homoeostasis. *Proceedings of the Nutrition Society*. 71(4): 456–462.

Hira, T., Nakajima, S., Eto, Y. and Hara, H., 2008. Calcium-sensing receptor mediates phenylalanine-induced cholecystokinin secretion in enteroendocrine STC-1 cells. *FEBS Journal*. 275(18): 4620–4626.

Jongwattanapisan, P., Suntornsaratoon, P., Wongdee, K., Dorkkam, N., Krishnamra, N. and Charoenphandhu, N., 2012. Impaired body calcium metabolism with low bone density and compensatory colonic calcium absorption in cecectomized rats. *American Journal of Physiology: Endocrinology and Metabolism*. 302(7): E852–E863.

Liou, A. P., Sei, Y., Zhao, X., Feng, J., Lu, X., Thomas, C., Pechhold, S., Raybould, H. E. and Wank, S. A., 2011. The extracellular calcium-sensing receptor is required for cholecystokinin secretion in response to L-phenylalanine in acutely isolated intestinal I cells. *American Journal of Physiology: Gastrointestinal and Liver Physiology*. 300(4): G538–G546.

Mace, O. J., Schindler, M. and Patel, S., 2012. The regulation of K- and L-cell activity by GLUT2 and the calcium-sensing receptor CasR in rat small intestine. *Journal of Physiology*. 590(Pt 12): 2917–2936.

MacLeod, R. J., Hayes, M. and Pacheco, I., 2007. Wnt5a secretion stimulated by the extracellular calcium-sensing receptor inhibits defective Wnt signaling in colon cancer cells. *American Journal of Physiology: Gastrointestinal and Liver Physiology*. 293(1): G403–G411.

Muramatsu, M., Hira, T., Mitsunaga, A., Sato, E., Nakajima, S., Kitahara, Y., Eto, Y. and Hara, H., 2014. Activation of the gut calcium-sensing receptor by peptide agonists reduces rapid elevation of plasma glucose in response to oral glucose load in rats. *American Journal of Physiology: Gastrointestinal and Liver Physiology*. 306(12): G1099–G1107.

Nakajima, S., Hira, T., Eto, Y., Asano, K. and Hara, H., 2010. Soybean beta 51–63 peptide stimulates cholecystokinin secretion *via* a calcium-sensing receptor in enteroendocrine STC-1 cells. *Regulatory Peptides*. 159(1–3): 148–155.

Nakajima, S., Hira, T. and Hara, H., 2012. Calcium-sensing receptor mediates dietary peptide-induced CCK secretion in enteroendocrine STC-1 cells. *Molecular Nutrition & Food Research*. 56(5): 753–760.

Ohsu, T., Amino, Y., Nagasaki, H., Yamanaka, T., Takeshita, S., Hatanaka, T., Maruyama, Y., Miyamura, N. and Eto, Y., 2010. Involvement of the calcium-sensing receptor in human taste perception. *Journal of Biological Chemistry*. 285(2): 1016–1022.

Pacheco, I. I. and Macleod, R. J., 2008. CaSR stimulates secretion of Wnt5a from colonic myofibroblasts to stimulate CDX2 and sucrase-isomaltase using Ror2 on intestinal epithelia. *American Journal of Physiology: Gastrointestinal and Liver Physiology*. 295(4): G748–G759.

Peiris, D., Pacheco, I., Spencer, C. and MacLeod, R. J., 2007. The extracellular calcium-sensing receptor reciprocally regulates the secretion of BMP-2 and the BMP antagonist Noggin in colonic myofibroblasts. *American Journal of Physiology: Gastrointestinal and Liver Physiology*. 292(3): G753–G766.

Ray, J. M., Squires, P. E., Curtis, S. B., Meloche, M. R. and Buchan, A. M., 1997. Expression of the calcium-sensing receptor on human antral gastrin cells in culture. *Journal of Clinical Investigation*. 99(10): 2328–2333.

Rey, O., Chang, W., Bikle, D., Rozengurt, N., Young, S. H. and Rozengurt, E., 2012. Negative cross-talk between calcium-sensing receptor and β-catenin signaling systems in colonic epithelium. *Journal of Biological Chemistry*. 287(2): 1158–1167.

Riccardi, D. and Brown, E. M., 2010. Physiology and pathophysiology of the calcium-sensing receptor in the kidney. *American Journal of Physiology: Renal Physiology*. 298(3): F485–F499.

Sheinin, Y., Kállay, E., Wrba, F., Kriwanek, S., Peterlik, M. and Cross, H. S., 2000. Immunocytochemical localization of the extracellular calcium-sensing receptor in normal and malignant human large intestinal mucosa. *Journal of Histochemistry and Cytochemistry*. 48(5): 595–602.

Tolhurst, G., Reimann, F. and Gribble, F. M., 2012. Intestinal sensing of nutrients. *Handbook of Experimental Pharmacology*. 209: 309–335.

Wang, M., Yao, Y., Kuang, D. and Hampson, D. R., 2006. Activation of family C G-protein-coupled receptors by the tripeptide glutathione. *Journal of Biological Chemistry*. 281(13): 8864–8870.

Wang, W., Lu, M., Balazy, M. and Hebert, S. C., 1997. Phospholipase A2 is involved in mediating the effect of extracellular Ca^{2+} on apical K^{+} channels in rat TAL. *American Journal of Physiology*. 273(3 Pt 2): F421–F429.

Wang, Y., Chandra, R., Samsa, L. A., Gooch, B., Fee, B. E., Cook, J. M., Vigna, S. R., Grant, A. O. and Liddle, R. A., 2011. Amino acids stimulate cholecystokinin release through the Ca^{2+}-sensing receptor. *American Journal of Physiology: Gastrointestinal and Liver Physiology*. 300(4): G528–G537.

CHAPTER 25

Taste Cells and Calcium Signaling

KATHRYN F. MEDLER*[a]

[a]Department of Biological Sciences, University at Buffalo, The State University of New York, Buffalo, Buffalo, NY 14260, USA
*E-mail: kmedler@buffalo.edu

25.1 Introduction

The ability to detect chemicals in our environment is required to identify nutrients, conspecifics, predators or potentially harmful conditions. Because these functions are crucial for survival, all organisms, including single-celled organisms, detect chemicals in their environment. While simple organisms have a single chemosensory system, vertebrates and many invertebrates utilize both olfactory and taste systems to detect chemicals in their environment. While the olfactory system is used in many social interactions, the sense of taste is used to discriminate between items that have potentially nutritional value and items that are potentially harmful to consume. Thus, our sense of taste is critical for our survival and well-being.

The chemical detectors of the taste system are the taste receptor cells (TRCs) that are found in specialized groupings of cells called taste buds (Figure 25.1). Taste buds consist of 50–150 TRCs and are found on the tongue and palate in the mouth. TRCs extend microvilli into the oral cavity where they contact the chemicals that are being released from the potential food item as it is masticated. TRCs convert the initial chemical signal into an

Food and Nutritional Components in Focus No. 10
Calcium: Chemistry, Analysis, Function and Effects
Edited by Victor R. Preedy

Published by the Royal Society of Chemistry, www.rsc.org

electrical signal that is sent to the brain for processing *via* the afferent gustatory neurons. Depending on the signal that is sent to the brain, the organism will either consume the food item or eject it.

TRCs are neuroepithelial cells because they have properties of both neurons and epithelial cells. Like neurons, TRCs are excitable cells, capable of firing action potentials and forming both conventional and nonconventional synapses with gustatory neurons. Like epithelial cells, TRCs are derived from epithelia precursors, express epithelial markers including epithelial ion channels and have a limited lifespan that ranges from two weeks to two months (Chaudhari and Roper, 2010; Perea-Martinez *et al.*, 2013). In mammals, taste buds are localized in specialized bumps and grooves called papillae. There are three types of papillae that house taste buds on the tongue: circumvallate (CV), foliate (Fol) and fungiform (Fun). The CV papillae are comprised of crypts that are lined with taste buds, are found on the back of the tongue and house hundreds of taste buds. The Fol papillae are slits found on the both sides of the tongue and also house hundreds of taste buds. The Fun papillae are small protrusions that house 1–2 taste buds each and are scattered across the anterior two-thirds of the tongue (Figure 25.1). Additional taste buds are found on the palate and throat, including a row of taste buds found between the hard and soft palate called the geschmacksstreifen. However, most taste buds are localized in the three papillae types that are found on the tongue. The taste buds within these different papillae are innervated by different cranial nerves, including the facial, glossopharyngeal and vagus nerves. A branch of the facial nerve called the chorda tympani nerve innervates the Fun papillae, while the glossopharyngeal nerve innervates the CV papillae. The Fol papillae are innervated by

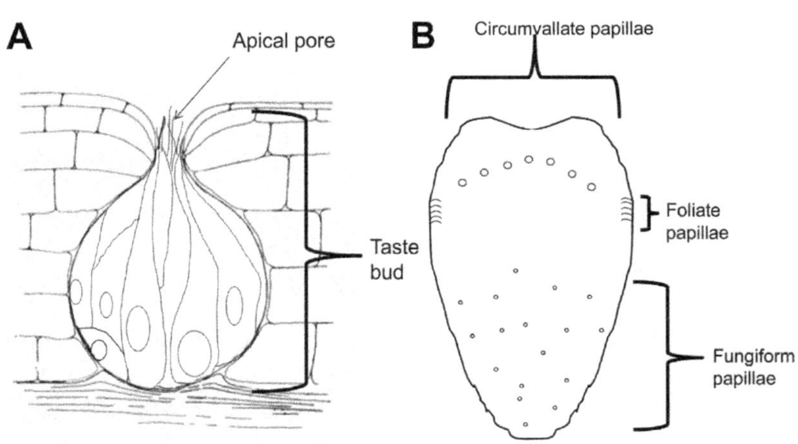

Figure 25.1 General taste bud and tongue morphology. (A) A schematic of a taste bud that is comprised of taste receptor cells (TRCs). Elongate TRCs extend microvilli into the apical pore where contact with taste stimuli in the oral cavity occurs. (B) Schematic of a mammalian tongue indicating the approximate location of the circumvallate, foliate and fungiform papillae.

both the chorda tympani and glossopharyngeal nerve, while the vagus nerve innervates the taste buds in the back of the throat. The palate taste buds are innervated by the greater superficial petrosal branch of the facial nerve.

Compared to other sensory systems, the taste system is unique in its ability to detect a large number of chemically diverse stimuli that vary in their molecular size, pH, charge and lipophilicity. In order to detect these different taste stimuli, TRCs use multiple transduction mechanisms but a particular TRC does not express all the signaling mechanisms and is only sensitive to a subset of different taste stimuli (Figure 25.2). This specificity allows the taste system to detect lots of different types of taste stimuli but still discriminate between them. There are five basic taste qualities that include sour, salty, bitter, sweet and umami which is the detection of amino acids, in particular glutamate. Salty taste is used to detect ions that are needed to maintain a

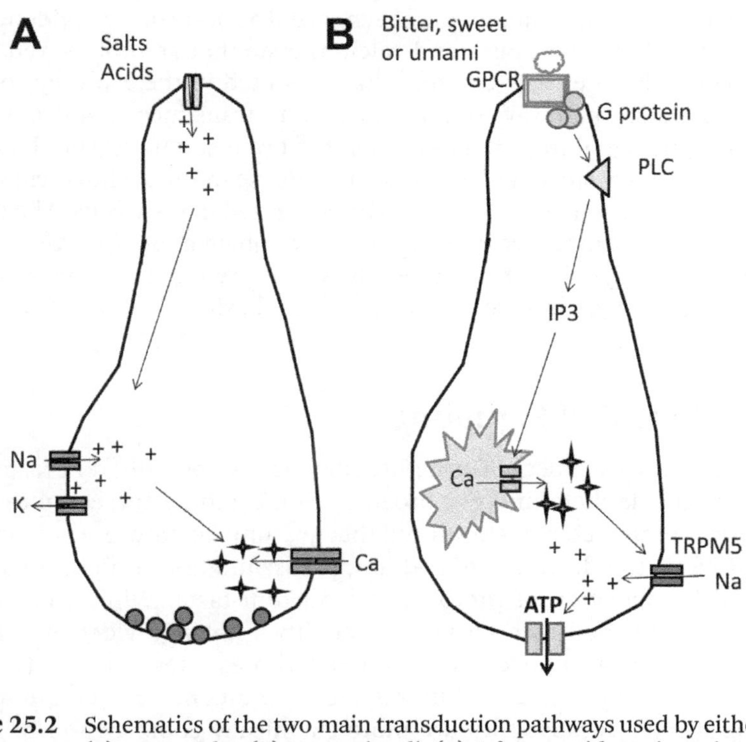

Figure 25.2 Schematics of the two main transduction pathways used by either ionic (A) or complex (B) taste stimuli. (A) Salts or acids activate ionotropic receptors to cause an influx of positively charged ions that depolarize the cell. This initial depolarization activates voltage-gated sodium and potassium channels that fire an action potential and open voltage-gated calcium channels. Calcium influx initiates vesicular release of neurotransmitter. (B) Bitter, sweet and umami stimuli activate a G-protein coupled receptor (GPCR) that initiates a signaling pathway. Activated G proteins turn on phospholipase C (PLC) that produces inositol triphosphate (IP_3) that binds to a receptor on the internal calcium stores to cause release of calcium into the cytosol. Calcium activates TRPM5 that depolarizes the cell and activates a hemichannel to release ATP.

normal ionic balance in our bodies. Sweet is used to detect carbohydrates such as sugars that are consumed to provide energy while umami, is used to detect amino acids in proteins. We also use our taste system to avoid potentially harmful compounds. Sour is used to detect acids in foods, which is a general indicator that the food has spoiled. Bitter compounds are often alkaloids or toxins that are harmful (Iwata *et al.*, 2014).

The sensitivity of a particular organism to these different taste qualities differs depending on their diet and nutritional needs. Studies have shown that strict carnivores do not have functional sweet receptors because they do not need to consume carbohydrates for energy but instead have a well-developed umami taste. Herbivores are less sensitive to bitter compounds since much of their diet can be somewhat bitter. Other organisms are primarily sensitive to bitter taste and will consume most things that are not bitter, while avoiding bitter compounds (Finger and Simon, 2000).

While these five taste qualities are recognized as basic tastes, evidence also suggests that there is fat taste and calcium taste that are not as well established. These divergent taste stimuli have resulted in the evolution of multiple transduction pathways that are needed to transduce chemical stimuli into an output signal from the TRCs that the brain can understand. Regardless of the signaling mechanism used, an increase in intracellular calcium is required for a normal output signal to be generated in taste cells. The goal of this review is to summarize what is currently known about how these taste-evoked calcium signals are generated, how they are regulated and how they appear to be changed by disease, in particular obesity.

25.2 Taste Cell Signaling

While there is a great deal of structural diversity across different taste stimuli, we can divide them into two broad categories: ionic and complex. Ionic stimuli are comprised of taste stimuli that are ionic in nature. These include salty, which is the detection of sodium, potassium and lithium ions; sour, which is the detection of protons; and calcium taste. While calcium has not been established as a basic taste quality, growing evidence indicates that calcium acts as a taste stimulus and activates a response in TRCs. All other taste stimuli, including bitter, sweet and umami, can be grouped as complex stimuli because they are more chemically complex and primarily activate a G-protein coupled receptor (GPCR) signaling pathway (DeFazio *et al.*, 2006). The detection of fat has also not been established as a basic taste but multiple studies demonstrate that free fatty acids are detected by TRCs, though its effects on cells are not yet completely understood. Some studies have demonstrated exceptions to the absolute dependence on calcium signaling in the ability of the TRCs to respond to some taste stimuli, however, each taste quality requires an increase in intracellular calcium to generate a normal output signal (Gilbertson and Boughter, 2003). The source of these calcium increases varies by taste quality.

25.2.1 Ionic Stimuli

Two established taste qualities, salty and sour, detect the presence of different ions in the oral cavity. Because ions have a simple structure and a positive charge, they are able to interact with ionotropic receptors localized on the microvilli of the TRCs. These ions can pass through the channels and directly depolarize the taste cells. Since ions can also pass through tight junctions between taste cells, they can also potentially interact with channels on the basolateral membrane of the TRCs. While the detection of calcium ions is not an established taste quality, there is ample evidence that taste receptor cells detect and respond to external calcium ions. Interestingly, these initial studies indicate that calcium is not activating an ion channel but is instead activating GPCRs (Iwata *et al.*, 2014).

25.2.1.1 Salty Taste

Terrestrial animals all detect salts that are needed to maintain a correct ionic balance within the body. For most organisms, low concentrations of salts, primarily sodium, are appetitive and are readily consumed. Very high concentrations of salts are generally aversive and are avoided. These differences likely evolved from the need to maintain an appropriate ionic balance but also to not disrupt the body's osmotic balance by consuming too much salt. Since salts are small charged ions, salt stimuli interact with apically located channels that open and allow sodium influx that initiates signaling inside the cell. Multiple channels have been identified that are salt sensitive. Primarily there are two types of salt receptors: (1) the epithelial sodium channel that is sensitive to the diuretic amiloride that blocks a resting sodium conductance in many taste cells and (2) a salt receptor that is not sensitive to amiloride. While multiple channels have been proposed to be responsible for the amiloride-insensitive component, the channel identity is still unclear. All organisms that have been studied to date have both amiloride-sensitive and amiloride-insensitive components to their salt taste, though the relative contribution of each varies across species. Regardless of the receptors, when salt is present, the taste receptor cell depolarizes, which activates VGCCs, leading to a large calcium influx and synaptic transmission (Medler and Kinnamon, 2004) (Figure 25.3A).

25.2.1.2 Sour Taste

Sour taste is used to detect protons that are usually a sign of spoiled food that should be avoided. There is still debate as to what constitutes a sour response since application of protons causes a general drop in intracellular pH in all cells. However, it is not clear that all taste cells transmit sour information to the brain. Multiple receptors have been identified as potential sour receptors but conflicting studies have prevented a clear identification (Huang *et al.*, 2006; Ishimaru *et al.*, 2006; Ishii *et al.*, 2009). Despite the lack of an

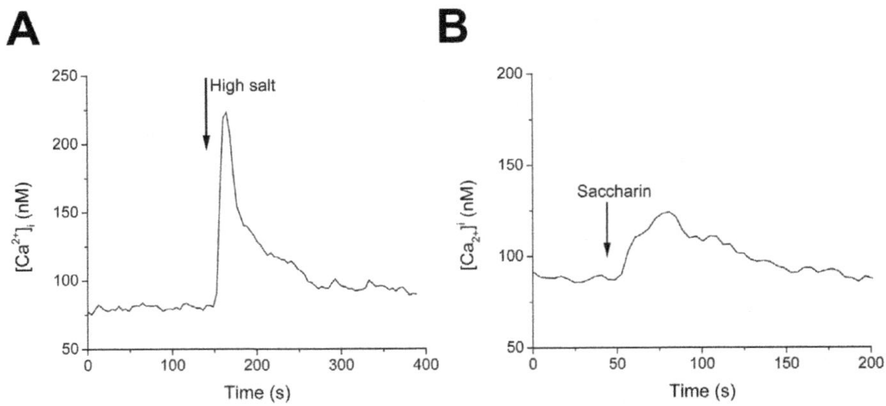

Figure 25.3 Different taste stimuli evoke distinct calcium signals in TRCs. Taste cells were loaded with Fura2-AM, a calcium-sensitive dye and the taste-evoked changes in cytosolic calcium were recorded. (A) An ionic stimulus, a salt (50 mM potassium chloride) evokes a large calcium influx in TRCs. (B) A more chemically complex stimulus such as the sweetener saccharin (2 mM) causes a slower, smaller cytosolic calcium increase.

established sour receptor, it is generally accepted that protons from the oral cavity enter the taste cell and cause an intracellular acidification of the cell. This acidification blocks constitutively active potassium channels that causes a depolarization that will activate voltage-gated calcium channels (VGCCs) and generate a large calcium influx that triggers vesicular fusion and neurotransmitter release. While not all TRCs have conventional chemical synapses that depend on calcium influx through VGCCs to initiate synaptic transmission, most of the TRCs that do have these synaptic specializations also respond to sour stimuli by generating a large calcium influx (Richter *et al.*, 2003).

25.2.1.3 Calcium Taste

Increasing evidence indicates that animals in need of calcium to maintain ionic homeostasis develop a "calcium appetite" that is controlled by the taste system. Animals deprived of calcium will actively ingest calcium salts, while animals not deprived of calcium do not (Tordoff, 2001). There are currently two candidate calcium receptors that have been identified in taste cells: the calcium-sensing receptor (CaSR) and the T1R3 receptor. Both of these candidate calcium receptors are GPCRs that activate second messenger signaling pathways and do not allow calcium ions to enter the cell directly. T1R3 is an established taste receptor that heterodimerizes with other members of the T1R family to form either a sweet or umami receptor. In studies using T1R3–KO mice, it was shown that mice lacking a functional T1R3 did not respond appropriately to calcium ions compared to wild type. It is still unclear which candidate receptor is the calcium receptor but perhaps both receptors are required and they form heterodimers. Regardless of the receptor, increases in extracellular calcium cause a phospholipase C (PLC)-dependent calcium signal that is generated in response to the production of inositol triphosphate (IP_3) by PLC, which causes the subsequent release of calcium from internal stores (Bystrova *et al.*, 2010; Tordoff *et al.*, 2012).

25.2.2 Chemically Complex Stimuli

Taste stimuli that comprise bitter, sweet and umami tastes are too chemically complex to efficiently activate or permeate ion channels and instead, primarily bind to GPCRs. These GPCRs activate a second messenger signaling pathway that is largely conserved between bitter, sweet and umami taste stimuli. Two families of GPCRs have been identified as taste receptors: the T2Rs that are bitter receptors and the T1Rs that are required for the detection of both sweet and umami stimuli. While T2Rs are well established as the primary family of bitter receptors and the T1Rs are important for the normal detection of sweet and umami stimuli, there is strong evidence that additional receptors are also contributing to the transduction of at least some of these stimuli. Both families of receptors activate a specific calcium-dependent second messenger pathway that is described below (Chandrashekar *et al.*, 2000; Nelson *et al.*, 2001, 2002). While this is the primary signaling pathway to transduce bitter, sweet or umami stimuli, other studies have shown that these taste stimuli can also affect cyclic nucleotide levels and potassium channels that would modulate membrane potential and cell excitability. A thorough discussion of these pathways is beyond the scope of this review, but it should be noted that not all of these taste responses solely depend on a calcium signal to generate a response.

25.2.2.1 Bitter, Sweet and Umami Stimuli

Bitter stimuli are detected by a family of GPCRs called the T2Rs (Chandrashekar *et al.*, 2000). When bound by a bitter stimulus, this GPCR activates a heterotrimeric G protein that may include an alpha G protein called gustducin. Gustducin is primarily found in taste cells and activates a phosphodiesterase to reduce cyclic cAMP levels that modulates the cell's responsiveness. Gustducin is coupled to its $\beta\gamma$ partners that activate a PLC pathway. When PLC is activated, it cleaves phosphatidylinositol 4,5-bisphosphate (PIP_2) to produce IP_3 and diacylglycerol (DAG). IP_3 binds to IP_3 receptors on internal calcium stores and causes a calcium release from these stores. This increase in intracellular calcium activates a transient receptor potential (TRP) channel called TRPM5. TRPM5 is a monovalent-selective TRP channel that opens and allows sodium influx. This sodium influx causes a membrane depolarization that generates an action potential and opens a hemichannel to release ATP onto the afferent gustatory neuron. Thus, neurotransmitter release from the taste cell is completely dependent on calcium release from stores (Huang and Roper, 2010). This differs from other neurons that all depend on calcium influx through VGCCs to release a chemical neurotransmitter.

Sweet taste is detected primarily by a heterodimer of two GPCRs called T1R2 and T1R3. This heterodimer is sensitive to multiple sweeteners, including sugars and artificial sweeteners. When bound by a sweet stimulus, this heterodimer activates the same PLC signaling pathway that was described for the bitter receptors. This signaling pathway appears to be the primary mechanism to transmit information about sweet stimuli to the afferent gustatory neurons that carry that information to the brain. There is some evidence of additional sweet transducing pathways, but these are not as well characterized (Nelson *et al.*, 2001).

Umami stimuli activate a heterodimer of two GPCRs called T1R1 and T1R3. This is the same T1R3 receptor that couples with T1R2 to form a sweet receptor. Like the pathways for bitter and sweet, the detection of umami stimuli results in the activation of the PLC signaling pathway to cause calcium release from stores, TRPM5 activation, and ultimately, ATP release from a hemichannel. There are multiple studies that also support a role for a truncated metabotropic glutamate receptor (mGluR4) in the detection of umami stimuli. When activated, this receptor also causes a release of calcium from internal stores, but the exact downstream effects of this receptor activation are not well known (Zhao *et al.*, 2003).

Thus, the receptors that detect these chemically complex, but divergent taste stimuli, all appear to activate the same signal-transduction pathway that relies on calcium release from internal stores to cause the release of a chemical neurotransmitter (Figure 25.3B). This is unique among neurons that primarily have conventional chemical synapses with neurotransmitter stored in vesicles that are released in response to a large calcium influx through VGCCs. While some taste cells do have the conventional chemical synapse machinery, the taste cells that are responsible for the detection of bitter, sweet and umami taste stimuli do not (Iwata *et al.*, 2014).

All of these taste qualities rely on ATP release from a hemichannel to transmit a signal to the brain. This unique signaling system is still not completely characterized and the identity of the hemichannel is the subject of conflicting studies. Some studies conclude that the hemichannel is a pannexin, while other studies point to connexin hemichannels (Huang *et al.*, 2007; Romanov *et al.*, 2007). A recent study suggests that the calcium homeostasis modulator 1 (CALHM1) may be the ATP release channel (Taruno *et al.*, 2013). Further studies are needed to better characterize this hemichannel.

Another interesting puzzle in this signaling pathway is the source of the ATP that acts as a neurotransmitter. These taste cells release ATP in response to bitter, sweet or umami stimuli, but they do not possess vesicles to store the ATP and the source of this ATP is unclear (Finger *et al.*, 2005). Anatomical studies have identified the presence of "atypical" mitochondria localized at the site of nerve contact in these taste cells (Royer and Kinnamon, 1988); however, the specialized function of these mitochondria, if any, is unknown. It is intriguing to speculate that these large mitochondria are producing ATP in response to an increase in intracellular calcium and are acting as the source of ATP that is released through the hemichannel.

25.2.2.2 Fat Taste

While the detection of fat in our food is not yet considered one of the basic tastes, a growing number of studies demonstrate that taste receptor cells are sensitive to and respond to free fatty acids, in particular, long-chain fatty acids (LCFAs). Some studies have shown that fatty acids can alter potassium-channel activity that will change the responsiveness of the taste cell. Other studies have identified two types of receptors that are expressed in TRCs and respond to fatty acids. One is called CD36, which is an integral membrane glycoprotein that binds to LCFAs with high affinity. CD36 is found in the CV and Fol papillae but is not significantly expressed in the Fun papillae. When CD36 is nonfunctional or absent, taste cells are not responsive to the free fatty acids that have been tested, suggesting an essential role for CD36 in fat taste. However, the downstream effects of CD36 are not known beyond the finding that CD36 positive cells respond to fatty acids with a large cytosolic calcium release that depends on calcium release from internal stores (Gilbertson and Khan, 2013). Further studies are needed to identify the key components of this pathway, including any potential involvement with other pathways. The other identified receptor type is a GPCR. Two GPCRs have been implicated as fat receptors, GPR120 and GPR40. Ablation of either of these GPCRs caused a reduction in the ability to detect LCFAs but GPR120 has much broader expression within the TRCs. While the downstream signaling pathway of the LCFA sensitive GPCRs has not been characterized, initial studies suggest that GPR120 activates a similar signaling pathway as seen for the chemically complex taste stimuli, in that it depends on activation of PLC to cause calcium release and activation of TRPM5 (Liu *et al.*, 2011; Gilbertson and Khan, 2013). TRPM5 has been reported to have a key role in the detection of lipids since TRPM5-null mice lose their preference for lipids (Sclafani *et al.*, 2007).

Another study identified a potential role for the stromal interaction molecule (STIM 1) in fat taste. STIM1 is the calcium sensor on the internal calcium stores that activates the store-operated channel, Orai1, which is located on the plasma membrane and replenishes the internal calcium stores after activation. While STIM1 is widely expressed, its role in taste transduction has not been studied with the exception of this potential role in fat taste. In taste cells, the STIM1-dependent activation of Orai channels causes a large calcium influx that is required for fat-taste perception (Dramane *et al.*, 2012). Based on the similarities between what is known about the signaling pathway that is activated by LCFAs and the signaling pathway for chemically complex taste stimuli, one could predict that STIM1 signaling is critical for bitter, sweet and umami taste stimuli as well. However, further studies are needed to identify the role of STIM1 in these transduction pathways.

25.3 Calcium Regulation in Taste Cells

Since calcium is a ubiquitous molecule that is an important second messenger in all cells, its concentration inside a cell must be tightly regulated. Taste cells are no exception and indeed, the transduction of all known taste stimuli depends primarily on calcium. It is somewhat surprising then, how little is currently known about the regulation of calcium in taste cells. Taste cells are especially intriguing in this respect compared to other excitable cells since they are constantly exposed to variations in their external environment. The ionic balance between the inside and outside of a cell is especially critical for excitable cells because their ability to respond to signals depends on this balance since this controls excitability. Taste cells are in physical contact with the external environment of the oral cavity and this environment changes every time the organism is consuming potential nutrients. Since taste cells are responding to chemicals that are introduced during consumption, it is absolutely critical that these cells maintain an appropriate ionic balance during feeding (Finger and Simon, 2000).

Our initial studies found that taste cells use multiple mechanisms to regulate their cytosolic calcium levels. The relative role of these calcium regulatory mechanisms depends on the calcium signal being generated which varies significantly between the ionic and complex taste stimuli (Szebenyi *et al.*, 2010). We have also found that taste cells actively regulate their cytosolic calcium, even in the absence of taste stimulation, keeping their baseline cytosolic calcium levels low by actively engaging mitochondrial buffering and sodium–calcium exchanger activity to regulate a constitutive calcium influx that can occur. The channel responsible for this constitutive calcium influx in the absence of cell stimulation has not been identified in taste cells but our initial study indicates that TRPV1 may be responsible (Hacker and Medler, 2008; Laskowski and Medler, 2009). However, since that research was initially published, Calhm1 has been identified in taste cells with a postulated role as the ATP release channel (Taruno *et al.*, 2013). Based on the known properties of this channel, it is possible that it may contribute to the regulation of basal calcium levels in taste cells. Since the pharmacology of the Calhm1 channel is not well characterized, the TRPV1 blocker used in our earlier study may also affect the activity of Calhm1. Further studies are needed to elucidate the channels contributing to this constitutive calcium influx. Upon stimulation, the calcium regulatory mechanisms also contribute to the reduction of the evoked calcium response (Szebenyi *et al.*, 2010). A more indepth discussion can be found in an earlier review (Medler, 2010).

25.4 Effects of Obesity on Taste Cell Signaling

Many studies have shown that obesity affects taste perception in both humans and rodents (Bartoshuk *et al.*, 2006; Kovacs and Hajnal, 2008). These studies are not always in agreement, which likely reflects the fact that obesity is a complex disease and the relationship between taste, appetite and satiety is not well characterized. Since food consumption is strongly influenced by appetite, the taste system is a potential target in appetite regulation (Shin *et al.*, 2008; Horio *et al.*, 2010). In humans, studies have shown that obesity causes an increased preference for sweet and umami taste stimuli, even though the subjects may not taste them as well as their lean counterparts (Bartoshuk *et al.*, 2006; Pepino *et al.*, 2010). Other studies have shown no differences in sweet preferences between obese and lean individuals (Anderson, 1995). Similar studies on obesity and fat taste have found a decreased ability to detect LCFAs in obese people. Thus, a higher concentration of LCFAs was needed for detection in people with a high body-mass index. Further studies have found genetic differences in the CD36 and GPR120 genes that are associated with a reduced ability to detect LCFAs in some obese subjects. These studies suggest that a relationship exists between the ability to detect taste stimuli and obesity; however, the cellular mechanisms responsible for this relationship are poorly understood.

Most studies have focused on the effects of obesity on behavioral responses or on the effects of obesity in the central taste system that controls perception. We recently published a study that focused on the taste-evoked calcium signals in the TRCs of diet-induced obese (DIO) mice compared to their lean littermates (Maliphol *et al.*, 2013). We performed this study to determine if these initial taste signals are different between the obese and normal mice. We found several highly significant differences between the two groups. Fewer taste cells were responsive to the taste stimuli tested in the obese mice (Figure 25.4). In addition, the taste-evoked responses that were generated were significantly smaller in the DIO mice. This effect was selective for the stimuli tested, the sweet stimuli were significantly reduced but the umami stimulus was not. The frequency of the responses to the bitter stimulus that we tested was not reduced, but the evoked signal was significantly reduced in size. Our behavioral studies on these mice also found a reduction in their ability to detect both sweet and bitter stimuli with no difference in their ability to detect umami compared to control (Figure 25.5). This is interesting since both sweet and umami are considered appetitive and are normally preferred

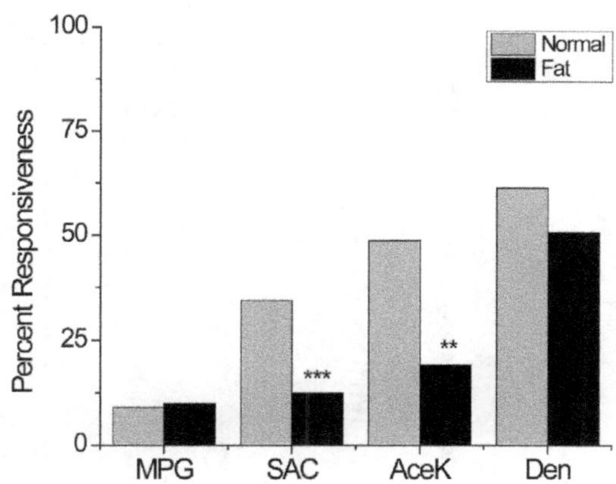

Figure 25.4 DIO reduces the frequency of taste responses to sweet stimuli. Bar graphs represent the percentages of taste-evoked calcium responses. The number of responsive cells for the sweet stimuli (2 mM Sac, 20 mM AceK) were significantly reduced for the obese mice (black bar) compared to controls (gray bar) (***, $p < 0.001$; **, $p < 0.01$; *, $p < 0.05$). The numbers of taste-evoked calcium responses to the umami stimulus (20 mM MPG) and the bitter stimulus (5 mM Den) were not significantly different. Reprinted from Maliphol *et al.* (2013).

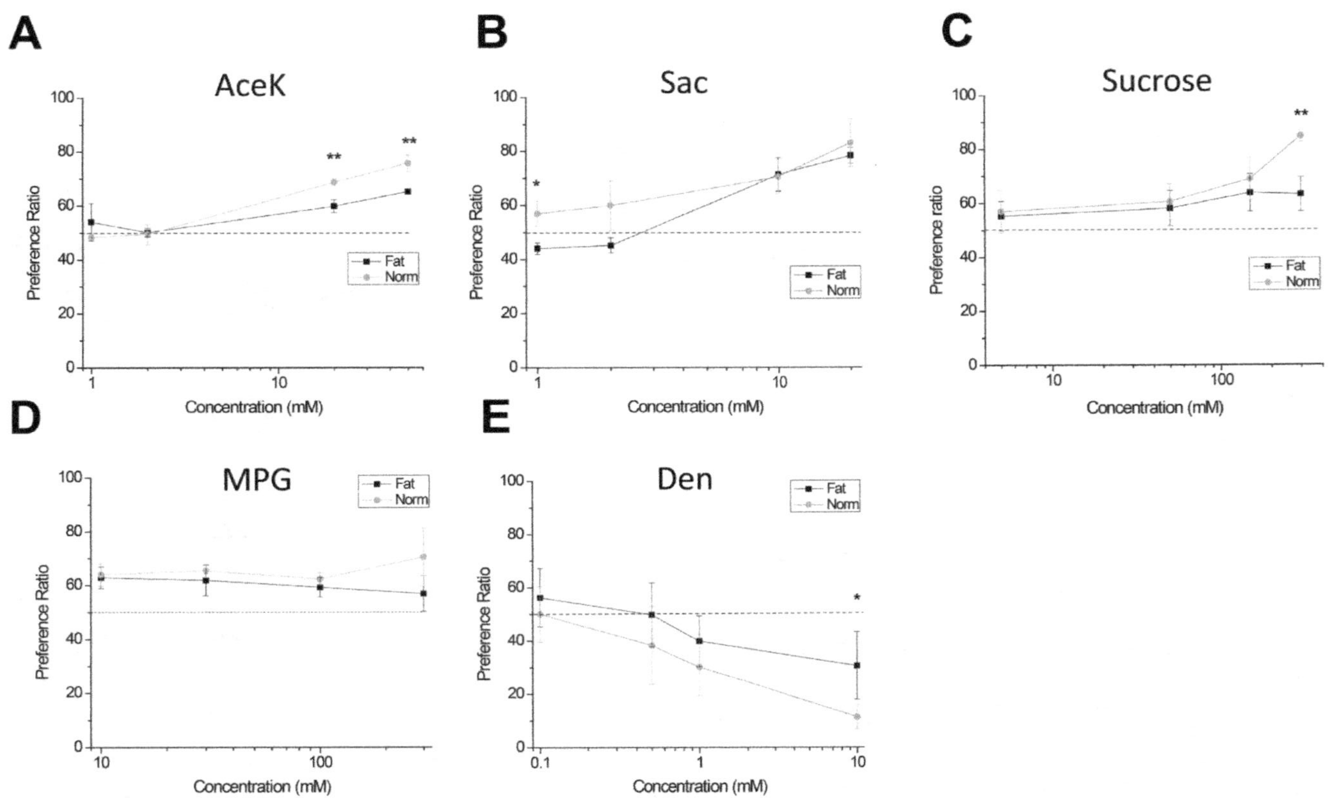

Figure 25.5 DIO alters the ability to detect some taste stimuli in mice. (A) The ability to detect the sweetener, AceK, at higher concentrations was impaired in obese mice compared to controls. (B) For saccharin, obese mice were significantly impaired in their ability to detect lower concentrations of the stimulus but were not significantly different from controls at higher concentrations. (C) Obese mice preferences were comparable to control mice for sucrose except for the highest concentration of sucrose tested (300 mM). At this concentration, control mice had a significantly higher preference for sucrose compared to the obese mice. (D) Control mice showed an increased preference for the umami stimulus, MPG, at the highest concentration but no significant differences between normal and obese mice were measured ($p = 0.056$). (E) The ability to avoid denatonium at the highest concentration was compromised in obese mice compared to littermate controls.

by the mice, while the bitter compound is aversive and actively avoided by the mice. Our data suggests that the responsiveness of the peripheral TRC is suppressed in DIO mice, which may contribute to overconsumption and weight gain (Maliphol *et al.*, 2013).

25.5 Conclusions

Calcium signaling in TRCs is an important and complex facet of taste perception. Regardless of the taste stimulus, a change in intracellular calcium is required to transmit signals to the brain that specific chemical groups have been detected. This leads to the decision to either consume the item or reject it. Different taste stimuli activate different signaling pathways within the taste cells, which generates distinct calcium signals. As a result, these calcium signals must be tightly regulated to ensure that the appropriate output response in produced. Obesity can adversely affect these calcium signals, which may ultimately alter consumption. Clearly, calcium signaling is a critical component of taste transduction and could potentially be affected by other diseases in addition to obesity. Further studies are needed to determine how taste is affected by other disease states.

Summary Points

- This chapter focuses on calcium signaling in the peripheral taste system.
- Taste receptor cells are located on the tongue and in the oral cavity and are responsible for the detection and identification of chemicals in our food.
- There are five established taste qualities: sour, salty, bitter, sweet and umami, which is the detection of amino acids/proteins.
- Different taste qualities have divergent chemical structures.
- Depending on the taste stimulus, it can either activate a channel that depolarizes the cell and activates voltage-gated calcium channels or a receptor that activates a second messenger signaling pathway.
- Calcium and fat stimuli can also activate taste receptor cells.
- Regardless of the activation mechanism, all taste stimuli cause an increase in cytosolic calcium to generate a signal that is sent to the brain.
- These calcium signals must be regulated and multiple mechanisms are present in taste cells to control cytosolic calcium levels.
- Diseases, such as obesity, can change how taste receptor cells respond to stimuli.

Key Facts of Taste

1. All organisms have a sense of taste and it was the first sensory system to evolve.
2. Taste is used to detect chemicals in the environment.

3. Taste is used to detect nutrients to ingest and harmful chemicals to avoid.
4. Taste receptor cells are derived from epithelial cells but have properties like neurons.
5. Taste cells are constantly replaced since they can be damaged by the environment.

Definitions of Words and Terms

Ionic stimuli. Taste stimuli that are comprised of ions and have a positive charge. Both salty and sour stimuli are ionic stimuli. These stimuli interact with ionotropic receptors to generate responses.

Complex stimuli. Taste stimuli that are more chemically complex and are too bulky to interact with ionotropic receptors. Instead these stimuli activate GPCRs and initiate second messenger signaling pathways. Bitter, sweet and umami are complex stimuli.

Salt taste. The detection of salts, including sodium, potassium, and lithium, by taste receptor cells. Consumption of these ions is needed to maintain an appropriate ionic environment within the body.

Sour taste. The detection of protons in acids. This taste quality is used to detect potentially harmful compounds or rotten food.

Calcium taste. The detection of calcium ions by taste receptor cells. This taste is used to maintain an appropriate calcium balance in our bodies by motivating the consumption of calcium.

Bitter taste. An aversive taste that motivates the organism to not consume a potential food item. This taste is often associated with toxins and alkaloids.

Sweet taste. The detection of sugars and carbohydrates that are consumed to provide energy needed for normal function.

Umami taste. The detection of amino acids, specifically glutamate in our diet.

Fat taste. The detection of fatty acids in our diet. This is an appetitive taste since fatty acids are used as an energy source.

Taste receptor cells. Neuroepithelial cells that are housed in taste buds and express the receptors that detect chemicals in our food. These cells convert the chemical signal into an output signal that the brain can understand.

Excitable cells. Cells that are able to produce and respond to electrical signals.

Taste papillae. Specialized epithelial bumps or grooves on the tongue that house the taste buds that are made of taste receptor cells.

Lipophilicity. The ability to dissolve in oils or fats.

Ionotropic receptor. A receptor that opens in response to a stimulus and allow ions to enter the cell directly.

G-protein coupled receptor. A receptor that activates a G-protein to initiate a signaling cascade inside a cell in response to binding a stimulus.

Long chain fatty acid. A chemical component of fat that consists of carboxylic acid and a long aliphatic chain of 13–20 carbons.

Stim1. A calcium sensor found on internal calcium stores in a cell that is activated when calcium is depleted from the internal stores.
Orai channel. A channel that gets activated by Stim1 and when open, allows for calcium influx to help refill the internal calcium stores.
Appetitive taste. A chemical stimulus that activates the taste system and generates a positive response that results in the consumption of the potential food item.
Aversive taste. A chemical stimulus that activates the taste system and generates a negative response that usually causes the organism to not consume the item.

List of Abbreviations

TRCs	Taste receptor cells
CV	Circumvallate papillae
Fol	Foliate papillae
Fun	Fungiform papillae
GPCR	G-protein coupled receptor
VGCC	Voltage-gated calcium channel
CaSR	Calcium-sensing receptor
PLC	Phospholipase C
PIP_2	Phosphatidylinositol 4,5-bisphosphate
IP_3	Inositol trisphosphate
DAG	Diacylglycerol
TRP	Transient receptor potential
CALHM1	Calcium homeostasis modulator 1
LCFA	Long-chain fatty acid
mGluR	Metabotropic glutamate receptor
TRPM5	Melastatin transient receptor potential 5
STIM1	Stromal interaction molecule
TRPV1	Vanilloid transient receptor potential 1
ATP	Adenosine triphosphate
DIO	Diet-induced obesity

References

Anderson, G. H., 1995. Sugars, sweetness, and food intake. *American Journal of Clinical Nutrition*. 62: 195S–201S, discussion 201S–202S.

Bartoshuk, L. M., Duffy, V. B., Hayes, J. E., Moskowitz, H. R. and Snyder, D. J., 2006. Psychophysics of sweet and fat perception in obesity: problems, solutions and new perspectives. *Philosophical Transactions of the Royal Society of London, Series B: Biological Sciences*. 361: 1137–1148.

Bystrova, M. F., Romanov, R. A., Rogachevskaja, O. A., Churbanov, G. D. and Kolesnikov, S. S., 2010. Functional expression of the extracellular-Ca^{2+}-sensing receptor in mouse taste cells. *Journal of Cell Science*. 123: 972–982.

Chandrashekar, J., Mueller, K. L., Hoon, M. A., Adler, E., Feng, L., Guo, W., Zuker, C. S. and Ryba, N. J., 2000. T2Rs function as bitter taste receptors. *Cell*. 100: 703–711.

Chaudhari, N. and Roper, S. D., 2010. The cell biology of taste. *Journal of Cell Biology*. 190: 285–296.

DeFazio, R. A., Dvoryanchikov, G., Maruyama, Y., Kim, J. W., Pereira, E., Roper, S. D. and Chaudhari, N., 2006. Separate populations of receptor cells and presynaptic cells in mouse taste buds. *Journal of Neuroscience*. 26: 3971–3980.

Dramane, G., Abdoul-Azize, S., Hichami, A., Vogtle, T., Akpona, S., Chouabe, C., Sadou, H., Nieswandt, B., Besnard, P. and Khan, N. A., 2012. STIM1 regulates calcium signaling in taste bud cells and preference for fat in mice. *Journal of Clinical Investigation*. 122: 2267–2282.

Finger, T. E., Danilova, V., Barrows, J., Bartel, D. L., Vigers, A. J., Stone, L., Hellekant, G. and Kinnamon, S. C., 2005. ATP signaling is crucial for communication from taste buds to gustatory nerves. *Science*. 310: 1495–1499.

Finger, T. E. and Simon, S. A., 2000. Cell biology of taste epithelium. In: Finger T. E., Silver W. L. and Restrepo D. (ed.) *The Neurobiology of Taste and Smell*. Wiley-Liss, New York, pp. 287–314.

Gilbertson, T. A. and Boughter, Jr, J. D., 2003. Taste transduction: appetizing times in gustation. *Neuroreport*. 14: 905–911.

Gilbertson, T. A. and Khan, N. A., 2013. Cell signaling mechanisms of oro-gustatory detection of dietary fat: advances and challenges. *Progress in Lipid Research*. 53C: 82–92.

Hacker, K. and Medler, K. F., 2008. Mitochondrial calcium buffering contributes to the maintenance of basal calcium levels in mouse taste cells. *Journal of Neurophysiology*. 100: 2177–2191.

Horio, N., Jyotaki, M., Yoshida, R., Sanematsu, K., Shigemura, N. and Ninomiya, Y., 2010. New frontiers in gut nutrient sensor research: nutrient sensors in the gastrointestinal tract: modulation of sweet taste sensitivity by leptin. *Journal of Pharmacological Sciences*. 112: 8–12.

Huang, A. L., Chen, X., Hoon, M. A., Chandrashekar, J., Guo, W., Trankner, D., Ryba, N. J. and Zuker, C. S., 2006. The cells and logic for mammalian sour taste detection. *Nature*. 442: 934–938.

Huang, Y. A. and Roper, S. D., 2010. Intracellular Ca^{2+} and TRPM5-mediated membrane depolarization produce ATP secretion from taste receptor cells. *Journal of Physiology*. 588: 2343–2350.

Huang, Y. J., Maruyama, Y., Dvoryanchikov, G., Pereira, E., Chaudhari, N. and Roper, S. D., 2007. The role of pannexin 1 hemichannels in ATP release and cell–cell communication in mouse taste buds. *Proceedings of the National Academy of Sciences of the United States of America*. 104: 6436–6441.

Ishii, S., Misaka, T., Kishi, M., Kaga, T., Ishimaru, Y. and Abe, K., 2009. Acetic acid activates PKD1L3-PKD2L1 channel – a candidate sour taste receptor. *Biochemical and Biophysical Research Communications*. 385: 346–350.

Ishimaru, Y., Inada, H., Kubota, M., Zhuang, H., Tominaga, M. and Matsunami, H., 2006. Transient receptor potential family members PKD1L3 and PKD2L1 form a candidate sour taste receptor. *Proceedings of the National Academy of Sciences of the United States of America*. 103: 12569–12574.

Iwata, S., Yoshida, R. and Ninomiya, Y., 2014. Taste transductions in taste receptor cells: basic tastes and moreover. *Current Pharmaceutical Design.* 20(16): 2684–2692.

Kovacs, P. and Hajnal, A., 2008. Altered pontine taste processing in a rat model of obesity. *Journal of Neurophysiology*. 100: 2145–2157.

Laskowski, A. I. and Medler, K. F., 2009. Sodium/calcium exchangers contribute to the regulation of cytosolic calcium levels in mouse taste cells. *Journal of Physiology*. 587(16): 4077–4089.

Liu, P., Shah, B. P., Croasdell, S. and Gilbertson, T. A., 2011. Transient receptor potential channel type m5 is essential for fat taste. *Journal of Neuroscience*. 31: 8634–8642.

Maliphol, A. B., Garth, D. J. and Medler, K. F., 2013. Diet-induced obesity reduces the responsiveness of the peripheral taste receptor cells. *PLoS One*. 8: e79403.

Medler, K. F., 2010. Calcium signaling in taste cells: regulation required. *Chemical Senses.* 35(9): 753–765.

Medler, K. F. and Kinnamon, S. C., 2004. Transduction mechanisms in taste cells. In: Stephan Frings J. B. (ed.) *Transduction Channels in Sensory Cells.* Wiley-VCH, Weinheim, Germany, pp. 153–177.

Nelson, G., Chandrashekar, J., Hoon, M. A., Feng, L., Zhao, G., Ryba, N. J. and Zuker, C. S., 2002. An amino-acid taste receptor. *Nature*. 416: 199–202.

Nelson, G., Hoon, M. A., Chandrashekar, J., Zhang, Y., Ryba, N. J. and Zuker, C. S., 2001. Mammalian sweet taste receptors. *Cell*. 106: 381–390.

Pepino, M. Y., Finkbeiner, S., Beauchamp, G. K. and Mennella, J. A., 2010. Obese women have lower monosodium glutamate taste sensitivity and prefer higher concentrations than do normal-weight women. *Obesity*. 18: 959–965.

Perea-Martinez, I., Nagai, T. and Chaudhari, N., 2013. Functional cell types in taste buds have distinct longevities. *PLoS One*. 8: e53399.

Richter, T. A., Caicedo, A. and Roper, S. D., 2003. Sour taste stimuli evoke Ca^{2+} and pH responses in mouse taste cells. *Journal of Physiology*. 547: 475–483.

Romanov, R. A., Rogachevskaja, O. A., Bystrova, M. F., Jiang, P., Margolskee, R. F. and Kolesnikov, S. S., 2007. Afferent neurotransmission mediated by hemichannels in mammalian taste cells. *EMBO Journal*. 26: 657–667.

Royer, S. M. and Kinnamon, J. C., 1988. Ultrastructure of mouse foliate taste buds: synaptic and nonsynaptic interactions between taste cells and nerve fibers. *Journal of Comparative Neurology*. 270(11–24): 58.

Sclafani, A., Zukerman, S., Glendinning, J. I. and Margolskee, R. F., 2007. Fat and carbohydrate preferences in mice: the contribution of alpha-gustducin and Trpm5 taste-signaling proteins. *American Journal of Physiology: Regulatory, Integrative and Comparative Physiology*. 293: R1504–R1513.

Shin, Y. K., Martin, B., Golden, E., Dotson, C. D., Maudsley, S., Kim, W., Jang, H. J., Mattson, M. P., Drucker, D. J., Egan, J. M. and Munger, S. D., 2008. Modulation of taste sensitivity by GLP-1 signaling. *Journal of Neurochemistry*. 106: 455–463.

Szebenyi, S. A., Laskowski, A. I. and Medler, K. F., 2010. Sodium/calcium exchangers selectively regulate calcium signaling in mouse taste receptor cells. *Journal of Neurophysiology*. 104: 529–538.

Taruno, A., Vingtdeux, V., Ohmoto, M., Ma, Z., Dvoryanchikov, G., Li, A., Adrien, L., Zhao, H., Leung, S., Abernethy, M., Koppel, J., Davies, P., Civan, M. M., Chaudhari, N., Matsumoto, I., Hellekant, G., Tordoff, M. G., Marambaud, P. and Foskett, J. K., 2013. CALHM1 ion channel mediates purinergic neurotransmission of sweet, bitter and umami tastes. *Nature*. 495: 223–226.

Tordoff, M. G., 2001. Calcium: taste, intake, and appetite. *Physiological Reviews*. 81: 1567–1597.

Tordoff, M. G., Alarcon, L. K., Valmeki, S. and Jiang, P., 2012. T1R3: a human calcium taste receptor. *Scientific Reports*. 2: 496.

Zhao, G. Q., Zhang, Y., Hoon, M. A., Chandrashekar, J., Erlenbach, I., Ryba, N. J. and Zuker, C. S., 2003. The receptors for mammalian sweet and umami taste. *Cell*. 115: 255–266.

CHAPTER 26

Intracellular Calcium Modulation of Gene Expression

MARIANA CASAS[a] AND ENRIQUE JAIMOVICH*[b]

[a]Center for Molecular Studies of the Cell, ICBM, Physiology and Biophysics Program, Facultad de Medicina, Universidad de Chile, Independencia 1027, Santiago, Chile; [b]Center for Molecular Studies of the Cell, ICBM, Facultad de Medicina, Universidad de Chile, Independencia 1027, Santiago, Chile
*E-mail: ejaimovi@med.uchile.cl

26.1 Introduction

Calcium ions are an essential element for the regulation of muscle contraction. Due to the presence of calcium-binding proteins in the contractile machinery of striated muscle, the free calcium concentration in the cytosol is able to regulate the interaction of actin and myosin filaments, thus controlling the "on" and "off" of muscle contraction. Electrical stimulation of the nerve elicits an action potential in the muscle fiber end plate that propagates along the sarcolemma and radially penetrates to the muscle fiber interior through the transverse tubule system. In the T-tubule, the action potential activates a voltage sensor (Cav1.1 channels or dihydropyridine receptors) that, in skeletal muscle is capable of interacting with ryanodine receptors located in the sarcoplasmic reticulum membrane, increasing the opening probability of these calcium release channels and eliciting a fast free calcium transient in the muscle fiber cytosol, responsible for the muscle fiber twitch (see Figure 26.1).

Food and Nutritional Components in Focus No. 10
Calcium: Chemistry, Analysis, Function and Effects
Edited by Victor R. Preedy

Published by the Royal Society of Chemistry, www.rsc.org

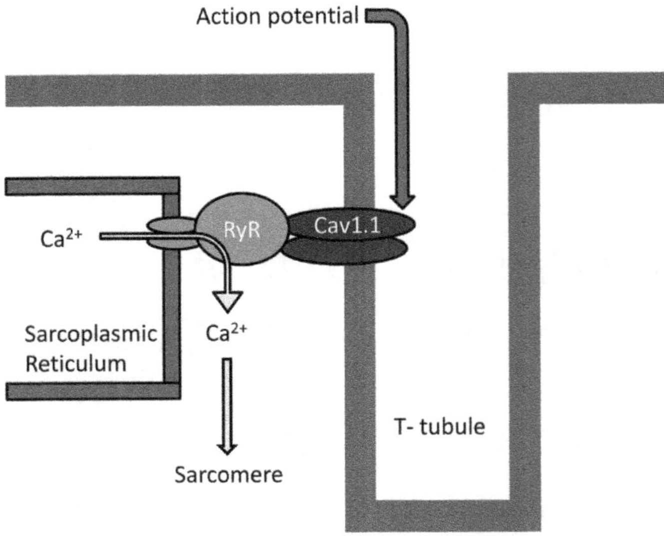

Figure 26.1 Schematic description of the excitation–contraction process in skeletal muscle cells. Electrical stimulation (the action potential) propagating along the sarcolemma of the muscle fiber propagates radially to the fiber interior through the transverse tubule membranes. Here, the voltage-gated calcium channel (Cav1.1) senses the membrane potential change and its conformational change impinges on RyR type 1 channels located in the adjacent sarcoplasmic reticulum membrane, to increase their open probability and release calcium to the cytoplasm. This calcium increase will trigger muscle contraction in the sarcomere.

Increased use, decreased use and injury are known to induce different forms of muscle adaptation, but the cellular mechanisms responsible for these changes remain unknown. The process of muscle adaptation is largely dependent on nerve activity and on nerve–muscle interaction and therefore constitutes a model for both plasticity and regeneration of excitable cells. Ca^{2+} is an intracellular messenger normally involved in processes that lead to gene regulation in many cells. Nevertheless, the important role of Ca^{2+} in muscle contraction has precluded the study of other possible roles for Ca^{2+} in skeletal muscle. Our group has found that the transient elevation in intracellular Ca^{2+} seen in muscle cells after tetanic stimulation can be modeled into two spatially, kinetically and pharmacologically distinct processes. Only the early component is related to contraction. The slow component, lasting tens of seconds, represents a different type of signal and has been linked to the excitation–transcription process.

26.2 Muscle-Cell Adaptations to Training

Skeletal muscle is a highly plastic tissue and a striking feature of muscle-cell plasticity is the specificity of the adaptive response to a given stimulus (Fluck and Hoppeler, 2003); the degree of loading and the number of muscular contractions appear to be the dominant stimuli for muscular adaptations. For instance, highly repetitive, low-load exercise training will cause differentiation of muscle fibers towards a fatigue-resistance phenotype with a high mitochondrial volume density. This change towards adaptation to aerobic metabolism is assisted by an increase in capillary density and may involve a shift of the contractile character of the fibers towards a slow type accompanied by changes in sarcomere components (Fluck and Hoppeler, 2003). This cellular specialization allows the recruited muscle fibers to sustain a high number of slow contractions. On the other hand, exercise protocols involving major loading will produce fiber hypertrophy and increased force as well as a shift toward glycolitic metabolism generating fibers adapted to perform short-lasting, high-force contractions. Importantly, maintenance of both skeletal muscle mass and oxidative capacity is dependent on the persistency of contractile stimuli, as shown by the dramatic deconditioning of muscle function with inactivity. Thus, the profile of muscle stimulation exerts an essential control over the muscle phenotype. Gene expressions are therefore an essential phase of processing for integration of exercise stimuli into the adjustments of muscle phenotype needed to adapt muscle function to changes in demand.

26.3 Gene Transcription Regulation by Calcium in Skeletal Muscle

Ca^{2+} is a ubiquitous intracellular signal that controls an enormous number of cellular processes. The different roles of Ca^{2+} as second messenger are ensured by spatiotemporal patterns of intracellular Ca^{2+} to generate both transient and oscillatory changes in Ca^{2+} concentrations.

We have described the complex Ca^{2+} release response induced by K^+ depolarization. This response involves at least two components: a first fast Ca^{2+} transient associated with the RyR and E–C coupling, and a second slower transient not related with contraction that are mediated by the IP_3 receptors (Estrada *et al.*, 2001). The DHPR voltage sensor involved in excitation–contraction coupling also plays a role as voltage sensor for the IP_3-mediated slow calcium signal (Jaimovich and Carrasco, 2002; Araya *et al.*, 2003).

Importantly, several works from our group linked the described slow Ca^{2+} signal to regulation of gene expression in skeletal muscle. We have shown that membrane depolarization induces ERK1/2 and CREB phosphorylation in a slow Ca^{2+}-signal-dependent way, as well as an increase in mRNA levels of the early genes c-fos, c-jun and egr-1 (Powell *et al.*, 2001; Carrasco *et al.*, 2003). Various other transcription factors and late genes are activated as well (Juretic *et al.*, 2006; Juretic *et al.*, 2007; Valdes *et al.*, 2007). Depolarization affects transcriptional activity of a limited number of genes, mainly associated with metabolism, cell communication and response to stress. A number of genes related to calcium signaling pathways are induced at 4 h, reinforcing the potential role of calcium in the early steps of signal transduction that leads to gene expression. Significant changes in the expression of molecules involved in muscle-cell structure were observed, as well as an increase in the specific muscle proteins troponin I and alpha actin mRNA levels in both differentiated C2C12 and rat skeletal muscle cells in primary culture (Juretic *et al.*, 2007). Also, a transient increase in interleukin 6 mRNA level upon membrane depolarization was evidenced in separate studies (Juretic *et al.*, 2006; Bustamante *et al.*, 2014). In adult muscle fibers, changes in interleukin 6 mRNA levels, after IP_3-dependent signals induced by ATP stimulation were also observed (Fernández-Verdejo *et al.*, 2014). Recent evidences point to expression of antiapoptotic genes after electrical stimulation of normal muscle fibers (Valladares *et al.*, 2013).

26.4 What We Do Know about the Mechanism from Work in Myotubes and Adult Muscle Fibers

The study of fluorescent calcium signals from cultured rat myotubes has provided important results in the past few years. Both K^+ depolarization and tetanic electrical stimulation were shown to produce slow Ca^{2+} signals, unrelated to contraction and associated to regulation of gene expression. A slow calcium signal first described in rat myotubes stimulated with K^+ depolarization, occurs several seconds after the fast calcium signal that is due to ryanodine receptor (RyR) opening and mediates E–C coupling. The former is refractory to ryanodine treatment and depends on IP_3 generation (Jaimovich *et al.*, 2000; Araya *et al.*, 2003). This slow calcium signal has a predominant nuclear component and its propagation between adjacent nuclei in the myotube can be frequently recorded (Jaimovich *et al.*, 2000). Besides the fact that the three types of IP_3 receptors (IP_3R) are located at different sites within rat myotubes, including the nuclear region, a functional calcium store in isolated nuclei has been described (Cardenas *et al.*, 2005). Treatment of these isolated nuclei with IP_3 induces calcium release to the nucleoplasm, regulating phosphorylation of the transcription factor cAMP response element-binding protein (CREB) (Cardenas *et al.*, 2005). The slow calcium signal appears to participate in the regulation of transcription factors and the modulation of gene expression mediated by membrane depolarization and calcium *via* an IP_3 signaling pathway (Powell *et al.*, 2001; Carrasco *et al.*, 2003; Cardenas *et al.*, 2004).

The molecular mechanism involved in phospholipase C (PLC) activation by K^+ depolarization in myotubes is still not completely understood, but both the IP_3 transient and the slow calcium signal were blocked by *pertussis* toxin (PTX), suggesting the participation of a PTX-sensitive heterotrimeric G protein in the signaling mechanism (Araya *et al.*, 2003). In several systems, G protein-dependent PLC activation has been described. Both $G\alpha q$ and $G\beta\gamma$ (released from heterotrimeric $G_{i/o}$) may directly activate $PLC\beta$ isoforms. Also, $G\beta\gamma$ may indirectly activate $PLC\gamma$ increasing phosphatidylinositol-3,4,5-trisphosphate (PIP_3) levels by activation of $PI3K\gamma$ (Rebecchi and Pentyala, 2000). When tetanic electrical field stimulation was used instead of K^+

depolarization, an enhanced IP_3-dependent slow calcium signal was observed and again the DHPR was necessary to sense the action potentials induced by the stimulation protocol, mimicking the K^+ depolarization results (Eltit *et al.*, 2004). We demonstrated that tetanic electrical stimulation of myotubes activates both G protein and PI3K and that both signaling pathways are required for the activation of PLC and the onset of the slow calcium signal (Eltit *et al.*, 2006). We presented evidence that both Gβγ and PI3K are activated by tetanic stimulation and are involved in the generation of the slow calcium signal.

In the work of Buvinic *et al.* (2009) we provided evidence that favors a mechanistic model for the origin of IP_3-dependent calcium transients in primary myotubes with ATP released by pannexin-1 channels as a key factor. Indeed, in these cells, after electrical stimulation, the pannexin-1 channel releases ATP to the extracellular medium; ATP is quickly degraded by ectonucleotidases to ADP, AMP and adenosine, but before disappearing, both ATP and ADP can interact with purinergic P2Y receptors (G protein-linked receptors). Combining this evidence with our previous studies we suggest that this interaction releases the G protein βγ subunit and activate a γ-type PI3 kinase. The product of this activation, PIP_3 in the membrane will recruit phospholipase C (mainly γ-type) to the membrane to actively produce both IP_3 and diacylglycerol. IP_3 will interact with its receptors located both in the sarcoplasmic reticulum and in the nuclear envelope to elicit calcium release, responsible for the slow calcium signal.

Importantly, in adult muscle fibers, we showed that Cav1.1, the voltage sensor of the sarcoplasmic membrane physically interact with pannexin-1 channels. Moreover, we showed that pharmacological inhibition of Cav1.1 impairs ATP release after electrical stimulation, suggesting a functional interaction between these two channels and pointing to Cav1.1 as the initial molecule triggering the whole cascade (Jorquera *et al.*, 2013).

The sequence of events leading from electrical stimulation to gene expression was schematized in Figure 26.2 and examples of genes regulated by electrical stimuli are provided in Table 26.1.

26.5 Fiber-Type Diversity

Muscle fibers can be classified into four main groups: slow fatigue resistant, fast fatigue resistant, fast with mixed metabolism and fast fatigable. Muscle fibers within a given group can be recognized by the specific molecular markers they express, like different myosin heavy-chain isoforms and a different set of metabolic enzymes (Schiaffino and Reggiani, 1994; Pette and Staron, 2002).

For instance, fast fatigable fibers express a particular form of myosin that produces force more effectively at rapid shortening velocities (MHC and MLC fast isoforms), and a set of metabolic enzymes to provide energy from their relatively large stores of glycogen (like glycerol-3-phosphate dehydrogenase, GPDH), while slow ones express a slow isoform of myosin (β-MHC) and a set of oxidative enzymes (like SDH).

These muscle-fiber properties are in agreement with the activity of motoneuron that innervates them. Indeed, sustained contractile activity of slow

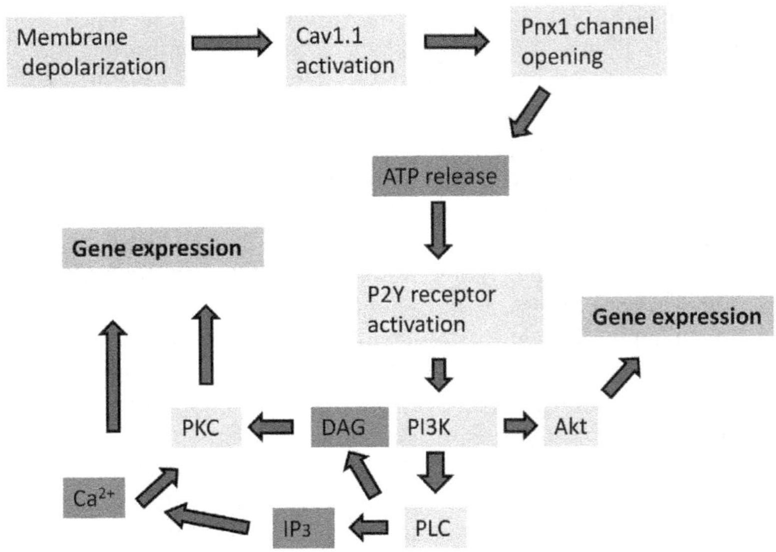

Figure 26.2 Schematic description of a plausible process of excitation–transcription coupling in skeletal muscle fibers. Action potentials driven by exercise will depolarize the T-tubule membrane and this membrane potential change will induce a conformational modification of the Cav1.1 voltage sensor. Coupling between the voltage sensor and pannexin-1 channels will induce opening of the latter and release of ATP to the extracellular medium. ATP and metabolites, acting on purinergic P2Y receptors will induce activation of PI3 kinase (PI3K), giving rise to several signaling events, among them activation of phospholipase C (PLC) with production of both diacylglycerol (DAG) and inositol 1,4,5-trisphosphate (IP_3). IP_3 will interact with its receptors located both in the sarcoplasmic reticulum membrane and in the nuclear envelope to release calcium ions to the cytosol and to the nucleoplasm. Calcium-dependent events will be responsible for expression of a number of genes. Additional signaling pathways leading to gene expression will originate from activation of protein kinase C isoforms sensitive both to calcium and DAG and from activation of protein kinase B (Akt) by phosphorylation reactions starting with PI3K.

fibers is the result of a pattern of low-frequency tonic stimulation proper to electrical activity of motoneurons innervating them. Conversely, the phasic high-frequency pattern characteristic of fast motoneurons matches with the contractile properties of fast muscle fibers.

Nerve activity also plays a major role in control of gene expression in skeletal muscle fibers. Indeed, fiber-type specification depends on both myoblast lineage and motoneuron innervation. The role of motoneuron activity in the specification and maintenance of fiber-type identity have been demonstrated by different experimental approaches, namely denervation, cross-reinnervation and external electrostimulation. Moreover, different firing patterns from different motoneuron subclasses allow establishment of specific transcriptional programs that control fiber-type identity and growth.

Table 26.1 Examples of genes regulated by short-term electrical stimuli and calcium signals in skeletal muscle cells. This table summarizes some of the genes that have been studied and whose expression have demonstrated to depend on electrical stimulation; likely depending on slow calcium signals.

Gene-protein	Stimulus and cell model	Identified elements	References
c-fos	Potassium depolarization (myotubes)	IP_3	Carrasco *et al.* (2003)
c-jun	Potassium depolarization (myotubes)	IP_3	Carrasco *et al.* (2003)
egr-1	Potassium depolarization (myotubes)	IP_3	Carrasco *et al.* (2003)
Interleukin-6	Potassium depolarization, electrical stimulation, ATP (myotubes and fibers)	ATP release, IP_3, calcium	Juretic *et al.* (2006), Bustamante *et al.* (2014), Fernández-Verdejo *et al.* (2014)
Slow troponin I	Electrical stimulation, ATP (adult muscle fibers)	Cav1.1, ATP release, IP_3, calcium	Jorquera *et al.* (2013)
Fast troponin I	Electrical stimulation, ATP (adult muscle fibers)	Cav1.1, ATP release, IP_3, calcium	Jorquera *et al.* (2013)
Heme oxygenase-1	Potassium depolarization (myotubes)	IP_3	Jorquera *et al.* (2009)
HSP-70	Potassium depolarization (myotubes)	IP_3	Jorquera *et al.* (2009)
Bax	Electrical stimulation, ATP (adult muscle fibers)	ATP release	Valladares *et al.* (2013)
Bim	Electrical stimulation, ATP (adult muscle fibers)	ATP release	Valladares *et al.* (2013)
PUMA	Electrical stimulation, ATP (adult muscle fibers)	ATP release	Valladares *et al.* (2013)
BCL-2	Electrical stimulation, ATP (adult muscle fibers)	ATP release	Valladares *et al.* (2013)
Neuregulin 1-β	Electrical stimulation (myotubes)	IP_3	Juretic *et al.* (2012)

Although the mechanisms by which the muscle fiber translates different firing patterns of motoneurons to induce specific gene transcription programs remains largely unknown, some molecular mechanisms involved in downstream differential gene expression induced by motoneuron electrical activity are beginning to be elucidated.

26.5.1 Ca^{2+}-Dependent Kinases and Phosphatases Involved in Muscle Phenotype Specification

As reviewed by Mellström *et al.* (2008), there are several ways in which calcium may control gene expression in a given cell. Some examples of transcription regulation are described in Figure 26.3. In this example (studied in T-cells,

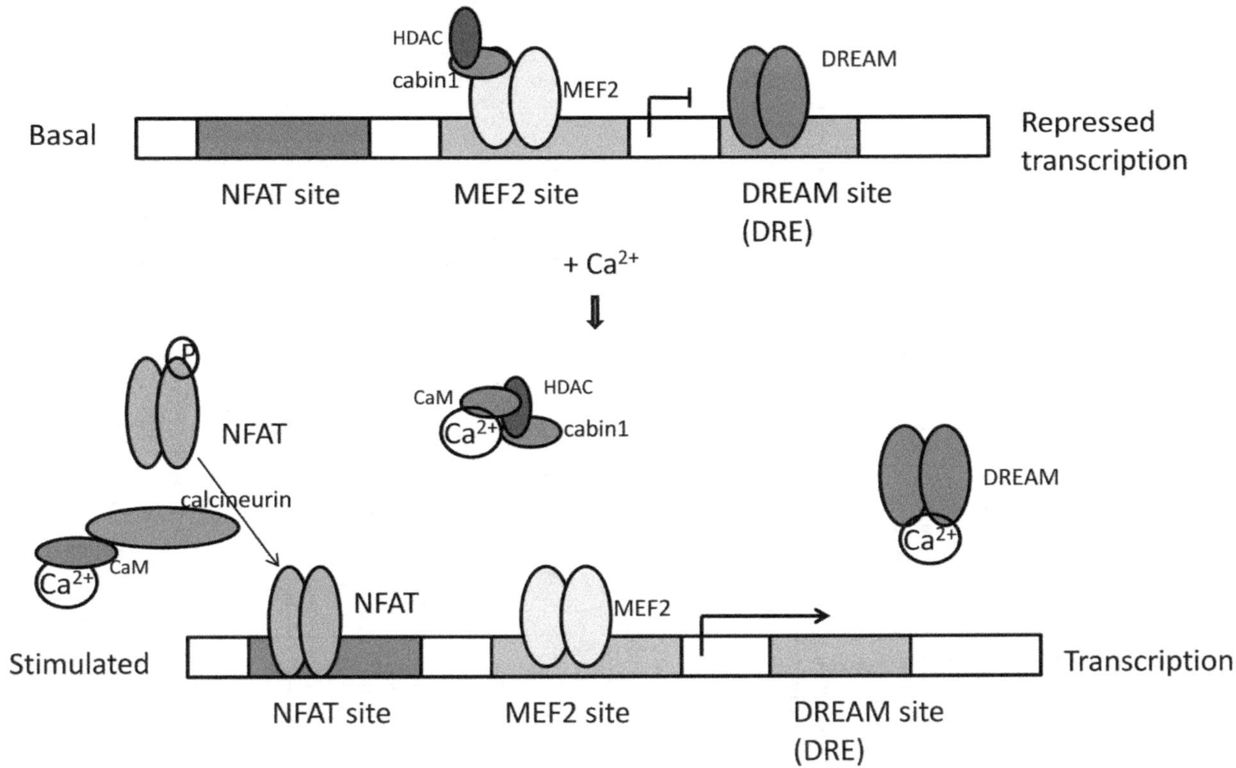

Figure 26.3 Examples of calcium-dependent processes regulating gene expression. Three DNA sites are schematically represented; the NFAT site, the MEF2 site and the DRE, DREAM site. The upper part of the figure reflects the resting condition in which all three sites are transcriptionally inactive. An increase in free calcium concentration will induce transcription initiation as schematized in the lower part of the figure. In order for NFAT transcription factor to bind its site (and initiate transcription) it will first have to be dephosphorylated by the calcium-calmodulin (CaM) sensitive phosphatase calcineurin. The MEF2 site will have the transcription factor bound in resting conditions but it will be inactive because of being complexed by the proteins cabin and HDAC; binding of Ca–CaM to HDAC will release the complex and free MEF2 to activate transcription. DREAM is a calcium-sensitive repressor that when bound to the DRE site will block transcription. Released by calcium binding, will allow transcription to proceed (modified from (Savignac *et al.*, 2007)).

Savignac *et al.*, 2007), transcription will be repressed in basal conditions in several sites, namely NFAT, because the transcription factor is kept out of the nucleus, MEF2, because the complex cabin1-HDAC will be acting as a repressor, avoiding MEF2 phosphorylation by MAP kinases and DRE, occupied by the transcriptional repressor DREAM. A calcium transient in the nucleus will allow Ca^{2+} ions to bind to DREAM, releasing the repressor action; the binding of Ca^{2+}–calmodulin (CaM) to cabin1 will release the repressor complex from the MEF2 site and by Ca^{2+}–CaM activation of the phosphatase calcineurin, NFAT will be dephosphorylated and will translocate to the nucleus to bind its specific site. Activation of transcription will proceed then under these conditions, for as long as the free calcium levels remain elevated.

Among transcriptional pathways present in skeletal muscle that can be activated by calcium signals are the calcineurin-NFAT, Ca^{2+}–calmodulin-dependent kinases II and IV (CaMKII and CaMKIV) and pathways. The Ca^{2+} participates in gene expression control by dephosphorylating specific substrates, including the nuclear factor of activated T cell (NFAT) family proteins. When dephosphorylated, NFAT translocates from the cytosol to the nucleus in order to, in cooperation with other transcription factors, activate target genes.

A growing amount of evidence, points to Cn activity requirement for the induction and maintenance of the slow muscle fiber phenotype. In 1998, Chin and colleagues have shown that activation of Cn in skeletal myotubes cultures upregulates slow-fiber-specific gene promoters and that inhibition of this phosphatase by cyclosporin A in intact animals promotes slow-to-fast fiber-type transformation (Chin *et al.*, 1998). In transgenic mice expressing activated Cn under the control of the muscle creatine kinase enhancer, there is an increase in slow muscle fibers (Naya *et al.*, 2000).

Different experiments show the relationship between transcription control by nerve activity and Cn pathway. In regenerating rat soleus muscle, upregulation of slow myosin heavy chain (MyHC) induced by slow motoneurons activity or low-frequency external electrostimulation can be prevented by the Cn inhibitors cyclosporine A, FK506 and cabin (cabin-1) peptide (Serrano *et al.*, 2001).

The NFAT gene family of transcription factors have been implicated as downstream effectors of Cn in nerve-activity-dependent slow-to-fast transformation. In cultured adult mouse skeletal muscle fibers expressing NFATc-GFP fusion proteins, electrical stimulation with patterns typical of slow-twitch muscle causes a cyclosporin A-sensitive NFATc-GFP nuclear translocation (Liu *et al.*, 2001). Moreover, in cultured rabbit myocites, nuclear translocation of endogenous NFATc and slow MyHCs transcription are also induced by low-frequency stimulation (Kubis *et al.*, 2002). Strong evidence for a major role of Cn-NFAT pathway in muscle-fiber-type specification came from the McCullagh works using a peptide (VIVIT) that inhibits Cn-dependent NFAT nuclear translocation without disrupting other Cn-dependent pathways. This peptide prevents the upregulation of MyHC-slow induced by nerve activity in regenerating soleus muscle (McCullagh *et al.*, 2004).

As expected by its role in Cn activation, intracellular Ca^{2+} increases induce a slow-oxidative phenotype. Indeed, intracellular Ca^{2+} levels are higher in slow muscle compared to fast ones, and can be increased in these later by mimicking the firing pattern of slow motoneurons by chronic low-frequency stimulation (Carroll *et al.*, 1999). Moreover, rising intracellular Ca^{2+} levels by addition of a Ca^{2+}ionophore in cultured muscle cells produce a reversible fast to slow transition. In these experiences, the fiber-type transition, as well as the NFATc1 nuclear translocation, can be prevented by cyclosporin A treatment (Meissner *et al.*, 2001). Important questions remain about the Ca^{2+} signal necessary to activate the Cn-NFAT signaling and its intracellular origin. In primary myotubes from rabbit, some studies suggest that Cn and slow MyHC gene are activated by repetition of high-amplitude calcium transients rather than a sustained elevation of resting Ca^{2+} concentration, for short-term stimulations (Kubis *et al.*, 2003). Concerning intracellular Ca^{2+} origin, important works from DiMario's group in innervated avian cultured muscle fibers, show that inhibition of inositol trisphosphate receptor 1(IP_3R1), responsible for calcium release from the SR as slow wave, induces a fast-to-slow fiber-type transition accompanied by a reduction of NFAT-dependent transcription and nuclear localization, suggesting that IP_3R1 activity can regulate transcription factor activity in skeletal muscle fibers (Jordan *et al.*, 2005) (Table 26.1).

Nerve activity-dependent Cn regulation of skeletal muscle fiber type is also reported to be mediated by MEF2 transcription factors and seems to be integrated with additional calcium-regulated signaling inputs, like calmodulin-dependent protein kinase (CaMK) activity. Moreover, CaMKs are also involved in regulation of skeletal muscle genes. During acute and prolonged muscular activity there is an increase in CaMKII activity (Fluck *et al.*, 2000), associated to activation of mitochondrial biogenesis and muscle hypertrophy. Interestingly, CaMKII is sensitive to Ca^{2+} frequency oscillations, a very important feature in order to decode different electrostimulation patterns.

The CaMKIV pathway has also a role in skeletal muscle gene transcription, specifically in expression of mitochondrial biogenesis and oxidative enzymes (Wu *et al.*, 2002). In CaMKIV overexpressing transgenic mice, there is an increase in mitochondria content that is correlated to an increase in PGC-1α expression, whose overexpression in muscle fibers can switch them to a slow phenotype (Lin *et al.*, 2002).

Some other nuclear genes encoding mitochondria proteins like cytochrome c are dependent on protein kinase C-dependent pathway in response to Ca^{2+} signals.

26.5.2 Ras-MAPK Pathway

The role of Ras in nerve-activity-dependent skeletal muscle gene regulation is supported by the fact that physical exercise is a potent inductor of the MAP kinase signaling pathway. In 1996 it was first shown that exercise activates ERK1/2, JNK and P38 signaling in skeletal muscle. It is now well established that physical exercise or muscle contractile activity can activate several of the MAP kinase signaling pathways (Sakamoto and Goodyear, 2002).

Ras-MAPK signaling has also been involved in promoting nerve-activity-dependent induction of the slow muscle fiber gene program in regenerating rat soleus muscles. In fact, a dominant negative form of Ras (RasN17) prevents the induction of slow MHC by slow motoneuron activity, whereas a Ras mutant that selectively stimulates the MAPK/ERK pathway is able to induce slow MHC expression even in the absence of innervations. Moreover, MAPK/ERK activity is increased in adult muscle fibers electrostimulated with firing patterns typical of slow motoneurons, and no changes in this activity were observed after stimulation with patterns that mimic firing patterns from motoneurons innervating fast fatigable muscle fibers (Murgia *et al.*, 2000).

26.5.3 IP_3-Dependent Ca^{2+} Signals and Muscle Plasticity

In adult skeletal muscle we saw a mosaic pattern of staining for IP_3 receptors type 1 and 3 (Casas *et al.*, 2010) consistent with a much higher level in fast fibers compared to slow type fibers for the type 1 receptor, as observed in avian muscle cultures (Jordan *et al.*, 2005); a plausible hypothesis is that IP_3-mediated calcium signals could be involved in determination or maintenance of fiber phenotype. Moreover, low-frequency electrical stimulation induced increases in slow isoforms of Troponin I mRNA levels and a decrease in those of the fast isoforms, a feature of a slow-to-fast muscle fiber-type transformation (Casas *et al.*, 2010). Importantly, the amplitude of IP_3-dependent Ca^{2+} signals was dependent on both, duration and frequency of electrical stimulation. In a later study (Jorquera *et al.*, 2013), we saw that ATP release trough pannexin-1 channels was also frequency dependent, taking place at low (below 60 Hz, maximum at 10–20 Hz) but not at high (over 75 Hz) frequencies of stimulation. Accordingly, as 20 Hz stimulation did, ATP stimulation induced a slow-to-fast muscle fiber-type transformation, suggesting that electrical stimulation needs ATP release to induce changes in gene expression. In this work we provided evidence that Cav1.1 triggers the whole cascade, physically interacting with the pannexin-1 channel, and probably activating it to release ATP at low frequencies of electrical stimulation (Jorquera *et al.*, 2013).

Summary Points

- This chapter focuses on calcium signals driving gene expression in skeletal muscle.
- Free calcium ion transients are essential to elicit muscle contraction.
- A different type of calcium transient (slower and located in particular regions of the cell) generates intracellular signals that end in regulation of gene expression.

- Both signals (excitation-contraction coupling (E–C) and excitation-transcription coupling (E–T)) are dependent on nerve activity.
- Low-frequency nerve activity elicit both calcium transients (E–C and E–T), while high frequency nerve activity elicits E–C signals only.
- In the muscle cell, both signals are initiated by Cav1.1, the voltage sensor being located in the transverse tubule membrane of skeletal muscle.
- For the E–T signal, Cav1.1 activates pannexin-1 channels to release ATP from the muscle fiber.
- ATP is degraded to ADP, AMP and adenosine by membrane ectonucleotidases.
- ATP and ADP act on purinergic P2Y receptors to activate a G-protein dependent cascade that includes PI3 kinase and phospholipase C.
- Phospholipase C release the intracellular messengers inositol 1,4,5-trisphosphate (IP_3) and diacylglycerol.
- IP_3 acts on receptors located both in the sarcoplasmic reticulum and in the nuclear envelope to release calcium to both the cytosol and the nucleoplasm.
- Nuclear and cytosolic calcium participate in signaling cascades that ultimately end in either transcription factors binding or transcription repressors release from DNA to drive gene transcription.

Key Facts for Muscle Calcium Signals and Gene Expression

1. All cells possess mechanisms that allow them to have extremely low free calcium concentrations in the cytosol in basal conditions.
2. As calcium gets accumulated in the endosarcoplasmic reticulum, organelle-located release channels as ryanodine receptors and IP_3 receptors have an important role in eliciting cytosolic calcium transients.
3. Calcium ions are essential to regulate muscle contraction by binding to troponin C and allowing interaction between actin and myosin filaments.
4. For calcium ions to be involved in signals other than contraction, such signals appear to be extremely small and highly localized within the cell.
5. Calcium ions regulate gene expression in many cells by interaction with calcium-binding proteins.
6. Some calcium-binding proteins involved in gene expression are calmodulin and CaMKinase II, the protein phosphatase calcineurin and the transcription repressor DREAM.
7. The nerve patterned activity regulates both the strength of contraction and the specific gene-expression programs that determine the muscle-fiber phenotype.
8. Activity-dependent gene expression is important for muscle plasticity and adaptation to energy and force requirements.

Definitions of Words and Terms

Adaptive response of skeletal muscle, also known as **muscle plasticity**, involves a series of changes in gene expression and muscle-fiber phenotype that occur in response to exercise, immobilization, and lack of gravity or, in general, conditions that imply increased or decreased muscle function.

Calcineurin, Cn. Calcium–calmodulin regulated protein phosphatase. It is important in the regulation of signals linked to NFAT transcription factor, which can translocate to the nucleus after dephosphorylation.

CREB. Cyclic AMP response element, is a transcription factor.

DHPR, Cav1.1. Dihydropyidine receptors, refers to voltage-dependent calcium channels (L-type) that in skeletal muscle act as voltage sensors for the process of E–C coupling.

ERK1/2 and P38 signaling are specific examples of MAPK.

Fiber phenotype. This refers to the skeletal muscle fiber type (see below).

IP_3. Inositol 1,4,5-trisphosphate is an intracellular messenger involved in a number of cell functions. It acts by binding to the calcium release channels known as IP_3 receptors (IP_3R).

MAPK. Mitogen-activated protein kinases; refers to a family of protein kinases involved in a number of intracellular signaling pathways.

ERK. Extracellular signal-regulated kinases.

Slow MHC. This refers to a subtype of the heavy chain of the myosin molecule, present in slow type muscle fibers.

Slow motoneuron activity. The muscle-fiber phenotype is determined by the type of motoneuron that innervates the muscle. Slow and fast motoneurons differ in a number of characteristics, among them the frequency pattern of stimuli, being low frequency (10–20 Hz) in motoneurons innervating slow muscle fibers and high frequency (90–150 Hz) in motoneurons innervating fast muscle fibers.

Ras is one of the so-called small G proteins.

PGC-1α. Peroxisome proliferator-activated receptor gamma coactivator-1 alpha; a transcriptional regulator of mitochondrial function.

CaMKII, CaMKIV. Members of a family of protein kinases that are activated when they bind to the protein calmodulin that in turn is regulated by calcium ions.

IP_3R subtypes. There are three subtypes of inositol 2,4,5 trisphosphate (IP_3R) known as type I, type II and type III, that are expressed in different proportions and locations in different cells.

NFAT. Nuclear factor of activated T-cells, is a family of transcription factors shown to be important in immune response and also in the development of the nervous system, cardiac and skeletal muscle.

Mef2. Myocyte enhancer factor-2 proteins, are a family of transcription factors that through control of gene expression are important regulators of cellular differentiation. In adult organisms, Mef2 proteins mediate the stress response in some tissues.

Slow MyHCβ-MHC. Myosin heavy-chain, slow fiber-type isoform.

Myoblast. Muscle precursor cell.

Fiber type diversity. Skeletal muscle fibers have been classified according to their main metabolism, twitch characteristics and specific sarcomeric proteins in type I (slow) and type II (fast) fibers. Type II fibers can in turn be classified in types IIa, IIb and IIx.

Gα and Gβγ are the subunits of the heterotrimeric G-protein.

Pertussis toxin (PTX) is a known inhibitor of G-protein mediated processes.

K^+ depolarization. As potassium ions are highly permeant through the cell membrane in resting conditions, the membrane potential depends largely on the K^+ intracellular and extracellular concentrations. Increasing extracellular K^+ will reduce membrane potential in a process known as K^+-induced depolarization.

Myotubes. Refers to a stage during myogenesis in which the fusion of myoblasts give origin to multinucleated, elongated structures that are not yet mature muscle fibers and are known as myotubes. This process can be mimicked in culture starting from muscle satellite or progenitor cells to get a pure myotube preparation.

Tetanic electrical stimulation. High-frequency electrical stimulation of a muscle, producing a maintained maximal tension.

Immediate early genes are genes that are activated transiently and rapidly in response to a wide variety of cellular stimuli. Among them are **c-fos, c-jun and egr-1**.

RyR. Ryanodine receptor is a large protein tetramer, located mainly in the endo/sarcoplasmic reticulum membrane, and functions as a calcium-release channel. This molecule is fundamental for striated muscle contraction.

E–C coupling. Excitation–contraction coupling refers to the mechanism that links the action potential to the contraction of the muscle fiber.

Muscle fiber twitch is the contractile response of a muscle fiber to a single action potential.

References

Araya, R., Liberona, J. L., Cardenas, J. C., Riveros, N., Estrada, M., Powell, J. A., Carrasco, M. A. and Jaimovich, E., 2003. Dihydropyridine receptors as voltage sensors for a depolarization-evoked, IP_3R-mediated, slow calcium signal in skeletal muscle cells. *Journal of General Physiology*. 121: 3–16.

Bustamante, M., Fernandez-Verdejo, R., Jaimovich, E. and Buvinic, S., 2014. Electrical stimulation induces IL-6 in skeletal muscle through extracellular ATP by activating Ca^{2+} signals and an IL-6 autocrine loop. *American Journal of Physiology: Endocrinology and Metabolism*. 306: E869–E882.

Buvinic, S., Almarza, G., Bustamante, M., Casas, M., Lopez, J., Riquelme, M., Saez, J. C., Huidobro-Toro, J. P. and Jaimovich, E., 2009. ATP released by electrical stimuli elicits calcium transients and gene expression in skeletal muscle. *Journal of Biological Chemistry*. 284: 34490–34505.

Cardenas, C., Liberona, J. L., Molgo, J., Colasante, C., Mignery, G. A. and Jaimovich, E., 2005. Nuclear inositol 1,4,5-trisphosphate receptors regulate local Ca^{2+} transients and modulate cAMP response element binding protein phosphorylation. *Journal of Cell Science*. 118: 3131–3140.

Cardenas, C., Muller, M., Jaimovich, E., Perez, F., Buchuk, D., Quest, A. F. and Carrasco, M. A., 2004. Depolarization of skeletal muscle cells induces phosphorylation of cAMP response element binding protein *via* calcium and protein kinase Calpha. *Journal of Biological Chemistry*. 279: 39122–39131.

Carrasco, M. A., Riveros, N., Rios, J., Muller, M., Torres, F., Pineda, J., Lantadilla, S. and Jaimovich, E., 2003. Depolarization-induced slow calcium transients activate early genes in skeletal muscle cells. *American Journal of Physiology: Cell Physiology*. 284: C1438–C1447.

Carroll, S., Nicotera, P. and Pette, D., 1999. Calcium transients in single fibers of low-frequency stimulated fast-twitch muscle of rat. *American Journal of Physiology*. 277: C1122–C1129.

Casas, M., Figueroa, R., Jorquera, G., Escobar, M., Molgó, J. and Jaimovich, E., 2010. IP_3-Dependent, post-tetanic calcium transients induced by electrostimulation of adult skeletal muscle fibers. *Journal of General Physiology*. 136: 455–467.

Chin, E. R., Olson, E. N., Richardson, J. A., Yang, Q., Humphries, C., Shelton, J. M., Wu, H., Zhu, W., Bassel-Duby, R. and Williams, R. S., 1998. A calcineurin-dependent transcriptional pathway controls skeletal muscle fiber type. *Genes & Development*. 12: 2499–2509.

Eltit, J. M., Hidalgo, J., Liberona, J. L. and Jaimovich, E., 2004. Slow calcium signals after tetanic electrical stimulation in skeletal myotubes. *Biophysical Journal*. 86: 3042–3051.

Eltit, J. M., Garcia, A. A., Hidalgo, J., Liberona, J. L., Chiong, M., Lavandero, S., Maldonado, E. and Jaimovich, E., 2006. Membrane electrical activity elicits inositol 1,4,5-trisphosphate-dependent slow Ca^{2+} signals through a Gbetagamma/phosphatidylinositol 3-kinase gamma pathway in skeletal myotubes. *Journal of Biological Chemistry*. 281: 12143–12154.

Estrada, M., Cardenas, C., Liberona, J. L., Carrasco, M. A., Mignery, G. A., Allen, P. D. and Jaimovich, E., 2001. Calcium transients in 1B5 myotubes lacking ryanodine receptors are related to inositol trisphosphate receptors. *Journal of Biological Chemistry*. 276: 22868–22874.

Fernández-Verdejo, R., Casas, M., Galgani, J. E., Jaimovich, E. and Buvinic, S., 2014. Exercise sensitizes skeletal muscle to extracellular ATP for IL-6 expression in mice. *International Journal of Sports Medicine*. 35: 273–279.

Fluck, M. and Hoppeler, H., 2003. Molecular basis of skeletal muscle plasticity – from gene to form and function. *Reviews of Physiology, Biochemistry and Pharmacology*. 146: 159–216.

Fluck, M., Waxham, M. N., Hamilton, M. T. and Booth, F. W., 2000. Skeletal muscle Ca^{2+}-independent kinase activity increases during either hypertrophy or running. *Journal of Applied Physiology*. 88: 352–358.

Jaimovich, E. and Carrasco, M. A., 2002. IP_3 dependent Ca^{2+} signals in muscle cells are involved in regulation of gene expression. *Biological Research*. 35: 195–202.

Jaimovich, E., Reyes, R., Liberona, J. L. and Powell, J. A., 2000. IP_3 receptors, IP_3 transients, and nucleus-associated Ca^{2+} signals in cultured skeletal muscle. *American Journal of Physiology: Cell Physiology*. 278: C998–C1010.

Jordan, T., Jiang, H., Li, H. and DiMario, J. X., 2005. Regulation of skeletal muscle fiber type and slow myosin heavy chain 2 gene expression by inositol trisphosphate receptor 1. *Journal of Cell Science*. 118: 2295–2302.

Jorquera, G., Altamirano, F., Contreras-Ferrat, A., Almarza, G., Buvinic, S., Jacquemond, V., Jaimovich, E. and Casas, M., 2013. Cav1.1 controls frequency-dependent events regulating adult skeletal muscle plasticity. *Journal of Cell Science*. 126: 1189–1198.

Jorquera, G., Juretic, N., Jaimovich, E. and Riveros, N., 2009. Membrane depolarization induces calcium-dependent up regulation of Hsp70 and Hmox-1 in skeletal muscle cells. *American Journal of Physiology: Cell Physiology*. 297(3): C581–C590.

Juretic, N., Garcia-Huidobro, P., Iturrieta, J. A., Jaimovich, E. and Riveros, N., 2006. Depolarization-induced slow Ca^{2+} transients stimulate transcription of IL-6 gene in skeletal muscle cells. *American Journal of Physiology: Cell Physiology*. 290: C1428–C1436.

Juretić, N., Jorquera, G., Caviedes, P., Jaimovich, E. and Riveros, N., 2012. Electrical stimulation induces calcium-dependent up-regulation of neuregulin-1β in dystrophic skeletal muscle cell lines. *Cellular Physiology and Biochemistry*. 29: 919–930.

Juretic, N., Urzua, U., Munroe, D. J., Jaimovich, E. and Riveros, N., 2007. Differential gene expression in skeletal muscle cells after membrane depolarization. *Journal of Cellular Physiology*. 210: 819–830.

Kubis, H. P., Scheibe, R. J., Meissner, J. D., Hornung, G. and Gros, G., 2002. Fast-to-slow transformation and nuclear import/export kinetics of the transcription factor NFATc1 during electrostimulation of rabbit muscle cells in culture. *Journal of Physiology*. 541: 835–847.

Kubis, H. P., Hanke, N., Scheibe, R. J., Meissner, J. D. and Gros, G., 2003. Ca^{2+} transients activate calcineurin/NFATc1 and initiate fast-to-slow transformation in a primary skeletal muscle culture. *American Journal of Physiology: Cell Physiology*. 285: C56–C63.

Lin, J., Wu, H., Tarr, P. T., Zhang, C. Y., Wu, Z., Boss, O., Michael, L. F., Puigserver, P., Isotani, E., Olson, E. N. *et al.*, 2002. Transcriptional co-activator PGC-1 alpha drives the formation of slow-twitch muscle fibres. *Nature*. 418: 797–801.

Liu, Y., Cseresnyes, Z., Randall, W. R. and Schneider, M. F., 2001. Activity-dependent nuclear translocation and intranuclear distribution of NFATc in adult skeletal muscle fibers. *Journal of Cell Biology*. 155: 27–39.

McCullagh, K. J., Calabria, E., Pallafacchina, G., Ciciliot, S., Serrano, A. L., Argentini, C., Kalhovde, J. M., Lomo, T. and Schiaffino, S., 2004. NFAT is a nerve activity sensor in skeletal muscle and controls activity-dependent myosin switching. *Proceedings of the National Academy of Sciences of the United States of America*. 101: 10590–10595.

Meissner, J. D., Gros, G., Scheibe, R. J., Scholz, M. and Kubis, H. P., 2001. Calcineurin regulates slow myosin, but not fast myosin or metabolic enzymes, during fast-to-slow transformation in rabbit skeletal muscle cell culture. *Journal of Physiology*. 533: 215–226.

Mellström, B., Savignac, M., Gomez-Villafuertes, R. and Naranjo, J. R., 2008. Ca^{2+}-operated transcriptional networks: molecular mechanisms and *in vivo* models. *Physiological Reviews*. 88: 421–449.

Murgia, M., Serrano, A. L., Calabria, E., Pallafacchina, G., Lomo, T. and Schiaffino, S., 2000. Ras is involved in nerve-activity-dependent regulation of muscle genes. *Nature Cell Biology*. 2: 142–147.

Naya, F. J., Mercer, B., Shelton, J., Richardson, J. A., Williams, R. S. and Olson, E. N., 2000. Stimulation of slow skeletal muscle fiber gene expression by calcineurin in vivo. *Journal of Biological Chemistry*. 275: 4545–4548.

Pette, D. and Staron, R. S., 2002. The adaptive potential of skeletal muscle fibers transitions of muscle fiber phenotypic profiles. *Canadian Journal of Applied Physiology*. 27: 423–448.

Powell, J. A., Carrasco, M. A., Adams, D. S., Drouet, B., Rios, J., Muller, M., Estrada, M. and Jaimovich, E., 2001. IP_3 receptor function and localization in myotubes: an unexplored Ca^{2+} signaling pathway in skeletal muscle. *Journal of Cell Science*. 114: 3673–3683.

Rebecchi, M. J. and Pentyala, S. N., 2000. Structure, function, and control of phosphoinositide-specific phospholipase C. *Physiological Reviews*. 80: 1291–1335.

Sakamoto, K. and Goodyear, L. J., 2002. Invited review: intracellular signaling in contracting skeletal muscle. *Journal of Applied Physiology*. 93: 369–383.

Savignac, M., Mellstrom, B. and Naranjo, J. R., 2007. Calcium-dependent transcription of cytokine genes in T lymphocytes. *Pflugers Archiv: European Journal of Physiology*. 454: 523–533.

Schiaffino, S. and Reggiani, C., 1994. Myosin isoforms in mammalian skeletal muscle. *Journal of Applied Physiology*. 77: 493–501.

Serrano, A. L., Murgia, M., Pallafacchina, G., Calabria, E., Coniglio, P., Lomo, T. and Schiaffino, S., 2001. Calcineurin controls nerve activity-dependent specification of slow skeletal muscle fibers but not muscle growth. *Proceedings of the National Academy of Sciences of the United States of America*. 98: 13108–13113.

Valdes, J. A., Hidalgo, J., Galaz, J. L., Puentes, N., Silva, M., Jaimovich, E. and Carrasco, M. A., 2007. NF-kappaB activation by depolarization of skeletal muscle cells depends on ryanodine and IP_3 receptor-mediated calcium signals. *American Journal of Physiology: Cell Physiology*. 292: C1960–C1970.

Valladares, D., Almarza, G., Contreras, A., Pavez, M., Buvinic, S., Jaimovich, E. and Casas, M., 2013. Electrical stimuli are anti-apoptotic in skeletal muscle *via* extracellular ATP. Alteration of this signal in Mdx mice is a likely cause of dystrophy. *PLoS One*. 8: e75340.

Wu, H., Kanatous, S. B., Thurmond, F. A., Gallardo, T., Isotani, E., Bassel-Duby, R. and Williams, R. S., 2002. Regulation of mitochondrial biogenesis in skeletal muscle by CaMK. *Science*. 296: 349–352.

CHAPTER 27

Mitochondrial Calcium Homeostasis and Implications for Human Health

TITO CALÌ[a], DENIS OTTOLINI[a], AND MARISA BRINI*[a]

[a]Department of Biology, University of Padova, Via Ugo Bassi 58/b, 35131 Padova, Italy
*E-mail: marisa.brini@unipd.it

27.1 Introductory Elements of Calcium Signaling

Calcium homeostasis is central for all cellular function. Its fundamental role of carrier of messages has been established since 1883 when the serendipitous observation by Sidney Ringer that the addition of Ca^{2+} to isolated rat hearts induced contraction. Ca^{2+} is now recognized as a ubiquitous second messenger involved in almost all physiological processes such as fertilization, muscle contraction, secretion, gene expression and cell death (Brini *et al.*, 2013).

The development of progressively sophisticated tools (both Ca^{2+} probes and technologies) to monitor intracellular Ca^{2+} fluxes in living cells, and in whole animals as well, has enormously improved our knowledge on the molecular characteristics of the transmission of the Ca^{2+} signal.

Cytosolic Ca^{2+} concentration can vary by several order of magnitude: under resting conditions it is maintained at about 100 nM by the coordinated

Food and Nutritional Components in Focus No. 10
Calcium: Chemistry, Analysis, Function and Effects
Edited by Victor R. Preedy

Published by the Royal Society of Chemistry, www.rsc.org

activity of the Ca^{2+} ATPases (both in the plasma membrane and in the endo/sarcoplasmic reticulum (ER/SR) and Golgi compartments), of the plasma membrane Na^{+}/Ca^{2+} exchanger, by the buffering activity of the cytosolic Ca^{2+} binding proteins and the mitochondria. Both the extracellular ambient and intracellular Ca^{2+} stores, that is, ER/SR and Golgi apparatus, are unlimited Ca^{2+} reservoir where Ca^{2+} concentration is in the mM range (Figure 27.1). The steep gradient against the cytosolic compartment is highly energy demanding, however, it guarantees the rapid generation of Ca^{2+} transients and the triggering of cascades of cellular events. Ca^{2+} signals are shaped with specific spatiotemporal patterns by the membrane transport proteins that generate cellular Ca^{2+} microdomains that may have distinct functional significance according to their restricted spatial localization (Berridge and Dupont, 1994; Berridge, 2006). The intracellular organelles, and in particular the mitochondria, contribute to precisely regulate the amplitude and duration of Ca^{2+} signals in a site-specific manner.

Ca^{2+} signals occur in two main forms: waves and oscillations. Propagating Ca^{2+} waves mainly arise from the release of Ca^{2+} from internal stores

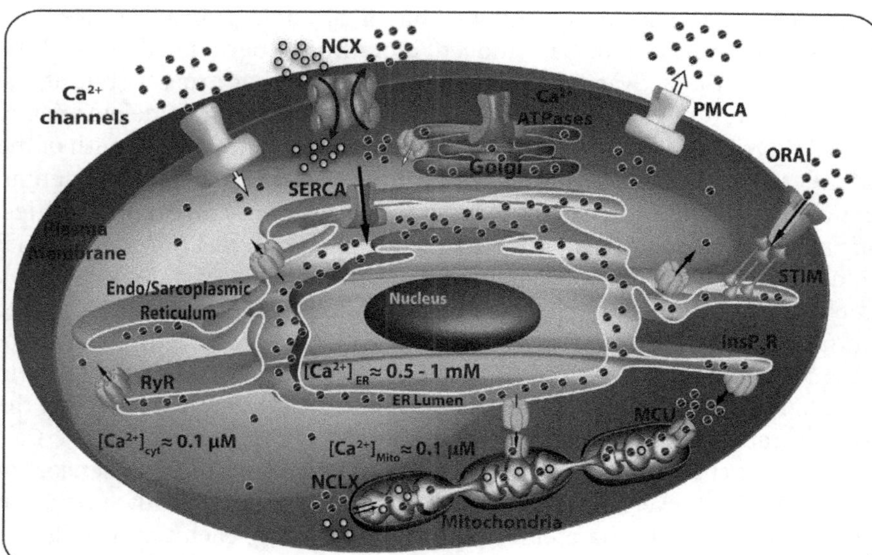

Figure 27.1 Schematic representation of intracellular Ca^{2+} homeostasis. The Ca^{2+}-transport proteins of the plasma membrane (PM) and the intracellular organelles, including mitochondria, endoplasmic reticulum (ER), Golgi apparatus, are indicated. PMCA, plasma membrane Ca^{2+} ATPase; NCX, Na^{+}/Ca^{2+} exchanger, SERCA, sarco/endoplasmic reticulum Ca^{2+} ATPase; $InsP_3R$, inositol 1,4,5-trisphosphate receptor; RyR, ryanodine receptor; MCU, mitochondrial Ca^{2+} uniporter, NCLX, mitochondrial Na^{+}/Ca^{2+} exchanger, STIM, sensors stromal interaction molecule, ORAI, calcium release-activated calcium modulator. Black and gray circles represent Ca^{2+} and Na^{+} ions, respectively.

through the opening of intracellular channels, the most important being the inositol 1,4,5-tris-phosphate receptor ($InsP_3R$) and the ryanodine receptor (RyR) in the ER/SR membranes. Two main types of intracellular waves are known: those mediated through the opening of $InsP_3Rs$ (most common) and those mediated by the regenerative activation of RyRs. Localized events of $InsP_3R$ opening, known as "puffs", are responsible for a local Ca^{2+} release that may act as a trigger to generate a global Ca^{2+} wave that involves sequential activation of $InsP_3Rs$ (and RyRs channel/receptor clusters). Regenerative activation of RyRs has been initially observed in cardiac myocytes (but it is now known to occur also in skeletal muscle and neurons), and depends on the spontaneous localized Ca^{2+} release events, often called "sparks", that result from the opening of clusters of RyRs in the SR by a local mechanism named calcium-induced calcium release (CICR). Ca^{2+} entry through voltage-gated plasma membrane Ca^{2+} channels can also modulate the generation of this type of wave, thus suggesting a role for extracellular Ca^{2+} as well.

The temporal regulation of the Ca^{2+} signal is determined by oscillations of Ca^{2+} concentration that generally are characterized by a constant amplitude and a variable frequency, which ranges from 5 to 60 s depending on the cell type, and on the nature and the strength of the stimulus that initiates them. The oscillatory pathway has a physiological advantage over the sustained elevation of Ca^{2+}. Indeed, it permits to periodically exceed the threshold for the activation of a high Ca^{2+}-demanding process, but at the same time prevents persistent Ca^{2+} overload, which would be deleterious to the cell (Berridge, 1990). The generation of Ca^{2+} hot spots at the mouth of the Ca^{2+} release/influx channels is another sophisticated way to compartmentalize Ca^{2+} signals and to selectively activate sensors with different affinities. Ca^{2+} microdomains are generated by the juxtaposition of ER-mitochondrial membranes, the so-called mitochondria-associated ER-membranes (MAMs), which are specialized regions where ER and mitochondria become physically and functionally tethered (Naon and Scorrano, 2014). Relevant Ca^{2+} microdomains are also those generated by the close contacts between the PM and the ER. They are responsible for the control of Ca^{2+} entering through the store-operated Ca^{2+} channels (SOCCs) activated by the depletion of the Ca^{2+} stores. These channels are controlled by STIM proteins, which are sensors of ER luminal Ca^{2+} levels. Their extrareticular portion by cyclically moving in and out of the ER-plasma membrane junctions during each Ca^{2+} oscillatory spike reversibly activates the channel pore subunit Orai1 and thus favouring Ca^{2+} entry (Luik *et al.*, 2006; Prakriya *et al.*, 2006).

27.2 Mitochondrial Calcium Handling

Mitochondria are organelles endowed with a double membrane that separates them from the cytoplasm (Figure 27.2). The outer mitochondrial membrane (OMM) and the inner mitochondrial membrane (IMM) undergo a continuous process of fission and fusion, that, together with organelles movements, control and guarantee proper mitochondrial function. Several

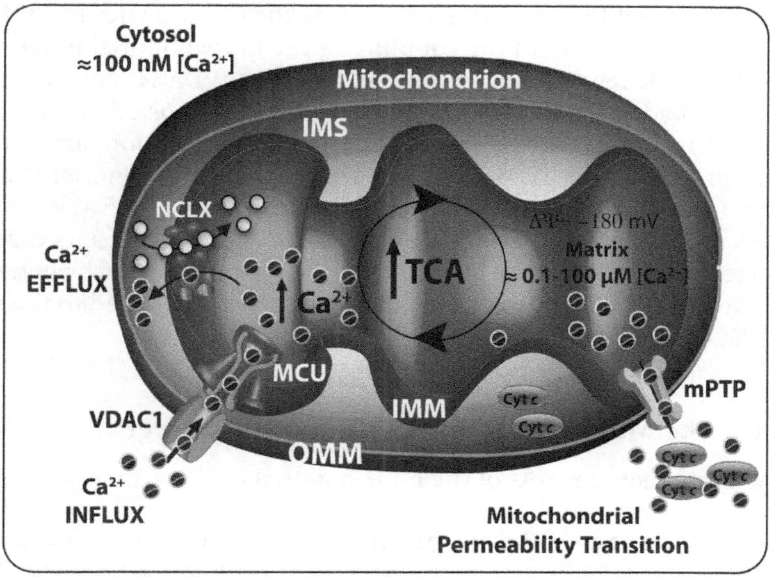

Figure 27.2 Schematic representation of mitochondrial Ca^{2+} transporters. Mitochondria accumulate Ca^{2+} in the matrix *via* an electrogenic Ca^{2+} uniporter (MCU) using the mitochondrial membrane potential ($\Delta\psi$) generated by the respiratory chain. Voltage-dependent anion channel (VDAC1) facilitates Ca^{2+} diffusion from intracellular stores through the outer mitochondrial membrane (OMM) by coupling $InsP_3$ receptor to MCU. A Na^+/Ca^{2+} (NCLX) exchanger operates as the Ca^{2+} efflux pathway. Mitochondrial Ca^{2+} overload induces the opening of the mitochondrial permeability transition pore (mPTP) and the release of cytochrome c (Cyt c) and cell death. IMM, inner mitochondrial membrane; IMS, intermembrane space. Black and gray circles represent Ca^{2+} and Na^+ ions, respectively. TCA, tricarboxylic acid cycle.

mitochondrial processes are controlled by Ca^{2+} within the organelle, *i.e.*, the activity of three dehydrogenases of the tricarboxylic acid (TCA) cycle and the release of proapoptotic factors following mitochondrial Ca^{2+} overload (Berridge, 2006). The IMM contains a specific Ca^{2+} transport machinery composed by a tetrameric pore-forming structure, the mitochondrial Ca^{2+} uniporter (MCU), plus several regulatory components, among which MICU1, MICU2, EMRE and MCUR1 have been recently characterized (see Section 27.2.1). It also contains a Na^+/Ca^{2+} extrusion system (NCLX), which operates in most cell types (in some cells a mitochondrial H^+/Ca^{2+} exchanger is instead present). The electrochemical gradient generated across the IMM by the chemiosmotic operation of the respiratory chain acts as a driving force for mitochondrial Ca^{2+} uptake. Ca^{2+} influx also critically depends on microdomains of high Ca^{2+} concentration generated by the opening of Ca^{2+} channels in the neighbouring ER that satisfy the low affinity of the uniporter (Rizzuto *et al.*, 1992; Rizzuto *et al.*, 1993). The OMM participates to the mitochondrial Ca^{2+} handling by favoring the Ca^{2+} transfer from the ER to mitochondria thanks to the coupling of the mitochondrial outer membrane voltage-dependent anion-channel protein 1 (VDAC1) and the $InsP_3R$ to MCU (Szabadkai *et al.*, 2006). Even if not directly involved in physiological Ca^{2+} transport, the mitochondrial permeability transition pore (mPTP), another protein of the IMM, is important for the mitochondrial function and Ca^{2+} signaling. The recent molecular identification of the ATP synthase complex (a subunit of it or the ATP synthase dimer *per se*) as its essential component further links the mitochondrial Ca^{2+} transport with energy balance. The mPTP will be discussed in Section 27.2.1 dedicated to the mitochondrial Ca^{2+} players.

27.2.1 The Molecular Identification of the Players

The molecular components of the mitochondrial Ca^{2+} machinery have been only recently identified.

Their discovery now offers the possibility to modify their expression level and to create cellular and animal models to directly investigate the consequences of their manipulation. This will permit us to deeply increase our knowledge on the importance of Ca^{2+} signaling and the consequences of its alteration for human health.

27.2.1.1 The Mitochondrial Calcium Uniporter Complex

Mitochondria Ca^{2+} transport has unique characteristic: Ca^{2+} accumulation in the mitochondrial matrix is mediated by a uniporter (MCU) and does not require ATP but utilizes the mitochondrial membrane potential ($\Delta\Psi m$, negative inside) maintained across the inner membrane by the respiratory chain. Ca^{2+} uptake is countered by Ca^{2+} efflux so that electrochemical gradient equilibrium is never attained. Pharmacological studies led to the identification of compounds able to either inhibit or activate it; however, the very low Ca^{2+} affinity of MCU appeared difficult to reconcile with its function in the extremely low cytosolic Ca^{2+} concentration. Thus, for a long period mitochondrial Ca^{2+} accumulation was considered to occur only after Ca^{2+}-overload conditions, even if it was evident that it had a direct role in the regulation of mitochondrial dehydrogenases, and consequently of the oxidative phosphorylation (Denton *et al.*, 1980). In 1992 the direct monitoring of Ca^{2+} transients in the mitochondrial matrix of living cells by the recombinant Ca^{2+}-sensitive photoprotein aequorin solved the discrepancies (Rizzuto *et al.*, 1992). The low affinity of the mitochondrial uptake system was overcome by the generation of localized microdomains at high Ca^{2+} concentration in close proximity to the mouth of the ER Ca^{2+} channels and to the mitochondrial Ca^{2+} uptake sites (Rizzuto *et al.*, 1993). The finding renewed interest in mitochondrial Ca^{2+} and new research aimed at investigating its physiological role and identifying the mitochondrial

Ca^{2+} transporters. In spite of patch clamp experiments on mitoplasts (swollen mitochondria without the OMM) showing that the MCU was a highly selective hardly saturable Ca^{2+} channel (Kirichok *et al.*, 2004), its molecular identity had remained elusive for a long time. Different proposals had been made for different candidate proteins, but none of them was completely convincing. Only in 2011, using a MitoCarta database, a uniporter component (MICU1), first (Perocchi *et al.*, 2010) and the MCU channel later (Baughman *et al.*, 2011; De Stefani *et al.*, 2011) were identified. MCU is an integral inner membrane protein of 40 kDa that includes in its N-terminal domain a mitochondrial targeting sequence. After mitochondrial import, this sequence was cleaved thus generating a 35 kDa form that represents the mature protein. MCU is ubiquitously expressed in mammals, but missing in yeast. It is predicted to have both the N- and C-terminal domain in the matrix, with two transmembrane regions connected by a short loop containing the DIME motif, which contains a cluster of negatively charged amino acids in proximity to the pore region. MCU is arranged in oligomers, possibly tetramers, in the inner membrane and forms a super-complex with a molecular weight around 480 kDa (Figure 27.3). MCU physically

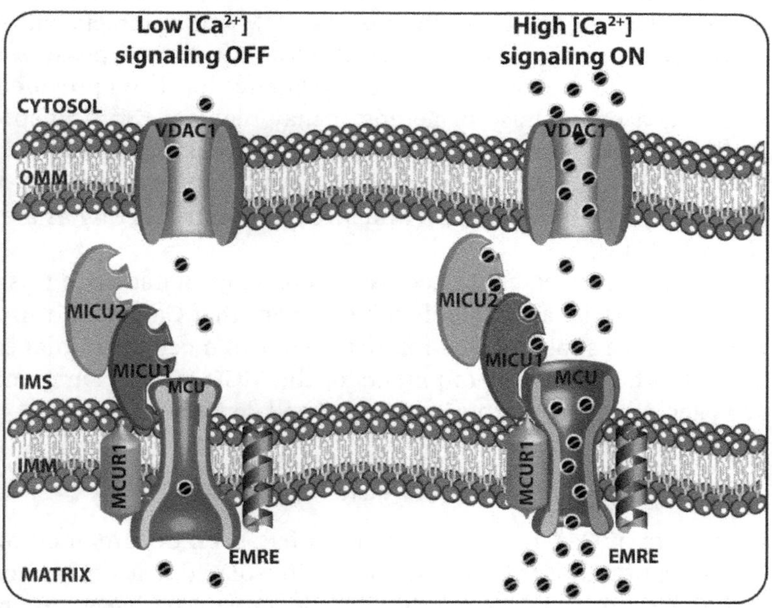

Figure 27.3 Schematic map of the mitochondrial calcium uniporter complex. The mitochondrial calcium uniporter complex is composed by the pore-forming tetrameric structure (MCU) and several regulators (MICU1, MICU2, MCUR1, EMRE) that control both the activity of the channel and the assembly of the multicomplex. MICU1 and 2, mitochondrial Ca^{2+} uptake 1 and 2, MCUR1, mitochondrial Ca^{2+} uptake regulator; EMRE, essential MCU regulator. Two opposite representative situations (signaling OFF and signaling ON) are schematized. VDAC1, voltage-dependent anion channel 1.

interacts with MICU1, a 54 kDa protein, associated with the mitochondrial inner membrane. MICU1 has one putative transmembrane domain and two canonical EF hands Ca^{2+} binding motifs that are probably faced to the intermembrane space to sense changes in cytosolic Ca^{2+} concentration and thus control the Ca^{2+} affinity of the MCU complex. In addition to MICU1, two other paralogs (MICU2 and MICU3) are expressed in most eukaryotic cells (Plovanich *et al.*, 2013). MICU1 and MICU2 heterodimerize and their protein levels are strictly related to the MCU complex stability. MICU1 has a double role in the regulation of MCU activity: it stimulates the mitochondrial Ca^{2+} uptake *via* MCU during the cytosolic Ca^{2+} rises, and inhibits it in the resting condition (Mallilankaraman *et al.*, 2012; Csordas *et al.*, 2013). Recently, it has been shown that the inhibitory role is rather played by the interacting MICU2 that has been defined the gatekeeper of MCU (Patron *et al.*, 2014). The existence of an isoform of MCU, the MCUb, further complicated the situation. It differs from MCU by some residues in the pore region and may act as a dominant negative subunit that, by assembling together with canonical MCU subunits, prevents Ca^{2+} transport into the matrix (Raffaello *et al.*, 2013). Another two proteins (*e.g.*, MCUR1 and EMRE) contribute to the structural and possibly functional arrangement of the MCU complex. MCUR1 is a 40 kDa protein of the IMM and interacts with MCU but not with MICU1. It is mainly exposed to the matrix and possesses two putative transmembrane domains, but no evidence has been provided that it could act as a pore, thus suggesting it may regulate Ca^{2+} transport by prevalently regulating the assembly of the MCU complex (Mallilankaraman *et al.*, 2012). EMRE is a 10 kDa single-pass transmembrane protein of the IMM that interacts both with MCU and MICU1, possibly acting as a bridge (Sancak *et al.*, 2013).

Despite further studies being necessary to deeply characterize this complex machinery and its activity, it is quite evident that Ca^{2+} import in mitochondria requires a sophisticated multistep control mechanism that can be achieved only when all the components of the MCU complex are perfectly adjusting each other to promptly respond to cell demand.

27.2.1.2 The Mitochondrial Na^+/Ca^{2+} Exchanger

The efflux route of Ca^{2+} from mitochondria has been documented following the observation that the addition of Na^+ to isolated heart mitochondria promoted the efflux of Ca^{2+}. Later, this Na^+-dependent pathway was characterized as an electrogenic Na^+/Ca^{2+} antiporter (NCLX) that transports Ca^{2+} against the Ca^{2+} electrochemical gradient. Together with the reversible formation of buffering Ca^{2+}-phosphate complexes it contributes to maintain nanomolar free Ca^{2+} levels inside the mitochondrial matrix. As the Ca^{2+} uptake pathway, the efflux pathway has been pharmacologically studied for years without having molecularly identified the transporter. The NCLX has been identified by Sekler and coworkers as a mammalian member of

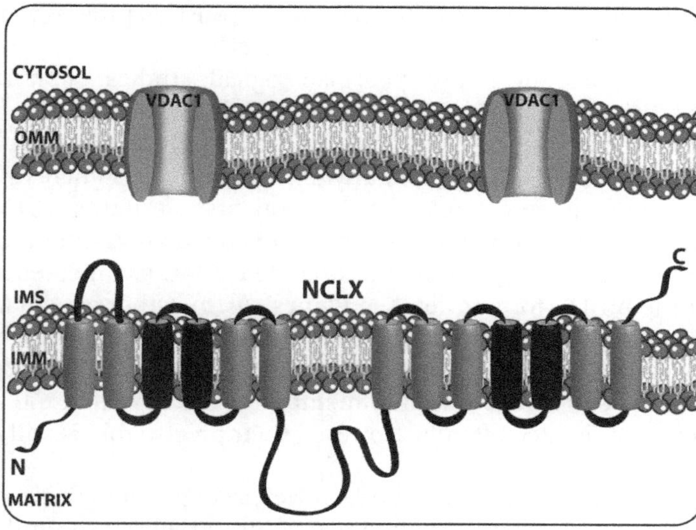

Figure 27.4 Putative topology model of the mitochondrial Na^+/Ca^{2+} exchanger. The Na^+/Ca^{2+} exchanger (NCLX) prevents the attainment of Ca^{2+} electrochemical equilibrium. On the basis of its amino acids sequence, NCLX has been predicted to have 9 transmembrane domains, which are shown as gray cylinders plus two α-repeat regions, indicated by black cylinders, which contain residues essential to cation binding and transport.

the phylogenetically ancestral Ca^{2+}/anion exchangers family that catalyses Na^+-dependent (but also Li^+) Ca^{2+} transport (Palty *et al.*, 2010) (Figure 27.4). As expected, it was found to be enriched in the mitochondrial cristae and was sensitive to the specific inhibitor chloro-5-(2-chlorophenyl)-1,5-dihydro-4,1-benzothiazepin-2(3H)-one (CGP 37157). In addition to the NCLX, a H^+/Ca^{2+} exchanger is also present in some tissues, *i.e.* in the liver. The transport is electroneutral and it has been characterized as a 1 Ca^{2+} for $2H^+$ exchanger.

27.2.1.3 The Mitochondrial Permeability Transition Pore

The mitochondrial permeability transition pore (mPTP) may play a role in the mitochondrial Ca^{2+} efflux, especially in conditions in which mitochondrial Ca^{2+} concentration in the matrix reaches threshold levels. The mPTP is a nonspecific channel, permeable to solutes up 1.5 kDa. Its opening is induced by high Ca^{2+} levels in the matrix, but pH, adenine nucleotides, free radicals and $\Delta\Psi m$ also modulate it. Mitochondrial Ca^{2+} overload and excessive reactive oxygen species (ROS) in the matrix cause permeabilization of the IMM, proton electrochemical gradient dissipation, ATP depletion, further ROS production and organelle swelling. These events are collectively termed "mitochondrial permeability transition" (MPT), a process that culminates in the release of cytochrome c and cell death.

Numerous biochemical and pharmacological studies supported the hypothesis that the mPTP was a multiprotein complex. The voltage-dependent anion channel (VDAC), the adenine nucleotide translocase (ANT) and the cyclophilin-D (CypD), a peptidyl-prolyl *cis–trans* isomerase (PPIase) were considered the most relevant components (Bernardi, 2013). More recently, however, genetic studies that have generated knockout models seriously questioned this suggestion and excluded VDAC as an essential component of the mPTP (for a review see Baines, 2010). Other candidates were proposed to participate to the mPTP complex formation: the mitochondrial phosphate carrier (PiC), the peripheral benzodiazepine receptor PBR/TSPO, a protein of the outer mitochondrial membrane, the hexokinase II and several members of the Bcl-2 family. However, their precise role is still debated (Bernardi, 2013).

In 2013, two different groups found a novel protagonist in the long way to the identification of the mPTP. Bonora *et al.* proposed the subunit c of the F_0F_1-ATPase as the major component of the mPTP (Bonora *et al.*, 2013). At the same time, Giorgio *et al.* hypothesized the mPTP was formed by dimers of ATP synthase protein (Giorgio *et al.*, 2013). Both possibilities perfectly fit with the previous studies that have considered ANT, CypD, and PiC as candidates, since ATP synthase interacts with them (Figure 27.5). Unfortunately, its ablation is incompatible with the life, thus genetic manipulation cannot be used to finally demonstrate the precise composition of the pore complex.

The finding that ATP synthase (or one of its subunits) could participate in MPT regulation is extremely interesting since unequivocally links metabolic pathway, *i.e.*, ATP production, to cell death. mPTP opening can result in cell death by apoptosis or necrosis, possibly depending on the residual ATP content of the cells. Apoptosis is an energy-dependent process and is activated by the release of proapoptotic factors from mitochondria; necrosis instead is linked to massive mPTP opening in a large fraction of the mitochondria population with consequent ATP depletion.

It is, however, important to underline that the opening of mPTP could not necessarily be linked to cell death. The channel indeed could be transiently activated, *i.e.* in the "flickering" mode, thus acting as a Ca^{2+} releasing channel to favor Ca^{2+} exit under high intramitochondrial Ca^{2+} concentration conditions and avoid Ca^{2+} overload. Moreover, the flickering transiently depolarizes mitochondria, thus reducing their ability to accumulate Ca^{2+} and preserving them from overload. In this light, mPTP may mediate a fast Ca^{2+}-induced Ca^{2+}-release mechanism and thus have a prosurvival role. However, the question is debated since the evidence for a role in Ca^{2+} extrusion is uniquely based on the fact that both CsA treatment and CypD ablation induce an increase in mitochondrial Ca^{2+} content. Very recently, by manipulating the level of expression of subunit c of ATP synthase, identified as a

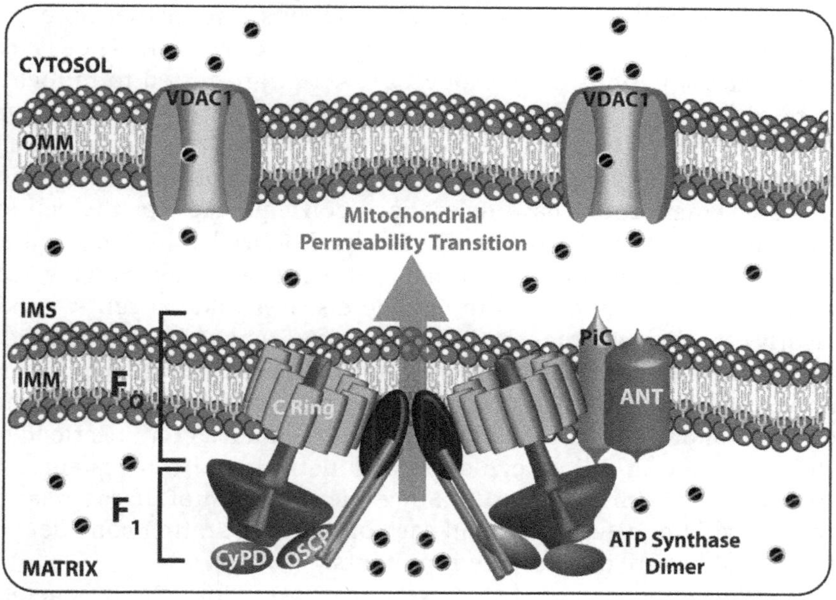

Figure 27.5 Schematic representation of the different components of the mitochondrial permeability transition pore (mPTP). The F_0F_1ATP synthase is the enzyme that synthesizes ATP. It is formed by the catalytic portion F_1 and the membrane-bound proton-translocating F_0 portion. Dimers of ATP synthase or its subunit c (c ring) have been recently proposed to constitute the pore-forming unit of the mPTP. Adenine nucleotide translocase (ANT), mitochondrial phosphate carrier (PiC) and cyclophilin-D (CypD) may participate in the pore composition or/and act as regulatory subunits. The scheme shows the different subunits of ATP synthase, the interactions with the proposed mPTP regulators. OSCP subunit of ATP synthase binds CypD. The arrow indicates the mitochondrial permeability transition process.

critical component of the mPTP, De Marchi *et al.* have shown that the mPTP does not directly participate in the Ca^{2+}-efflux mechanism in physiological conditions, but it is rather implicated in the dissipation of mitochondrial membrane potential (De Marchi *et al.*, 2014).

27.3 Mitochondrial Calcium in Human Health

The ability of mitochondria to take up substantial amounts of Ca^{2+} is important for two principal reasons: it enables mitochondria to shape the spatiotemporal profile of cytosolic Ca^{2+} signals and it simultaneously transmits these events into the mitochondrial matrix. In the matrix, Ca^{2+} upregulates the activity of Ca^{2+}-sensitive dehydrogenases of the Krebs cycle and the F_0F_1-ATP synthase, thereby controlling the rate of ATP production. Interestingly, the intensity of activation is directly proportional to the amplitude of mitochondrial Ca^{2+} transients, and furthermore to its spatiotemporal characteristics, since imaging studies on single cells have shown that oscillations of cytosolic Ca^{2+} were transmitted to mitochondria, resulting in the prolonged activation of the matrix enzymes, thus sustaining ATP production for times longer than those of the Ca^{2+} transients (Hajnoczky *et al.*, 1995). Any dysregulation in the control of mitochondrial Ca^{2+} signals could have deleterious consequences for the cell and the organism (Figure 27.6). Impaired mitochondrial function has been associated with a plethora of human pathological states ranging from neurodegenerative conditions to cardiac diseases and cancer. It is hard to clearly establish whether the link is causative or whether mitochondria impairments are rather the consequences of different defective pathways. However, consistent evidence is emerging on the possibility that defective mitochondrial Ca^{2+} uptake as well as mitochondrial Ca^{2+} overload can both be involved in cell deregulation with detrimental consequences for human health (Brini *et al.*, 2013). Since the spectrum of diseases is very broad, we will not discuss in detail each of them, but rather consider general aspects that are common determinants.

The evidence that connects Ca^{2+} dyshomeostasis with neurodegenerative disease like Parkinson's (PD), Alzheimer's (AD), Huntington's diseases (HD)

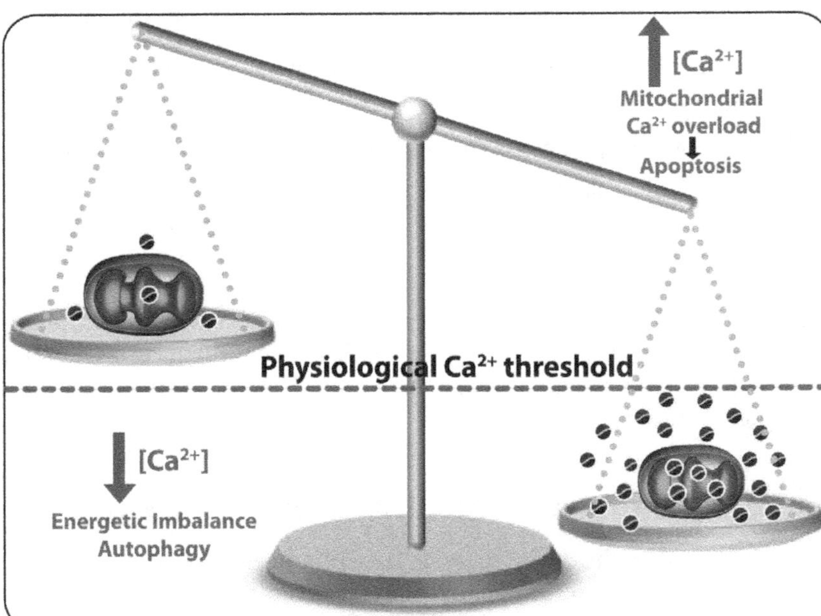

Figure 27.6 Balanced mitochondrial Ca^{2+} levels are essential to guarantee cell wellness. Mitochondrial Ca^{2+} overload as well as defective mitochondrial Ca^{2+} uptake can result in cell-death pathways, *i.e.*, apoptosis and autophagy. Both them are, however, also essential to the normal development of the organisms, thus their impairment impacts human health.

and amyotrophic lateral sclerosis (ALS) is abundant (Brini *et al.*, 2014). Multiple mitochondrial defects have also been detected. Mitochondrial function is particularly relevant for high energy demanding cells such as neurons. Moreover, neurons are uniquely exposed to glutamate excitotoxicity, which can lead to massive Ca^{2+} influx and to the activation of detrimental enzymes like phospholipases, endonucleases and proteases (*e.g.* calpains). Mitochondria exert a double action: they move along axons and supply the energy necessary to the pumps for the extrusion of Ca^{2+} (and other) ions, and also temporarily counteract the effects of cytosolic Ca^{2+} overload by buffering the excessive level of Ca^{2+} in the cytosol. When mitochondrial Ca^{2+} buffering capacity is overwhelmed, mitochondrial Ca^{2+} overload assumes a deleterious role. Another aspect, which is relevant for human health, is directly related to the necessity for the cells to have a proper Ca^{2+} transfer from the ER to mitochondria.

The proper communication between ER and mitochondria is a critical element for cell wellness: it is not only crucial to generate, in stimulated cells, microdomains of high [Ca^{2+}] in the proximity of ER Ca^{2+} release channels and thus permit that Ca^{2+} can be taken up by the MCU, but also, under basal conditions, to sustain cell metabolism and bioenergetics. The interaction between the two compartments is bifunctional: alterations in the ER–mitochondria juxtaposition determine mitochondrial dysfunctions, but also compromise ER-based processes such as lipid metabolism, protein synthesis and folding (Cali *et al.*, 2013).

Apoptosis is a process of programmed cell death that occurs in all multicellular organisms and that is essential for normal tissue development. Its dysregulation has very deleterious effects: a lot of pathological conditions like neurodegeneration or cancer find their genesis in a defective apoptosis. The process can be induced through different pathways, among which that mediated by Bcl-2 family proteins is Ca^{2+} dependent (Giorgi *et al.*, 2012). It has been convincingly demonstrated that the antiapoptotic action of Bcl-2 protein is strictly associated to its ability to reduce ER Ca^{2+} content/release through the $InsP_3R$, thus preventing mitochondria Ca^{2+} sensitization to apoptotic stimuli. However, it has been recently recognized that the complete absence of Ca^{2+} transfer from the ER to the mitochondria could be responsible for the activation of another cell-death pathway, namely autophagy. Similarly to apoptosis, dysfunctional autophagy has been associated with the pathogenesis of several diseases. Having this in mind, the finding that the silencing of the $InsP_3R$ by siRNA, or its pharmacological inhibition, induced autophagy is particularly intriguing (Cardenas *et al.*, 2010). This underlines the close relationship between the autophagic and apoptotic pathways, since both of them are strictly related to the concentration of luminal ER Ca^{2+} and to the amount that can be released by it and sensed by mitochondria.

Following the molecular identification of mitochondrial Ca^{2+} players, gene-inactivation studies are now possible and the generation of animal models as well as the identification of pathological conditions associated to specific gene mutations will permit novel insights into the specific role of mitochondrial Ca^{2+} in human health to be reached. The finding that a MICU1 loss of function mutation, which severely compromises mitochondria Ca^{2+} handling by enhancing the uptake in resting conditions, was found in a patient with proximal myopathy, learning difficulties and a progressive extrapyramidal movement disorder provides direct evidence for the crucial role of mitochondrial Ca^{2+} in human health (Logan *et al.*, 2014). Interesting results have been obtained by the characterization of MCU knockout mice (Pan *et al.*, 2013), even if the finding that MCU ablation not only was compatible with survival but also generated a very mild phenotype was quite surprising. As expected, mice have alteration in metabolic function, particularly in those tissues with high metabolic rate such as skeletal muscle and under conditions of increased ATP demand. Pyruvate dehydrogenase activity was clearly compromised in skeletal muscle and cardiac mitochondria, but no difference in oxygen consumption was observed. MCU KO mitochondria did not show Ca^{2+}-induced mPTP opening, but, unexpectedly, the absence of MCU did not confer protection against ischemic-reperfusion damage in the heart, thus suggesting the presence of compensating mechanisms, even if no mitochondrial Ca^{2+} uptake was observed. Further studies are necessary to clarify the relevance of MCU in cell physiology and pathology.

This is just the beginning of a new "golden period" for mitochondrial Ca^{2+}, as documented by the appearance of a very recent study based on a loss-of-function genetic screen that identified $InsP_3R$ and MCU (and the absence of ER-mitochondria Ca^{2+} transfer) as novel signaling molecules for the control of senescence (Wiel *et al.*, 2014). Thus, even if further investigations are required to dissect the multifaceted role of mitochondrial Ca^{2+} in human health, we must admit that the evidence that has emerged so far suggests that mitochondrial Ca^{2+} machinery could be thought as a new therapeutical target for many diseases and numerous studies are now exploring this field.

Summary Points

- This chapter focuses on mitochondrial Ca^{2+} handling.
- Mitochondria accumulate Ca^{2+} in the matrix by the mitochondrial Ca^{2+} uniporter (MCU) using the electrochemical gradient generated by the respiratory chain.
- Mitochondria extrude Ca^{2+} by the Na^+/Ca^{2+} exchanger, which operates against Ca^{2+} gradients.
- Mitochondrial Ca^{2+} accumulation modulates the bioenergetics metabolism.
- Mitochondrial Ca^{2+} accumulation contributes to the shaping of cytosolic Ca^{2+} signals.
- Defects in mitochondrial Ca^{2+} handling have detrimental consequences for cell physiology.

- Mitochondrial Ca^{2+} overload can culminate in cell death.
- Mitochondrial Ca^{2+} dyshomeostasis is involved in numerous human pathologies such as neurodegenerative conditions, cardiac diseases and cancer.

Key Facts

Key Facts of Calcium

1. Ca^{2+} ion is a universal carrier of signals.
2. Ca^{2+} controls a large array of processes that may be general to all cells (fertilization, gene transcription, differentiation, cell death, metabolic pathways, cell motility and migration) or cell specific (secretion, contraction/relaxation of skeletal muscles and heart).
3. Ca^{2+} transport across membranes is performed by Ca^{2+} transport proteins: channels, ATPases, exchangers, in which Ca^{2+} is exchanged for another ion (usually Na^{+}).
4. In the cytosol, Ca^{2+}-binding proteins modulate the Ca^{2+} signal spatially and temporally.
5. Ca^{2+}-binding proteins act as pure Ca^{2+} buffers, *e.g.*, parvalbumin, calbindins, and calreticulin, or as Ca^{2+} sensors, since they, in addition to buffering Ca^{2+}, also process its signal.
6. The most important and versatile Ca^{2+} sensor is the ubiquitous protein calmodulin.
7. Intracellular organelles participate in the regulation of Ca^{2+} homeostasis.

Key Facts of Mitochondrial Ca^{2+} History

1. 1961–1962: Vasington and Murphy and De Luca and Engstrom found that isolated mitochondria accumulated Ca^{2+} using either respiration or ATP hydrolysis.
2. 1974: Carafoli and coworkers found that Na^{+} stimulates heart mitochondria to release Ca^{2+}.
3. 1980: Denton and McCormack demonstrated that three matrix dehydrogenases were Ca^{2+} sensitive and could be regulated by the Ca^{2+} transport into the mitochondria.
4. 1992: Rizzuto and Pozzan introduced the recombinant Ca^{2+} indicator targeted to the mitochondrial matrix and demonstrated that, in living cells, mitochondria are able to accumulate Ca^{2+} following cell stimulation.
5. 1993: Rizzuto and Pozzan documented that microdomains of high Ca^{2+} concentration generated at the mouth of the Ca^{2+} release channels of the endoplasmic reticulum membrane are sensed by neighboring mitochondria.
6. 1998: Rizzuto and coworkers visualized close contact sites between endoplasmic reticulum and mitochondria by using targeted GFP and high resolution imaging techniques.

Key Facts of the Molecular Identification of the Mitochondrial Ca^{2+} Players

1. 2004: Clapham and coworkers, based on electrophysiological measurements on mitoplast, demonstrated that the mitochondrial calcium uniporter (MCU) is a Ca^{2+} specific ion channel.
2. 2010: Palty and Sekler identified the mitochondrial Na^{+}/Ca^{2+} exchanger (NCLX).
3. 2011: MCU was identified by screening the Mito-Carta (an inventory of nuclear and mitochondrial genes that encode mitochondrially localized proteins) by the two independent groups of Rizzuto and Mootha.
4. 2010–2014: Several groups have established that the mitochondrial Ca^{2+} uptake system is a multicomplex composed by different proteins that are necessary to regulate and stabilize the complex.
5. 2013: Mitochondrial permeability transition pore (mPTP) was identified by two independent groups. Pinton and coworkers proposed that the subunit c of ATP synthase is its essential pore component; Bernardi and coworkers proposed that the pore is composed by ATP synthase dimers.

Key Facts of ATP

1. The adenosine triphosphate (ATP) molecule is the chemical energy for cell metabolism.
2. ATP is mostly produced through a process called oxidative phosphorylation.
3. Oxidative phosphorylation is operated at the mitochondrial inner membrane by the respiratory chain complexes (I, II, III, IV) that are responsible for the transfer of electrons to different substrates and the pumping of protons in the intermembrane space.
4. The transport of protons across the inner membrane is responsible for the generation of an electrochemical gradient.
5. ATP synthase, that is the complex V of the respiratory chain, catalyzed the reaction of ATP synthesis.
6. ATP synthase uses the electrochemical gradient of protons across the inner mitochondrial membrane to generate ATP from ADP and a molecule of inorganic phosphate.

Key Facts of Apoptosis

1. The term apoptosis originates from a Greek word meaning "falling leaves".
2. It was initially proposed in 1972 to describe a form of natural cell death occurring in multicellular organisms as different as nematodes, insects and mammals.
3. Apoptosis can be induced through two different principal pathways that selectively involve specific proteins and intracellular compartments.

4. The extrinsic pathway engages plasma membrane receptors belonging to a subgroup of the TNF-receptor superfamily known as death receptors and the activation of the death-inducing caspase cascade.
5. The intrinsic pathway engages Bcl-2-related proteins upon cellular insults.
6. Mitochondria are key effectors in the intrinsic apoptosis pathway.

Key Facts of Autophagy

1. Autophagy is the general term used to define a cellular process responsible for the delivery of proteins or organelles to lysosomes.
2. Autophagy has different biological significance according to the specific cell conditions.
3. During metabolic stress, it represents a way to recycle macromolecules to satisfy the energy demands.
4. During defective protein homeostasis it selectively eliminates unfolded or aggregate proteins.
5. Under nonstress conditions, low levels of autophagy guarantee normal cellular homeostasis by removing dysfunctional proteins or even organelles.
6. Mitophagy is a type of autophagy specifically deputed to remove dysfunctional mitochondria thus preserving the functional pool and guaranteeing its functionality.
7. Autophagy is not just a way for the cells to die, but it is a sophisticated mechanism for cell quality control.

Definitions of Words and Terms

Ca^{2+} microdomains. Localized domains of Ca^{2+} that regulate specific cellular processes in different regions of the cell.
Ca^{2+} oscillations. Periodic and transitory changes of cytosolic Ca^{2+} concentration characterized by specific amplitude and time frame.
Ca^{2+} waves. Propagation of the cytosolic Ca^{2+} transients originated in a specific region of the cell to the whole cell or in different cell regions.
Mitochondrial associated ER membranes (MAMs). A specialized cell compartment where endoplasmic reticulum and mitochondrial membranes become structurally and functionally into close contact. Several specific proteins are found to be enriched in the MAMs, among them Mitofusin 2 is responsible for the physical tethering between the two membranes.
Mitochondrial fission. A process that allows the division of mitochondria after their fragmentation, their division during cell duplication and the segregation of damaged mitochondria.
Mitochondrial fusion. A process that permits mitochondria to join. Together with fission, this process allows the continuous remodeling of mitochondrial network.

Uniporter. A protein that acts as a carrier for the selective transport of ions (or molecules) across the cellular membranes according to their concentration gradient.

Exchanger. A protein that mediates the transport across the cellular membranes of a chemical species against it concentration gradient in exchange of another species (usually Na^+), which is transported according to its concentration gradient.

Respiratory chain. A chain of oxidoreductive reactions that couples electrons transfer from specific compounds with the transfer of protons (H^+). The process generates an electrochemical proton gradient across the mitochondrial inner membrane and drives the synthesis of adenosine tris-phosphate (ATP) by the F_0F_1 ATP synthase.

Apoptosis. Originally called "programmed cell death". It is characterized by blebbing, cell shrinkage, extensive chromatin condensation and DNA degradation and naturally occurs to regulate cells number and organism development.

Autophagy. A cellular process responsible for the delivery of proteins or organelles to lysosomes for their degradation.

List of Abbreviations

ANT	Adenine nucleotide translocase
ATP	Adenosine trisphosphate
CICR	Ca^{2+}-induced Ca^{2+} release
CypD	Cyclophilin-D
CsA	Cyclosporine-A
EMRE	Essential MCU regulator
ER	Endoplasmic reticulum
$\Delta\Psi m$	Mitochondrial membrane potential
$InsP_3$	Inositol 1,4,5-trisphosphate,
$InsP_3R$	Inositol 1,4,5-trisphosphate receptor
IMM	Inner mitochondrial membrane
IMS	Intermembrane space
MAMs	Mitochondria-associated ER membranes
MCU	Mitochondrial Ca^{2+} uniporter
MCUb	Mitochondrial Ca^{2+} uniporter isoform b
MICU1	Mitochondrial calcium uptake 1
MICU2	Mitochondrial calcium uptake 2
MICU3	Mitochondrial calcium uptake 3
MCUR1	Mitochondrial calcium uptake regulator
MPT	Mitochondrial permeability transition
mPTP	Mitochondrial permeability transition pore
NCX	Na^+/Ca^{2+} exchanger
NCLX	Mitochondrial Na^+/Ca^{2+} exchanger
OMM	Outer mitochondrial membrane

Orai1	Calcium release-activated calcium modulator 1
Pi	Inorganic phosphate
PiC	Mitochondrial phosphate carrier
PMCA	Plasma membrane Ca^{2+}-ATPase
RyR	Ryanodine receptor
ROS	Reactive oxygen species
SERCA	Sarco/endoplasmic reticulum Ca^{2+}-ATPase
SOCCs	Store-operated Ca^{2+} channels
SR	Sarcoplasmic reticulum
STIM	Sensors stromal interaction molecule
TCA	Tricarboxylic acid
PBR/TSPO	Peripheral benzodiazepine receptor
VDAC	Voltage-dependent anion-channel protein

Acknowledgements

The original work by the authors has been supported over the years by grants from the Italian Ministry of University and Research (PRIN 2003, 2005 and 2008), the Telethon Foundation (Project GGP04169), the Italian National Research Council (Agenzia 2000, CNR), from the University of Padova (Progetto di Ateneo 2008 CPDA082825) to M.B. Tito Calì is supported by the University of Padova (Progetto Giovani GRIC128SP0, Bando 2012).

References

Baines, C. P., 2010. *Annual Review of Physiology*. 72: 61–80.

Baughman, J. M., Perocchi, F., Girgis, H. S., Plovanich, M., Belcher-Timme, C. A., Sancak, Y., Bao, X. R., Strittmatter, L., Goldberger, O., Bogorad, R. L., Koteliansky, V. and Mootha, V. K., 2011. *Nature*. 476: 341–345.

Bernardi, P., 2013. *Frontiers in Physiology*. 4: 95.

Berridge, M. J., 1990. *Journal of Biological Chemistry*. 265: 9583–9586.

Berridge, M. J., 2006. *Cell Calcium*. 40: 405–412.

Berridge, M. J. and Dupont, G., 1994. *Current Opinion in Cell Biology*. 6: 267–274.

Bonora, M., Bononi, A., De Marchi, E., Giorgi, C., Lebiedzinska, M., Marchi, S., Patergnani, S., Rimessi, A., Suski, J. M., Wojtala, A., Wieckowski, M. R., Kroemer, G., Galluzzi, L. and Pinton, P., 2013. *Cell Cycle*. 12: 674–683.

Brini, M., Cali, T., Ottolini, D. and Carafoli, E., 2013. *Metal Ions in Life Sciences*. 12: 119–168.

Brini, M., Ottolini, D., Cali, T. and Carafoli, E., 2013. *Metal Ions in Life Sciences*. 13: 81–137.

Brini, M., Cali, T., Ottolini, D. and Carafoli, E., 2014. *Cellular and Molecular Life Sciences*. 71: 2787–2814.

Cali, T., Ottolini, D. and Brini, M., 2013. *DNA and Cell Biology*. 32: 140–146.

Cardenas, C., Miller, R. A., Smith, I., Bui, T., Molgo, J., Muller, M., Vais, H., Cheung, K. H., Yang, J., Parker, I., Thompson, C. B., Birnbaum, M. J., Hallows, K. R. and Foskett, J. K., 2010. *Cell*. 142: 270–283.

Csordas, G., Golenar, T., Seifert, E. L., Kamer, K. J., Sancak, Y., Perocchi, F., Moffat, C., Weaver, D., de la Fuente Perez, S., Bogorad, R., Koteliansky, V., Adijanto, J., Mootha, V. K. and Hajnoczky, G., 2013. *Cell Metabolism*. 17: 976–987.

Denton, R. M., McCormack, J. G. and Edgell, N. J., 1980. *Biochemical Journal*. 190: 107–117.

De Marchi, E., Bonora, M., Giorgi, C. and Pinton, P., 2014. *Cell Calcium*. 56: 1–13.

De Stefani, D., Raffaello, A., Teardo, E., Szabo, I. and Rizzuto, R., 2011. *Nature*. 476: 336–340.

Giorgi, C., Baldassari, F., Bononi, A., Bonora, M., De Marchi, E., Marchi, S., Missiroli, S., Patergnani, S., Rimessi, A., Suski, J. M., Wieckowski, M. R. and Pinton, P., 2012. *Cell Calcium*. 52: 36–43.

Giorgio, V., von Stockum, S., Antoniel, M., Fabbro, A., Fogolari, F., Forte, M., Glick, G. D., Petronilli, V., Zoratti, M., Szabo, I., Lippe, G. and Bernardi, P., 2013. *Proceedings of the National Academy of Sciences USA*. 110: 5887–5892.

Hajnoczky, G., Robb-Gaspers, L. D., Seitz, M. B. and Thomas, A. P., 1995. *Cell*. 82: 415–424.

Kirichok, Y., Krapivinsky, G. and Clapham, D. E., 2004. *Nature*. 427: 360–364.

Logan, C. V., Szabadkai, G., Sharpe, J. A., Parry, D. A., Torelli, S., Childs, A. M., Kriek, M., Phadke, R., Johnson, C. A., Roberts, N. Y., Bonthron, D. T., Pysden, K. A., Whyte, T., Munteanu, I., Foley, A. R., Wheway, G., Szymanska, K., Natarajan, S., Abdelhamed, Z. A., Morgan, J. E., Roper, H., Santen, G. W., Niks, E. H., van der Pol, W. L., Lindhout, D., Raffaello, A., De Stefani, D., den Dunnen, J. T., Sun, Y., Ginjaar, I., Sewry, C. A., Hurles, M., Rizzuto, R., Duchen, M. R., Muntoni, F. and Sheridan, E., 2014. *Nature Genetics*. 46: 188–193.

Luik, R. M., Wu, M. M., Buchanan, J. and Lewis, R. S., 2006. *Journal of Cell Biology*. 174: 815–825.

Mallilankaraman, K., Cardenas, C., Doonan, P. J., Chandramoorthy, H. C., Irrinki, K. M., Golenar, T., Csordas, G., Madireddi, P., Yang, J., Muller, M., Miller, R., Kolesar, J. E., Molgo, J., Kaufman, B., Hajnoczky, G., Foskett, J. K. and Madesh, M., 2012. *Nature Cell Biology*. 14: 1336–1343.

Mallilankaraman, K., Doonan, P., Cardenas, C., Chandramoorthy, H. C., Muller, M., Miller, R., Hoffman, N. E., Gandhirajan, R. K., Molgo, J., Birnbaum, M. J., Rothberg, B. S., Mak, D. O., Foskett, J. K. and Madesh, M., 2012. *Cell*. 151: 630–644.

Naon, D. and Scorrano, L., 2014. *Biochimica et Biophysica Acta*. 1843: 2184–2194.

Palty, R., Silverman, W. F., Hershfinkel, M., Caporale, T., Sensi, S. L., Parnis, J., Nolte, C., Fishman, D., Shoshan-Barmatz, V., Herrmann, S., Khananshvili, D. and Sekler, I., 2010. *Proceedings of the National Academy of Sciences USA*. 107: 436–441.

Pan, X., Liu, J., Nguyen, T., Liu, C., Sun, J., Teng, Y., Fergusson, M. M., Rovira, I. I, Allen, M., Springer, D. A., Aponte, A. M., Gucek, M., Balaban, R. S., Murphy, E. and Finkel, T., 2013. *Nature Cell Biology*. 15: 1464–1472.

Patron, M., Checchetto, V., Raffaello, A., Teardo, E., Vecellio Reane, D., Mantoan, M., Granatiero, V., Szabo, I., De Stefani, D. and Rizzuto, R., 2014. *Molecular Cell*. 53: 726–737.

Perocchi, F., Gohil, V. M., Girgis, H. S., Bao, X. R., McCombs, J. E., Palmer, A. E. and Mootha, V. K., 2010. *Nature*. 467: 291–296.

Prakriya, M., Feske, S., Gwack, Y., Srikanth, S., Rao, A. and Hogan, P. G., 2006. *Nature*. 443: 230–233.

Plovanich, M., Bogorad, R. L., Sancak, Y., Kamer, K. J., Strittmatter, L., Li, A. A., Girgis, H. S., Kuchimanchi, S., De Groot, J., Speciner, L., Taneja, N., Oshea, J., Koteliansky, V. and Mootha, V. K., 2013. *PLoS One*. 8: e55785.

Raffaello, A., De Stefani, D., Sabbadin, D., Teardo, E., Merli, G., Picard, A., Checchetto, V., Moro, S., Szabo, I. and Rizzuto, R., 2013. *EMBO Journal*. 32: 2362–2376.

Rizzuto, R., Simpson, A. W., Brini, M. and Pozzan, T., 1992. *Nature*. 358: 325–327.

Rizzuto, R., Brini, M., Murgia, M. and Pozzan, T., 1993. *Science*. 262: 744–747.

Sancak, Y., Markhard, A. L., Kitami, T., Kovacs-Bogdan, E., Kamer, K. J., Udeshi, N. D., Carr, S. A., Chaudhuri, D., Clapham, D. E., Li, A. A., Calvo, S. E., Goldberger, O. and Mootha, V. K., 2013. *Science*. 342: 1379–1382.

Szabadkai, G., Bianchi, K., Varnai, P., De Stefani, D., Wieckowski, M. R., Cavagna, D., Nagy, A. I., Balla, T. and Rizzuto, R., 2006. *Jornal of Cell Biology*. 175: 901–911.

Wiel, C., Lallet-Daher, H., Gitenay, D., Gras, B., Le Calve, B., Augert, A., Ferrand, M., Prevarskaya, N., Simonnet, H., Vindrieux, D. and Bernard, D., 2014. *Nature Communications*. 5: 3792.

CHAPTER 28

Pregnancy and Calcium

BELAL ALSHAIKH[a] AND KAMRAN YUSUF*[b]

[a]Department of Pediatrics, Section of Neonatal-Perinatal Medicine, Department of Community Health Sciences, University of Calgary, Room 780, Foothills Medical Centre, 1403 29th St NW, Calgary, Alberta T2N 2T9, Canada; [b]Department of Pediatrics, Section of Neonatal-Perinatal Medicine, Cummings School of Medicine, University of Calgary, Rm 2273, Health Sciences Center, 3330 Hospital Drive NW, Calgary, Alberta T2N 4N1, Canada
*E-mail: kyusuf@ucalgary.ca

28.1 Introduction

Calcium is fundamental to many cellular processes and biological pathways. In addition to being the principal component of bone, calcium is involved in muscle contraction, nerve transmission, blood coagulation, hormone secretion and second messenger formation. In the human circulation, calcium comprises three fractions – ionized, complexed and albumin bound, of which the ionized fraction, corresponding to 50% of the total, is the most important physiologically and clinically. During normal pregnancy, total serum calcium declines. This, however, is due to expansion of plasma volume by almost 45% during pregnancy and lower serum albumin leading to a decreased albumin bound fraction. *Ionized calcium remains unaltered and should be the fraction measured to assess calcium status during pregnancy*. Serum phosphate and magnesium also remain unchanged during pregnancy.

Food and Nutritional Components in Focus No. 10
Calcium: Chemistry, Analysis, Function and Effects
Edited by Victor R. Preedy

Published by the Royal Society of Chemistry, www.rsc.org

28.2 Maternal Adaptations during Pregnancy

During normal intrauterine life, the human fetus accumulates 30 g of calcium, 80% of it in the third trimester (Trotter and Hixon, 1974). To meet this extravagant demand of the fetus for calcium, several maternal adaptations of calcium metabolism take place, of which increased intestinal absorption of calcium is the most important.

28.2.1 Intestinal Calcium Absorption

Intestinal absorption of calcium doubles during pregnancy, starting as early as the 12th week of gestation (Cross *et al.*, 1995). This early increase in absorption of calcium allows the mother to store calcium for the third trimester when fetal demands are greater. In addition, the intestinal absorption of calcium is further enhanced in mothers whose dietary intake of calcium is low. Women with a normal calcium intake absorb almost 57% in the first trimester and 72% during the third trimester. This number can be as high as 87% in women with deficient calcium intake (Hacker *et al.*, 2012). The increase in calcium absorption is mediated by 1,25-dihydroxyvitamin D_3 (calcitriol). While levels of 25-hydroxyvitamin D remain unchanged during pregnancy, levels of 1,25-dihydroxyvitamin D_3 double in the first trimester and remain so during the rest of the pregnancy. 1,25-Dihydroxyvitamin D_3 augments the synthesis of vitamin D-dependent intracellular calcium-binding proteins (CaBP) including CaBP-D9K, enhancing the transfer of calcium across the intestinal epithelium (Delorme *et al.*, 1979). Although the placenta has 1-α-hydroxylase activity, most of the conversion of 25-hydroxyvitamin to 1,25-dihydroxyvitamin D_3 takes place in the kidneys. During pregnancy parathyroid hormone (PTH) levels do not rise and animal studies also suggest that the increase in 1,25-dihydroxyvitamin D_3 and intestinal absorption of calcium in pregnancy is PTH independent. Although animal studies have demonstrated that prolactin and placental lactogen stimulate renal 1-α-hydroxylase and also increase intestinal absorption of calcium independent of 1,25-dihydroxyvitamin D_3, their role in calcium metabolism during human pregnancy remains to be defined (Kovacs and Kronenberg, 1997; Quan-Sheng and Miller, 1989; Tolaymat *et al.*, 1994).

In summary, upregulation of intestinal absorption of calcium is mediated by several factors including calcitriol, prolactin, placental lactogen, and probably other factors.

28.2.2 Renal Excretion of Calcium

Total calcium renal excretion per day is increased in pregnant women as early as the 12th week of gestation, rising by 43% in the third trimester compared to prepregnancy levels. However, the fasting urine calcium values are normal or low and rise postprandial, indicating that the rise in 24 h urine calcium results from the increased intestinal absorption rather than increase in renal loss (Gertner *et al.*, 1986). For women with decreased calcium intake, renal excretion of calcium may be lower in the third trimester as compared to the first trimester.

28.2.3 Skeletal Calcium Metabolism

The evidence whether bone-mineral content undergoes resorption or slightly increases during pregnancy is conflicting (Kalkwarf and Specker, 2002). In the only study of its kind, bone-biopsy specimens from pregnant women with elective termination of pregnancy in the first trimester, showed increased bone resorption as compared to nonpregnant controls (Purdie *et al.*, 1988). Biochemical markers of bone turnover start to rise early during pregnancy to reach their highest levels by the third trimester (Bezerra *et al.*, 2004). However, measuring bone turnover during pregnancy using these markers has several limitations such as the high variability level between individual pregnant women, the hemodilution effect of pregnancy and the diurnal variation of these markers. In addition, contributions from the placenta and fetus cannot be ignored. Generally, most studies show that bone markers do not confirm significant resorption in pregnant women.

Imaging studies have tried to estimate the bone-mineral density. Investigators have used dual-energy X-ray absorptiometry (DXA) scan to compare pregnant women before conception and after delivery. Bone densities were less by 3.5–4.5% when the study was done between 4 and 6 weeks after delivery, but no change occurred when DXA was done within 2 weeks of delivery (Ritchie *et al.*, 1998; Black *et al.*, 2000; Naylor *et al.*, 2000). Therefore, it is speculated that puerperium and lactation but not pregnancy reduce the bone density. However, recent quantitative ultrasound studies of the calcaneus in pregnant women have shown a mild decrease in their bone density (Javaid *et al.*, 2005). Also, peripheral quantitative computed tomography (pQCT) showed a similar decrease in trabecular bone (spine and pelvis) density, the bone most likely to be mobilized in the adult skeleton during pregnancy (Wisser *et al.*, 2005). There is also data indicating cortical thickness of the femur increases during pregnancy, with an increase in markers of bone formation during the third trimester (Smith, 1967). It has been speculated that resorption of trabecular bone helps to meet the calcium demands of the fetus, while an increase in cortical thickness may help to store calcium for use during lactation. However, the net calcium balance remains zero in pregnant women with an adequate intake of calcium. This balance may, however, be altered in women with low calcium intake, adolescents, multiple fetuses, and short intervals between pregnancies.

Most epidemiological studies show no association between the increase in number of pregnancy and postmenopausal osteoporosis (Mendoza-Romo *et al.*, 2013). This may be explained partially by the ability to restore the bone minerals loss in the time free of pregnancy.

28.3 Calcium Transfer from Mother to Fetus

Calcium is actively transported across the placenta, allowing the fetus to maintain ionized calcium levels higher than maternal levels and emphasizing the importance of calcium to the fetus. This transfer is estimated to start around the end of first semester and peak at 36 weeks gestation when the fetus starts to accumulate 250–350 mg per day (Forbes, 1976). Investigators have shown that this active transfer is largely mediated by a complicated model of transport proteins located on the syncytiotrophoblast of the placenta involving three steps (Martin *et al.*, 2007). The first step involves voltage-dependent calcium channels on the syncytiotrophoblast brush border at the placental-maternal interface. Although many channels have been described, TRPV (transient receptor potential vanilloid) 6 is considered the most important as its expression increases 14-fold during the third trimester, the time period of maximum calcium transfer to the fetus. TRPV 6 is also well expressed in intestinal epithelial cells, while TRPV 5 is the major transporter of calcium in the kidneys. The second step involves transporting calcium from the placental–maternal border to the basolateral membrane at the placental/fetal interface using intracellular calcium binding proteins. Of these, CaBP-D9K is the most significant as its expression increases rapidly as gestational age advances. The third step involves calcium transfer across the basolateral membrane at the placental/fetal interface using plasma membrane Ca^{2+} ATPase (PMCA) proteins (Lafond and Simoneau, 2006) (Figure 28.1).

Hormones regulating calcium transfer across the placenta to the fetus are the subject of intensive investigation using animal models. Current evidence indicates that both PTH and vitamin D are not involved. Although calcitonin levels are higher in the fetal circulation compared to the maternal circulation, this hormone also does not play any role in placental calcium transfer. Data on PTHrP is mixed with studies showing both increased and decreased calcium transfer in mice lacking PTHrP (Bond *et al.*, 2008; Mitchell and Juppner, 2010) (Table 28.1).

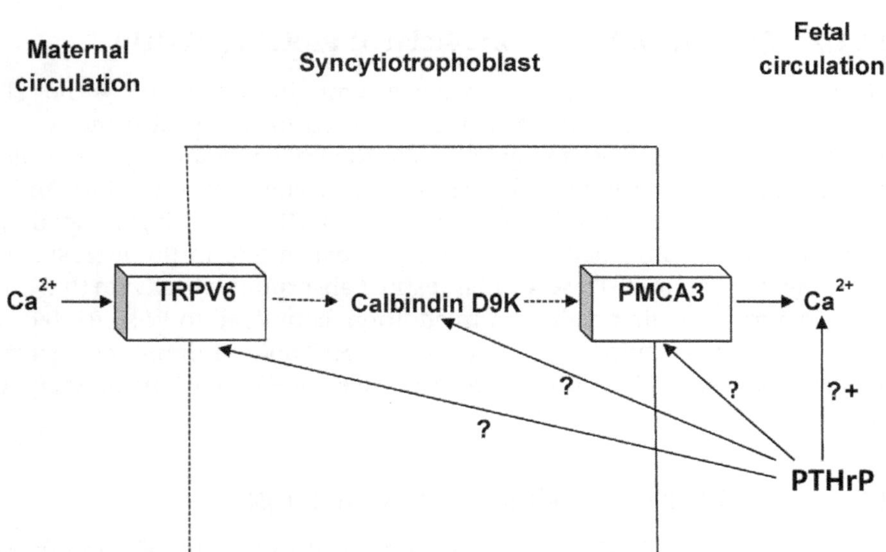

Figure 28.1 Calcium transport across the placenta.

Table 28.1 Placental transfer of calcium and hormones involved in calcium homeostasis.

	Placental transfer
Calcium	Active transport
25-Hydroxyvitamin D	Yes
1,25-Dihydroxyvitamin D_3	No
Phosphorus	Active transport
Calcitonin	No
Parathyroid hormone	No
Parathyroid hormone-related peptide	No

28.4 Calcitropic Hormones – Changes during Pregnancy

Many hormones are altered during pregnancy. Some of these hormones that are known to affect calcium metabolism are well studied, but the roles of other pregnancy-related hormones such as steroid hormones and placental lactogen are largely undetermined (Table 28.2).

28.4.1 Parathyroid Hormone (PTH)

Women who consume diets with adequate calcium have normal or a slight decline in PTH in the first trimester followed by an increase towards midnormal levels by the end of third trimester. In contrast, pregnant women who consume very low amounts of calcium have frequently higher PTH levels.

28.4.2 Parathyroid Hormone-Related Protein (PTHrP)

PTHrP is one of the mediators of hypercalcemia in malignancy. Although undetectable in adults, this protein is increased in the third trimester of pregnancy. It derives from many tissues in the body including parathyroid gland, breast and placenta and is commonly secreted in the breast milk during the lactation period. The exact role of PTHrP during pregnancy is still not well established. It is unlikely to play a role in the increase in 1,25-dihydroxyvitamin D levels and intestinal absorption of calcium that are in place during the first trimester. In addition, compared to PTH, PTHrP is less effective in activating 1-α-hydroxylase. PTHrP does inhibit bone resorption and may have role in preventing excessive bone loss during pregnancy (Kovacs, 2011).

28.4.3 1,25-Dihydroxyvitamin D_3 (Calcitriol)

1,25-Dihydroxyvitamin D_3 is the hormonally active form of vitamin D and levels double during pregnancy. Although the placenta may produce 1,25-dihydroxyvitamin D_3, maternal kidneys are the leading source during

Table 28.2 Changes in calcium and calciotropic hormones during pregnancy.

	Pregnancy trimester		
	First	Second	Third
Ionized calcium	Unchanged	Unchanged	Unchanged
Total calcium	Decreased	Decreased	Decreased
Phosphate	Unchanged	Unchanged	Unchanged
Parathyroid hormone	Low normal	Low normal	Normal
Calcitonin	Unchanged	Increased	Increased
25-Hydroxyvitamin D	Unchanged	Unchanged	Unchanged
1,25-Dihydroxyvitamin D	Increased	Increased	Increased
PTHrP	Increased	Increased	Increased

pregnancy. The total 1,25-dihydroxyvitamin D_3 rises quickly in the first trimester, but the free 1,25-dihydroxyvitamin D_3 fraction rises later in the third trimester, as initially during pregnancy vitamin D binding proteins also rise. Levels of 25-hydroxyvitamin D remain constant during pregnancy. Unlike 1,25-dihydroxyvitamin D_3, 25-hydroxyvitamin D readily crosses the placenta.

28.4.4 Calcitonin

Calcitonin is derived mainly from thyroidal C cells during pregnancy, with a small proportion produced from breast and placenta. Calcitonin levels start to rise early during pregnancy. The exact role of this rise in not well understood, as the absence of calcitonin during pregnancy has no effect on maternal skeleton in animal studies (Woodrow *et al.*, 2006).

28.5 Maternal Calcium Metabolism and Fetal Health

Fetal-bone mineralization appears to be affected by maternal calcium metabolism in addition to fetal regulation. A recent study demonstrated that unlike phosphate and alkaline phosphatase, calcium increases steadily with gestational age (Fenton *et al.*, 2011). Furthermore, markers of bone turnover such as osteocalcin, bone-specific alkaline phosphatase, and carboxy terminal collagen crosslinks (CTX) are higher in cord blood compared to the maternal blood (Yamaga *et al.*, 1999). In contrast, ionized calcium levels in pregnant women with essential hypertension or pre-eclampsia are lower in umbilical-cord blood when compared with normotensive controls (Varner *et al.*, 1983; Yusuf *et al.*, 2012), see Figures 28.2 and 28.3.

This is consistent with results from Hache *et al.* who revealed a 60% reduction in calcium transport in syncytiotrophoblasts from pre-eclamptic mothers (Hache *et al.*, 2011). In addition, placental proteins that play a key role in placental calcium transport such as TRPV6, CaBP-9K and PMCA1 were found to be decreased.

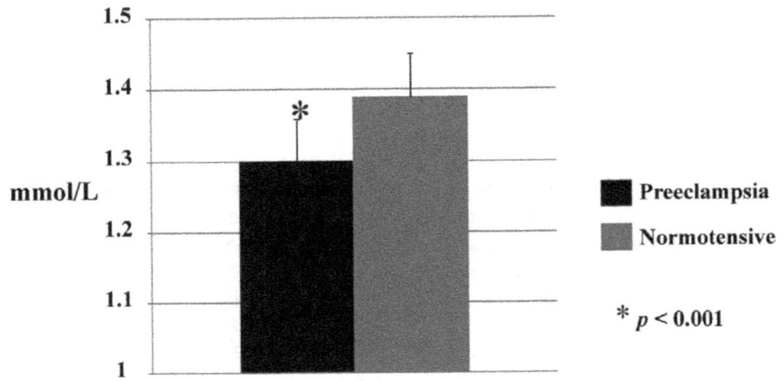

Figure 28.2 Ionized calcium levels in umbilical cord arterial blood in pre-eclampsia and normotensive pregnancies.

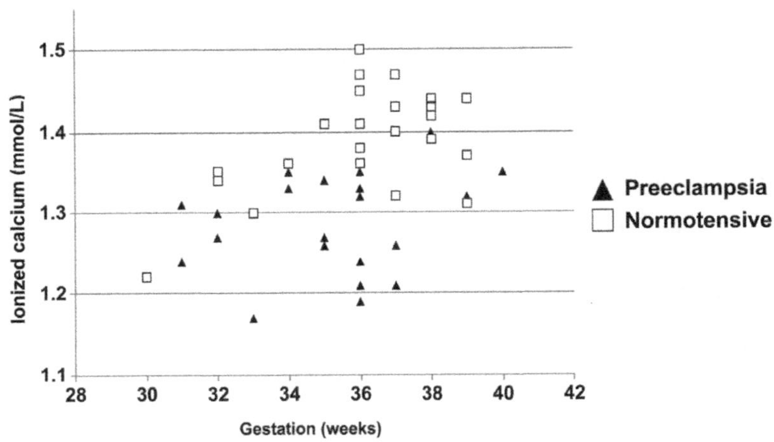

Figure 28.3 Scattergram of umbilical-cord arterial blood ionized calcium and gestational age in pre-eclampsia and normotensive pregnancies.

The effect of maternal calcium intake on fetal bone is uncertain. Although pregnant mothers may be advised to increase their intake of calcium, current recommendations by the Institute of Medicine for calcium intake during pregnancy is the same as for nonpregnant women. There is, however, also evidence that there may be a positive relationship between maternal dietary calcium intake and fetal-bone mineralization (Ganpule *et al.*, 2006). In contrast, the results of oral calcium supplementation in pregnancy have shown inconsistent benefits. Measurement of fetal femur length using ultrasound in pregnant African-American women have shown better femur growth in mothers consuming the highest dairy intake (>3 serving per day) compared with lowest dairy intake (<2 serving per day) (Chang *et al.*, 2003). Nevertheless, the bone density of infants born to women with low dietary calcium intake who were assigned to oral supplements has ranged between slight reduction to higher measure in women consuming the lowest amount of calcium at baseline (Raman *et al.*, 1978; Jarjou *et al.*, 2006).

28.6 Maternal Calcium Metabolism and Infant Growth

Infant length, weight, mid-upper arm circumference, and total body weight calcium have all been reported to positively correlate with higher maternal calcium intake (Hacker *et al.*, 2012). This, however, is an inconsistent finding and several studies have failed to confirm this association. This raises the question again as to whether supplementation benefits only infants of mothers with deficient calcium in their diet or infants of any pregnant women.

The offspring's long-term outcome in relation to maternal calcium status is infrequently described in the literature. This is due to confounding by lifestyle and nutrition on the bone development during childhood. Javaid *et al.* have described a positive correlation between pregnant women consuming higher calcium intake and whole body bone area and bone-mineral content at 9 years of age (Javaid *et al.*, 2006).

28.7 Impact of Pregnancy on Disorders of Calcium Metabolism

28.7.1 Pre-eclampsia

Pre-eclampsia is a pregnancy-specific disorder characterized by onset of hypertension and proteinuria after 20 weeks of gestation. It is a major cause of maternal and neonatal mortality and morbidity including preterm birth. Several changes in calcium metabolism have been described in pregnant women with pre-eclampsia. These include lower levels of serum and urinary calcium, 25-hydroxy vitamin D, and 1,25-dihydroxyvitamin D. The last two forms of vitamin D are also reported to be lower in umbilical-cord blood of infants born to mothers with pre-eclampsia.

Adequate calcium intake has revealed effectiveness in reducing the risk of pre-eclampsia and preterm delivery in pregnant women whose dietary intake of calcium is low. A Cochrane review, assessing 13 trials, has reported calcium supplementation in women may reduce the risk of preterm birth by 24%, and approximately halve the risk of pre-eclampsia (Hofmeyr *et al.*, 2010). The highest effect of calcium supplementation was in women with a low baseline calcium intake at the beginning of their pregnancy. A previous study showed that supplementation with additional calcium (1500 mg per day) to pregnant women who regularly consumed less than 600 mg of calcium per day decreases the risk of preterm delivery, but without decreasing the risk of pre-eclampsia (Villar *et al.*, 2006). However, supplementation did reduce its severity, and reduce neonatal mortality.

Fetal calcium metabolism has long been thought to be independent of maternal or fetal vitamin D levels. However, recent studies suggest that placental calcium transport and fetal bone development might largely be affected by maternal and fetal vitamin D (Mitchell and Juppner, 2010). This complicated relationship can explain the findings of Mahon *et al.* who have used three-dimensional ultrasonography to detect rickets-like changes in fetal bones as early as 18 weeks of gestation in mothers with vitamin D deficiency (Mahon *et al.*, 2010). All these changes in calcium homeostasis support the recent findings of lower ionized calcium seen in the umbilical cord blood of neonates of pre-eclamptic mothers (Yusuf *et al.*, 2012).

28.7.2 Gestational Diabetes

Gestational diabetes is a common complication of pregnancy. In gestational diabetes, maternal levels of total calcium, ionized calcium and phosphorus do not differ between diabetic and nondiabetic women. In contrast, serum magnesium levels are significantly lower in diabetic than in normal pregnant women. These low levels of maternal magnesium are thought to suppress fetal parathyroid gland resulting in transient neonatal hypoparathyroidism after birth. Thus, infants of diabetic mothers are prone to have low concentrations of serum calcium and low levels of parathyroid hormone. Neonatal hypocalcemia might result in early seizure in these infants (Banerjee *et al.*, 2003).

28.7.3 Hyperparathyroidism

Hypercalcemia during pregnancy is uncommon, and mostly associated with primary hyperparathyroidism. It can result in significant maternal and fetal adverse outcomes such as abortions and severe intrauterine growth retardation. Symptoms of hypercalcemia during pregnancy, like nausea, vomiting and fatigue, can overlap with those of hyperemesis gravidarum. Therefore, most cases of hypercalcemia are diagnosed during routine laboratory evaluation for a calcium profile or when rare conditions such as hypercalcemic crisis or acute pancreatitis occur. As large amounts of calcium transfer to the fetus during pregnancy, severe maternal hypercalcemia rarely occurs. Nonetheless, it may present shortly after delivery due to further stimulation of parathyroid glands by lactation (Kort *et al.*, 1999).

High calcium transfer to the fetus results in suppression of fetal parathyroid glands. Once the fetus is born, calcium levels start to drop and the suppressed parathyroids may not be able to respond adequately. This can lead to severe hypocalcemic seizures in the first 3 days of life. Moreover, mild and moderate maternal hyperparathyroidism have been frequently diagnosed after delivery when neonatal hypocalcemia is evaluated.

28.7.4 Hypoparathyroidism

Total calcium levels are usually low during normal pregnancy, therefore diagnosis of hypoparathyroidism for the first time in pregnancy is problematic. However, ionized calcium that remains normal during pregnancy and albumin-corrected serum calcium may help in confirming the diagnosis. Nevertheless, most patients are known to have it before pregnancy. The physiologic adaptations in pregnancy include many mechanisms that increase calcium levels without involving PTH and have been discussed above. The available case reports suggest the two possibilities in that some women with hypoparathyroidism symptoms may improve with lower requirement for calcium and vitamin D, while others required higher doses of both (Parkes *et al.*, 2013). Thus, close monitoring of calcium levels in pregnant women with a history of hypoparathyroidism is essential to prevent adverse fetal consequences.

28.7.5 Pseudohypoparathyroidism

Pseudohypoparathyroidism (PHP) is a rare condition characterized by genetic resistance to PTH that results in hypocalcemia, hyperphosphatemia and high PTH. As in hypoparathyroidism, pregnancy has been shown to alter calcium levels in pregnant women with pseudohypoparathyroidism in two different ways in the literature (Kovacs, 2011). In some cases, women became normocalcemic and do not require therapeutic doses of calcium or calcitriol. Other women have shown increased requirements of these medications. The mechanism of how pseudohypoparathyroidism might be improved during pregnancy is unknown, but may involve the effect of pregnancy-related hormones on 1α hydroxylase. Neonates of women with pseudohypoparathyroidism are prone to secondary hyperparathyroidism, which may result in subsequent hypocalcemia. The management goal is to prevent maternal hypocalcemia in these women in order to protect neonates from secondary hyperparathyroidism.

Summary Points

- This chapter focuses on calcium metabolism during pregnancy.
- Calcium is fundamental to many cellular processes and biological pathways.
- The fetus accumulates 30 g of calcium during the course of a normal pregnancy, most of it during the third trimester. The fetus maintains calcium levels higher than the mother, emphasizing the importance of calcium to fetal growth.
- Total serum calcium has a physiologic decline as a result of reduction of serum albumin levels. However, ionized calcium stays unchanged during pregnancy and should be the fraction measured to assess calcium status during pregnancy.
- Maternal adaptations to meet the large calcium demand of the fetus include increased intestinal absorption and bone resorption.
- The increased intestinal absorption of calcium during pregnancy is mediated mainly by 1,25-dihydroxyvitamin D_3, levels of which double during pregnancy. During early pregnancy resorption of trabecular bone takes place and there is evidence to suggest that increased bone deposition takes place in long bones in the latter half of pregnancy. Although there is increased renal excretion of calcium during pregnancy, this is mostly postprandial and net calcium balance during pregnancy remains zero.

- Levels of PTH and 25-hydroxyvitamin D remain unchanged during pregnancy. The roles of PTHrP and calcitonin in calcium metabolism during pregnancy remain to be defined.
- The active transport of calcium across the placenta is mediated by an intricate system of proteins, the regulation of which is poorly understood. However, PTHrP may have a role.
- Calcium supplementation during pregnancy may prevent pre-eclampsia.
- Both hypo- and pseudo-hypoparathyroidism may either get better or worse during pregnancy, requiring careful monitoring.
- Pregnancy is not a risk factor for osteoporosis later in life.

Key Features of Pregnancy and Calcium

1. Due to the extravagant fetal demand for calcium during pregnancy, calcium is actively transported across the placenta.
2. This active transport is mediated by a complicated model of transport voltage dependent proteins located on the syncytiotrophoblast.
3. Transient receptor potential vanilloid 6 (TRPV6) is considered the most important pathway in calcium transport because its expression on the maternal/placental interface increases 14-fold during the third trimester.
4. TRPV6 is a specialized epithelial Ca^{2+} channel responsible for the critical Ca^{2+} entry step in transcellular Ca^{2+} absorption in the intestine and transplacental transfer during pregnancy.
5. 1,25-Dihydroxyvitamin D_3 is the most important and biologically active metabolite of vitamin D, promoting the active intestinal absorption of calcium and phosphorus and playing a key role in maintaining appropriate serum calcium and phosphorus concentrations.
6. Bone and intestine are the main target organs of 1,25-dihydroxyvitamin D action; however, receptors have been recognized in other tissues not directly involved in calcium metabolism such as the hematopoietic cells and reproductive tissues.
7. The combined action of PTH and 1,25-dihydroxyvitamin D is to stimulate calcium mobilization from bones.
8. Decreased serum concentrations of ionized calcium increase serum parathyroid hormone synthesis and secretion, which sequentially stimulates the hydroxylation of 25-hydroxyvitamin D to 1,25-dihydroxyvitamin D in the renal cortex.

Definitions of Key Terms

1-α-Hydroxylase. A cytochrome P450 enzyme that is located in the proximal tubule of the kidney and a variety of other tissues including skin, immune cells and bone. 1-α-Hydroxylase catalyses the hydroxylation of 25-hydroxyvitamin D to the active form of the vitamin, 1,25-dihydroxyvitamin D_3, also called calcitriol.

Calbindin. These include several calcium binding proteins. Calbindin D9K is found mainly in the placenta and intestinal epithelial cells. Calbindin D28K is expressed mainly in the kidneys.

Calcitonin. A 32-amino acid polypeptide hormone that is produced primarily by the parafollicular cells (also known as C-cells) of the thyroid. It acts to reduce blood calcium (Ca^{2+}), opposing the effects of parathyroid hormone (PTH).

Calcitriol. The hormonally active form of vitamin D with three hydroxyl groups (also known as 1,25-dihydroxyvitamin D_3).

Carboxy-terminal collagen crosslinks (CTX). A serum biomarker used to measure the rate of bone turnover.

Dual-energy X-ray absorptiometry (DXA). DXA or as previously known DEXA, is a means of measuring bone-mineral density. Two X-ray beams with different energy levels are intended at the patient's bones. When soft-tissue absorption is subtracted, the bone-mineral density can be determined from the absorption of each beam by bone.

Gestational diabetes is a condition in which women who have not been diagnosed with diabetes develop high glucose levels during pregnancy.

Hyperparathyroidism. Hyperparathyroidism is overactivity of the parathyroid glands resulting in increased levels of parathyroid hormone.

Hypoparathyroidism. Hypoparathyroidism is decreased function of the parathyroid glands resulting in lower levels of parathyroid hormone.

Parathyroid hormone (PTH). A polypeptide hormone produced in the parathyroid gland and helps regulate the blood levels of calcium and phosphate.

Plasma Membrane Calcium ATPase proteins (PCMA). Transport proteins on the plasma membrane of cells that help to remove calcium from cells.

Pre-eclampsia. A pregnancy-related condition characterized by high blood pressure and significant amounts of protein in the urine.

Pseudohypoparathyroidism. This is a condition in which there is resistance to the action of parathyroid hormone.

Quantitative computed tomography (QCT). A medical technique that measures bone-mineral density using a standard X-ray computed tomography (CT) scanner with a calibration standard to convert Hounsfield Units (HU) of the CT image to bone-mineral density values. Quantitative CT scans are primarily used to evaluate bone-mineral density at the lumbar spine and hip.

Peripheral QCT (pQCT) is a type of QCT, used for making measurements of the bone-mineral density in a peripheral part of the body, such as the forearms or legs as opposed to QCT that measures bone-mineral density at the hip and spine.

Syncytiotrophoblast. These are multinucleated, terminally differentiated cells which line the highly vascular placental villi. They are in direct contact with maternal blood and nutrients from the mother cross these cells to reach the fetus.

TRPV (transient receptor potential vanilloid). A family of transient receptor potential (TRP) ion channels that are selective for calcium and magnesium over sodium ions.

Trophoblast. These are specialized cells of the placenta that are essential for embryo implantation and transfer of nutrients from the mother to the fetus.

Vitamin D-dependent calcium binding proteins (CaBP). Proteins that bound calcium in the micromolar range. CaBP exist in two distinct sizes with a molecular weight of approximately 9 kDa and 28 kDa. CaBP-D9k is found in mammalian intestine, while CaBP-D28k is in avian intestine and in mammalian kidney.

List of Abbreviations

DXA	Dual-energy X-ray absorptiometry
TRPV	Transient receptor potential vanilloid
CaBP	Calcium-binding proteins
PMCA	Plasma membrane Ca^{2+} ATPase
pQCT	Peripheral quantitative computed tomography
PTH	Parathyroid hormone
PTHrP	Parathyroid hormone-related protein
CTX	Carboxy terminal collagen crosslinks
Calcitriol	1,25-Dihydroxyvitamin D_3

References

Banerjee, S., Mimouni, F. B., Mehta, R., Llanos, A., Bainbridge, R., Varada, K. and Sheffer, G., 2003. Lower whole blood ionized magnesium concentrations in hypocalcemic infants of gestational diabetic mothers. *Magnesium research*. 16(2): 127–130.

Bezerra, F. F., Mendonca, L. M., Lobato, E. C., O'Brien, K. O. and Donangelo, C. M., 2004. Bone mass is recovered from lactation to postweaning in adolescent mothers with low calcium intakes. *American Journal of Clinical Nutrition*. 80(5): 1322–1326.

Black, A. J., Topping, J., Durham, B., Farquharson, R. G. and Fraser, W. D., 2000. A detailed assessment of alterations in bone turnover, calcium homeostasis, and bone density in normal pregnancy. *Journal of Bone and Mineral Research*. 15(3): 557–563. DOI: 10.1359/jbmr.2000.15.3.557.

Bond, H., Dilworth, M. R., Baker, B., Cowley, E., Requena Jimenez, A., Hugh Boyd, R.D., Husain, S. M., Ward, B. S., Sibley, C. P. and Glazier, J. D., 2008. Increased maternofetal calcium flux in parathyroid hormone-related protein-null mice. *Journal of Physiology*. 586(7): 2015–2025.

Chang, S. C., O'Brien, K. O., Nathanson, M. S., Caulfield, L. E., Mancini, J. and Witter, F. R., 2003. Fetal femur length is influenced by maternal dairy intake in pregnant African American adolescents. *American Journal of Clinical Nutrition*. 77(5): 1248–1254.

Cross, N. A., Hillman, L. S., Allen, S. H., Krause, G. F. and Vieira, N. E., 1995. Calcium homeostasis and bone metabolism during pregnancy, lactation, and postweaning: a longitudinal study. *American Journal of Clinical Nutrition*. 61(3): 514–523.

Delorme, A. C., Marche, P. and Garel, J. M., 1979. Vitamin D-dependent calcium-binding protein. Changes during gestation, prenatal and postnatal development in rats. *Journal of Developmental Physiology*. 1(3): 181–194.

Fenton, T. R., Lyon, A. W. and Rose, M. S., 2011. Cord blood calcium, phosphate, magnesium, and alkaline phosphatase gestational age-specific reference intervals for preterm infants. *BMC Pediatrics*. 11: 76. DOI: 10.1186/1471-2431-11-76.

Forbes, G. B., 1976. Letter: calcium accumulation by the human fetus. *Pediatrics*. 57(6): 976–977.

Ganpule, A., Yajnik, C. S., Fall, C. H., Rao, S., Fisher, D. J., Kanade, A. and Joglekar, C., 2006. Bone mass in Indian children – relationships to maternal nutritional status and diet during pregnancy: the Pune Maternal Nutrition Study. *Journal of Clinical Endocrinology and Metabolism*. 91(8): 2994–3001. DOI: 10.1210/jc.2005-2431.

Gertner, J. M., Coustan, D. R., Kliger, A. S., Mallette, L. E., Ravin, N. and Broadus, A. E., 1986. Pregnancy as state of physiologic absorptive hypercalciuria. *American Journal of Medicine*. 81(3): 451–456.

Hache, S., Takser, L., LeBellego, F., Weiler, H., Leduc, L., Forest, J. C. and Lafond, J., 2011. Alteration of calcium homeostasis in primary preeclamptic syncytiotrophoblasts: effect on calcium exchange in placenta. *Journal of Cellular and Molecular Medicine*. 15(3): 654–667. DOI: 10.1111/j.1582-4934.2010.01039.x.

Hacker, A. N., Fung, E. B. and King, J. C., 2012. Role of calcium during pregnancy: maternal and fetal needs. [Review]. *Nutrition Reviews*. 70(7): 397–409. DOI: 10.1111/j.1753-4887.2012.00491.x.

Hofmeyr, G. J., Lawrie, T. A., Atallah, A. N. and Duley, L., 2010. Calcium supplementation during pregnancy for preventing hypertensive disorders and related problems. *The Cochrane database of systematic reviews*. (8): CD001059. DOI: 10.1002/14651858.CD001059.pub3.

Jarjou, L. M., Prentice, A., Sawo, Y., Laskey, M. A., Bennett, J., Goldberg, G. R. and Cole, T. J., 2006. Randomized, placebo-controlled, calcium supplementation study in pregnant Gambian women: effects on breast-milk calcium concentrations and infant birth weight, growth, and bone mineral accretion in the first year of life. *American Journal of Clinical Nutrition*. 83(3): 657–666.

Javaid, M. K., Crozier, S. R., Harvey, N. C., Gale, C. R., Dennison, E. M. and Boucher, B. J., 2006. Maternal vitamin D status during pregnancy and childhood bone mass at age 9 years: a longitudinal study. *Lancet*. 367(9504): 36–43. DOI: 10.1016/S0140-6736(06)67922-1.

Javaid, M. K., Crozier, S. R., Harvey, N. C., Taylor, P., Inskip, H. M., Godfrey, K. M. and Southampton Women's Survey Study Group, 2005. Maternal and seasonal predictors of change in calcaneal quantitative ultrasound during pregnancy. *Journal of Clinical Endocrinology and Metabolism*. 90(9): 5182–5187. DOI: 10.1210/jc.2005-0195.

Kalkwarf, H. J. and Specker, B. L., 2002. Bone mineral changes during pregnancy and lactation. *Endocrine*. 17(1): 49–53. DOI: 10.1385/ENDO:17:1:49.

Kort, K. C., Schiller, H. J. and Numann, P. J., 1999. Hyperparathyroidism and pregnancy. *American Journal of Surgery*. 177(1): 66–68.

Kovacs, C. S., 2011. Calcium and bone metabolism disorders during pregnancy and lactation. *Endocrinology and Metabolism Clinics of North America*. 40(4): 795–826. DOI: 10.1016/j.ecl.2011.08.002.

Kovacs, C. S. and Kronenberg, H. M., 1997. Maternal-fetal calcium and bone metabolism during pregnancy, puerperium and lactation. *Endocrine Reviews*. 18(6): 832–872.

Lafond, J. and Simoneau, L., 2006. Calcium homeostasis in human placenta: role of calcium-handling proteins. *International Review of Cytology*. 250: 109–174. DOI: 10.1016/S0074-7696(06)50004-X.

Mahon, P., Harvey, N., Crozier, S., Inskip, H., Robinson, S., Arden, N. and Group, S. W. S. S., 2010. Low maternal vitamin D status and fetal bone development: cohort study. *Journal of Bone and Mineral Research*. 25(1): 14–19. DOI: 10.1359/jbmr.090701.

Martin, R., Harvey, N. C., Crozier, S. R., Poole, J. R., Javaid, M. K., Dennison, E. M. and SWS Study Group, 2007. Placental calcium transporter (PMCA3) gene expression predicts intrauterine bone mineral accrual. *Bone*. 40(5): 1203–1208. DOI: 10.1016/j.bone.2006.12.060.

Mendoza-Romo, M. A., Ramirez-Arriola, M. C., Velasco-Chavez, J. F., Rivera-Martinez, J. G., Nieva-de Jesus, R. N. and Valdez-Jimenez, L. A., 2013. Parity and menarche as risk factors of osteoporosis in postmenopausal Mexican women. *Ginecologia y Obstetricia de Mexico*. 81(3): 133–139.

Mitchell, D. M. and Juppner, H., 2010. Regulation of calcium homeostasis and bone metabolism in the fetus and neonate. *Current Opinion in Endocrinology, Diabetes and Obesity*. 17(1): 25–30. DOI: 10.1097/MED.0b013e328334f041.

Naylor, K. E., Iqbal, P., Fledelius, C., Fraser, R. B. and Eastell, R., 2000. The effect of pregnancy on bone density and bone turnover. *Journal of Bone and Mineral Research*. 15(1): 129–137. DOI: 10.1359/jbmr.2000.15.1.129.

Parkes, I., Schenker, J. G. and Shufaro, Y., 2013. Parathyroid and calcium metabolism disorders during pregnancy. *Gynecological Endocrinology*. 29(6): 515–519.

Purdie, D. W., Aaron, J. E. and Selby, P. L., 1988. Bone histology and mineral homeostasis in human pregnancy. *British Journal of Obstetrics and Gynaecology*. 95: 849–854.

Quan-Sheng, D. and Miller, S. C., 1989. Calciotrophic hormone levels and calcium absorption during pregnancy in rats. *American Journal of Physiology*. 257(1 Pt 1): E118–E123.

Raman, L., Rajalakshmi, K., Krishnamachari, K. A. and Sastry, J. G., 1978. Effect of calcium supplementation to undernourished mothers during pregnancy on the bone density of the bone density of the neonates. *American Journal of Clinical Nutrition*. 31(3): 466–469.

Ritchie, L. D., Fung, E. B., Halloran, B. P., Turnlund, J. R., Van Loan, M. D., Cann, C. E. and King, J. C., 1998. A longitudinal study of calcium homeostasis during human pregnancy and lactation and after resumption of menses. *American Journal of Clinical Nutrition*. 67(4): 693–701.

Smith, R. W., 1967. Dietary and hormonal factors in bone loss. *Federation Proceedings*. 26(6): 1737–1746.

Tolaymat, A., Sanchez-Ramos, L., Yergey, A. L., Vieira, N. E., Abrams, S. A. and Edelstein, P., 1994. Pathophysiology of hypocalciuria in preeclampsia: measurement of intestinal calcium absorption. *Obstetrics and Gynecology*. 83(2): 239–243.

Trotter, M. and Hixon, B. B., 1974. Sequential changes in weight, density, and percentage ash weight of human skeletons from an early fetal period through old age. *Anatomical Record*. 179(1): 1–18. DOI: 10.1002/ar.1091790102.

Varner, M. W., Cruikshank, D. P. and Pitkin, R. M., 1983. Calcium metabolism in the hypertensive mother, fetus, and newborn infant. *American Journal of Obstetrics and Gynecology*. 147(7): 762–765.

Villar, J., Abdel-Aleem, H., Merialdi, M., Mathai, M., Ali, M. M., Zavaleta, N. and World Health Organization Calcium Supplementation for the Prevention of Preeclampsia Trial, Group, 2006. World Health Organization randomized trial of calcium supplementation among low calcium intake pregnant women. *American Journal of Obstetrics and Gynecology*. 194(3): 639–649. DOI: 10.1016/j.ajog.2006.01.068.

Wisser, J., Florio, I., Neff, M., Konig, V., Huch, R., Huch, A. and von Mandach, U., 2005. Changes in bone density and metabolism in pregnancy. *Acta Obstetricia et Gynecologica Scandinavica*. 84(4): 349–354. DOI: 10.1111/j.0001-6349.2005.00766.x.

Woodrow, J. P., Sharpe, C. J., Fudge, N. J., Hoff, A. O., Gagel, R. F. and Kovacs, C. S., 2006. Calcitonin plays a critical role in regulating skeletal mineral metabolism during lactation. *Endocrinology*. 147(9): 4010–4021. DOI: 10.1210/en.2005-1616.

Yamaga, A., Taga, M., Hashimoto, S. and Ota, C., 1999. Comparison of bone metabolic markers between maternal and cord blood. *Hormone Research*. 51(6): 277–279.

Yusuf, K., Kamaluddeen, M., Hasan, S. U., Al-Awad, E., Finch, R. A. and Akierman, A. R., 2012. Ionized calcium levels in umbilical cord blood of women with preeclampsia and normotensive pregnancies. *Journal of Maternal-Fetal & Neonatal Medicine*. 25(2): 203–205. DOI: 10.3109/14767058.2011.561895.

CHAPTER 29

Calcium Supplementation during Pregnancy and Lactation: Implications for Maternal and Infant Bone Health

FLÁVIA FIORUCI BEZERRA[b] AND CARMEN MARINO DONANGELO*[a]

[a]Escuela de Nutrición, Universidad de la República, Paysandú 843, Montevideo, Uruguay; [b]Instituto de Nutrição, Universidade do Estado do Rio de Janeiro, Rua São Francisco Xavier 524, Rio de Janeiro, Brazil
*E-mail: cmarino@nutricion.edu.uy

29.1 Introduction

It is widely recognized that pregnancy and lactation are periods of high maternal calcium demands for fetal and infant skeletal growth and mineralization (Olausson *et al.*, 2012). During pregnancy, the fetus accumulates a total of 20–30 g of calcium at a rate up to 250–300 mg per day in the third trimester. During lactation, the rate of maternal calcium transfer to the mammary gland for breast milk secretion is 200–300 mg per day, thus providing a total of 35–50 g of calcium to the infant during 6 months of breastfeeding.

Food and Nutritional Components in Focus No. 10
Calcium: Chemistry, Analysis, Function and Effects
Edited by Victor R. Preedy

Published by the Royal Society of Chemistry, www.rsc.org

Considering that the skeleton of an adult woman contains 800–900 g of calcium, total maternal calcium losses during pregnancy and 6 months of lactation represent 3 and 5%, respectively, of total maternal bone calcium. In theory, mobilization of maternal bone calcium could supply the calcium required for the developing fetus and the lactating infant. In fact, temporary loss of maternal bone mass is a well-established physiological adaptation during human pregnancy and lactation.

Several stage-specific physiological adaptations are known to contribute to meet the increased calcium requirements for pregnancy and lactation. These adaptations have been extensively reviewed and recently summarized elsewhere (Olausson *et al.*, 2012). Briefly, during pregnancy, additional calcium is obtained primarily from increased efficiency of intestinal calcium absorption, and to some extent from increased maternal bone turnover, particularly during the last trimester. During lactation, conservation of renal calcium plays a role but the primary adaptation is mobilization of maternal bone calcium resulting in maternal bone loss that is recovered after weaning.

It is well accepted that in women with calcium intakes close to current recommendations (National Research Council, 2011), the physiologic adaptations for providing calcium to the fetus and infant are largely independent of calcium intake (Olausson *et al.*, 2012). In these women, there is no apparent bone benefit for the mother or infant of using calcium supplements. However, in women consuming low-calcium diets, maternal bone adaptations during reproduction have been found to respond to increased calcium intake or supplementation, although not always as expected. Moreover, fetal and infant skeletal development has been found to be positively associated with increased maternal calcium intake/supplementation during pregnancy in some but not all studies. These apparently inconsistent results are possibly due to the complex interactions between genetics, diet composition, environment and lifestyle on maternal and infant bone responses during pregnancy and lactation.

In this chapter, studies evaluating the effect of maternal calcium intake during pregnancy and lactation, from the diet and from supplements, on maternal bone outcomes and on fetal and infant bone growth are reviewed. The effects of other factors on bone outcomes and the possible implications for the maternal and infant bone health are considered.

29.2 Calcium Supplementation and Maternal Bone Outcomes

When interpreting maternal bone responses to calcium intake or supplementation, it is important to consider that bone outcomes during pregnancy and lactation are affected by multiple factors and that bone responses are difficult to fully evaluate because of limitations in study design, sample size and bone measurements. These limitations have been thoroughly discussed in a recent review (Olausson *et al.*, 2012). Measurement of BMC and BMD prior to pregnancy and lactation is the ideal baseline but it is challenging to achieve in practice. Because of this difficulty only a few studies have used this approach (Ritchie *et al.*, 1998; Olausson *et al.*, 2008). Changes in bone mass and density are not generally measured during pregnancy due to poor precision of the ultrasound technique suitable for use in pregnant women, or to avoid unnecessary exposure of the fetus to ionizing radiation when using DXA. Bone measurements 1–3 weeks after delivery are generally assumed to reflect bone changes during pregnancy even in breastfeeding mothers. Serial bone measurements after delivery during breastfeeding, and after a period of time since cessation of breastfeeding are assumed to reflect bone responses to lactation and weaning, respectively. Comparisons over time with an appropriate control group (nonlactating postpartum or nonpregnant nonlactating) have been done only in few studies. In postweaning assessment, the time elapsed since onset of weaning may have not been sufficient to ensure that bone has reached a steady state. In many studies, bone outcomes are assessed indirectly by measuring changes in bone-turnover markers during pregnancy and lactation, assuming that changes in the dynamics of bone metabolism reflect changes in bone status. Confounding factors considered in some, but not all studies, include maternal age, parity, changes in body weight, vitamin D status, breastfeeding practice, hormone contraception, return of menses, and the residual effect of a previous recent pregnancy and lactation.

29.2.1 Effects of Calcium Supplementation during Pregnancy

There are several longitudinal studies relating maternal dietary calcium intake during pregnancy to maternal bone outcomes during pregnancy and postpartum although only a few studies have tested the effect of calcium supplementation during pregnancy in randomized placebo-controlled trials (Table 29.1).

Among observational studies, no effect (Olausson *et al.*, 2008) and positive effects (O'Brien *et al.*, 2003; Zeni *et al.*, 2003; O'Brien *et al.*, 2006; Avendaño-Badillo *et al.*, 2009) of dietary calcium intake during pregnancy on maternal bone have been described. No effect was found in a study in UK adult women with mean calcium intake ≥1000 mg per day, with comparison over time between pregnant and nonpregnant nonlactating groups (Olausson *et al.*, 2008). In this study, pregnancy was associated with substantial decreases in whole body (−1.7%) and regional (spine, −3.03% and total hip, −1.87%) BMC. Increase in body weight was a significant predictor of the skeletal changes but calcium intake did not affect bone responses to pregnancy in these women. On the other hand, in a study of adolescent women from the third trimester of pregnancy to early postpartum, with mean calcium intake of 1200 mg per day, lumbar spine *Z* scores postpartum were significantly associated with calcium intake during pregnancy ($R^2 = 0.355$, $P < 0.02$) (O'Brien *et al.*, 2003). Although dietary calcium was close to the 1300 mg per day reference intake for adolescents (National Research Council, 2011), a higher intake during pregnancy appeared to be protective against loss of trabecular bone in these young mothers.

Table 29.1 Studies relating dietary calcium or calcium supplementation during pregnancy and maternal bone outcomes.[a]

References	Type of study	Country, maternal age and study groups (*n*)	Mean dietary Ca mg per day	Ca supplement mg per day	Maternal bone outcomes	Overall effect of calcium intake/supplementation during pregnancy on maternal bone
Zeni *et al.* (2003)	Prospective cohort	Argentina, 17–30 years, Pregnant, 12 to 38 weeks (39)	790	—	Bone resorption markers increased during pregnancy; changes between the 3rd and 2nd trimester were negatively correlated with Ca intake	Positive
		NPNL women (30)	530	—		
O'Brien *et al.*, 2003	Prospective cohort	US, 13–18 years, Pregnant, 3rd tri (23) to 1 month PP (15)	1200	—	LS BMD *Z* scores at postpartum positively associated with Ca intake during the third trimester of pregnancy	Positive
O'Brien *et al.* (2006)	Prospective cohort	Brazil, 21–34 years, Pregnant, 10–12 weeks pregnancy to 7–8 weeks PP (10)	460	—	Net balance in bone calcium turnover positively associated with dietary calcium intake during pregnancy and early lactation	Positive
Avendaño-Badillo *et al.* (2009)	Prospective cohort	Mexico, 15–43 years, Pregnant, 12 to 34 weeks (206)	998	—	Inverse association between dietary calcium intake and NTx (bone resorption marker) during pregnancy	Positive
Olausson *et al.* (2008)	Prospective cohort	UK, 23–37 years, Pregnant, prepreg. to 2 weeks PP (34)	1008–1345	—	Decrease in adjusted BMD and BMC at whole-body, spine and hip in the pregnant women, independent of Ca intake during pregnancy	No effect
		NPNL (84)	1001			

(*continued*)

Table 29.1 (*continued*)

References	Type of study	Country, maternal age and study groups (*n*)	Mean dietary Ca mg per day	Ca supplement mg per day	Maternal bone outcomes	Overall effect of calcium intake/supplementation during pregnancy on maternal bone
O'Brien *et al.* (2012)	Prospective cohort	Three cohorts Pre/early preg. to 3–10 weeks PP US, 13–18 years (23) US, 25–34 years (13) Brazil, 20–35 years (10)	 1200 1200 450	—	At late pregnancy, inverse association between rate of bone calcium deposition and 1,25-$(OH)_2D$, particularly in the adolescents and low calcium cohorts. No association with calcium intake	No effect
Janakiraman *et al.* (2003)	RCO	Mexico, 15–43 years Pregnant, 25–35 weeks, treated during 20 days Calcium-Placebo (16) Placebo-Calcium (15)	 1031 959	1200	14% decrease in NTx (bone resorption marker) in response to Ca supplementation compared to placebo (multivitamin)	Positive
Liu *et al.* (2011)	RCT	China, 24–31 years Pregnant, 20 weeks preg. to 6 weeks PP Control (12) Milk suppl. (12) Milk suppl. + Ca suppl. (11)	 480 479 486	 — 350 950	Higher total and LS BMD in women with calcium and milk powder supplementation than in those in the control group. Serum osteocalcin increased only in the calcium/milk intervention groups	Positive

Diogenes *et al.* (2013)	RCT	Brazil, 13–19 years Pregnant, treatment 26 weeks preg. to delivery			At 20 weeks PP, higher BMC (13.9%), BA (6.2%) and BMD (10.6%) at the LS in the supplemented group. Less pronounced decrease in femoral neck BMD from 5 to 20 weeks PP in the supplemented (3.0%) than in the placebo (4.5%) groups	Positive
		Ca suppl. (30)	500	600 + vit D (200 IU)		
		Placebo (26)	743	—		
Jarjou *et al.* (2010)	RCT	The Gambia, 27.4 ± 7.5 years Pregnant, treated 20 weeks preg. to delivery			Lower size-adjusted BMC, BA, and BMD at the hip and greater decreases in BMC at LS and distal radius throughout 12 months lactation in the Ca-supplemented group	Negative
		Ca suppl. (61)	355	1500		
		Placebo (64)	355	—		
Jarjou *et al.* (2013)	RCT	The Gambia, 29 ± 8 years Treated 20 weeks to delivery in a previous pregnancy			After 5 years, the lower maternal BMC in the calcium-supplemented group from the previous trial persisted over time irrespective of subsequent pregnancy and lactation	Negative
		NPNL				
		Ca suppl. (31)	329	1500		
		Placebo (28)		—		
		Lact.				
		Ca suppl. (24)	330	1500		
		Placebo (20)		—		

[a]Overall effect of calcium intake/supplementation: positive when associated with preservation of bone, and negative with loss of bone. RCT, randomized controlled trial; RCO, randomized crossover trial; PP, postpartum; NPNL, nonpregnant nonlactating women; BMC, bone-mineral content; BMD, bone-mineral density; BA, bone area; LS, lumbar spine; UD, ultradistal.

Studies in different populations with dietary calcium between 500 and 1000 mg per day indicate that increased calcium intake during pregnancy reduces maternal bone turnover and possibly bone calcium mobilization. In a longitudinal study in Argentina (Zeni *et al.*, 2003), a negative correlation was found between increase from the second to third trimester of pregnancy of different biochemical markers of bone turnover (NTx, βCTx, bone alkaline phosphatase) and calcium intake at late pregnancy. Similarly, an inverse association between dietary calcium, particularly from dairy products, and urinary levels of NTx was found during pregnancy in a study in Mexico (Avendaño-Badillo *et al.*, 2009). Consistent with these results, a crossover trial (Janakiraman *et al.*, 2003) showed that short-term use of calcium supplements during the third trimester of pregnancy decreased urinary NTx, a marker of bone resorption. Also, net balance in bone calcium turnover (deposition minus resorption) was positively associated with dietary calcium in a kinetic study during pregnancy and lactation in Brazilian women (O'Brien *et al.*, 2006). When three different cohorts of women were compared during pregnancy and postpartum using kinetic modeling (O'Brien *et al.*, 2012), elevated 1,25-$(OH)_2D$ was associated with decreased rates of bone calcium deposition during late pregnancy, particularly in the adolescent US cohort and in the Brazilian adult cohort with low calcium intake.

Among randomized controlled trials studies, both positive (Liu *et al.*, 2011; Diogenes *et al.*, 2013) and negative (Jarjou *et al.*, 2010; Jarjou *et al.*, 2013) maternal bone outcomes in response to calcium supplementation during pregnancy have been described.

Positive effects have been found irrespective of maternal age. A dose-dependent relationship between calcium intake from midpregnancy to 6 weeks postpartum (usual diet supplemented with milk powder and calcium carbonate) and maternal BMD at postpartum was observed in adult Chinese women habitually consuming <500 mg per day dietary calcium. Maternal bone responses were site specific, being significant at the whole-body and spine, but not at the hip. Milk/calcium supplementation suppressed postpartum bone resorption (NTx) but serum osteocalcin (a marker of bone formation and bone calcium retention) was increased by supplementation in these women (Liu *et al.*, 2011). In a study of Brazilian adolescent mothers with habitual low calcium diets (≈600 mg per day), supplementation with calcium (600 mg per day) plus vitamin D (200 IU per day) during the third trimester of pregnancy resulted in higher lumbar spine bone mass at 5 and 20 weeks postpartum, and reduced rate of bone loss at the femoral neck during the first 20 weeks of lactation, compared to placebo (Diogenes *et al.*, 2013).

In contrast, calcium supplementation (1500 mg per day) during the second half of pregnancy of Gambian women with very low calcium diets (≈350 mg per day) resulted in lower BMC, BA and BMD at the hip throughout 12 months of lactation, compared to placebo. The supplemented group also had greater decreases in BMC at the lumbar spine and radius, and had biochemical changes consistent with greater bone mineral mobilization (Jarjou *et al.*, 2010). Serum 1,25-$(OH)_2D$ decreased from 20 weeks pregnancy to 13 weeks lactation irrespective of treatment but to a larger extent in the calcium-supplemented women. A follow-up study conducted 5 years later in a subset from the same cohort scanned after a subsequent pregnancy and lactation indicated that the lower BMC in the calcium supplemented group from the previous trial persisted over time (Jarjou *et al.*, 2013). It was suggested that in these women accustomed to very low calcium intakes, the use of calcium supplementation during pregnancy disrupted the physiologic adaptation to conserve calcium for fetal and breast milk needs resulting in lower long-term maternal bone calcium. Besides habitual calcium intake, differences in diet composition, vitamin D status, breastfeeding practice, duration of lactation amenorrhea, physical activity, exposure to environmental pollutants, and genetics may explain differences in response to calcium supplementation between women from Gambia and from other countries, but further research is needed to confirm these hypotheses.

29.2.2 Effects of Calcium Supplementation during Lactation

Prospective cohort studies and randomized controlled trials have been done to test associations between calcium intake/supplementation during lactation and maternal bone outcomes (Table 29.2).

In general, studies in adult women consuming close to 1000 mg per day or higher dietary calcium indicate that there is no relationship between calcium intake and maternal bone outcomes during lactation and postweaning (Sowers *et al.*, 1993; Sowers *et al.*, 1995; Laskey *et al.*, 1998; Laskey *et al.*, 2011). It appears that in well-nourished adult women the physiological maternal bone loss during lactation and recovery after weaning is explained mostly by factors such as duration of breastfeeding, total breast-milk output, hormonal changes, resumption of menses, and changes in body weight, but not by different amounts of calcium intake close or higher than reference intakes during these periods. However, some studies provide evidence of positive effects of increasing calcium intake on maternal bone outcomes under certain conditions (Chan *et al.*, 1987; Krebs *et al.*, 1997; O'Brien *et al.*, 2012).

In a longitudinal study comparing three groups of well-nourished lactating women, adolescents and adults, from 4 weeks predelivery until 16 weeks postpartum (Chan *et al.*, 1987), BMC decreased during lactation in the adolescent group consuming 900 mg per day but not in the adult and adolescent groups consuming ≥1500 mg per day calcium. These results suggest that bone loss during lactation in adolescent mothers may be prevented by increased dietary calcium intake. Consistent with these results, in a kinetic study comparing three cohorts of women from different populations (O'Brien *et al.*, 2012), a higher calcium intake postpartum was associated with higher rates of bone-calcium deposition, particularly in the adolescent cohort and in the adult cohort accustomed to a low-calcium diet.

In a study up to 7 months postpartum of lactating and nonlactating adult women consuming 1400 mg per day calcium (Krebs *et al.*, 1997), BMD at the lumbar spine in the lactating women was positively associated with the ratio

Table 29.2 Studies relating dietary calcium or calcium supplementation during lactation and maternal bone outcomes.[a]

Reference	Type of study	Country, maternal age and study groups (*n*)	Mean dietary Ca mg per day	Ca supplement mg per day	Maternal bone outcomes	Overall effect of calcium intake/ supplementation during lactation on maternal bone
Sowers *et al.* (1993)	Prospective cohort	US, 20–40 years Lactation and weaning 1 to 18 months PP (98)	>1500	—	5% bone loss at the spine and femoral neck at 6 months PP during lactation and complete recovery after weaning. No relationship with dietary calcium intake	No effect
Sowers *et al.* (1995)	Prospective cohort	US, 20–40 years Lactation and weaning 1 to 18 months PP (112)	1526–1830	—	Higher markers of bone turnover with longer duration of breastfeeding. No relationship with dietary calcium intake	No effect
Krebs *et al.* (1997)	Prospective cohort	US, 31 ± 4 years 0.5 to 7 months PP Lactating (27) NLPP (8)	 1400 875	—	↓LSBMD only in the lactating women (4.0%); LSBMD positively associated with the ratio of calcium to protein intake	Positive
Laskey *et al.* (1998)	Prospective cohort	UK, 20–40 years 0.5 to 3 months PP Lactating (47) NLPP (11) NPNL (22)	 1390 1180 960	—	↓BMC at spine (3.96%), femoral neck (2.39%), total hip (1.51%) and whole body (0.86%) only in the lactating group. No relationship with calcium intake	No effect
Laskey *et al.* (2011)	Prospective cohort	UK, 32 ± 4 years 2 weeks to 12 months PP Lactating (48) NPNL (23)	 1254 904	—	Lactation was associated with significant but transient changes in hip BMD and measures of structural geometry. Bone changes were not associated with Ca intake	No effect

O'Brien *et al.* (2012)	Prospective cohort	Three cohorts Pre/early preg. to 3–10 weeks PP US, 13–18 years (23) US, 25–34 years (13) Brazil, 20–35 years (10)	 1200 1200 450	—	During the postpartum period, positive association between rate of bone calcium deposition and dietary calcium	Positive
Chan *et al.* (1987)	Prospective cohort	US, Lactating 4 weeks prepartum to 16 weeks PP 19–35 years (12) 15–18 years (21) 15–18 years (15)	 1500 >1600 900	—	10% decrease in BMC in the adolescent group consuming daily 900 mg Ca. No decrease in the high Ca groups. Positive correlation between dietary calcium intake and BMC in all adolescents	Positive
Cross *et al.* (1995)	RCT	US, 28 ± 1 years Lactating 2 weeks PP to 3 months and PW Ca suppl. (7) Placebo (8)	950	1000	Lactation: ↓LSBMD in both groups (Ca sup, −6.3%; placebo, −4.3%); ↑UDradius BMD (5.7%) only in the calcium group. After weaning compared to baseline: ↓UDradius BMD (−5.2%) only in the placebo group; no significant change in LSBMD in either group	Positive (slight)
Kalkwarf *et al.* (1997)	RCT	US, 21–40 years Lactation: 16 days to 6 months PP Lact. (97) NLPP (99) Weaning: 6 to 12 months PP Lact. (95) NLPP (92)	 821 650 706 683	1000 + vit D (400 IU) or placebo	Lactation study: ↓LSBMD only in the lactating women; slightly lower reduction with calcium supplementation: Ca sup, 4.2%; placebo, 4.9%. No effect in forearm BMD Weaning study: ↑LSBMD in both groups; slightly higher increase with calcium supplementation: Ca sup, 5.9%; placebo, 4.4%. No effect in forearm BMD	Positive (slight)

(*continued*)

Table 29.2 (*continued*)

Reference	Type of study	Country, maternal age and study groups (*n*)	Mean dietary Ca mg per day	Ca supplement mg per day	Maternal bone outcomes	Overall effect of calcium intake/supplementation during lactation on maternal bone
Kalkwarf *et al.* (1999)	RCT	US, 21–40 years			Biomarkers of bone turnover higher in lactating than in nonlactating women during lactation and post weaning. No effect of Ca supplementation	No effect
		Lactation: 16 days to 6 months PP		1000 + vit D (400 IU) or placebo		
		Lact. (97)	860			
		NLPP (99)	699			
		Weaning: 6 to 12 months PP				
		Lact (95)	739			
		NLPP (92)	711			
Prentice *et al.* (1995)	RCT	The Gambia, 16–41 years			No significant differences in fore-arm BMC between Ca supple-mented and placebo groups at any stage of lactation	No effect
		Lactating, 2 to 52 weeks PP				
		Ca suppl. (30)	275	714		
		Placebo (30)	288	—		
Prentice *et al.* (1998)	RCT	The Gambia, 16–41 years			Increased bone turnover markers and decreased serum PTH and 1,25-$(OH)_2$D during the first months of lactation; no effect of Ca supplementation	No effect
		Lactating, 1.5 to 78 weeks PP				
		Ca suppl. (30)	278	714		
		Placebo (30)	288	—		

[a]Overall effect of calcium intake/supplementation: positive when associated with preservation of bone, and negative with loss of bone. RCT, randomized controlled trial; PP, postpartum; PW, postweaning; NPNL, nonpregnant nonlactating; NLPP, nonlactating postpartum women; BMC, bone-mineral content; BMD, bone-mineral density; BA, bone area; LS, lumbar spine; UD, ultradistal.

of calcium to protein intake in the diet, suggesting that maternal bone loss during lactation may be attenuated by an increased intake of calcium relative to protein in populations with habitually high protein intake. It was hypothesized that in women with low calcium intakes, the effects of lactation on BMD could be attenuated by the fact that their habitual diet may be also low in protein (Krebs *et al.*, 1997).

Randomized controlled studies indicate that calcium supplementation during lactation has no effect on maternal bone outcomes during lactation and postweaning, both in women accustomed to dietary calcium ≥800 mg per day (Cross *et al.*, 1995; Kalkwarf *et al.*, 1997; Kalkwarf *et al.*, 1999) and in women on very low dietary calcium (≈300 mg per day) (Prentice *et al.*, 1995; Prentice *et al.*, 1998). However, subtle positive effects have been described in some of these studies (Kalkwarf *et al.*, 1997; Prentice *et al.*, 1998).

When well-nourished Caucasian women received 1000 mg per day calcium (together with vitamin D, 400 IU per day) or placebo during 6 months postpartum, the decrease in lumbar spine BMD seen in the lactating women was slightly but significantly lower in the calcium supplemented (−4.2%) than in the placebo (−4.9%) group (Kalkwarf *et al.*, 1997). In the same study, when cohorts were treated during weaning, lumbar spine BMD increased slightly more in the calcium supplemented (+5.9%) than in the placebo (+4.4%) group. Therefore, calcium supplementation did not prevent bone loss during lactation in these women but slightly reduced loss and enhanced regain in bone density after weaning.

In lactating Gambian women, providing calcium supplement to increase calcium intake to about 1000 mg per day proved to be of no benefit for maternal BMC at the radius midshaft and wrist (Prentice *et al.*, 1995), neither modified the increased bone turnover (Prentice *et al.*, 1998) during several months of lactation. But, at 52 weeks postpartum bone alkaline phosphatase was significantly lower in the calcium-supplemented group compared to placebo, suggesting a reduction in bone turnover during lactation by use of the supplemental calcium. On the other hand, at 13 weeks postpartum, and irrespective of calcium supplementation, Gambian mothers had higher serum PTH, 1,25-$(OH)_2D$, and osteocalcin than British lactating women with greater habitual dietary calcium (>1000 mg per day), suggesting differences in the magnitude of adaptive mechanisms between these two populations. As previously mentioned, ethnic, genetic, life style, and environmental differences could be underlying factors.

29.3 Calcium Supplementation and Fetal/Infant Bone Growth

It is well accepted that calcium homeostasis during pregnancy and lactation is overwhelmingly in favor of the fetus/neonate (Prentice, 2003). Nevertheless, fetus and neonate are dependent on maternal calcium and it is plausible to expect that changes in maternal calcium intake may affect intrauterine and/or postnatal bone development through changes in placental calcium transfer and/or breast-milk calcium content. Studies evaluating the influence of maternal calcium intake from diet or from supplements during pregnancy on fetal and infant skeletal development are summarized in Table 29.3. Studies of calcium supplementation during lactation have been mainly focused on breast-milk calcium.

Increasing maternal calcium intake during pregnancy, through diet or supplements, has been shown to positively affect newborn bone mineral mass in some (Koo *et al.*, 1999; Chang *et al.*, 2003; Chan *et al.*, 2006; Young *et al.*, 2012), but not all studies (Jarjou *et al.*, 2006; Abdel-Aleem *et al.*, 2009; Abalos *et al.*, 2010). Many factors possibly contribute to different observations, such as differences in study design, methods used for fetal/infant bone assessment, timing of fetal/infant bone evaluation, genetic background, maternal age, maternal gestational weight gain, overall maternal nutritional state, and more specifically maternal habitual calcium intake.

29.3.1 Fetal Bone Growth

Fetal growth rate is the highest throughout lifespan, even greater than during puberty. The fetus typically accumulates about 30 g of calcium during intrauterine life, of which 80% is in the third trimester. These means that an average daily calcium transfer of 200 mg from mother to fetal skeleton is needed during this period and may reach 330 mg per day at 35 weeks of gestation (Prentice, 2003).

In the final third of pregnancy, calcium transfer through the placenta occurs at a rapid rate by active transport (Abrams, 2011). Multiple calcium-binding proteins are involved in this process, but hormonal regulation is still unclear. Parathyroid hormone-related protein (PTHrP), which is produced in several fetal tissues and the placenta, appears to be the main determinant of fetal calcium levels. Furthermore, it is possible that vitamin D increases the synthesis of various calcium-binding proteins (Abrams, 2011).

It appears that efficient mechanisms for fetal calcium conservation operate in late gestation, including fetal intestinal absorption of calcium present in amniotic fluid that is predominantly originated from fetal urine and available for reuse (Done, 2012). As observed in extrauterine life, 1,25-$(OH)_2D$, produced by both the placenta and the fetus, exerts positive influence in calcium intestinal absorption and may play an important role in fetal calcium reutilization (Done, 2012). Moreover, it appears that maternal vitamin D status is an important factor affecting fetal bone development.

29.3.2 Evaluation of Fetal and Infant Bone Outcomes

Measuring multiple fetal ultrasound parameters is considered an effective way for evaluation of fetal growth. Selection of a single biometric parameter depends on the timing and purpose of the measurement. Biparietal diameter (BPD) better correlates with gestational age; abdominal circumference is the

Table 29.3 Studies relating maternal calcium from diet or supplements during pregnancy and fetal/infant bone-growth outcomes.[a]

Reference	Study type	Country, maternal age and study groups (n)	Mean dietary Ca mg per day	Ca supplement mg per day	Fetal/infant outcome measurements	Overall effect of Ca intake/ supplementation on fetal/infant bone growth
Raman *et al.* (1978)	RCT	India, 16–32 years placebo (38) Ca suppl. G1 (24) Ca suppl. G2 (25)	Not informed	300 (G1) or 600 (G2) from 18–22 weeks gestation until parturition	In both 300 and 600 mg Ca-supplemented groups, neonate densities of ulna, radio, tibia and fibula bones were higher than those of the neonates born to nonsupplemented mothers	Positive
Koo *et al.* (1999)	RCT	US, 19.5 ± 0.4 years Placebo (48) Ca suppl. (43)	1035 1010	2000 from ~22 weeks gestation until parturition	No differences between treatment groups in birth weight or length, and in TB or LS BMC at 1st weeks postpartum Higher TB BMC at 1st weeks postpartum in infants born to Ca-supplemented mothers in the lowest quintile (<600 mg per day) of dietary Ca intake	Positive
Chang *et al.* (2003)	Retrospective cohort	US African American, ≤17 years Groups by dairy intake: Low (180) Medium (86) High (84)	Dairy intake (servings per day): Low – <2 Medium – 2 to 3 High – >3	—	Dairy intake not associated with: biparietal diameter, abdominal circumference and head circumference at 20–34 weeks of gestation High dairy intake associated with greater femur length after adjustment for gestational age, maternal age, maternal height, prepregnancy BMI, and biparietal diameter	Positive

(*continued*)

Table 29.3 (*continued*)

Reference	Study type	Country, maternal age and study groups (*n*)	Mean dietary Ca mg per day	Ca supplement mg per day	Fetal/infant outcome measurements	Overall effect of Ca intake/ supplementation on fetal/infant bone growth
Chan *et al.* (2006)	RCT	US, 15–17 years		~1200 during pregnancy	Infants in the dairy group were heavier and had higher total body Ca at birth than infants in the control and orange juice plus calcium groups	Positive
		Control (23)	835			
		Ca fortified orange juice (24)	856			
		Dairy (25)	817			
Jarjou *et al.* (2006)	RCT	The Gambia, 27.4 ± 7.6 years			No differences between the 2 groups for birth weight and infant weight, body length and head circumference at 2 weeks postpartum	Negative
		Placebo (64)	363	1500		
		Ca suppl. (61)	345	from ~20 weeks gestation until parturition	Slower rate of increase in infant BMC and bone area from 2 to 52 weeks in the Ca supplement group compared to placebo	
Abdel-Aleem *et al.* (2009)	RCT	Egyptian, 20.5 ± 3.0 years	Estimated <600	1500 from ~20 weeks gestation until parturition	No differences between placebo and supplemented groups for: biparietal diameter, femur length, humerus length and abdominal circumference from 20 to 36 weeks gestation	No effect
		Placebo (48)			No differences between placebo and supplemented groups for: weight, length, head circumference and abdominal circumference at birth	
		Ca suppl. (43)				

Abalos *et al.* (2010)	RCT	Argentina, 20.5 ± 4.6 years Placebo (230) Ca suppl. (231)	Estimated <600	1500 from ~20 weeks gestation until parturition	No differences between placebo and supplemented groups for: head circumference, biparietal diameter, abdominal circumference, femoral diaphysis length, humeral diaphysis length from 20 to 36 weeks gestation No differences between placebo and supplemented groups for: weight, length, head circumference and abdominal circumference at birth	No effect
Young *et al.* (2012)	Prospective cohort	US, ≤18 years (171)	917	—	Mothers consuming >1050 mg Ca per day had fetuses with higher femur and humerus *Z* scores at 34 weeks than those consuming <1050 mg Ca per day Maternal dietary Ca intake and 25-(OH)D status interacted to affect bone-length outcomes Mothers consuming >1100 mg Ca per day delivered longer neonates compared to mothers consuming <1100 mg Ca per day	Positive

[a]Overall effect of calcium supplementation: positive when associated with increased bone growth; negative, with decreased bone growth. RCT, randomized control trial; BMC, bone mineral content; BMI, body-mass index; TB, total.

most useful dimension to evaluate overall fetal growth, and femur length is the best parameter in the evaluation of skeletal development (Degani, 2001). This set of information has been used to evaluate patterns of intrauterine growth and response to interventions. However, it is important to note that ultrasound measurements are often subjected to observer error, and systematic variations in measurement accuracy may exist between different studies. For this and other reasons, when investigating the effect of nutrient intervention during pregnancy on the offspring, a number of studies evaluate only birth or neonatal outcomes that are also known to reflect fetal growth and development (Done, 2012).

More sophisticated methods for the assessment of infant bone accrued during pregnancy include radiograph and DXA, both restricted to postpartum measurement. DXA has been used for the evaluation of bone mineral content, area and density in infants and, if the assessment is done during the first days postpartum, it reflects bone accretion during gestation. The main limitation appears to be the difficulty in preventing the infant from moving during measurement in order to obtain images that are technically satisfactory without movement artifacts.

29.3.3 Effects of Maternal Calcium Supplementation during Pregnancy on Fetal and Infant Bone Growth

The influence of maternal calcium intake or calcium supplementation during pregnancy on fetal growth has been approached with different methodologies. The first study evaluating the effect of calcium supplementation during pregnancy on bone measurements of neonates was done in an underprivileged population (Raman *et al.*, 1978). In this study, bone densities (evaluated by X-ray) at ulna, radius, tibia and fibula of neonates from mothers supplemented with 300 or 600 mg daily calcium from 18–22 weeks of gestation until parturition were higher than bone densities of neonates from nonsupplemented mothers. More recently, the effect of calcium supplementation (2000 mg Ca per day) during pregnancy on neonatal bone mass at the first weeks postpartum was tested in overall well-nourished mothers, in a placebo-controlled study (Koo *et al.*, 1999). Habitual maternal calcium intake was on average close to 1000 mg per day at midpregnancy and all women received a prenatal supplement containing 400 IU D_2 during pregnancy. Maternal calcium supplementation was found to be associated with increased DXA measurements of total body BMC only in newborns whose mothers were in the lowest quintile (<600 mg) of habitual daily calcium intake during pregnancy. Considered together, the results of both studies suggest that bone growth of neonates benefits from additional calcium when their mothers consume low-calcium diets.

On the other hand, the effect of calcium supplementation (1500 mg per day) during pregnancy on fetal and infant growth that was investigated in Gambian mothers accustomed to very low calcium intakes (300–400 mg per day) produced different results (Jarjou *et al.*, 2006). In this randomized, placebo-controlled trial, there was a trend to a slightly higher BMC and bone area in neonates (~2 weeks postpartum) whose mothers received calcium supplements during pregnancy. However, this trend was not sustained over time and it was actually reversed, resulting in a lower bone accretion rate over the first year of life. Therefore, it appears that the low maternal calcium intake was not the primary determining factor of infant calcium accretion in this population.

The effect of prenatal calcium supplementation was investigated in two other studies resulting in no beneficial effect for fetal growth (Abdel-Aleem *et al.*, 2009; Abalos *et al.*, 2010). These studies were subsets of a WHO multicenter randomized trial of calcium supplementation among low-calcium intake pregnant women for the prevention of pre-eclampsia and preterm delivery. Mothers received a calcium supplement (1500 mg) or placebo daily from ~20 weeks of pregnancy until parturition. Fetal growth was monitored by five ultrasound examinations at 20, 24, 28, 32 and 36 weeks. In both studies, no differences in fetal biometric measurements (BPD, humerus length, abdominal circumference and femur length) at any stage of gestation were observed between fetuses of women who were supplemented with calcium and those who were not (Abdel-Aleem *et al.*, 2009; Abalos *et al.*, 2010). Also, neonatal outcomes (birth weight, length, head circumference, abdominal and thigh circumferences) were similar between the groups, regardless of the supplementation status.

Interestingly, studies focusing on pregnant adolescents are not controversial and provide consistent results irrespective of study design. The effect of maternal dairy intake at early pregnancy on fetal femur development during gestation was investigated in a retrospective cohort study of African-American adolescents (Chang *et al.*, 2003). Dairy intake was considered to be high in those pregnant adolescents consuming 3 or more servings per day, medium in those consuming 2 to 3 servings per day and low if consumption was less than 2 servings per day. High maternal dairy intake at entry into prenatal care was associated with greater fetal femur length evaluated between 20 and 34 weeks of pregnancy, after adjustment for gestational age, maternal age, maternal height, prepregnancy BMI, and fetal BPD. Pregnant adolescents with high dairy intake had fetuses with significantly longer femurs than did those with low dairy intake, suggesting a dose–response relation.

Another study focusing in pregnant adolescents in the US tested the effect of calcium intervention through fortified orange juice or dairy products during the second half of pregnancy on newborn anthropometric data and total bone calcium content, estimated by DXA (Chan *et al.*, 2006). Calcium intake was increased to >1200 mg per day in the supplemented groups compared to ≈860 mg per day in the control group. Newborns from mothers receiving dairy products during pregnancy were heavier at birth and had higher total bone calcium content than those born from mothers receiving fortified orange juice or placebo. The hypothesis was that consumption of dairy products increased also the intake of vitamin D, that in turn may have contributed to a better utilization of the extra calcium provided, and probably promoted a higher fetal calcium accretion.

More recently, maternal vitamin D status and calcium intake were found to interact in the effect on fetal skeletal growth in pregnant adolescents (Young *et al.*, 2012). In this prospective longitudinal designed study, fetal femur and fetal humerus *Z* scores were singly associated with maternal calcium intake and serum 25-(OH)D, with higher scores at maternal calcium intake ≥1050 mg per day or maternal serum 25-(OH)D > 50 nmol L^{-1}. When the interaction between the two nutrients was evaluated, calcium intake remained associated with fetal outcomes only when maternal 25-(OH)D was <50 nmol L^{-1}. Similarly, maternal 25-(OH)D was associated with fetal measurements only when maternal calcium intake was <1050 mg per day. These results suggest that, in adolescent mothers, both adequate maternal calcium intake or sufficient 25-(OH)D status have positive effects on fetal skeletal development and there is a capacity for compensation when intake or status of the other nutrient is limited. The associations and interactions observed *in utero* remained evident at delivery, as indicated by significant differences in neonatal birth length. Adolescents consuming higher amounts of calcium delivered neonates that were longer at birth, especially when maternal 25-(OH)D was suboptimal (Young *et al.*, 2012).

29.3.4 Effects of Maternal Calcium Supplementation on Breast-Milk Calcium Concentration

Most reports on breast-milk calcium concentration and its relationship with mother's calcium intake were published several years ago. In general, these studies suggest that there is no influence of maternal calcium intake during lactation, through diet or supplements, on breast-milk calcium concentrations, even in women consuming very low calcium diets (Prentice *et al.*, 1995; Kalkwarf *et al.*, 1997; Laskey *et al.*, 1998; Jarjou *et al.*, 2006).

It has long been hypothesized that calcium intake during pregnancy, rather than during lactation, may influence breast-milk calcium concentration and hence the calcium intake of breastfed infants. This hypothesis was tested in Gambian mothers that received calcium supplementation (1500 mg per day) from 20 weeks of gestation until delivery (Jarjou *et al.*, 2006). No significant differences between supplemented and nonsupplemented mothers were observed in breast milk concentrations of calcium or calcium-to-phosphorus ratio at 2, 13 and 52 weeks of lactation. Also, no effect of calcium supplementation (1000 mg per day) from 28 weeks gestation until parturition was observed on breast-milk calcium concentrations in Iranian mothers (Karandish *et al.*, 2007). However, in a longitudinal study from the third trimester of pregnancy to 40 days lactation done in Spain (Ortega *et al.*, 1998) it was found that mothers consuming less than 1100 mg per day calcium during pregnancy had significantly lower calcium concentration in mature milk than those consuming >1100 mg per day, suggesting that breast fed babies of mothers with lower calcium intakes during pregnancy may receive less calcium from mother's milk.

In contrast to most studies in adult mothers, teenage motherhood was found to consistently affect calcium concentration in breast milk. Moreover, it was suggested that lower maternal calcium intake results in lower calcium concentrations in the milk secreted by adolescent mothers, but not by adult mothers (Vítolo *et al.*, 2004).

29.4 Conclusions

Several potential interacting factors may explain different results between studies testing the effects of maternal calcium supplementation on maternal and infant bone responses during pregnancy and lactation. Maternal age merits special attention considering that the adolescent pregnant body may compete with her fetus for the nutrients required for optimal bone mineralization. Therefore, adolescent pregnant women and their infants may be particularly benefited by a higher maternal calcium intake during pregnancy and lactation. Other aspects requiring attention when examining the effects of maternal calcium supplementation include limited or inexistent data on dietary habits and vitamin D status. This is an important issue since responses to a calcium-supplement intervention certainly depend on the initial nutritional and bone status of the mother to which her body is already adapted. Additionally, in populations with restricted access to a variety of foods, the concomitant deficiencies of calcium and other nutrients need to be considered. Besides calcium, dietary intake of several nutrients and food components affecting bone health should be examined when evaluating the maternal bone responses to pregnancy and lactation. Finally, interpretation of results depends on the study design and techniques used for assessment of maternal and infant bone mass, and fetal and neonatal bone growth, each one with specific limitations.

The available evidence indicates that use of calcium supplementation, particularly during pregnancy, may have a protective effect on maternal bone mass and may benefit fetal and neonatal bone growth in certain (but not all) populations with habitual calcium intakes <1000 mg per day, and in adolescent mothers. Bone benefits appear to be site specific and the responsiveness of bone sites varies among different populations. Additional research is required before making public-health recommendations for use of calcium supplementation during pregnancy and lactation aiming at maternal and infant bone health. It is likely that recommendations may vary for different population groups.

Summary Points

- It is well accepted that in women with calcium intakes close to current recommendations there is no apparent bone benefit for the mother or infant of using calcium supplements.
- However, in women consuming low-calcium diets, and in adolescent mothers, maternal bone adaptations during reproduction have been found to respond to increased calcium intake or supplementation, although not always as expected.

- Fetal and infant skeletal development has been found to be positively associated with increased maternal calcium intake or supplementation during pregnancy in some but not all studies.
- Inconsistent results across studies of maternal and infant bone responses to calcium supplementation may be due to complex interactions between genetics, diet composition, calcium intake, vitamin D status, environment, and lifestyle.
- Bone effects of calcium supplementation during pregnancy and lactation appear to vary for different population groups.

Key Facts of Bone

1. Adult bone is composed of 32–36% calcium deposited as mixed-phosphate salts within a protein matrix.
2. Bone mineral content and density can be measured by dual-energy-X-ray absorptiometry.
3. Bone serves two main body functions: to provide structural support and to contribute to calcium homeostasis.
4. Bone tissue is continuously renewed during a lifetime through processes of resorption and deposition known as bone turnover.
5. Bone turnover can be indirectly measured by biochemical markers such as osteocalcin and crosslinked *N*-telopeptide of type-I collagen.
6. Bone turnover is increased during pregnancy and lactation.

Definitions of Key Terms

1,25-$(OH)_2$D – 1,25-Dihydroxyvitamin D. Hormonal form of vitamin D that becames active after two sequential hydroxylations at liver (25C position) and kidneys (1C position).

25-(OH)D – 25-Hydroxyvitamin D. Main circulating form of vitamin D, considered the best marker of nutritional status.

Bone mineral content. Total amount of mineral matter that is present in bones, expressed in g.

Bone-mineral density. Amount of bone mineral per unit of bone area scanned, expressed in g cm^{-2}.

Bone turnover. A continuous process of bone remodeling that consists in removal of old bone and replacement with new bone matrix and minerals.

DXA – dual-energy X-ray absorptiometry. X-ray based image technique considered the gold standard for bone-mineral density evaluation.

Fetal biometry. Measurements obtained by ultrasound examination often used to estimate gestational age, fetal growth and skeletal development. Main measurements include abdominal circumference, biparietal diameter and femoral length.

PTH – parathyroid hormone. Polypeptide hormone secreted by the parathyroid glands that has a critical role in calcium and bone homeostasis.

PTHrP – parathyroid hormone-related peptide. Polypeptide hormone produced by tumors and some tissues under specific conditions that exerts biological actions similar to PTH.

***Z* score.** Indicates the number of standard deviations a single measurement distances from a given mean of an age-matched population.

List of Abbreviations

$1,25\text{-}(OH)_2D$	1,25-Dihydroxyvitamin D
25-(OH)D	25-Hydroxyvitamin D
BA	Bone area
BPD	Biparietal diameter
BMC	Bone-mineral content
BMD	Bone-mineral density
BMI	Body-mass index
DXA	Dual-energy X-ray absorptiometry
LS	Lumbar spine
NLPP	Non lactating postpartum women
NPNL	Nonpregnant nonlactating women
NTx	Crosslinked *N*-telopeptide of type-I collagen
PP	Postpartum
PTH	Parathyroid hormone
PTHrP	Parathyroid hormone-related protein
PW	Post weaning
RCO	Randomized crossover trial
RCT	Randomized controlled trial
TB	Total body
UD	Ultradistal
WHO	World Health Organization
βCTx	Carboxyterminal telopeptide of type-I collagen

References

Abalos, E., Merialdi, M., Wojdyla, D., Carroli, G., Campodónico, L., Yao, S. E., Gonzalez, R., Deter, R., Villar, J. and Van Look, P., 2010. Effects of calcium supplementation on fetal growth in mothers with deficient calcium intake: a randomized controlled trial. *Paediatric and Perinatal Epidemiology*. 24: 53–62.

Abdel-Aleem, H., Merialdi, M., Elsnosy, E. D., Elsedfy, G. O., Abdel-Aleem, M. A. and Villar, J., 2009. The effect of calcium supplementation during pregnancy on fetal and infant growth: a nested randomized controlled trial within WHO calcium supplementation trial. *Journal of Maternal-Fetal and Neonatal Medicine*. 22: 94–100.

Abrams, S. A., 2011. What are the risks and benefits to increasing dietary bone minerals and vitamin D intake in infants and small children? *Annual Review of Nutrition*. 2011(31): 285–297.

Avendaño-Badillo, D., Hernandez-Avila, M., Hernandez-Cadena, L., Rueda-Hernandez, G., Solano González, M., Ibarra, L. G., Hu, H. and Téllez-Rojo, M. M., 2009. High dietary calcium intake decreases bone mobilization during pregnancy in humans. *Salud Pública de México*. 51: S100–S107.

Chan, G. M., McElligott, K., McNaught, T. and Gill, G., 2006. Effects of dietary calcium intervention on adolescent mothers and newborns. *Obstetrics and Gynecology*. 108: 565–571.

Chan, G. M., McMurry, M., Westover, K., Engelbert-Fenton, K. and Thomas, M. R., 1987. Effects of increased dietary calcium intake upon the calcium and bone mineral status of lactating adolescent and adult women. *American Journal of Clinical Nutrition*. 46: 319–323.

Chang, S., O'Brien, K. O., Nathanson, M. S., Caulfield, L. E., Mancini, J. and Witter, F. R., 2003. Fetal femur length is influenced by maternal dairy intake in pregnant African American adolescents. *American Journal of Clinical Nutrition*. 77: 1248–1254.

Cross, N. A., Hillman, L. S., Allen, S. H., Krause, G. F. and Vieira, N. E., 1995. Calcium homeostasis and bone metabolism during pregnancy, lactation, and postweaning: a longitudinal study. *American Journal of Clinical Nutrition*. 61: 514–523.

Degani, S., 2001. Fetal biometry: clinical, pathological, and technical considerations. *Obstetrical and Gynecologycal Survey*. 56: 159–167.

Diogenes, M. E. L., Bezerra, F. F., Rezende, E. P., Taveira, M. F., Pinhal, I. and Donangelo, C. M., 2013. The effect of calcium plus vitamin D supplementation during pregnancy in Brazilian adolescent mothers: a randomized placebo-controlled trial. *American Journal of Clinical Nutrition*. 98: 82–91.

Done, S. L., 2012. Fetal and neonatal bone health: update on bone growth and manifestations in health and disease. *Pediatric Radiology*. 42: S158–S176.

Janakiraman, V., Ettinger, A., Mercado-Garcia, A., Hu, H. and Hernandez-Avila, M., 2003. Calcium supplements and bone resorption in pregnancy. A randomized crossover trial. *American Journal of Preventive Medicine*. 24: 260–264.

Jarjou, L. M. A., Laskey, M. A., Sawo, Y., Goldberg, G. R., Cole, T. J. and Prentice, A., 2010. Effect of calcium supplementation in pregnancy on maternal bone outcomes in women with a low calcium intake. *American Journal of Clinical Nutrition*. 92: 450–457.

Jarjou, L. M. A., Prentice, A., Sawo, Y., Laskey, M. A., Bennet, J., Goldberg, G. R. and Cole, T. J., 2006. Randomized, placebo-controlled, calcium supplementation study in pregnant Gambian women: effects in breast-milk calcium concentrations and infant birth weight, growth, and bone mineral accretion in the first year of life. *American Journal of Clinical Nutrition*. 83: 657–666.

Jarjou, L. M. A., Sawo, Y., Goldberg, G. R., Laskey, M. A., Cole, T. J. and Prentice, A., 2013. Unexpected long-term effects of calcium supplementation in pregnancy on maternal bone outcomes in women with a low calcium intake: a follow up study. *American Journal of Clinical Nutrition*. 98: 723–730.

Kalkwarf, H. J., Specker, B. L., Bianchi, D. C., Ranz, J. and Ho, M., 1997. The effect of calcium supplementation on bone density during lactation and after weaning. *New England Journal of Medicine*. 337: 523–528.

Kalkwarf, H. J., Specker, B. L. and Ho, M., 1999. Effects of calcium supplementation on calcium homeostasis and bone turnover in lactating women. *Journal of Clinical Endocrinology and Metabolism*. 84: 464–470.

Karandish, M., Djazayery, A., Michaelsen, K. F., Rashidi, A., Mohammadpour-Ahranjani, B., Behrooz, A. and Mølgaard, C., 2007. Does supplementation with calcium during pregnancy affect the mineral concentration in mature breast-milk? *International Journal of Endocrinology and Metabolism*. 4: 134–141.

Koo, W. W. K., Walters, J. C., Esterlitz, J., Levine, R. J., Bush, A. J. and Sibai, B., 1999. Calcium supplementation and fetal bone mineralization. *Obstetrics and Gynaecology*. 94: 577–582.

Krebs, N. F., Reidinger, C. J., Robertson, A. D. and Brenner, M., 1997. Bone mineral density changes during lactation: maternal, dietary, and biochemical correlates. *American Journal of Clinical Nutrition*. 65: 1738–1746.

Laskey, M. A., Prentice, A., Hanratty, L. A., Jarjou, L. M. A., Dibba, B., Beavan, S. R. and Cole, T. J., 1998. Bone changes after 3 months of lactation: influence of calcium intake, breast-milk output, and vitamin D-receptor genotype. *American Journal of Clinical Nutrition*. 67: 685–692.

Laskey, M. A., Price, R. I., Khoo, B. C. and Prentice, A., 2011. Proximal femur structural geometry changes during and following lactation. *Bone*. 48: 755–759.

Liu, Z., Qiu, L., Chen, Y. and Su, Y., 2011. Effect of milk and calcium supplementation on bone density and bone turnover in pregnant Chinese women: a randomized controlled trial. *Archives of Gynaecology and Obstetrics*. 283: 205–211.

National Research Council, 2011. Dietary Reference Intakes for Calcium and Vitamin D. In: Ross, A. C., Taylor, C. L., Yaktine, A. L. and Del Valle, H. B. (ed.) National Academies Press, Washington, DC, vol. 5, pp. 345–402.

O'Brien, K. O., Nathanson, M. S., Mancini, J. and Witter, F. R., 2003. Calcium absorption is significantly higher in adolescents during pregnancy than in the early postpartum period. *American Journal of Clinical Nutrition*. 78: 1188–1193.

O'Brien, K. O., Donangelo, C. M., Zapata, C. L. V., Abrams, S. A., Spencer, E. M. and King, J. C., 2006. Bone calcium turnover during pregnancy and lactation in women with low calcium diets is associated with calcium intake and circulating insulin-like growth factor 1 concentrations. *American Journal of Clinical Nutrition*. 83: 317–323.

O'Brien, K. O., Donangelo, C. M., Ritchie, L. D., Gildengorin, G., Abrams, S. and King, J. C., 2012. Serum 1,25-dihydroxyvitamin D and calcium intake affect rates of bone calcium deposition during pregnancy and the early postpartum period. *American Journal of Clinical Nutrition*. 96: 64–72.

Olausson, H., Goldberg, G. R., Laskey, M. A., Schoenmakers, I., Jarjou, L. M. A. and Prentice, A., 2012. Calcium economy in human pregnancy and lactation. *Nutrition Research Reviews*. 25: 40–67.

Olausson, H., Laskey, M. A., Goldberg, G. R. and Prentice, A., 2008. Changes in bone mineral status and bone size during pregnancy and the influence of body segue and calcium intake. *American Journal of Clinical Nutrition*. 88: 1032–1039.

Ortega, R. M., Martínez, R. M., Quintas, M. E., López-Sobaler, A. M. and Andrés, P., 1998. Calcium levels in maternal milk: relationships with calcium intake during the third trimester of pregnancy. *British Journal of Nutrition*. 79: 501–507.

Prentice, A., Jarjou, L. M., Cole, T. J., Stirling, D. M., Dibba, B. and Fairweather-Tait, S., 1995. Calcium requirements of lactating Gambian mothers: effects of a calcium supplement on breast milk calcium concentration, maternal bone mineral content, and urinary calcium excretion. *American Journal of Clinical Nutrition*. 62: 58–67.

Prentice, A., Jarjou, L. M. A., Stirling, D. M., Buffenstein, R. and Fairwether-Tait, S., 1998. Biochemical markers of calcium and bone metabolism during 18 months of lactation in Gambian women accustomed to a low calcium intake and in those consuming a calcium supplement. *Journal of Clinical and Endocrinology Metabolism*. 83: 1059–1066.

Prentice, A., 2003. Micronutrients and the bone mineral content of mother, fetus and newborn. *Journal of Nutrition*. 133: 1693S–1699S.

Raman, L., Rajalakshmi, K., Krishnamachari, K. A. and Sastry, J. G., 1978. Effect of calcium supplementation to undernourished mothers during pregnancy on the bone density of the bone density of the neonates. *American Journal of Clinical Nutrition*. 31: 466–469.

Ritchie, L. D., Fung, E. B., Halloran, B. P., Turnlund, J. R., Van Loan, M. D., Cann, C. E. and King, J. C., 1998. A longitudinal study of calcium homeostasis during human pregnancy and lactation and after resumption of menses. *American Journal of Clinical Nutrition*. 67: 693–701.

Sowers, M., Corton, G., Shapiro, B., Jannausch, M. L., Crutchfield, M., Smith, M. L., Randolph, J. F. and Hollis, B., 1993. Changes in bone density with lactation. *Journal of the American Medical Association*. 269: 3130–3135.

Sowers, M., Eyre, D., Hollis, B. W., Randolph, J. F., Shapiro, B., Jannausch, M. L. and Crutchfield, M., 1995. Biochemical markers of bone turnover in lactating and nonlactating postpartum women. *Journal of Clinical Endocrinology and Metabolism*. 80: 2210–2216.

Vítolo, M. R., Valente Soares, L. M., Carvalho, E. B. and Cardoso, C. B., 2004. Calcium and magnesium concentrations in mature human Milk: influence of calcium intake, age and socioeconomic level. *Archivos Latinoamericanos de Nutrición*. 54: 118–122.

Young, B. E., McNanley, T. J., Cooper, E. M., McIntyre, A. W., Witter, F., Harris, Z. L. and O'Brien, K. O., 2012. Maternal vitamin D status and calcium intake interact to affect fetal skeletal growth in utero in pregnant adolescents. *American Journal of Clinical Nutrition*. 95: 1103–1112.

Zeni, S. N., Ortela Soler, C. R., Lazzari, A., López, L., Suarez, M., Gregorio, S. D., Somoza, J. I. and Portela, M. L., 2003. Interrelationship between bone turnover markers and dietary calcium intake in pregnant women: a longitudinal study. *Bone*. 33: 606–613.

CHAPTER 30

The Effects of Dietary Calcium on Hypertension

FILIPPO ROSSI*[a]

[a]Istituto di Scienze degli Alimenti e della Nutrizione, Facoltà di Scienze Agraria, Alimentari e Ambientali, Via Emilia Parmense 84, 29122 Piacenza, Italy
*E-mail: filippo.rossi@unicatt.it

30.1 Introduction

Sodium is considered one of the most important dietary factors affecting blood pressure and the First National Health and Nutrition Examination Survey (NHANES I) showed that alcohol and Na had the strongest correlations with blood pressure in American adults (Gruchow *et al.*, 1985). However, in the same study, there also emerged an inverse relationship between Ca intake and blood pressure, although limited to nonwhite men, a subgroup with a low Ca intake compared to the white population (Sempos *et al.*, 1986). This finding has stimulated more research aimed at a better understanding of the relationship between Ca and blood pressure.

Food and Nutritional Components in Focus No. 10
Calcium: Chemistry, Analysis, Function and Effects
Edited by Victor R. Preedy

Published by the Royal Society of Chemistry, www.rsc.org

30.2 Does Calcium Really Affect Blood Pressure?

As recently reviewed by Nguyen *et al.* (2013) research investigating the relationship between Ca intake and blood pressure does not always give positive results. In general, we have two different types of study:

- observational studies, focused on the relationship between Ca intake and blood pressure;
- intervention studies, where volunteers received Ca supplements.

In the first type of study, many authors reported an inverse relationship between blood pressure and Ca intake (McCarron *et al.*, 1984; Reed *et al.*, 1985; Whitteman *et al.*, 1989; Cutler and Brittain, 1990; Ascherio *et al.*, 1992; Dwyer *et al.*, 1996; Wang *et al.*, 2008).

However, in many other studies limited or no significant effect of dietary Ca on blood pressure were detected (Strazzullo *et al.*, 1986; Cappuccio *et al.*, 1987; Ascherio *et al.*, 1996). Several factors could explain the limited effect of Ca supplementation, one of which is the different amount of Ca ingested. Cappuccio *et al.* (1987) administered 160 mg per day of the mineral, while in the experiment of Strazzullo *et al.* (1986) 1 g per day of Ca was given to volunteers. Interestingly, in this work a relevant effect of Ca supplementation on the fall in blood pressure was observed in a subgroup of volunteers with a high urinary excretion of Ca. The high urinary Ca loss could determine a compensatory increase in PTH and vitamin D_3 that, in turn, could suppress renin activity and consequently angiotensin II synthesis (Chun Li *et al.*, 2004). Ascherio *et al.* (1996) reported an inverse relationship between Ca intake and the prevalence of hypertension in the 4 year follow-up of their epidemiological study, although they did not observe any significant effect of Ca on blood pressure.

Four meta-analyses have assessed the role of Ca intake on the regulation of blood pressure, and the results are summarized in Table 30.1.

The work of van Mierlo *et al.* (2006) in agreement with Gruchow *et al.* (1985), reports a larger impact of Ca supplementation on systolic blood pressure (SBP) in subjects having a Ca intake equal to or lower than the recommended daily allowance (RDA): 800 mg per day. The mean Ca supplementation in this work was 1200 mg per day. It should be remembered that the Ca requirement has been increased and has now been set at 900–1000 mg per day for adult (but not elderly) subjects.

Another subpopulation showing a better Ca intake response is represented by hypertensive subjects, who obtained a reduction in SBP almost double that of normotensive subjects Allender *et al.* (1996).

To explain the reasons why dietary Ca can modify blood pressure, several hypotheses have been made:

- the effects of intracellular Ca on vascular smooth muscle contraction (Smajilovic and Jacob Tfelt-Hansen, 2007);
- Ca increases urinary excretion of Na;

Table 30.1 Summary of meta-analyses assessing the effect of dietary Ca on blood pressure.[a]

Authors	No. of studies	Results	*P* value
van Mierlo *et al.* (2006)	40 studies	-1.86 mm Hg SBP	*
	2492 subjects	-0.99 mm Hg DBP	*
		-2.63 mm Hg SBP Ca intake <800 mg per day	*
		-1.30 mm Hg DBP Ca intake <800 mg per day	*
Dickinson *et al.* (2006)	13 studies	-2.5 mm Hg SBP	*
	485 subjects	-0.8 mm Hg DBP	NS
Griffith *et al.* (1999)	42	-1.44 mm Hg SBP	*
		-0.68 mm Hg DBP	*
Allender *et al.* (1996)	28 studies	-0.89 mm Hg SBP all subjects	*
	1231 subjects	-0.18 mm Hg DBP all subjects	NS
		-0.53 mm Hg SBP normotensive	*
		-1.68 mm Hg SBP hypertensive	*
Bucher *et al.* (1996)	33 studies	-1.27 mm Hg SBP	*
	2412 subjects	-0.24 mm Hg DBP	NS

[a]*$P < 0.05$; SBP = systolic blood pressure; DBP = diastolic blood pressure.

- interrelationship between Ca intake and some hormones involved in blood-pressure regulation;
- modulation of the sympathetic nervous system by Ca.

Smajilovic and Jacob Tfelt-Hansen (2007) excellently discussed the role of Ca in the regulation of blood vessel contraction. They upheld the role of Ca as first messenger, through a calcium sensing receptor (CaR) that has been shown to be present in several tissues, in rat arteries (Bukoski *et al.*, 1997) and also in vascular smooth muscle cells (VSMC) of rat aorta (Smajilovic and Jacob Tfelt-Hansen, 2007). Evidence for the presence of the CaR in human aortic endothelial cells has been provided by Ziegelstein *et al.* (2006). Interestingly, when stimulated by Ca, this receptor determines the synthesis of nitric oxide (NO) which is a potent vasodilator (see Figure 30.1).

According to Smajilovic and Jacob Tfelt-Hansen (2007) extracellular Ca can reduce blood pressure acting in three different parts of the blood vessel. At the level of tunica adventitia the CaR, present in perivascular nerves, is activated by extracellular Ca determining relaxation of blood vessels (Bukoski *et al.*, 1997).

In VSMC, constituting tunica media, the link of Ca with CaR generates membrane hyperpolarization, due to the opening of Ca-sensitive K channels and consequently vessel relaxation.

The third mechanism by which Ca can induce a reduction in blood pressure is the production of NO in endothelial cells (Smajilovic and Jacob Tfelt-Hansen, 2007; Ohashi *et al.*, 2012). Ca bound to calmodulin reacts with L-arginine in a reaction catalysed by eNOS to produce NO that exerts a vasodilation activity (Figure 30.2).

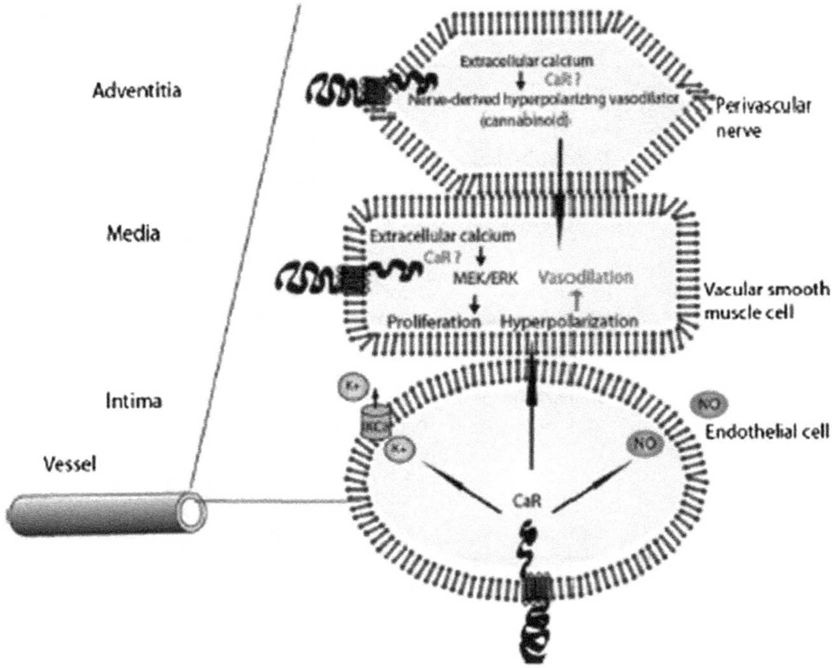

Figure 30.1 Effects of Ca on blood vessel diameter. Black text and arrows signify known effects, vasodilation is a proposed effects. (With permission from Smajilovic and Jacob Tfelt-Hansen (2007)).

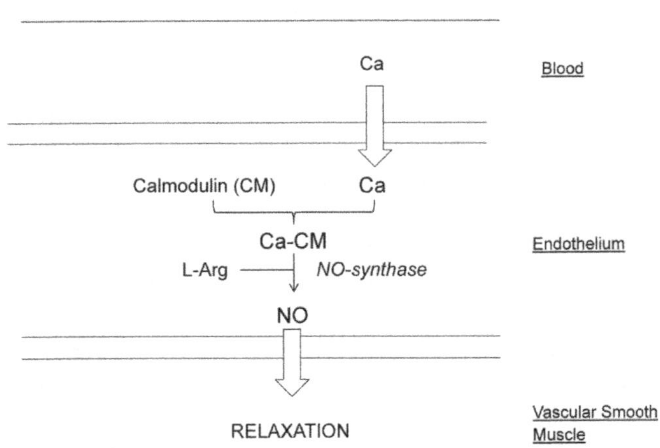

Figure 30.2 Mode of action of Ca on renin–angiotensin system in salt-sensitive subjects. The increase in dietary Ca intake determines an increase in osmotic pressure into the juxta-glomerular cells (JGA). The result is a suppression of renin synthesis and consequently a lower production of angiotensin II, a vasoconstrictor molecules the action of which is mediated by Ca release from intracellular stores. Intracellular Ca (Ca_i) does not change because an higher Ca intake leads to an increase in Ca uptake, that counterbalance the suppression of Ca release from endoplasmic reticulum.

30.3 Ca in the Regulation of the Renin–Angiotensin System

Resnick intensively investigated the role of the renin–angiotensin system and hormones involved in Ca metabolism, in the genesis of hypertension and on the effect of dietary Ca supplementation on blood pressure.

According to his theory the two main categories of hypertensive subjects: salt sensitive and salt insensitive, are the results of an altered regulation of Ca and Na metabolism. In salt-sensitive subjects, there is a high Ca uptake linked to low renin levels or low renin activity, while in salt-insensitive subjects, renin levels are normal or high and intracellular Ca is high, due to increased release from the endoplasmic reticulum (Resnick, 1999).

A confirmation of the role of Ca transport across cell membranes, in hypertensive salt-sensitive subjects, comes from the work of Resnick *et al.* (1989). The authors treated with a Ca channel blocker a cohort of hypertensive volunteers with different degrees of responsiveness to sodium intake. They observed a greater drug-induced lowering of blood pressure in the salt-sensitive compared to salt-insensitive subjects. The first cohort of volunteers also had low initial levels of serum ionized Ca and higher levels of vitamin D_3 induced by low serum Ca. Two other effects of Ca channel blocker were an increase in extracellular Ca and a reduction in vitamin D_3 plasma levels.

This hypothesis confirms the observation of the NHANES I study of an inverse relationship between Ca intake and blood pressure in black men, with a low Ca intake. This population group has high levels of 1,25-dihydroxyvitamin D and parathyroid hypertensive factor. Both of these hormones promote the transport of extracellular Ca in the cell and have a suppressive effect on renin activity.

In salt-sensitive subjects, an increase in dietary Ca determines a reduction in synthesis of vitamin D_3 and parathyroid hypertensive factor, restoring the activity of renin, otherwise suppressed. If renin activity is not restored, the increase in serum Na does not activate the compensatory metabolic process (lowering of angiotensinogen synthesis, increased urinary Na excretion) that prevents blood pressure from increasing.

According to Resnick (1999), the interrelationship between extracellular Ca, cytosolic free Ca (Ca_i) and stored intracellular Ca, plays an important role in the regulation of blood pressure (see Figure 30.3).

In this hypothesis there are three important points:

- blood-vessel contraction is directly proportional to Ca_i;
- angiotensin II increases Ca_i by means of Ca release from intracellular stores;
- vitamin D_3 increases Ca_i, through an increment in Ca uptake from external space into cells.

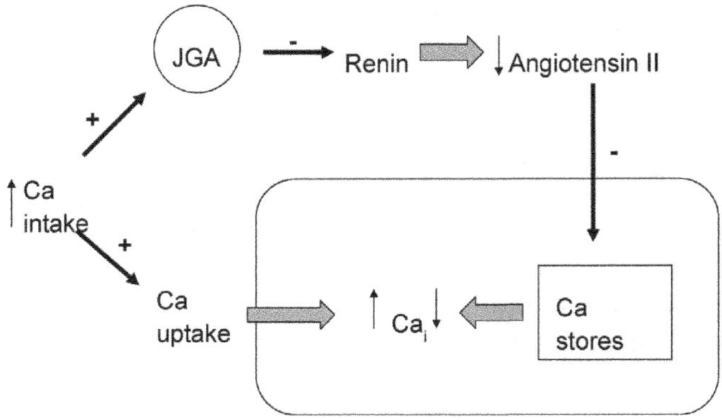

Figure 30.3 Metabolic pathway of relationship between Ca and nitric oxide (NO) synthesis in blood vessel.

In subjects with high Ca intake, we observe an increased Ca uptake that leads to an increase of Ca_i that in turn suppresses the Ca release from cellular stores mediated by angiotensin II, so there is no change on Ca_i and smooth muscles are not stimulated to contract. On the other hand, the increase of Ca_i in cells of juxtaglomerular apparatus, determines a reduction in renin production, which results in a suppression of angiotensin II synthesis, with a consequent secondary suppressive effect in angiotensin-mediated vasoconstriction activity.

In salt-sensitive subjects PTH and vitamin D_3 levels are high, while renin activity is low, probably because vitamin D_3 suppresses renin activity and sensitivity to Na (Chun Li *et al.*, 2004). In this population group, an increase in Ca intake reduces the blood levels of Ca-regulating hormones and, on the one hand, this increases renin sensitivity, but on the other this does not determine angiotensin II synthesis because the increase of blood Ca is detected in the juxtaglomerular apparatus determining a suppression of renin synthesis.

30.4 Conclusion

At present there no is clear evidence of a benefit of Ca supplementation for the prevention or treatment of hypertension. In recent years, the daily requirements (RDA) for Ca has been increased (at least in Italy), but this is related to the role of Ca in bone metabolism in young and elderly people, not to the prevention of hypertension. In subjects with a low Ca intake, it is useful to suggest the consumption of dairy products, which are excellent sources of Ca and also of ACE-inhibiting peptides – molecules that can reduce blood pressure. Obviously, as suggested more than 30 years ago by the DASH Project, low-fat dairy products should be the first dietary choice to fulfil Ca requirements without excessively increasing energy intake.

Summary Points

- This chapter focuses on the relationship between Ca intake (diet plus supplements) and blood pressure.
- Hypertension is an illness of growing importance, the therapy of which is based on drugs and some dietary and lifestyle recommendations (low salt intake, low-fat diet, weight loss).
- Any nonpharmacological tool that can prevent the development of hypertension, reduces costs for the National Health Service.
- Hypertensive subjects can be divided into two groups: salt sensitive (an increase in Na is reflected on blood pressure) and salt insensitive (Na intake does not affect blood pressure).
- Blood pressure is regulated by the renin–angiotensin system. Renin is an enzyme, produced by the kidney that converts angiotensinogen, secreted from the liver, in angiotensin I. At pulmonary level angiotensin I is converted to angiotensin II by angiotensin-converting enzyme (ACE). Angiotensin II determines blood-vessel contraction and an increase in blood pressure.
- Calcium can reduce blood pressure in salt-sensitive subjects by restoring the dis-regulated set-point of the renin–angiotensin system. In this population group, the renin–angiotensin system does not recognize the increase in Na blood level and consequently does not reduce the synthesis of renin.
- Another mode of action of Ca on blood pressure is the stimulation of synthesis of NO in the endothelium. NO is a strong vasodilator.

Key Facts

The key facts of blood pressure and hypertension.

1. Systolic blood pressure (SBP) is the maximum pressure exerted by the blood on arteries and vessels when the heart is beating, which occurs at the end of ventricle contraction. The normal values for adults range from 90 to 120 mm Hg.
2. Diastolic blood pressure (DBP) represents the minimum pressure exerted by the blood on arteries, which occurs when the ventricles are relaxed. The normal values for adults range from 60 to 80 mm Hg.
3. Hypertension is defined as blood pressure (BP) $\geq$ 140 mm Hg for SBP and $\geq$90 mm Hg for DBP, hypertensive subjects can be categorized into 3 classes of increasing severity of the illness:

   - grade I 140–159 mm Hg for SBP and 90–99 mm Hg for DBP;
   - grade II 160–179 mm Hg for SBP and 100–109 mm Hg for DBP;
   - grade III $\geq$180 mm Hg for SBP and $\geq$110 mm Hg for DBP.

Optimal values of SBP are <130 mm Hg, while people with SBP ranging from 130 to 139 mm Hg are classified as normal subjects. It has recently been proposed to consider people with SBP from 120 to 139 mm Hg and with DBP from 80 to 89 mm Hg, as belonging to the new class of prehypertension. In western countries, although some dietary factors (salt intake, excess energy intake, cholesterol consumption) have been demonstrated to affect SBP, hypertension increases with age. Several epidemiologic studies have demonstrated a positive relationship of high blood pressure with cardiovascular and kidney diseases. Therefore, if the severity and prevalence of hypertension could be reduced, then many associated diseases could be prevented.

Definition and Explanation of Key Terms

Systolic blood pressure (SBP). This is the maximum value reached by blood pressure during ventricle contraction.

Diastolic blood pressure (DBP). This is the minimum value reached by blood pressure during ventricle relaxation.

Blood vessels. Blood vessels are tubes of different diameter, components of the circulatory system, where blood flows and reaches the whole body. According to their function they can be classified as arteries (which carry the blood from the heart to the tissues), capillaries (with a small diameter and the site where gas and nutrient exchange between blood and tissue takes place) and veins (which carry blood from the capillaries to the heart).

Blood-vessel structure. In arteries and veins we can find 3 layers: tunica intima (a single layer of endothelial cells); tunica media, made of connective tissue and containing vascular smooth muscle cells, involved in the regulation of vessel diameter; tunica adventitia exclusively made of connective tissue. Capillaries have a small caliber and are formed of one layer of endothelial cells (tunica intima) surrounded by connective tissue.

Vasoconstriction. This is the narrowing of blood vessels, a physiological activity operated by vascular smooth muscle cells. The result of vasoconstriction is an increase in blood pressure.

Vasodilation. This is the widening of blood vessels. The relaxation of vascular smooth muscle cells results in vasodilation and in a decrease of blood pressure.

Salt-sensitive subjects. Salt sensitivity means the responsiveness of blood pressure to Na intake. In salt sensitive subjects, any increase in Na intake results in a corresponding increase in blood pressure.

Salt-insensitive subjects. In salt-insensitive subjects, blood pressure is not responsive to changes in dietary Na.

Angiotensin II. Angiotensin II is a peptide hormone derived from enzymatic cleavage of angiotensin I operated by angiotensin-converting enzyme (ACE), whose biological action results in an increase of blood pressure. In fact, angiotensin II is a vasoconstrictor agent, that also increases Na retention by aldosterone biosynthesis.

Nitric oxide. Nitric oxide, chemical formula NO, is a multipurpose signaling molecule, part of the regulatory mechanism of blood pressure, which acts as a vasodilator factor that leads to a reduction in blood pressure.

List of Abbreviations

NHANES I	First National Health and Nutrition Examination Survey
SBP	Systolic blood pressure
DBP	Diastolic blood pressure
RDA	Recommended daily allowance
CaR	Calcium sensing receptor
VSMC	Vascular smooth muscle cells
NO	Nitric oxide
eNOS	Endothelial nitric oxide synthase

References

Allender, P. S., Cutler, J. A., Follmann, D., Cappuccio, F. P., Pryer, J. and Elliott, P., 1996. Dietary calcium and blood pressure: a meta-analysis of randomized clinical trials. *Annals of Internal Medicine*. 124: 825–831.

Ascherio, A., Rimm, E. B., Giovannucci, E. L., Colditz, G. A., Rosner, B., Willett, W. C., Sacks, F. and Stampfer, M. J., 1992. A prospective study of nutritional factors and hypertension among US men. *Circulation*. 86: 1475–1484.

Ascherio, A., Hennekens, C., Willett, W. C., Sacks, F., Rosner, B., Manson, J. A., Witteman, J. and Stampfer, M. J., 1996. Prospective study of nutritional factors, blood pressure, and hypertension among US women. *Hypertension*. 27: 1065–1072.

Bucher, H. C., Cook, R. J., Guyatt, G. H., Lang, J. D., Cook, D. J., Hatala, R. and Hunt, D. L., 1996. Effects of dietary calcium supplementation on blood pressure. A meta-analysis of randomized controlled trials. *JAMA, the Journal of the American Medical Association*. 275: 1016–1022.

Bukoski, R. D., Bian, K., Wang, Y. and Mupanomunda, M., 1997. Perivascular sensory nerve Ca^{2+} receptor and Ca^{2+}-induced relaxation of isolated arteries. *Hypertension*. 30: 1431–1439.

Cappuccio, F. P., Markandu, N. D., Singer, D. R. J., Smith, S. J., Shore, A. C. and MacGregor, G. A., 1987. Does oral calcium supplementation lower high blood pressure? A double blind study. *Journal of Hypertension*. 5: 67–71.

Chun Li, Y., Qiao, G., Uskokovic, M., Xiang, W., Zheng, W. and Kong, J., 2004. Vitamin D: a negative endocrine regulator of the renin–angiotensin system and blood pressure. *Journal of Steroid Biochemistry and Molecular Biology*. 89–90(2004): 387–392.

Cutler, J. A. and Brittain, E., 1990. Calcium and blood pressure. An epidemiologic perspective. *American Journal of Hypertension*. 3: 137S–146S.

Dickinson, H. O., Nicolson, D. J., Cook, J. V., Campbell, F., Beyer, F. R., Ford, G. A. and Mason, J., 2006. Calcium supplementation for the management of primary hypertension in adults. *Cochrane Database of Systematic Review*. 19: CD004639.

Dwyer, J. H., Li, L., Dwyer, K. M., Curtin, L. R. and Feinleib, M., 1996. Dietary calcium, alcohol, and incidence of treated hypertension in the NHANES I epidemiologic follow-up study. *American Journal of Epidemiology*. 144(9): 828–838.

Griffith, L. E., Guyatt, G. H., Cook, R. J., Bucher, H. C. and Cook, D. J., 1999. The influence of dietary and nondietary calcium supplementation on blood pressure: an updated metaanalysis of randomized controlled trials. *American Journal of Hypertension*. 12(1 Pt 1): 84–92.

Gruchow, H. W., Sobocinski, K. A. and Barboriak, J. J., 1985. Alcohol, nutrient intake and hypertension in US adults. *JAMA, the Journal of the American Medical Association*. 253: 1567–1570.

McCarron, D. A., Morris, C. D., Henry, H. J. and Stanton, J. L., 1984. Blood pressure and nutrient intake in the United States. *Science*. 224: 1392–1398.

Nguyen, H., Odelola, O. A., Janani Rangaswami, J. and Amanullah, A., 2013. A review of nutritional factors in hypertension management. *International Journal of Hypertension*. DOI: 10.1155/2013/698940.

Ohashi, J., Sawada, A., Nakajima, S., Noda, K., Takaki, A. and Shimokawa, H., 2012. Mechanisms for enhanced endothelium-derived hyperpolarizing factor-mediated responses in microvessels in mice. *Circulation Journal*. 76: 1768–1779.

Reed, D., McGee, D., Yano, K. and Hankin, J., 1985. Diet, blood pressure, and multicollinearity. *Hypertension*. 7: 405–410.

Resnick, L. M., 1999. The role of dietary calcium in hypertension. A hierarchal overview. *American Journal of Hypertension*. 12: 99–112.

Resnick, L. M., Nicholson, J. P. and Laragh, J. H., 1989. The effects of calcium channel blockade on blood pressure and calcium metabolism. *American Journal of Hypertension*. 2: 927–930.

Strazzullo, P., Siani, A., Guglielmi, S., Di Carlo, A., Galletti, F., Cirillo, M. and Mancini, M., 1986. Controlled trial of long-term oral calcium supplementation in essential hypertension. *Hypertension*. 8: 1084–1088.

Sempos, C., Cooper, R., Kovar, M., Johnson, C., Drizd, T. and Yettey, E., 1986. Dietary calcium and blood pressure in National Health and Nutrition Examination Surveys I and II. *Hypertension*. 8: 1067–1074.

Smajilovic, S. and Jacob Tfelt-Hansen, J., 2007. Calcium acts as a first messenger through the calcium-sensing receptor in the cardiovascular system. *Cardiovascular Research*. 75: 457–467.

van Mierlo, L. A., Arends, L. R., Streppel, M. T., Zeegers, M. P., Kok, F. J., Grobbee, D. E. and Geleijnse, J. M., 2006. Blood pressure response to calcium supplementation: a meta-analysis of randomized controlled trials. *Journal of Human Hypertension*. 20: 571–580.

Wang, L., Manson, J. E., Buring, J. E., Lee, I. M. and Sesso, H. D., 2008. Dietary intake of dairy products, calcium, and vitamin D and the risk of hypertension in middle-aged and older women. *Hypertension*. 51: 1073–1079.

Witteman, J. C., Willett, W. C., Stampfer, M. J., Colditz, G. A., Sacks, F. M., Speizer, F. E., Rosner, B. and Hennekens, C. H., 1989. A prospective study of nutritional factors and hypertension among US women. *Circulation*. 80: 1320–1327.

Ziegelstein, R. C., Xiong, Y., He, C. and Hu, Q., 2006. Expression of a functional extracellular calcium-sensing receptor in human aortic endothelial cells. *Biochemical and Biophysical Research Communications*. 342: 153–163.

CHAPTER 31

Interaction of Dietary Calcium with Genes of Transporters, Receptors and Enzymes Involved in Cholesterol Metabolism

XIAOBO WANG[a], LIN LEI[a], YUWEI LIU[a], KA YING MA[a], JINGNAN CHEN[b], RUI JIAO[c], YU HUANG[d], AND ZHEN-YU CHEN*[a]

[a]School of Life Sciences, The Chinese University of Hong Kong, Shatin, NT, Hong Kong, China; [b]Lipids Technology and Engineering, School of Food Science and Engineering, Henan University of Technology, Lianhua Road, Zhengzhou, Henan, China; [c]Deparment of Food Science and Engineering, Jinan University, Guangzhou, China; [d]School of Biomedical Sciences, Chinese University of Hong Kong, Shatin, NT, Hong Kong, China
*E-mail: zhenyuchen@cuhk.edu.hk

31.1 Introduction

It has been demonstrated that calcium supplementation decreases plasma cholesterol concentrations in both humans and animals. Jacqmain *et al.* (2003) examined the association of daily calcium intake with plasma total cholesterol (TC), low-density lipoprotein (LDL) cholesterol and high-density

Food and Nutritional Components in Focus No. 10
Calcium: Chemistry, Analysis, Function and Effects
Edited by Victor R. Preedy

Published by the Royal Society of Chemistry, www.rsc.org

lipoprotein (HDL) cholesterol in both men and women, finding that daily calcium intake was negatively correlated with plasma TC, LDL cholesterol, and TC : HDL cholesterol ratio. A similar observation was made in a study on menopausal women given a calcium supplement, finding that calcium supplement increased HDL-C concentration and decreased LDL-C : HDL-C ratio (Reid *et al.*, 2002). It appears that dietary calcium is more effective in reducing plasma cholesterol in women than in men (Jacqmain *et al.*, 2003; Ditscheid *et al.*, 2005). In a placebo-controlled, double-blind, crossover study with a supplement of calcium phosphate (1 g), calcium supplement increased the fecal excretion of bile acids. When supplemented together with vitamin D, calcium has been shown to enhance the beneficial effect of body-weight loss on the lipid and lipoprotein cholesterol profile in overweight and obese women with usual low daily calcium intake (Major *et al.*, 2007). In rabbits, a calcium-deficient diet had raised serum TC and LDL-cholesterol concentrations 2-fold higher than a high-Ca diet (Hsu and Culley, 2006). In hamsters, addition of calcium into regular milk fat lowered the LDL cholesterol : HDL cholesterol ratio (Pellizzon *et al.*, 2005). All the evidence suggests that calcium in diet favorably affects blood lipoprotein profile. This chapter summarizes the findings of recent studies on the underlying mechanisms by which dietary calcium reduces plasma TC and interacts with the genes of transporters, receptors and enzymes involved in cholesterol absorption and metabolism.

31.2 Cholesterol Absorption

The mechanism by which enterocytes absorb cholesterol is only partially known. Evidence suggests that cholesterol absorption through a simple passive diffusion into the membrane of enterocytes is very low. Instead, strong evidence indicates that cholesterol is absorbed *via* a transporter-facilitated mechanism (Figure 31.1). It is estimated that a total 1200–1700 mg of free cholesterol and cholesteryl esters (CE) enter the lumen of the small intestine with 300–500 mg coming from the diet and the remainder coming from bile each day (Grundy and Metzger, 1972; Grundy, 1993). CE is subjected to the hydrolysis of pancreatic cholesterol esterase (PCE) and also possibly hormone-sensitive lipase (HSL) (Howles *et al.*, 1996; Grober *et al.*, 2003). Only free cholesterol can be absorbed in the intestine.

Recently, a newly discovered sterol transporter namely Niemann-Pick C1 Like Protein 1 (NPC1L1), which is localized in the brush-border membrane of absorptive enterocytes in the small intestine, has been shown to play a major role in cholesterol absorption (Davis and Altmann, 2009). It was evident that NPC1L1 knockout mice decreased intestinal cholesterol absorption by 77% and ezetimibe, a NPC1L1 inhibitor drug, could decrease plasma TC by 20% (Davis *et al.*, 2007). Once cholesterol enters the absorptive enterocytes, acyl coenzyme A: cholesterol acyltransferase 2 (ACAT-2) esterifies cholesterol to produce CE. ACAT2 plays a critical role in cholesterol absorption because it catalyzes the esterification reaction converting free cholesterol into CE, which can then be assembled into the core of chylomicrons (CMs) particles.

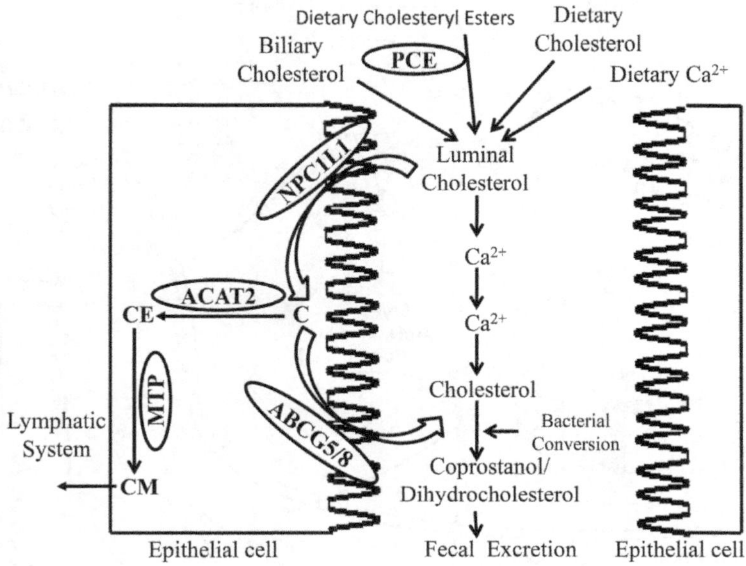

Figure 31.1 Cholesterol-absorption pathway. Cholesterol is from three sources, bile fluid, free cholesterol from diet and hydrolysis of cholesteryl esters (CE) by pancreatic cholesterol esterase. Niemann-Pick C1 like Protein 1 (NPC1L1) is responsible for the uptake of cholesterol. ATP binding cassette transporters (ABCG5 and ABCG8) return the unabsorbed cholesterol to the lumen of the intestine. Cholesterol acyltransferase 2 (ACAT2) esterifies cholesterol to form cholesteryl ester (CE), which is packed with microsomal triacylglycerols protein (MTP) into chylomicrons (CM) and transferred into blood through the lymphatic system. Some of the unabsorbed cholesterol is converted to coprostanol or dihydrocholesterol by colon bacteria.

Microsomal triacylglycerol transport protein (MTP) assembles CE and triacylglycerols (TG) into CMs, which is subsequently transferred into blood through the lymphatic system. Both ACAT-2 and MTP are considered a promising therapeutic target in reducing plasma TC concentration and the risk of heart diseases. Overall cholesterol absorption is around 50–70% in the intestine. In this connection, intestinal ATP binding cassette transporters (ABCG5 and ABCG8) serve to return the unabsorbed cholesterol to the lumen of the intestine for excretion and elimination (Figure 31.1).

31.3 Cholesterol in Circulation

There are five major lipoprotein particles namely CMs, very-low density lipoprotein (VLDL), intermediate-density lipoprotein (IDL), LDL, and HDL in circulation responsible for transporting TG and cholesterol in the aqueous solution of bloodstream. The functions of CMs are to deliver exogenous TG to the muscle and adipose tissue while to carry exogenous cholesterol to the liver (Figure 31.2). In brief, CMs consists of mainly TG, phospholipids (PL),

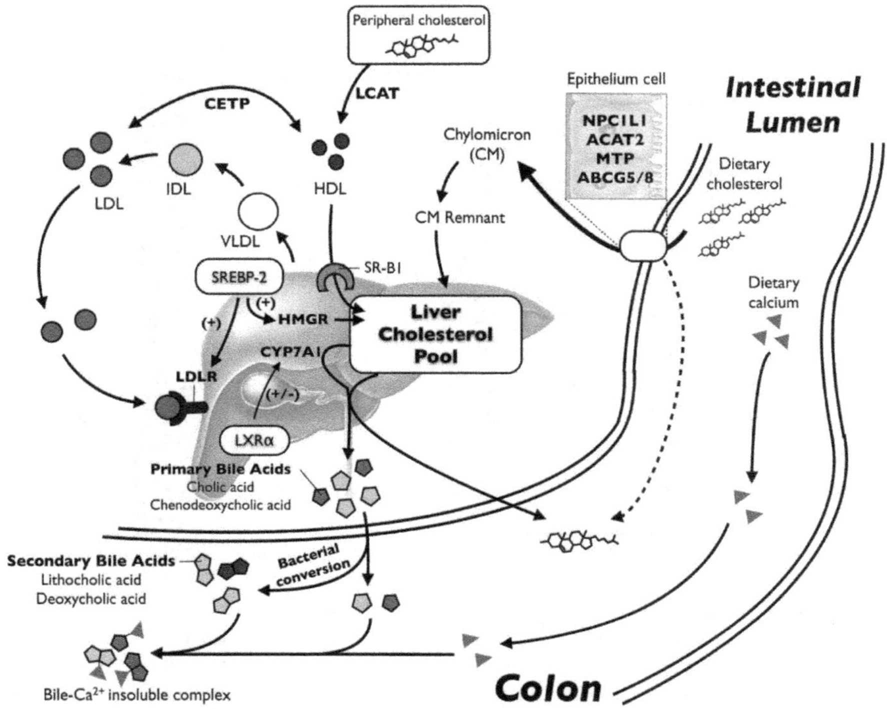

Figure 31.2 Cholesterol homeostasis is governed by coordination of sterol regulatory element binding protein 2 (SREBP2), liver X receptor (LXR), farnesoid X receptor (FXR), cholesteryl ester transport protein (CETP), lecithin-cholesterol acyltransferase (LCAT), LDL receptor (LDL-R), scavenger receptor B class 1 (SR-B1), 3-hydroxy-3-methylglutaryl-CoA reductase (HMGR), and cholesterol-7α-hydroxylase (CYP7A1). SREBP-2 regulates the activation of LDL receptor and HMG-CoA reductase, while LXRα and FXR control the transcription of CYP7A1. LDL receptor mediates the removal of LDL-cholesterol from blood. HMG-CoA reductase is a key enzyme in cholesterol synthesis, while CYP7A1 encoding cholesterol-7α-hydroxylase is a regulatory enzyme in bile-acid synthesis. CETP transfers a cholesteryl ester (CE) from HDL to LDL or VLDL with an exchange of the equivalent triacylglycerol (TG). LCAT esterifies the free cholesterol and removes cholesterol from peripheral tissue. SR-B1, a HDL receptor, mediates the delivery of HDL cholesteryl esters to the liver and steroidogenic organs.

cholesterol, CE and proteins. They circulate to the capillaries of skeletal muscle and adipose tissue where mainly TG is hydrolyzed by lipoprotein lipase, releasing two fatty acids and one monoacylglycerol, followed by using two fatty acids and monoacylglycerol to resynthesize TG stored in adipocytes or muscle cells as energy reserve. The remaining parts of CMs containing cholesterol and CE, called chylomicron remnants, re-enter the circulation and deliver cholesterol and CE to the liver. The function of VLDL is to transport the endogenous TG and cholesterol from the liver to the other organs where they are needed. When VLDL circulates to come into contact with lipoprotein lipase in the capillary beds in the adipose tissue and muscle, lipoprotein lipase will remove TG for energy storage or utilization. The remaining part of VLDL containing cholesterol and CE is called IDL, which becomes LDL after it obtains some CE from HDL *via* cholesteryl ester transport protein (CETP). The function of LDL is to carry cholesterol to other tissues that require additional cholesterol. This delivery of cholesterol to other tissues is mediated by an LDL receptor, which is synthesized and inserted on the plasma membrane of tissues. The LDL particle binds to its receptor and is subjected to endocytosis, delivering cholesterol into the cell. The function of HDL is to mainly remove cholesterol from the cells of other tissues back to the liver for excretion or reuse mediated by scavenger receptor B1 (Figure 31.2).

31.4 Cholesterol Homeostasis

Humans synthesize about 1000 mg cholesterol each day while dietary cholesterol is around 300 mg. To maintain cholesterol homeostasis, three transcriptional factors, sterol regulatory element binding protein-2 (SREBP-2), liver X receptor alpha (LXRα) and farnesoid X receptor (FXR) act in a coordinated manner (Figure 31.2) (Davis *et al.*, 2007). SREBP-2 regulates the activation of the transcription for LDL receptor and 3-hydroxy-3-methylglutaryl-CoA reductase (HMGR), a key enzyme in cholesterogenesis. In contrast, LXRα governs the transcription of CYP7A1 encoding cholesterol-7α-hydroxylase, which is a key enzyme in bile-acid synthesis. Regarding FXR, it is a bile-acid receptor and also controls the bile-acid synthesis because bile acids, reabsorbed *via* the enterohepatic circulation, activate FXR and downregulate CYP7A1 (Figure 31.2) (Lee *et al.*, 2004).

31.5 Effect of Dietary Calcium on Plasma Lipoprotein in Ovariectomized Hamsters

It is known that dietary calcium decreases plasma TC, particularly in menopausal women (Reid *et al.*, 2002; Jacqmain *et al.*, 2003; Ditscheid *et al.*, 2005; Major *et al.*, 2007). To investigate the underlying mechanisms why dietary calcium decreases favorably plasma TC in menopausal women, ovariectomized hamster as an animal model is the choice because the cholesterol metabolism in hamsters is similar or close to that in humans (Zhang *et al.*, 2009). In one of our previous studies, forty-eight ovariectomized hamsters were divided into four groups and fed one of the four diets containing 0, 2, 6 and 8 g calcium per kg (Ma *et al.*, 2011). Accompanied by a dose-dependent increase in plasma calcium concentrations, plasma TC, TG, and LDL-C were decreased, whereas HDL-C was increased with the increasing dietary calcium in a dose-dependent manner (Figure 31.3). It was also noticed that hepatic cholesterol was also dose-dependently decreased with the increasing dietary calcium (Ma *et al.*, 2011). Data from this study on ovariectomized hamsters confirm the observation in menopausal women that calcium supplementation favorably modulates the lipoprotein profile.

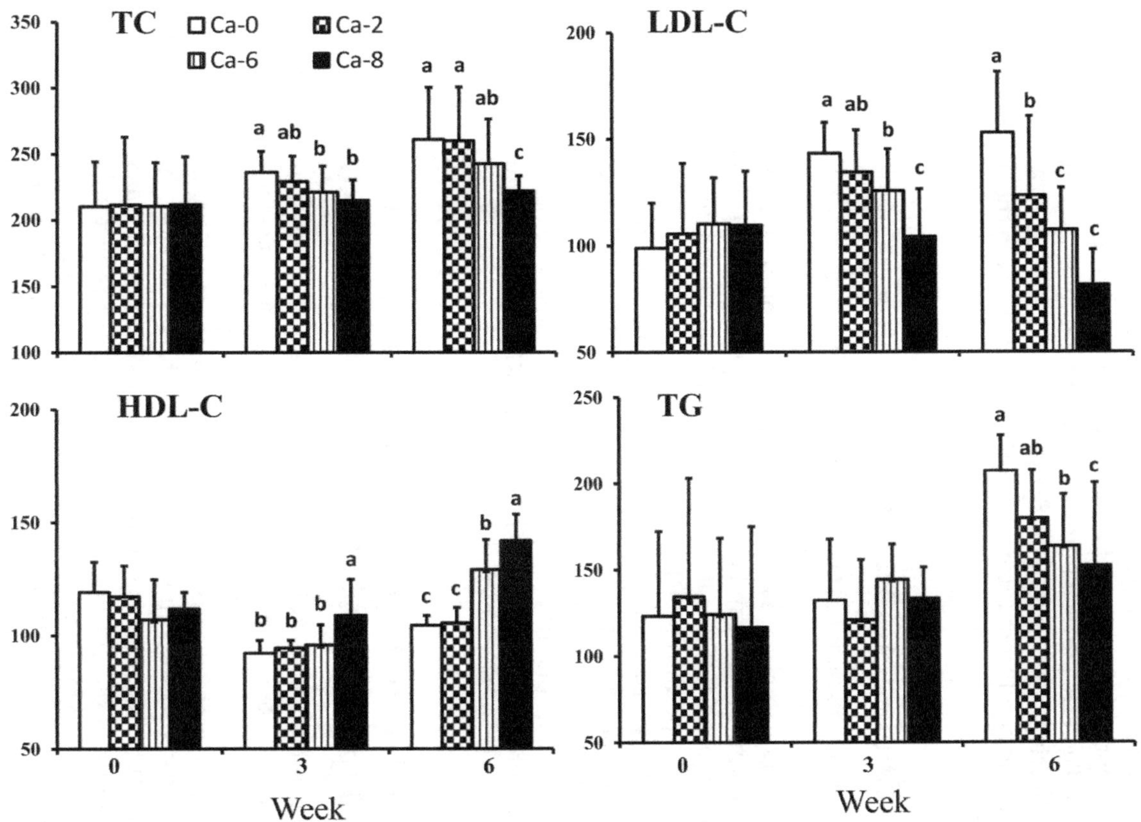

Figure 31.3 Changes in plasma total cholesterol (TC), low-density lipoprotein cholesterol (LDL-C), high-density lipoprotein cholesterol (HDL-C) and total triacylglycerols (TG) in ovariectomized hamsters fed a Ca deficient diet (Ca-0) and three experimental diets supplemented with 2 g (Ca-2), 6 g (Ca-6) and 8 g (Ca-8) Ca per kg diet (source: modified from Ma *et al.*, 2011).

31.6 Interaction of Dietary Calcium with Genes of Transporters, Receptors and Enzymes Involved in Cholesterol Metabolism

Despite extensive research on the link between dietary calcium and plasma lipoprotein profile, the underlying mechanism is only partially known. We have examined the effect of dietary calcium on the gene expression of intestinal NPC1L1, ABCG 5 and 8, ACAT2 and MTP as well as hepatic SREBP-2, LXRα, HMGR, LDL receptor, and CYP7A1 (Ma *et al.*, 2011). RT-PCR analysis found that dietary calcium could upregulate the production of mRNA ABCG5/8 and downregulate the mRNA MTP and NPC1L1 (Figure 31.4). While it had no effect on the mRNA ACAT-2, calcium supplementation suppressed intestinal ACAT activity (Ma *et al.*, 2011). Accompanied with these changes, dietary calcium increased the fecal excretion of total neutral sterols, suggesting calcium supplementation could decrease the cholesterol absorption (Figure 31.5).

Regarding its effect on cholesterol homeostasis, calcium supplementation could upregulate hepatic mRNA CYP7A1 in a dose-dependent manner, while it had no effect in mRNA abundance of SREBP-2, HMGR, and LDL receptor (Figure 31.4). Upregulation of CYP7A1 was accompanied by greater excretion

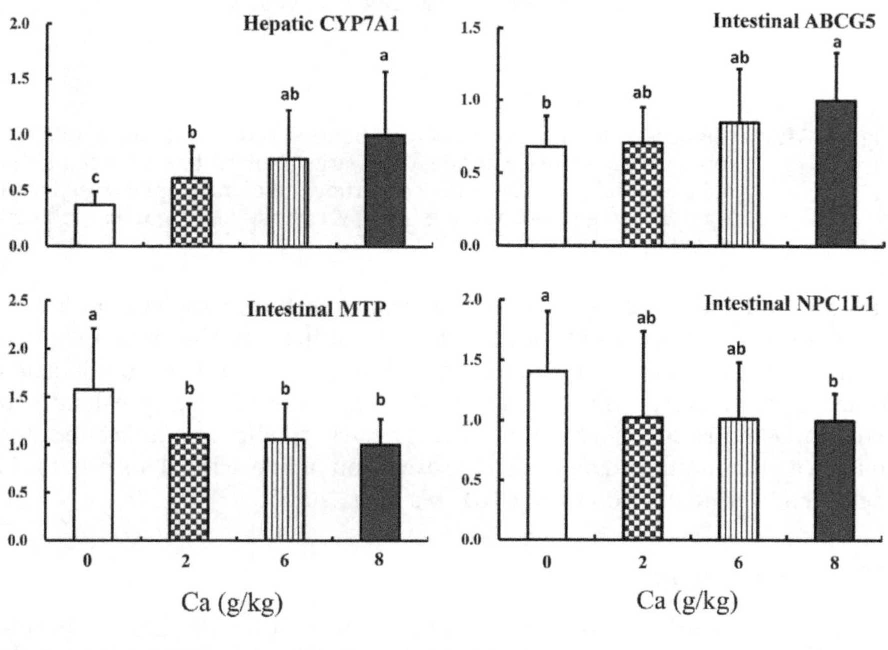

Figure 31.4 Effect of calcium supplementation on mRNA levels of hepatic cholesterol-7α-hydroxylase (CYP7A1) and intestinal Niemann-Pick C1 like 1 (NPC1L1), microsomal triacylglycerol transport protein (MTP), and ATP binding cassette transporter 5 (ABCG5) in ovariectomized hamsters fed a Ca deficient diet (Ca-0) and three experimental diets supplemented with 2 g (Ca-2), 6 g (Ca-6) and 8 g (Ca-8) Ca per kg diet (source: modified from Ma *et al.*, 2011).

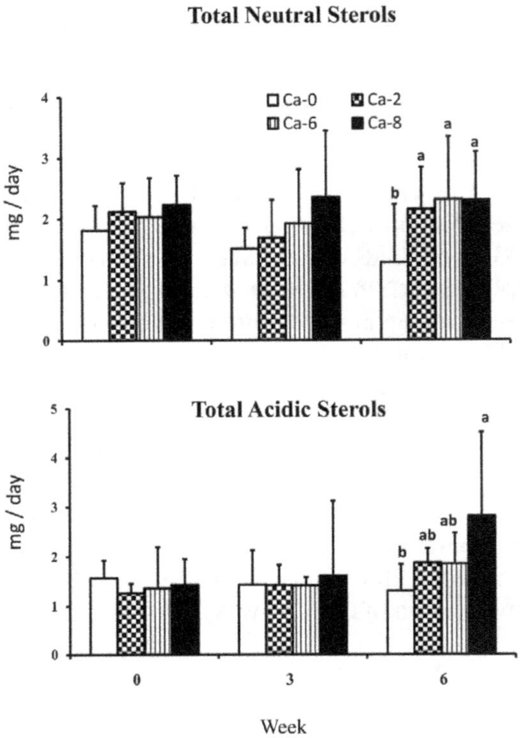

Figure 31.5 Changes in fecal excretion of total neutral sterols and total acidic sterols (mg per hamster per day) in ovariectomized hamsters fed a Ca-deficient diet (Ca-0) and three experimental diets supplemented with 2 g (Ca-2), 6 g (Ca-6) and 8 g (Ca-8) Ca per kg diet (source: modified from Ma *et al.*, 2011).

of fecal total acidic sterols with the increasing calcium supplement. Excessive cholesterol from either diet or *de novo* synthesis is eliminated *via* two mechanisms, namely incorporation in bile fluid or conversion to bile acids followed by excretion into feces. The driving force of dietary calcium to reduce plasma TC level starts in the intestine, where calcium binds bile acids to form a precipitate and increases the excretion of bile acids. This process is mediated by upregulation of CYP7A1 (Ma *et al.*, 2011).

31.7 Conclusion

Evidence has clearly demonstrated that calcium supplementation favorably modifies plasma lipoprotein profile by decreasing plasma TC in humans, particularly menopausal women. A study on ovariectomized hamsters has found that calcium supplements decrease plasma TC, LDL cholesterol, and TC/HDL cholesterol ratio. These changes are mediated by inhibition on intestinal ACAT activity, upregulation of mRNA CYP7A1, downregulation of mRNA NPC1L1 and MTP as well as upregulation of intestinal mRNA ABCG 5.

Summary Points

- Evidence shows dietary calcium is inversely related to plasma cholesterol.
- This chapter discusses why dietary calcium decreases plasma cholesterol.
- Calcium increases the bile-acid excretion due to its binding interaction with bile acids in the intestine.
- Dietary calcium upregulates the gene expression of cholesterol-7α-hydroxylase (CYP7A1), a key enzyme in converting cholesterol to bile acids in the liver.
- Dietary calcium also increases the cholesterol excretion, mediated by downregulation of the gene expression of intestinal Niemann-Pick C1 like protein 1 (NPC1L1) and acyl coenzyme A: cholesterol acyltransferase 2 (ACAT-2), both of which are responsible for cholesterol absorption.
- Two mechanisms by which dietary calcium decreases plasma cholesterol are (1) to decrease the cholesterol absorption and (2) to increase the excretion of bile acids.

Key Facts of Cholesterol Absorption and Elimination

1. Cholesterol is an essential compound for humans. It is used to (1) synthesize hormones, oestrogen and androgen; (2) synthesize vitamin D; (3) synthesize the bile acids for fat absorption; and (4) maintain the integrity of cell membranes. However, high cholesterol in blood can cause artery walls to thicken (atherosclerosis) as a result of the accumulation of fatty materials in the inner surface of the arteries.
2. Humans daily consume about 400 mg cholesterol and synthesize about 1000 mg cholesterol. The synthesis of cholesterol mainly occurs in the liver. Low-density lipoprotein (LDL) carries cholesterol from the liver to peripheral tissues where cholesterol is needed, while high-density lipoprotein (HDL) carries excessive cholesterol from the peripheral tissues back to the liver for elimination. LDL cholesterol is called bad cholesterol, while HDL cholesterol is called good cholesterol.
3. Dietary cholesterol increases plasma cholesterol. Cholesterol absorption is a very complex process and requires several proteins, enzymes and transporters NPC1L1, ACAT 2, MTP and ABCG 5 and 8.
4. Cholesterol in the liver is converted to bile acids, which are transported to the intestine and needed for fat absorption. The conversion is also the major way to eliminate the excessive cholesterol in the body.
5. Dietary calcium increases the excretion of bile acids and decreases the cholesterol absorption, leading to reduction in plasma cholesterol.

Definition of Words and Terms

ABCG5 and ABCG8. These are abbreviations for ATP binding cassette transporters. They serve to return the unabsorbed cholesterol from enterocytes to the lumen of the intestine for excretion and elimination.

ACAT2. This is an abbreviation for Acyl-CoA: cholesterol acyltransferase 2, which esterifies cholesterol to produce cholesteryl ester (CE) and plays a critical role in cholesterol absorption.

CYP7A1. This is an abbreviation for cholesterol-7α-hydroxylase, which a key enzyme in converting cholesterol to bile acids in the liver.

HDL. This is an abbreviation for the high-density lipoprotein, which is responsible for removing excessive cholesterol from the peripheral tissues back to the liver.

LDL. This is an abbreviation for low-density lipoprotein, which is responsible for carrying cholesterol from the liver to other peripheral tissues.

MTP. This is an abbreviation for microsomal triacylglycerol transport protein that packs cholesteryl esters and triacylglycerols (TG) into chylomicrons (CM) before absorption.

NPC1L1. This is an abbreviation for a sterol Niemann-Pick C1 like 1 transporter, which is responsible for transporting sterols from the lumen of the intestine to the enterocytes.

List of Abbreviations

ABCG5/8	ATP-binding cassette transporters subfamily G member 5 and 8
ACAT 2	Acyl-CoA: cholesterol acyltransferase 2
C	Cholesterol
CE	Cholesteryl ester
CM	Chylomicrons
HDL-C	High-density lipoprotein cholesterol
LDL-C	Low-density lipoprotein cholesterol
MTP	Microsomal triacylglycerol transport protein
LDL-C	Nonhigh-density lipoprotein cholesterol
NPC1L1	Niemann-Pick C1 like 1
PCE	Pancreatic cholesterol esterase
TC	Total cholesterol
TG	Triacylglycerols

Acknowledgement

This project was supported by Hong Kong GRF grants (Project Number CUHK 461112 and 462813).

References

Davis Jr, H. R., Hoos, L. M., Tetzloff, G., Maguire, M., Zhu, L. J., Graziano, M. P. and Altmann, S. W., 2007. Deficiency of Niemann-Pick C1 like 1 prevents atherosclerosis in ApoE−/− mice. *Arteriosclerosis, Thrombosis, and Vascular Biology*. 27: 841–8499.

Davis Jr, H. R. and Altmann, S. W., 2009. Niemann-Pick C1 like 1 (NPC1L1) an intestinal sterol transporter. *Biochimica et Biophysica Acta*. 1791: 679–683.

Ditscheid, B., Keller, S. and Jahreis, G., 2005. Cholesterol metabolism is affected by calcium phosphate supplementation in humans. *Journal of Nutrition*. 135: 1678–1682.

Grober, J., Lucas, S., Sorhede-Winsell, M., Zaghini, I., Mairal, A., Contreras, J. A., Besnard, P., Holm, C. and Langin, D., 2003. Hormone-sensitive lipase is a cholesterol esterase of the intestinal mucosa. *Journal of Biological Chemistry*. 278: 6510–6515.

Grundy, S. M., 1993. Absorption and metabolism of dietary cholesterol. *Annual Review of Nutrition*. 3: 71–96.

Grundy, S. M. and Metzger, A. L., 1972. A physiological method for estimation of hepatic secretion of biliary lipids in man. *Gastroenterology*. 62: 1200–1217.

Howles, P. N., Carter, C. P. and Hui, D. Y., 1996. Dietary free and esterified cholesterol absorption in cholesterol esterase (bile salt-stimulated lipase) gene-targeted mice. *Journal of Biological Chemistry*. 271: 7196–7202.

Hsu, H. H. T. and Culley, N. C., 2006. Effects of dietary calcium on atherosclerosis, aortic calcification, and icterus in rabbits fed a supplemental cholesterol diet. *Lipids in Health and Disease*. 5: 16.

Jacqmain, M., Doucet, E., Despres, J. P., Bouchard, C. and Tremblay, A., 2003. Calcium intake, body composition, lipoprotein–lipid concentration in adults. *American Journal of Clinical Nutrition*. 77: 1448–1452.

Lee, J. Y., Mitmesser, S. H. and Carr, T. P., 2004. Regulation of cellular cholesterol. In: Zempleni, J. and Daniel, H. (ed.) *Molecular Nutrition*. CABI Publishing, Wallingford, Oxon, UK, pp. 309–319.

Ma, K. Y., Yang, N., Jiao, R., Peng, C., Guan, L., Huang, Y. and Chen, Z. Y., 2011. Dietary calcium decreases plasma cholesterol by down-regulation of intestinal NPC1L1 and MTP and up-regulation of CYP7A1 and ABCG 5/8 in hamsters. *Molecular Nutrition & Food Research*. 55: 247–258.

Major, G. C., Alarie, F., Dore, J., Phouttama, S. and Tremblay, A., 2007. Supplementation with calcium + D enhances the beneficial effect of weight loss on plasma lipid and lipoprotein concentrations. *American Journal of Clinical Nutrition*. 85: 54–59.

Pellizzon, M., Ana, J. S., Buison, E., Martin, J., Buison, A. and Jen, K. L., 2005. Effect of a modified milk fat and calcium on cholesterol metabolism in hamsters. *Lipids*. 39: 441–448.

Reid, I. R., Mason, B., Horne, A., Ames, R., Clearwater, J., Bava, U., Orr-Walker, B., Wu, F., Evans, M. C. and Gamble, G. D., 2002. Effects of calcium supplementation on serum lipid concentration in normal older women: a randomized controlled trial. *American Journal of Medicine*. 112: 343–347.

Zhang, Z. S., Wang, H., Jiao, R., Peng, C., Wong, Y. M., Yeung, V. S. Y., Huang, Y. and Chen, Z. Y., 2009. Choosing hamsters but not rats as a model for studying plasma cholesterol-lowering activity of functional foods. *Molecular Nutrition & Food Research*. 53: 921–930.

CHAPTER 32

Bone Health: The Independent and Combined Effects of Calcium, Vitamin D and Exercise in Children and Adolescents

DIMITRIS VLACHOPOULOS*[a], LUIS GRACIA-MARCO[a], ALAN R. BARKER[a], INGE HUYBRECHTS[b], LUIS A. MORENO[c], AND THEODORA MOURATIDOU[c]

[a]Children's Health and Exercise Research Centre, Sport and Health Sciences, University of Exeter, Heavitree Road, EX1 2LU, Exeter, UK; [b]Dietary Exposure Assessment Group, International Agency for Research on Cancer, 150 Cours Albert Thomas, F-69372 Lyon Cedex 08, France; [c]Growth, Exercise, Nutrition and Development (GENUD) Research Group, University of Zaragoza, c/ Pedro Cerbuna 12, 50009 Zaragoza, Spain
*E-mail: dv231@exeter.ac.uk

32.1 Introduction

Bone-mass acquisition during growth is a key determinant of osteoporosis risk later in life (Marshall *et al.*, 1996). During the last two decades, increasing evidence suggests that young population groups fail to attain adequate

Food and Nutritional Components in Focus No. 10
Calcium: Chemistry, Analysis, Function and Effects
Edited by Victor R. Preedy

Published by the Royal Society of Chemistry, www.rsc.org

bone mass; this is alarming given the raised concerns that inadequate bone mass may result in fragile bones and increased fracture risk both in adolescence and in adulthood (Rizzoli *et al.*, 2010). Childhood and adolescence are considered crucial periods of bone-mass acquisition, with up to 50% of total body (TB) bone mass achieved during these life stages (Perez-Lopez *et al.*, 2010). Peak bone mass (PBM) attainment typically occurs between the second and third decade of life, with 80–90% acquired by late adolescence (second decade of life), although this is skeletal site dependent (Baxter-Jones *et al.*, 2011). While up to 80% of bone mass is determined by genetics (Davies *et al.*, 2005), lifestyle factors, such as diet (*i.e.* calcium, vitamin D) and physical activity (PA) have been shown to contribute to skeletal development.

Dietary calcium is an essential component of bone formation in young people as it influences the development of the skeleton during growth. For this reason, adequate calcium levels are necessary for the optimal attainment of PBM (Huncharek *et al.*, 2008). Increasing dietary calcium to optimal levels through consumption of dairy products (fortified or not) may increase bone mass during growth, but the magnitude and the maintenance of the gain in bone mass remain controversial (Huybrechts *et al.*, 2011). Another substantial nutrient in bone mineralization is vitamin D, which plays an important role in calcium absorption from the skeleton and in the attainment of PBM. In addition, vitamin D deficiency may be linked to rickets in children (Pettifor and Prentice, 2011). Therefore, it is important to fully understand the impact of vitamin D on bone development and its association with enhanced bone mass in children and adolescents.

PA contributes to the development of bone mass in young people due to its association with increases in lean mass as explained by the mechanostat theory "*bigger muscles exert higher tensile forces on the bones they attach*" (Rauch *et al.*, 2004). The importance of PA during childhood is dependent on the ability of the skeleton to adapt to mechanical loading after exercise and appears to elicit the greatest bone-accrual response in the growing skeleton (Bielemann *et al.*, 2013). The intensity, frequency and the type of PA are important factors to consider in order to understand the impact of PA on the paediatric skeleton; due to the ground reaction forces applied at different sites (Vicente-Rodriguez, 2006). In addition, specific PA levels associated with optimal bone health have been published (Gracia-Marco *et al.*, 2011).

Calcium, vitamin D and PA have been shown to independently influence bone-mineral accrual in young people. However, there is evidence suggesting that exercise interacts with calcium intake (Courteix *et al.*, 2005) and vitamin D levels (Valtuena *et al.*, 2012) to improve bone status. Understanding these interactions during childhood and adolescence is a public health priority. Therefore, this chapter will discuss the independent and interactive effects of calcium, vitamin D and exercise on bone status in children and adolescents (Figure 32.1).

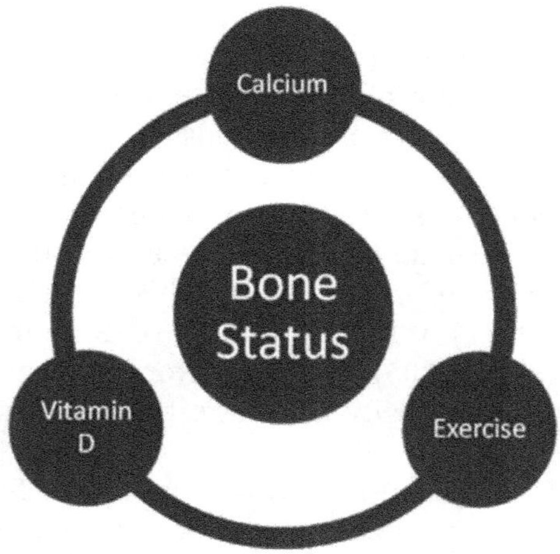

Figure 32.1 Combined effects of calcium, vitamin D and exercise on bone status in children and adolescents. The association of the nutritional factors (calcium and vitamin D) and exercise in bone development.

32.2 Bone Mass Increase during Childhood and Adolescence

Childhood and adolescence is characterized by profound changes in the morphology of the skeleton and bone development determined through the dynamic process of bone formation and resorption. Skeletal growth depends, among other factors, on bone remodeling from osteoblast (formation) and osteoclast (resorption) cells where the ratio of the activity of these cells determines the rate of bone mineral accrual. During the period of bone growth the activity of osteoblasts is higher than the activity of osteoclasts until the time of PBM (Deng *et al.*, 2008). Several other factors are shown to contribute to differences in bone status in children and adolescents such as genetics, sexual maturation, body composition, nutrition and PA.

Pubertal status is associated with bone growth both in size and density, and evidence shows that boys and girls acquire bone mass at similar rates before puberty, whereas after puberty males tend to acquire a higher bone mass than females (Davies *et al.*, 2005). Findings from a longitudinal study monitoring bone mass suggest that the highest rate of bone-mass accrual in females and males occurs at 12–15 years and 14–16 years, respectively. In the same study, bone mass increase rate decreased by the age of 16–18 years in girls and 17–20 years in boys (Bailey *et al.*, 1999). During the years of puberty, girls acquired ~40% of their PBM and ~90% by the age of 18 (Theintz *et al.*, 1992). Maturation onset is another important component of bone-mass acquisition due to the relationship between the pubertal timing and the PBM accrual. The Saskatchewan Paediatric Bone Mineral Accrual Study reported that PBM was achieved 6 months later than peak height

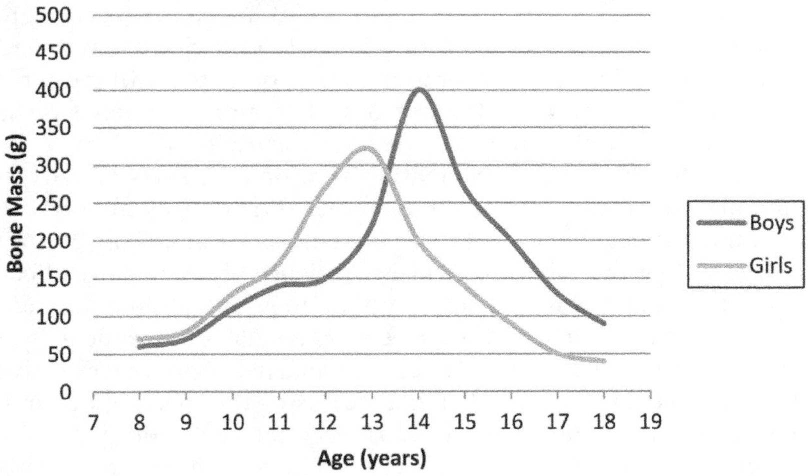

Figure 32.2 Increases in bone mass during childhood and adolescence. Bone mass increases by chronological age are shown for males and females from childhood to adulthood. Peak bone mass increase values achieved ~2 years earlier for girls compared with boys. Adapted from: (Bailey *et al.*, 1999).

velocity (PHV), where boys and girls obtained 90% of total height at PHV but only 57% of total BMC (Bailey *et al.*, 1999). The inverse association of PHV and PBM accrual showed that children who reach PHV earlier tend to have higher BMC than those with a late PHV. In addition, children with a higher stature have greater BMC and BMD z-scores than shorter peers (Zemel, 2013) (Figure 32.2).

32.3 Calcium, Vitamin D and Bone Status in Children and Adolescents

The development of the human skeleton requires an adequate supply of many nutrients, including calcium. Ninety-nine per cent of calcium is concentrated in bone and explains 1–5% of the variance in childhood bone mass and later in PBM attainment (Levine, 2012). Adequate calcium levels are defined as 800 mg per day for children aged 4–8 years old and 1300 mg per day for adolescents (Baker *et al.*, 1999). Adequate calcium levels are crucial during the pubertal growth spurt due to their relationship with peak bone acquisition. Peak calcium accretion occurs during puberty at ~12.5 years in females and ~14 years in males (Bailey *et al.*, 2000).

Increasing dietary calcium to optimal levels through diet rich in calcium and dairy products promotes bone growth in the skeleton. A double-blind placebo-controlled trial in 149 girls aged 7.9 years has shown that 850 mg per day of calcium coming from fortified foods, taken for one year, significantly increased BMD at the femoral and radial sites by 7–12 mg cm^{-2} per year and by 2 mg cm^{-2} per year in the lumbar spine (LS) compared to a placebo group. Furthermore, a 7 year longitudinal study in adolescent girls showed that calcium supplementation and milk products were correlated with greater BMC on the hip and forearm sites. However, only dairy products were associated with an increase of BMD in the spine (Matkovic *et al.*, 2004).

In contrast to the above, a systematic review on the effects of dairy products and dietary calcium on bone mass revealed that only 9 out of 27 RCTs reported modest benefits in children's bone mass ranging from 1% to 6%. This suggests that there is no conclusive evidence of the positive effects of dairy products and dietary calcium on bone mass in children and adolescents (Lanou *et al.*, 2005). A more recent meta-analysis included 21 RCTs (study duration 1–2 years) with 3821 participants (83% girls) aged between 4 and 17 years. The findings revealed that increasing the consumption of milk products and calcium intake improved TB BMC by 2 g. In addition, children with calcium intakes below 800 mg per day had a greater increase in BMC (49 g), and vitamin D added to the milk supplement improved BMC at the LS by 35 g. Calcium/milk supplement groups alone showed a nonsignificant improvement of 2.1 g in TB BMC compared to control groups (Huncharek *et al.*, 2008).

Vitamin D plays an important role in calcium homeostasis and is an essential nutrient for the bone mineralization process. This is done via its participation in the absorption of calcium and phosphorous in the skeleton through calcification of the growth plate and mineralization of osteoid on trabecular and cortical bone surfaces. The natural source of vitamin D in the body is sunlight accounting for 95%, with the remaining 5% being dependent on dietary intake. Vitamin D appears in the body in the form of vitamin D_2 and D_3 and is then converted in the liver into 25-hydroxyvitamin D (25(OH)D); the latter is the most appropriate biochemical marker to assess vitamin D availability by measuring serum 25(OH)D concentration in the blood (Perez-Lopez *et al.*, 2010). Table 32.1 shows the recommended dietary intakes of vitamin D for children and adolescents aged 4–18 years (15 μg per day).

Table 32.1 Dietary reference intakes for vitamin D in children and adolescents across the world. Childhood is defined as the age between 4 and 8 years and adolescence as the age between 9 and 18 years. Adapted from: (Ross, 2011).

	Vitamin D (mg per day)		
Age group	Estimated average requirement	Recommended dietary allowance	Upper level intake
Childhood (4–8 years)			
	10	15	75
Adolescence (8–18 years)			
Boys	10	15	100
Girls	10	15	100

Clinical vitamin D (in)sufficiency in young people includes the following categories: *severe deficiency* of vitamin D when 25(OH)D levels are below 27.5 nmol L^{-1}, *deficiency* between 27.5 nmol L^{-1} and 50 nmol L^{-1}, *insufficiency* between 50 nmol L^{-1} and 75 nmol L^{-1} and *sufficiency* above 75 nmol L^{-1} (Gonzalez-Gross *et al.*, 2012). It should be noted that the definition of sufficient 25(OH)D levels in adolescents is controversial and therefore predefined cut-offs are used to define optimal levels ≥75 mmol L^{-1} and insufficiency <75 mmol L^{-1} (Valtuena *et al.*, 2012). Depending on the magnitude and the skeletal site, deficiency in vitamin D may lead to rickets and increase fragility by reducing the absorption of calcium in the skeleton (Pettifor, 2005). Insufficient vitamin D status has been observed in 1006 European adolescents from 9 different countries, aged 12.5 to 17.5 participating in the Healthy Lifestyle in the Europe by Nutrition in Adolescence study. After adjusting by age, gender and weight status, it was revealed that 15% of adolescents were severely deficient, 27% were deficient and 39% insufficient for their 25(OH)D levels (Gonzalez-Gross *et al.*, 2012). The daily dose of vitamin D in children and adolescence is a matter of debate, indicating that optimal vitamin D levels should be maintained throughout the year.

32.4 Impact of Physical Activity on Bone Status during Growth

32.4.1 The Role of Physical Activity on Bone Mass

PA is also considered an important determinant of the accrual and maintenance of bone mass. Weight-bearing PA can increase bone mass due to the ability of the skeleton to adapt to the loads under which it is placed. According to Wolf's law, when the loading on a particular bone increases, the bone remodels itself over time to become stronger to resist that loading (Frost, 1994). Based on the previous theory, studies in animals and humans have shown that the growing skeleton adapts to the mechanical stimuli to a greater extent than mature bone due to a higher osteoblast activity and bone formation (Robling *et al.*, 2002). Thus, engaging in PA is particularly important during childhood and adolescence.

A review of 19 cohort studies from childhood to adulthood highlighted the importance of PA during the growth period and its link with bone mass later in life. The positive association was higher in males than in females at the femoral neck (FN) and LS sites (Bielemann *et al.*, 2013). The results of the six-year Saskatchewan Pediatric Bone Mineral Accrual study in adolescent boys and girls showed that both girls and boys classified in the highest PA quartile acquired 17% and 9% higher TB BMC, respectively compared with their inactive peers one year after peak BMC velocity. At the LS site, active adolescents displayed 18% higher BMC (adjusted for weight and height) one year after peak BMC velocity. In addition, there was a difference of 7% and 11% in active and inactive boys and girls, respectively, at the FN (Bailey *et al.*, 1999).

Table 32.2 Physical activity and bone health during adolescence. Adjusted bone mineral content by height and weight covariates; significant difference set at $p < 0.05$. Adapted from: (Baxter-Jones *et al.*, 2008).

Significant differences in bone mineral content (g) at total body, lumbar spine, total hip and femoral neck at 3 activity groups			
Gender (age at peak height velocity)	Inactive	Average	Active
Males (14.1)	Compared to active group: total body, lumbar spine and total hip	Compared to active group: total body and total hip	Compared to both groups: total body and total hip
Females (11.6)	Compared to active group: total body and lumbar spine	Compared to active group: lumbar spine	Compared to both groups: lumbar spine

In a 7 year study, including 72 males and 82 females, the authors examined whether the positive effects of PA on BMC acquired in adolescence preserved into the second and third decade of life. It was reported that both active males and females during adolescence remained active during adulthood. Furthermore, active boys showed 8–10% greater adjusted BMC for weight, height, calcium intake, PA and maturation status at TB, FN and total hip (TH) in adulthood. In addition, active females had 9–10% higher adjusted BMC at TH and FN in young adulthood (Baxter-Jones *et al.*, 2008). These results highlight the importance of PA for bone accrual during childhood and adolescence and its relation with bone status in adulthood. Therefore, it is important to understand the type of exercise and the optimal "dose" (*i.e.* frequency, intensity, duration) that is required to maximize increases in BMC and BMD during childhood and adolescence (Table 32.2).

32.4.2 Weight-Bearing Exercise and Bone Mass

Weight-bearing exercise such as jumping, running and weight training can enhance bone mass due to the high ground reaction forces (GRF) applied in the developing skeleton of children and adolescents. Evidence supports the beneficial effects of weight-bearing activities in bone accrual by applying GRF of 3.5 to 5 times body weight, which represents a moderate exercise intensity intervention (McKay *et al.*, 2005). The majority of the evidence so far comes from studies conducted in a school-based environment and includ jumping exercises to improve bone health at different stages of growth.

Findings from a 4 year longitudinal study of 205 prepubertal children aged 10 ± 1 years suggested that a 7 month jumping intervention significantly increased the adjusted BMC for age, sexual maturation and tissue mass by 7.9% at LS, 8.4% at TH, 7.7% at FN and 7.3% at TB. Three years after the end of the intervention the effects decreased but remained significant at the above sites accounting for 2.3%, 3.2%, 4.4% and 2.9% higher BMC, respectively (Gunter *et al.*, 2008). A review of RCTs showed that weight-bearing activities (*i.e.* games, dance, resistance training and jumping exercise) eliciting GRF of 2–9 times body weight can effectively improve bone mass by 0.9–4.9% in prepubertal children, 1.1–5.5% in early pubertal and 0.3–1.9% in pubertal children (Hind and Burrows, 2007). These findings are in agreement with a recent meta-analysis of 27 studies addressing the impact of weight-bearing activities (*i.e.* body weight, resistance training machines or both) on BMC and BMD of 2985 boys and girls aged 10.3 ± 2.7 years. The results revealed trivial effect sizes of 0.17 (0.05–0.29) for BMC and 0.26 (0.02–0.49) for BMD. Interestingly, calcium intake explained 35% of the variance and the changes in BMC were greater during the prepubertal years (Tanner stage 1) when compared with mid- and late pubertal years (Tanner stages 2–5) (Behringer *et al.*, 2014). A meta-analysis of school-based exercise interventions, such as jumping, aerobic and circuit training in children and adolescents confirmed the small but beneficial effect of exercise on BMC and added the negative effect on fat mass. Effect sizes presented that BMC improved at LS by 0.38 (standardized mean difference), at FN by 0.29, at TB by 0.48 and fat mass decreased by 0.25 (Nogueira *et al.*, 2014).

These studies show the positive impact of weight-bearing activities, mainly jumping interventions, on BMC and BMD but, it should be noted that the optimal type, intensity and duration of exercise is not clear yet (Bailey and Brooke-Wavell, 2008). Most of the studies involved exercise interventions lasting from 3 to 20 min per day with a frequency of 3 times per week. This is in agreement with the findings of a recent meta-analysis in girls after controlling for the important confounding factors of maturation stage, exercise mode, intervention strategy, duration and frequency of exercise, design and duration of the study (Ishikawa *et al.*, 2013).

The intensity of exercise may contribute to the changes of BMC and BMD among children and adolescents, and there is evidence to suggest that greater differences are observed when interventions include participation in different sports (Boreham and McKay, 2011). The positive long-term effects of sport participation on BMD were demonstrated in a 27 year follow-up study of 154 Belgian boys aged 13 at baseline. Engagement in high impact sports (GRF > 4 times body weight) during adolescence and adulthood resulted in significantly higher BMD (1.12 g cm^{-2}) at LS compared to participants that stopped high impact sports in adulthood (BMD = 1.01 g cm^{-2}) or did not engaged in high-impact sports during adolescence and in adulthood (BMD = 0.99 g cm^{-2}) (Van Langendonck *et al.*, 2003).

In contrast, other sports, such as swimming and cycling, may not have an effect or have a negative impact on bone status. Cycling training of 10 h per week for 2 to 7 years in adolescent males (<17 years of age) resulted in 10% lower BMC in legs, adjusted for total lean mass and height, but 8% higher in the hip area compared with a control group (Olmedillas *et al.*, 2011). Furthermore, a comparison between adolescent female football players and swimmers that engaged in their sport for 10 h per week, reported significantly higher BMD at all sites for footballers. Also, swimmers had lower values of

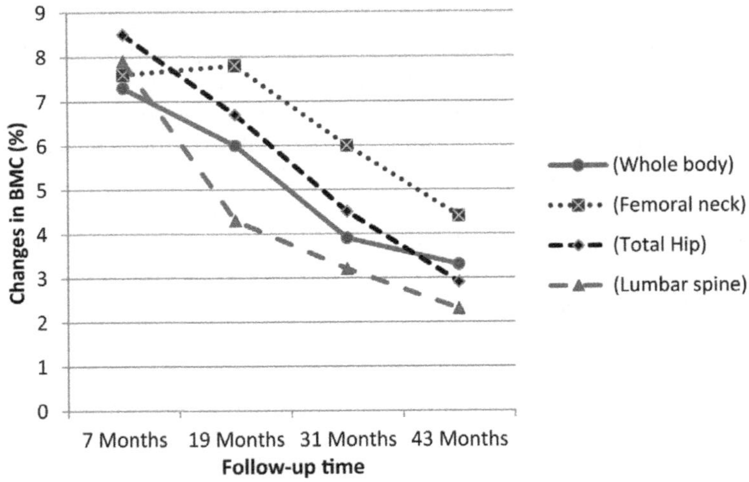

Figure 32.3 Changes (%) in bone-mineral content after a jumping intervention. Percentage of change in bone-mineral content at four sites of the skeleton (total body, femoral neck, total hip, lumbar spine) after a jumping intervention programme. The duration of the school-based intervention was 7 months and the participants were followed up for 3 years. Adapted from: (Gunter *et al.*, 2008).

bone geometry compared with the controls (Ferry *et al.*, 2011). Prolonged participation in weight-bearing activities, such as jumping and osteogenic sports (*e.g.* football, handball, gymnastic), may elicit greater improvements in BMC and BMD compared to nonweight-bearing/osteogenic sports, such as cycling and swimming, which may decrease or not affect the bone development in children and adolescents (Hind and Burrows, 2007) (Figure 32.3).

32.5 Associations of Calcium, Vitamin D and Exercise with Bone Mass of Children and Adolescents

Combined actions of calcium, vitamin D and exercise can determine bone development during growth by taking into account the interactions between hormonal changes and GRF placed on the skeleton. Evidence shows that adequate calcium and vitamin D levels are necessary to acquire bone mass and these nutrients interact with exercise to enhance bone growth (Valtuena *et al.*, 2012). However, the magnitude of the benefits in boys and girls differ at various sites of the skeleton and may depend on the baseline levels of calcium and vitamin D, on the dose of the supplementation and on previous loading experience (Specker and Vukovich, 2007).

The impact of calcium and exercise was investigated in 113 prepubertal females for 12 months. Girls were separated into four groups: two exercise

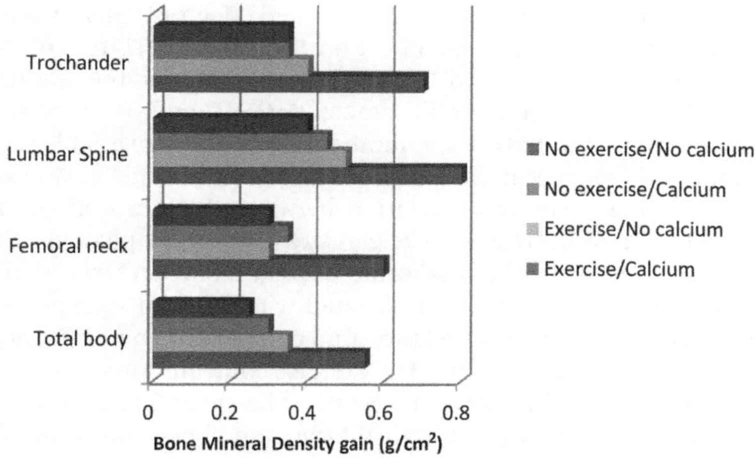

Figure 32.4 Bone-mineral density gain after 12 months of calcium supplementation and exercise intervention. Bone-mineral density gain at total body, femoral neck, lumbar spine and trochanter sites after calcium supplementation and exercise intervention period of 12 months in four groups. Adapted from: (Courteix *et al.*, 2005).

groups (7 ± 4 h of sport participation per week) that received 800 mg of calcium phosphate (exercise/calcium) or placebo (exercise/placebo) and two sedentary groups (1 ± 1 h of sport participation per week) that received 800 mg of calcium phosphate (sedentary/calcium) or placebo (sedentary/placebo) (Courteix *et al.*, 2005). The findings showed that in the exercise/calcium group there was a 6.3% greater BMD at the TB, 11% at the LS, 8.2% at FN and 9.3 at Ward's triangle compared to other groups. It should be mentioned that exercise might increase the calcium intake requirements by 422 mg due to calcium excretion through sweat following exercise at intensities. This was observed in adults but there is no evidence for this in young people (Bullen *et al.*, 1999) (Figure 32.4).

The positive combined associations between calcium, vitamin D and bone status have been shown in adolescents at different sites of the skeleton and at different maturation stages (Neville *et al.*, 2002). The relationship between PA and 25(OH)D concentration was investigated in 100 adolescents aged 12.5–17.5 years after controlling for calcium and vitamin D intakes, maturation, age, gender, season, PA and fitness level. Findings revealed that the interaction between 25(OH)D levels and PA was significantly related to BMC at TB and legs. Furthermore, it was found that the active group of adolescents had increased total BMC when the vitamin D levels were sufficient (>75 nmol L^{-1}) (Valtuena *et al.*, 2012). The beneficial effects of PA have been also demonstrated in participants with vitamin D deficiency. It has been reported that adjusted BMD for age, BMI, parathyroid hormone and bone turnover markers was positively related to PA at TB, FN and LS site in 166 adolescent girls with vitamin D deficiency (Constantini *et al.*, 2010).

However, there is evidence that the combined effects of calcium and vitamin D may not have a beneficial effect on BMC and BMD in adolescents. A study of 101 adolescents aged 12.5–17.5 years examined the relationship between calcium, vitamin D, milk intake, 25(OH)D and bone status after adjusting the BMC and BMD for a number of cofounding variables such as height, family affluence scale, maturation status, season and PA (Mouratidou *et al.*, 2013). There was no association between calcium and vitamin D intake and BMC and BMD at any skeletal site. However, a positive relationship was found between milk intake and BMD and BMC in boys. Also, BMD for TB, head and right arm was associated with calcium intakes. In girls there was a significant positive relationship of 25(OH)D and BMC and BMD at TB, subtotal, left and right arm. The exercise contribution to the relationship of bone status and dietary nutrients has been verified by a study that investigated the association of physical fitness and body composition with 25(OH)D levels in 1006 adolescents (Valtuena *et al.*, 2013). Physical fitness was positively related to 25(OH)D levels in boys and upper limp muscular strength was associated with 25(OH)D levels in girls. In addition, boys with higher fitness levels and lower BMI had significantly higher 25(OH)D levels compared to those with lower fitness levels.

32.6 Conclusions

Calcium, vitamin D and exercise are recognized as important factors for bone health in children and adolescences. In this chapter, we discussed the potential independent and combined effects of calcium, vitamin D and exercise on bone health in children and adolescents. The beneficial effects of calcium in relation to vitamin D and exercise were found to be greater during prepubertal years for both boys and girls. Therefore, future interventions should focus on this period of life in order to elicit greater improvements in bone mass that may translate to an increased PBM in adulthood. Current evidence suggests bone development is enhanced to a greater extent when optimal calcium and vitamin D levels combined with weight-bearing exercise. Overall, the findings suggest that regular exercise is considered the most important modifiable factor to enhance bone development during childhood and adolescence. Further longitudinal research is needed to provide valuable information about the continuation of the impact of calcium in combination with vitamin D and exercise from childhood and adolescence into adulthood.

Summary Points

- Peak bone mass increase occurs approximately 2 years earlier in girls compared with boys.
- Boys gain greater amounts of bone mass compared to girls during adolescence.

- Calcium intake between 700 and 1000 mg per day is recommended for normal bone growth in children and 1300 mg per day in adolescents.
- Sufficient vitamin D intake of 15 mg per day or 25-hydroxyvitamin D blood levels above 75 nmol L^{-1} is recommended for children and adolescents.
- Exercise stimulates bone mineral content and bone-mineral density increase and active children and adolescents have better bone status compared to their nonactive peers.
- Exercise seems to be the most important contributing factor to bone development compared to calcium and vitamin D.

Key Facts of Bone Health during Growth

1. Bone-mass acquired during youth is an important determinant of the risk of osteoporotic fracture during later life. The higher the peak bone mass, the lower the risk of osteoporosis.
2. From birth to the onset of the physical maturation, the bone mass is the same in girls as in boys.
3. During puberty, the speed of bone development in the spine and hip increases by approximately five times.
4. By the end of the teen years, approximately 90% of adult bone mass is in place.
5. During growth the gain in bone mineral mass is mainly due to an increase in bone size with very little change in bone tissue within the bones.
6. Young children who engage in 40 min of normal vigorous activity each day have significantly stronger bones than their less active peers.
7. It is estimated that a 10% increase of peak bone mass in children reduces the risk of an osteoporotic fracture during adult life by 50%.

Definitions and Explanations of Key Terms

Bone-mineral content. Amount of bone mineral mass in grams per centimeter.

Bone-mineral density. Amount of mineral mass in grams per square centimeter.

Ground reaction forces. The forces exerted by the ground on a body in contact with it.

Peak bone mineral accrual. The period of most rapid gain in bone mineral.

Peak-height velocity. Maximum growth in height.

Tanner maturity stages. A five-stage scale system based on breast and pubic hair development.

Peak bone mass. Attainment of maximum bone mass that occurs between the second and third decade of life.

25-Hydroxyvitamin D. Biochemical marker to assess vitamin D availability in the blood.

Osteogenic/nonosteogenic sports. Sports that elicit or not improvements in bone mass.

Weight-bearing exercise. Type of exercise that enhance bone status.

List of Abbreviations

25(OH)D	25-Hydroxyvitamin D
BMC	Bone-mineral content
BMD	Bone-mineral density
Ca	Calcium
FN	Femoral neck
GRF	Ground reaction forces
LS	Lumbar spine
PA	Physical activity
PBM	Peak bone mass
PHV	Peak height velocity
RCT	Randomized controlled trial
TB	Total body
TH	Total hip

References

Bailey, C. A. and Brooke-Wavell, K., 2008. Exercise for optimising peak bone mass in women. *Proceedings of the Nutrition Society*. 67: 9–18.

Bailey, D. A., Martin, A. D., McKay, H. A., Whiting, S. and Mirwald, R., 2000. Calcium accretion in girls and boys during puberty: a longitudinal analysis. *Journal of Bone and Mineral Research*. 15: 2245–2250.

Bailey, D. A., McKay, H. A., Mirwald, R. L., Crocker, P. R. and Faulkner, R. A., 1999. A six-year longitudinal study of the relationship of physical activity to bone mineral accrual in growing children: the university of Saskatchewan bone mineral accrual study. *Journal of Bone and Mineral Research*. 14: 1672–1679.

Baker, S. S., Cochran, W. J., Flores, C. A., Georgieff, M. K., Jacobson, M. S., Jaksic, T. and Krebs, N. F., 1999. American Academy of Pediatrics. Committee on Nutrition. Calcium requirements of infants, children, and adolescents. *Pediatrics*. 104: 1152–1157.

Baxter-Jones, A. D., Faulkner, R. A., Forwood, M. R., Mirwald, R. L. and Bailey, D. A., 2011. Bone mineral accrual from 8 to 30 years of age: an estimation of peak bone mass. *Journal of Bone and Mineral Research*. 26: 1729–1739.

Baxter-Jones, A. D., Kontulainen, S. A., Faulkner, R. A. and Bailey, D. A., 2008. A longitudinal study of the relationship of physical activity to bone mineral accrual from adolescence to young adulthood. *Bone*. 43: 1101–1107.

Behringer, M., Gruetzner, S., McCourt, M. and Mester, J., 2014. Effects of weight-bearing activities on bone mineral content and density in children and adolescents: a meta-analysis. *Journal of Bone and Mineral Research*. 29: 467–478.

Bielemann, R. M., Martinez-Mesa, J. and Gigante, D. P., 2013. Physical activity during life course and bone mass: a systematic review of methods and findings from cohort studies with young adults. *BMC Musculoskeletal Disorders*. 14: 77.

Boreham, C. A. and McKay, H. A., 2011. Physical activity in childhood and bone health. *British Journal of Sports Medicine*. 45: 877–879.

Bullen, D. B., O'Toole, M. L. and Johnson, K. C., 1999. Calcium losses resulting from an acute bout of moderate-intensity exercise. *International Journal of Sport Nutrition*. 9: 275–284.

Constantini, N. W., Dubnov-Raz, G., Chodick, G., Rozen, G. S., Giladi, A. and Ish-Shalom, S., 2010. Physical activity and bone mineral density in adolescents with vitamin D deficiency. *Medicine & Science in Sports & Exercise*. 42: 646–650.

Courteix, D., Jaffre, C., Lespessailles, E. and Benhamou, L., 2005. Cumulative effects of calcium supplementation and physical activity on bone accretion in premenarchal children: a double-blind randomised placebo-controlled trial. *International Journal of Sports Medicine*. 26: 332–338.

Davies, J. H., Evans, B. A. and Gregory, J. W., 2005. Bone mass acquisition in healthy children. *Archives of Disease in Childhood*. 90: 373–378.

Deng, Z. L., Sharff, K. A., Tang, N., Song, W. X., Luo, J., Luo, X., Chen, J., Bennett, E., Reid, R., Manning, D., Xue, A., Montag, A. G., Luu, H. H., Haydon, R. C. and He, T. C., 2008. Regulation of osteogenic differentiation during skeletal development. *Frontiers in Bioscience*. 13: 2001–2021.

Ferry, B., Duclos, M., Burt, L., Therre, P., Le Gall, F., Jaffre, C. and Courteix, D., 2011. Bone geometry and strength adaptations to physical constraints inherent in different sports: comparison between elite female soccer players and swimmers. *Journal of Bone and Mineral Metabolism*. 29: 342–351.

Frost, H. M., 1994. Wolff's Law and bone's structural adaptations to mechanical usage: an overview for clinicians. *Angle Orthodontist*. 64: 175–188.

Gonzalez-Gross, M., Valtuena, J., Breidenassel, C., Moreno, L. A., Ferrari, M., Kersting, M., De Henauw, S., Gottrand, F., Azzini, E., Widhalm, K., Kafatos, A., Manios, Y., Stehle, P. and Grp, H. S., 2012. Vitamin D status among adolescents in Europe: the healthy lifestyle in Europe by nutrition in adolescence study. *British Journal of Nutrition*. 107: 755–764.

Gracia-Marco, L., Moreno, L. A., Ortega, F. B., Leon, F., Sioen, I., Kafatos, A., Martinez-Gomez, D., Widhalm, K., Castillo, M. J. and Vicente-Rodriguez, G., 2011. Levels of physical activity that predict optimal bone mass in adolescents: the HELENA study. *American Journal of Preventive Medicine*. 40: 599–607.

Gunter, K., Baxter-Jones, A. D., Mirwald, R. L., Almstedt, H., Fuller, A., Durski, S. and Snow, C., 2008. Jump starting skeletal health: a 4-year longitudinal study assessing the effects of jumping on skeletal development in pre and circum pubertal children. *Bone*. 42: 710–718.

Hind, K. and Burrows, M., 2007. Weight-bearing exercise and bone mineral accrual in children and adolescents: a review of controlled trials. *Bone*. 40: 14–27.

Huncharek, M., Muscat, J. and Kupelnick, B., 2008. Impact of dairy products and dietary calcium on bone-mineral content in children: results of a meta-analysis. *Bone*. 43: 312–321.

Huybrechts, I., Lin, Y., De Keyzer, W., Sioen, I., Mouratidou, T., Moreno, L. A., Slimani, N., Jenab, M., Vandevijvere, S., De Backer, G. and De Henauw, S., 2011. Dietary sources and sociodemographic and economic factors affecting vitamin D and calcium intakes in Flemish preschoolers. *European Journal of Clinical Nutrition*. 65: 1039–1047.

Ishikawa, S., Kim, Y., Kang, M. and Morgan, D. W., 2013. Effects of weight-bearing exercise on bone health in girls: a meta-analysis. *Sports Medicine*. 43: 875–892.

Lanou, A. J., Berkow, S. E. and Barnard, N. D., 2005. Calcium, dairy products, and bone health in children and young adults: a reevaluation of the evidence. *Pediatrics*. 115: 736–743.

Levine, M. A., 2012. Assessing bone health in children and adolescents. *Indian Journal of Endocrinology and Metabolism*. 16: S205–S212.

Marshall, D., Johnell, O. and Wedel, H., 1996. Meta-analysis of how well measures of bone mineral density predict occurrence of osteoporotic fractures. *British Medical Journal*. 312: 1254–1259.

Matkovic, V., Landoll, J. D., Badenhop-Stevens, N. E., Ha, E. Y., Crncevic-Orlic, Z., Li, B. and Goel, P., 2004. Nutrition influences skeletal development from childhood to adulthood: a study of hip, spine, and forearm in adolescent females. *Journal of Nutrition*. 134: 701S–705S.

McKay, H., Tsang, G., Heinonen, A., MacKelvie, K., Sanderson, D. and Khan, K. M., 2005. Ground reaction forces associated with an effective elementary school based jumping intervention. *British Journal of Sports Medicine*. 39: 10–14.

Mouratidou, T., Vicente-Rodriguez, G., Gracia-Marco, L., Huybrechts, I., Sioen, I., Widhalm, K., Valtuena, J., Gonzalez-Gross, M. and Moreno, L. A., 2013. Associations of dietary calcium, vitamin D, milk intakes, and 25-hydroxyvitamin D with bone mass in Spanish adolescents: the HELENA study. *Journal of Clinical Densitometry*. 16: 110–117.

Neville, C. E., Robson, P. J., Murray, L. J., Strain, J. J., Twisk, J., Gallagher, A. M., McGuinness, M., Cran, G. W., Ralston, S. H. and Boreham, C. A., 2002. The effect of nutrient intake on bone mineral status in young adults: the Northern Ireland young hearts project. *Calcified Tissue International*. 70: 89–98.

Nogueira, R. C., Weeks, B. K. and Beck, B. R., 2014. Exercise to improve pediatric bone and fat: a systematic review and meta-analysis. *Medicine & Science in Sports & Exercise*. 46: 610–621.

Olmedillas, H., Gonzalez-Aguero, A., Moreno, L. A., Casajus, J. A. and Vicente-Rodriguez, G., 2011. Bone related health status in adolescent cyclists. *PLoS One*. 6: e24841.

Perez-Lopez, F. R., Chedraui, P. and Cuadros-Lopez, J. L., 2010. Bone mass gain during puberty and adolescence: deconstructing gender characteristics. *Current Medicinal Chemistry*. 17: 453–466.

Pettifor, J. M., 2005. Rickets and vitamin D deficiency in children and adolescents. *Endocrinology and Metabolism Clinics of North America*. 34: 537–553.

Pettifor, J. M. and Prentice, A., 2011. The role of vitamin D in paediatric bone health. *Best Practice & Research, Clinical Endocrinology & Metabolism*. 25: 573–584.

Rauch, F., Bailey, D. A., Baxter-Jones, A., Mirwald, R. and Faulkner, R., 2004. The 'muscle-bone unit' during the pubertal growth spurt. *Bone*. 34: 771–775.

Rizzoli, R., Bianchi, M. L., Garabedian, M., McKay, H. A. and Moreno, L. A., 2010. Maximizing bone mineral mass gain during growth for the prevention of fractures in the adolescents and the elderly. *Bone*. 46: 294–305.

Robling, A. G., Hinant, F. M., Burr, D. B. and Turner, C. H., 2002. Improved bone structure and strength after long-term mechanical loading is greatest if loading is separated into short bouts. *Journal of Bone and Mineral Research*. 17: 1545–1554.

Ross, A. C., 2011. The 2011 report on dietary reference intakes for calcium and vitamin D. *Public Health Nutrition*. 14: 938–939.

Specker, B. and Vukovich, M., 2007. Evidence for an interaction between exercise and nutrition for improved bone health during growth. *Medicine and Sport Science*. 51: 50–63.

Theintz, G., Buchs, B., Rizzoli, R., Slosman, D., Clavien, H., Sizonenko, P. C. and Bonjour, J. P., 1992. Longitudinal monitoring of bone mass accumulation in healthy adolescents: evidence for a marked reduction after 16 years of age at the levels of lumbar spine and femoral neck in female subjects. *Journal of Clinical Endocrinology and Metabolism*. 75: 1060–1065.

Valtuena, J., Gracia-Marco, L., Huybrechts, I., Breidenassel, C., Ferrari, M., Gottrand, F., Dallongeville, J., Sioen, I., Gutierrez, A., Kersting, M., Kafatos, A., Manios, Y., Widhalm, K., Moreno, L. A. and Gonzalez-Gross, M., 2013. Cardiorespiratory fitness in males, and upper limbs muscular strength in females, are positively related with 25-hydroxyvitamin D plasma concentrations in European adolescents: the HELENA study. *QJM*. 106: 809–821.

Valtuena, J., Gracia-Marco, L., Vicente-Rodriguez, G., Gonzalez-Gross, M., Huybrechts, I., Rey-Lopez, J. P., Mouratidou, T., Sioen, I., Mesana, M. I., Martinez, A. E., Widhalm, K. and Moreno, L. A., 2012. Vitamin D status and physical activity interact to improve bone mass in adolescents. The HELENA Study. *Osteoporosis International*. 23: 2227–2237.

Van Langendonck, L., Lefevre, J., Claessens, A. L., Thomis, M., Philippaerts, R., Delvaux, K., Lysens, R., Renson, R., Vanreusel, B., Vanden Eynde, B., Dequeker, J. and Beunen, G., 2003. Influence of participation in high-impact sports during adolescence and adulthood on bone mineral density in middle-aged men: a 27-year follow-up study. *American Journal of Epidemiology*. 158: 525–533.

Vicente-Rodriguez, G., 2006. How does exercise affect bone development during growth? *Sports Medicine*. 36: 561–569.

Zemel, B., 2013. Bone mineral accretion and its relationship to growth, sexual maturation and body composition during childhood and adolescence. *World Review of Nutrition and Dietetics*. 106: 39–45.

CHAPTER 33

Dietary Calcium and Osteoprotegerin

DIANJUN SUN*[a], JUN YU*[a], YANHUI GAO*[a], AND YUANYUAN LI*[a]

[a]Center for Endemic Disease Control, Chinese Center for Disease Control and Prevention, Harbin Medical University, 157 Baojian Road, Nangang District, Harbin, Heilongjiang Province 150081, China
*E-mail: hrbmusdj@163.com, 6yujun@126.com, gaoyh411@163.com, Liyuanyuan418@163.com

33.1 Osteoprotegerin in Health: A Brief Overview

OPG was discovered by two laboratories at the same time in 1997 (Simonet *et al.*, 1997; Yamaguchi *et al.*, 1998), and was named osteoclastogenesis inhibitory factor (OCIF). The tumor necrosis factor receptor-like molecule-1 (TNFR-1) and follicular dendritic cell receptor-1 (FDCR-1), which are new protein molecules discovered by other researchers from different cell lines, are encoded by the same gene after DNA sequencing and amino acid sequence analysis. On 8th February 2000, cytokines discovered previously, such as OPG, OCIF, TNFR-1 and FDCR-1, were determined as a homonymous protein, and named OPG by the American Society for Bone and Mineral Research President's Committee on Nomenclature who proposed the standard nomenclature for the new tumor necrosis factor (TNF) family member involved in the regulation of bone resorption. OPG gene is located on human chromosome 8q23-24.

Food and Nutritional Components in Focus No. 10
Calcium: Chemistry, Analysis, Function and Effects
Edited by Victor R. Preedy

Published by the Royal Society of Chemistry, www.rsc.org

33.1.1 Molecular Structure

OPG is a soluble secreted glycoprotein, a growth factor receptor, which belongs to the TNF receptor family. The OPG molecule consists of three structural domains influencing the biological function. The N-terminal part is a cysteine-rich domain important for dimerization and osteoclastogenesis, whereas the C-terminal contains a death domain and also responsible for heparin binding (Yamaguchi *et al.*, 1998). OPG is synthesized as a monomer (60 kDa) and assembled as a homodimer within the cell, and then secreted mainly as a disulfide-linked homodimer into the circulation (Simonet *et al.*, 1997; Yamaguchi *et al.*, 1998). Differentiation, activation and signal transduction of osteoclast is negatively regulated through OPG. OPG acts as a decoy receptor of RANKL, and has a high affinity with RANKL. OPG blocks RANK and RANKL binding by competitive binding with RANKL, thereby inhibiting osteoclast formation and bone resorption (Figures 33.1–33.3).

33.1.2 Expression and Regulation

OPG is highly expressed in liver, heart, lung, kidney, stomach, small intestine, skin, brain, spinal cord and bone among other tissues (Simonet *et al.*, 1997). OPG is a hormone regulating bone metabolism and a cytokines regulator. The expression of OPG is also adjusted by hormones and cytokines, such as parathyroid hormone (PTH), platelet-derived growth factor, Ca^{2+} and epinephrine (Halladay *et al.*, 2001; Olesen *et al.*, 2005). PTH, a potent inducer of osteoclast formation, suppresses OPG mRNA expression *in vitro* and *in vivo* (Halladay *et al.*, 2001). *In vitro* experiments, human smooth muscle cells (SMCs) produce large amounts of OPG (up to 30 times more than endothelial cells) after stimulation with TNF-α, whereas insulin was found to decrease the production. The effects were related to the same changes in mRNA expression that indicating a transcriptional regulation (Olesen *et al.*, 2005). Plasma OPG in growth-hormone deficiency patients increased significantly after 6 months of growth-hormone replacement therapy and this increase was significantly correlated with the increase of osteocalcin and deoxypyridinoline values (Lanzi *et al.*, 2003). Legumes have a beneficial effect on bone through modulation of OPG and RANKL

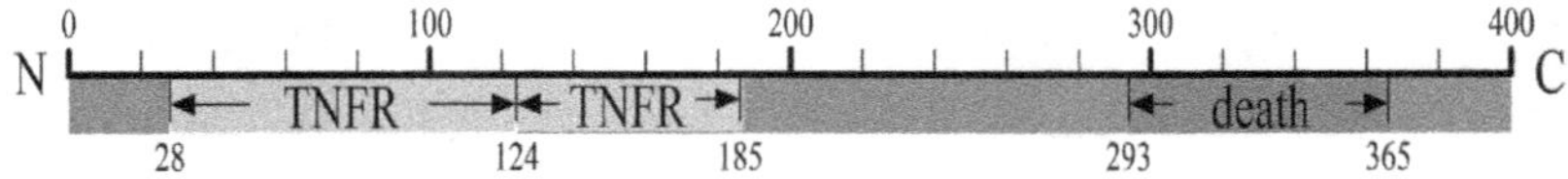

Figure 33.1 Osteoprotegerin (OPG) molecular structure. Note: the length of OPG protein is 401 bp, which is TNFR domain in the fragment of 28 to 124 and 124 to 185, and is the death domain in the fragment of 293 to 365. The N-terminal part is a cysteine-rich domain important for dimerization and osteoclastogenesis whereas the C-terminal contains a death domain and is also responsible for heparin binding.

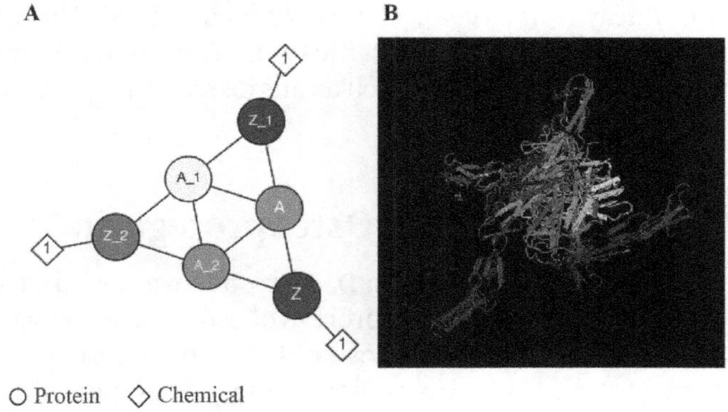

Figure 33.2 Crystal structure of human rankl complexed with its decoy receptor osteoprotegerin. (A) tumor necrosis factor ligand superfamily member 11, soluble3 form; tumor necrosis factor receptor superfamily member 11b; *N*-acetyl-D-glucosamine. (B) is 3D crystal structure of human rankl complexed with its decoy receptor osteoprotegerin.

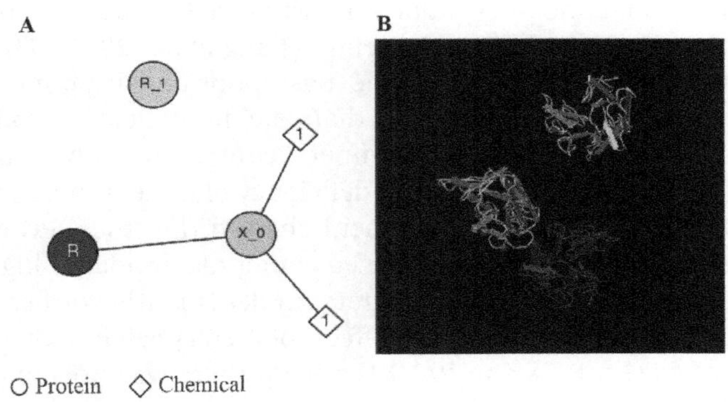

Figure 33.3 Crystal structure of mouse rankl–opg complex. (A) tumor necrosis factor ligand superfamily member 11, soluble3 form; tumor necrosis factor receptor superfamily member 11b; Cl (chloride ion). (B) is 3D crystal structure of mouse rankl-opg complex.

expression in ovariectomized rats, which could help to compensate for an estrogen-deficiency by preventing bone loss induced by ovarian hormone deficiency (Park *et al.*, 2013).

33.1.3 Function

Transgenic mice (OPG+/+) with OPG overexpression had the phenotype of bone sclerosis. On the other hand, OPG gene knockout (OPG−/−) mice who lacked OPG, emerged with severe osteoporosis and arterial calcification before mature (Bucay *et al.*, 1998). The use of recombined OPG therapy can prevent osteoporosis and arterial calcification. The relationships between OPG and metabolic bone disease as well as angiosteosis have caused increasing attention (Asanuma *et al.*, 2007).

33.2 Calcium Intake and Osteoprotegerin

Calcium, in combination with vitamin D, is an effective way to treat of osteoporosis. Calcium would stimulate the process of osteoporosis since bone mineralization occurs concurrently from osteoblast to osteocyte transition. The effect of calcium (1.8–11.8 mM) on human primary osteoblast differentiation was tested *in vitro*. Treatment with calcium resulted in a striking dose- and time-dependent increase in cell-associated mineralization. The expression of osteoclast inhibitor, OPG, was dramatically enhanced by calcium. Studies on the effects of calcium intake level on OPG expression in bone show that low calcium intake decreased the expression of OPG (Liu *et al.*, 2009; Jiang *et al.*, 2013). Low dietary calcium has resulted in weaker eggshells and poorer bone quality than that from hens fed with the control diet. High dietary calcium increased serum calcium concentration, OPG mRNA, osteocalcin and inhibited serum alkaline phosphatase activity and decreased its mRNA compared with low or control dietary calcium (Jiang *et al.*, 2013). The calcium-deficient ducks showed decreased bone mass, bone density, mineral content and OPG expression. It is concluded that calcium-deficiency could inhibit the secretion of OPG, decrease bone mineral content and inhibit the synthesis of bone matrix, which may induce deficiency of bone mineralization (Liu *et al.*, 2009). Another animal experiment showed that the OPG expression was increased in the 5/6 nephrectomized young rats model in high-calcium diet compared with normal calcium rats model (Rianthavorn *et al.*, 2010). However, in another study about the effects of dietary calcium on the expression levels of OPG mRNA and RANKL mRNA, there was no significant difference in the distal femur joints of the experimental rats fed with purified rodent diets containing either 1000 mg calcium per kg, 5000 mg calcium per kg, or 20 000 mg calcium per kg (Yu *et al.*, 2013).

33.3 Calcium and Osteoprotegerin and Bone Diseases

33.3.1 Calcium and Bone Diseases

Bone has developed as the storage of calcium as well as a supporting tissue in vertebrates. Bone is a complex tissue in which resorption and formation takes place throughout life. This process is called bone remodeling. Osteotrophic hormones, such as 1,25-dihydroxy-vitamin D_3, PTH and calcitonin, maintain serum calcium homeostasis within a narrow range of 9 to 10 mg dL^{-1} by regulating intestinal absorption of calcium and bone remodeling (Suda, 2002).

Hypoenergetic diets higher in dairy foods, dietary calcium, and protein with daily exercise, favorably affected important bone health biomarkers *vs.* diets with less of these bone-supporting nutrients (Josse *et al.*, 2012). A narrow serum calcium level is essential for many metabolic processes.

There are still controversies in the relationship between calcium supplementation and bone-mineral density (BMD). A number of early epidemiological data showed that BMD was higher in women having higher dietary calcium intake (>1000 mg per day) than those who have lower dietary calcium intake (<500 mg per day). Osteopenia and increased risk of fractures in inflammatory bowel disease (IBD) are caused by several factors. Inadequate calcium intake is found in one third of IBD patients and represents a reversible risk factor for osteoporosis, suggesting the need for tailored nutritional advice in IBD (Vernia *et al.*, 2014).

Habitually high dietary calcium intake can reduce the risk of hip fracture in the elderly. Animal experiments also found that the ovariectomized rats with an adequate-calcium diet showed significantly higher BMD of the lumbar spine compared to the low-calcium-diet group (Bae and Kim, 2010). However, Feskanich *et al.* disagreed with the hypothesis that intake of mass milk or other food containing calcium can reduce the risk of bone fracture (Feskanich *et al.*, 1997). The results of their study showed that the calcium intake level had no relation with the bone fracture in women. Calcium supplement is only helpful to the women with normally low calcium intake. Neither higher estrogen level nor higher calcium intake level would increase the BMD to the maximum limit. Only maintaining adequate calcium intake and adequate estrogen in the whole life contributed to higher BMD in postmenopausal women. To perimenopause women, the beneficial effect on the bone mass, especially for collum femoris and hip, can be achieved if their dietary calcium intake reached 851.7 mg per day.

People attach great importance to the relationship of calcium and osteoporosis. Animal experimental and epidemiological data showed that long-term insufficiency of calcium intake is hazardous to osteoporosis. For the osteoporosis, its real aetiological agent is the accommodative dysfunction of hormonal regulation that induces the endocrine-metabolic anomaly. And the reason for bone lesion caused by metabolic disorder is the decreased ability on calcium intake, absorption and utilization, which could not be corrected through oral calcium supplement. Proper calcium supplement is able to increase bone calcium, BMD and bone mineral substance content. Increasing the dose of calcium supplement had no more effects, and adverse effects of high-calcium diets may occur at intakes greater than 2000 mg per day (Whiting and Wood, 1997). Calcium supplement cannot decrease the incidence of osteoporotic fracture. On the contrary, after high calcium intake, the risk of hip fracture increased 50% in patient with osteoporosis. The prominent side-effects of excessive calcium intake were illustrated in many animal experiments. The animals suffered osteochondrosis, renal failure and death when they were fed with high doses of calcium. Because excessive calcium intake disturbs the secretion of calcium-regulation hormones, such as vitamin D_3 and calcitonin, excessive calcium intake induces the occurrence of osteochondrosis. The other side-effect of excessive calcium intake includes soft-tissue calcification, vascular wall calcification, arthrosis periphery calcification, cartilage calcification, kidney calcification, conjunctiva calcification and tympanic membrane calcification, and *et al.* The calcification may induce functional lesions of corresponding organs. Genistein plus calcium and vitamin D_3 as part of a healthy diet showed positive effects to modulate bone turnover in osteopenic and postmenopausal women, and improved sRANKL-OPG balance after 2 years of treatment (Marini *et al.*, 2008) in a cohort study. Breast-cancer patients usually have low dietary calcium intake and high bone turnover. Treatments to correct calcium insufficiency and/or treatments with antiresorptive agents, such as OPG, may be of benefit in the adjuvant as well as palliative setting (Zheng *et al.*, 2007).

33.3.2 Osteoprotegerin and Bone Diseases

Osteoclast-mediated bone resorption plays an essential role in calcium homeostasis. Bone resorption by osteoclasts is coupled with bone formation by osteoblasts, and this balanced process is continuously remodeled for the skeleton. RANKL has been identified as an essential cytokine for the formation and activation of osteoclasts. The effects of RANKL are physiologically counterbalanced by the decoy receptor OPG (Hofbauer and Schoppet, 2004). OPG, RANK and RANKL system regulate the balance between osteoclasts–osteoblasts. The expression of these genes affects the maturation and function of osteoblasts–osteoclasts and bone remodeling.

In bone, OPG is derived from osteoblast-lineage cells such as osteoblasts and osteocytes. As the decoy receptor, OPG competitively combined with RANKL, inhibits its biological effect, and eliminates the support function of RANKL on all stages of osteoclast. The expression level of RANKL and OPG differs from the differentiation stages. Under physiological status, the normal bone remodeling from bone resorption to bone construction (or *vice versa*) is regulated by OPG/RANK/RANKL system. The ratio of RANKL/OPG is responsible for bone transition status and the net bone mass (Figure 33.4). Under various kinds of pathologic conditions, such as inflammatory reaction, cancer, pausimenia and disuse, the expression level of RANKL increased significantly, the bone reconstitution is excessively activated (Figure 33.5). The bone loss cannot be blocked by OPG. Increased osteoclast-mediated bone resorption exceeds the rate of bone formation. The unbalanced ratio of RANKL to OPG may increase bone loss or bone formation. This bone metabolic disorder phenomenon is illustrated in the hyperparathyroidism, renal osteopathy and osteomalacia. Hormones, cytokines and medicine that exert the regulation effects on the bone metabolism by all means are changing the ratio of RANKL to OPG. The synovial membrane of early untreated rheumatoid arthritis is characterized by a high RANKL/OPG ratio that can be reversed by methotrexate (Revu *et al.*, 2013).

The serum OPG level reflects the bone metabolic status. The bone formation slows down when the serum OPG decreases, indicating bone resorption exceeds formation. Most cases of juvenile Paget's disease (JPD), affected infants

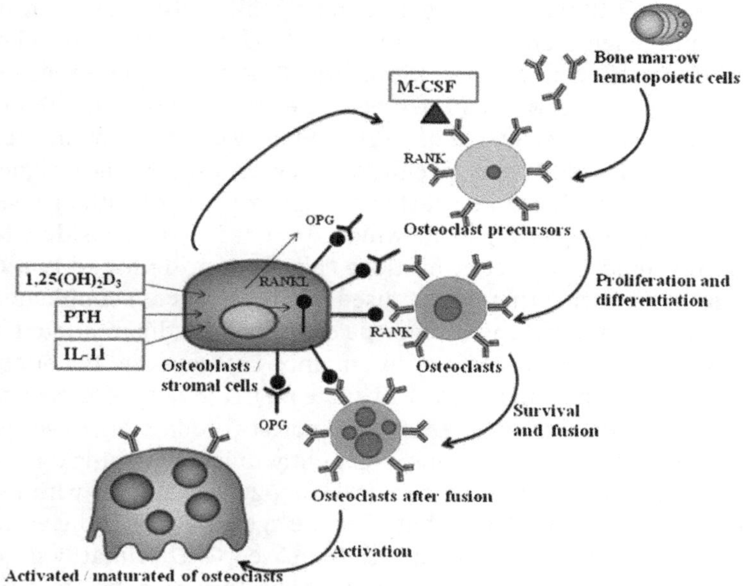

Figure 33.4 Osteoclasts formation. Abbreviation: macrophage colony-stimulating factor (M-CSF); 1,25-dihydroxyvitamin D_3 (1,25(OH)$_2$D$_3$); parathyroid hormone (PTH); interleukin-11 (IL-11).

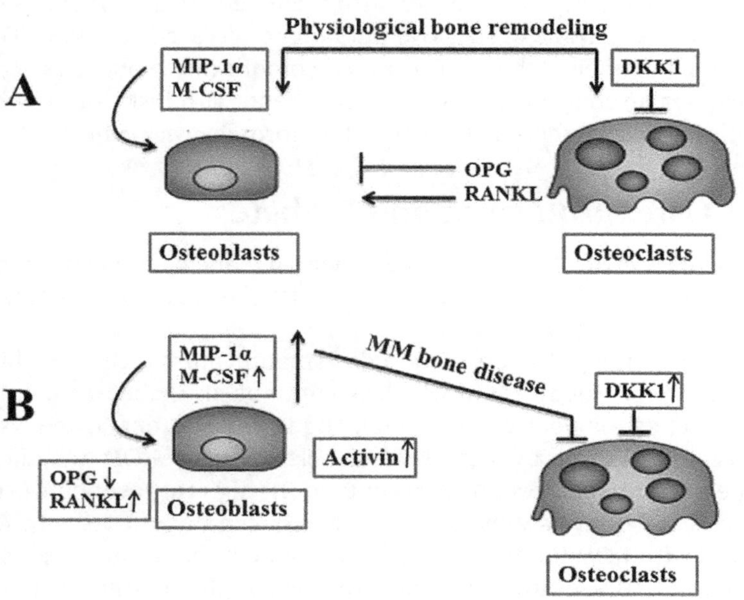

Figure 33.5 Biological characteristics of multiple myeloma bone disease. (A) Bone homeostasis under normal circumstances; (B) The biological basis of rebuilding myeloma bone. Abbreviation: multiple myeloma bone disease (MM bone disease); macrophage colony-stimulating factor (M-CSF); macrophage inflammatory protein 1α (MIP-1α); Dickkopf-1 (DKK1).

and children typically are suffering from bone pain, fractures and deformities, deafness, and macrocranium, are caused by OPG deficiency due to homozygous loss-of-function mutations within the TNFRSF11B gene that encodes OPG. Those patients' serum OPG levels were extremely low and RANKL levels were markedly elevated (Saki *et al.*, 2013). Mogelvang and Omland reported that the OPG level is gender specific, *i.e.* women usually have higher OPG levels than men. Furthermore, OPG is strongly associated with age (Omland *et al.*, 2007; Mogelvang *et al.*, 2013), which also needs to be considered before defining threshold or risk stratification. Palumbo and Li found that OPG is a pro-osteogenic factor that can be used as an osteogenic supplement in a growth medium to prime the osteogenic capacity of undifferentiated human mesenchymal stem cells (hMSCs) to enhance osteogenesis for bone-tissue engineering (Palumbo and Li, 2013). An increase in the ratio of RANKL to OPG, cannot explain the observed increase in osteoclast production, as serum OPG levels increase with age in humans, apparently to levels insufficient to counteract bone loss and prevent fracture (Matsuo, 2013). Patients with osteoporosis showed higher OPG levels, but after the age adjustment the correlation no longer existed (Pobeha *et al.*, 2014). Increased OPG primarily in adrenal insufficiency might be reflected in a compensatory response to enhanced bone resorption due to exogenous steroid excess and/or by a deficit in adrenal androgens. RANKL levels remain within the normal range on standard steroid replacement (Fichna *et al.*, 2012). Up to the present, findings indicate that osteoporosis of the hip is associated with increased circulatory levels of OPG in patients with chronic obstructive pulmonary disease (COPD). OPG might serve as a biomarker of this COPD-related comorbidity (Pobeha *et al.*, 2014). The mean serum concentration of OPG in patients with cystic fibrosis was significantly lower than those in the control (Ambroszkiewicz *et al.*, 2012).

33.4 Osteoprotegerin and Diabetes

Recently, the role of OPG and its mechanism in diabetes receives more attention (Figure 33.6). The reason for elevated OPG in diabetes patients is not clear (Rashtchizadeh *et al.*, 2012).

Osteogenic differentiation of vascular smooth muscle cells could result in medial artery calcification, which is common in diabetes mellitus, but the pathogenesis is poorly understood. High-turnover renal osteodystrophy (ROD) is characterized by increased osteoclast activity. ROD was diagnosed as a type-II (normal or low turnover) or type-III (high turnover plus osteoidosis) disease. OPG levels were significantly reduced in patients with ROD III compared with ROD II and correlated with bone mineralization density distribution (BMDD). Serum OPG concentrations might be used to distinguish among forms of ROD (Haas *et al.*, 2002). Thirty per cent higher serum OPG levels were found in a group of elderly women with diabetes mellitus as compared to healthy individuals (Browner *et al.*, 2001). A significant correlation between HbA1c and OPG was shown in patients with type-1 diabetes (T1D) or type-2 diabetes (T2D) (Knudsen *et al.*, 2003). Serum OPG is associated with

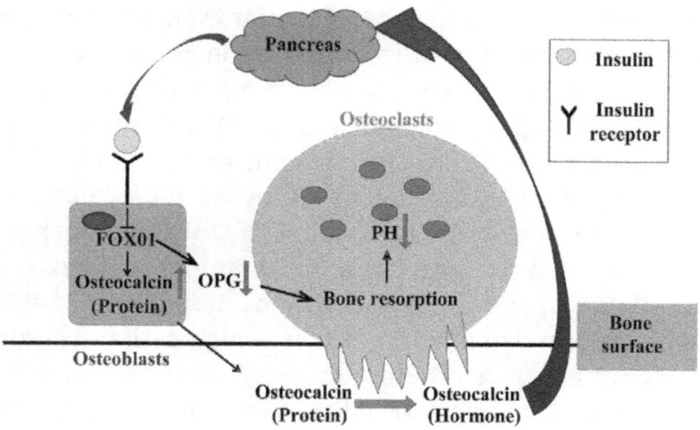

Figure 33.6 The role of osteoprotegerin (OPG) and insulin in bone metabolism. Insulin signaling inhibits transcription factor Fox01, thereby increasing the expression of osteocalcin (the form of protein), and reducing the expression of osteoprotegerin. Osteoprotegerin can stimulate the osteoclast to resorption of bone tissue. pH value decreases, which will cause osteocalcin (the form of protein) converted into the form of hormone osteocalcin. This signal of hormone osteocalcin results in the pancreas secreting more insulin. Abbreviation: forkhead box transcription factor-1 (FOX01); pH value or hydrogen ion concentration (pH).

the development and progression of diabetic complications in a large group of adults with T1D. OPG is an independent predictor of cardiovascular (CV) complications (Gordin *et al.*, 2013). In addition, patients with renal impairment had elevated OPG levels compared to T1D patients who did not have overt kidney disease. Jorsal found that OPG was not only an independent marker of mortality, but increased OPG levels also predicted the severity of diabetic nephropathy (Jorsal *et al.*, 2008). Similarly, Knudsen found increased OPG levels in T2D patients with microvascular complications compared to those without complications. Elevated OPG levels predicted CV events in patients with uncomplicated T2D and increased OPG levels in patients with uncomplicated well-controlled T2D (Knudsen *et al.*, 2003).

33.5 Osteoprotegerin and Cardiovascular Diseases

Many experiments were carried out to test the hypothesis that arterial calcification is linked to bone resorption by determining whether the selective inhibition of bone resorption with osteoprotegerin will inhibit arterial calcification (Price *et al.*, 2001). OPG has been implicated to be involved in various inflammations and linked to diabetes and poor glycemic control.

Bucay *et al.* showed that OPG knockout mice developed spontaneous arterial calcification in aorta and renal artery, *etc.*, except the severe osteoporosis and bone fracture. In the calcification artery, RANK and RANKL were obviously expressed, which did not exist in normal artery, thus OPG appears to be protective against vascular calcification (Bucay *et al.*, 1998). Therefore, the elevated OPG levels may prevent cardiovascular events in humans. However, the protective role of OPG found in animal models has not been observed in humans. In fact, increased OPG levels have consistently been associated with the severity of cardiovascular disease (CVD). Browner *et al.* found an association between high OPG levels in plasma and increased CVD mortality in a cohort of 490 women older than 65 (Browner *et al.*, 2001). Several associations between OPG levels and traditional CVD risk factors have now been reported, including positive correlations with smoking, fasting glucose levels, older age, diabetes, and renal impairment. In addition, OPG concentrations are positively correlated with coronary calcification, vascular stiffness and the presence of unstable atherosclerotic plaques (Nybo and Rasmussen, 2008). Logstrup *et al.* recently showed that increased serum OPG levels on admission for acute myocardial infarct are associated with decreased microcirculation after revascularization (Logstrup *et al.*, 2013). In population-based cohort studies, increased OPG levels were found to be associated with future risk of myocardial infarct, ischemic stroke, and CVD mortality (Mogelvang *et al.*, 2013). Recently, circulating OPG level has emerged as an independent biomarker of CVD in the population.

Vascular calcification (VCs) is a common pathobiological process that occurs in the elderly population and in patients with diabetes, chronic kidney disease (CKD) and/or after kidney transplantation (KTx). OPG has recently emerged as an important regulator of the development of vascular calcification. Meneghini *et al.* evaluated the prevalence of VCs at the abdominal aortic site in KTx patients at the time of transplantation and 1 year after, and showed that high levels of serum calcium and OPG are significantly associated with the progression of VCs. Whether these associations are based on a causal relationship and whether their correction might impact on the calcification process could be verified by prospective interventional studies (Meneghini *et al.*, 2013).

Arterial thickening contributes to elevated cardiovascular risk in patients with maintenance renal replacement therapy. Sixty-seven patients with chronic kidney disease were observed who aged 53 ± 13 years (range, 19–75 years) and were treated with peritoneal dialysis for 30 ± 24 months. In a multiple regression analysis, age, osteocalcin, and log-transformed OPG values remained significantly associated with the common carotid artery intima-media thickness (CCA-IMT). The highest CCA-IMT value was observed in patients with OPG concentrations in the upper tertiles. OPG concentrations strongly and positively correlated with the duration of dialysis treatment. Elevated serum OPG levels may be useful as a prognostic marker of cardiovascular risk in dialyzed patients (Janda *et al.*, 2013).

OPG plays a major role in the development of transplantation osteoporosis because of the immunosuppressive therapy. The serum OPG levels decline consistently in 57 cardiac transplant recipients following initiation of immunosuppressive therapy and were independently correlated with changes in bone density. Bone loss at the lumbar spine and femoral neck after 6 months was correlated to the decrease in serum OPG at 6 months as well as 3 months after cardiac transplantation (Fahrleitner *et al.*, 2003).

33.6 Clinical Application of Osteoprotegerin

The discovery of OPG may establish a new way to treat several metabolic bone diseases caused by abnormal bone formation and resorption. Osteoporosis is a common age-related disease, which has become a growing public-health problem. This degenerative bone disease primarily affects postmenopausal women, but men may also suffer from reduced bone-mineral density. The development of prophylactic treatments and medications of osteoporosis has become an urgent issue due to the increasing proportion of the elderly in the population. A chimeric form of osteoprotegerin (cOPG) inhibits hypercalcemia and bone resorption induced by IL-1beta, TNF-alpha, PTH, PTHrP, and $1,25(OH)_2D_3$, which demonstrates that cOPG may be useful in the treatment of diseases including hyperparathyroidism, humoral hypercalcemia of malignancy, osteoporosis, and inflammatory bone disease, which are characterized, in part, by increases in osteoclastic bone resorption (Morony *et al.*, 1999).

33.6.1 Osteoprotegerin and Hypercalcemia

Animal model studies showed that after injection of OPG, serum calcium concentrations significantly decreased in both normal mice and hypercalcemic nude mice carrying tumors associated with humoral hypercalcemia of malignancy. Urinary excretion of Ca, P, and creatinine did not change significantly after injection of OPG (Akatsu *et al.*, 1998). Yamamoto also found that OPG decreased serum calcium rapidly and dramatically when it was administered into rats with established hypercalcemia, but OPG did not increase urinary calcium excretion (Yamamoto *et al.*, 1998). These results suggest that OPG decreases serum Ca through its inhibitory effect on bone resorption. OPG has therapeutic potential for the treatment of hypercalcemic conditions such as malignancy-associated hypercalcemia. In tumor-bearing mice, OPG treatments reduced osteoclast activity from approximately 2-fold above normal to the subphysiological range but had no effects on tumor size, tumor-induced cachexia, or PTHrP levels. The potent effects of OPG in this humoral hypercalcemia of malignancy model also suggest a potential therapeutic role for OPG in the prevention and treatment of this disorder (Capparelli *et al.*, 2000). A xenograft model showed that OPG inhibits osteoclastic osteolysis and tumor survival when administered to severe combined immunodeficiency (SCID) mice injected within the tibia with a human prostate cancer cell line (Zhang *et al.*, 2001). Bisphosphonates is the standard treatment for cancer hypercalcemia. Success is achieved in more than 90% of the cases. Other classes of compounds are being tested for their antiosteoclastic activity, and animal studies suggest that OPG could be as efficient but act faster than bisphosphonates (Body, 2000). Long-term growth hormone (GH) treatment increases the accumulation of insulin-like growth factor I (IGF-I) and OPG in cortical and trabecular bone in patients with GH deficiency, and in turn this may lead to an increase in bone mass and improve skeletal biomechanical competence (Ueland *et al.*, 2002). The bone anabolic effect of GH replacement therapy could in part be mediated by a positive bone balance at each remodeling unit due to the inhibitory action of OPG on osteoclastogenesis (Lanzi *et al.*, 2003).

33.6.2 Osteoprotegerin and Bone-Tissue Engineering

OPG is a pro-osteogenic factor that can be used as an osteogenic supplement in a growth medium to prime the osteogenic capacity of undifferentiated hMSCs to enhance osteogenesis for bone-tissue engineering (Palumbo and Li, 2013). Mesenchymal stem cells can differentiate into osteoblasts and hold promise for applications of bone regeneration such as bone tissue engineering. Palumbo profiled soluble factors secreted from MG-63 cells using a comparative protein array and found that OPG, known as a potent antiosteoclastogenic protein, was at the highest relative level among 507 soluble molecules by the array detection. Furthermore, treating human mesenchymal stem cells (hMSCs) with OPG before osteogenic induction significantly increased the expression of osteocalcin mRNA transcription and the production of calcium deposits compared to the untreated control cells, suggesting that OPG is capable of priming undifferentiated hMSCs for the enhancement of subsequent osteogenesis. Furthermore, the nuclear factor-kappa B (NF-κB) was activated by OPG in undifferentiated hMSCs and that blocking NF-κB activation before osteogenic induction decreased osteogenesis of OPG-pretreated cells upon receiving osteogenic stimuli.

To summarize, the effect of dietary calcium on OPG is still unclear. Being a member of TNF receptor family, OPG is expressed extensively in the body. In skeletal system, OPG blocks the binding of RANK and its ligand to inhibit bone loss and increase bone-mineral density. The function of OPG in the pathogenesis of VCs needs further study. Only adequate dietary calcium is beneficial to people, excessive calcium cannot correct the disorder of bone metabolism induced by osteoporosis. The therapeutic potential of OPG may be widely used in the future.

Summary Points

- Osteoprotegerin is a decoy receptor for osteoclast differentiation factor RANKL. OPG not only inhibits the formation of osteoclasts but also affects the function and/or survival of mature osteoclasts.
- Osteoprotegerin regulates bone homeostasis, and its inactivation in mice results in severe osteoporosis.
- Osteoprotegerin can be used in the treatment of osteoporosis associated with increased osteoclast activity.
- The discovery and characterization of RANKL, RANK, OPG and subsequent studies have changed the concepts of bone and calcium metabolism, and have led to more detailed understanding of the pathogenesis of metabolic bone diseases, and may form the basis of innovative therapeutic strategies.

- Vascular calcification is a potent predictor of plaque instability and cardiac events. Osteoprotegerin, the well-known vascular calcification mediator, is a signaling molecule involved in bone remodeling, which has been involved in the regulation of vascular calcification and atherogenesis.

Key Facts

Inadequate calcium intake is a common phenomenon around the world. In order to avoid suffering the diseases related to low calcium intake, people should eat calcium-rich products to meet their calcium needs. OPG plays an important role in the bone metabolism. OPG can decrease serum calcium rapidly and dramatically. Although the relationship of dietary calcium and OPG is not clear, the clinical application of OPG in bone-tissue engineering may provide a novel method to treat bone disease.

Definitions of Words and Terms

Osteoprotegerin is known as an osteoclastogenesis inhibitor, which works primarily by binding osteoprotegerin ligand. OPG not only inhibits osteoclast formation, but also inhibits osteoclasts resorption.

Osteoporosis is a very common skeletal disorder characterized by reduced bone mass and altered trabecular microarchitecture that leads to bone fragility and fractures.

Juvenile Paget's disease is an extremely rare, yet painful and debilitating bone disease with onset occurring during early childhood. Increased bone turnover and lack of bone modeling lead to severe deformities, frequent fractures, short stature, and loss of hearing.

Chronic kidney disease is characterized by numerous metabolic, nutritional and cardiovascular complications including loss of muscle mass, osteoporosis and vascular calcification contributing to the high mortality among these patients.

Denosumab is the therapeutic RANKL-targeted antibody, a kind of recombinant RANKL antagonist.

Vascular calcification is a common pathological manifestation that usually occurs among the elder population and in patients with atherosclerosis, hypertension, diabetes vascular lesion, and chronic kidney disease. The main manifestation is the increased stiffness and the decreased compliance of vascular wall that is liable to induce myocardial ischemia, left ventricular hypertrophy and cardiac failure.

List of Abbreviations

OPG	Osteoprotegerin
TNF	Tumor necrosis factor
RANK	Receptor activator of nuclear factor-κ B

TRAIL	TNF-related apoptosis-inducing ligand
RANKL	Receptor activator of nuclear factor-κ B ligand
OCIF	Osteoclastogenesis inhibitory factor
TNFR-1	Tumor necrosis factor receptor-like molecule-1
MCSF	Macrophage-colony stimulating factor
BMDD	Bone-mineralization density distribution
BMD	Bone-mineral density
CKD	Chronic kidney disease
CV	Cardiovascular
CVD	Cardiovascular disease
VCs	Vascular calcifications
KTx	Kidney transplantation
JPD	Juvenile Paget's disease
CCA-IMT	Common carotid artery intima-media thickness
ROD	Renal osteodystrophy
T1D	Type-1 diabetes
T2D	Type-2 diabetes
IBD	Inflammatory bowel disease
hMSCs	Human mesenchymal stem cells
NF-κB	Nuclear factor-kappa B
SCID	Severe combined immunodeficiency
GH	Growth hormone
IGF-I	Insulin-like growth factor
SMCs	Smooth muscle cells

References

Akatsu, T., Murakami, T., Ono, K., Nishikawa, M., Tsuda, E., Mochizuki, S. I., Fujise, N., Higashio, K., Motoyoshi, K., Yamamoto, M. and Nagata, N., 1998. Osteoclastogenesis inhibitory factor exhibits hypocalcemic effects in normal mice and in hypercalcemic nude mice carrying tumors associated with humoral hypercalcemia of malignancy. *Bone*. 23(6): 495–498.

Ambroszkiewicz, J., Gajewska, J., Sands, D., Chełchowska, M., Ołtarzewski, M. and Laskowska-Klita, T., 2012. Assessment of selected bone metabolism marker concentrations in children with cystic fibrosis. *Medycyna wieku rozwojowego*. 16(2): 117–123.

Asanuma, Y., Chung, C. P., Oeser, A., Solus, J. F., Avalos, I., Gebretsadik, T., Shintani, A., Raggi, P., Sokka, T., Pincus, T. and Stein, C. M., 2007. Serum osteoprotegerin is increased and independently associated with coronary artery atherosclerosis in patients with rheumatoid arthritis. *Atherosclerosis*. 195(2): 135–141.

Bae, Y. J. and Kim, M. H., 2010. Calcium and magnesium supplementation improves serum OPG/RANKL in calcium-deficient ovariectomized rats. *Calcified Tissue International*. 87(4): 365–372.

Body, J. J., 2000. Current and future directions in medical therapy: hypercalcemia. *Cancer*. 88(suppl. 12): 3054–3058.

Bucay, N., Sarosi, I., Dunstan, C. R., Morony, S., Tarpley, J., Capparelli, C., Scully, S., Tan, H. L., Xu, W., Lacey, D. L., Boyle, W. J. and Simonet, W. S., 1998. Osteoprotegerin-deficient mice develop early onset osteoporosis and arterial calcification. *Genes & Development*. 12(9): 1260–1268.

Browner, W. S., Lui, L. Y. and Cummings, S. R., 2001. Associations of serum osteoprotegerin levels with diabetes, stroke, bone density, fractures, and mortality in elderly women. *Journal of Clinical Endocrinology and Metabolism*. 86(2): 631–637.

Capparelli, C., Kostenuik, P. J., Morony, S., Starnes, C., Weimann, B., Van, G., Scully, S., Qi, M., Lacey, L. and Dunstan, C. R., 2000. Osteoprotegerin prevents and reverses hypercalcemia in a murine model of humoral hypercalcemia of malignancy. *Cancer Research*. 60(4): 783–787.

Feskanich, D., Willett, W. C., Stampter, M. J. and Colditz, G. A., 1997. Milk, dietary calcium, and bone fractures in woman: a 12-year prospective study. *American Journal of Public Health*. 87(6): 992–997.

Fichna, M., Zurawek, M., Fichna, P., Gryczynska, M., Nowak, J. and Ruchala, M., 2012. Increased serum osteoprotegerin in patients with primary adrenal insufficiency receiving conventional hydrocortisone substitution. *Journal of Physiology and Pharmacology*. 63(6): 677–682.

Fahrleitner, A., Prenner, G., Leb, G., Tscheliessnigg, K. H., Piswanger-Sölkner, C., Obermayer-Pietsch, B., Portugaller, H. R., Berghold, A. and Dobnig, H., 2003. Serum osteoprotegerin is a major determinant of bone density development and prevalent vertebral fracture status following cardiac transplantation. *Bone*. 32(1): 96–106.

Gordin, D., Soro-Paavonen, A., Thomas, M. C., Harjutsalo, V., Saraheimo, M., Bjerre, M., Forsblom, C., Flyvbjerg, A. and Groop, P. H., 2013. Osteoprotegerin is an independent predictor of vascular events in Finnish adults with type 1 diabetes. *Diabetes Care*. 36(7): 1827–1833.

Hofbauer, L. C. and Schoppet, M., 2004. Clinical implications of the osteoprotegerin/RANKL/RANK system for bone and vascular diseases. *Journal of the American Medical Association*. 292(4): 490–495.

Haas, M., Leko-Mohr, Z., Roschger, P., Kletzmayr, J., Schwarz, C., Domenig, C., Zsontsich, T., Klaushofer, K., Delling, G. and Oberbauer, R., 2002. Osteoprotegerin and parathyroid hormone as markers of high-turnover osteodystrophy and decreased bone mineralization in hemodialysis patients. *American Journal of Kidney Diseases*. 39(3): 580–586.

Halladay, D. L., Miles, R. R., Thirunavukkarasu, K., Chandrasekhar, S., Martin, T. J. and Onyia, J. E., 2001. Identification of signal transduction pathways and promoter sequences that mediate parathyroid hormone 1–38 inhibition of osteoprotegerin gene expression. *Journal of Cellular Biochemistry*. 84(1): 1–11.

Janda, K., Krzanowski, M., Chowaniec, E., Kuśnierz-Cabala, B., Dumnicka, P., Kraśniak, A., Podolec, P. and Sułowicz, W., 2013. Osteoprotegerin as a marker of cardiovascular risk in patients on peritoneal dialysis. *Polskie Archiwum Medycyny Wewnetrznej*. 123(4): 149–155.

Jiang, S., Cui, L., Shi, C., Ke, X., Luo, J. and Hou, J., 2013. Effects of dietary energy and calcium levels on performance, egg shell quality and bone metabolism in hens. *Veterinary Journal*. 198(1): 252–258.

Josse, A. R., Atkinson, S. A., Tarnopolsky, M. A. and Phillips, S. M., 2012. Diets higher in dairy foods and dietary protein support bone health during diet- and exercise-induced weight loss in overweight and obese premenopausal women. *Journal of Clinical Endocrinology and Metabolism*. 97(1): 251–260.

Jorsal, A., Tarnow, L., Flyvbjerg, A., Parving, H. H., Rossing, P. and Rasmussen, L. M., 2008. Plasma osteoprotegerin levels predict cardiovascular and all-cause mortality and deterioration of kidney function in type 1 diabetic patients with nephropathy. *Diabetologia*. 51(11): 2100–2107.

Liu, J. D., Liu, H. X., Gu, J. H., Duan, X. J., Wang, J. and Liu, Z. P., 2009. Effects of calcium-deficiency diet on bone histomorphometry and expression of osteoprotegerin in Shan-ma laying ducks. *Chinese Journal of Veterinary Science*. 29(2): 217–223.

Knudsen, S. T., Foss, C. H., Poulsen, P. L., Andersen, N. H., Mogensen, C. E. and Rasmussen, L. M., 2003. Increased plasma concentrations of osteoprotegerin in type 2 diabetic patients with microvascular complications. *European Journal of Endocrinology*. 149(1): 39–42.

Logstrup, B. B., Hofsten, D. E., Christophersen, T. B., Moller, J. E., Bjerre, M., Flyvbjerg, A., Botker, H. E. and Egstrup, K., 2013. Microvascular dysfunction is associated with plasma osteoprotegerin levels in patients with acute myocardial infarction. *Coronary Artery Diseases*. 24(6): 487–492.

Lanzi, R., Losa, M., Villa, I., Gatti, E., Sirtori, M., Dal Fiume, C. and Rubinacci, A., 2003. GH replacement therapy increases plasma osteoprotegerin levels in GH-deficient adults. *European Journal of Endocrinology*. 148(2): 185–191.

Matsuo, K., 2013. Osteoimmunologic regulation of aging. *Clinical Calcium*. 23(1): 59–64.

Meneghini, M., Regalia, A., Alfieri, C., Barretta, F., Croci, D., Gandolfo, M. T., Vettoretti, S., Rastaldi, M. P. and Messa, P., 2013. Calcium and osteoprotegerin levels predict the progression of the abdominal aortic calcifications after kidney transplantation. *Transplantation*. 96(1): 42–48.

Marini, H., Minutoli, L., Polito, F., Bitto, A., Altavilla, D., Atteritano, M., Gaudio, A., Mazzaferro, S., Frisina, A., Frisina, N., Lubrano, C., Bonaiuto, M., D'Anna, R., Cannata, M. L., Corrado, F., Cancellieri, F., Faraci, M., Marini, R., Adamo, E. B., Wilson, S. and Squadrito, F., 2008. OPG and sRANKL serum concentrations in osteopenic, postmenopausal women after 2-year genistein administration. *Journal of Bone and Mineral Research*. 23(5): 715–720.

Mogelvang, R., Haahr-Pedersen, S., Bjerre, M., Frystyk, J., Iversen, A., Galatius, S., Flyvbjerg, A. and Jensen, J. S., 2013. Osteoprotegerin improves risk detection by traditional cardiovascular risk factors and hsCRP. *Heart*. 99(2): 106–110.

Morony, S., Capparelli, C., Lee, R., Shimamoto, G., Boone, T., Lacey, D. L. and Dunstan, C. R., 1999. A chimeric form of osteoprotegerin inhibits hypercalcemia and bone resorption induced by IL-1beta, TNF-alpha, PTH, PTHrP, and 1,25$(OH)_2D_3$. *Journal of Bone and Mineral Research*. 14(9): 1478–1485.

Nybo, M. and Rasmussen, L. M., 2008. The capability of plasma osteoprotegerin as a predictor of cardiovascular disease: a systematic literature review. *European Journal of Endocrinology*. 159(5): 603–608.

Omland, T., Drazner, M. H., Ueland, T., Abedin, M., Murphy, S. A., Aukrust, P. and de Lemos, J. A., 2007. Plasma osteoprotegerin levels in the general population: relation to indices of left ventricular structure and function. *Hypertension*. 49(6): 1392–1398.

Olesen, P., Ledet, T. and Rasmussen, L. M., 2005. Arterial osteoprotegerin: increased amounts in diabetes and modifiable synthesis from vascular smooth muscle cells by insulin and TNF-alpha. *Diabetologia*. 48(3): 561–568.

Park, Y., Moon, H. J., Paik, D. J. and Kim, D. Y., 2013. Effect of dietary legumes on bone-specific gene expression in ovariectomized rats. *Nutrition Research and Practice*. 7(3): 185–191.

Palumbo, S. and Li, W. J., 2013. Osteoprotegerin enhances osteogenesis of human mesenchymal stem cells. *Tissue Engineering*. 19(19–20): 2176–2187.

Pobeha, P., Petrasova, D., Tkacova, R. and Joppa, P., 2014. Circulatory osteoprotegerin is related to osteoporosis of the hip in patients with COPD. *Respiratory Medicine*. 108(4): 621–627.

Price, P. A., June, H. H., Buckley, J. R. and Williamson, M. K., 2001. Osteoprotegerin inhibits artery calcification induced by warfarin and by vitamin D. *Arteriosclerosis Thrombosis and Vascular Biology*. 21(10): 1610–1616.

Rashtchizadeh, N., Ghorbanihaghjo, A., Argani, H., Mahmoudi Meimand, S., Safa, J., Vatankhahan, H. and Shahidi, M., 2012. Serum receptor activator of nuclear factor-κ B ligand, osteoprotegrin, and intact parathyroid hormone in hemodialysis and renal transplant patients. *Therapeutic Apheresis and Dialysis*. 16(6): 600–604.

Revu, S., Neregård, P., Af Klint, E., Korotkova, M. and Catrina, A. I., 2013. Synovial membrane immunohistology in early-untreated rheumatoid arthritis reveals high expression of catabolic bone markers that is modulated by methotrexate. *Arthritis Research & Therapy*. 15(6): R205.

Rianthavorn, P., Ettenger, R. B., Salusky, I. B. and Kuizon, B. D., 2010. Trabecular bone volume and osteoprotegerin expression in uremic rats given high calcium. *Pediatric Nephrology*. 25(11): 2311–2319.

Saki, F., Karamizadeh, Z., Nasirabadi, S., Mumm, S., McAlister, W. H. and Whyte, M. P., 2013. Juvenile paget's disease in an Iranian kindred with vitamin D deficiency and novel homozygous TNFRSF11B mutation. *Journal of Bone and Mineral Research*. 28(6): 1501–1508.

Simonet, W. S., Lancey, D. L., Dunstan, C. R., Kelley, M., Chang, M. S., Lüthy, R., Nguyen, H. Q., Wooden, S., Bennett, L., Boone, T., Shimamoto, G., DeRose, M., Elliott, R., Colombero, A., Tan, H. L., Trail, G., Sullivan, J., Davy, E., Bucay, N., Renshaw-Gegg, L., Hughes, T. M., Hill, D., Pattison, W., Campbell, P., Sander, S., Van, G., Tarpley, J., Derby, P., Lee, R. and Boyle, W. J., 1997. Osteoprotegerin: a novel secreted protein involved in the regulation of bone density. *Cell*. 89(2): 309–319.

Suda, T., 2002. How is bone formed and resorbed? – Molecular mechanisms of bone formation and resorption. *Rinsho Byori*. 50(3): 267–272.

Ueland, T., Bollerslev, J., Flyvbjerg, A., Hansen, T. B., Vahl, N. and Mosekilde, L., 2002. Effects of 12 months of GH treatment on cortical and trabecular bone content of IGFs and OPG in adults with acquired GH deficiency: a double-blind, randomized, placebo-controlled study. *Journal of Clinical Endocrinology and Metabolism*. 87(6): 2760–2763.

Vernia, P., Loizos, P., Di Giuseppantonio, I., Amore, B., Chiappini, A. and Cannizzaro, S., 2014. Dietary calcium intake in patients with inflammatory bowel disease. *Journal of Crohn's & Colitis*. 8(4): 312–317.

Whiting, S. J. and Wood, R. J., 1997. Adverse effects of high calcium diets in humans. *Nutrition Reviews*. 55: 1–9.

Yamamoto, M., Murakami, T., Nishikawa, M., Tsuda, E., Mochizuki, S., Higashio, K., Akatsu, T., Motoyoshi, K. and Nagata, N., 1998. Hypocalcemic effect of osteoclastogenesis inhibitory factor/osteoprotegerin in the thyropara thyroidectomized rat. *Endocrinology*. 139(9): 4012–4015.

Yamaguchi, K., Kinosaki, M., Goto, M., Kobayashi, F., Tsuda, E., Morinaga, T. and Higashio, K., 1998. Characterization of structural domains of human osteoclastogenesis inhibitory factor. *Journal of Biological Chemistry*. 273(9): 5117–5123.

Yu, J., Gao, Y. H. and Sun, D. J., 2013. Effect of fluoride and low versus high levels of dietary calcium on mRNA expression of osteoprotegerin and osteoprotegerin ligand in the bone of rats. *Biological Trace Element Research*. 152(3): 387–395.

Zhang, J., Dai, J., Qi, Y., Lin, D. L., Smith, P., Strayhorn, C., Mizokami, A., Fu, Z., Westman, J. and Keller, E. T., 2001. Osteoprotegerin inhibits prostate cancer-induced osteoclastogenesis and prevents prostate tumor growth in the bone. *Journal of Clinical Investigation*. 107(10): 1235–1244.

Zheng, Y., Zhou, H., Modzelewski, J. R., *et al.*, 2007. Accelerated bone resorption, due to dietary calcium deficiency, promotes breast cancer tumor growth in bone. *Cancer Research*. 67(19): 9542–9548.

CHAPTER 34

Inadequate Calcium Intake and Body Fat in Adults

DIANE K. TIDWELL*[a] AND MATTHEW J. MCALLISTER[b]

[a]Department of Food Science, Nutrition and Health Promotion, Mississippi State University, Box 9805, 206 Herzer, 945 Stone Blvd, Mississippi State, MS 39762, USA; [b]Department of Kinesiology, Mississippi State University, Box 9575, 124 McCarthy Gym, Mississippi State, MS 39762, USA
*E-mail: dtidwell@fsnhp.msstate.edu

34.1 Introduction

As the most abundant mineral in the body, calcium is metabolically involved in numerous physiological functions including fat metabolism. Many analytical and observational studies deduced that calcium affects adiposity (Heaney *et al.*, 2002; Zemel, 2009) with perhaps more than one mechanism of action (Major *et al.*, 2008). Due to the strong relationship between calcium and the calcitrophic hormone, 1,25-dihydroxyvitamin D_3 (1,25-D_3) or calcitriol, investigations have also reported that both calcium and vitamin D affect fat metabolism (Zemel, 2009; Rosenblum *et al.*, 2012; Zhu *et al.*, 2013). Although the relationship between dietary calcium and body fat was not investigated for years after the serendipitous reporting of an inverse relationship between dietary calcium and body weight by McCarron *et al.* (1984), a large body of evidence has since appeared in the literature. Zemel, a pioneer investigator in calcium and obesity research, referred to calcium as having an antiobesity effect (Zemel, 2002). Zemel *et al.* (2000) published one of the

Food and Nutritional Components in Focus No. 10
Calcium: Chemistry, Analysis, Function and Effects
Edited by Victor R. Preedy

Published by the Royal Society of Chemistry, www.rsc.org

first studies that investigated calcium's role in adiposity. Furthermore, Zemel (2009) has investigated the association between adiposity and whey protein, which contains calcium, branched-chain amino acids and small peptides. Research concerning calcium and its involvement with obesity continues with investigators such as He *et al.* (2012); Laraichi *et al.* (2013), and Marotte *et al.* (2013) making significant strides in identifying complex interrelated mechanisms associated with calcium and adiposity. Additionally, numerous cross-sectional, prospective, longitudinal, clinical, and intervention studies have reported inverse relationships between calcium intake and body fat, and calcium and body-mass index (BMI).

34.2 Randomized Clinical Trials Investigating Calcium and Body Fat in Adults

Several randomized controlled trials (RCTs) have reported positive effects of adequate calcium intakes and positive weight-management outcomes and decreases in excess body fat. Zhou *et al.* (2010) reported from a double-blind, placebo RCT with 870 postmenopausal women that the calcium intervention groups gained less trunk fat and maintained more trunk lean mass compared to the placebo group over four years. A 16 week RCT with 90 premenopausal overweight and obese women, who were randomized into three groups, determined that a reduction in visceral adipose tissue was significantly correlated with calcium and protein intakes (Josse *et al.*, 2011). The group consuming the high-dairy, high-protein diet with a mean calcium intake of 1840 mg per day lost more visceral adipose tissue than the group consuming adequate protein and low dairy, which contained a much lower calcium intake of 299 mg per day; all women had low calcium intakes prior to the study (Josse *et al.*, 2011). In two 16 week parallel double-blind, placebo RCTs with 171 overweight or obese men and women aged 18 to 65 years, Rosenblum *et al.* (2012) reported calcium and/or vitamin D supplementation significantly contributed to a reduction in visceral adipose tissue. Additionally, the RCT conducted by Shahar *et al.* (2010) with 322 men and women with a mean age of 52 years and BMI of 31 kg m^{-2} concluded that higher levels of dietary calcium and increased serum levels of vitamin D were associated with increased weight loss over a 2 year period.

A large randomized, double-blind, placebo RCT involving 36 282 postmenopausal women concluded that calcium plus 1,25-D_3 supplementation had a small effect on preventing weight gain over three years; these results were primarily observed in women who reported prior low calcium intakes (Caan *et al.*, 2007). A recent RCT conducted by Zhu *et al.* (2013) concluded that calcium plus vitamin D supplementation for 12 weeks facilitated body fat and visceral fat loss in low calcium-consuming college students during energy restriction. Fifty-three participants aged 18 to 25 years were randomly assigned to receive either an energy-restricted diet (500 kcal less per day) supplemented with 600 mg of elemental calcium and 125 International units of vitamin D_3 or only energy restriction. The supplemented group lost more fat mass compared to the control group, 2.8 ± 1.3 kg (standard deviation) *vs.* 1.8 ± 1.3 kg, respectively (Zhu *et al.*, 2013). Results from a 15 week double blind, placebo RCT with 63 overweight or obese women (mean age 43 years, BMI 32) indicated that participants in a weight-reduction program with calcium and vitamin D supplementation, and were low calcium consumers of 600 mg per day or less, lost more body weight and fat mass compared to the placebo group (Major *et al.*, 2009). Zemel *et al.* (2004) conducted a 24 week placebo RCT with 32 obese adults consuming energy-restricted diets of 500 kcal per day deficit. Increasing dietary calcium with 800 mg per day of calcium supplementation significantly augmented weight and fat loss secondary to energy restriction and a greater effect was seen in the high dairy group (Zemel *et al.*, 2004).

A meta-analysis with 14 RCTs representing 883 adults aged 18 to 85 years reported that including dairy products, which contributed to higher amounts of calcium, in energy-restricted weight-loss diets caused a significant decrease in body weight, body-fat mass, and waist circumference, and a gain in lean body mass compared to the controls, which received energy-restricted diets without increased dairy consumption (Abargouei *et al.*, 2012). Another meta-analysis with seven RCTs evaluating the use of calcium supplements concluded that calcium supplementation for at least six months resulted in significant weight loss in obese and overweight people (Onakpoya *et al.*, 2011). The mean ages of participants in the seven trials ranged from 36.6 to 61.6 years and the number of participants ranged from 43 to 87 individuals. Total calcium intakes ranged from 1600 to over 2500 mg per day (Onakpoya *et al.*, 2011), which exceeded the Institute of Medicine's (2011) recommended dietary allowances of 1000 to 1300 mg per day.

There are also RCTs that do not support the hypothesis that calcium has an effect on fat metabolism. Boon *et al.* (2005) reported that altering the calcium content using calcium carbonate supplements did not influence fat metabolism in 12 healthy men (age range 20–40 years, BMI range 21.2–27.9) in three 7 day dietary interventions in a randomized crossover design. Bortolotti *et al.* (2008) reported that calcium supplementation in 10 overweight or obese men and women with habitually low calcium intakes failed to alter fat metabolism in a placebo-controlled crossover 5 week trial. Barr (2003) reviewed calcium supplementation on changes in body weight and body composition in 17 placebo RCTs with interventions lasting three months to four years. Only one trial observed greater weight loss in the group receiving 1200 mg of calcium supplements per day compared to the control group over a 4 year period. Overall, Barr (2003) concluded that increased dairy product or calcium intake provided little evidence to support the hypothesis that calcium, or other components of dairy, have a measurable impact on body weight in generally healthy people.

Lanou and Barnard (2008) also reported that the overall evidence from 49 RCTs does not support the hypothesis that calcium or dairy consumption aided in weight or fat loss. Twenty of the RCTs were conducted in adults without energy restriction and included the use of calcium supplements or dairy products; one RCT reported a small reduction in the rate of weight gain with supplemented calcium and another RCT observed greater weight loss in the supplemented group. The other 18 trials with adults did not report significant differences between the randomized groups except two trials observed an increase in weight gain with an increase in dairy-product consumption (Lanou and Barnard, 2008). Chen *et al.* (2012) also conducted a meta-analysis with 29 RCTs and concluded that increasing dairy consumption did not support a beneficial effect on body weight and fat loss in long-term studies or studies without energy restriction; however, dairy products may have modest benefits with weight loss in short-term or energy-restricted trials.

Trowman *et al.* (2006) conducted a meta-analysis with 13 RCTs and concluded that calcium supplementation and a reduction in body weight were not significantly associated after adjusting for differences in baseline body weights between the control and intervention groups. The 13 trials lasted from 12 weeks to 36 months and consisted of 23 to 223 participants (men and women with mean ages of 23.7 to 72 years); it was noted that only one RCT conducted by Zemel *et al.* (2004) was specifically designed and powered to examine the effect of calcium supplements or dairy products on body weight. The majority of RCTs were investigating bone mass; however, body weight and calcium-supplementation information was reported and available for review (Trowman *et al.*, 2006). This has been a criticism in calcium and obesity research with some trials reporting calcium and body fat results when the studies were not specifically designed for that intention or outcome. However, Heaney and Rafferty (2009) stated that clinical trials are less suited for evaluating the effects of nutrients and more applicable for evaluating the effects of drugs, and also, clinical trials have limited generalizability. Evidence can be critically appraised with sound conclusions when RCTs, which can determine cause and effect relationships, are in concordant with observational studies that contribute to generalizability (Heaney and Rafferty, 2009).

34.3 Observational Studies Investigating Calcium and Body Fat in Adults

Numerous observational studies have reported significant inverse relationships between calcium intake and body fat or dietary calcium and BMI. Using data from the National Health and Nutrition Examination Survey III, Zemel *et al.* (2000) observed that for any given level of energy intake and physical activity, diets with lower amounts of calcium favored increased adipose tissue and the converse was true for diets with higher amounts of calcium. Jacqmain *et al.* (2003) examined calcium intake and body composition in 470 men and women aged 20 to 65 years and concluded that a low daily calcium intake was strongly associated with adiposity, especially in women. Women with calcium intakes less than 600 mg per day ($n = 52$) had a mean BMI of 31.8 and body fat of 37.3%, which was significantly greater than for women consuming

600 to 1000 mg per day (BMI 27.0, body fat 31.3%, n = 113). Women consuming greater than 1000 mg of calcium per day (n = 70) had the lowest BMI of 25.2 and 28.9% body fat (Jacqmain *et al.*, 2003). Tidwell and Valliant (2011) reported similar results; women who consumed less than 600 mg of calcium per day (n = 40) had an obese mean BMI of 35.4 and 42.4% body fat, which was greater than for women (n = 40) who consumed 600 to 1000 mg of calcium per day (BMI 28.0, body fat 36.4%). Women consuming more than 1000 mg of calcium per day (n = 20) had a significantly lower BMI of 22.5 and body fat of 27.0% compared to the other groups of women (Tidwell and Valliant, 2011).

A recent study reported that women with lower intakes of calcium had higher values for BMI, waist circumference, waist : height ratio, and percentage of body fat (Da Silva Ferreira *et al.*, 2013). Women (mean age 31.4 years, BMI 26.2, n = 76) with high calcium intakes had a lower odds ratio for prevalent overweight, obesity, and abdominal obesity (Da Silva Ferreira *et al.*, 2013). A study with 63 year-old men (n = 301) reported an inverse association between calcium intake and sagittal abdominal obesity (Rosell *et al.*, 2004). Mirmiran *et al.* (2005) reported that as dairy consumption, and consequently calcium intake, increased in men (38 ± 15 years old, n = 223) and women (32 ± 13 years old, n = 239), the number of men and women with a BMI less than 25 increased and the number of obese people, BMI of 30 or greater, significantly decreased. A large retrospective study with 10 591 men and women aged 53 to 57 years concluded that calcium supplements taken for other reasons, such as osteoporosis prevention, may have a small beneficial effect on reducing weight gain especially in women during midlife (Gonzalez *et al.*, 2006). Ochner and Lowe (2007) concluded from their longitudinal study that dietary calcium may oppose weight regain by reducing the effect of greater energy intake; their study was comprised of a 22 week weight loss program with 103 overweight or obese women who had 6 month and 18 month follow-up sessions. However, Samara *et al.* (2013) reported higher calcium consumption was associated with a lower 5 year increase in BMI and waist circumference in men (n = 288) but not in the women (n = 300).

Murakami *et al.* (2006) did not observe a significant relationship between calcium intake and BMI in a cross-sectional study with 1905 young, lean, Japanese women aged 18 to 20 years with a BMI of 20.8. They concluded that intakes of calcium, and also dairy products, may not be associated with BMI in young Japanese women who were not only lean, but also had rather low intakes of calcium. However, a study with 2975 American Indians aged 47 to 79 years concluded that dietary calcium was not associated with BMI and body fat and it was noted that the population has a high prevalence of obesity (Eilat-Adar *et al.*, 2007). Venti *et al.* (2005) also reported nonsignificant associations between calcium intake and body weight, body fat, or BMI in Pima Indians, which is an obesity-prone population. The study contained 65 men and women aged 33.0 ± 7.0 years with a BMI of 33.3 ± 8.1 and body fat percentage of 40.0 ± 10.0%. Mean daily calcium intake was close to the recommended intake of 1000 mg per day; however, mean energy intake was approximately 3000 kcal per day and daily fat intake was 40%, which exceeded recommended amounts.

The authors concluded that the high prevalence of obesity and the high-energy, high-fat diet may have masked a small influence of calcium intake on body weight (Venti *et al.*, 2005). In a large cross-sectional study by Kamycheva *et al.* (2002) with 9252 men (mean age 43.9 years, BMI 25.6) and 9662 women (mean age 42.8 years, BMI 24.4), it was observed, after correcting for variables such as age in the multiple regression analysis, that BMI and calcium intake were positively related in men but there was no association in women.

34.4 The Biological Plausibility of Calcium's Effect on Body Fat

Determining the specific effect that calcium has on body fat in humans has been challenging due to the complex nature that diet and other life-style factors have on an individual's body fat status, in addition to the complexity of the endocrine system and factors that affect absorption and bioavailability of calcium. Research has focused on calcium and definitive mechanisms to explain dietary calcium's ability to modulate body fat. Zemel *et al.* (2000) observed that a low-calcium (0.4%), high-fat, and high-sucrose diet increased adipocyte lipogenesis, inhibited lipolysis, and accelerated increases in body weight and adipose tissue mass in agouti transgenic mice. This potential relationship is demonstrated in Figure 34.1. The high

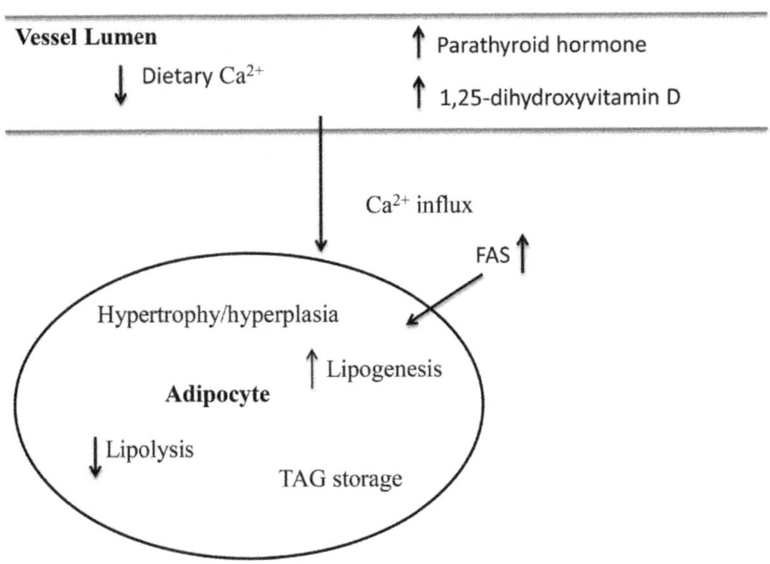

Figure 34.1 Dietary calcium's influence on adipocytes. Low levels of dietary calcium (Ca^{2+}) increase 1,25-dihydroxyvitamin D and parathyroid hormone levels, which stimulate an influx of Ca^{2+} into adipocytes. This effect is known to be associated with an inhibition of lipolysis and promotion of lipogenesis by increasing fatty acid synthase (FAS) activity. Elevated FAS activity promotes triglyceride (TAG) storage and possibly adipocyte hypertrophy/hyperplasia.

calcium (1.2%) diet reduced lipogenesis by 51% and stimulated lipolysis that resulted in 26% to 39% reductions in body weight and adipose tissue (Zemel *et al.*, 2000). It has also been shown that low dietary calcium can increase calcitriol levels, increase cortisol production, which is associated with visceral adipose accumulation, and suppress uncoupling protein 2 (UCP2) *via* action of the nuclear vitamin D receptor (nVDR) through the action of 1,25-D_3-membrane activated rapid response to steroids (1,25-D_3-MARRS) (Zemel and Sun, 2008), which can affect adipose apoptosis, lipolysis, and lipogenesis (Figure 34.2). Low dietary calcium may also affect additional hormones secreted from white adipose tissue including resistin, adiponectin (De Angel *et al.*, 2009), interleukins, and tumor necrosis factor-alpha (TNF-α) (Zemel and Sun, 2008), which can affect metabolic health (Figure 34.3).

Zemel *et al.* (2000) confirmed that increasing dietary calcium suppresses 1,25-D_3 and parathyroid hormone, and thereby reduces adipocyte intracellular calcium and lipid storage. High-calcium diets inhibit adipocyte fatty acid synthase (FAS) activity and stimulate lipolysis (Zemel *et al.*, 2000). Also, high intakes of calcium reduce calcitriol turnover through its action on parathyroid hormone suppression and conversely, low dietary calcium increases calcitriol. These actions affect lipid metabolism through FAS

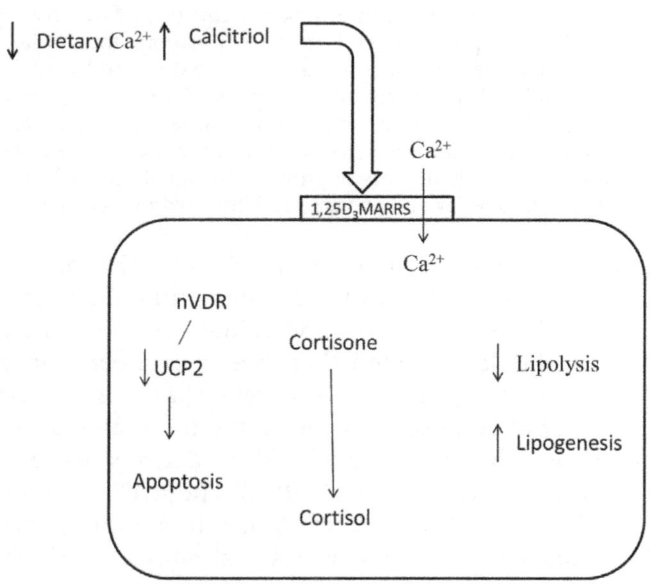

Figure 34.2 Dietary calcium's effect on selected cellular functional outcomes. Low dietary calcium (Ca^{2+}) increases calcitriol (also known as 1,25-dihydroxyvitamin D) and through the action of 1,25-D_3-membrane activated rapid response to steroids (1,25-D_3-MARRS), suppresses uncoupling protein 2 (UCP2) *via* the nuclear vitamin D receptor (nVDR). These events can affect adipose apoptosis, decrease lipolysis, increase lipogenesis, and increase cortisol production, which is associated with visceral adipose accumulation.

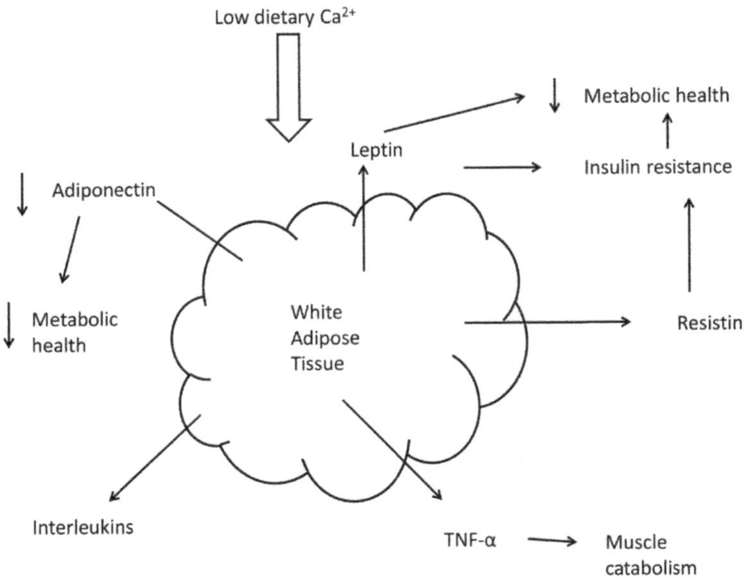

Figure 34.3 Low dietary calcium and possible metabolic occurrences. Low dietary calcium (Ca^{2+}) may affect additional hormones and adipokines secreted from white adipose tissue including, but not limited to, resistin, adiponectin, interleukins, and tumor necrosis factor alpha (TNF-α). Adiponectin is known to promote metabolic health and tends to be secreted in lower amounts in obese individuals. Inflammatory molecules such as interleukins and TNF-α play important physiological roles; however, they can also be potentially dangerous and contribute to atherosclerosis. Elevated resistin secretion is strongly associated with insulin resistance. Elevated leptin levels are associated with obesity and elevated risk for cardiovascular disease.

activity and other possible mechanisms (Figure 34.4) that provide evidence that an increase in serum calcitriol is a contributing factor in excess adiposity. High calcitriol levels occur in individuals with low calcium intakes (Zemel, 2009). It should be noted that excessive visceral adipose tissue accumulation is potentially dangerous for several reasons. In terms of metabolic health, visceral adipose cells contribute to inflammation and cytokine production, which are associated with oxidative stress and excessive production of reactive oxygen species (ROS). The ROS are associated with cardiovascular disease and insulin resistance. In addition, glucocorticoid secretion is associated with excessive visceral adiposity, which can exacerbate the aforementioned. The potential dangers of calcium deficiency on metabolic health, as well as visceral adipose tissue accumulation are depicted in Figure 34.5.

Additionally, higher intakes of calcium may increase fat excretion. Bendsen *et al.* (2008) reported that increasing the intake of calcium from low-fat dairy products by 1600 mg per day for seven days doubled the excretion of fecal fat from 5.4 to 11.5 g per day. Saturated, monounsaturated, and polyunsaturated

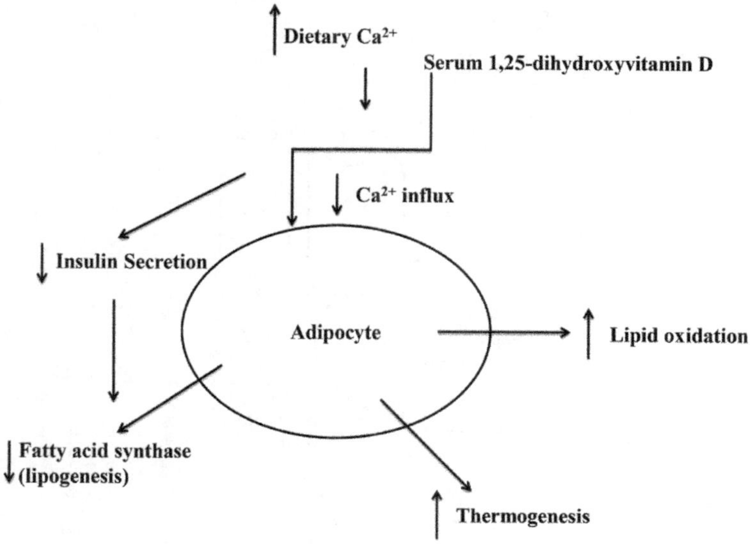

Figure 34.4 Increased dietary calcium and decreased obesity risk. High dietary calcium (Ca^{2+}) suppresses serum 1,25-dihydroxyvitamin D (also known as calcitriol) levels that is associated with properly maintained insulin regulation. Improper insulin regulation, such as chronic hyperinsulinemia, is associated with elevated fatty acid synthase (FAS) activity and hence, optimal maintenance of this hormone is associated with decreased FAS activity. Increased dietary Ca^{2+} is also known to promote adipocyte thermogenesis and β-oxidation. Each of these metabolic responses can potentially result in weight loss or decreased obesity risk.

fatty acids were all excreted in larger amounts on the diet with high calcium in a randomized crossover trial with 11 healthy participants (five men and six women) with a mean age of 33 years and BMI of 27.6 (Bendsen *et al.*, 2008). Lorenzen and Astrup (2011) also reported higher fecal-fat excretion during high-calcium diet periods compared to low-calcium diet periods in nine healthy men (mean age 32.8 years, BMI 28.1) in a randomized crossover design study. In a meta-analysis by Christensen *et al.* (2009) with 13 RCTs, it was determined that increasing dairy calcium by 1241 mg per day resulted in increased fecal fat of 5.2 g per day. These findings are significant since this mechanism could contribute to increased fecal-fat excretion to an extent that could be relevant for preventing weight gain or weight regain (Christensen *et al.*, 2009).

Another mechanism to explain dietary calcium's ability to modulate body fat is the effect calcium has on fat oxidation. Lower calcium intakes are related to reduced rates of fat oxidation, which has been shown to be a predictor of weight gain (Melanson *et al.*, 2003). Fat oxidation was significantly increased in 35 healthy adults (mean age 31.6 years, BMI 23.7, body fat 21.4%) in energy-deficit conditions on a high-calcium diet based on dairy products compared to those on a low-dairy, low-calcium diet (Melanson

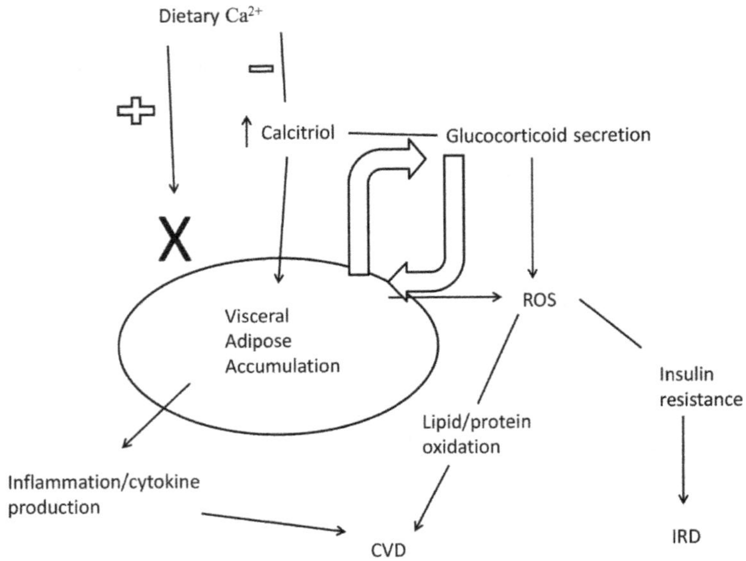

Figure 34.5 Potential effect of high or low dietary calcium on metabolic and cardiovascular health. An increased level of dietary calcium (Ca^{2+}) potentially blunts visceral adipose tissue accumulation. Concurrently, excessive visceral adipose tissue accumulation is associated with low dietary calcium. Visceral adipose is well known to contribute to elevated oxidative stress and elevated cardiovascular disease (CVD) risk *via* reactive oxygen species (ROS) and adipokine production. Oxidative stress is also associated with insulin resistance that can contribute to the development of insulin resistant diabetes (IRD). Chronic glucocorticoid, *i.e.* cortisol, secretion has been shown to result from (and potentially contribute to) visceral adipose tissue accumulation. Chronic hypercortisolemia is also associated with elevated CVD risk, insulin resistance, and oxidative stress.

et al., 2003). Additionally, Ping-Delfos and Soares (2011) reported that higher calcium and vitamin D intakes at a breakfast meal increased postprandial fat oxidation rates and diet-induced thermogenesis and reduced energy intake in the subsequent 24 h in a randomized, within-subject comparison of 11 men and women with a mean age of 54 years.

Evidence by Tardoff (2001) suggests there is an effect of calcium homeostasis on variables influenced by appetite control; inadequate intakes of calcium may lead to increased intakes in fat and energy. Calcium may influence appetite and when a person is deficient in calcium, or other mineral, the person may innately seek out sources of that mineral and hunger is increased (Major *et al.*, 2008). Calcium supplements of 1200 mg per day for three months reduced food cravings in women usually suffering from cravings such as sweets or salt. A 15 week calcium plus vitamin D double blind RCT using *ad libitum* buffet-type meals determined that women with initial calcium intakes of 600 mg per day or less (n = 7) had a significant decrease in lipid intake when taking supplements as opposed to the placebo group

(n = 6) that did not decrease their lipid intake (Major *et al.*, 2009). Astrup *et al.* (2010) stated that there is evidence suggesting that inadequate intakes of calcium during energy-restricted, weight-loss programs may trigger hunger and impair dietary compliance. Research also indicates that humans have taste receptors for calcium in the gastrointestinal tract and signaling may be linked to appetite regulation (Astrup *et al.*, 2010).

Calcium probably affects body composition and fat mass by several physiologic mechanisms with 1,25-D_3 interacting in a complex system of absorptive and endocrine functions. Furthermore, bioactive compounds in whey protein must be considered when examining the influence of consuming dairy products and body fat. A high proportion of branched-chain amino acids occurs in dairy products. Zemel *et al.* (2005) stated that branched-chain amino acids, and especially leucine, have specific metabolic roles as energy substrates and muscle protein synthesis. Also, cortisol production is stimulated by 1,25-D_3 in human adipocytes (Zemel *et al.*, 2005). Since high-calcium diets suppress 1,25-D_3, Zemel *et al.* (2005) proposed that loss of central adiposity in African-American adults who consumed high-dairy diets may be attributable, in part, to suppression of 1,25-D_3 levels and a consequent reduction in cortisol production by visceral adipocytes. Consuming dairy products, specifically dairy proteins, has also been shown to suppress short-term food intake, increase satiety, and stimulate food-intake regulatory mechanisms known to signal satiation and satiety (Major *et al.*, 2008).

34.5 Summary

Although calcium is an important nutrient for adipocyte metabolism, energy intake is important. When we consume more energy than we expend, fat stores will increase regardless of calcium intake. Teegarden (2003) iterated that energy balance is the most important factor in obesity; however, calcium may play a substantial contributing role in reducing the incidence of obesity. Major *et al.* (2008) concluded that the scientific evidence supports that inadequate intakes of calcium may increase the risk of positive energy balance. The effect of calcium on the regulation of body weight appears to be exerted mainly during energy restriction and in persons with low habitual calcium intakes. This seems to be the key in calcium's relationship to body fat; chronic or habitually low levels of dietary calcium may contribute to adiposity and obesity.

Boon *et al.* (2005) deduced from their research that there is a calcium-intake threshold of approximately 800 mg per day, and higher amounts had no additional beneficial effect on body composition. Heaney and Rafferty (2009) stated that "many nutrients (perhaps most) exhibit threshold behavior" and a low intake clearly below the threshold (control group) is needed to allow a nutrient effect to be detected by the high-intake group. Heaney (2008) presented functional outcomes of interactions between calcium and vitamin D, and concluded that vitamin D status and the benefits associated with vitamin D appear to be dependent on, or augmented by, calcium intakes at or above currently recommended amounts. However, many individuals are not consuming adequate amounts of calcium or vitamin D and their health may be adversely affected. Although there may not be scientific consensus on calcium's role in body composition, Heaney *et al.* (2002) stated that improving calcium intake has the potential to cause a substantial reduction in the prevalence of obesity. Also, Heaney and Rafferty (2009) stated that while adult calcium intakes in the United States are generally below recommended amounts and obesity rates are not improving, advocating for increased calcium intake could be a safe and inexpensive strategy for improving body composition. When examining all the evidence for calcium and its effect or influence on adiposity and obesity, it seems reasonable that promoting adequate intakes of calcium is warranted.

Summary Points

- This chapter focuses on calcium intake and body fat in adults.
- Calcium has a role in fat metabolism.
- Complex biological mechanisms may influence fat accumulation in the presence of inadequate intakes of calcium, which can contribute to the development of obesity.
- Numerous analytical and observational studies have strongly associated inadequate dietary calcium and obesity in adults.
- Therefore, the intake of calcium may be an important factor in obesity prevention.

Key Facts

Listed here are interesting facts about calcium intake and body fat including how the area of calcium and obesity research came into fruition and continues today.

1. Inadequate calcium intake and obesity has a relatively long history of association.
2. The association between inadequate dietary calcium and obesity began in the 1980s by McCarron *et al.* (1984) but was not expounded upon until the seminal publication by Zemel *et al.* (2000) appeared in the literature, which sparked a deluge in calcium and obesity research.
3. Michael Zemel is a pioneer in calcium and obesity research but began his career primarily in the area of calcium and its effect on hypertension.
4. Zemel has promoted the consumption of dairy products, which contain calcium in addition to other nutritious components that may have a greater positive effect on calcium and body fat compared to calcium supplements.
5. Research investigating the specific roles that calcium plays regarding the accumulation of body fat continues today.
6. An adequate intake of calcium is important for overall health and could possibly help decrease the prevalence of obesity.

Definitions and Explanations of Key Terms

1,25-Dihydroxyvitamin D_3. The hormonally active form of vitamin D. The structure contains three hydroxyl groups. It is also called calcitriol or 1,25-dihydroxycholecalciferol.

Adipocyte. A cell specialized in storing fat. Adipocytes contain adipose (fat) and store energy as fat.

Adiposity. The accumulation of fat in adipose tissue.

Body-mass index. An index measurement obtained by dividing a person's weight in kilograms by the square of the person's height in meters (kg m^{-2}).

Calcium-sensing receptor. A protein-coupled receptor that senses and affects extracellular levels of calcium ion.

Calcitriol. The hormonally active form of vitamin D. It is also called 1,25-dihydroxyvitamin D_3 or 1,25-dihydroxycholecalciferol.

Fatty-acid synthase. A multienzyme protein that catalyzes the synthesis of fatty acids.

Hormone sensitive lipase. An enzyme capable of hydrolyzing tri-, di-, and monoacylglyerol and other ester compounds. The main function of hormone sensitive lipase is to mobilize stored fat by catalyzing the release of fatty acids from adipose tissue.

Lipogenesis. The metabolic production of fat. The formation of fatty acids from acetyl coenzyme A from carbohydrate or protein, which is stored in adipocytes as triacylglyerols. It is also called adipogenesis.

Lipolysis. The metabolic breakdown of fat. The breakdown of fat involves the hydrolysis of adipose tissue compounds such as triacylglyerols into glycerol and fatty acids.

Obesity. The condition of having an excess amount of body fat, which can have a negative impact on health. Adults are considered obese when their body-mass index is 30 kg m^{-2} or higher.

White adipose tissue. Tissue containing white adipocytes, the other type is brown adipose tissue. A white adipocyte contains a large lipid (fat) droplet, cytoplasm, and a flattened nucleus that is located on the periphery of the adipocytes. White adipose tissue secretes proteins such as adiponectin, leptin, and resistin. Adipocytes increase in size with positive energy balance and when they reach maximum capacity, they may replicate to allow more fat storage.

List of Abbreviations

1,25-D_3	1,25-Dihydroxyvitamin D
1,25-D_3-MARRS	1,25-Dihydroxyvitamin D_3-membrane activated rapid response to steroids
BMI	Body-mass index
FAS	Fatty-acid synthase
IU	International units
kg m^{-2}	Kilograms divided by squared meters

mg	Milligram
nVDR	Nuclear vitamin D receptor
ROS	Reactive oxygen species
TNF-α	Tumor necrosis factor-alpha
UCP2	Uncoupling protein 2

References

Abargouei, A. S., Janghorbani, M., Salehi-Marzijarani, M. and Esmaillzadeh, A., 2012. Effect of dairy consumption on weight and body composition in adults: a systematic review and meta-analysis of randomized controlled clinical trials. *International Journal of Obesity*. 36: 1485–1493.

Astrup, A., Chaput, J.-P., Gilbert, J.-A. and Lorenzen, J. K., 2010. Dairy beverages and energy balance. *Physiology & Behavior*. 100: 67–75.

Barr, S., 2003. Increased dairy product or calcium intake: is body weight or composition affected in humans? *Journal of Nutrition*. 133: 245S–248S.

Bendsen, N. T., Hother, A.-L., Jensen, S. K., Lorenzen, J. K. and Astrup, A., 2008. Effect of dairy calcium on fecal-fat excretion: a randomized crossover trial. *International Journal of Obesity*. 32: 1816–1824.

Boon, N., Koppes, L. L. J., Saris, W. H. M. and Van Mechelen, W., 2005. The relation between calcium intake and body composition in a Dutch population. *American Journal of Epidemiology*. 162: 27–32.

Bortolotti, M., Rudelle, S., Schneiter, P., Vidal, H., Loizon, E., Tappy, L. and Acheson, K. J., 2008. Dairy calcium supplementation in overweight or obese persons: its effect on markers of fat metabolism. *American Journal of Clinical Nutrition*. 88: 877–885.

Caan, B., Neuhouser, M., Aragaki, A., Lewis, C. B., Jackon, R., LeBoff, M. S., Margolis, K. L., Powell, L., Uwaifo, G., Whitlock, E., Wylie-Rosett, J. and LaCroix, A., 2007. Calcium plus vitamin D supplementation and the risk of postmenopausal weight gain. *Archives of Internal Medicine*. 167: 893–902.

Chen, M., Pan, A., Malik, V. S. and Hu, F. B., 2012. Effects of dairy intake on body weight and fat: a meta-analysis of randomized controlled trials. *American Journal of Clinical Nutrition*. 96: 735–747.

Christensen, R., Lorenzen, J. K., Svith, C. R., Bartels, E. M., Melanson, E. L., Saris, W. H., Tremblay, A. and Astrup, A., 2009. Effect of calcium from dairy and dietary supplements on fecal-fat excretion: a meta-analysis of randomized controlled trials. *Obesity Reviews*. 10: 475–486.

De Angel, R. E., Berrigan, D., Nunez, N. P., Hursting, S. D. and Perkins, S. N., 2009. Dietary calcium source influences body composition, glucose metabolism and hormone levels in a mouse model of postmenopausal obesity. In Vivo. 23: 527–536.

Da Silva Ferreira, T., Torres, M. R. S. G. and Sanjuliani, A. F., 2013. Dietary calcium intake is associated with adiposity, metabolic profile, inflammatory state and blood pressure, but not with erythrocyte intracellular calcium and endothelial function in healthy pre-menopausal women. *British Journal of Nutrition*. 110: 1079–1088.

Eilat-Adar, S., Xu, J., Loria, C., Mattil, C., Goldbourt, U., Howard, B. V. and Resnick, H. E., 2007. Dietary calcium is associated with body mass index and body fat in American Indians. *Journal of Nutrition*. 137: 1955–1960.

Gonzalez, A. J., White, E., Kristal, A. and Littman, A. J., 2006. Calcium intake and 10 year weight change in middle-aged adults. *Journal of the American Dietetic Association*. 106: 1066–1073.

He, Y.-H., He, Y., Liao, X.-L., Niu, Y.-C., Wang, G., Zhao, C., Wang, L., Tian, M.-J., Li, Y. and Sun, C. H., 2012. The calcium-sensing receptor promotes adipocyte differentiation and adipogenesis through PPARγ pathway. *Molecular and Cellular Biochemistry*. 361: 321–328.

Heaney, R. P., 2008. Vitamin D and calcium interactions: functional outcomes. *American Journal of Clinical Nutrition*. 88: 541S–544S.

Heaney, R. P., Davies, K. M. and Barger-Lux, J., 2002. Calcium and weight: clinical studies. *Journal of the American College of Nutrition*. 21: 152S–155S.

Heaney, R. P. and Rafferty, K., 2009. Preponderance of the evidence: an example from the issue of calcium intake and body composition. *Nutrition Reviews*. 67: 32–39.

Institute of Medicine, 2011. *Dietary Reference Intakes for Calcium and Vitamin D*. The National Academies Press, Washington, DC.

Jacqmain, M., Doucet, E., Despres, J. P., Bouchard, C. and Tremblay, A., 2003. Calcium intake, body composition, and lipoprotein–lipid concentrations in adults. *American Journal of Clinical Nutrition*. 77: 1448–1452.

Josse, A. R., Atkinson, S. A., Tarnopolsky, M. A. and Phillips, S. M., 2011. Increased consumption of dairy foods and protein during diet- and exercise-induced weight loss promotes fat mass loss and lean mass gain in overweight and obese premenopausal women. *Journal of Nutrition*. 141: 1626–1634.

Kamycheva, E., Joakimsen, R. M. and Jorde, R., 2002. Intakes of calcium and vitamin D predict body mass index in the population of Northern Norway. *Journal of Nutrition*. 132: 102–106.

Lanou, A. J. and Barnard, N. D., 2008. Dairy and weight loss hypothesis: an evaluation of the clinical trials. *Nutrition Reviews*. 66(5): 272–279.

Laraichi, S., Parra, P., Zamanillo, R., Amarti, A. E., Palou, A. and Serra, F., 2013. Dietary supplementation of calcium may counteract obesity in mice mediated by changes in plasma fatty acids. *Lipids*. 48: 817–826.

Lorenzen, J. K. and Astrup, A., 2011. Dairy calcium intake modifies responsiveness of fat metabolism and blood lipids to a high-fat diet. *British Journal of Nutrition*. 105: 1823–1831.

Major, G. C., Chaput, J.-P., Ledoux, M., St-Pierre, S., Anderson, G. H., Zemel, M. B. and Tremblay, A., 2008. Recent developments in calcium-related obesity research. *Obesity Reviews*. 9: 428–445.

Major, G. C., Alarie, F. P., Doré, J. and Tremblay, A., 2009. Calcium plus vitamin D supplementation and fat mass loss in female very low-calcium consumers: potential link with a calcium-specific appetite control. *British Journal of Nutrition*. 101: 659–663.

Marotte, C., Weisstaub, A., Bryk, G., Olguin, M. C., Posadas, M., Lucero, D., Schreier, L., de Portela, M. L. P. M. and Zeni, S. N., 2013. Effect of dietary calcium (Ca) on body composition and Ca metabolism during growth in genetically obese (β) male rats. *European Journal of Nutrition*. 52: 297–305.

McCarron, D. A., Morris, C. D., Henry, H. J. and Stanton, J. L., 1984. Blood pressure and nutrient intake in the United States. *Science*. 224: 1392–1398.

Melanson, E. L., Sharp, T. A., Schneider, J., Donahoo, W. T., Grunwald, G. K. and Hill, J. O., 2003. Relation between calcium intake and fat oxidation in adult humans. *International Journal of Obesity*. 27: 196–203.

Mirmiran, P., Esmaillzadeh, A. and Azizi, F., 2005. Dairy consumption and body mass index: an inverse relationship. *International Journal of Obesity*. 29: 115–121.

Murakami, K., Okubo, H. and Sasaki, S., 2006. No relation between intakes of calcium and dairy products and body mass index in Japanese women aged 18 to 20 y. *Nutrition*. 22: 490–495.

Ochner, C. N. and Lowe, M. R., 2007. Self-reported changes in dietary calcium and energy intake predict weight regain following a weight loss diet in obese women. *Journal of Nutrition*. 137: 2324–2328.

Onakpoya, I. J., Perry, R., Zhang, J. and Ernst, E., 2011. Efficacy of calcium supplementation for management of overweight and obesity: systematic review of randomized clinical trials. *Nutrition Reviews*. 69: 335–343.

Ping-Delfos, W. C. S. and Soares, M., 2011. Diet induced thermogenesis, fat oxidation and food intake following sequential meals: influence of calcium and vitamin D. *Clinical Nutrition*. 30: 376–383.

Rosell, M., Johansson, G., Berglund, L., Vessby, B., De Faire, U. and Hellenius, M.-L., 2004. Associations between the intake of dairy fat and calcium and abdominal obesity. *International Journal of Obesity*. 28: 1427–1434.

Rosenblum, J. L., Castro, V. M., Moore, C. E. and Kaplan, L. M., 2012. Calcium and vitamin D supplementation is associated with decreased abdominal visceral adipose tissue in overweight and obese adults. *American Journal of Clinical Nutrition*. 95: 101–108.

Samara, A., Herbeth, B., Ndiaye, N. C., Fumeron, F., Billod, S., Siest, G. and Visvikis-Siest, S., 2013. Dairy product consumption, calcium intakes, and metabolic syndrome-related factors over 5 years in the STANISLAS study. *Nutrition*. 2013(29): 519–524.

Shahar, D. R., Schwarzfuchs, D., Fraser, D., Vardi, H., Thiery, J., Fiedler, G. M., Bluher, M., Stumvoll, M., Stampfer, M. J., Shai, I. and The DIRECT Group, 2010. Dairy calcium intake, serum vitamin D, and successful weight loss. *American Journal of Clinical Nutrition*. 92: 1017–1022.

Tardoff, M. G., 2001. Calcium: taste, intake, appetite. *Physiological Reviews*. 81: 1567–1597.

Teegarden, D., 2003. Calcium intake and reduction in weight or fat mass. *Journal of Nutrition*. 133: 249S–251S.

Tidwell, D. K. and Valliant, M. W., 2011. Higher amounts of body fat are associated with inadequate intakes of calcium and vitamin D in African American woman. *Nutrition Research.* 31: 527–536.

Trowman, R., Dumville, J. C., Hahn, S. and Torgerson, D. J., 2006. A systematic review of the effects of calcium supplementation on body weight. *British Journal of Nutrition.* 95: 1033–1038.

Venti, C. A., Tataranni, A. and Salbe, A. D., 2005. Lack of relationship between calcium intake and body size in an obesity-prone population. *Journal of the American Dietetic Association.* 105: 1401–1407.

Zemel, M. B., 2002. Regulation of adiposity and obesity risk by dietary calcium: mechanisms and implications. *Journal of the American College of Nutrition.* 21: 146S–151S.

Zemel, M. B., 2009. Proposed role of calcium and dairy food components in weight management and metabolic health. *Physician and Sportsmedicine.* 37(2): 29–39.

Zemel, M. B., Richards, J., Mathis, S., Milstead, A., Gebhardt, L. and Silva, E., 2005. Dairy augmentation of total and central fat loss in obese subjects. *International Journal of Obesity.* 29: 391–397.

Zemel, M. B., Shi, H., Greer, B., Dirienzo, D. and Zemel, P. C., 2000. Regulation of adiposity by dietary calcium. *FASEB Journal.* 14: 1132–1138.

Zemel, M. B. and Sun, X., 2008. Calcitriol and energy metabolism. *Nutrition Reviews.* 66(suppl. 2): S139–S146.

Zemel, M. B., Thompson, W., Milstead, A., Morris, K. and Campbell, P., 2004. Calcium and dairy acceleration of weight and fat loss during energy restriction in obese adults. *Obesity Research.* 12: 582–590.

Zhou, J., Zhao, L.-J., Watson, P., Zhang, Q. and Lappe, J. M., 2010. The effect of calcium and vitamin D supplementation on obesity in postmenopausal women: secondary analysis for a large-scale, placebo controlled, double-blind, 4-year longitudinal clinical trial. *Nutrition and Metabolism.* 7: 62. Available at: http://www.nutritionandmetabolism.com/content/7/1/62. Accessed 13 February 2014.

Zhu, W., Cai, D., Wang, Y., Lin, N., Hu, Q., Ma, S. and Amarasekara, S., 2013. Calcium plus vitamin D_3 supplementation facilitated fat loss in overweight and obese college students with very-low calcium consumption: a randomized controlled trial. *Nutrition Journal.* 12: 1–8. Available at: http://www.nutritionj.com/content/12/1/8. Accessed 20 January 2014.

CHAPTER 35

Calcium in Critical Care

DIMA YOUSSEF[a] AND KARIN AMREIN*[b]

[a]Department of Internal Medicine, Division of Infectious Diseases, East Tennessee State University, Box 70622, VAMC Bldg. 1, Dogwood Ave., Johnson City, TN 37614, USA; [b]Medical University of Graz, Department of Internal Medicine, Division of Endocrinology and Metabolism, Auenbruggerplatz 15, 8036 Graz, Austria

*E-mail: estecina@hotmail.com

35.1 Introduction

Calcium is the most abundant divalent cation in the body. The majority (98%) of calcium is in the form of hydroxyapatite in the bone and teeth, and only a small amount is extracellular. Plasma and extracellular calcium play a vital role in neuromuscular coupling and contraction. Plasma calcium is regulated by hormones, vitamins, and organ function and also depends on phosphorus and magnesium regulation. In severe illness, hypocalcemia is highly prevalent and associated with adverse outcomes. This may be caused by the underlying disease, accompanying medication and other factors. Severe hypercalcemia is rare and due primarily to pre-existing comorbidities such as malignant disorders, hyperparathyroidism, sarcoidosis, and others.

Calcium in serum is bound to proteins, principally albumin (*ca.* 40% of total calcium, ratio of 0.8 mg dL^{-1} calcium per 1.0 g dL^{-1} of albumin). Approximately 15% is bound to anions such as sulfate, phosphate, lactate, and citrate, the rest (*ca.* 45%) circulates as physiologically active ionized calcium.

Food and Nutritional Components in Focus No. 10
Calcium: Chemistry, Analysis, Function and Effects
Edited by Victor R. Preedy

Published by the Royal Society of Chemistry, www.rsc.org

Serum calcium varies with serum albumin, acid–base balance and may vary with hydration status. Total calcium alone can be misinterpreted, since it can change without altering the ionized calcium (Ladenson *et al.*, 1978), and *vice versa.*

Change in total but not ionized calcium: Hypoalbuminemia, hyperalbuminemia, and multiple myeloma can change the total serum calcium without altering the ionized calcium.

- Hypoalbuminemia: is associated with hypocalcemia. →Equation for correction of serum calcium in case of hypoalbuminemia:
  Serum calcium falls by 0.8 mg dL^{-1} for each 1.0 g dL^{-1} fall in the serum albumin.
  Corrected [Ca] = measured total [Ca] + (0.8 × (4.5 − [alb])) (Fukugawa and Kurokawa, 2002).
- Hyperalbuminemia: can cause high serum calcium, and this can be due to extracellular volume depletion or by fluid shift out of the vascular space.
- Multiple myeloma: causes pseudohypercalcemia when a monoclonal myeloma protein binds calcium with high affinity. Here, the ionized fraction is normal and there is no need to treat.

Change in ionized fraction but not total calcium: Physiologically alterations in the ionized calcium may occur without changes in the total serum calcium.

Acid–base changes: may affect ionized calcium because of changes in binding affinity. Alkalosis enhances calcium binding to albumin, thus decreasing ionized calcium. In acute respiratory alkalosis, ionized calcium falls approximately 0.16 mg dL^{-1} for each 0.1 unit increase in pH, as demonstrated in tetany accompanying psychogenic hyperventilation. Chronic respiratory alkalosis decreases the ionized calcium through relative hypoparathyroidism and renal resistance to PTH, with hypercalciuria. In chronic metabolic acidosis, ionized calcium is high due to less albumin binding without affecting the total serum calcium.

Hyperphosphatemia: Acute hyperphosphatemia reduces the ionized calcium by binding to circulating calcium. Calcium-phosphate precipitates, and deposits in soft tissues.

CKD: In reduced renal function, total serum calcium does not predict the ionized calcium. Measuring the ionized calcium is preferable to measuring the total calcium.

In summary, in a number of clinical situations that often occur in intensive care, total serum calcium therefore may not well reflect the physiologically important ionized calcium. Measuring ionized calcium is therefore preferable in this setting. The normal ionized calcium ranges between 4.65 and 5.25 mg dL^{-1}.

Serum calcium measured is usually reported in units of mg dL^{-1} in the United States, in mmol L^{-1} in other countries, and in meq. L^{-1} in some laboratories.

1 mg dL^{-1} is equivalent to 0.25 mmol L^{-1} or to 0.5 meq. L^{-1}. The normal total serum calcium range is 8.8 to 10.3 mg dL^{-1} (equivalent to 2.2 to 2.6 mmol L^{-1} or 4.4 to 5.2 meq. L^{-1}).

35.1.1 Recommended Intake

The recommended intake is 1000 mg a day for adults. A dose of 1200 mg of calcium and 800 IU of vitamin D daily for most postmenopausal women with osteoporosis is recommended (Institute of Medicine, 2010). In pre-menopausal women or in men with osteoporosis, 1000 mg of calcium and sufficient amounts of vitamin D daily are recommended. It is preferable that the major part comes from dietary sources, since supplements may cause side effects, particularly kidney stones. Moreover, adverse effects of calcium supplements on cardiovascular risk are under debate. The total intake of calcium should not exceed 2000 mg per day.

35.1.2 Sources of Calcium

The NIH recommends getting the needed calcium from foods before turning to supplements. Additionally to milk, there are a variety of foods that contain calcium. Dairy products, especially yogurt, are the top source of calcium. Cheese also provides plenty of calcium. Fortified cereals, soy milk, sardines, tofu, pink salmon, collard greens and molasses are good source of calcium as well. Table 35.1 summarizes the main sources of calcium in normal diet.

In critical care, often adequate oral feeding is not possible. Enteral nutrition products typically contain ~60–80 mg calcium/100 mL and parenteral nutrition products normally do not contain calcium.

35.1.3 Calcium Supplements

Usually, calcium carbonate and calcium citrate are used. Calcium carbonate is cheaper and contains 40% elemental calcium (*e.g.*: 1250 mg of calcium carbonate contains 500 mg of elemental calcium, whereas there are 211 mg elemental Ca per gram of calcium citrate).

Table 35.1 Main dietary sources of calcium.

Dairy foods	Milk, yogurt, cheese
Leafy green vegetables	Broccoli, spinach, kale
Fruits	Oranges
Beans and peas	Tofu, peas, peanuts, baked beans, black beans
Fish	Salmon, sardines
Others	Sesame seeds, blackstrap molasses, corn tortillas, almonds *In ICU*: enteral nutrition formulas, supplements

35.1.4 Calcium Absorption

Calcium is absorbed in the small intestine *via* both active and passive transport. Intestinal transport of calcium is primarily regulated by 1,25-$(OH)_2D$. If calcium intake exceeds 1 to 2 g daily, significant amounts of calcium may be absorbed passively. Calcium carbonate absorption is better than calcium citrate when taken with meals, especially in achlorhydria; calcium citrate is better absorbed in the fasting state. Proton pump inhibitors or H_2 blockers may decrease the absorption of calcium carbonate, and so calcium citrate is preferable in such situations.

35.1.5 Side Effects of Calcium Supplements

Nephrolithiasis: Calcium supplements have been associated with an increased risk of kidney stones.

Cardiovascular disease: Controversial, under debate. Calcium supplements, high dietary intake of calcium, or vitamin D supplements have not been shown to increase all-cause mortality and actually calcium plus vitamin D was associated with decreased all-cause mortality.

Gastrointestinal side effects: High calcium intake may cause dyspepsia and constipation, and may decrease the absorption of iron and thyroid hormone.

35.1.6 Special Situations

Gastrointestinal disease: Decreased calcium absorption can be due to decreased gastric acidity, impaired pancreatic secretions, decreased gut transit time, steatorrhea, and vitamin D deficiency.

Primary hyperparathyroidism: Adequate dietary calcium (800 to 1000 mg daily) and vitamin D, not restriction, is recommended in primary hyperparathyroidism.

Chronic kidney disease: Excess calcium supplementation in patients with eGFR <30 mL min^{-1} may be associated with an increase in greater arterial calcification and cardiovascular disease.

Diuretic therapy: Loop diuretics increase calcium excretion, whereas thiazide diuretics are hypocalciuric and protect against calcium stones and bone loss.

Granulomatous diseases: Chronic glucocorticoid therapy carries a risk for osteoporosis. Extrarenal production of calcitriol by macrophages and increased calcium intestinal absorption may cause hypercalcemia and hypercalciuria.

35.2 Hypocalcemia

Hypocalcemia is common in critical illness and correlates with poor outcomes (Carlstedt *et al.*, 1998; Carlstedt and Lind, 2001; Hästbacka and Pettilä, 2003; Kelly and Levine, 2013).

Decreased plasma calcium can have serious consequences in critically ill patients, potentiating arrhythmias by QT prolongation and seizures. When hypocalcemia occurs, there is failure of normal plasma calcium regulation. Calcium can leave the extracellular space when driven by reactions that deposit calcium in the bones and soft tissues or when there is insufficient PTH or 1,25-$(OH)_2D$ to mobilize calcium from the bone. Besides the usual causes of hypocalcemia found in other populations (disorders of PTH or vitamin D and disorders that decrease serum ionized calcium by binding of calcium within the vascular space or by its deposition in tissues), another important pathophysiologic mechanism has been proposed in the critically ill patient. In a rat model, intraperitoneal injection of interleukin 6 reduced parathyroid hormone, 1,25-dihydroxyvitamin D and calcium levels within hours *via* increased expression of the calcium-sensing receptor (CaSR) at the parathyroid, thyroid, and kidney cells (Canaff *et al.*, 2008). Other ICU-specific factors that may also lead to hypocalcemia include drugs, hyperphosphatemia, and insufficient calcium infusion during citrate dialysis.

35.2.1 Causes of Hypocalcemia

Table 35.2 represents the main causes of hypocalcemia seen in critical care.

Often, hypoalbuminemia found in severe illness, fluid overload, malnutrition or nephrotic syndrome may cause low total calcium but normal ionized calcium. Also, acute and chronic changes in acid–base status may

Table 35.2 Causes of hypocalcemia in ICU patients.

General causes
Malabsorption of calcium or vitamin D
Decreased intake of calcium/malnutrition
Steatorrhea
Vitamin D deficiency
Chronic renal insufficiency
Decreased PTH or decreased PTH effectiveness
Hypoparathyroidism, postsurgical and other etiologies
Pseudohypoparathyroidism
Hypomagnesemia
Acute pancreatitis
Rhabdomyolysis
Tumor lysis syndrome
Acute hyperphosphatemia
Treatment of hypercalcemia
Loop diuretics, aminoglycosides
Specific causes in critical illness
Systemic inflammatory response syndrome/Sepsis, possibly *via* IL-6
Transfusion, especial fresh frozen plasma (citrate-mediated)
Insufficient calcium replacement during continuous renal replacement therapy when citrate is used as anticoagulant

affect calcium status as described above. Serum calcium is under strict control by PTH. Through the calcium-sensing receptor (CaSR), calcium inhibits PTH and stimulates renal calcium excretion in the loop of Henle. PTH raises calcium by decreasing urinary calcium excretion and stimulation of calcium reabsorption in the distal tubule, increasing calcium absorption in the intestine by increased renal production of 1,25-$(OH)_2D$ (calcitriol), or increased bone resorption. It is thus useful to characterize hypocalcemia broadly as associated with low or high PTH levels (Goltzman and Cole, 2006).

35.2.1.1 Hypocalcemia with Low PTH

Hypoparathyroidism may occur as a complication after surgical procedures at the neck (*e.g.* thyroid, parathyroid, or other radical neck surgery). Other causes of hypoparathyroidism include autoimmune disease, irradiation and infiltrative diseases. In critically ill patients, hypocalcemia also may be due to a decreased effect of PTH action. Functional hypoparathyroidism may be mediated by magnesium. Hypomagnesemia has an important effect of decreasing PTH release from parathyroids, contributing to hypocalcemia. Furthermore, hypomagnesemia decreases the action of PTH on bone. Causes for hypomagnesemia include severe malnutrition, alcoholism, diuretics, drugs such as cisplatin, malabsorption and aminoglycosides. Severe hypermagnesemia can also cause hypocalcemia by suppressing PTH secretion. Serum magnesium can be >5 meq. L^{-1}, which is usually seen only when given in large doses to women with eclampsia. Hypocalcemia has also been described to occur during magnesium treatment in aneurysmal subarachnoid hemorrhage (Desai *et al.*, 1988).

Other, rare disorders that usually present in childhood and affect calcium and parathyroid metabolism through abnormal parathyroid gland development or altered PTH regulation include X-linked or autosomal recessive hypoparathyroidism and congenital syndromes like DiGeorge syndrome besides CaSR mutations.

35.2.1.2 Hypocalcemia with High PTH

- Acute pancreatitis: Hypocalcemia occurs due to deposition in the form of calcium soaps (*i.e.*, poorly soluble salts of Ca^{2+} and fatty acids) in the abdominal cavity (Dettelbach *et al.*, 1990). PTH is typically elevated in response to hypocalcemia, and there is resistance to PTH. Hypocalcemia in acute pancreatitis is linked to poor outcome.
- Sepsis and severe illness: Hypocalcemia frequently occurs in critically ill or postsurgical patients (80–90%) (Zivin *et al.*, 2001), sepsis and severe burns. This is due to impaired secretion of PTH and reduced calcitriol production and end-organ resistance to PTH, hypomagnesemia,

cytokines effects on the parathyroid glands, kidneys, and bone, and high serum calcitonin. Hypocalcemia is common in toxic-shock syndrome (Sperber *et al.*, 1990). This may also affect patients after prolonged resuscitation (Cairns *et al.*, 1991).

- Transfusion: Hypocalcemia may happen during and after surgery, mainly with blood transfusion (citrate used for anticoagulating blood products chelates calcium). Here, total calcium is normal while ionized calcium is reduced. As a rule of thumb, per 2 units of fresh frozen plasma, 1 vial of intravenous calcium is needed. Critically ill patients with impaired liver or renal function and who receive blood transfusion, may have a fall in ionized calcium but rarely become symptomatic. This can also happen with massive amounts of citrate given rapidly, such as in plasma exchange, plasma- or leukapheresis, or massive blood transfusion (Bolan *et al.*, 2002).
- Surgery: Hypocalcemia can also occur during and after surgery in patients not having transfusions (Lepage *et al.*, 1999). Volume expansion and hypoalbuminemia cause hypocalcemia. PTH increases in case of low ionized calcium. These changes subside within hours after surgery.
- Continuous renal replacement therapy using citrate as anticoagulant requires regular calcium measurement and replacement. If this is inadequate, hypocalcemia accompanied by secondary hyperparathyroidism may develop rapidly when ionized calcium levels fall below 4.5 mg dL^{-1} (1.12 mmol L^{-1}) (Raimundo *et al.*, 2013). Each 0.1 mmol L^{-1} increase in ionized calcium is associated with a 30% decrease in PTH. Citrate anticoagulation may be especially problematic in impaired citrate metabolism (*e.g.* liver cirrhosis).
- Profound vitamin D deficiency or resistance: Vitamin D deficiency is common in critically ill patients and caused by poor intake or malabsorption, decreased sun exposure, decreased 25-hydroxylation of vitamin D to form calcidiol (25-hydroxyvitamin D) in the liver, increased metabolism to inactive metabolites, decreased 1-hydroxylation of calcidiol to 1,25-$(OH)_2D$ in the kidney, and decreased calcitriol action.
- Chronic kidney disease: Causes a decrease in the renal production of 1,25-$(OH)_2D$. In end-stage CKD, hyperphosphatemia occurs and so does hypocalcemia. In a rat model, it has also been shown that uremia leads to reduced hepatic synthesis (Michaud *et al.*, 2010).
- PTH resistance: In pseudohypoparathyroidism, target organs are unresponsive to PTH. Hypocalcemia, hyperphosphatemia and elevated PTH are the consequence.
- Hyperphosphatemia with and without calcium deposition in soft tissues: may be seen in ICU patients with acute renal failure (ARF), tumor lysis syndrome, tissue breakdown or excessive phosphorus repletion. In CKD, calcitriol synthesis is impaired and thus leads to worsening hypocalcemia. Hypocalcemia can also occur due to calcium deposition in calciphylaxis.

35.2.1.3 Hypocalcemia Induced by Drugs

For ICU patients, review of medications that may affect calcium should be undertaken. Such medications include loop-acting diuretics like furosemide, phenytoin, calcium-lowering drugs such as osteoporosis or cancer therapy (particularly intravenous bisphosphonates), and phosphate therapy, calcium chelators like citrate, lactate, sodium EDTA and foscarnet. Chemotherapy (such as cisplatin, 5-fluorouracil, and leucovorin), denosumab, bisphosphonates, fluoride poisoning, and cinacalcet can all induce hypocalcemia.

35.2.2 Clinical Manifestations of Hypocalcemia

Symptoms are often absent but if present, may range from mild to life-threatening seizures, refractory heart failure, or laryngospasm depending on the degree of hypocalcemia and the time in which it developed. Table 35.3 summarizes these manifestations.

Acute manifestations: Central and peripheral nervous system effects are the most common symptoms, among them altered mental status, coma, tetany, paresthesias, hyperreflexia, and muscle weakness.

Tetany: This occurs when ionized calcium falls below 4.3 mg dL^{-1}. Acid–base status, hypomagnesemia, and potassium balance also affect the frequency and severity of symptoms (Navarro *et al.*, 1991). It starts with perioral and acral paresthesias, and may be followed by hyperventilation that causes respiratory alkalosis, worsening the paresthesias. Stiffness, myalgia, and muscle spasms and cramps as well as carpopedal spasm may happen. In severe cases, spasm of the respiratory muscles and of the glottis can cause cyanosis. Associated diaphoresis, bronchospasm, and biliary colic may occur. Acute hypocalcemia directly increases peripheral neuromuscular irritability. Hyperexcitability of peripheral neurons is probably the most important

Table 35.3 Clinical manifestations of acute hypocalcemia.

Acute manifestations
Neuromuscular irritability (tetany) • Paresthesias (perioral, extremities) • Muscle twitching • Carpopedal spasm • Trousseau's sign • Chvostek's sign • Seizures • Laryngospasm • Bronchospasm
Papilledema QT prolongati on Arrhythmia Impaired coagulation

pathophysiologic effect of hypocalcemia. Neuromuscular irritability due to latent tetany can be detected by Trousseau's and Chvostek's signs, but they are nonspecific. Trousseau's sign is characterized by carpopedal spasm through inflation of a sphygmomanometer above systolic blood pressure for three minutes. Chvostek's sign consists of contraction of the facial muscles upon tapping the facial nerve anterior to the ear. The contraction varies between twitching of the lip to spasm of facial muscles and depends upon the degree of hypocalcemia.

Seizures may be focal or generalized and may be the sole presenting symptom (Armelisasso *et al.*, 2004). An electroencephalogram shows spikes and bursts of high-voltage paroxysmal slow waves. Hypocalcemia may also complicate a known seizure disorder.

Cardiovascular: Hypotension may be present especially when hypocalcemia is caused either by EDTA or transfusion of citrated blood (Ghent *et al.*, 1994). Decreased myocardial performance and congestive heart failure can also occur (Wong *et al.*, 1990; Kazmi and Wall, 2007). Myocardial dysfunction is reversible with calcium repletion. On the ECG, hypocalcemia causes prolongation of the QT interval (Meyer *et al.*, 2007). Prolongation of the QT interval is associated with early after-repolarizations and this may trigger dangerous arrhythmias such as torsades de pointes and ventricular fibrillation.

Papilledema is seen only in severe cases and usually improves with reversal of hypocalcemia.

Psychiatric symptoms may include anxiety, confusion, hallucinations, and frank psychosis.

Hypocalcemia may also impair coagulation as it is an essential factor in the coagulation cascade.

35.2.3 Diagnostic Approach to Hypocalcemia

In the ICU, we recommend using primarily ionized serum calcium.

35.2.3.1 Clinical History (If Feasible)

A family history of hypocalcemia suggests a genetic cause. Chronic hypocalcemia may be seen with hypoparathyroidism of any cause, a mutation of the CaSR and in pseudohypoparathyroidism.

35.2.3.2 Laboratory Workup

Laboratory workup should include intact PTH, serum creatinine, magnesium, phosphate, alkaline phosphatase, amylase, urinary calcium and magnesium excretion, 25-(OH)D and 1,25-$(OPH)_2$D.

Table 35.4 Laboratory findings in hypocalcemia.

	PTH	Corrected serum calcium	Phosphorus	Magnesium	25-(OH)D
Hypoparathyroidism	Low	Low	High	Normal	Normal
Activating mutation CaSR	Low to normal	Low	High	Normal	Normal
Hypomagnesemia	Low to normal	Low	Normal	Low	Normal
Chronic kidney disease	High	Low	High	High	Normal
Pseudohypoparathyroidism	High	Low	High	Normal	Normal
Vitamin D deficiency	High	Low to normal	Low to normal	Normal	Low

Table 35.4 simplifies the findings on lab work seen in the different causes of hypocalcemia.

- **Serum PTH**: A low or normal PTH in hypocalcemia suggests hypoparathyroidism. It is elevated in acute or chronic kidney disease, vitamin D deficiency, and pseudohypoparathyroidism. It is normal or low in hypomagnesemia or autosomal dominant hypocalcemia, where there is mutation in the CaSR gene.
- **Magnesium**: Hypomagnesemia (magnesium < 0.8 meq. L^{-1}) may induce hypocalcemia by inducing PTH resistance or deficiency. The calcium level should go back to normal soon after correction of hypomagnesemia.
- **Phosphate**: Hyperphosphatemia with hypocalcemia indicates hypoparathyroidism or pseudohypoparathyroidism. High serum phosphate in kidney disease or tissue breakdown is due to lack of PTH effect on urinary phosphate excretion. However, if phosphate is low, then PTH is high, meaning secondary hyperparathyroidism or low dietary phosphate intake.
- **Vitamin D**: Vitamin D deficiency (defined by a low serum 25-(OH)D) increases PTH secretion by causing hypocalcemia and eliminating the normal inhibitory effect of calcitriol on PTH production. Phosphate absorption in the intestines is diminished.
- **Alkaline phosphatase**: An elevated alkaline phosphatase is common in osteomalacia and osteoblastic bone metastases that can cause hypocalcemia due to rapid deposition of calcium in metastases.
- **Amylase**: Serum amylase is elevated in acute pancreatitis but only slightly in chronic pancreatitis.
- **Urinary calcium**: Low urinary calcium is seen in untreated hypoparathyroidism or vitamin D deficiency.
- **Urinary magnesium**: May be helpful in hypomagnesemia. A high value reflects renal loss.

35.2.4 Treatment of Hypocalcemia

Patients with hypocalcemia and symptoms like carpopedal spasm, tetany, or seizures, and arrhythmias, and cases of acute drop in corrected calcium to ≤7.5 mg dL^{-1} need intravenous calcium supplementation. Intravenous calcium (1 to 2 g of calcium gluconate) can be infused over 10 to 20 min (not as a bolus dose due to the risk of serious cardiac dysfunction, including systolic arrest) (Tohme and Bilezikian, 1996). This dose will last for only two or three hours. For persistent severe hypocalcemia, continuous intravenous calcium gluconate can be given as 8–12 mg kg^{-1} of calcium over 6–8 h. During treatment, calcium should be regularly checked, along with phosphorus and magnesium. Physical examination and ECG (QT interval) may guide when treatment should be slowed or changed to oral supplementation.

However, there are no studies that have evaluated the association between parenteral calcium supplementation in critically ill ICU patients and important clinical outcomes (Forsythe *et al.*, 2008).

Patients with milder symptoms of neuromuscular irritability and corrected calcium concentrations > 7.5 mg dL^{-1}, oral calcium supplementation can be initiated. Treatment consists of 1500 to 2000 mg of elemental calcium given as calcium carbonate or calcium citrate daily, in divided doses, as such a large dose cannot be resorbed at once.

Patients with pancreatitis and rhabdomyolysis may have transient hypocalcemia followed by release of calcium back into the extracellular space. Close monitoring for normocalcemia or even rebound hypercalcemia is recommended. When vitamin D deficiency or hypoparathyroidism is the cause of hypocalcemia, management should include administration of calcium and active and/or native vitamin D.

Patients with chronic renal failure with hypocalcemia usually have calcitriol deficiency caused by reduced renal conversion of 25-(OH)D and active vitamin D supplementation should be initiated. Several options are available in this situation:

*__Calcitriol__: (1,25-dihydroxyvitamin D). This is the most active metabolite of vitamin D. Calcitriol does not need activation, has a rapid onset of action, and a biologic half-life of about four to six hours. It is also is available for parenteral administration for rapid replenishment of stores in case of critical illness.

*__Alfacalcidiol__: (1-α-hydroxyvitamin D_3) is a synthetic analog of vitamin D that is converted in the liver to the active metabolite 1,25-$(OH)_2$D. It has a rapid onset of action and relatively short half-life.

Other active vitamin D analogs are available.

Additionally, 25-(OH)D status needs to be replete (at least <20 ng mL^{-1} or 50 nmol L^{-1}), which is often not the case. Native vitamin D (ergo- or cholecalciferol) also needs to be supplemented in these situations.

Accompanying hypomagnesemia, because of multiple effects contributing to hypocalcemia, is also a treatment priority and usually can be corrected with little risk of complications, except in renal insufficiency. 2 g of magnesium sulfate should be infused as a 10 percent solution over 20 min, followed by 1 g in 100 mL of fluid per hour. Persistent hypomagnesemia requires supplementation with oral magnesium, typically 300 to 400 mg daily divided into three doses.

Presence of hyperphosphatemia: Hypocalcemia with severe hyperphosphatemia is problematic. Raising plasma calcium in this setting may cause calcium phosphate deposition. Only enough calcium to prevent or reverse cardiovascular complications should be given. Plasma ionized calcium may help guide treatment. Acute hemodialysis to lower plasma phosphorus may be needed.

35.3 Hypercalcemia

Most often, hypercalcemia is accidentally identified by a routine plasma calcium level (>10.5 mg dL^{-1}). Hypercalcemia is the result of failure of the regulatory mechanisms for calcium. Hypercalcemia occurs with accelerated bone resorption, increased intestinal absorption, or decreased renal calcium excretion. Primary hyperparathyroidism and malignancy are the most common causes. Primary hyperparathyroidism is favored in asymptomatic chronic hypercalcemia or known multiple endocrine neoplasia. Hypercalcemia of malignancy is more symptomatic, and often comes with higher and rapid increases in calcium.

Hypercalcemia may be a cause of admission to the ICU. Severe hypercalcemia with altered mental status is a medical emergency. Patients may be admitted to the ICU especially for close monitoring of intravascular fluid therapy.

35.3.1 Causes of Hypercalcemia

Table 35.5 contains the essential causes of hypercalcemia in ICU.

Table 35.5 Differential diagnosis of hypercalcemia.

Primary hyperparathyroidism
Vitamin A or D intoxication
Milk-alkali syndrome
Adrenal insufficiency
Paget disease of bone
Familial hypocalciuric hypercalcemia
Granulomatous diseases: sarcoidosis, tuberculosis, fungal infections
Malignancy
Thiazide diuretics
Thyrotoxicosis
Immobilization
Over-supplementation with (primarily intravenous) calcium

35.3.1.1 *Malignancy*

Malignancy is the most frequent cause of severe hypercalcemia (>13 mg dL^{-1}). The most important mechanism is release by the tumor of a peptide that has structural homology with PTH, called parathyroid hormone-related peptide (PTHrP). PTHrP increases plasma calcium and decreases plasma phosphorus. Malignancy also causes bony metastases with bone destruction. Cytokines including tumor necrosis factor and interleukin-1 stimulate the differentiation of osteoclast precursors into mature osteoclasts. In lymphoma, extrarenal production of calcitriol from calcidiol by activated mononuclear cells cause hypercalcemia.

35.3.1.2 *Primary Hyperparathyroidism*

This is mostly due to an adenoma or hyperplasia. It causes hypercalcemia through activation of osteoclasts (bone resorption), and increased intestinal calcium absorption. Serum calcium is slightly elevated (<11 mg dL^{-1}), or at high-normal values with intermittent hypercalcemia.

35.3.1.3 *Secondary and Tertiary Hyperparathyroidism*

CKD and secondary hyperparathyroidism may develop hypercalcemia. This is due to a reduction in the bone uptake of calcium. In renal failure, hypercalcemia is due to autonomous overproduction of PTH, representing tertiary hyperparathyroidism.

35.3.1.4 *Thyrotoxicosis*

Slight hypercalcemia may occur here due to increased bone resorption. It usually resolves after resolution of hyperthyroidism.

35.3.1.5 *Increased Calcium Intake*

This rarely causes hypercalcemia, since the initial rise in calcium inhibits PTH and the synthesis of calcitriol. However, the combination of high calcium intake and low urine calcium excretion may cause hypercalcemia, and is seen in CKD and milk-alkali syndrome.

35.3.1.6 *Hypervitaminosis D*

Vitamin D intoxication is rare and usually caused by accidental long-term overdosing of vitamin D supplements, *e.g.* in pharmaceutical errors in dosing. 25-(OH)D (calcidiol) or 1,25-$(OH)_2$D (calcitriol) increase calcium absorption and bone resorption, and if high, may cause hypercalcemia. Hypercalcemia due to calcitriol lasts one to two days because of its short half-life. Hypercalcemia due to vitamin D or calcidiol lasts longer. Up to 10 000 IU per day of native vitamin D are safe; hypercalcemia usually occurs when 25-(OH)D levels excess 150 ng mL^{-1}.

35.3.1.7 Granulomatous Diseases

Macrophages within granulomas may synthesize 1,25-$(OH)_2D_3$. Sarcoidosis, tuberculosis, berylliosis, and fungal diseases are associated with hypercalcemia. Hypercalciuria may occur because high calcium and vitamin D inhibit PTH release, thus increasing renal calcium excretion. Both Hodgkin's and non-Hodgkin's lymphoma can produce 1,25-$(OH)_2D_3$.

35.3.1.8 Lithium

Chronic lithium intake may induce hypercalcemia, by increased secretion of PTH related to an increase in the set point at which calcium suppresses PTH. Stopping lithium should reverse the hypercalcemia.

35.3.1.9 Thiazide Diuretics

Thiazides may cause mild hypercalcemia, by lowering urinary calcium excretion, in patients with increased bone resorption.

35.3.1.10 Pheochromocytoma

Hypercalcemia results from production of PTH-related protein. Alpha-adrenergic blockers can decrease this production.

35.3.1.11 Adrenal Insufficiency

Hypercalcemia occurs in occasional patients with Addisonian crisis, due to increased bone resorption, volume contraction and increased proximal tubular calcium reabsorption, hemoconcentration. Cortisol administration reverses the hypercalcemia.

35.3.1.12 Rhabdomyolysis and Acute Renal Failure

Hypercalcemia may happen during the diuretic phase of acute renal failure, mainly in rhabdomyolysis (Akmal *et al.*, 1986). Here, it is secondary to the mobilization of the deposited calcium in the injured muscle. Hypercalcemia can result from the correction of hyperphosphatemia, mild secondary hyperparathyroidism, and increased serum calcitriol.

35.3.1.13 *Acute Pancreatitis*

Patients that are recovering can have a rebound of plasma calcium levels as it gets released back into the extracellular space.

35.3.1.14 *Theophylline Toxicity*

This may present with mild hypercalcemia that usually resolves with administration of a beta-adrenergic antagonist.

35.3.1.15 *Other Causes*

These include immobilization, Paget disease of bone, administration of estrogen or an antiestrogen to patients with breast cancer and extensive skeletal metastases, hypervitaminosis A or the administration of retinoic acid (*via* the mechanism of increased bone resorption).

35.3.2 Clinical Manifestations of Hypercalcemia

Mild hypercalcemia (calcium <12 mg dL^{-1}) may be asymptomatic, or non-specific symptoms, such as constipation, fatigue, and depression may exist.

Calcium of 12 to 14 mg dL^{-1} may be well tolerated chronically, while an acute rise may cause symptoms, like polyuria, polydipsia, dehydration, anorexia, nausea, muscle weakness, and changes in sensorium.

In severe hypercalcemia (calcium >14 mg dL^{-1}), there is usually progression of symptoms.

Table 35.6 gathers the clinical findings seen in hypercalcemia.

Mental status changes: These include anxiety, depression, and cognitive dysfunction. Lethargy, confusion, stupor, and coma may happen with severe hypercalcemia and in rapid calcium increase.

Gastrointestinal symptoms: These include constipation, anorexia, and nausea. Pancreatitis and peptic ulcer disease may happen.

Renal tubular acidosis: Chronic hypercalcemia infrequently causes type 1 (distal) renal tubular acidosis.

Nephrogenic diabetes insipidus: This is manifested by polyuria and polydipsia.

Nephrolithiasis: This results from chronic hypercalciuria, and increased calcitriol production.

Renal insufficiency: This depends on the degree and duration of hypercalcemia.

Cardiovascular: Acute hypercalcemia causes a shortened QT interval. ST-segment elevation mimicking myocardial infarction has been reported (Turhan *et al.*, 2005; Nishi *et al.*, 2006).

Musculoskeletal: Mild muscle weakness may occur, but improves after parathyroidectomy. Bone pains can occur.

Table 35.6 Clinical manifestations of hypercalcemia.

Renal
Nephrocalcinosis, nephrolithiasis
Nephrogenic diabetes insipidus
Acute renal failure
Polyuria, polydipsia
Distal renal tubular acidosis
Gastrointestinal
Anorexia, nausea, vomiting, constipation
Pancreatitis
Peptic ulcer disease
Musculoskeletal
Muscle weakness, bone pain
Neurologic
Confusion
Fatigue
Stupor, coma
Cardiovascular
Shortening of QT interval on ECG
Bradycardia
Hypertension

Physical findings: There are no specific physical findings of hypercalcemia other than those related to the cause. Band keratopathy, a reflection of subepithelial calcium phosphate deposits in the cornea, is a very rare finding. Patients may have absent deep tendon reflexes and muscle weakness. Besides that, bowel sounds are often decreased.

35.3.3 Diagnostic Approach to Hypercalcemia

A plasma calcium concentration over 10.5 mg dL^{-1} or high ionized calcium levels above the normal range (1.10–1.26 mmol L^{-1}) makes the diagnosis of hypercalcemia.

Hypercalcemia in the outpatient setting is usually due to primary hyperparathyroidism while malignancy is more often responsible for hypercalcemia in hospitalized patients.

Interpretation of serum calcium: Hypercalcemia represents increased ionized calcium. As 40–45% of calcium is bound to protein, an increased binding increases the total calcium without affecting the ionized calcium. This is pseudohypercalcemia. Also, hypoalbuminemia elevates ionized calcium, while total calcium may be normal. In general, primary hyperparathyroidism is associated with mild hypercalcemia (calcium <11 mg dL^{-1}), whereas malignancy-associated hypercalcemia is severe (calcium >13 mg dL^{-1}).

PTH level is essential in the workup of hypercalcemia. It is essential to differentiate parathyroid hormone (PTH)-mediated hypercalcemia (primary hyperparathyroidism and familial hyperparathyroid syndromes) from non-PTH mediated hypercalcemia (malignancy, vitamin D intoxication, granulomatous disease, *etc.*).

PTH: A high PTH with hypercalcemia usually reflects primary hyperparathyroidism. PTH may still be in the upper end of normal; urinary calcium should be measured to rule out familial hypocalciuric hypercalcemia. A low intact PTH (<20 pg mL^{-1}) is most consistent with non-PTH-mediated hypercalcemia. In this case, PTHrP and vitamin D metabolites should be measured to assess for hypercalcemia of malignancy and vitamin D intoxication. If PTHrP and vitamin D metabolites are also low, then other hypercalcemia causes should be sought.

PTH-rP: Malignancy is mostly causative of non-PTH-mediated acute hypercalcemia. In chronic hypercalcemia and low PTH, non-PTH related disorders other than malignancy are the culprit.

Hypercalcemia of malignancy shows an elevated PTHrP while PTH and 1,25-$(OH)_2$D are suppressed.

Vitamin D metabolites: Calcidiol and calcitriol should be measured if there is no obvious malignancy and neither PTH nor PTHrP levels are elevated. A high 25-(OH)D is significant of vitamin D intoxication (Jacobus *et al.*, 1992). A high 1,25-$(OH)_2$D may be due to increased intake, granulomatous diseases or lymphoma, or increased renal production. Pulmonary evaluation for malignancy or sarcoidosis is needed.

Other tests: Serum and urinary protein electrophoresis, TSH, vitamin A are needed to evaluate diseases like multiple myeloma, thyrotoxicosis, immobilization, or vitamin A toxicity and milk-alkali syndrome.

Serum phosphate and urinary calcium can be helpful. Hypophosphatemia is present in hyperparathyroidism and hypercalcemia of malignancy. Urinary calcium excretion is usually raised in these conditions, but is low in milk-alkali syndrome, thiazide diuretics, and familial hypocalciuric hypercalcemia.

35.3.4 Treatment of Hypercalcemia

In an acute setting, nonspecific treatment as listed below is warranted. Depending on the very heterogeneous reasons for hypercalcemia as discussed above, specific treatment may be necessary to ameliorate calcium levels in the long term.

- **Mild hypercalcemia:** (calcium <12 mg dL^{-1}) does not require treatment. Avoidance of thiazides, volume depletion, and high-calcium diet is recommended. Good hydration is advisable.
- **Moderate hypercalcemia:** Asymptomatic or mildly symptomatic chronic moderate hypercalcemia (12–14 mg dL^{-1}) may not need immediate therapy. Mental status changes may happen in acute rise to such levels, and then saline hydration and bisphosphonates may be used.
- **Severe hypercalcemia:** (calcium >14 mg dL^{-1}), should be treated even if asymptomatic. Treatment aims at increasing renal excretion and decreasing mobilization from bones through.

1. Volume expansion with isotonic saline. Loop diuretics are not recommended.

Calcitonin used to be used for the treatment of hypercalcemia. However, a panel of advisers to the Food and Drug Administration recommended that calcitonin no longer be used as a treatment for osteoporosis in postmenopausal women, because the benefit of calcitonin salmon products in treating bone-thinning associated with osteoporosis is outweighed by a potential increase in the risk of cancer. Also, European regulators recommended that long-term use of calcitonin be curtailed after a review found evidence of a small increased risk of cancer with long-term use of the drugs.

2. Intravenous bisphosphonates such as zoledronic acid or pamidronate were used concomitantly with calcitonin.

The use of calcitonin and saline reduces calcium within 12 to 48 h. The effect of bisphosphonate starts by the second to fourth day, thus maintaining control.

a. *Increased excretion of calcium*: Expansion of extracellular volume with saline decreases passive proximal reabsorption of calcium. Loop diuretics avoid fluid overload and inhibit active sodium and passive calcium reabsorption in the loop of Henle. Normal saline should be administered at a rate of 200–300 mL h^{-1}. The elderly and cardiac patients may not tolerate this massive expansion, and so central venous pressure or pulmonary artery wedge pressure measurements may be necessary.

Since calcium absorption is coupled with sodium reabsorption in the ascending loop of Henle, furosemide increases calcium excretion. It maintains diuresis while giving large volumes of isotonic saline. Hypokalemia and hypomagnesemia may develop. Thiazide diuretics inhibit renal calcium excretion in the distal tubule and should be avoided. Serum calcium usually declines within a few hours with these measures.

b. *Decreased mobilization of calcium from bone*: Plicamycin blocks bone resorption of calcium. It can be given to any patient with hypercalcemia but mostly in hypercalcemia of malignancy. The usual dose is 25 μg kg^{-1} in 50 mL D5W intravenous over 3–6 h. Plicamycin should not be used in the presence of severe renal or liver failure or thrombocytopenia. It is much less often used than bisphosphonates.

Bisphosphonates: These bind to bone hydroxyapatite and inhibit osteoclast activity. They are more potent than calcitonin and saline in moderate or severe hypercalcemia (Stewart, 2005). Their action onset is within 2 days and maximum effect at about 1 week. They are given with saline and/or calcitonin, as their maximal effects start at 4 days. They prevent hypercalcemia and fractures in metastatic cancer to bone. Intravenous zoledronic acid or pamidronate are the most used. ZA is more potent than pamidronate. With repetitive use, there is a risk of jaw osteonecrosis in multiple myeloma or metastatic bone disease. They have low toxicity, but may cause flu-like illness, hypocalcemia, hypophosphatemia, ocular inflammation, osteonecrosis of the jaw, and renal impairment.

- **Pamidronate:** This is given as a single dose of 60 or 90 mg intravenously over 24 h, with 60–100% chances of normal calcium 10 days later. In renal insufficiency, smaller dose of pamidronate is given. Fever is the main side effect.
- **Zoledronic acid:** Hypercalcemia of malignancy, due to autonomous production of PTHrP, responds better to ZA. It requires a shorter time of administration (15 min *versus* 2 h).
- **Ibandronate:** This effectively treats hypercalcemia of malignancy, and is as effective as pamidronate.

Glucocorticoids: Used in hypercalcemia related to high vitamin D (in granulomatous disorders) and in multiple myeloma. They decrease calcitriol production by the activated mononuclear cells. They may decrease calcium absorption from the gut.

Gallium nitrate: Inhibits osteoclastic bone resorption by inhibition of PTH secretion. It is effective in both PTHrP-mediated, and non-PTHrP-mediated hypercalcemia. It's more potent than etidronate, pamidronate and calcitonin alone, but is nephrotoxic, and needs a continuous infusion over five days.

Calcimimetics: As for cinacalcet, they reduce calcium in severe hypercalcemia due to parathyroid carcinoma and in hemodialysis patients.

Dialysis: This may be indicated in severe malignancy-associated hypercalcemia and inadequate renal function or heart failure. It may be used in hypercalcemia without renal failure or in intolerance to diuresis.

Summary Points

- This chapter focuses on disturbances of calcium in critical illness and gives an overview of the available data in this field.
- Serum calcium levels are tightly regulated and both hyper- and hypocalcemia may be detrimental to many physiologic functions.
- In critically ill patients, the measurement of ionized calcium is preferred because total serum calcium may not reliably represent biologically active calcium concentrations.
- Hypocalcemia is commonly encountered in the ICU and often caused by ICU-specific factors besides those found in other, healthier populations. It is associated with poor outcomes including excess mortality. However, there are no data if correction of calcium levels is beneficial.

- Symptoms are often absent but range from mild to life-threatening symptoms. This depends mainly on the degree of hypocalcemia and how fast it developed. Rapid intravenous calcium is needed in severely symptomatic patients and in acute decreases. Hypercalcemia is seldom an issue in critical illness. Primary hyperparathyroidism and malignancy are the most common causes of hypercalcemia.
- Asymptomatic hypercalcemia and moderate hypercalcemia (12 to 14 mg dL^{-1}) may not require treatment. Calcium >14 mg dL^{-1} requires treatment, regardless of symptoms. Also serum ionized calcium levels that are out of proportion (4.65 to 5.25 mg dL^{-1}) need to be taken into consideration. Intravenous saline hydration is needed for immediate management of hypercalcemia. Bisphosphonates are also used. Glucocorticoids are effective in hypercalcemia of lymphomas, sarcoidosis, or other granulomatous diseases. Dialysis may be used in severe, otherwise refractory hypercalcemia.

Key Facts for Calcium in Critical Care

1. Calcium is the most abundant divalent cation in the body.
2. It is tightly regulated by vitamin D and parathyroid hormone.
3. The majority (98%) of calcium is bound in bone, and only a small amount is extracellular.
4. Calcium plays an essential role in many cellular processes.
5. Total plasma calcium is unreliable in critical illness because a large proportion is bound to albumin and other proteins. In addition, changes in acid–base status affect binding.
6. Ionized calcium levels are recommended to assess calcium status in critical illness.
7. Hypocalcemia is common in the ICU and is associated with adverse outcomes.

Definitions of Words and Terms

Normocalcemia. Total calcium range between 8.5 and 10.5 mg dL^{-1} or ionized calcium between 4.65 and 5.25 mg dL^{-1}.

Hypoparathyroidism. Low PTH levels or impaired PTH action.

Pseudohypoparathyroidism. Characterized by targeted organ (kidney and, perhaps, bone) unresponsiveness to PTH.

Chvostek sign. Tapping facial nerve leads to grimace.

Trousseau sign. Inflating blood pressure cuff causes carpopedal spasm of outstretched hand.

Pseudohypocalcemia. Describes low total serum calcium with normal ionized calcium.

Pseudohypercalcemia. Reflects high total serum calcium with normal ionized fraction.

List of Abbreviations

ARF	Acute renal failure
BMD	Bone-mineral density
CaSR	Calcium-sensing receptor
CKD	Chronic kidney disease
ECG	Electrocardiogram
EDTA	Ethylenediaminetetraacetic acid
GFR	Glomerular filtration rate
IU	International units
IV	Intravenous
Lab	Laboratory
NIH	National Institute of Health
PTH	Parathyroid hormone
PTHrP	PTH-related peptide
25-(OH)D	25-Hydroxyvitamin D, calcidiol
1,25-$(OH)_2$D	1,25-Dihydroxyvitamin D, calcitriol

References

Akmal, M., Bishop, J. E., Telfer, N., Norman, A. W. and Massry, S. G., 1986. Hypocalcemia and hypercalcemia in patients with rhabdomyolysis with and without acute renal failure. *Journal of Clinical Endocrinology & Metabolism*. 63: 137.

Bolan, C. D., Cecco, S. A., Wesley, R. A., Horne, M., Yau, Y. Y., Remaley, A. T., Childs, R. W., Barrett, A. J., Rehak, N. N. and Leitman, S. F., 2002. Controlled study of citrate effects and response to i.v. calcium administration during allogeneic peripheral blood progenitor cell donation. *Transfusion*. 42: 935.

Cairns, C. B., Niemann, J. T., Pelikan, P. C. and Sharma, J., 1991. Ionized hypocalcemia during prolonged cardiac arrest and closed-chest CPR in a canine model. *Annals of Emergency Medicine*. 20: 1178.

Canaff, L., Zhou, X. and Hendy, G. N., 2008. The proinflammatory cytokine, interleukin-6, up-regulates calcium-sensing receptor gene transcription *via* Stat1/3 and Sp1/3. *Journal of Biological Chemistry*. 283: 13586–13600.

Carlstedt, F. and lind, L., 2001. Hypocalcemic syndromes. *Critical Care Clinics*. 17: 139.

Carlstedt, F., Lind, L., Rastad, J., Stiernström, H., Wide, L. and Ljunghall, S., 1998. Parathyroid hormone and ionized calcium levels are related to the severity of illness and survival in critically ill patients. *European Journal of Clinical Investigation*. 28: 898–903.

Dettelbach, M. A., Deftos, L. J. and Stewart, A. F., 1990. Intraperitoneal free fatty acids induce severe hypocalcemia in rats: a model for the hypocalcemia of pancreatitis. *Journal of Bone and Mineral Research*. 5: 1249.

Desai, T. K., Carlson, R. W. and Geheb, M. A., 1988. Prevalence and clinical implications of hypocalcemia in acutely ill patients in a medical intensive care setting. *American Journal of Medicine*. 84: 209.

Forsythe, R. M., Wessel, C. B., Billiar, T. R., Angus, D. C. and Rosengart, M. R., 2008. Parenteral calcium for intensive care unit patients. *The Cochrane Database of Systematic Reviews*. CD006163.

Fukugawa, M. and Kurokawa, K. L., 2002. Calcium homeostasis and imbalance. *Nephron*. 92: 41.

Ghent, S., Judson, M. A. and Rosansky, S. J., 1994. Refractory hypotension associated with hypocalcemia and renal disease. *American Journal of Kidney Diseases*. 23: 430.

Goltzman, D. and Cole, D. E. C., 2006. Hypoparathyroidism. In primer on the metabolic bone diseases and disorders of bone metabolism. *American Society of Bone and Mineral Research*. 6: 216.

Hästbacka, J. and Pettilä, V., 2003. Prevalence and predictive value of ionized hypocalcemia among critically ill patients. *Acta Anaesthesiologica Scandinavica*. 47: 1264–1269.

Institute of Medicine, 2010. Report at a Glance, Report Brief: Dietary Reference Intakes for Calcium and Vitamin D. http://www.iom.edu/Reports/2010/Dietary-Reference-Intakes-for-Calcium-and-Vitamin-D/Report-Brief.aspx.

Jacobus, C. H., Holick, M. F., Shao, Q., Chen, T. C., Holm, I. A., Kolodny, J. M., Fuleihan, G. E. and Seely, E. W., 1992. Hypervitaminosis D associated with drinking milk. *New England Journal of Medicine*. 326: 1173.

Kazmi, A. S. and Wall, B. M., 2007. Reversible congestive heart failure related to profound hypocalcemia secondary to hypoparathyroidism. *American Journal of Medical Sciences*. 333: 226.

Kelly, A. and Levine, M. A., 2013. Hypocalcemia in the critically ill patient. *Journal of Intensive Care Medicine*. 28: 166–177.

Ladenson, J. H., Lewis, J. W. and Boyd, J. C., 1978. Failure of total calcium corrected for protein, albumin, and pH to correctly assess free calcium status. *Journal of Clinical Endocrinology and Metabolism*. 46: 986.

Lepage, R., Légaré, G., Racicot, C., Brossard, J. H., Lapointe, R., Dagenais, M. and D'Amour, P., 1999. Hypocalcemia induced during major and minor abdominal surgery in humans. *Journal of Clinical Endocrinology and Metabolism*. 84: 2654.

Meyer, T., Ruppert, V., Karatolios, K. and Maisch, B., 2007. Hereditary long QT syndrome due to autoimmune hypoparathyroidism in autoimmune polyendocrinopathy-candidiasis-ectodermal dystrophy syndrome. *Journal of Electrocardiology*. 40: 504.

Michaud, J., Naud, J., Demers, C., Petit, J. L., Leblond, F. A., Bonnardeaux, A., Gascon-Barré, M. and Pichette, V., 2010 Sep. Reduced hepatic synthesis of calcidiol in uremia. *Journal of the American Society of Nephrology*. 2: 1488–1497.

Navarro, J., Oster, J. R., Gkonos, P. J., Ruiz, J. P., Rhamy, R. K. and Perez, G. O., 1991. Tetany induced on separate occasions by administration of potassium and magnesium in a patient with hungry-bone syndrome. *Mineral and Electrolyte Metabolism*. 17: 340.

Nishi, S. P., Barbagelata, N. A., Atar, S., Birnbaum, Y. and Tuero, E., 2006. Hypercalcemia-induced ST-segment elevation mimicking acute myocardial infarction. *Electrocardiology*. 39: 298.

Raimundo, M., Crichton, S., Lei, K., Sanderson, B., Smith, J., Brooks, J., Ng, J., Lemmich Smith, J., McKenzie, C., Beale, R., Dickie, H. and Ostermann, M., 2013. Maintaining normal levels of ionized calcium during citrate-based renal replacement therapy is associated with stable parathyroid hormone levels. *Nephron Clinical Practice*. 124: 124–131.

Sperber, S. J., Blevins, D. D. and Francis, J. B., 1990. Hypercalcitoninemia, hypocalcemia, and toxic shock syndrome. *Reviews of Infectious Diseases*. 12: 736.

Stewart, A. F., 2005. Clinical practice. Hypercalcemia associated with cancer. *New England Journal of medicine*. 352: 373.

Tohme, J. F. and Bilezikian, J. P., 1996. Diagnosis and treatment of hypocalcemic emergencies. *The Endocrinologist*. 6: 10.

Turhan, S., Kilickap, M. and Kilinc, S., 2005. ST segment elevation mimicking acute myocardial infarction in hypercalcaemia. *Heart*. 91: 999.

Wong, C. K., Lau, C. P., Cheng, C. H., Leung, W. H. and Freedman, B., 1990. Hypocalcemic myocardial dysfunction: short- and long-term improvement with calcium replacement. *American Heart Journal*. 120: 381.

Zivin, J. R., Gooley, T., Zager, R. A. and Ryan, M. J., 2001. Hypocalcemia: a pervasive metabolic abnormality in the critically ill. *American Journal of Kidney Diseases*. 37: 689.

CHAPTER 36

Hypercalcemia: An Overview of its Pathology

FRANCO LUMACHI*[a] AND STEFANO M. M. BASSO[b]

[a]Department of Surgery, Oncology & Gastroenterology, University of Padua, School of Medicine, Via Giustiniani 2, 35128 Padova, Italy; [b]Surgery 1, S. Maria degli Angeli Hospital, Via Montereale 24, 33170 Pordenone, Italy
*E-mail: flumachi@unipd.it

36.1 Calcium Distribution in the Human Body

Calcium is an alkaline-earth metal that is essential for several human physiological processes, including the function of cellular membranes, nerves and muscles, enzymatic activities, and blood clotting. Calcium also plays a role in the control of growth and proliferation of both normal and cancer cells (Landowski *et al.,* 2011).

Bones and teeth are the major storage site for calcium containing more than 99% of the calcium in the body, mainly in the form of hydroxyapatite [$Ca_{10}(PO_4)_6(OH)_2$] (Mundy and Guise, 1999). The total skeletal calcium is relatively stable, but there is a dynamic process of bone turnover through a balance between the process of bone deposition and reabsorption (O'Toole, 2011). The remaining 1% of the total-body calcium resides within the body fluids.

The calcium absorbed in the gut diffuses throughout the body and is then distributed in the extracellular and intracellular compartments. The level of extracellular calcium is approximately 10 000-fold higher than intracellular

Food and Nutritional Components in Focus No. 10
Calcium: Chemistry, Analysis, Function and Effects
Edited by Victor R. Preedy

Published by the Royal Society of Chemistry, www.rsc.org

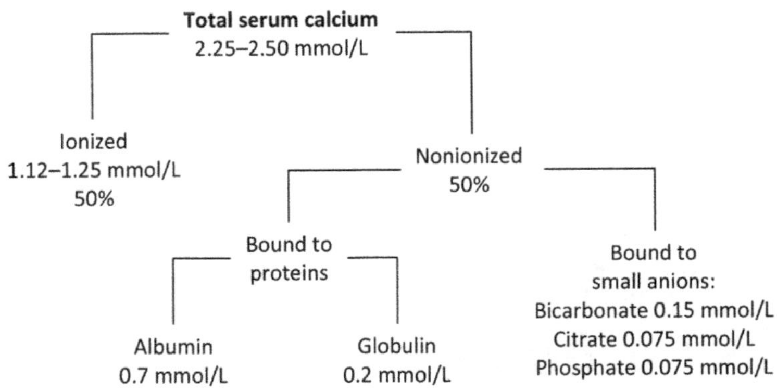

Figure 36.1 Distribution of serum calcium fractions (modified from Lumachi and Basso (2010)). Ionized calcium (Ca^{2+}) represents the active fraction of calcium, corresponding to approximately 50% of total serum calcium, and is tightly regulated, ranging from 2.25 to 2.50 mmol L^{-1}. Approximately 40% of serum calcium is bound to proteins (primarily to albumin and globulin), and 10% is complex bound with small anions, such as carbonate, citrate, and phosphate.

free cytosolic calcium, which acts as a second messenger in the regulation of cell division, membrane trafficking, and muscle contractility. Extracellular calcium has several physiological functions, including the synaptic transmission of skeletal and heart muscles, coagulation, and platelets aggregation (Shoback *et al.*, 2001).

Ionized calcium (Ca^{2+}) represents the active fraction of calcium, corresponding to approximately 50% of total serum calcium and is tightly regulated at 1.25 ± 0.07 mmol L^{-1}. Approximately, 40% of serum calcium is bound to proteins (primarily to albumin), and 10% is complex bound with small anions, such as carbonate, citrate, and phosphate (Figure 36.1). Thus, fluctuations in plasma protein levels do not affect Ca^{2+} levels. The normal reference ranges of calcium fractions are reported in Table 36.1.

Ca^{2+} concentration exhibits a relationship with blood pH because hydrogen ions compete with calcium for binding sites on serum proteins, and it has been shown that calcium entry channels in the plasma membrane play a major role in maintaining cellular Ca^{2+} levels (Landowski *et al.*, 2011). The endocrine control of Ca^{2+} is achieved by a homeostatic mechanism involving (i) parathyroid hormone (PTH); (ii) 1,25-dihydroxyvitamin D ($1,25(OH_2)D_3$, also named calcitriol), the active form of vitamin D; and (iii) calcitonin, which plays only a secondary role. However, other factors, such as fibroblast growth factor-23 (FGF-23), the transient receptor potential vanilloid type 5 (TRPV5), and a transmembrane protein named mKlotho, regulate calcium metabolism (Medici *et al.*, 2008; Lumachi and Basso, 2010). FGF-23 is produced by osteocytes, regulates osteogenesis and kidney function by promoting urinary phosphate excretion, and inhibiting the synthesis of $1,25(OH)_2D_3$, whereas Klotho is an ageing suppressor gene of restricted expression (*i.e.*, kidneys,

Table 36.1 Calcium distribution in the body fluids (modified from Shoback *et al.* (2001)).

Calcium fractions	Normal reference range
Ionized (Ca^{2+})	1.1–1.3 mmol L^{-1}
Protein-bound	0.9–1.1 mmol L^{-1}
Complexed	0.1–0.2 mmol L^{-1}
Total serum	2.2–2.6 mmol L^{-1}
Intracellular free	180–190 nmol L^{-1}

brain, parathyroid glands) encoding mKlotho, which acts as a coreceptor in FGF-23 signaling (Medici *et al.*, 2008; Schmid *et al.*, 2013). Thus, both FGF-23 and mKlotho are secretory proteins that regulate mineral-ion metabolism, and there is functional evidence that transcellular epithelial calcium uptake is essential for bone formation (Vanoevelen *et al.*, 2011).

36.2 Regulatory Mechanisms of Calcium Metabolism

The recommended intake of calcium in adults (>18 years old) ranges from 1.0 to 1.3 g per day, according to age, gender and breastfeeding status (Lumachi *et al.*, 2011). An estimated 800 mg per day of calcium is eliminated in the feces. Vitamin D and PTH play a major role in the mechanism of calcium absorption in the gut, which occurs *via* either TRPV5-mediated active transcellular or passive paracellular pathways (O'Toole, 2011). From 8 g to 10 g of calcium is filtered daily at the glomerulus, but 98–99% is then reabsorbed by the renal tubule, mainly at the level of the proximal tubule and thick ascending limb of the loop of Henle (TALH) (Figure 36.2).

The recommended dietary allowance (RDA) for dietary protein in adults is 0.8 g kg^{-1} body weight, and a protein intake above the RDA reduces the incidence of bone fracture by increasing bone-mineral density (BMD), especially in the elderly (Cao and Nielsen, 2010). A relationship between calcemia and the risk of cardiovascular diseases has been hypothesized, but calcium intake does not increase the incidence of vascular calcifications, and in patients with primary hyperparathyroidism (PHPT) serum calcium and PTH levels do not correlate with arterial blood pressure (Lumachi *et al.*, 2002).

Vitamin D is synthesized in the skin from 7-dehydrocholesterol to vitamin D_3 (cholecalciferol) by a nonenzymatic ultraviolet sunlight-mediated conversion, but there is also a clinically important dietary source of the vitamin (Shoback *et al.*, 2001). Vitamin D_3 is then hydroxylated in the liver by CYP2R1 to 25(OH)D_3 and further 1α-hydroxylated in the proximal convoluted tubule of the kidney by CYP27B1 (both members of the cytochrome P450 family) to 1,25$(OH)_2D_3$ (Zhu *et al.*, 2013). Figure 36.3 illustrates these steps. This activated form of vitamin D increases the intestinal calcium absorption *via* a transcellular pathway that involves the nuclear vitamin D receptor VDR (O'Toole, 2011). The production of 1,25$(OH)_2D_3$ is inhibited by FGF-23, hypercalcemia

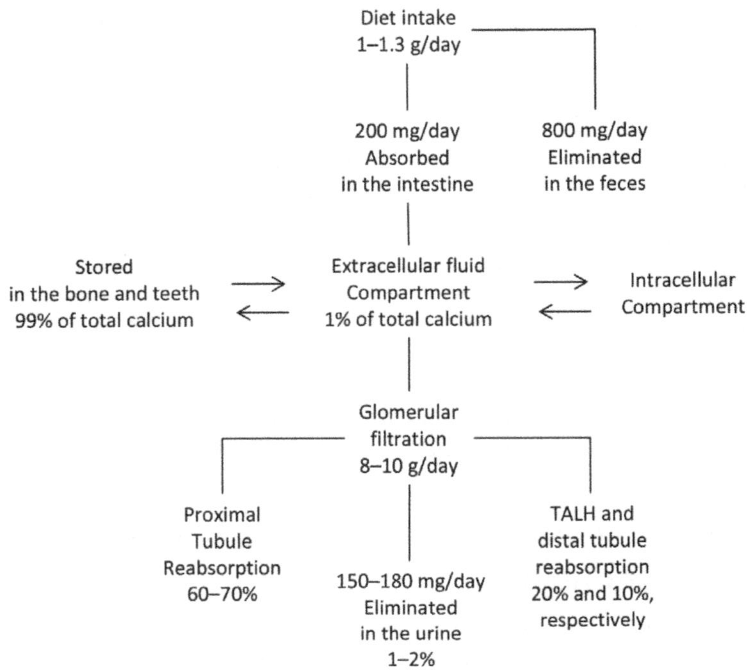

Figure 36.2 Calcium metabolism in the human body. The intake of calcium in adults (>18 years old) ranges from 1.0 to 1.3 g per day, according to age, gender and breastfeeding status. An estimated 800 mg per day of calcium is eliminated in the feces. Bones and teeth are the major sites for calcium, containing more than 99% of the calcium in the body. The remaining 1% of total-body calcium resides within the body fluids. From 8 g to 10 g of calcium is filtered daily at the glomerulus, but 98–99% is then reabsorbed by the renal tubule, at the level of the proximal tubule (60–70%), distal tubule (20%), and thick ascending limb of the loop of Henle (TALH). Approximately 1–2% (150–180 mg per day) of filtered calcium is eliminated in the urine.

and hyperphosphatemia in favor of 24,25$(OH)_2D_3$, and stimulated by hypocalcemia, PTH, and insulin-like growth factor-1 (IGF-1).

Vitamin D deficiency is common, especially in the elderly. Furthermore, more than 95% of morbidly obese patients present with vitamin D deficiency and elevated serum PTH levels, that gradually resolve after sleeve gastrectomy to return within the normal range, without any impact on serum calcium levels (Ruiz-Trovar *et al.*, 2012).

Parathyroid hormone is a straight-chain polypeptide containing 84 amino acids that is synthesized and released by the parathyroid glands (Lumachi and Basso, 2010). Serum PTH derives from either the release of stored PTH within the parathyroid glands or the increased synthesis of pre/pro-PTH, the intracellular precursor of PTH (Shoback *et al.*, 2001). The response of parathyroid chief cells to changes in serum Ca^{2+} is sensed by the calcium-sensing receptor (CaSR), localized on the cells surface, which plays a crucial role in

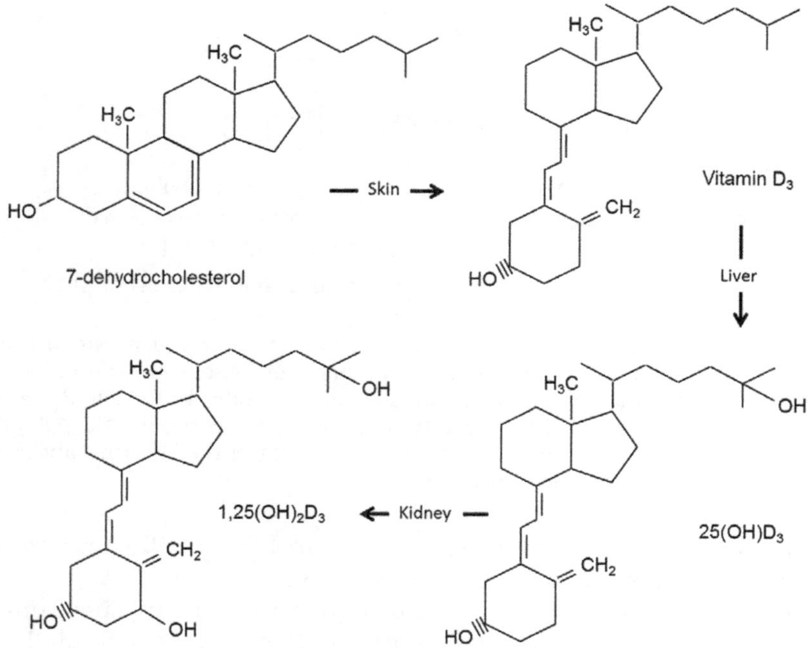

Figure 36.3 Main steps of synthesis and activation of vitamin D. Vitamin D is synthesized in the skin from 7-dehydrocholesterol to vitamin D_3 (cholecalciferol) by a nonenzymatic ultraviolet sunlight-mediated conversion, and then hydroxylated in the liver by CYP2R1 to $25(OH)D_3$ and further 1α-hydroxylated in the proximal convoluted tubule of the kidney by CYP27B1 (both members of the cytochrome p450 family) to $1,25(OH)_2D_3$ (calcitriol), the active form of vitamin D. The production of $1,25(OH)_2D_3$ is inhibited by FGF-23, hypercalcemia and hyperphosphatemia in favor of $24,25(OH)_2D_3$, and stimulated by hypocalcemia, PTH, and insulin-like growth factor-1 (IGF-I).

the regulation of Ca^{2+} reabsorption in the kidney, and modulates the effect of $1,25(OH)_2D_3$ on PTH release. PTH and the CaSR should be considered the key regulators of calcium homeostasis, as CaSRs are also present in the osteoblasts, renal glomerulus, cortical TALH, and proximal and distal tubules (O'Toole, 2011). Calcium-sensing receptors work by activating a guanylyl nucleotide binding protein (G protein) and the phospholipase C pathway, thus increasing intracellular Ca^{2+} concentration (Lumachi and Basso, 2010). The biological actions of PTH include the following (Mundy and Guise, 1999; Bikle *et al.*, 2002):

- The stimulation of osteoclasts, bone resorption and the release of calcium and phosphate from bone, mediated by the by the receptor-activated nuclear factor-κ (RANK)–RANK ligand (RANKL) system.
- The stimulation of the renal production of $1,25(OH)_2D_3$, which increases the intestinal absorption of calcium and phosphate.

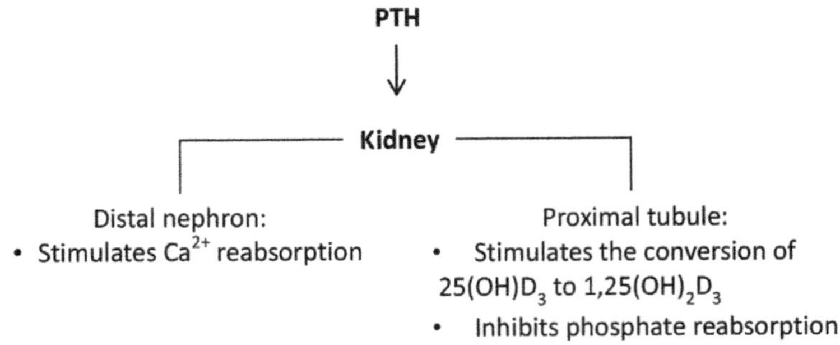

Figure 36.4 Biological actions of PTH on distal nephron and proximal tubule. Ca^{2+}: ionized calcium. PTH inhibits phosphate reabsorption from the proximal tubule, and stimulates calcium reabsorption and the conversion of $25(OH)D_3$ to $1,25(OH)_2D_3$ in the distal tubule. $1,25(OH)_2D_3$ is the active form of vitamin D, which increases the intestinal absorption of calcium and phosphate.

- The direct effect on the activity of the osteoblast, a cell known to have PTH receptors, stimulating bone formation.
- The inhibition of phosphate reabsorption from the proximal tubules and the stimulation of calcium reabsorption in the distal tubule and TALH (Figure 36.4).

In addition, PTH stimulates the function of mature osteoblasts, also promoting the differentiation and proliferation of osteoblast precursors (Bikle *et al.*, 2002).

36.3 Evaluation of Hypercalcemia

Circulating total calcium is maintained within a range between 2.25 and 2.50 mmol L^{-1} (9–10 mg dL^{-1}), and hypercalcemia can be defined as a serum calcium concentration that is 2 or more standard deviations above the average value of the reference interval of the population (Lumachi *et al.*, 2011). In adults, the upper limit of the normal range for total calcium and serum Ca^{2+} is usually set at 2.65 and 1.32 mmol L^{-1}, respectively, but so-called pseudohypercalcaemia can also occur (Lumachi *et al.*, 2012). Because the concentration of plasma proteins may alter total calcemia, its adjustment is required; or alternatively the free calcium (Ca^{2+}) concentration should be measured. A commonly used formula is the following (Clase *et al.*, 2000):

$$\text{Corrected calcium (mg dL}^{-1}) = \text{total calcium} - [0.707 \times (\text{albumin} - 3.4)]$$

Preanalytical causes of pseudohypercalcaemia include venous stasis and subsequent changes in albumin concentrations in the blood, dehydration, hemoconcentration, and the use of certain drugs (*i.e.*, acetylcysteine, paracetamol, salicylates), but also the pH changes may affect serum Ca^{2+} levels, that increase by 0.05 mmol L^{-1} for each pH decline of 0.1 (Lumachi *et al.*, 2012).

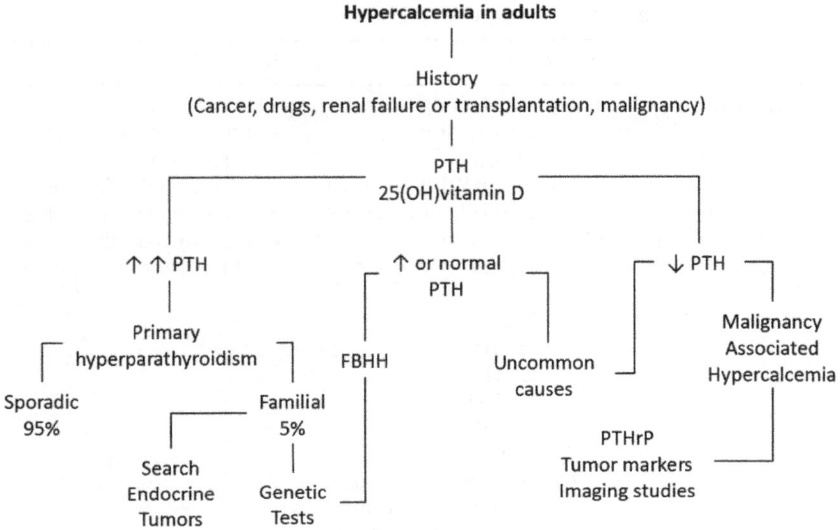

Figure 36.5 Suggested diagnostic algorithm for hypercalcemia. FBHH: familial benign hypocalciuric hypercalcemia; PTHrP: parathyroid hormone-related protein. Once a true hypercalcemia is confirmed by either repeated measurements of total calcium or Ca^{2+}, the intact PTH and 25(OH)vitamin D assay are required for further investigation. Patients with elevated PTH serum levels usually have sporadic or familial primary hyperparathyroidism. When PTH levels are normal or low, malignancy-associated hypercalcemia should be considered.

Once a true hypercalcemia is confirmed by either repeated measurements of total calcium or Ca^{2+}, the intact PTH and 25(OH) vitamin D assay are required for further investigation. A suggested diagnostic algorithm for hypercalcemia is presented in Figure 36.5. Primary hyperparathyroidism is the most common cause of benign hypercalcemia in nonhospitalized patients, accounting for at least 80% of cases, whereas malignancy-associated hypercalcemia (MAH) is more frequent in hospitalized patients (Table 36.2).

36.4 Primary Hyperparathyroidism and Other Benign Diseases

There are considerable geographical differences in the incidence of PHPT, ranging from 5 to 20 cases per 100 000 in recent series, with an estimated prevalence up to 4 per 1000 in women over 60 years (Shoback *et al.*, 2001; Abood and Vestergaard, 2013). Today, 80–90% of patients with PHPT are asymptomatic, and thus, radiological bone findings, such as *osteitis fibrosa cystica*, bone cysts, and brown tumors, observed in the past, are now extremely uncommon. However, the prolonged catabolic and anabolic effects of PTH on cortical and cancellous bone may still lead to several deleterious effects on the skeleton, such as osteopenia and osteoporosis, which are revealed using dual-energy X-ray

Table 36.2 Main causes of hypercalcemia.[a]

Frequent	Infrequent	Rare
• Sporadic primary hyperparathyroidism • Cancer-induced: PTHrP Local osteolysis Parathyroid cancer Paraneoplastic: PTH $1,25(OH)_2D_3$	• Drug induced: Thiazide diuretics Lithium Vitamin A intoxication Vitamin D intoxication • Familial primary hyperparathyroidism: MEN1 MEN2A FBHH FIPH HPT-JT NSHPT • Tertiary hyperparathyroidism: Chronic renal failure Postrenal transplantation	• Sarcoidosis • Endocrine diseases: Thyrotoxicosis Addison's disease • Prolonged immobilization • Rabdomyolysis • Other more rare causes

[a]PTHrP: parathyroid hormone-related protein; MEN: multiple endocrine neoplasia; FBHH: familial benign hypocalciuric hypercalcemia; FIPH: familial isolated primary hyperparathyroidism; HPT-JT: hyperparathyroidism-jaw tumor; NSHPT: neonatal severe primary hyperparathyroidism.

absorptiometry (DXA) to obtain BMD measurements. Elevated PTH levels and chronic uncontrolled hypercalcemia also increase tubular calcium reabsorption, but the increased filtered load causes final hypercalciuria that may lead to a risk of nephrolithiasis, progressive loss of renal function, and midterm cardiovascular damages. However, the relationship between arterial blood pressure and either serum PTH or calcium levels is unclear (Lumachi and Basso, 2010).

Surgery should be the treatment of choice for all patients with confirmed PHPT because parathyroidectomy always results in improvements in BMD and cognitive functions, and a decreased incidence of fracture risk and kidney stones (Lumachi *et al.*, 2007; Udelsman *et al.*, 2009). However, recent guidelines have been proposed that suggest surgical treatment for patients with sporadic asymptomatic PHPT (Table 36.3). The serum Ca^{2+} level and 24 h urinary calcium excretion are not considered useful criteria (Bilezikian *et al.*, 2009).

In approximately 80–85% of patients with sporadic PHPT, a solitary monoclonal parathyroid adenoma can be demonstrated, 12–15% exhibit multiple adenomas or parathyroid gland hyperplasia, and parathyroid carcinoma is rare (1% of cases) and often has a poor outcome (Schulte *et al.*, 2012). Mutations in the type-1 multiple endocrine neoplasia (MEN1) and hyperparathyroidism type-2 (HRPT2/CDC73) suppressor genes, or in cyclin-dependent kinase inhibitor (CDKI) gene, and alterations of the cyclin D1 (CCND1) and parathyroid adenoma 1 (PRAD1) oncogenes, have been found to be related to several parathyroid tumors (Lumachi and Basso, 2010).

Table 36.3 Parameters suggesting surgical treatment for patients with asymptomatic primary hyperparathyroidism (modified from Bilezikian *et al.* (2009)).[a]

Parameter	Characteristics
Age of the patient	<50 years
Total serum calcium	>0.25 mmol L^{-1} above the upper limit of normal range
eGFR	<60 mL min^{-1}
BMD	>2.5 SD below the young normal mean (T-score) at any site

[a]eGFR: estimated glomerular filtration ratio; BMD: bone-mineral density; SD: standard deviation.

Several familial syndromes, such as MEN1, multiple endocrine neoplasia type-2A (MEN2A), familial isolated primary hyperparathyroidism (FIPH), familial benign hypocalciuric hypercalcemia (FBHH), hyperparathyroidism-jaw tumor (HPT-JT) syndrome, and neonatal severe primary hyperparathyroidism (NSHPT) may cause PHPT.

MEN1 and *MEN2A* are both autosomal dominant syndromes, and patients with MEN-related and familial non-MEN PHPT typically exhibit parathyroid hyperplasia. MEN1 is characterized by the combination of PHPT, pituitary adenomas, and pancreatic endocrine tumors (*i.e.*, gastrinoma, insulinoma), whereas MEN2A is characterized by medullary thyroid carcinoma, PHPT and pheochromocytomas, associated with a cluster of germline gain-of-function mutations of the rearranged during transfection (RET) proto-oncogene on chromosome 10q11.2 (Peppa *et al.*, 2008).

FBHH is a rare autosomal dominant disease caused by loss-of-function of the CaSR and is characterized by mild lifelong hypercalcemia, slightly elevated PTH, and low 24 h urinary calcium (Lumachi and Basso, 2010). Thus, inactivating mutations in CaSR may lead to FBHH, but additional loci (*i.e.*, on chromosomes 19p13.3 and 19q13) for this disease have been identified (O'Toole, 2011).

FIPH is another autosomal dominant disorder, with reduced genetic penetrance, affecting young patients that can be considered an early stage of either MEN1 or HPT-JT syndromes, and is associated with germline inactivating mutations in the MEN1 and HRPT2 tumor-suppressor genes (Lumachi and Basso, 2010).

HPT-JP syndrome is a rare familial cancer syndrome that results from germline inactivation of HRPT2/CDC73, a putative tumor-suppressor gene that encodes for parafibromin, a component of the RNA polymerase-associated factor-1 complex (PAF1C) that regulates transcription, with homology to the tumor suppressor Cdc73 (Panicker *et al.*, 2010).

NSHPT is a very rare type of familial PHPT occurring in infancy, characterized by severe hypercalcemia and precocious bone changes, due to a homozygous loss-of-function of CaSR mutation.

MEN1 gene mutation is responsible for at least 70% of cases of parathyroid carcinomas, whereas germline mutations of the HRPT2 gene (CDC73)

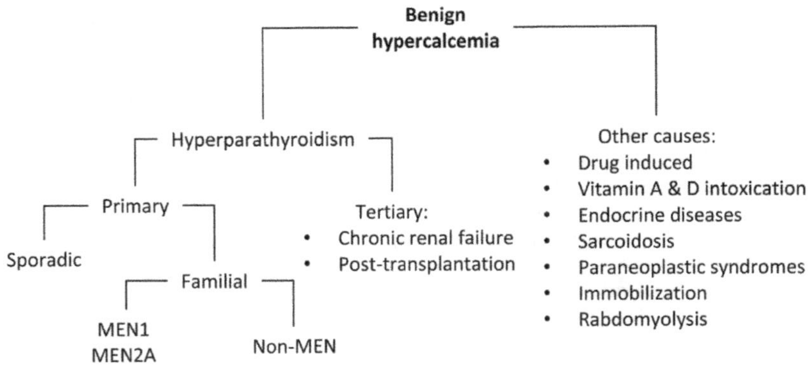

Figure 36.6 Main causes of benign hypercalcemia. MEN: multiple endocrine neoplasia; FBHH: familial benign hypocalciuric hypercalcemia; FIPH: familial isolated primary hyperparathyroidism; HPT-JT: hyperparathyroidism-jaw tumor. Primary hyperparathyroidism (HPT) is the most common cause of benign hypercalcemia. Several familial syndromes, such as multiple endocrine neoplasia type 1 (MEN1) and 2A (MEN2A), and non-MEN syndromes, such as familial benign hypocalciuric hypercalcemia (FBHH), familial isolated primary hyperparathyroidism (FIPH), hyperparathyroidism-jaw tumor (HPT-JT) syndrome, and neonatal severe primary hyperparathyroidism (NSHPT), may cause primary HPT. Tertiary HPT can be observed as the end result of long-standing secondary HPT and chronic kidney failure, and in approximately 25% of patients who underwent kidney transplantation. Other rare causes of benign hypercalcemia are reported.

coding for parafibromin have been identified also in families with HPT-JT and FIPH. CaSR is also expressed in several benign and malignant tumors, often at different expression levels with respect to their healthy counterparts, such as breast and prostate cancer, and in those cancers (*i.e.*, parathyroid and colon carcinomas) originating from organs involved in Ca^{2+} homeostasis, in which they can have a suppressive role (Saidak *et al.*, 2009).

Mild hypercalcemia may also occur in patients who are thiazide diuretics or lithium users, and in up to 10% of patients with granulomatous diseases, such as sarcoidosis, due to the inappropriate production of $1,25(OH)_2D_3$. Other causes of mild hypercalcemia are thyrotoxicosis (observed in up to 50% of patients), increased bone resorption subsequent to a prolonged immobilization (*i.e.*, in the elderly), vitamin A or D intoxication, and rhabdomyolysis, rarely observed in patients with acute renal failure, in whom mobilization of calcium from muscles may result (Figure 36.6).

36.5 Malignancy-Associated Hypercalcemia

Malignancy-associated hypercalcemia is the most common metabolic disorder in malignancy that occurs in up to 20–25% of cancer patients, exhibiting an estimated incidence of 15 cases per 100 000 persons per year (Strewler,

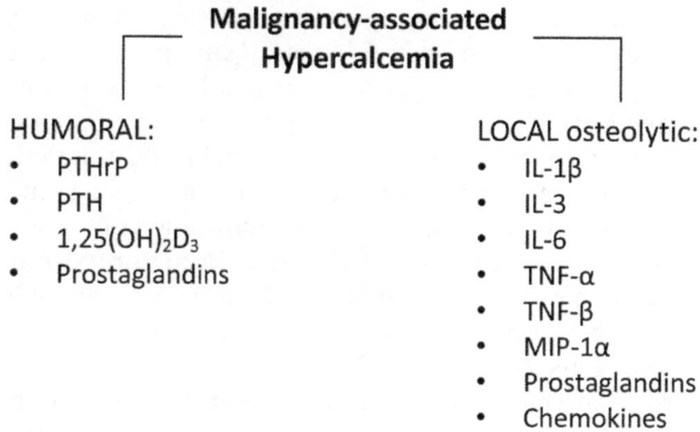

Figure 36.7 There are two different mechanisms that may lead to malignancy-associated hypercalcemia: the so-called humoral hypercalcemia, and local osteolytic hypercalcemia. Humoral hypercalcemia is largely caused by the secretion of parathyroid hormone-related protein (PTHrP) by the malignant cells or, rarely, PTH itself, whereas in very rare cases the disorder may be caused by ectopically activated 1,25$(OH)_2D_3$, or prostaglandins. Local osteolytic hypercalcemia is induced directly by cancer cells invading the bone and promoting the degradation of mineral matrix. Tumor-driven osteoclastic bone resorption is activated by various cytokines, including interleukin (IL)-1β, IL-3, IL-6, tumor necrosis factor (TNF)-α, TNF-β, macrophage inflammatory protein (MIP)-1α, prostaglandins, and chemokines, locally produced by the tumor cells.

2005). MAH is a serious complication that is observed in 10–15% of cancer patients, especially in those with advanced cancer, usually indicating a poor prognosis (Hu *et al.*, 2013). In patients with MAH, osteoblastic activity is suppressed, with the subsequent tendency towards metabolic alkalosis and increased 1,25(OH)2D3 serum levels, whereas those with PHPT typically experience hyperchloremic acidosis, with low circulating vitamin D (Mundy and Edwards, 2008). In these patients, osteolysis is the result of a disruption of the balance between osteoblast and osteoclast activity of in the bone marrow (Vallet *et al.*, 2010). Tumor cells may activate osteoclasts through several pathways, and all three processes regulating calcemia, such as the mobilization of Ca^{2+} from the bone, its renal tubular resorption, and intestinal absorption, can be involved in the pathogenesis of hypercalcemia (Basso *et al.*, 2011). Two different mechanisms may lead to MAH (Figure 36.7).

The so-called *humoral hypercalcemia* is largely caused by the secretion of parathyroid hormone-related protein (PTHrP) by the malignant cells or, rarely, PTH itself, whereas in very rare cases the disorder may be caused by ectopically activated 1,25$(OH)_2D_3$, with a subsequent increased intestinal calcium absorption and moderate effects of the kidney and bone (Stewart, 2005).

Local osteolytic hypercalcemia is induced directly by cancer cells invading the bone, leading to the degradation of mineral matrix (Lumachi *et al.*, 2008). The bone marrow represents an ideal site for the growth of cancer cells, such as malignant plasma cells, leukocytes and metastatic cells from solid tumors, and the complex network of paracrine and autocrine pathways (*vicious cycle*) between these cells, osteoclasts and osteoblasts ultimately stimulates tumor cells growth itself (Basso *et al.*, 2011). Tumor-driven osteoclastic bone resorption is activated by various cytokines, including interleukin (IL)-1β, IL-3, IL-6, tumor necrosis factor (TNF)-α, TNF-β, macrophage inflammatory protein (MIP)-1α, prostaglandins, and chemokines, locally produced by the tumor cells (Kremer *et al.*, 2011; Hu *et al.*, 2013). Both normal and malignant bone marrow cells interact together with this network of cytokines, locally increased RANKL, and other mediators, leading to osteoclast activation adjacent to the cancer cells (Lumachi *et al.*, 2009).

The *parathyroid hormone-related protein* was first discovered in the 1980s in patients with MAH, but PTHrP expression is present in cancer tissue even in normocalcemic patients (Kremer *et al.*, 2011). PTHrP is a circulating polypeptide of 139 to 173 amino acids (compared with the 84-amino acids of the PTH molecule), with the amino-terminal fragment (1–34) exhibiting PTH-like activity, and is produced by several cancers. These include solid tumors, such as lung (50%), breast (20%), head and neck (8%) or renal and urinary tract (8%) carcinomas, and multiple myeloma (10%) or lymphoma (3%) (Strewler, 2005). The expression of PTHrP is under the control of growth factors, such as tumor growth factor (TGF)-β and platelet-derived growth factor (PDGR), and angiogenic factors, such as vascular endothelial growth factor (VEGF) (Kremer *et al.*, 2011).

In humans, the *PTHrP* and the *PTH* genes are located in the short arm of chromosomes 12 and 11, respectively, but share the same G-protein coupled receptor (GCPR), named PTH1R, expressed in specific target cells, such as renal tubular cells, osteoblasts and possibly osteoclasts (Kremer *et al.*, 2011). However, other domains exhibit unique amino acid sequences, directing interactions with specific receptors that recognize nonhomologous parts of PTHrP (Basso *et al.*, 2011).

The physiological role of PTHrP is related to bone development during endochondral bone formation. In the embryonic and early postnatal life PTHrP regulates cartilage growth and long bones developing, stimulates the proliferation of chondrocytes in the growth plate and delays the mineralization of hypertrophic cartilage (Mundy and Edwards, 2008; Kremer *et al.*, 2011).

Up to 80% of patients with MAH exhibit elevated PTHrP serum levels, which is usually observed in those with advanced solid cancer, thus predicting short survival, whereas patients with multiple myeloma or metastatic breast cancer may exhibit temporary remission of the disease (Strewler, 2005; Lumachi *et al.*, 2009). In patients with advanced and metastatic cancer, PTHrP has a multifunctional role, regulating the progression of osteolytic bone metastases, cancer cells differentiation and growth, and the interference between cancer cells survival factors and apoptosis signaling pathways (Kremer *et al.*, 2011).

Macrophage colony-stimulating factor (M-CSF)-dependent macrophages differentiate into osteoclasts in the presence of receptor RANKL, which binds to RANK, a TNF receptor superfamily type-I transmembrane receptor that is expressed in mature osteoclasts (Tanaka *et al.*, 2005). Osteoprotegerin (OPG), another TNF superfamily member, acts as a competitive receptor of RANKL and inhibits RANKL–RANK interactions by binding to RANKL, ultimately suppressing osteoclast differentiation (Basso *et al.*, 2011). RANKL expression in bone-marrow stromal cells and osteoblasts is stimulated by PTHrP and cytokines (*i.e.*, TNF-α, and IL-1) resulting in osteoclast formation in the bone microenvironment, and RANK signaling also stimulates the epithelial–mesenchymal transition of cancer cells (Tanaka *et al.*, 2005; Kukita and Kukita, 2013). Thus, RANKL–RANK-related osteoclastogenesis activates downstream signaling pathways, leading to further osteoclastic bone destruction and Ca^{2+} mobilization (Yoneda *et al.*, 2013).

RANK and RANKL expression are observed in several human cancer cell lines, and both breast and prostate cancer preferentially metastasize to bone (Kukita and Kukita, 2013). Approximately 75% of patients with advanced breast cancer will develop osteolytic bone metastases, whereas 90% of those dying of advanced prostate cancer develop blastic or mixed blastic and lytic metastases (Saidak *et al.*, 2009).

36.6 Conclusions

Hypercalcemia is a relatively common disorder largely related to PHPT and MAH that requires early diagnosis to prevent the development of serious mid- and long-term complications. These complications are now uncommon in patients with PHPT, in whom effective parathyroidectomy leads to complete recovery, but they can be observed in cancer patients with cancer-related skeletal events.

Cinacalcet, a calcimimetic agent that inhibits PTH and is commonly used in patients with secondary hyperparathyroidism, is not useful in those with PHPT because it does not provide symptom relief and at midterm can be associated with chronic bone loss, due to persistently elevated PTH serum levels (Norman *et al.*, 2012). The drugs capable of inhibiting osteoclast action, such as bisphosphonates, may lead to a reduction in progressive bone resorption. Because RANKL is a potent stimulator of osteoclastogenesis by interacting with its receptor RANK, drugs such as the monoclonal antibody denosumab, mimicking the effects of OPG, significantly decrease osteoclast-related bone resorption and subsequent hypercalcemia in most patients with MAH. Other drugs are under study and will be developed in the near future.

Summary Points

- This chapter focuses on hypercalcemia and its causes.
- Parathyroid hormone (PTH) is synthesized and released by the parathyroid glands, and regulates ionized calcium (Ca^{2+}) levels in the blood.
- Primary hyperparathyroidism is the most common cause of benign hypercalcemia in nonhospitalized patients, accounting for approximately 80% of cases.

- Malignancy-associated hypercalcemia (MAH) is the most common metabolic disorder in malignancy that occurs in up to 20–25% of cancer patients.
- Two different mechanisms may lead to MAH. The so-called humoral hypercalcemia is largely caused by the secretion of parathyroid hormone-related protein (PTHrP) by the malignant cells. On the contrary, local osteolytic hypercalcemia is induced directly by cancer cells invading the bone, leading to the degradation of mineral matrix.
- Hypercalcemia is a relatively common disorder largely related to primary hyperparathyroidism and MAH that requires early diagnosis and treatment to prevent the development of serious mid- and long-term complications.

Key Facts

Key Facts of Calcium Functions and Control of Ionized Calcium

1. Calcium is essential for several human physiological processes, including the function of cellular membranes, nerves and muscles, enzymatic activities, and blood clotting. It also plays a role in the control of growth and proliferation of both normal and cancer cells.
2. Bones and teeth are the major storage site for calcium containing more than 99% of the calcium in the body, and the remaining 1% resides within the body fluids.
3. Ionized calcium (Ca^{2+}) represents the active fraction of calcium, corresponding to approximately 50% of total serum calcium.
4. The endocrine control of Ca^{2+} is achieved by a homeostatic mechanism involving parathyroid hormone (PTH), 1,25-dihydroxyvitamin D, and calcitonin, but other factors, such as fibroblast growth factor-23 (FGF-23), the transient receptor potential vanilloid type 5 (TRPV5), and a transmembrane protein named mKlotho, regulate calcium metabolism.

Key Facts of Parathyroid Hormone Actions

1. Serum parathyroid hormone (PTH) derives from either the release of stored PTH within the parathyroid glands or the increased synthesis of pre/pro-PTH, the intracellular precursor of PTH.
2. The response of parathyroid chief cells to changes in serum Ca^{2+} is sensed by the calcium-sensing receptor (CaSR), localized on the cells surface, which plays a crucial role in the regulation of Ca^{2+} reabsorption in the kidney, and modulates the effect of $1,25(OH)_2D_3$ on PTH release.
3. The main biological actions of PTH include: (i) the stimulation of osteoclasts, bone resorption and the release of calcium and phosphate from bone, mediated by the receptor-activated nuclear factor-κ (RANK)–RANK ligand (RANKL) system; (ii) the stimulation of the renal production of

$1,25(OH)_2D_3$; (iii) a direct effect on the activity of the osteoblast stimulating bone formation; and (iv) the inhibition of phosphate reabsorption from the proximal tubules and the stimulation of calcium reabsorption in the distal tubule and thick ascending limb of the loop of Henle.

Key Facts of Benign Hypercalcemia

1. Primary hyperparathyroidism (PHPT) is the most common cause of hypercalcemia in nonhospitalized patients, accounting for approximately 80% of cases.
2. Today, 80–90% of patients with PHPT are asymptomatic, but the prolonged catabolic and anabolic effects of PTH on cortical and cancellous bone may still lead to several deleterious effects on the skeleton, such as osteopenia and osteoporosis. The radiological bone findings observed in the past are now extremely uncommon.
3. In approximately 80–85% of patients with sporadic PHPT, a solitary monoclonal parathyroid adenoma can be demonstrated, 12–15% exhibit multiple adenomas or parathyroid gland hyperplasia, and parathyroid carcinoma is rare (1% of cases).
4. Surgery should be the treatment of choice for all patients with confirmed PHPT because parathyroidectomy always results in improvements in BMD and cognitive functions, and a decreased incidence of fracture risk and kidney diseases. Recent guidelines have been proposed that suggest surgical treatment for patients with sporadic asymptomatic PHPT.
5. Several rare (<10% of cases) familial syndromes, such as multiple endocrine neoplasia type 1 (MEN1), MEN2A, familial isolated primary hyperparathyroidism (FIPH), familial benign hypocalciuric hypercalcemia (FBHH), hyperparathyroidism-jaw tumor (HPT-JT) syndrome, and neonatal severe primary hyperparathyroidism (NSHPT) may cause PHPT.

Malignancy-Associated Hypercalcemia

1. Malignancy-associated hypercalcemia (MAH) is the most common paraneoplastic syndrome and metabolic disorder in malignancy, that can be observed in 10–15% of cancer patients.
2. Two different mechanisms may lead to MAH. The so-called humoral hypercalcemia is largely caused by the secretion of parathyroid hormone-related protein (PTHrP) by the malignant cells. On the contrary, local osteolytic hypercalcemia is induced directly by cancer cells invading the bone and promoting the degradation of mineral matrix.
3. Up to 80% of patients with MAH exhibit elevated PTHrP serum levels, which is usually observed in those with advanced solid cancer, thus predicting short survival, whereas patients with multiple myeloma or metastatic breast cancer may exhibit temporary remission of the disease.

4. PTHrP is a circulating polypeptide of 139 to 173 amino acids (compared with the 84-amino acids of the PTH molecule), with the amino-terminal fragment (1–34) exhibiting PTH-like activity, and is produced by several cancers.
5. The PTHrP stimulation of osteoclasts, bone resorption and release of calcium and phosphate from bone is mediated by the receptor-activated nuclear factor-κ (RANK)–RANK ligand (RANKL) system.
6. Osteoprotegerin acts as a soluble decoy receptor of RANKL and inhibits RANKL–RANK interactions by binding competitively to RANKL, ultimately suppressing osteoclast differentiation.
7. Drugs mimicking the effects of OPG (*i.e.*, denosumab), significantly decrease osteoclast-related bone resorption and subsequent hypercalcemia in most patients with MAH.

Definitions and Explanations of Key Terms

1,25(OH$_2$)D$_3$. 1,25-Dihydroxyvitamin D, also named calcitriol, is the active form of vitamin D.

CaSR. The calcium-sensing receptors are localized on the cells surface, play a crucial role in the regulation of Ca^{2+} reabsorption in the kidney, and modulate the effect of 1,25(OH)$_2$D$_3$ on PTH release.

FGF-23. Fibroblast growth factor-23 is produced by osteocytes, regulates osteogenesis and kidney function by promoting urinary phosphate excretion, and inhibiting the synthesis of 1,25(OH)$_2$D$_3$.

Hypercalcemia. Hypercalcemia can be defined as a serum calcium concentration that is two or more standard deviations above the average value of the reference interval of the population. In adults, the upper limit of the normal range for total calcium and serum Ca^{2+} is usually set at 2.65 and 1.32 mmol L^{-1}, respectively.

Ionized calcium. Ionized calcium (Ca^{2+}) represents the active fraction of calcium, corresponding to approximately 50% of total serum calcium. It is tightly regulated at 1.25 ± 0.07 mmol L^{-1}.

Klotho. Kloto is an ageing suppressor gene of restricted expression (*i.e.*, kidneys, brain, parathyroid glands) encoding mKlotho, which acts as a coreceptor in FGF-23 signaling.

MAH. Malignancy-associated hypercalcemia is the most common metabolic disorder in malignancy that occurs in up to 20–25% of cancer patients, especially in those with advanced cancer.

PHPT. Primary parathyroid glands hyperfunction (hyperparathyroidism) is the most common cause of benign hypercalcemia in nonhospitalized patients, accounting for at least 80% of cases.

PTH. Parathyroid hormone is a straight-chain polypeptide containing 84 amino acids that is synthesized and released by the parathyroid glands.

PTHrP. Parathyroid hormone-related protein is a circulating polypeptide of 139 to 173 amino acids (compared with the 84-amino acids of the PTH molecule), with the amino-terminal fragment (1–34) exhibiting PTH-like activity, and is produced by several cancers.

RANK–RANKL system. The receptor-activated nuclear factor-κ (RANK)–RANK ligand (RANKL) system mediates the stimulation of osteoclasts, bone resorption and the release of calcium and phosphate from bone.

List of Abbreviations

$1,25(OH)_2D_3$	1,25-Dihydroxyvitamin D
BMD	Bone-mineral density
Ca^{2+}	Ionized calcium
CaSR	Calcium-sensing receptor
CCND1	Cyclin D1 oncogene
CDKI	Cyclin-dependent kinase inhibitor
DXA	Dual-energy X-ray absorptiometry
FBHH	Familial benign hypocalciuric hypercalcemia
FGF-23	Fibroblast growth factor-23
FIPH	Familial isolated primary hyperparathyroidism
GCPR	G-protein coupled receptor
HPT-JT	Hyperparathyroidism-jaw tumor
IGF-1	Insulin-like growth factor-1
IL	Interleukin
MAH	Malignancy-associated hypercalcemia
M-CSF	Macrophage colony-stimulating factor
MEN1	Multiple endocrine neoplasia type 1
MEN2A	Multiple endocrine neoplasia type 2A
MIP	Macrophage inflammatory protein
NSHPT	Neonatal severe primary hyperparathyroidism
OPG	Osteoprotegerin
PAFIC	Polymerase-associated factor-1 complex
PDGR	Platelet-derived growth factor
PHPT	Primary hyperparathyroidism
PRAD1	Parathyroid adenoma 1 oncogene
PTH	Parathyroid hormone
PTHrP	Parathyroid hormone-related protein
RANK	Receptor-activated nuclear factor-κ
RANKL	Receptor-activated nuclear factor-κ ligand
RET	Rearranged during transfection proto-oncogene
TALH	Thick ascending limb of the loop of Henle
TGF	Tumor growth factor
TNF	Tumor necrosis factor
TRPV5	Transient receptor potential vanilloid type 5
VDR	Vitamin D receptor
VEGF	Vascular endothelial growth factor

References

Abood, A. and Vestergaard, P., 2013. Increasing incidence of primary hyperparathyroidism in Denmark. *Danish Medical Journal*. 60: A4567.

Basso, U., Maruzzo, M., Roma, A., Camozzi, V., Luisetto, G. and Lumachi, F., 2011. Malignant hypercalcemia. *Current Medicinal Chemistry*. 18: 3462–3467.

Bikle, D. D., Sakata, T., Leary, C., Elalieh, H., Ginzinger, D., Rosen, C. J., Beamer, W., Majumdar, S. and Halloran, B. P., 2002. Insulin-like growth factor I is required for the anabolic actions of parathyroid hormone on mouse bone. *Journal of Bone and Mineral Research*. 17: 1570–1578.

Bilezikian, J. P., Khan, A. A. and Potts, Jr, J. T., 2009. Guidelines for the management of asymptomatic primary hyperparathyroidism: summary statement from the third international workshop. *Journal of Clinical Endocrinology and Metabolism*. 94: 335–339.

Cao, J. J. and Nielsen, F. H., 2010. Acid diet (high-meat protein) effects on calcium metabolism and bone health. *Current Opinion in Clinical Nutrition and Metabolic Care*. 13: 698–702.

Clase, C. M., Norman, G. L., Beecroft, M. L. and Churchill, D. N., 2000. Albumin-corrected calcium and ionized calcium in stable haemodialysis patients. *Nephrology, Dialysis, Transplantation*. 15: 1841–1846.

Hu, M. I., Glezerman, I., Leboulleux, S., Insogna, K., Gucalp, R., Misiorowski, W., Yu, B., Ying, W. and Jain, R. K., 2013. Denosumab for patients with persistent or relapsed hypercalcemia of malignancy despite recent bisphosphonate treatment. *Journal of the National Cancer Institute*. 105: 1417–1420.

Kremer, R., Li, J., Camirand, A. and Karaplis, A. C., 2011. Parathyroid hormone related protein (PTHrP) in tumor progression. *Advances in Experimental Medicine and Biology*. 720: 145–160.

Kukita, A. and Kukita, T., 2013. Multifunctional properties of RANKL/RANK in cell differentiation, proliferation and metastasis. *Future Oncology*. 11: 1609–1622.

Landowski, C. P., Bolanz, K. A., Suzuki, Y. and Hediger, M. A., 2011. Chemical inhibitors of the calcium entry channel TRPV6. *Pharmaceutical Research*. 28: 322–330.

Lumachi, F., Ermani, M., Luisetto, G., Nardi, A., Basso, S. M. M. and Camozzi, V., 2002. Relationship between serum parathyroid hormone, serum calcium and arterial blood pressure in patients with primary hyperparathyroidism: results of a multivariate analysis. *European Journal of Endocrinology*. 146: 643–647.

Lumachi, F., Camozzi, V., Ermani, M., Nardi, A. and Luisetto, G., 2007. Lumbar spine bone mineral density changes in patients with primary hyperparathyroidism according to age and gender. *Annals of the New York Academy of Sciences*. 1117: 362–366.

Lumachi, F., Brunello, A., Roma, A. and Basso, U., 2008. Medical treatment of malignancy-associated hypercalcemia. *Current Medicinal Chemistry*. 15: 415–421.

Lumachi, F., Brunello, A., Roma, A. and Basso, U., 2009. Cancer-induced hypercalcemia. *Anticancer Research*. 29: 1551–1556.

Lumachi, F. and Basso, S. M. M. (ed.), 2010. *Hypercalcemia. Pathophysiology and Treatment*. Bentham Science Publisher, Sharya, UAE, 105 pp.

Lumachi, F., Motta, R., Cecchin, D., Ave, S., Camozzi, V., Basso, S. M. and Luisetto, G., 2011. Calcium metabolism and hypercalcemia in adults. *Current Medicinal Chemistry*. 18: 3529–3536.

Lumachi, F., Cappelletti, P., Tozzoli, R., Basso, S. M., Luisetto, G. and Camozzi, V., 2012. Diagnosis of alterations of serum calcium metabolism. *Medicinal Chemistry*. 8: 551–555.

Medici, D., Razzaque, M. S., Deluca, S., Rector, T. L., Hou, B., Kang, K., Goetz, R., Mohammadi, M., Kuro-O, M., Olsen, B. R. and Lanske, B., 2008. FGF-23-Klotho signaling stimulates proliferation and prevents vitamin D-induced apoptosis. *Journal of Cell Biology*. 182: 459–465.

Mundy, G. R. and Guise, T. A., 1999. Hormonal control of calcium homeostasis. *Clinical Chemistry*. 45: 1347–1352.

Mundy, G. R. and Edwards, J. R., 2008. PTH-related peptide (PTHrP) in hypercalcemia. *Journal of the American Society of Nephrology*. 19: 672–675.

Norman, J., Lopez, J. and Politz, D., 2012. Cinacalcet (Sensipar) provides no measurable clinical benefits for patients with primary hyperparathyroidism and may accelerate bone loss with prolonged use. *Annals of Surgical Oncology*. 19: 1466–1471.

O'Toole, J. F., 2011. Disorders of calcium metabolism. *Nephron Physiology*. 118: 22–27.

Panicker, L. M., Zhang, J. H., Dagur, P. K., Gastinger, M. J. and Simonds, W. F., 2010. Defective nucleolar localization and dominant interfering properties of a parafibromin L95P missense mutant causing the hyperparathyroidism-jaw tumor syndrome. *Endocrine-related Cancer*. 17: 513–524.

Peppa, M., Boutati, E., Kamakari, S., Pikounis, V., Peros, G., Panayiotides, I. G., Economopoulos, T., Raptis, S. A. and Hadjidakis, D., 2008. Multiple endocrine neoplasia type 2A in two families with the familial medullary thyroid carcinoma associated G533C mutation of the RET proto-oncogene. *European Journal of Endocrinology*. 159: 767–771.

Ruiz-Tovar, J., Oller, I., Tomas, A., Llavero, C., Arroyo, A., Calero, A., Martinez-Blasco, A. and Calpena, R., 2012. Mid-term effects of sleeve gastrectomy on calcium metabolism parameters, vitamin D and parathormone (PTH) in morbid obese women. *Obesity Surgery*. 22: 797–801.

Saidak, Z., Mentaverri, R. and Brown, E. M., 2009. The role of the calcium-sensing receptor in the development and progression of cancer. *Endocrine Reviews*. 30: 178–195.

Schmid, C., Neidert, M. C., Tschopp, O., Sze, L. and Bernays, R. L., 2013. Growth hormone and Klotho. *Journal of Endocrinology*. 219: R37–R57.

Schulte, K. M., Gill, A. J., Barczynski, M., Karakas, E., Miyauchi, A., Knoefel, W. T., Lombardi, C. P., Talat, N., Diaz-Cano, S. and Grant, C. S., 2012.

Classification of parathyroid cancer. *Annals of Surgical Oncology*. 19: 2620–2628.

Shoback, D., Marcus, R., Bikle, D. and Strewler, G., 2001. Mineral metabolism & metabolic bone diseases. In: Greenspan, F. S. and Gardner, D. G. (ed.) *Basic & Clinical Endocrinology*. Lange Medical Books/McGraw-Hill, New York, USA, pp. 273–333.

Stewart, A. F., 2005. Clinical practice. Hypercalcemia associated with cancer. *New England Journal of Medicine*. 352: 373–379.

Strewler, G. J., 2005. Hypercalcemia of malignancy and parathyroid hormone-related protein. In: Clark, O. H., Duh, Q.-I. and Kebebew, E. (ed.) *Textbook of Endocrine Surgery*. 2nd edn. Elsevier-Saunders, Philadelphia, USA, pp. 536–542.

Tanaka, S., Nakamura, K., Takahasi, N. and Suda, T., 2005. Role of RANKL in physiological and pathological bone resorption and therapeutics targeting the RANKL–RANK signaling system. *Immunological Reviews*. 208: 30–49.

Udelsman, R., Pasieka, J. L., Sturgeon, C., Young, J. E. and Clark, O. H., 2009. Surgery for asymptomatic primary hyperparathyroidism: proceedings of the third international workshop. *Journal of Clinical Endocrinology and Metabolism*. 94: 366–372.

Vallet, S., Mukherjee, S., Vaghela, N., Hideshima, T., Fulciniti, M., Pozzi, S., Santo, L., Cirstea, D., Patel, K., Sohani, A. R., Guimaraes, A., Xie, W., Chauhan, D., Schoonmaker, J. A., Attar, E., Churchill, M., Weller, E., Munshi, N., Seehra, J. S., Weissleder, R., Anderson, K. C., Scadden, D. T. and Raje, N., 2010. Activin A promotes multiple myeloma-induced osteolysis and is a promising target for myeloma bone disease. *Proceedings of the National Academy of Sciences of the United States of America*. 107: 5124–5129.

Vanoevelen, J., Janssens, A., Huitema, L. F., Hammond, C. L., Metz, J. R., Flik, G., Voets, T. and Schulte-Merker, S., 2011. Trpv5/6 is vital for epithelial calcium uptake and bone formation. *FASEB Journal*. 25: 3197–3207.

Yoneda, T., Tanaka, S. and Hata, K., 2013. Role of RANKL/RANK in primary and secondary breast cancer. *World Journal of Orthopedics*. 4: 178–185.

Zhu, J. G., Ochalek, J. T., Kaufmann, M., Jones, G. and Deluca, H. F., 2013. CYP2R1 is a major, but not exclusive, contributor to 25-hydroxyvitamin D production in vivo. *Proceedings of the National Academy of Sciences of the United States of America*. 110: 15650–15655.

Subject Index

Illustrations and figures are in **bold**. Tables are in *italics*.

absorption 8–9, 10, 15, 25
definition 218
isotope studies 70–1
speciation 47, **48**
see also bioavailability; intestinal absorption
ACAT2 (acyl-CoA: cholesterol acyltransferase 2), definition 528
acetylcholine, calcium function 4
acid–base equilibrium 24
acidosis, definition 233
acrylamide, definition 286
active transport 25
definition 344
acute myocardial infarction (AMI), definition 359
adaptive responses, definition 443
adenosine triphosphate (ATP), transcellular calcium transport 9, 15
adipocyte, definition 577
adiposity, definition 577
adolescence
calcium systemic balance 10
definition 218
adolescents, dairy consumption 34, 36
African-Americans, dairy consumption 33
African diet 5, 36, 38–9
age-related bone loss 76–7
calcium systemic balance 10
alcohol intake, calcium bioavailability 13, 51
alginates 55
alkaline-earth metals, properties 22, 72
alkalinity, definition 359
allotropes 68
allotropy, definition 72
amino acids
calcium bioavailability 49
definition 233
3-aminopropionamide (3-APA), definition 286
analytical determination methods
animal feeds 26
biological tissues 26
metabolic studies 71
angiotensin II, definition 516
animal feeds 22
anticonstipation formula milk 55–6
antinutritional factor (ANF), definition 270
antireflux formula milk 55
apatite 22
apoptosis, definition 464
appetitive taste, definition 427
arabinose, calcium bioavailability 49–50
arginine, calcium bioavailability 49
ascorbic acid 57
Asian diets 33–8

atomic absorption
spectrophotometry 26
atomic mass *68*, 69
atomic number *68*
atomic radius *68*
ATP binding cassette transporters, definition 527
Australia, dairy consumption 33
autophagy, definition 464
aversive taste, definition 427
azomethine ylide, definition 287

baked goods 59
bakery products, calcium content *6*
Bangladesh 37
barfi 33
beans 56
beryllium 72
beta decay, definition 72
beverages 59–60
bioaccessibility, definition 252
bioavailability *see* beverages
- carbohydrates effect 49–50
- definition 47, 60–1, 72, 196, 218, 252, 323, 359
- determination 47, 61
- dietary fat effect 50
- dietary fiber effect 11, 16, 50–1
- food types *11*, 12–13, 15–16, 51–60
- isotope studies 70–1
- oxalic acid effect 12, 50
- phytic acid effect 50
- protein effect 49
- solubility 11, 69

biofilm, definition 378
biomarkers, definition 378
biosynthesis
- 1,25-dihydroxyvitamin D (1,25D) 79, **80**
- 25-hydroxyvitamin D 79, **80**
- vitamin D_3 79

bisglycinocalcium, solubility *68*
bisphosphonates, definition 122
bitter taste, definition 426
blood coagulation cascade 76
blood levels
- regulation 25, 27
- *see also* serum levels

blood pressure, definition 516
blood-vessel structure, definition 516
blood vessels, definition 516
body-mass index (BMI), definition 577
boiling point *68*
bone
- age-related calcium loss 10, 76–7
- calcium content 4, 14, 23, 76
- composition 67
- as dietary source 38
- *see also* skeleton

bone mineral content, definition 504, 541
bone-mineral density
- definition 234, 504, 541
- menopause 77
- protein effect 49

bone remodeling, definition 16
bone resorption 76
bone turnover
- definition 504
- double stable isotope studies 71
- rate 11

bovine milk 52, *53*
Bragg's law, definition 141
branched-chain, definition, definition 305
brassicas, calcium bioavailability 57, 58
Bravais lattices, definition 141–2
bread
- calcium bioavailability 59
- calcium content *6*, *58*

breadmaking, definition 305
breakfast cereals *58*, 59
Britain, dairy consumption 31
broccoli *57*
brown rice, definition 270

brush border membranes 69
Buddhism 37
buffalo milk 53
Bulgaria, dairy consumption 32
buttermilk 32
calcium content 30

C-terminal telopeptide (CTX), definition 359
Ca^{2+} microdomains, definition 463
Ca^{2+} oscillations, definition 463
Ca^{2+} waves, definition 463
caffeine, calcium bioavailability 51
caffeine intake, calcium excretion 13
calbindin
definition 392, 479
vitamin D 47, **48**
calcemia, definition 233
calcineurin, definition 443
calciotropic hormones, definition 305
calcitonin
calcium transport regulation 25, 27
definition 16, 27, 392, 479
calcitriol
calcium homeostasis 9
definition 16, 479, 577
protein effect 49
transcellular calcium transport 9, 15, 47, **48**
see also 1,25-dihydroxyvitamin D
calcium
abundance 3, 15, 22, 27
functional overview 4, 14, 23, 46, 67, 76 *see also specific functions*
historical use 68
isolation 21, 68
nomenclature 67–8
physiological distribution 4, 14, 23, 27
properties 21–2, 23, 27, 67–70
sources 3, 14, 22
calcium acetate 51
calcium carbonate
absorption 13, **48**
eggshells 38
historical use 68
side effects 7
solubility *68*, 69
sources 3, 22
supplements 7, 13
calcium channels (TRPV6) 47, **48**
calcium chloride, solubility 22
calcium citrate
absorption 12, 13, **48**
solubility *68*, 69
supplements 7, 13
calcium citrate malate
absorption 12, **48**
solubility *68*
calcium fluoride, sources 3
calcium formate
absorption 13, **48**
supplements 13, 47
calcium glycerophosphate, solubility *68*, 69
calcium lactate, absorption **48**
calcium oxalate 12
solubility *68*, 69
calcium oxide, Roman concrete 68
calcium palmitate 53
calcium phosphate
solubility *68*, 69
sources 3
calcium phytate 50
solubility *68*
calcium propionate, solubility 22
calcium-sensing receptors (CaSR)
definition 577, 620
expression 77
functional overview 77
calcium species, definition 107
calcium sulfate
historical use 68
industrial uses 22
solubility 22, *68*
sources 3, 22
calcium taste, definition 426

calcium transporters, vitamin D 47, **48**
calciuria, definition 233
calculus 24, 27
calibration curve, definition 107
calmodulin, definition 392
CaMKII, definition 443
CaMKIV, definition 443
Canada, dairy consumption 33
cancer, colon 25
capillary electophoresis (CE), definition 107
carbohydrates, calcium bioavailability 49–50
carbonate apatite, definition 122
carbonated drinks 61
carboxy-terminal collagen crosslinks (CTX), definition 479
cardiovascular disease (CVD)
  definition 359
  research history 25
  supplements 8
casein
  buffalo milk 53
  calcium bioavailability 49, 52, 54–5
  human milk 52
  iron interactions 55
casein phosphopeptides (CPPs), definition 196
β-catenin, definition 408
cauliflower *57*
cecocolonectomy, definition 344
celery *57*
cell culture, definition 197
cells, calcium content 4
cellular signaling, definition 61
cellulose, calcium-binding 50–1
cereal products
  Asian consumption 36
  calcium bioavailability 58–9
  calcium content *6*, *7*, *58*
chai 33
cheese
  calcium bioavailability 54
  consumption levels 31, 32, 33
  lactose 31, 40
chelation, definition 287
chickpeas 56
children
  dairy consumption 31, 33, 34, 35
  grains intake 35
  recommended dietary allowances (RDA) *77*
  seafood intake 35, 37
  soy consumption 37
  vegetable intake 35, 37
China
  dietary sources 34, 35–7
  postpartum diet 38
Chinese Food Pagoda 36
cholecalciferol *see* vitamin D_3
cholecystokinin (CCK), definition 408
cholesterol, calcium intake levels 25
chondroitin polysulfate (CPS), definition 122
chronic kidney disease (CKD), definition 559
Chvostek sign, definition 601
citric acid, calcium bioavailability 51, 57, 58
coeliac disease 50
  definition 61
coffee 59
cola 61
colon cancer 25
colonic fermentation, fiber 51, 58
colvalent radius *68*
complex stimuli, definition 426
computed tomography, quantitative (QCT) 479
concrete 68
constipation 24, 27
consumption, definition 323

cooking methods
legumes 56
vegetables 58
coral 3
cottage cheese 32
cotyledons, definition 253
cow's milk *see* bovine milk
CREB (cyclic AMP response element), definition 443
critical pH, definition 378
cruciferous vegetables, Asian consumption 36–7
cuisine, definition 40
CYP2R1 (vitamin D-25 hydroxylase) 79
CYP7A1 (cholesterol-7α-hydroxylase), definition 528
CYP24 (25-hydroxyvitamin D-24-hydroxylase), functional overview 79
CYP27B1 (25-hydroxyvitamin D-1 alpha hydroxylase) 79, **80**
definition 88
cystinuria, definition 359
cytoplasm, definition 107

daidzein, definition 323
daily intake 10
dairy consumption 30–1
economic status 34
dairy, definition 40
dairy products
absorption 12
calcium bioavailability *11*, 13, 54
calcium content *5*
consumption levels 30–6, 39
definition 40
fermented 31
Western diet 5
see also milk
DASH eating plan 25
Davy, Sir Humphrey 21, 68
deficiency
bioavailability issues 46–7
carbonated drinks effect 61
minerals generally 24
physiological effects 24, 27, 76
vitamin D 78
dehulling, definition 253
7-dehydrocholesterol 79
demineralization, definition 378
denosumab, definition 559
dental caries, definition 378
dental pellicle, definition 378
detection limit (LOD), definition 107
developmental stages *see* life stages
diabetes mellitus
calcium intake levels 25
definition 378
dialysis, definition 107, 253
diarrhoea, secretory 408
diastolic blood pressure (DBP), definition 516
dietary cation– anion balance (DCAB) 24–5
dietary fiber *see* fiber, dietary
dietary sources 4–8, 14, 39
Africa 36, 38–9
Americas 33, 39
Asia 33–8
Australia 33
bones 38
eggshells 38
Europe 31–2, 39
insects 38–9
Middle East 32–3
speciation 47
vegans 27
dietary supplements *see* supplements
diets
African 5, 36, 38–9
Asian 33–8
DASH eating plan 25
fasting 76
Western 5, 31–2, 33, 35, 39
digest *see* soluble fraction
digestibility, definition 218
dihydropyidine receptors, definition 443

24,25-dihydroxyvitamin D 79
1,25-dihydroxyvitamin D (1,25D)
autocrine actions 78, 89
biosynthesis 79, **80**
definition 88, 504, 577, 620
endocrine actions 78, 89
functional overview **80**
paracrine actions 78, 89
see also calcitriol
disaccharide, definition 344
double stable isotope studies 71
dough (beverage) 32
drinks *see* beverages
dual-energy X-ray absorptiometry (DXA), definition 479, 504
duodenum 9, 25

E–C coupling, definition 444
economic status 34
eggshells as dietary source 38
Egypt 68
electrolytes, definition 378
electron affinity *68*
electron capture, definition 72
electronegativity *68*
electronic configuration 68
element speciation, definition 107
emulsifiers 54
enamel, definition 379
enriched foods *see* fortified foods
enterocytes 69
definition 393
ergocalciferol *see* vitamin D_2
ERK1/2 signaling
definition 443
see also extracellular-signal-regulated kinases (ERKs)
estrogens, definition 323
ethnicity, definition 40
Europe, dairy consumption 31–2, 39
exchanger, definition 464
excitable cells, definition 426
excretion 9, 15
caffeine intake effect 13, 51
calcium content 11
isotope studies 70–1
obligatory loss 76
potassium intake effect 12
protein intake effect 12, 49
sodium intake effect 12, 50
exercise, weight-bearing 542
extracellular fluid
calcium content 4
calcium functional overview 23, 27
calcium turnover 11
extracellular-signal-regulated kinases (ERKs)
definition 443
see also ERK1/2 signaling

fasting 76
fat, dietary
calcium bioavailability 50
infant formula supplementation 55
fat taste, definition 426
fatty-acid synthase (FAS), definition 577
fatty acids
calcium absorption 9
milk 52–3
fermentation
colonic 51, 58
definition 323
fermented dairy products 31, 40
feta cheese 31
fetal biometry, definition 504
fiber, dietary
calcium bioavailability 11, 16, 50–1
definition 197, 253
fortified foods 54
legumes 57
vegetables 58
wheat 58–9
fiber type diversity, definition 444

fibroblast growth factor-23 (FGF-23), definition 620
fish 37
  Asian consumption 37
  calcium content *6*
flame atomic absorption and emission spectrometry (FAAES), definition 141
flour
  calcium bioavailability 59
  calcium content *58*
  fortified *7*
fluorescence, definition 122
fluorite *see* calcium fluoride
food fermentation, definition 61
food-frequency questionnaires (FFQ) 160
food supplementation, definition 305
fortified foods
  absorption 13
  bioavailability factor 61
  calcium content *7*
  definition 197, 323
  dietary intake 6
  fiber, dietary 54
  fructo-oligosaccharides (FOS) 53
fracture prevention, supplements 8
fracture risk
  deficiency 24
  definition 234
  protein intake 49
  supplementation studies 78
fructo-oligosaccharides (FOS)
  calcium bioavailability 51
  definition 344
  food types 53
  infant formula supplementation 53
  see also 1-kestose
fructose, calcium bioavailability 49
fruit
  calcium bioavailability *11*
  calcium content *5*
Fulani 36

G-protein coupled receptor (GPCR), definition 407, 426
galacto-oligosaccharide, definition 344
gastrin, definition 407
gastrointestinal tract, pH effects 69
genistein, definition 323
geographical variations, vegetable calcium bioavailability 58
germinated rice, definition 270
germination, definition 253
gestational diabetes, definition 479
geung cho 38
Giza, Great Pyramid 68
glucagon-like peptide-1 (GLP-1), definition 408
glucose, calcium absorption 9
gluten, definition 305
gluten-free diet, definition 305
gluten sensitivity, definition 305
goat milk 36, 53
grains intake
  Korea 35, 37
  see also cereal products
Greece, dairy consumption 31, 32
green vegetables
  African diet 38
  Asian diets 36–7
  calcium bioavailability *11*, 12, 57–8
  calcium content *5*
  children's intake 34
  wild compared to cultivated 38
ground reaction forces, definition 541
growth hormone, definition 218

guar gum 55
gulab jamun 33
gut hormones, definition 407
gypsum *see* calcium sulfate

HCl extractability, definition 253
hemicellulose, calcium bioavailability 11, 50, 58
herbs, Africa 36
hexosans 58
high-density lipoprotein (HDL), definition 528
high density lipoproteins (HDL), definition 323
Hinduism 34
homeostasis 9, 14, 15
  see also systemic balance
hormone sensitive lipase, definition 577
human milk, calcium bioavailability 51–2, 53
hydrocolloids, definition 305
hydroxyapatite
  bone composition 67
  bone matrix 23
  definition 122
  solubility *68*
hydroxylapatite, definition 379
1-α-hydroxylase, definition 478
5-hydroxymethyl-2-furfural (HMF), definition 287
25-hydroxyvitamin D
  biosynthesis 79, **80**
  definition 89, 504, 542
hypercalcemia
  definition 620
  symptoms 76
hypercalciuria 50
  definition 359
  phosphorus treatment 50
hyperoxaluria, definition 359
hyperparathyroidism, definition 479, 620
hypertension 25
hyperuricosuria, definition 359
hyphenated technique, definition 107
hypocalcaemia
  infant milk types consumed 53
  symptoms 76
hypocitraturia, definition 359
hypoparathyroidism, definition 479, 601

ileum 9
image-analyzing software, definition 122
immediate early genes, definition 444
incretin, definition 408
India, dairy consumption 33–4
inductively coupled plasma optical emission spectrometry (ICP-MS)
  definition 107
  metabolic studies 71
infancy
  calcium systemic balance 10
  milk comparison 53
infant cereals *58*
infant formula
  calcium bioavailability 54–6
  calcium content *7*
  fructo-oligosaccharides (FOS) 53
  special diet formulas 55–6
ingestion, definition 323
inositol hexaphosphoric acid *see* phytic acid
inositol hexaphosphoric acid (IP6) 50
  see also phytic acid
insects as dietary source 38–9
insulin-like growth factor-1 (IGF-1), protein effect 49
intact parathyroid hormone (iPTH), definition 359
intestinal absorption
  calcium solubility 69
  definition 233

passive diffusion 9, 25, 51
reabsorption 11
intestine, definition 393
intracellular calcium, functional overview 23, 24, 76
intracellular signaling 76
inulin, definition 344
ionic stimuli, definition 426
ionization energy, definition 72
ionization potential *68*
ionized calcium, definition 620
IP_3R subtypes, definition 443
Iran, dairy consumption 32
Iraq, dairy consumption 32
iron, absorption interference 9, 24, 50, 55
isotonicity 52
isotopes
metabolic studies 70–1
properties 69–70
Israel, dairy consumption 32
Italy, dairy consumption 31

Japan
dairy consumption 35
fish consumption 37
tofu 38
jejunum 9
Jewish cuisine 32–3
Jordan, dairy consumption 32
juvenile Paget's disease (JPD), definition 559

kale *57*, 58
kashk 32
kefir 31
1-kestose 53
khir 33
kidney
calcium reabsorption 9
1,25-dihydroxyvitamin D (1,25D) biosynthesis 79, **80**
obligatory excretion 76
reabsorption 11, 76
kidney stones *see* calculus
Klotho, definition 620
Korea 37
dairy consumption 35
tofu 38
kosher dietary laws 32
kulfi 33

labneh 32
α-lactalbumin 52
lactase, definition 40
lactase persistence
definition 39–40
population differences 31
lactose
absorption 40
calcium absorption 9
calcium bioavailability 12, 16, 49, 52
dairy food differences 31
definition 40, 345
lactose-free formula milk 55
lactose intolerance
definition 40
ethnicity 31, 35, 40
symptoms 40
lactulose, calcium bioavailability 52
large intestine, definition 393
Lebanon, dairy consumption 32
leben 32
legumes
calcium bioavailability 56–7
calcium content *5*, *56*
lentils 56, 57
ready-to-eat 61
Lewis's acids 23
life stages
calcium intake levels 26
calcium systemic balance 10
intestinal absorption 9
recommended dietary allowances (RDA) *77*
supplements 9
lime *see* calcium oxide
limestone *see* calcium carbonate
lipogenesis, definition 577
lipolysis, definition 577
lipophilicity, definition 426

liver, vitamin D 79, **80**
long chain fatty acid, definition 426
low density lipoproteins (LDL), definition 323, 528
lumen, definition 393
lysine, calcium bioavailability 49
lysosome, definition 393

Maasai 36
magnesium
  absorption interference 9, 24, 50
  properties 72
  soil 22
Maillard reaction, definition 287
Maillard reaction products (MRPs), definition 197
Maillard reactions 56
  definition 305
maize 39
malic acid, calcium bioavailability 51, 58
malignancy-associated hypercalcemia (MAH), definition 620
malting, definition 61
MAPK (mitogen-activated protein kinases), definition 443
mate (beverage) 60
meat, calcium content *6*
Mediterranean diet, definition 219
Mef2 (myocyte enhancer factor-2), definition 443
melting point *68*
membrane channels, calcium functional overview 23
membrane permeability, calcium function 4
men, recommended dietary allowances (RDA) *77*
menopause
  bone-mineral density 77
  calcium systemic balance 10
  oestrogen effects 77
menstrual cycle, calcium systemic balance 10
mercury oxide 68
messenger function *see* signal transduction
meta-analyses, protein intake 49
metabolism
  isotope studies 70–1
  system overview **10**, 25, **26**
Mexico, tortillas 39
Middle East, dairy consumption 32–3
milk
  calcium bioavailability *11*, 13, 51–3
  calcium content *5*, 30, 51–2
  composition *53*
  consumption levels 30–6, 51
  fortification 54
  see also dairy foods
Miller indices, definition 142
mineral deficiencies 24
mineral water 59
mitochondrial associated ER membranes (MAMs), definition 463
mitochondrial fission, definition 463
mitochondrial fusion, definition 463
Moldova, dairy consumption 32
Mongolia, dairy consumption 34
MTP (microsomal triacylglycerol transport protein), definition 528
mugwort 37
muscle fiber twitch, definition 444
muscle plasticity, definition 443
myo-inositol 1,2,3,4,5,6 hexakis dihydrogen phosphate *see* phytic acid
myoblast, definition 443
myotubes, definition 444

nanocalcium additive, definition 141
nanotechnology, definition 141
National Heart, Lung and Blood Institute (NHLBI) 25

NCX *see* sodium–calcium exchanger
nephrolithiasis, definition 359
NFAT (nuclear factor of activated T-cells), definition 443
NHANES (National Health and Nutrition Examination Survey)
definition 16
Korea 35, 37
supplement intake 7
nitric oxide, definition 516
nixtamilizaion 39
nomad, definition 40
nomenclature 67–8
nonosteogenic sports, definition 542
nonradiometric fluorescence, definition 28
normocalcemia, definition 601
NPC1L1 (Niemann-Pick C1 like 1 transporter), definition 528
nuclear spin, definition 72
nucleophilicity, definition 73
nuts
calcium bioavailability *11*
calcium content *6*

obesity, definition 577
oestrogen, total-body calcium levels 77–8
oligosaccharides
definition 345
milk 52
orai channel, definition 427
osteoblasts
calcium-sensing receptors (CaSR) 77
definition 17, 89, 233
osteoclast function, definition 122
osteoclasts
definition 16–17, 89, 233
resorption 25
osteocytes, definition 17, 89
osteogenic sports, definition 542
osteomalacia 78
definition 88
osteoporosis
definition 27, 122, 218, 305, 559
oestrogen status 77
risk factors 26
osteoprotegerin (OPG), definition 559
ovariectomized (OVX), definition 345
overconsumption
minerals generally 24
physiological effects 24, 27
oxalate, calcium absorption 9
oxalic acid
calcium-binding 12, 16
calcium bioavailability 50, 56, 57, 58
oxidation state, definition 73
oxidation states *68*
oxygen affinity 23

P38 signaling, definition 443
paddy, definition 270
palmitic acid 53, 55–6
pancreatitis 50
definition 61
paneer 33
parathyroid gland 25
parathyroid hormone (PTH)
calcium homeostasis 9
calcium transport regulation 25, 27
definition 16, 28, 89, 359, 393, 408, 479, 504, 620
protein effect 49
parathyroid hormone-related peptide (PTHrP), definition 505, 620–1
parboiled rice, definition 270
parotid saliva, definition 379
passive diffusion 9, 15, 25
definition 345
dietary fiber effect 51
phosphate 13
pastoralists 36, 40
definition 40

PCMA (plasma membrane Ca^{2+} ATPase) 47, **48**
peak bone mass
  age 26
  definition 218, 541
peak bone mineral accrual, definition 541
peak-height velocity, definition 541
Pearson's classification 23
pectin 11, 57
pentosans 58
periodontitis, definition 379
peripheral QCT (pQCT), definition 479
pertussis toxin (PTX), definition 444
PGC-1α (peroxisome proliferator-activated receptor gamma coactivator-1 alpha), definition 443
pH
  absorption effect 25
  dietary fiber effect 51
  solubility effects 9, 15, 69
phosphate
  absorption 12–13
  salivary buffer 27
  soft drinks 13
phosphopeptides, calcium bioavailability 49, 52
phosphoric acid 61
phosphorus
  absorption interference 50
  calcium absorption 9, 53
  human milk 53
photoisomerization, definition 393
phytase, definition 270
phytases 59
phytates, definition 61, 197, 270
phytic acid
  calcium-binding 11
  calcium bioavailability 50, 56, 57, 58
  definition 270
  fructo-oligosaccharides (FOS) 51
  see also inositol hexaphosphoric acid (IP6)
phytoestrogens, definition 323
pickled foods 38
placenta 79
plants 27
  calcium functional overview 23
  calcium levels 3–4
plasma
  calcium content 4
  calcium homeostasis 75–6, **80**
  double stable isotope studies 71
plasma ionized calcium, definition 88
plasma membrane calcium ATPase proteins (PCMA), definition 479
plaster of Paris *see* calcium sulfate
polyanion, definition 122
polysaccharides, definition 345
Portland cement 22
potassium intake, calcium systemic balance 12
pre-eclampsia, definition 479
prebiotic, definition 345
prebiotics, milk supplements 53, 56
pregnancy, 1,25-dihydroxyvitamin D (1,25D) biosynthesis 79
pressure cooking, definition 253
probiotics, definition 324, 345
proofing, definition 305
protein, definition 233
protein intake
  calcium bioavailability 12
  calcium excretion 12
protein quality, definition 233
proteins, calcium bioavailability 49
pseudohypercalcemia, definition 601
pseudohypocalcemia, definition 601
pseudohypoparathyroidism, definition 479, 601

quantitation limit (LOQ), definition 107

quantitative computed tomography (QCT), definition 479

radioisotopes
 atomic mass *70*
 decay mode *70*
 definition 73
 half-life *70*
 metabolic studies 70
radish leaves 37
RANK-RANKL system, definition 621
Ras, definition 443
ratiometric fluorescence, definition 28
recommended dietary allowances (RDA)
 bioavailability factor 47, 61
 historical research 75
 life stages 76–7
 vitamin D 79
red blood cell, definition 107
relative standard deviation, definition 107
religion, dietary rules 32, 34, 37
remineralization, definition 379
respiratory chain, definition 464
retention
 definition 197, 218
 protein effect 12
rhubarb *57*
rice
 calcium content 58
 types 270
rice milling, definition 270
rice noodle, definition 270
rice pudding 33
rickets 78
 definition 88
Romania, dairy consumption 32
Romans 68
Russia, dairy consumption 32–3
RyR (ryanodine receptor), definition 444
saliva, definition 379
salivary buffers 27
salivary flow rate, definition 379
salivary glands, definition 379
salt-insensitive subjects, definition 516
salt-sensitive subjects, definition 516
salt taste, definition 426
Scandinavians 31
Schiff bases, definition 287
seafood
 Asian consumption 37
 calcium content 6
secretory diarrhoea, definition 408
Serbia, dairy consumption 32
serum levels
 calcium content 23
 homeostasis 9
sheep milk 53
shells 3
short-bowel syndrome 50
 definition 61
short-chain fatty acid (SCFA), definition 345
signal transduction
 calcium functional overview 23, 24
 definition 73
 see also membrane channels
sintered CaP, definition 122
skeleton
 calcium content age effects 76–7
 as calcium store 75–6
 see also bone
skin, obligatory loss 76
slow MHC, definition 443
slow motoneuron activity, definition 443
slow MyHCβ-MHC, definition 443
slurry, definition 122
small intestine, definition 393
soaking, definition 253
sodium–calcium exchanger (NCX) 47, **48**

sodium intake, calcium systemic balance 12
soft drinks, fracture risk 13
soil
  calcium form 3
  fertility 22
  magnesium 22
solubility
  absorption effect 12, 69
  calcium salts 22, *68*, 69
  definition 253
  pH effects 9, 15, 69
soluble fraction, definition 197
sour taste, definition 426
sources
  calcium carbonate 3
  calcium fluoride 3
  calcium in Earth's crust 3, 22
  calcium phosphate 3
  calcium sulfate 3
  see also dietary sources
soy-based formula milk 55
soybean products 57
  Asian consumption 37, 37–8
soybeans 56
soymilk, definition 324
Spain, dairy consumption 31
speciation, definition 107
spices, calcium content *7*
spinach *57*, 58
  bioavailability *11*, 13, 16
  oxalic acid 12, 16
stable isotopes method, definition 233
steatorrhoea 50
  definition 61
stim1, definition 427
stimulated saliva, definition 379
stomach, calcium solubility 9, 15
sublingual saliva, definition 379
submandibular saliva, definition 379
sugar dehydration, definition 287
supplementation, bioavailability factor 61
supplements
  absorption 8–9, 13
  calcium sources 3
  cardiovascular disease (CVD) 8, 25
  forms 7–8, 47
  fracture risk 8, 78
  life stage benefits 8
  menopause 77–8
  side effects 7
  usage increase 6–7
sweat, calcium content 11
sweet taste, definition 426
synbiotics, definition 345
syncytiotrophoblast, definition 479
Syria, dairy consumption 32
systemic balance 9–11, 15
  definition 14
  see also homeostasis
systolic blood pressure (SBP), definition 516

Taiwan, dairy consumption 35
Tanner maturity stages, definition 541
tartrate-resistant acid phosphatase (TRAP), definition 122
taste papillae, definition 426
taste receptor cells, definition 426
tea 33
teenagers, recommended dietary allowances (RDA) *77*
teeth, calcium content 4, 14
tetanic electrical stimulation, definition 444
thyroid, calcitonin secretion 25
tight junctions 76
tofu 57
  calcium content 37–8
tooth decay 27
tortillas 39
toxicity, vitamin D 79
transcellular calcium transport, dietary fiber effect 51

transcellular transport 9, 15
transport
definition 197
transcellular 9, 15
tricalcium phosphate, absorption **48**
1,24,25-trihydroxyvitamin D 79
trophoblast, definition 480
Trousseau sign, definition 601
TRPV (transient receptor potential vanilloid), definition 480
turnover 11

Ukraine, dairy consumption 32
ultrasonic dialysis, definition 107
umami taste, definition 426
uniporter, definition 464
United Kingdom *see* Britain
United States of America, dairy consumption 33
unstimulated saliva, definition 379
uptake *see* retention
urine
calcium excretion 9
double stable isotope studies 71
uronic acids, calcium-binding 11, 50, 57

Van der Waals radius *68*
vascular calcification 4
definition 559
vasoconstriction, definition 516
vasodilation, definition 516
vegans 27
vegetables
Asian consumption 36–7
calcium bioavailability *11*, 57–8
calcium content *5*
geographical variations 58
see also green vegetables
vegetarianism
religious dietary rules 32, 37
see also vegans
Vietnam, dairy consumption 34–6
vitamin D
alcohol consumption 51
calcium bioavailability 47, **48**
calcium homeostasis 9, 75–6, **80**
calcium transport regulation 9, 15, 25, 27
deficiency 78
definition 393, 504
fracture risk studies 78
recommended dietary allowances (RDA) 79
sources 79
tolerable upper limit 79
total-body calcium levels 75–6
toxicity 79
see also calcitriol; 1,25-dihydroxyvitamin D
vitamin D binding protein (DBP) 79
vitamin D-dependent calcium binding proteins (CaBP), definition 480
vitamin D receptor (VDR) 9, 15, 77, 78
vitamin D_2, sources 79
vitamin D_3
biosynthesis 79
definition 16

water
hardness 359
mineral water 59
weaning 40
weight-bearing exercise, definition 542
Western diet 5, 31–2, 33, 35, 39
wheat products
calcium bioavailability 58–9
calcium content *6*, *58*
white adipose tissue (WAT), definition 577

white rice, definition 270
whole saliva,
  definition 379
Wnt signal pathways,
  definition 408
women, recommended dietary
  allowances (RDA) *77*

X-ray diffraction, definition 141

yoghurt
  calcium bioavailability 54
  consumption levels 31, 32, 33
  fiber content 54

Z score, definition 505
zao dan 38
zinc, absorption interference 9, 24
zwitterion, definition 287